Earthworms and Ecological Processes

Yahya Kooch • Yakov Kuzyakov
Editors

Earthworms and Ecological Processes

Springer

Editors
Yahya Kooch
Faculty of Natural Resources & Marine Sciences
Tarbiat Modares University
Noor, Iran

Yakov Kuzyakov
Department of Agricultural Soil Science
University of Göttingen
Göttingen, Germany

ISBN 978-3-031-64509-9 ISBN 978-3-031-64510-5 (eBook)
https://doi.org/10.1007/978-3-031-64510-5

This Springer imprint is published by the registered company Springer Nature Switzerland AG
The registered company address is: Gewerbestrasse 11, 6330 Cham, Switzerland

Contents

Part I
Global Distribution of Earthworms and Ecological Processes

Invasive Earthworms and Ecosystem Processes in North America

Maryam Nouri-Aiin and Josef H. Görres

Abstract Earthworms significantly influence soil structure and plant growth in agricultural ecosystems, but their invasions in North America, particularly by species within the Megascolecidae family, have raised ecological concerns. While earthworms are generally considered indicators of soil health, invasive species can negatively impact ecosystems by altering soil structure. These invasions, facilitated by human activities, disrupt natural soil processes and nutrient cycles, contributing to environmental changes on a large scale. Effective management and control measures, including quarantine protocols and biological control agents, are crucial to mitigate the spread and impact of invasive earthworm species.

1 Introduction

Earthworms play a crucial role in shaping the soil, and their actions have a beneficial impact on plant growth, especially in agricultural ecosystems (van Groenigen et al. 2014). The general high regard in which they are held by the public and agronomists means that invasions may be socially cryptic. Earthworm invasions have only been of concern for a few decades in North America. Any of the negative effects of invasive earthworms, such as increased greenhouse gas emissions (Lubbers et al. 2013), increased leaching of agrochemicals (Shuster et al. 2003), and decreased biodiversity (Ferlian et al. 2018) are not generally recognized and have not changed the positive perception of these organisms. The presence of earthworms is indeed still one of the positive indicators of soil quality in the USDA soil quality set (Ditzler and Tugel 2002). However, over the past decade, the invasion of North American forests and horticultural ecosystems by a group of cosmopolitan species within the Megascolecidae family, also known as pheretimoids, is causing a shift in the favorable perception of at least this family of earthworms within the US

M. Nouri-Aiin (✉) · J. H. Görres
University of Vermont, Burlington, VT, USA
e-mail: maryam.nouri-aiin@uvm.edu; jgorres@uvm.edu

Y. Kooch, Y. Kuzyakov (eds.), *Earthworms and Ecological Processes*,
https://doi.org/10.1007/978-3-031-64510-5_1

American gardening and forestry communities (Johnson et al. 2021). Although these earthworms invaded after European lumbricids (Chang et al. 2021), they are more visible than invasive lumbricids because of their behavior, peak densities, and loose castings.

The broad impacts that invasive earthworms have in ecosystems, particularly on forested systems, begin with the engineering activities of earthworms that alter soil structure and decomposition rates (Bohlen et al. 2004b, Hale et al. 2005). Langmaid (1964) noted that when earthworms invaded podzols in Canada, the organic horizon and the albic horizon characteristic of podzols were lost within a couple of years and replaced by a dark A horizon. Soil structure changes of these magnitudes impact other ecosystem processes. Ecological cascades set off by these soil modifications affect both soil ecology at the invasion site and ecological and environmental processes of societal importance at much larger scale (Frelich et al. 2019).

While much research has focused on the impact of earthworms on North American forested ecosystems (Bohlen et al. 2004a, 2004b; Frelich et al. 2006; Hale 2008; Hale et al. 2008), earthworm invasions are a global issue (Tiunov et al. 2006; Brown et al. 2006; Kim et al. 2015; Prasanth Narayanan et al. 2016; Janion-Scheepers and Griffiths 2020). The areas of the globe that were glaciated in the last ice age were devoid of earthworms until they were introduced largely by humans. These areas encompass the northern latitudes of Eurasia and North America (Tiunov et al. 2006), although some small ice sheets also covered parts of the southern hemisphere (Hulton et al. 2002). Success of the earthworm invasions in previously glaciated regions may be due to earthworms finding unoccupied niches (Wardle et al. 2011). But exotic earthworms are also introduced into and become established in ecosystems south of the maximum extent of the last glaciation (Brown et al. 2006; Callaham et al. 2006) or into established native earthworm communities elsewhere (Kim et al. 2015). Here they are interacting with native earthworm communities and can strongly affect the trajectory of the plant and decomposer communities.

Biological invasions refer to the introduction of organisms into a new habitat and their subsequent successful establishment. The ability of new species to persist within an already established community has been an important research topic since Wallace (1911) introduced the question. Elton (1958) discussed some essential and necessary elements of what might determine a successful invasion in terms of the ecological knowledge of the time. He suggested that complex food webs and species-rich environments might resist invasions and that the absence of predators may promote an invasion. But there are more barriers to invasions such as geographic obstacles and various abiotic impedances. Conceptual models of invasion recognize several stages of biological invasions (Fig. 1), each of which can be associated with a barrier to success (Blackburn et al. 2011). In a first step, an organism has to overcome geographic barriers and be moved from its original to its new range (dispersal or transport). Geographic barriers may operate at different scales. This is followed by the introduction to or colonization of the new range where the invasive organism has to survive. This step may be associated with acclimation to a new region under artificial conditions (greenhouse, gardens, etc.). But this step may be bypassed when the organism was introduced to the new environment from one with

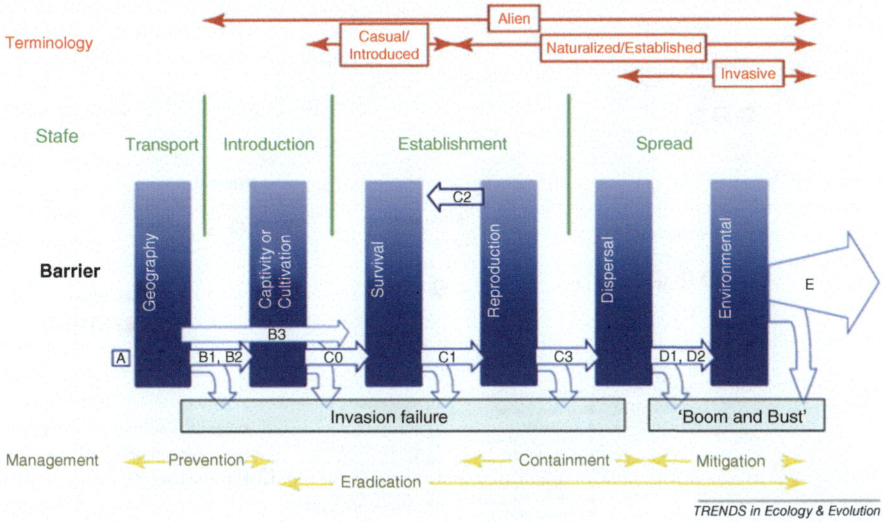

Fig. 1 The four stages of a successful biological invasion. Note the introduction step (acclimation in an artificial ecosystem) may be redundant (from Blackburn et al. 2011). The capital letters A, B1, B3, C1, C2, C3, D1, D2, and E represent different categories of populations within the framework, which correspond to specific stages and barriers in the invasion process: A, not moved outside its original habitat range; B1, moved outside its original habitat range and kept in captivity or quarantine; B2, moved outside its original habitat range and cultivated; B3, moved outside its original habitat range and released directly into a new environment; C0, released into the wild in a new area, but unable to survive for a long time; C1, surviving in the wild in a new area, but not reproducing; C2, surviving in the wild in a new area, reproducing, but not maintaining a stable population; C3, surviving in the wild in a new area, reproducing, and maintaining a stable population; D1, stable population in the wild, with individuals living a significant distance from the original introduction point; D2, a population that can sustain itself in the wild, with individuals surviving and reproducing at a considerable distance from the original introduction point, E: Fully invasive species, with individuals spreading, surviving, and reproducing at multiple locations across a wide range of habitats. Yellow arrows at the bottom of the figure show the level of management possible to reduce the severity of the invasion

similar conditions. Establishment requires the organism to persist as a population in the new range, i.e., it reproduces reliably under the new conditions and the offspring survives. Finally, the last step requires further spread from the point of introduction which once again requires that the invader has a way to move from the point where the population was established (Blackburn et al. 2011). Each successful invader will have to undergo at least the first three stages of the invasion. As the invader establishes in a new habitat, it restructures the endemic community through competition, predation, or engineering the environment. The distribution of invasive species across the globe is of a heightened importance now, given concerns for biodiversity and the current ongoing episode of global species extinctions.

The "natural" distribution of earthworm families has been linked to six biogeographical realms (hereinafter referred to as bioregions) vaguely associated with the tectonic plates. Earthworms are ubiquitous and occur on all continents except

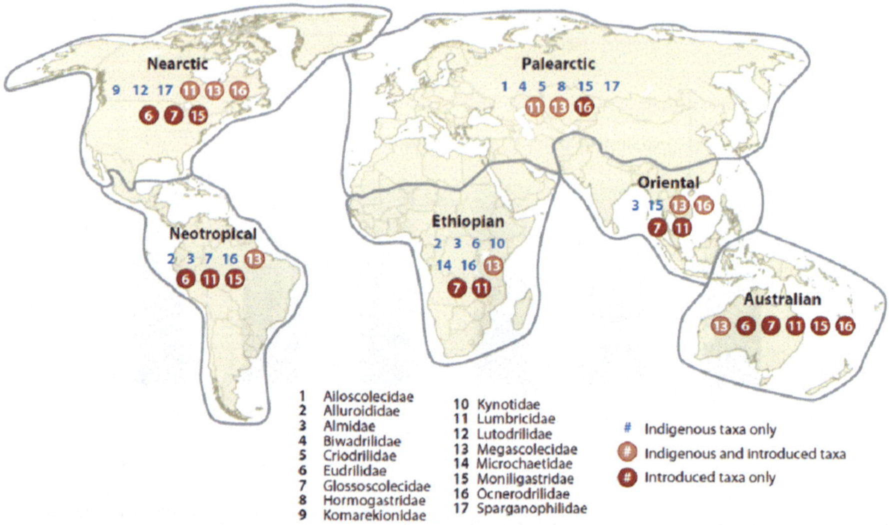

Fig. 2 The distribution of endemic and exotic earthworm families across the six bioregions of the Earth (Hendrix et al. 2008; Reynolds 1994). Included on the map are families indigenous to the continent (blue symbols), families with taxa that are either indigenous or introduced (orange symbols), and families whose taxa were introduced (red symbols). All continental-scale bioregions have species that were introduced from other bioregions

Antarctica, although some are found on sub-Antarctic islands (Smith and Steenkamp 1992). Figure 2 shows the current distribution of families across the biogeographical areas (Hendrix et al. 2008; Reynolds 1994).

The global-scale partitioning of earthworm taxa shows a very coarse and simple geographic pattern suggesting an initial distribution at the family level. However, earthworms have habitat preferences which superimpose current climate and edaphic conditions on their distribution. In addition, long-term climate variations such as episodes of glaciation also structured the biogeography of earthworms, leaving northern Eurasia and northern North America without earthworms when the glaciers receded.

The coarse distribution map of earthworm families begs the question what an invasive species is and what the geographic scale is at which this is determined. Continental boundaries may be overly broad and horizons in time may be artificial limits. American conservation biologists take the arrival of Europeans in North America as a threshold that determines the status of species as exotic or native in the Nearctic region (Stahl 2009). However, the Vikings reached North America earlier than the Iberian colonizers (Kuitems et al. 2022), yet large contiguous areas of North America were icebound until about 10,000 years ago and there is evidence that some North American species began to colonize these areas shortly after the retreat of the glaciers (McCay et al. 2017). What are the spatial and temporal boundaries for these postglacial colonization events in the Palearctic and Nearctic regions?

Answering this question may come down to semantics and the meaning of "invasive." Blakemore (2009) distinguishes several components of an earthworm community: native endemics (indigenous), introduced exotics (invasive), translocated natives, and neo-endemics (naturalized). Native endemics are the original fauna of a region. Introduced exotics and translocated endemics are the invasive species that we are addressing in this chapter. Introduced exotics were introduced recently and may not yet be in equilibrium with their new habitat. Translocated endemics are species transported into a new range within a bioregion. Neoendemic species are naturalized species introduced long enough ago to have undergone speciation in their new range. Frequently, an invasive species is given the attributes of a pest, i.e., they do damage in the range into which they were introduced. There are a small number of earthworm species that are considered cosmopolitan. These are introduced species with a broad global distribution which is strongly determined by human activities.

Although these definitions seem clear-cut, classifying earthworms into these categories is not straightforward for a number of reasons. James and Hendrix (2004) argued that in order to know whether a species is in a new habitat, the term "original range" needs to be defined and that phylogeny of the worms needs to be known. Phylogenetic studies can overturn preconceptions on the origins of a species. For example, the species *Dendrodrilus rubidus* and *Allolobophoridella eiseni* that were formerly regarded as Palearctic (Eurasian) in origin are now included in the Nearctic genus *Bimastos* (Czuzdi et al. 2017). Furthermore, bioregions are large. Within each of the continental-scale regions, there are ecoregions differentiated by climate, soils, vegetation, and topography and the time since the last glaciation, factors aligning with Jenny's factors of soil formation. In North America, there are 182 Level III Ecoregions and over 900 Level IV Ecoregions (Omernik and Griffith 2014; Reynolds 2017; Reynolds 2020). Geographic aggregation at various spatial scales is likely to place the same earthworm species into different categories of an earthworm community (sensu Blakemore 2009). Reynold's earthworm species checklists are often aggregated either by state or by Level III Ecoregion (e.g., Reynolds 2016; Reynolds 2017).

An example of a translocated endemic taxon in North America is the Nearctic species *Eisenoides lonnbergi* which is endemic south of the extent of the Wisconsin glaciation, but can now be found in wetland habitats further north (McCay et al. 2017). Its distribution may have been facilitated by stream flow in north-draining sub-watersheds of Lake Ontario (Fig. 3). Other examples are from arctic Europe where earthworms in the *Lumbricus* and *Aporrectodea* genera were moved by human activity from their southern range into arctic ecosystems (Tiunov et al. 2006; Wackett et al. 2018; Blume-Werry et al. 2020). These examples blur the simple notion that invasiveness is associated with continental-scale bioregions. Wackett et al. (2018) make the case that these newcomers are significantly impacting the formerly glaciated Fennoscandian arctic ecosystems and should be considered introduced species, even though they are translocated endemics at the continental bioregion scale (sensu Blakemore 2009). This introduces one more dimension used in the definition of an invasive earthworm species: they considerably alter an ecosystem.

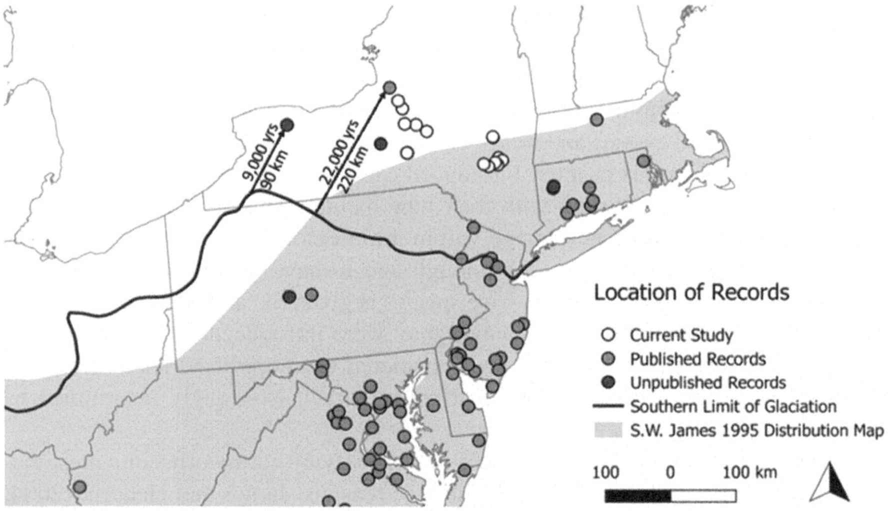

Fig. 3 Expansion of *E. lonnbergi* in the northeastern USA beyond the bounds of the last glaciation with travel times from the glacial limit to the most northerly observations (modified from McCay et al. 2017)

2 The Invasion Mechanisms

2.1 *Dispersal (Transport)*

In many areas across the globe where earthworm invasions have occurred, it is difficult to find the mode of transport and the point and time of introduction either because they occurred long ago or multiple, widespread introductions confound investigations. However, in places where earthworm invasions are not yet at advanced stages, means of introductions are more easily discerned. In remote northern Canada, Cameron et al. (2007) identified four ways by which transport may occur: flowing water, fishing bait, vehicle tire treads, and large animals. Costello et al. (2011) found that forest logging (vehicle traffic) and transport by water were predictors of earthworm abundance and diversity in Alaska. These investigations highlight the two classes of processes that Schwert (1980) suggested as the main, and generally accepted, long-distance transport conduits: anthropochory, i.e., moved by human activity, and hydrochory, i.e., moved by water. In contrast to the anthropochorous rate of dispersal, natural earthworm locomotion rates are of the order of 10 m per year on land (Marinissen and van den Bosch 1992; Dymond et al. 1997; Hale et al. 2005; Eijsackers 2011). It would take thousands of years for earthworms to move even 100 km that way. At 10 m per year, travel times for *Eisenoides lonnbergi* from the line of the last glaciation to the their most northerly reported location in the US state of New York (Fig. 3) would be between 9000 and 22,000 years.

It is generally accepted that the most important long-distance mode of earthworm dispersal between and within continents, by and large, is through human activities. Molecular evidence of human-mediated introductions of *Dendrobaena octaedra* along Canadian roadways, probably by vehicle traffic, underscores the reliance of earthworm invaders on human activity to facilitate long-distance travel (Cameron et al. 2008). Trade in particular is responsible for transporting earthworms between continents and within continents. In North America, early post-Columbian introductions of lumbricid earthworms from Europe into North America were likely by ship ballast comprising earth materials that included earthworms and their cocoons (Hendrix and Bohlen 2002; Hendrix et al. 2008). Similarly, pheretimoids were probably introduced into North America by trade (Chang et al. 2021). A close association of earthworms and their transport with the horticultural trade was reported by Gates (1963) who found 19 earthworm species in Maine greenhouses that are now regarded as exotic species in the Nearctic bioregion. The eight species that Gates found exclusively in greenhouses of the northeastern USA at the time are now found in the wild (Chang et al. 2018; Nouri-Aiin et al. 2022). Schwert (1980) also makes a connection between an isolated population of *Aporrectodea icterica* at an arboretum in southern Ontario and horticulture, suggesting that the organism was first introduced there with trees imported from England. Speculatively, the introduction of jumping worms in the pheretimoid clade to the East Coast of the USA is linked to the gift of cherry trees by Japan to the Washington, DC, area in the early parts of the twentieth century. A study on the distribution of the spider *Atypus karschi* (Donitz) introduced from East Asia in the Philadelphia area suggests that their co-occurrence with pheretimoids may be due to both being transported with potted plant materials in the nineteenth century and a gift of ornamental plants from Japan to Philadelphia in 1926 (Řezáč et al. 2022). Improved phytosanitary regulations that prohibit the importation of soils and other growing media into the USA and Canada without permits may decrease the number of introductions into North America (e.g., USDA-APHIS 2022). Recently, US regulations in effect since October 2022 require import permits and strict adherence to rearing protocols to prevent transport of plant and animal diseases with earthworms (USDA-APHIS 2022). Interestingly, the import of *Lumbricus terrestris* as fishing bait from Canada is not regulated if the earthworms were collected from Canadian soil. Also, there are few phytosanitary restrictions for plant trade from the USA into Canada, even when plants are moved with growing media, unless pests are present in the region where the product originated. Earthworms are however not listed as pests in Canada (Moore et al. 2018) and thus permitting in Canada is not a concern for horticultural importers. New records from Canada report pheretimoid jumping worm populations in Ontario, New Brunswick, and Quebec (Reynolds and McTavish 2021; McAlpine et al. 2022). The populations in Ontario were reported mainly from private gardens in the Toronto and Niagara Falls areas.

The current distribution of introduced earthworms depends not only on trade at the intercontinental scale but also on trade within a continent. While trade in the modern era is commonly associated with the expansion of invasive species, earlier trade is often not considered. In the Eurasian, Palearctic context, there is a long history of trade that may have moved earthworms along trade routes within continents. Ancient trade including those between Rome and Scandinavia between 0 and 500 BP (Hansen 2000), the Silk Road, and those of the Mediterranean (Blakemore 2009) could have distributed earthworms throughout Eurasia. It is not as clear how early trade among North American indigenous nations would have transported plant materials from the unglaciated South into areas previously covered by the ice shield of the last glaciation. However, there is evidence that the indigenous Cahokian people controlled trade between southern regions along the Gulf Coast, where there were pre-Columbian populations of earthworms, and areas near the Great Lakes before colonization by Europeans (Belshaw 2020). Migration of Mississippian indigenous farmers to southern Manitoba and Alberta (Belshaw 2020) could also have been a vector of earthworms from the unglaciated South to areas in which the ice age had extirpated earthworms. However, there is no evidence of earthworms native to the Nearctic bioregion in the Prairie ecozone that encompasses southern Manitoba, Saskatchewan, and Alberta (Reynolds 2020). However, there are several Nearctic species found in Canada (Reynolds 2020). First, several *Bimastos* species, including the newly reclassified *B. rubidus*, are found in most provinces. Reynolds (2020) also lists *Sparganophilus tamesis* for Ontario, New Brunswick, Newfoundland, and Quebec.

Travel in stream flow was proposed for *E. lonnbergi* by McCay et al. (2017) which would accelerate their dispersal. Transport by flowing water has been proven to be a viable potential transport mode for earthworms, although there have not been any studies that directly link one population upstream to one further downstream. However, earthworms and their cocoons have been found in streamwater. Schwert (1979) recovered earthworm cocoons from streamwater. Various researchers have demonstrated that earthworms can hatch and survive submerged in water for extended periods of time (Roots 1956; Edwards and Lofty 1980). Overland flow can also transport earthworms down gradient (Schwert 1979; Bouche 1977, Görres et al. 2014). In one study, the downslope transport was about 100 m during a single, 1-day storm event, but further movement with runoff was bounded by a wetland at the bottom of the hill (Görres et al. 2014), suggesting that there are natural barriers to this faster mode of transport. However, transport of cocoons and earthworms into streams may cause an even more rapid dispersal. Some aquatic and salt-tolerant species, e.g., *Pontoscolex litoralis*, may travel long distances as part of flotsam on ocean currents or with wooden boats (Blakemore 2009; Chen et al. 2021). Wackett et al. (2018) hypothesized that the epigeic species *Dendrobaena octaedra* and *D. rubidus* (syn. *Bimastos rubidus*) might have colonized northern Scandinavia traveling north into the arctic with brackish water as the glaciers of the last ice age receded, although their distribution could also be explained by early activities of the nomadic indigenous people of the region, the Sami. The wide distribution of the corticolous *B. rubidus* and the limnicolous species *Sparganophilus tamesis* in Canada (Reynolds 2020) from southern Canadian ecoregions into Newfoundland (B. *rubidus*) and the Maritime Provinces (*S. tamesis*) potentially indicates hydrochorous transport. However, the rivers along which *S. tamesis* is found were used by the indigenous Mi'kmaqs and Maliseet people for transport and food (Reynolds 2020). As with *B. rubidus* of the Fennoscandian arctic, it is not clear whether the mode of transport into these northerly, previously glaciated areas was by indigenous people or by hydrochory.

The recasting of *D. rubidus*, a Palearctic species, as *B. rubidus*, a Nearctic species, brings into focus the paradox of their presence in the Fennoscandian realm. If these earthworms were early arrivals in lands inhabited by the Sami, when and how did they make the jump across the Atlantic or across the Arctic Ocean? Could *D. rubidus* (syn. *B. rubidus*) have arrived in Fennoscandia much later than suggested by Wackett et al. (2018)?

2.2 Introduction (Colonization)

The second step of earthworm invasion is introduction or colonization. For some invasive earthworm species, this step occurs in protected environments such as gardens and greenhouses where Gates (1963) found many exotics. These horticultural installations may be environmental refuges for invasive earthworms where they can weather inclement climate until conditions in the wild are more suitable for establishment. In gardens, mulching and irrigation would keep soil temperatures and moisture high, thus affording the earthworms more optimal environmental conditions and improved chances to establish a population by providing a longer season to reach reproductive maturity. In addition, mulching and organic amendments will provide resources for earthworms (Bellitürk et al. 2015). Horticultural habitats also have well-controlled soil pH and nutrient availability, often providing high organic substrates to sustain earthworms. Many earthworms require calcium to buffer gut pH (Piearce 1972) and exchangeable soil base cation concentrations are important (Moore et al. 2013). In horticultural systems, base cations are carefully regulated by the addition of liming agents, for example, calcitic and dolomitic lime or wood ash, among others. This is not the case in natural ecosystems where the probability of acidification and loss of exchangeable cations is much greater than the reverse.

Some earthworms are parthenogenetic, which means that theoretically a single earthworm can find a colony. In the northeastern USA and Canada, the pheretimioids *Amynthas agrestis*, *Amynthas tokionesis* and *Metaphire hilgendorfi* are of concern. All three of these worms are thought to be parthenogenetic, although they may have a small number of hermaphroditic morphs in their populations (Gates 1956; Shen et al. 2011). One expects clonal populations, but interestingly, populations at least in Vermont (USA) are genetically diverse (Nouri-Aiin et al. 2022) suggesting multiple introductions or co-introductions of several clonal lineages together. Whether the genetic diversity is due to occasional sexual reproduction or aggregation of different lineages in the horticultural supply chain is not known. It is not clear whether acclimation in horticulture is necessary for their establishment since the climate in their original range is similar to that in northeastern North America (Tecimen et al. 2021; Chang et al. 2021). In regions where the conditions are optimal, these earthworms could be introduced directly to the wild. This may be true for fishing bait that is often introduced directly to riparian areas when it is discarded by anglers at fishing sites (Hale 2008). However, much of the fishing bait sold commercially is extracted from populations already existing at sites similar to those to which they are introduced. In North America, the dew worm or night crawler (*L. terrestris*) is a common bait species, and most of these are collected in Canada for use in the USA (Keller et al. 2007; Kilian et al. 2012; Steckley 2023). Soil and climate conditions in Canada and the northern tier of US states are similar and acclimation is not necessary. The establishment of earthworm populations around remote fishing spots in Minnesota and elsewhere is a good example of this step being redundant in some cases (Hale et al. 2006). Cameron et al. (2008) suggests multiple introductions of *Dendrobaena octaedra* into the wild along roads in northern Canada, presumably by traffic.

2.3 Establishment

The distribution of some earthworms is global, suggesting a generalist strategy of these cosmopolitan or peregrine earthworms. Some species are found across several climate zones and in vastly different soils. Some cosmopolitan lumbricids such as *Aporrectodea* species and megascolecids such as *Amynthas* species inhabit andic soils along the Pacific Ring of Fire (Blakemore 2003; Kim et al. 2015) as well as postglacial inceptisols with varying mineralogies in North America and Europe (Tiunov et al. 2006; Wackett et al. 2018; Reynolds 2020; Keen et al. 2022).

The observation of pheretimoid earthworms in the northeastern USA, which has cold long winters, was initially in horticultural ecosystems. Observations in the wild are more recent and individuals may have been able to escape the horticultural systems with climate change increasing their chances to establish populations in the wild. Görres et al. (2016) and Görres and Melnichuk (2012) estimated that the pheretimoid species of concern in Vermont can reach reproductive maturity in approximately 90 days measured from the observation of the first hatchling. The frost-free period would thus need to be at least that long. In the northeastern USA, the frost-free period has increased from 1970 to 2000 by 7 days with an average increase in temperature of 2.1°C. The forecast predicts a lengthening of the frost-free period by 29–40 days by the end of the twenty-first century. Based on the maturation period of 90 days, Moore et al. (2018) produced a map of the potential range where pheretimoids could become established based on average frost-free periods (Fig. 4). The northernmost areas that may be affected in North America include Labrador and regions just south of Hudson Bay in Quebec and Ontario at current climate conditions, but their potential range based on climate will undoubtedly expand north. The cold and drought hardiness of their cocoons (see below) may also offer further expansion northward beyond the range shown in Fig. 4.

For invasive earthworms to establish in the wild, soil conditions will also have to be within environmental tolerances of the earthworm species in question. While

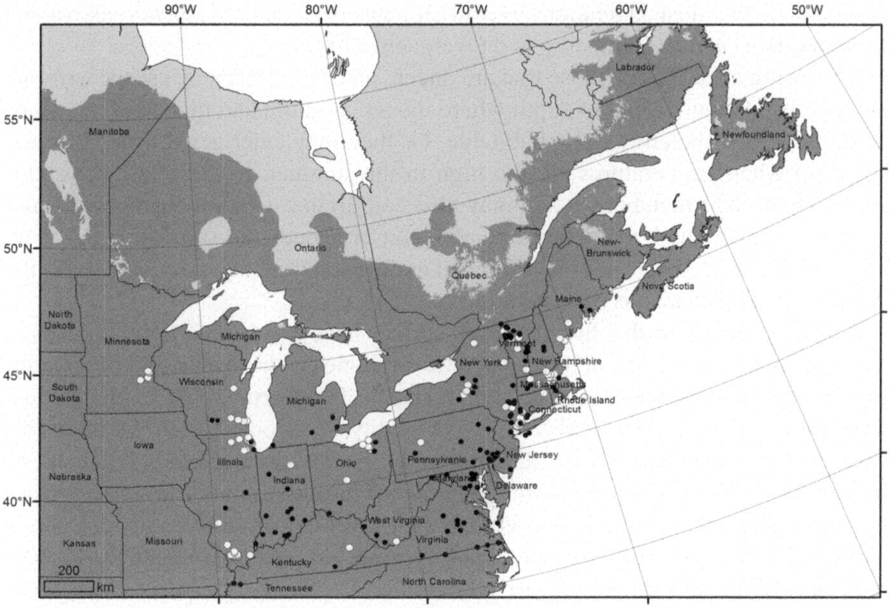

Fig. 4 This map shows the potential range of jumping worms (Megascolecidae) determined as the area where they can reach reproductive maturity based on an average frost-free period of 90 days (from Moore et al. 2018). White and black dots are where jumping worms were confirmed. In 2022–2023, jumping worms have also been found in Nova Scotia, New Brunswick, and Ontario

conditions in horticulture are maintained for crop production, in the wild, soil acidification occurs due to acid deposition that causes the loss of base cations. Moore et al. (2013) and Bernard et al. (2009) suggested that pheretimoid earthworms require base cations to thrive. For this reason, in areas such as the Adirondack Mountains of northern New York, where acid deposition was high, earthworms are not found frequently (Bernard et al. 2009).

Tolerance of climate and its variability at different timescales are also important for establishment and persistence in new territories. Resources, along with weather conditions, vary in time. The cyclical seasonal variations such as litter availability in forests and temperature and moisture are somewhat predictable. However, stochastic variations of the climate system, such as drought, flooding, and weather extremes, require special adaptations. Earthworm species differ in their temperature and moisture tolerances and have various adaptations to environmental conditions (Singh et al. 2019). Some of the differences and adaptations can be explained by the ecological groups to which species belong because their habitats, stratified by soil depth, expose them to different amplitudes of temperature and moisture variations. Earthworms have developed a variety of adaptations to extend the range of environmental conditions they encounter in their habitat. In particular, at the northern edge of their invaded range, adaptation to frost is important. An exhaustive summary of physical and chemical adaptations to cold climates is given in Meshcheryakova and Berman (2014) and in Holmstrup and Overgaard (2007).

Burrowing species, i.e., those that are anecic and endogeic, may escape extreme temperatures by burrowing to depth where the temperature and moisture variations are dampened sufficiently to be habitable. Or they may enter aestivation (summer diapause) when temperatures are too high in the summer. In this state, the earthworms lower their metabolism and may also coil up in estivation chambers deeper in the soil to avoid high temperatures and to conserve body moisture (Jimenez et al. 2001). Tilikj and Novo (2022) investigated expression of genes associated with the physiological mechanisms of aestivation.

Epigeic populations that live at the soil surface have to rely on different means of survival. One of the most important survival strategies is the production of cocoons (egg casings) that are freeze and drought resistant (Holmstrup 2002; Eggleton et al. 2009; Moreau-Valancogne et al. 2013). The ability of cocoons to survive these two extremes are related and rely on dehydration of the cocoon to avoid cell damage in the embryos. This has important implications for the expansion of invasive species into areas where they could not survive in their juvenile or adult life stage. Cocoon dehydration interrupts the development of the embryos, so that emergence of hatchlings can be spread over a long period of time. *Amynthas agrestis* and *Amynthas tokioensis* hatchlings can emerge throughout the year, even during the winter when temperatures are temporarily above freezing (Görres et al. 2018). Cocoons of these species also form a cocoon bank from which juveniles may hatch over several years (Nouri-Aiin and Görres 2019) which probably allows these invaders to persist not only over short episodes of unfavorable conditions but also over longer periods of drought and frost. This may in fact extend their range northward where the interannual weather conditions may vary to allow colonization and persistence as cocoons during years when the frost-free periods are too short for the vermiform to reach reproductive maturity.

Another important issue is whether invasive earthworms encounter competition from indigenous earthworms. Kim et al. (2015) reported that in New Zealand exotic lumbricids coexist in agricultural systems with native megascolecids. However, in the regions of North America that were covered by ice sheets, pheretimoids are replacing lumbricid earthworm invaders (Laushman et al. 2018) that are thought to be the first wave of earthworm invaders in North America (Chang et al. 2021). Hendrix et al. (2008) also found that incoming invaders can coexist with native earthworms potentially facilitated by different temperature tolerances. Competitive exclusion is by no means an inevitable fate of native earthworms, nor an insurmountable barrier for the invaders.

3 Environmental Impacts of Earthworm Invasions

3.1 Soil Organic Matter and Greenhouse Gas Emission

Invasive earthworms alter soil structure (Langmaid 1963; Bohlen et al. 2004a; Hale et al. 2005; Frelich et al. 2006), and at a very basic level, the earthworm's engineering activities accelerate organic matter decomposition (Haimi and Huhta 1990; Bohlen et al. 2004b; Zhang et al. 2013) and the cycling of other elements (Lemtiri et al. 2014; Resner et al. 2015; Čapek et al. 2021). Inevitably this will include the production of climate change gases, CO_2 and N_2O (Don et al. 2008; Lubbers et al. 2013).

Carbon mineralization may have important consequences where earthworm species invade new territories that were previously free of earthworms. The introduction of exotic earthworms into tundra and taiga ecosystems can have devastating consequences for the vast carbon stocks contained in their soils. Model simulations of earthworm effects on carbon storage in these northern systems projected that forest-floor carbon would be reduced by between 50 and 94%, mostly in the first 40 years after earthworm invasion (Cameron et al. 2015). The rapid consumption of organic materials by invasive earthworms may be a temporary effect, called worm aging: C emissions are high early in the invasion but then settle to lower levels when the most accessible C stocks are exhausted.

The ecological group of an earthworm species may determine its effect on the fate of soil and litter carbon (Huang et al. 2020). Differential effects of two invasive taxa of earthworms belonging to different ecological groups on carbon budgets (*Amynthas* spp., epi-endogeic species and *P. corethrurus*, an endogeic species) have been documented in southern subtropical China. *P. corethrurus* stabilized organic carbon by repackaging soil into large macroaggregates, whereas *Amynthas* spp. increased litter decomposition rate through the rapid consumption of organic residues and the stimulation of microbial biomass resulting in a net carbon loss from the soil (Wu et al. 2017).

However, earthworms contribute not only to the consumption of carbon but also its redistribution in soils at different spatial scales. At the pedon scale, Langmaid (1963) described the rapid changes that resulted from earthworm invasions of Canadian podzols. Langmaid noted the loss of the organic horizon, which was replaced by a granulated layer of earthworm casts. Similar transformations and translocations of organic matter have been reported from surface horizons to lower depths of the soil by anecic and epi-endogeic earthworms (Bossuyt et al. 2005; Bossuyt et al. 2006; Don et al. 2008). At lower depths, organic matter is less susceptible to mineralization. But these changes may not be universal across soil taxonomic units (Lejoly et al. 2021). At the much smaller aggregate scale, carbon can be protected in vermigenic aggregates (Bossuyt et al. 2004; Bossuyt et al. 2005; Zhang et al. 2013; Knowles et al. 2016). This is due to incorporation of organic matter into micro-aggregates that form during gut passage. The paradox of simultaneous carbon mineralization and physical protection is explored by Lubbers et al. (2017). The long-term effects of earthworms on carbon loss from soil are not fully established (Angst et al. 2019; Lubbers et al. 2017). It is likely dependent on the balance between the rate at which C is protected within aggregates and the C mineralization rate. These in turn depend on the age of an invasion and the ecological group of earthworms. It is estimated that net C loss becomes insignificant after 200 days of incubation in mesocosm studies (Borken et al. 2000; Lubbers et al. 2013; Zhang et al. 2013, Garnier et al. 2022). Aggregate size in earthworm casts might also contribute to the protection of soil C, as CO_2 emissions were found to be significantly lower from small aggregates (<250 μm) compared to large aggregates (>500 μm) (Jouquet et al. 2011, Thomas et al. 2020).

Closely related to the effect of earthworms on soil organic matter is the emission of greenhouse gases. When earthworms invade new environments with high carbon stocks, they would, at least initially, raise CO_2 emissions over those observed in a similar system where earthworms have been present for a long time. However, after the available soil carbon stocks have been either metabolized or stabilized, emissions would settle at a lower rate and be influenced by the production of litter. A meta-analysis of lumbricid earthworms showed a 33% increase in CO_2 emission (Lubbers et al. 2013). They predicted earthworms will significantly increase soil greenhouse gas fluxes over time due to the range of expansion of earthworms into higher latitudes due to climate change, something that is already seen in arctic soils (e.g., Wackett et al. 2018). Lubbers et al. (2013) concluded that earthworms increase the net global warming potential of soil by 16%.

Earthworms can have a wide range of impacts on N_2O emission depending on the soil type, C/N ratio, and earthworm functional group (Svensson et al. 1986; Zhang et al. 2013; Lv et al. 2018). In one study, earthworm presence increased N_2O emissions from forest soil by 57% in the first 6 weeks and then decreased significantly after 10 weeks (Borken et al. 2000). Although earthworm effects on N_2O emission rate is not significant in forest soil (Borken and Brumme 1997), in agricultural soils, the application of organic residues or fertilizers results in an increase of N_2O emission by as much as 43% (Giannopoulos et al. 2010; Lubbers et al. 2013). This increase occurred 30 days after application, when soils with high organic matter provided a C source for denitrifying bacteria after a decline in the decomposition rate (Rizhiya et al. 2007; Drake and Horn 2006; Speratti and Whalen 2008; Lubbers et al. 2013). Ammonification of organic matter may be a rate-limiting step that might delay the production of N_2O (Lubbers et al. 2013). However, when NH_4 was added to soil amended by organic residues, it triggered N_2O emissions in the presence of earthworms within a few days (Amador and Avizinis 2013). Fertilizer additions to agricultural land are not the only source of NH_4 additions. For example, the experimental forests at Hubbard Brook and Harvard Forest received a quarter to more than half of atmospheric nitrogen deposition as NH_4 (Aber et al. 1997). Global atmospheric NH_4 deposition increased from 1986 to 2012 by about 30% (Ackerman et al. 2019).

Few studies have analyzed the contribution of pheretimoids to greenhouse gases from newly invaded soils. The epi-endogeic pheretimoid *A. agrestis* caused greater emissions of N_2O from soils than the epi-endogeic lumbricids earthworm *Lumbricus rubellus* (Tecimen et al. 2021; Melnichuk 2016). A mesocosm study that simulated typical Vermont, USA, forest soils, examined C and N dynamics in soil invaded by two epi-endogeic invasive earthworm species, *L. rubellus* (Lumbricidae) and *A. agrestis* (Melnichuk 2016). Peak N_2O emission occurred within 15 days for *L. rubellus* and 30 days for *A. agrestis*. Interestingly, over the long term, N_2O emission from soils with *L. rubellus* remained flat after reaching the peak, but for *A. agrestis*, it increased again after several weeks. Cumulative N_2O emissions in the absence or presence of an A-horizon exceeded a control soil by 37 and 44%, respectively, when *A. agrestis* was present, but only by 19 and 21%, respectively, when *L. rubellus* was present (Melnichuk 2016). A flattening of N_2O emissions was observed in a forested riparian buffer ecosystem in Quebec monitored over a single year (Ejack et al. 2021), suggesting that the effect is not long term. However, studies do not usually span several years which may be important in deciduous forested ecosystems where annual cycles determine the availability of litter resources. Longer monitoring periods are needed to ascertain any cyclical variations in N_2O and CO_2 emissions coupled to cyclical litter availability.

An interesting feedback between increased plant growth stimulated by earthworm invasions in cold climates suggests greater greenhouse gas emissions. In a field study in northern Sweden (Blume-Werry et al. 2020; Klaminder et al. 2023), invasive earthworms significantly boost plant nitrogen uptake in a shrub-dominated (heath) tundra. Active plants prevent the soil from freezing which facilitates faster thawing of permafrost in summer, thereby also increasing greenhouse gas emissions. More vermigenic macroporosity in tundra soils results in a dramatic reduction in soil moisture (Klaminder et al. 2023), which in turn promotes plant growth and nitrification which can cause N_2O emissions (Lubbers et al. 2013; Ferlian et al. 2020).

3.2 Water Quality

Most studies on water quality effects by earthworms have been conducted in areas with neoendemic or endemic communities. However, hydrologic processes are universal and also valid in ecosystems invaded by earthworms. In agricultural systems, anecic earthworms are believed to have a positive effect on soil porosity and water infiltration (Clements et al. 1991; Pitkänen and Nuutinen 1998, Bertrand et al. 2015) which benefits crop productivity. However, the trade-off of improved infiltration rates is that macropores engineered by anecic earthworm allow water and agrochemicals from the surface to bypass the biologically active soil matrix without further biochemical transformations. In this way, anecic earthworms increase the probability of nutrient, salt, and pesticide leaching (Villholth et al. 2000; Wang et al. 2016; Schneider et al. 2018). In agriculture this is particularly important where anecic earthworm burrows connect the soil surface to subsurface tile drains (Shipitalo et al. 2004) which subsequently increases the export of agricultural amendments. There are some soils in the upland of the UK with natural soil pipes (large macropores) that are important for runoff on hill slopes with large humus accumulations (Wilson and Smart 1984). These may act similarly to drainage pipes installed in agriculture. Nutrient losses by macropore flow may be exacerbated by accelerated nutrient cycling in the presence of earthworms. For example, anecic earthworms are associated with increases in dissolved organic nitrogen losses (Shuster et al. 2003; Van Kessel et al. 2009). Earthworms in other ecological groups such as epi-endogeic earthworms are known to compact soils by producing small castings that promote surface water runoff and soil erosion (Blanchart et al. 1999; Görres et al. 2001; Hale et al. 2005; Görres et al. 2019). Furthermore, dead earthworm tissues release a significant amount of nutrients such as ammonium and phosphorus quickly into soil or water sources (Trap et al. 2021).

A study in northern Germany by Plum and Filser (2005) evaluated earthworm effects on nutrient mobilization. Earthworm-free microcosms were compared with microcosms of three species of annelid worms (the endogenic earthworm *Octolasion cyaneum*, the epigeic earthworm *L. rubellus*, and the enchytraeid *Cognettia glandulosa*). They found that ammonium and total phosphorus mobilization was significant with *O. cyaneum* presence even at a low density (200 individuals m^{-2}). As was indicated by others, mobilization of ammonium and phosphorus was considerably high in dead tissues of both *O. cyaneum and L. rubellus* (Plum and Filser 2005).

3.3 Biodiversity

Forest ecosystems in northern North America evolved in the absence of earthworms and typically have a thick layer of organic matter due to slow decomposition rates (Hendrix et al. 2008). This layer is integral to these forests and their biodiversity. The introduction of earthworms into these forests alters below- and aboveground biodiversity through depleting this layer and changing the physical structure and chemical properties in the soil profile (Fig. 5). Moreover, introduction of earthworms of different ecological groups into ecological niches has complex effects on other soil biota (Bouché 1977; Hale et al. 2005).

There is still a dearth of information on the effect of earthworms on microorganisms in newly invaded forests (McLean et al. 2006). The consumption of organic matter by epigeic earthworms and the mixing with mineral soil by epi-endogeic earthworms deprives microorganisms of the rich and variable habitat of the organic horizon (Vidal et al. 2019). The loss of spatial heterogeneity, both in terms of fewer horizons and within horizons, is thought to cause a decrease in fungal diversity. The repackaging of organic matter in more stable organo-mineral complexes by anecic and endogeic earthworms limits microbial growth (McLean et al. 2006). Generally, the gut microbiome is affected, in order of decreasing importance, by the earthworm ecological group, habitat characteristics, and earthworm species (Thakuria et al. 2010). This suggests that invasive earthworms may bring with them microorganisms from their original range. Whether these bacterial strains persist is not known, but natural selection may favor the indigenous species.

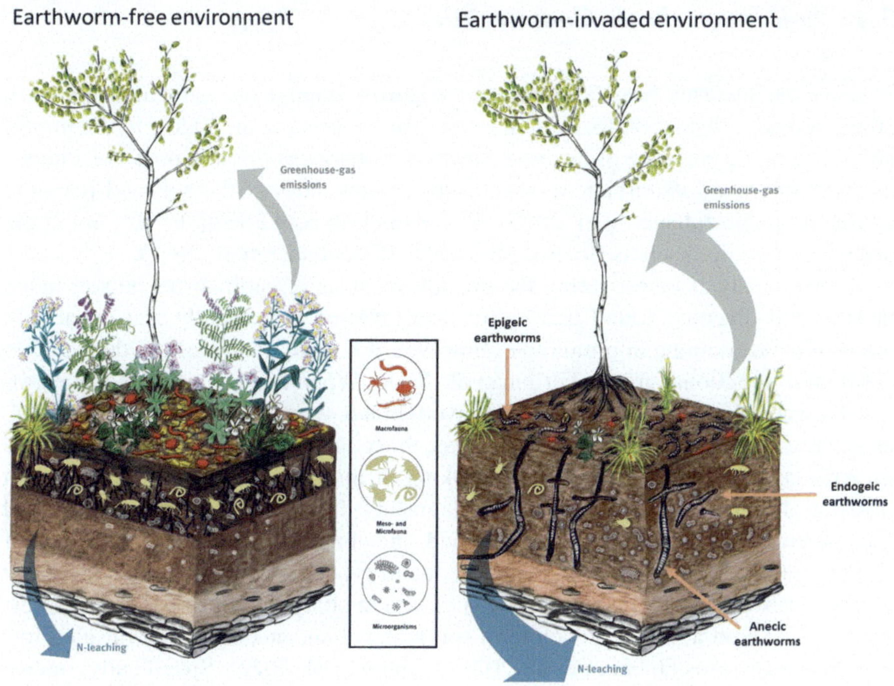

Fig. 5 Earthworm ecological groups and the effect of invasive earthworms on soil profiles and nitrogen fluxes (diagram from Ferlian et al. 2018)

Earthworm invasions affect the vertical distribution of microorganisms because anecic and epigeic earthworms translocate processed organic matter to lower depths where it is further modified by endogeic earthworms (Bohlen et al. 2004a, 2004b; McLean et al. 2006; Frelich et al. 2006; Eisenhauer et al. 2007). Earthworms make nutrients available to microorganisms through the mixing of soil layers (vertical transfer of organic matter to deeper layer) in addition to producing nutrient hotspots in the casts and burrows (McLean et al. 2006; Blouin et al. 2013; Groffman et al. 2015). Conversely, they can also negatively impact microbial biomass through habitat destruction and resource competition, which take place mostly in the organic layer (Eisenhauer et al. 2007).

Earthworms may have different effects on bacterial, fungal, and mycorrhizal networks in soil. Meta-analyses found that invasive earthworms had no significant effects on fungi but did significantly decrease mycorrhizal colonization (Paudel et al. 2016; Ferlian et al. 2018). Though the impact of earthworms on general bacterial communities is neutral, increases in denitrifier bacterial communities have been found to be significant when earthworms are present (Medina-Sauza et al. 2019). Despite the identification of parasite species such as *Apolocystis bosanqueti* and *Monocystis perplexa* in *A. tokioensis* and *A. agrestis*, as well as *Monocystis* sp. in L. terrestris, their impact on the health of these invasive earthworm species remains unknown (Field and Michiels 2005; Schall 2021; Schall et al. 2023; Keller and Schall 2024).

3.4 Soil Invertebrate Communities

Invasive earthworms have a significant negative impact on soil fauna diversity (Bohlen et al. 2004a, 2004b; Ferlian et al. 2018; Ferlian et al. 2020). Reduction in soil fauna in the presence of invasive earthworms happens through physical, chemical, and microbial disturbances, increasing compactness of the soil and resource competition (Eisenhauer et al. 2007). This reduction could be up to 90% for some arthropod families such as Brachychthoniidae (Eisenhauer et al. 2007).

A meta-analysis investigating the impacts of invasive earthworms on soil fauna density and diversity found that earthworm invasion consistently and negatively alters soil invertebrate communities; however, the scale of this effect depends on earthworm functional group (Ferlian et al. 2018). When endogeic and anecic earthworms were present, densities of soil invertebrates were significantly diminished, especially in the organic layer; nevertheless, the reduction was also significant in the mineral layer (Ferlian et al. 2018). Looking at more specific effects, earthworm invasion did not affect Astigmata and Gamasida. In contrast the analysis showed that prostigmatid mites' density increased and that the density and diversity of Collembola and oribatid mites decreased due to synergistic effects of endogeic and anecic earthworms (Ferlian et al. 2018). A recent study on North American forest fauna discovered a 33–45% decline in soil fauna abundance in highly invaded forests by earthworms (Jochum et al. 2021; Jochum et al. 2022). Specifically, microfauna biomass reduction (−38%) in the presence of earthworms is more significant than mesofauna (−14%) and macrofauna (−7%) (Jochum et al. 2021). Invasion intensity is significantly correlated with a decline in soil macrofauna abundance reduction (Burtis et al. 2014; McCay and Scull 2019; Jochum et al. 2022).

3.5 Plants

Invasive earthworms induce physiochemical changes in nutrient and water dynamics, mycorrhizal associations, and microhabitats for seed germination, which on a broader scale result in a changed plant community (Edwards and Bohlen 1996; Bohlen et al. 2004a, 2004b; Hale et al. 2006; Paudel et al. 2016). These effects are exacerbated by earthworms' direct feeding on seeds and seedlings resulting in changes to seed bank dynamics (Griffith et al. 2013; Nuzzo et al. 2015). In North America, shifts in plant species composition and diversity have been reported in invaded forests (Holdsworth et al. 2007; Craven et al. 2017; Fleri et al. 2021). Different plant functional groups respond to earthworm invasion differently, since they have specific requirements for germination and establishment including nutrient and water retention, access to mycorrhizal symbionts, and temperature buffering (Lawrence et al. 2003; Groffman et al. 2015; Hale et al. 2006).

Some plants are seen as indicators of earthworm invasion in North American forests (Corio et al. 2009). Earthworms increase certain herbaceous, sedge, grass, and generalist fern species abundance (Eisenhauer et al. 2007; Eisenhauer et al. 2009; Loss et al. 2012; Greiner et al. 2012; Melnichuk 2016; Bowe et al. 2020). Often a reduction in maple seedling is observed (Corio et al. 2009). Generally, in northern North American forests, the biomass of earthworms present at a site is negatively related to native plant diversity (Hale et al. 2005; Nuzzo et al. 2009; Resner et al. 2015; Bal et al. 2018; Craven et al. 2017) and positively to the abundance of exotic plant species (Frelich et al. 2006; Dobson and Blossey 2015; Craven et al. 2017). These changes in plant community affect the belowground and aboveground food web over long periods of time (Jochum et al. 2022; Eisenhauer et al. 2007), suggesting complex cascading effects of earthworm invasions (Frelich et al. 2019).

3.6 *Vertebrate Community*

At the forest-floor scale, various taxa such as invertebrates, salamanders, and ground-nesting birds are affected by the presence of invasive earthworms (Bohlen et al. 2004a; Ransom 2011; Loss et al. 2012; Bethke and Midgley 2020). Litter-dwelling animals that rely on leaf litter and invertebrates are significantly impacted by accelerated reductions in litter depth when earthworms are present. For example, salamander populations are threatened by declines in small arthropod abundance (Maerz et al. 2009). However, in the absence of a leaf litter layer, salamanders may benefit from earthworm burrows as an additional habitat, as well as from directly feeding on earthworm juveniles (Bohlen et al. 2004a, 2004b; Cáceres-Charneco and Ransom 2010; Ransom 2012).

Invasive earthworm abundance, as novel food sources, is directly correlated to the distribution of some bird species such as American robin (*Turdus migratorius*) (Cameron and Bayne 2012; Frelich et al. 2019). Although earthworms are a food source for some bird species and other wildlife taxa, they reduce the probability of survival of ground-nesting birds (Loss and Blair 2011). Litter depletion results in reduced nest concealment, increased nest predation, and declines in arthropod abundance resulting in local declines of forest songbird populations (Mattsson and Niemi 2006; Loss and Blair 2011). Nonetheless, the presence of invasive earthworms is shown to have wide-ranging impacts on different species of ground-nesting birds. For example, invasion of earthworms has no significant effect on the hermit thrush (*Catharus guttatus*), black-and-white warbler (*Mniotilta varia*), and Veery (*Catharus fuscescens*). However, invasion significantly reduces ovenbird density (*Seiurus aurocapilla*), especially in sugar maple woodlands (Loss et al. 2012).

3.7 Control

European (Lumbricid) earthworms are already highly dispersed in many forests in North America, and controlling their presence in landscapes is probably not feasible. However, management of pheretimoid species strongly associated with the horticultural trade may still be possible. Preventing jumping worms from being introduced to new habitats is the most effective method to protect earthworm-free, critical ecosystems (Hendrix and Bohlen 2002; Callaham et al. 2006; Chen et al. 2021).

Horticultural settings and commercial nurseries seem to be the main sources of pheretimoids (Görres et al. 2016; Nouri-Aiin et al. 2022). Organic and chemical vermicides that were used on golf courses to reduce earthworm castings are no longer certified for this use due to their broad target range and high toxicity (Potter et al. 2010; Potter et al. 2011). Physical methods such as prescribed fire have been tested to manage invasive earthworm populations, but are unlikely to be applied on a large scale. For instance, prescribed fire has been shown to reduce cocoon viability; however, it did not significantly affect the number of pheretimoid earthworms (Ikeda et al. 2015). Heating soil to 40 °C for at least 3 days could eliminate both cocoons and earthworms; however, this method is mainly applicable to compost products (Johnston and Herrick 2019). Using soil steamers has not been tried as an earthworm control. Although there is no biological control measure available to apply in the wild (McCay et al. 2020), entomopathogenic fungi such as *Beauveria bassiana and Metarhizium anisopliae* have shown efficacy against two *Amynthas* species (Nouri-Aiin and Görres 2021; Nouri-Aiin 2022).

4 Summary

While earthworms are generally seen as beneficial organisms that can improve soil fertility, their presence in certain parts of the world poses a threat to ecosystem functioning, leading to widespread consequences. Invasive earthworms present systemic challenges for ecosystems, particularly in formerly glaciated temperate and subarctic regions, where they can significantly diminish biodiversity and impact carbon storage in these vast ecosystems. These issues are of global importance, and efforts should focus on preventing the introduction of exotic species into these critical environments. Although controlling wild populations may present challenges, there are available tools to at least hinder their spread through human activity, including quarantine measures for materials from infested areas and/or their heat treatment.

References

Aber JD, Ollinger SV, Driscoll CT (1997) Modeling nitrogen saturation in forest ecosystems in response to land use and atmospheric deposition. Ecol Model 101(1):61–78

Ackerman D, Millet DB, Chen X (2019) Global estimates of inorganic nitrogen deposition across four decades. Glob Biogeochem Cycles 33(1):100–107

Amador JA, Avizinis EJ (2013) Response of nitrous oxide flux to addition of anecic earthworms to an agricultural field

Angst G, Mueller CW, Prater I, Angst Š, Frouz J, Jílková V, Peterse F, Nierop KG (2019) Earthworms act as biochemical reactors to convert labile plant compounds into stabilized soil microbial necromass. Commun Biology 2(1):1–7

Bal TL, Storer AJ, Jurgensen MF (2018) Evidence of damage from exotic invasive earthworm activity was highly correlated to sugar maple dieback in the upper Great Lakes region. Biol Invasions 20(1):151–164

Bellitürk K, Görres JH, Kunkle J, Melnichuk RDS (2015) Can commercial mulches be reservoirs of invasive earthworms? Promotion of ligninolytic enzyme activity and survival of Amynthas agrestis (Goto and Hatai, 1899). Appl Soil Ecol 87:27–31

Belshaw JD (2020) 2.4 the millennia before contact. Canadian History: Pre-Confederation-2nd Edition

Bernard MJ, Neatrour MA, McCay TS (2009) Influence of soil buffering capacity on earthworm growth, survival, and community composition in the Western Adirondacks and Central New York. Northeast Nat 16(2):269–284

Bertrand M, Barot S, Blouin M, Whalen J, de Oliveira T, Roger-Estrade J (2015) Earthworm services for cropping systems. A Review. *Agron Sustain Dev* 35(2):553–567

Bethke PG, Midgley MG (2020) Amynthas spp. impacts on seedlings and forest soils are tree species-dependent. Biol Invasions 22(10):3145–3162

Blackburn TM, Pyšek P, Bacher S, Carlton JT, Duncan RP, Jarošík V, Wilson JR, Richardson DM (2011) A proposed unified framework for biological invasions. Trends Ecol Evol 26(7):333–339

Blakemore RJ (2003) Japanese earthworms (Annelida: Oligochaeta): a review and checklist of species. Org Divers Evol 3(3):241–244

Blakemore RJ (2009) Cosmopolitan earthworms–a global and historical perspective. In: Annelids in modern biology, pp 257–283

Blanchart E, Albrecht A, Alegre J, Duboisset A, Gilot C, Pashanasi B, Lavelle P, Brussaard L (1999) Effects of earthworms on soil structure and physical properties. In: Earthworm Management in Tropical Agroecosystems, vol 5, pp 149–171

Blouin M, Hodson ME, Delgado EA, Baker G, Brussaard L, Butt KR et al (2013) A review of earthworm impact on soil function and ecosystem services. Eur J Soil Sci 64:161–182

Blume-Werry G, Krab EJ, Olofsson J et al (2020) Invasive earthworms unlock arctic plant nitrogen limitation. Nat Commun 11:1766. https://doi.org/10.1038/s41467-020-15568-3

Bohlen PJ, Groffman PM, Fahey TJ, Fisk MC, Suarez E, Pelletier DM, Fahey RT (2004a) Ecosystem consequences of exotic earthworm invasion of north temperate forests. Ecosystems 7:1e12. https://doi.org/10.1007/s10021-003-0126-z

Bohlen PJ, Scheu S, Hale CM, McLean MA, Migge S, Groffman PM, Parkinson D (2004b) Non-native invasive earthworms as agents of change in northern temperate forests. Front Ecol Environ 2:427e435

Borken W, Brumme R (1997) Liming practice in temperate forest ecosystems and the effects on CO_2, N_2O and CH4 fluxes. Soil Use Manag 13:251–257

Borken W, Gründel S, Beese F (2000) Potential contribution of Lumbricus terrestris L. to carbon dioxide, methane and nitrous oxide fluxes from a forest soil. Biol Fertil Soils 32(2):142–148

Bossuyt H, Six J, Hendrix PF (2004) Rapid incorporation of carbon from fresh residues into newly formed stable microaggregates within earthworm casts. Eur J Soil Sci 55(2):393–399

Bossuyt H, Six J, Hendrix PF (2005) Protection of soil carbon by microaggregates within earthworm casts. Soil Biol Biochem 37(2):251–258

Bossuyt H, Six J, Hendrix PF (2006) Interactive effects of functionally different earthworm species on aggregation and incorporation and decomposition of newly added residue carbon. Geoderma 130(1–2):14–25

Bouché MB (1977) Strategies lombriciennes. Ecol Bull 25:122–132

Bowe A, Dobson A, Blossey B (2020) Impacts of invasive earthworms and deer on native ferns in forests of northeastern North America. Biol Invasions 22(4):1431–1445

Brown GG, James SW, Pasini A, Nunes DH, Benito NP, Martins PT, Sautter KD (2006) Exotic, peregrine, and invasive earthworms in Brazil: diversity, distribution, and effects on soils and plants. Caribb J Sci 42(3):339

Burtis JC, Fahey TJ, Yavitt JB (2014) Impact of invasive earthworms on Ixodes scapularis and other litter-dwelling arthropods in hardwood forests, Central New York state, USA. Appl Soil Ecol 84:148–157

Cáceres-Charneco RI, Ransom TS (2010) The influence of habitat provisioning: use of earthworm burrows by the terrestrial salamander, Plethodon cinereus. Popul Ecol 52(4):517–526

Callaham MA, González G, Hale CM, Heneghan L, Lachnicht SL, Zou X (2006) Policy and management responses to earthworm invasions in North America. In: Biological invasions belowground: earthworms as invasive species. Springer, Dordrecht, pp 117–129

Cameron EK, Bayne EM (2012) Invasion by a non-native ecosystem engineer alters distribution of a native predator. Divers Distrib 18(12):1190–1198

Cameron EK, Bayne EM, Clapperton MJ (2007) Human-facilitated invasion of exotic earthworms into northern boreal forests. Ecoscience 14(4):482–490

Cameron EK, Bayne EM, Coltman DW (2008) Genetic structure of invasive earthworms Dendrobaena octaedra in the boreal forest of Alberta: insights into introduction mechanisms. Mol Ecol 17(5):1189–1197

Cameron EK, Shaw CH, Bayne EM, Kurz WA, Kull SJ (2015) Modelling interacting effects of invasive earthworms and wildfire on forest floor carbon storage in the boreal forest. Soil Biol Biochem 88:189–196

Čapek P, Choma M, Tahovská K, Kaňa J, Kopáček J, Šantrůčková H (2021) Coupling the resource stoichiometry and microbial biomass turnover to predict nutrient mineralization and immobilization in soil. Geoderma 385:114884

Chang CH, Johnston MR, Görres JH, Dávalos A, McHugh D, Szlavecz K (2018) Co-invasion of three Asian earthworms, *Metaphire hilgendorfi*, *Amynthas agrestis* and *Amynthas tokioensis* in the USA. Biol Invasions 20(4):843–848

Chang CH, Bartz ML, Brown G, Callaham MA, Cameron EK, Dávalos A, Dobson A, Görres JH, Herrick BM, Ikeda H, James SW (2021) The second wave of earthworm invasions in North America: biology, environmental impacts, management and control of invasive jumping worms. Biol Invasions 23(11):3291–3322

Chen SY, Hsu CH, Soong K (2021) How to cross the sea: testing the dispersal mechanisms of the cosmopolitan earthworm Pontodrilus litoralis. R Soc Open Sci 8(8):202297

Clements RO, Murray PJ, Sturdy RG (1991) The impact of 20 years absence of earthworms and 3 levels of N-fertilizer on a grassland soil. Agric Ecosyst Environ 36:75–85. https://doi.org/10.1016/0167-8809(91)90037-X

Corio K, Wolf A, Draney M, Fewless G (2009) Exotic earthworms of great lakes forests: a search for indicator plant species in maple forests. For Ecol Manag 258(7):1059–1066

Costello DM, Tiegs SD, Lamberti GA (2011) Do non-native earthworms in Southeast Alaska use streams as invasional corridors in watersheds harvested for timber? Biol Invasions 13(1):177–187

Craven D, Thakur MP, Cameron EK, Frelich LE, Beauséjour R, Blair RB, Blossey B, Burtis J, Choi A, Dávalos A, Fahey TJ (2017) The unseen invaders: introduced earthworms as drivers of change in plant communities in North American forests (a meta-analysis). Glob Chang Biol 23(3):1065–1074

Czuzdi C, Chang CH, Pavlícek T, Szederjesi T, Esopi D, Szlávecz K (2017) Molecular phylogeny and systematics of native North American lumbricid earthworms (Clitellata: Megadrili). PLoS One 12(8):e0181504
Ditzler CA, Tugel AJ (2002) Soil quality field tools: experiences of USDA-NRCS soil quality institute. Agron J 94(1):33–38
Dobson A, Blossey B (2015) Earthworm invasion, white-tailed deer and seedling establishment in deciduous forests of north-eastern North America. J Ecol 103(1):153–164
Don A, Steinberg B, Schöning I, Pritsch K, Joschko M, Gleixner G, Schulze ED (2008) Organic carbon sequestration in earthworm burrows. Soil Biol Biochem 40(7):1803–1812
Drake HL, Horn MA (2006) Earthworms as a transient heaven for terrestrial denitrifying microbes: a review. Eng Life Sci 6(3):261–265
Dymond P, Scheu S, Parkinson D (1997) Density and distribution of Dendrobaena octaedra (Lumbricidae) in aspen and pine forests in the Canadian Rocky Mountains (Alberta). Soil Biol Biochem 29(3–4):265–273
Edwards CA, Lofty JR (1980) Effects of earthworm inoculation upon the root growth of direct drilled cereals. J Appl Ecol:533–543
Edwards CA, Bohlen PJ (1996) Biology and ecology of earthworms, vol 3. Springer Science and Business Media
Eggleton P, Inward K, Smith J, Jones DT, Sherlock E (2009) A six year study of earthworm (Lumbricidae) populations in pasture woodland in southern England shows their responses to soil temperature and soil moisture. Soil Biol Biochem 41(9):1857–1865
Eijsackers H (2011) Earthworms as colonizers of natural and cultivated soil environments. Appl Soil Ecol 50:1–13
Eisenhauer N, Partsch S, Parkinson D, Scheu S (2007) Invasion of a deciduous forest by earthworms: changes in soil chemistry, microflora, microarthropods and vegetation. Soil Biol Biochem 39(5):1099–1110
Eisenhauer N, Straube D, Johnson EA, Parkinson D, Scheu S (2009) Exotic ecosystem engineers change the emergence of plants from the seed bank of a deciduous forest. Ecosystems 12(6):1008–1016
Ejack L, Kernecker ML, Prieto R, Chen C, Gul S, Bradley RL, Whalen JK (2021) Earthworms did not increase long-term nitrous oxide fluxes in perennial forage and riparian buffer ecosystems. Pedobiologia 85:150727
Elton CS (1958) The ecology of invasions by animals and plants. Chicago University Press, Chicago, IL, USA
Ferlian O, Eisenhauer N, Aguirrebengoa M, Camara M, Ramirez-Rojas I, Santos F, Tanalgo K, Thakur MP (2018) Invasive earthworms erode soil biodiversity: a meta-analysis. J Anim Ecol 87(1):162–172
Ferlian O, Thakur MP, Castañeda González A, San Emeterio LM, Marr S, da Silva Rocha B, Eisenhauer N (2020) Soil chemistry turned upside down: a meta-analysis of invasive earthworm effects on soil chemical properties. Ecology 101(3):e02936
Field SG, Michiels NK (2005) Parasitism and growth in the earthworm Lumbricus terrestris: fitness costs of the gregarine parasite Monocystis sp. Parasitology 130(4):397–403
Fleri JR, Martin TG, Rodewald AD, Arcese P (2021) Non-native earthworms alter the assembly of a meadow plant community. Biol Invasions 23(8):2407–2415
Frelich LE, Hale CM, Reich PB, Holdsworth AR, Scheu S, Heneghan L, Bohlen PJ (2006) Earthworm invasion into previously earthworm-free temperate and boreal forests. In: Biological invasions belowground: earthworms as invasive species. Springer, Dordrecht, pp 35–45
Frelich LE, Blossey B, Cameron EK, Dávalos A, Eisenhauer N, Fahey T, Ferlian O, Groffman PM, Larson E, Loss SR, Maerz JC (2019) Side-swiped: ecological cascades emanating from earthworm invasions. Front Ecol Environ 17(9):502–510
Garnier P, Makowski D, Hedde M, Bertrand M (2022) Changes in soil carbon mineralization related to earthworm activity depend on the time since inoculation and their density in soil. Sci Rep 12(1):1–10

Gates GE (1956) Reproductive organ polymorphism in earth-worms of the oriental megascolecine genus Pheretima Kinberg 1867. Evolution 10(2):213–227

Gates GE (1963) Miscellanea Megadrilogica VII. Greenhouse earthworms. Proc Biol Soc Wash 76(99):9–18

Giannopoulos G, Pulleman MM, Van Groenigen JW (2010) Interactions between residue placement and earthworm ecological strategy affect aggregate turnover and N_2O dynamics in agricultural soil. Soil Biol Biochem 42(4):618–625

Görres JH, Savin MC, Amador JA (2001) Soil micropore structure and carbon mineralization in burrows and casts of an anecic earthworm (Lumbricus terrestris). Soil Biol Biochem 33(14):1881–1887

Görres JH, Melnichuk RD (2012) Asian invasive earthworms of the genus Amynthas Kinberg in Vermont. Northeastern Nat 19(2):313–322

Görres JH, Bellitürk K, Keller E (2014) Failure of an Amynthas agrestis (Goto and Hatai, 1899) (Oligochaeta: Megascolecidae) population to expand its range within a sugar maple (Acer saccharum) stand. Megadrilogica 17(1):7–13

Görres JH, Bellitürk K, Melnichuk RD (2016) Temperature and moisture variables affecting the earthworms of genus Amynthas Kinberg, 1867 (Oligochaeta: Megascolecidae) in a hardwood forest in the Champlain Valley, Vermont, USA. Appl Soil Ecol 104:111–115

Görres JH, Connolly ST, Chang CH, Carpenter NR, Keller EL, Nouri-Aiin M, Schall JJ (2018) Winter hatching in New England populations of invasive pheretimoid earthworms *Amynthas agrestis* and *Amynthas tokioensis*: a limit on population growth, or aid in peripheral expansion? Biol Invasions 20(7):1651–1655

Görres JH, Martin C, Nouri-Aiin M, Bellitürk K (2019) Physical properties of soils altered by invasive pheretimoid earthworms: does their casting layer create thermal refuges? Soil Syst 3(3):52

Greiner HG, Kashian DR, Tiegs SD (2012) Impacts of invasive Asian (Amynthas hilgendorfi) and European (Lumbricus rubellus) earthworms in a North American temperate deciduous forest. Biol Invasions 14:2017–2027

Griffith B, Türke M, Weisser WW, Eisenhauer N (2013) Herbivore behavior in the anecic earthworm species Lumbricus terrestris L.? Eur J Soil Biol 55:62–65

Groffman PM, Fahey TJ, Fisk MC, Yavitt JB, Sherman RE, Bohlen PJ, Maerz JC (2015) Earthworms increase soil microbial biomass carrying capacity and nitrogen retention in northern hardwood forests. Soil Biol Biochem 87:51–58

Haimi J, Huhta V (1990) Effect of earthworms on decomposition processes in raw humus forest soil: a microcosm study. Biol Fertil Soils 10:178–183

Hale CM (2008) Evidence for human-mediated dispersal of exotic earthworms: support for exploring strategies to limit further spread

Hale CM, Frelich LE, Reich PB, Pastor J (2005) Effects of European earthworm invasion on soil characteristics in northern hardwood forests of Minnesota, USA. Ecosystems 8(8):911–927

Hale CM, Frelich LE, Reich PB (2006) Changes in hardwood forest understory plant communities in response to European earthworm invasions. Ecology 87(7):1637–1649

Hale CM, Frelich LE, Reich PB, Pastor J (2008) Exotic earthworm effects on hardwood forest floor, nutrient availability and native plants: a mesocosm study. Oecologia 155(3):509–518

Hansen UL (2000) Chieftains and traders looking east-the late Roman and early Germanic period contacts between the eastern Baltic and the southern Scandinavian area. Archaeologia Baltica 4:81–88

Hendrix PF, Bohlen PJ (2002) Exotic earthworm invasions in North America: ecological and policy implications: expanding global commerce may be increasing the likelihood of exotic earthworm invasions, which could have negative implications for soil processes, other animal and plant species, and importation of certain pathogens. Bioscience 52(9):801–811

Hendrix PF, Callaham MA Jr, Drake JM, Huang CY, James SW, Snyder BA, Zhang W (2008) Pandora's box contained bait: the global problem of introduced earthworms. Annu Rev Ecol Evol Syst 39(1):593–613

Holdsworth AR, Frelich LE, Reich PB (2007) Effects of earthworm invasion on plant species richness in northern hardwood forests. Conserv Biol 21(4):997–1008

Holmstrup M (2002) Strategies for cold and drought tolerance in permeable soil invertebrates. Ministry of Environment. National Environmental Research Institute

Holmstrup M, Overgaard J (2007) Freeze tolerance in Aporrectodea caliginosa and other earthworms from Finland. Cryobiology 55(1):80–86

Huang W, González G, Zou X (2020) Earthworm abundance and functional group diversity regulate plant litter decay and soil organic carbon level: a global meta-analysis. Appl Soil Ecol 150:103473

Hulton NRJ, Purves RS, McCulloch RD, Sugden DE, Bentley MJ (2002) The last glacial maximum and deglaciation in southern South America. Quat Sci Rev 21:233–241

Ikeda H, Callaham MA Jr, O'Brien JJ, Hornsby BS, Wenk ES (2015) Can the invasive earthworm, Amynthas agrestis, be controlled with prescribed fire? Soil Biol Biochem 82:21–27

James SW, Hendrix PF (2004) Invasion of exotic earthworms into North America and other regions. Earthworm Ecol 441:75–88

Janion-Scheepers C, Griffiths CL (2020) Alien terrestrial invertebrates in South Africa. Biol Invasions in South Africa 14:185

Jimenez JJ, Brown GG, Decaëns T, Feijoo A, Lavelle P (2001) Chapter 9: differences in the timing of diapause and patterns of aestivation in tropical earthworms. Jimenez, JJ and RJ Thomas

Jochum M, Ferlian O, Thakur MP, Ciobanu M, Klarner B, Salamon JA, Frelich LE, Johnson EA, Eisenhauer N (2021) Earthworm invasion causes declines across soil fauna size classes and biodiversity facets in northern North American forests. Oikos 130(5):766–780

Jochum M, Thouvenot L, Ferlian O, Zeiss R, Klarner B, Pruschitzki U, Johnson EA, Eisenhauer N (2022) Aboveground impacts of a belowground invader: how invasive earthworms alter aboveground arthropod communities in a northern North American forest. Biol Lett 18(3):20210636

Johnson DM, Gale KM, Dobson AM, McCay TS (2021) Public reporting and perception of invasive Pheretimoid "jumping worms" in the northeastern United States. Northeast Nat 2(3):383–396

Johnston MR, Herrick BM (2019) Cocoon heat tolerance of pheretimoid earthworms *Amynthas tokioensis* and *Amynthas agrestis*. Am Midl Nat 181(2):299–309

Jouquet P, Thi PN, Hong HN, Henry-des-Tureaux T, Chevallier T, Duc TT (2011) Laboratory investigation of organic matter mineralization and nutrient leaching from earthworm casts produced by Amynthas khami. Appl Soil Ecol 47(1):24–30

Keen SC, Wackett AA, Willenbring JK, Yoo K, Jonsson H, Clow T, Klaminder J (2022) Non-native species change the tune of tundra soils: novel access to soundscapes of the Arctic earthworm invasion. Sci Total Environ 838:155976

Keller RP, Cox AN, Van Loon C, Lodge DM, Herborg LM, Rothlisberger J (2007) From bait shops to the forest floor: earthworm use and disposal by anglers. Am Midl Nat 158(2):321–328

Keller EL, Schall JJ (2024) Variation and trade-offs in life history traits of the protist parasite Monocystis perplexa (Apicomplexa) in its earthworm host Amynthas agrestis. PeerJ 12:e17161

Kilian JV, Klauda RJ, Widman S, Kashiwagi M, Bourquin R, Weglein S, Schuster J (2012) An assessment of a bait industry and angler behavior as a vector of invasive species. Biol Invasions 14(7):1469–1481

Kim YN, Robinson B, Boyer S, Zhong HT, Dickinson N (2015) Interactions of native and introduced earthworms with soils and plant rhizospheres in production landscapes of New Zealand. Appl Soil Ecol 96:141–150

Klaminder J, Krab EJ, Larsbo M, Jonsson H, Fransson J, Koestel J (2023) Holes in the tundra: invasive earthworms alter soil structure and moisture in tundra soils. Sci Total Environ 859:160125

Knowles ME, Ross DS, Görres JH (2016) Effect of the endogeic earthworm Aporrectodea tuberculata on aggregation and carbon redistribution in uninvaded forest soil columns. Soil Biol Biochem 100:192–200

Kuitems M, Wallace BL, Lindsay C et al (2022) Evidence for European presence in the Americas in ad 1021. Nature 601:388–391. https://doi.org/10.1038/s41586-021-03972-8

Langmaid KJ (1963) The Sea raiders. Jarrolds

Langmaid, K.K., 1964. Some effects of earthworm invasion in virgin podzols. Can. J. Soil Sci. 44, 34–37. https://doi.org/10.4141/cjss64-005

Laushman KM, Hotchkiss SC, Herrick BM (2018) Tracking an invasion: community changes in hardwood forests following the arrival of *Amynthas agrestis* and *Amynthas tokioensis* in Wisconsin. Biol Invasions 20(7):1671–1685

Lawrence B, Fisk MC, Fahey TJ, Suárez ER (2003) Influence of nonnative earthworms on mycorrhizal colonization of sugar maple (Acer saccharum). New Phytol 157:145–153

Lejoly J, Quideau S, Laganière J (2021) Invasive earthworms affect soil morphological features and carbon stocks in boreal forests. Geoderma 404:115262

Lemtiri A, Colinet G, Alabi T, Cluzeau D, Zirbes L, Haubruge É, Francis F (2014) Impacts of earthworms on soil components and dynamics. A review. Biotechnologie, Agronomie, Société et Environnement 18

Loss SR, Blair RB (2011) Reduced density and nest survival of ground-nesting songbirds relative to earthworm invasions in northern hardwood forests. Conserv Biol 25(5):983–992

Loss SR, Niemi GJ, Blair RB (2012) Invasions of non-native earthworms related to population declines of ground-nesting songbirds across a regional extent in northern hardwood forests of North America. Landsc Ecol 27(5):683–696

Lubbers IM, Van Groenigen KJ, Fonte SJ, Six J, Brussaard L, Van Groenigen JW (2013) Greenhouse-gas emissions from soils increased by earthworms. Nat Clim Chang 3(3):187–194

Lubbers IM, Pulleman MM, Van Groenigen JW (2017) Can earthworms simultaneously enhance decomposition and stabilization of plant residue carbon? Soil Biol Biochem 105:12–24

Lv B, Zhang D, Cui Y, Yin F (2018) Effects of C/N ratio and earthworms on greenhouse gas emissions during vermicomposting of sewage sludge. Bioresour Technol 268:408–414

Maerz JC, Nuzzo VA, Blossey B (2009) Declines in woodland salamander abundance associated with non-native earthworm and plant invasions. Conserv Biol 23(4):975–981

Marinissen JCY, Van den Bosch F (1992) Colonization of new habitats by earthworms. Oecologia 91(3):371–376

Mattsson BJ, Niemi GJ (2006) Factors influencing predation on ovenbird (Seiurus aurocapilla) nests in northern hardwoods: interactions across spatial scales. Auk 123(1):82–96

McAlpine DF, Reynolds JW, Manzer LR, Elton K (2022) First reports of invasive pheretimoid earthworms (Oligochaeta: Megascolecidae) of Asian origin in Atlantic Canada. Bioinvasions Rec 11(4):830–838

McCay TS, Scull P (2019) Invasive lumbricid earthworms in northeastern north American forests and consequences for leaf-litter fauna. Biol Invasions 21(6):2081–2093

McCay TS, Pinder RA, Alvarado E, Hanson WC (2017) Distribution and habitat of the endemic earthworm Eisenoides lonnbergi (Michaelsen) in the northeastern United States. Northeast Nat 24(3):239–248

McCay TS, Brown G, Callaham MA Jr, Chang CH, Dávalos A, Dobson A, Görres JH, Herrick BM, James SW, Johnston MR, McHugh D (2020) Tools for monitoring and study of peregrine pheretimoid earthworms (Megascolecidae). Pedobiologia 83:150669

McLean MA, Migge-Kleian S, Parkinson D (2006) Earthworm invasions of ecosystems devoid of earthworms: effects on soil microbes. Biol Invasions 8:1257–1273

Medina-Sauza RM, Álvarez-Jiménez M, Delhal A, Reverchon F, Blouin M, Guerrero-Analco JA, Cerdán CR, Guevara R, Villain L, Barois I (2019) Earthworms building up soil microbiota, a review. Front Environ Sci 7:81

Melnichuk RDS (2016) Earthworms in Vermont forest soils: a study of nutrient, carbon, nitrogen and native plant responses. The University of Vermont and State Agricultural College

Meshcheryakova EN, Berman DI (2014) Cold hardiness and geographic distribution of earthworms (Oligochaeta, Lumbricidae, Moniligastridae). Entomol Rev 94(4):486–497

Moore JD, Ouimet R, Bohlen PJ (2013) Effects of liming on survival and reproduction of two potentially invasive earthworm species in a northern forest Podzol. Soil Biol Biochem 64:174–180

Moore JD, Görres JH, Reynolds JW (2018) Exotic Asian pheretimoid earthworms (Amynthas spp., Metaphire spp.): potential for colonisation of South-Eastern Canada and effects on forest ecosystems. Environ Rev 26(2):113–120

Moreau-Valancogne P, Bertrand M, Holmstrup M, Roger-Estrade J (2013) Integration of thermal time and hydrotime models to describe the development and growth of temperate earthworms. Soil Biol Biochem 63:50–60

Nouri-Aiin M (2022) Elements of biocontrol strategies for Pheretimoid earthworms

Nouri-Aiin M, Görres JH (2019) Earthworm cocoons: the cryptic side of invasive earthworm populations. Appl Soil Ecol 141:54–60

Nouri-Aiin M, Görres JH (2021) Biocontrol of invasive pheretimoid earthworms using Beauveria bassiana. PeerJ 9:e11101

Nouri-Aiin M, Connolly S, Keough C, Smigelsky AJ, Wen Y, Howland J, Schall JJ, Görres JH (2022) Genetic population structure and reproductive system of two invasive Asian earthworms, *Amynthas tokioensis* and *Amynthas agrestis*. PeerJ 10:e13622

Nuzzo VA, Maerz JC, Blossey B (2009) Earthworm invasion as the driving force behind plant invasion and community change in northeastern north American forests. Conserv Biol 23(4):966–974

Nuzzo V, Dávalos A, Blossey B (2015) Invasive earthworms shape forest seed bank composition. Divers Distrib 21(5):560–570

Omernik JM, Griffith GE (2014) Ecoregions of the conterminous United States: evolution of a hierarchical spatial framework. Environ Manag 54(6):1249–1266

Paudel S, Longcore T, MacDonald B, McCormick MK, Szlavecz K, Wilson GWT, Loss SR (2016) Belowground interactions with aboveground consequences: invasive earthworms and arbuscular mycorrhizal fungi. Ecology 97:605–614

Piearce TG (1972) The calcium relations of selected Lumbricidae. J Anim Ecol 41:167–188

Pitkänen J, Nuutinen V (1998) Earthworm contribution to infiltration and surface runoff after 15 years of different soil management. Appl Soil Ecol 9:411–415. https://doi.org/10.1016/S0929-1393(98)00098-5

Plum NM, Filser J (2005) Floods and drought: response of earthworms and potworms (Oligochaeta: Lumbricidae, Enchytraeidae) to hydrological extremes in wet grassland. Pedobiologia 49(5):443–453

Potter DA, Redmond CT, Meepagala KM, Williams DW (2010) Managing earthworm casts (Oligochaeta: Lumbricidae) in turfgrass using a natural byproduct of tea oil (Camellia sp.) manufacture. *Pest management science: formerly*. Pestic Sci 66(4):439–446

Potter DA, Redmond CT, Williams DW (2011) The worm turns: earthworm cast reduction on golf courses. Golf Course Manag 79(9):86–96

Prasanth Narayanan S, Sathrumithra S, Christopher G, Thomas AP, Julka JM (2016) Current distribution of the invasive earthworm Pontoscolex corethrurus (Müller, 1857) after a century of its first report from Kerala state, India. *Opuscula Zoologica (Budapest)* 47(1):101–107

Ransom TS (2011) Earthworms, as ecosystem engineers, influence multiple aspects of a salamander's ecology. Oecologia 165(3):745–754

Ransom TS (2012) Comparison of direct, indirect, and ecosystem engineering effects of an earthworm on the red-backed salamander. Ecology 93(10):2198–2207

Resner K, Yoo K, Hale C et al (2015) Invasive earthworms deplete key soil inorganic nutrients (Ca, Mg, K and P) in a northern hardwood forest. Ecosystems 18:89–102

Reynolds JW (1994) Earthworms of the world. Global Biodiver 4(1):11–16

Reynolds JW (2016) Earthworms (Oligochaeta: lumbricidae and megascolecidae) in the central basin and range ecoregion (13), USA. Megadrilogica 21(9):170–175

Reynolds JW (2017) A summary of the status of earthworms (Annelida: Oligochaeta) in ecoregions of the United States

Reynolds JW (2020) Earthworms (Annelida: Oligochaeta) in Canadian ecozones. Megadrilogica 26(1):1–35

Reynolds JW, McTavish MJ (2021) New Asian Pheretimoid "jumping earthworm" records (Clitellata: Oligochaeta: Megascolecidae) In Canada
Řezáč M, Tessler S, Heneberg P, Herrera IMÁ, Gloríková N, Forman M, Řezáčová V, Král J (2022) Atypus karschi Dönitz, 1887 (Araneae: Atypidae): an Asian purse-web spider established in Pennsylvania, USA. PLoS One 17(7):e0261695
Rizhiya E, Bertora C, van Vliet PC, Kuikman PJ, Faber JH, van Groenigen JW (2007) Earthworm activity as a determinant for N_2O emission from crop residue. Soil Biol Biochem 39(8):2058–2069
Roots BI (1956) The water relations of earthworms: II. Resistance to desiccation and immersion, and behaviour when submerged and when allowed a choice of environment. J Exp Biol 33(1):29–44
Schall JJ (2021) Stomatocystis goerresi, a new species of gregarine parasite (Apicomplexa, Monocystidae) from the invasive Japanese earthworm Amynthas tokioensis (Megascolecidae), with a description of the parasite's life cycle. Folia Parasitol 68
Schall JJ, Nouri-Aiin M, Görres J (2023) Apolocystis bosanqueti n. sp.(Apicomplexa: Eugregarinorida) from the invasive earthworm Amynthas agrestis (Annelida: Megascolecidae), with significance for the monophyly of the family Monocystidae. J Parasitol 109(2):56–64
Schneider AK, Hohenbrink TL, Reck A, Zangerle A, Schroeder B, Zehe E, van Schaik L (2018) Variability of earthworm-induced biopores and their hydrological effectiveness in space and time. Pedobiologia 71:8–19
Schwert DP (1979) Description and significance of a fossil earthworm (Oligochaeta: Lumbricidae) cocoon from postglacial sediments in southern Ontario. Can J Zool 57(7):1402–1405
Schwert DP (1980) Active and passive dispersal of lumbricid earthworms. *Int Soil Zool Colloquium* 7(29):182–189
Shen HP, Tsai CF, Fang YP, Chen JH (2011) Parthenogenesis, polyploidy and reproductive seasonality in the Taiwanese mountain earthworm Amynthas catenus (Oligochaeta, Megascolecidae). Pedobiologia 54(2):133–139
Shipitalo MJ, Nuutinen V, Butt KR (2004) Interaction of earthworm burrows and cracks in a clayey, subsurface-drained, soil. Appl Soil Ecol 26(3):209–217
Shuster WD, Shipitalo MJ, Subler S, Aref S, McCoy EL (2003) Earthworm additions affect leachate production and nitrogen losses in typical midwestern agroecosystems. J Environ Qual 32(6):2132–2139
Singh J, Schädler M, Demetrio W, Brown GG, Eisenhauer N (2019) Climate change effects on earthworms-a review. Soil Org 91(3):114
Smith VR, Steenkamp M (1992) Soil macrofauna and nitrogen on a sub-Antarctic Island. Oecologia 92:201–206
Speratti AB, Whalen JK (2008) Carbon dioxide and nitrous oxide fluxes from soil as influenced by anecic and endogeic earthworms. Appl Soil Ecol 38(1):27–33
Stahl PW (2009) Adventive vertebrates and historical ecology in the pre-Columbian Neotropics. Diversity 1(2):151–165
Steckley J (2023) Nightcrawler commodities: how capital subsumes nature and labour in Ontario's underground bait worm industry (Doctoral dissertation, University of Toronto (Canada))
Svensson BH, Boström U, Klemedtson L (1986) Potential for higher rates of denitrification in earthworm casts than in the surrounding soil. Biol Fertil Soils 2(3):147–149
Tecimen HB, Gorres JH, Melnichuk RD (2021) Effect of Lumbricus rubellus and Amynthas agrestis earthworms on soil biogeochemistry at the aggregate scale in northern hardwood forests. J Sustain For 40(1):83–98
Thakuria D, Schmidt O, Finan D, Egan D, Doohan FM (2010) Gut wall bacteria of earthworms: a natural selection process. ISME J 4(3):357–366
Thomas E, Prabha VS, Kurien VT, Thomas AP (2020) The potential of earthworms in soil carbon storage: a review. Environ Exp Biol 18(2):61–75
Tilikj N, Novo M (2022) How to resist soil desiccation: transcriptional changes in a Mediterranean earthworm during aestivation. Comp Biochem Physiol Part A: Mol Integr Physiol 264:111112

Tiunov AV, Hale CM, Holdsworth AR, Vsevolodova-Perel TS (2006) Invasion patterns of Lumbricidae into the previously earthworm-free areas of northeastern Europe and the western Great Lakes region of North America. In: Biological invasions belowground: earthworms as invasive species. Springer, Dordrecht, pp 23–34. https://doi.org/10.1007/s10530-006-9018-4

Trap J, Blanchart E, Ratsiatosika O, Razafindrakoto M, Becquer T, Andriamananjara A, Morel C (2021) Effects of the earthworm Pontoscolex corethrurus on rice P nutrition and plant-available soil P in a tropical Ferralsol. Appl Soil Ecol 160:103867

USDA-Aphis (2022) Earthworm import policy. https://www.aphis.usda.gov/aphis/newsroom/stakeholder-info/stakeholder-messages/plant-health-news/earthworm-import-policy-postpone-october. Accessed 22 Dec 2022

US-USDA (2022) Plants for planting manual. https://www.aphis.usda.gov/importexport/plants/manuals/ports/downloads/plants_for_planting.pdf. Accessed 21 Dec 2022

Van Groenigen JW, Lubbers IM, Vos HM, Brown GG, De Deyn GB, Van Groenigen KJ (2014) Earthworms increase plant production: a meta-analysis. Sci Rep 4(1):1–7

Van Kessel C, Clough T, van Groenigen JW (2009) Dissolved organic nitrogen: an overlooked pathway of nitrogen loss from agricultural systems? J Environ Qual 38(2):393–401

Vidal A, Watteau F, Remusat L, Mueller CW, Nguyen Tu TT, Buegger F, Derenne S, Quenea K (2019) Earthworm cast formation and development: a shift from plant litter to mineral associated organic matter. Front Environ Sci 7:55

Villholth KG, Jarvis NJ, JacobsenOH, de Jonge H (2000) *Field investigations and modeling of particle-facilitated pesticide transport in macroporous soil* (Vol. 29, no. 4, pp. 1298–1309). American Society of Agronomy, Crop Science Society of America, and Soil Science Society of America

Wackett AA, Yoo K, Olofsson J, Klaminder J (2018) Human-mediated introduction of geoengineering earthworms in the Fennoscandian arctic. Biol Invasions 20(6):1377–1386

Wallace AR (1911) The world of life. Рипол Классик

Wang Y, Chen J, Gu W, Xu Y, Gu J, Tao J (2016) Earthworm activities increase the leaching of salt and water from salt-affected agricultural soil during the wet–dry process under simulated rainfall conditions. Biol Fertil Soils 52(3):323–330

Wardle DA, Bardgett RD, Callaway RM, Van der Putten WH (2011) Terrestrial ecosystem responses to species gains and losses. Science 332(6035):1273–1277

Wilson CM, Smart PL (1984) Pipes and pipe flow process in an upland catchment, Wales. Catena 11(1):145–158

Wu J, Li H, Zhang W, Li F, Huang J, Mo Q, Xia H (2017) Contrasting impacts of two subtropical earthworm species on leaf litter carbon sequestration into soil aggregates. J Soils Sediments 17(6):1672–1681

Zhang W, Hendrix PF, Dame LE, Burke RA, Wu J, Neher DA, Li J, Shao Y, Fu S (2013) Earthworms facilitate carbon sequestration through unequal amplification of carbon stabilization compared with mineralization. Nat Commun 4(1):1–9

Earthworms in Canadian Forest Regions Revisited

John Warren Reynolds

Abstract Approximately 30% of the world's boreal forests and 10% of all forests worldwide are found in Canada, which has 43% of its area covered by forests. Canada is divided into eight woodland zones in addition to tundra and grasslands. Earthworm diversity, function, and a description will be provided for each forest area. A summary table will list the earthworm species with ecological type, forest regions, dominant tree species, and soil type in both the Canadian and American classification systems. A second table will illustrate the earthworm species with just the forest regions. Provided are a soil order map of Canada and a table indicating the different earthworm species prevalent in different forest soils. The appendix describes and pictures soil type profiles where earthworms have been recorded. Migration of arctic earthworms and climate change are topics of discussion.

RÉSUMÉ Les forêts canadiennes couvrent 43% de sa masse continentale, ce qui représente 10% des forêts mondiales et 30% des forêts boréales mondiales. Il y a huit régions forestières au Canada, plus les prairies et la toundra. Chaque région forestière contiendra une description, une carte, la diversité et la fonction des vers de terre. Un tableau récapitulatif listera les espèces de vers de terre avec le type écologique, les régions forestières, les espèces d'arbres dominantes, le type de sol dans les systèmes de classification canadien et américain. Un deuxième tableau illustrera les espèces de vers de terre avec seulement les régions forestières. Une carte des ordres pédologiques au Canada et un tableau illustrant les espèces de vers de terre dans divers sols forestiers. L'annexe décrit et illustre les profils de type de sol où les vers de terre ont été enregistrés. La discussion inclut la migration des vers de terre arctiques et le changement climatique.

J. W. Reynolds (✉)
Oligochaetology Laboratory, Kitchener, ON, Canada

New Brunswick Museum, Saint John, NB, Canada

Y. Kooch, Y. Kuzyakov (eds.), *Earthworms and Ecological Processes*,
https://doi.org/10.1007/978-3-031-64510-5_2

1 Introduction

Forests play a crucial role in Canada. Encompassing 43% of Canada's geographical area, these regions play a substantial role in the nation's economy, as well as its environmental and social welfare. Canadian woods account for 10% of the world's forest cover, 30% of the world's boreal forest, and 23% of the planet's temperate rainforest. Earthworms possess the capacity to alter basic characteristics of soil, resulting in significant repercussions that extend to ecosystem biodiversity and functioning. The extensive infiltration of non-native earthworms into Canada's woods could consequently have worldwide ramifications (Addison 2009).

A substantial geographical belt or area that exhibits widespread consistency in both the physiography and the composition of the main tree species is referred to as a forest area. There are eight regions of forest in Canada. Although they are not genuine forests, tundra and grasslands occasionally appear as the nine and ten forest zones.

Most earthworms in North America are thought to have been introduced relatively recently, with the majority living in soils in Canada and the northern United States. The majority are thought to have perished during the late Wisconsin stage of the Pleistocene glacier (Gates 1970; Reynolds 1994). Soil, vegetation, and carbon and nutrient pools in Canada have all formed without earthworms, with the possible exception of a few small sites where remnant populations of native earthworms have persisted.

2 Canadian Forest Regions

There are eight major forest regions in Canada: Acadian, Boreal (three subtypes), Coast, Columbian, Deciduous/Carolinian, Great Lakes-St. Lawrence, Montane, and Subalpine (Fig. 1).

Prior to recent times, the majority of studies conducted in Canada on earthworms have concentrated on farming areas, where they are believed to have a broadly positive impact on the structure and fertility of the soil (e.g., Tomlin and Fox 2003; Clapperton et al. 1997). It has been verified by biogeographical publications—mostly developed by Reynolds and his associates—that foreign earthworms are spreading extensively throughout Canada. Nearly half a century ago, Reynolds (1976b) recorded the existence of non-native earthworms in North American woodlands, including several types of forests found in Canada, e.g., Acadian (Acadienne), Boreal (Boréale), Deciduous/Carolinian, Great Lakes-St. Lawrence (Caduque Septentrionale), Coast and Columbian (Côtière et Columbia), and Montane and Subalpine (Montagnarde).

In subsequent years, I have continued to study the earthworms in various Canadian Forest Regions: Acadian Forest (Reynolds 1975a, 1975d, 1976a, 2001; Reynolds et al. 2015; McAlpine et al. 2022), Boreal Forest (Reynolds 2000a, 2000b,

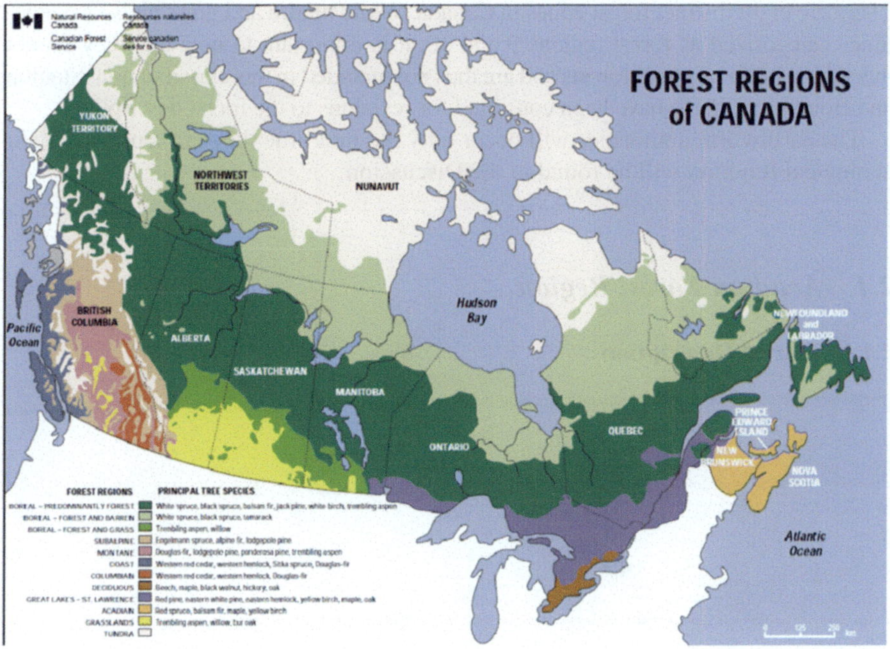

Fig. 1 Forest regions of Canada (after Rowe 1972; Natural Resources Canada 2007)

1975b, 1975c; Moore and Reynolds 2003; Moore et al. 2009; Reynolds et al. 2014, 2019), Deciduous/Carolinian Forest (Reynolds 2011a), and the Great Lakes-St. Lawrence Forest (Reynolds 1975b, 1976a, 1977a, 2007; Reynolds and Mayville 1994; Reynolds and Reynolds 1992).

Recently, ecologists and regulatory organizations have begun to show greater interest in the distribution and activity of non-native earthworms, particularly those that are encroaching onto native habitats.

There have been a number of other colleagues who have surveyed earthworms in various Canadian Forest Regions. Addison (2009) presented an excellent summary about earthworms in forest types, but her paper was metadata relying heavily on numerous surveys by Reynolds and colleagues (Moore and Reynolds 2003; Reynolds 1977a, 2000a, 2000b, 2001, 2002, 2003; Reynolds and Clapperton 1996; Reynolds and Khan 1999; Reynolds and Moore 1996; Reynolds and Reynolds 1992).

In western Canada, studies examining earthworms in forest regions were conducted by: McKey-Fender and Fender (1982), Spiers et al. (1986), McKey-Fender et al. (1994), Dymond et al. (1997), McLean and Parkinson (1997, 2000a, 2000b), Scheu and McLean (1993), Scheu and Parkinson (1994), and Cameron et al. (2007).

In eastern Canada, studies examining earthworms in forest regions not involving Reynolds were conducted by: Bohlen et al. (2004), Whalen (2004), Wironen and Moore (2006), and Sackett et al. (2013).

A forest zone is a significant geographical area distinguished by consistent characteristics in both the physical landscape and the types of trees that are most prevalent.

There are eight distinct forest zones in Canada. While tundra and grasslands are sometimes categorized as forest regions 9 and 10, it is important to note that they do not meet the criteria of actual forests and are thus not included in this classification. Stanton and Bourchier (2015) have been consulted for revisions to the forest descriptions.

The earthworm authorities will occur only the first time they are reported. Their ecological function will be found in the Discussion.

2.1 *Acadian Forest Region*

2.1.1 Forest Description

This area is limited to Nova Scotia, Prince Edward Island, and a significant portion of New Brunswick. It is closely associated with the Great Lakes-St Lawrence Forest area. The following trees are frequently encountered: sugar maple (*Acer saccharum* Marshall), red spruce (*Picea rubens* Sargent), balsam fir (*Abies balsamea* (L.) Miller). The following trees are found in wide distribution: black spruce (*Picea mariana* (Miller) Britton, Sterns, and Poggenburg), white and gray birch (*Betula papyrifera* Marshall and *Betula populifolia* Marshall, respectively), red oak (*Quercus rubra* L.), white elm (*Ulmus americana* L.), black ash (*Fraxinus nigra* Marshall), beech (*Fagus grandifolia* Ehrhart), red maple (*Acer rubrum* L. 1753), balsam poplar (*Populus balsamifera* L.), and Ulmus americana L.

2.1.2 Earthworm Diversity and Function

The earthworms in the Acadian Forest Region were presented in a paper on the Atlantic Maritime Ecoregion, which included areas outside the Acadian Forest Region (Reynolds 2010a). This woodland zone is located in the Magdalene Islands of Quebec as well as three provinces: Prince Edward Island, New Brunswick, and Nova Scotia. In all three provinces only introduced European lumbricids make up the earthworm's populations (McAlpine and Reynolds 2018, 2019, 2020, McAlpine et al. 2020; Reynolds, 1975a, 1975c, 1975d, 1976a, 2001, 2010b, 2014b; Reynolds et al. 2015).

The following earthworm species were found in forests in all four jurisdictions: *Aporrectodea trapezoides* (Dugès 1828), *Aporrectodea tuberculata* (Eisen 1874),

Aporrectodea turgida (Eisen 1874), *Dendrobaena octaedra* (Savigny 1826), *Dendrodrilus rubidus* (Savigny 1826), *Lumbricus castaneus* (Savigny 1826), and *Lumbricus rubellus* Hoffmeister, 1843. *Aporrectodea rosea* (Savigny 1826) and *Lumbricus terrestris* Linnaeus, 1758 were collected in the Acadian Forest of Prince Edward Island and the Magdalene Islands. In the forests of New Brunswick and Nova Scotia, two additional species were collected: *Octolasion cyaneum* (Savigny 1826) and *Octolasion tyrtaeum* (Savigny 1826). *Dendrobaena lusitana* Graff, 1957 and *Lumbricus festivus* (Savigny 1826) were only found in the Acadian Forest of New Brunswick. There are five species found in some of the jurisdictions, but not in the Acadian Forests: *Allolobophora chlorotica* (Savigny 1826), *Eisenia fetida* (Savigny 1826), *Eiseniella tetraedra* (Savigny 1826), *Octolasion cyaneum* (Savigny 1826), and *Sparganophilus tamesis* (Benham 1892)(= Sp. *eiseni*). You can find Reynolds' distributions in detail (2010a, 2014b). The earthworm authorities will occur only the first time they are reported. Their ecological function will be found in the Discussion.

2.2 *Boreal/Taiga Forest Region*

2.2.1 Forest Description

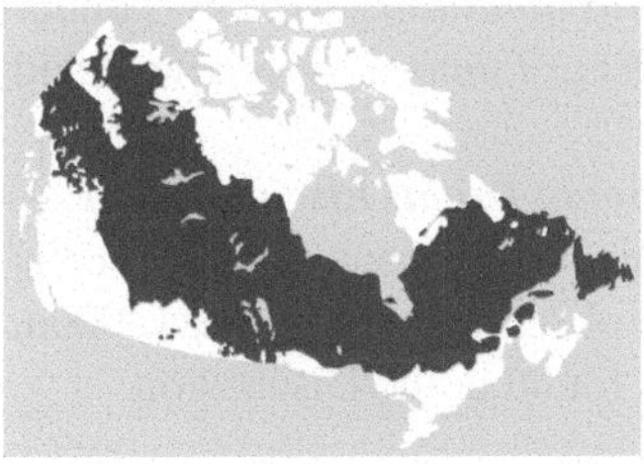

The vast Boreal Forest Region, which stretches south from the Mackenzie River Delta and Alaskan border to northeast British Columbia, across northern Alberta and Saskatchewan, through Manitoba, Ontario, and Québec, and terminates in northern Newfoundland on the Labrador Sea coastlines, contains roughly 80% of all of Canada's forest-covered land. The sparse forest that characterizes the northern boreal zone is where trees are smaller and more widely spread out as they grow toward the tundra, where only trees of lesser size can survive.

In the southern boreal area, there is a dense forest. Toward the southwest border in the prairie provinces (Alberta, Manitoba, and Saskatchewan), this forest transitions into a zone where poplar trees dominate. This section of the forest, referred to as the aspen grove, gradually becomes less dense and transforms into a vast plain with very few trees. The main species of the predominant coniferous boreal forest are white and black spruce (*Picea glauca* (Moench) Voss and *P. mariana* (Miller) Britton, Sterns, and Poggenburg, accordingly). However, other conifers, such as balsam fir (*Abies balsamea* (L.) Miller), jack pine (*Pinus banksiana* Lambert), and tamarack (*Larix laricina* (Du Roi) K. Koch), are also widely distributed. The area is characterized by a diverse assortment of deciduous trees, such as white birch (*Betula papyrifera* Marshall), balsam poplar (*Populus balsamifera* L.), and the widespread trembling aspen (*Populus tremuloides* Michaux).

2.2.2 Earthworm Diversity and Function

Earthworms in the Boreal Forest Zone were the subject of many publications: Cameron et al. (2007, 2015), Dymond et al. (1997), McLean and Parkinson (1997, 2000a, 2000b); Moore et al. (2009, 2018, 2022), Parkinson et al. (2004). Reynolds (1976b, 1977a), Reynolds and Reynolds (1992), Reynolds et al. (2014, 2019), Scheu and McLean (1993). This forest region is located in seven provinces—Alberta, British Columbia, Manitoba, Newfoundland and Labrador, Ontario, Quebec, and Saskatchewan—as well as three territories— Nunavut, Northwest, and Yukon.

In the Boreal Forest, the following earthworm species are frequently identified: *Aporrectodea tuberculata* (7/3)[1] and *Dendrobaena octaedra* (7/2) and *Dendrodrilus rubidus*, *Lumbricus rubellus* and *Lumbricus terrestris* (7/1). The next group of species are commonly encountered in this forest: *Aporrectodea trapezoides* (6/0), *Aporrectodea rosea*, *Aporrectodea turgida*, and *Octolasion tyrtaeum* (5/0). There are six species rarely found in the Boreal Forest: *Aporrectodea longa* (Ude 1885) (3/0) and *Lumbricus castaneus* (3/1) and *Arctiostrotus fontinalis* McKey-Fender et al. 1994 and *Bimastos parvus* (Eisen 1874) (0/1), *Lumbricus festivus* and *Octolasion cyaneum* (1/0).

Four species, namely *Allolobophora chlorotica, Eisenia fetida, Eiseniella tetraedra*, and *Sparganophilus tamesis*, are present in some jurisdictions yet not in the Boreal Forest. Their comprehensive distributions can be observed in Reynolds (2000a, 2014a, 2014b, 2015). Their ecological function will be found in the Discussion.

2.3 Coast Forest Region

2.3.1 Forest Description

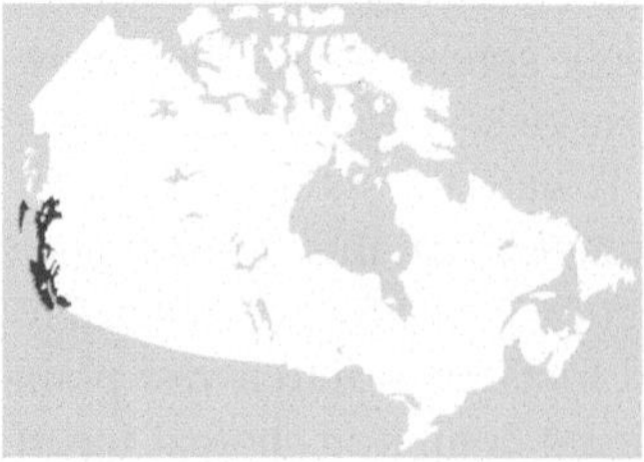

[1] Number of provinces/territories present in forest soils.

The forested zone reaches the coastal islands and the lower seaward slopes of the Coast Mountains in British Columbia. The distinctive species in this area are western hemlock (*Tsuga heterophylla* (Rafinesque) Sargent), Douglas fir (*Pseudotsuga menziesii* (Mirbel) Franco), western red cedar (*Thuja plicata* Donn ex D. Don), and Sitka spruce (*Picea sitchensis* (Bongard) Carrière). These trees are widely recognized for their high value in wood production. In contrast, the broad-leaved trees in the area, such as black cottonwood (*Populus trichocarpa* Torrey and Asa Gray ex. Hooker), red alder (*Alnus rubra* Bongard), big-leaf maple (*Acer macrophyllum* Pursh 1813), have a restricted range and are of little significance in terms of their economic value.

2.3.2 Earthworm Diversity and Function

Since the Coast Forest Region is only found in British Columbia in Canada, the most common earthworm species present are: *Aporrectodea trapezoides, Aporrectodea tuberculata, Aporrectodea turgida, Dendrobaena octaedra, Dendrodrilus rubidus, Lumbricus rubellus, Lumbricus terrestris, Octolasion cyaneum, Octolasion tyrtaeum, Arctiostrotus perrieri* (Benham 1892), and *Arctiostrotus vancouverensis* McKey-Fender et al. 1994.

Less frequently collected species, with five or fewer sites were *Aporrectodea longa, Bimastos lawrenceae* Fender, 1994, and *Toutellus oregonensis* (Smith, 1937).

Their ecological function will be found in the Discussion.

There are seven earthworm species found in British Columbia, but not in the Coast Forest soils, or forest locations: *Allolobophora chlorotica, Aporrectodea limicola* Michaelsen, 1890, *Aporrectodea longa, Aporrectodea rosea, Dendrobaena attemsi* Michaelsen, 1902, *Eisenia fetida, Eiseniella tetraedra, Lumbricus castaneus,* and *Lumbricus festivus.*

Reynolds (2015) provides an extensive explanation of their distributions.

2.4 *Columbia Forest Region*

2.4.1 Forest Description

This forested area is located in the southeastern part of British Columbia, stretching between the Rocky Mountains and the central plateau. It extends across the subalpine region, following the course of river valleys and lakes. The forest habitat in the inner damp belt has a significant resemblance to that of the coastal zone, but with a lower species diversity. Western red cedar (*Thuja plicata* Donn) and western hemlock (*Tsuga heterophylla* (Rafinesque) Sargent) are illustrative trees. The blue Douglas fir, scientifically known as *Pseudotsuga menziesii* var. glauca (Mayr) Franco, has a wide distribution. In the southern regions, it is accompanied by other tree species such as western white pine (*Pinus monticola* Douglas), western larch (*Larix occidentalis* Nuttall), grand fir (*Abies grandis* (Douglas ex D. Don) Lindley), and western yew (*Taxus brevifolia* Nuttall). The upper Fraser Valley is home to Engelmann spruce (*Picea engelmannii* Parry), which is also sporadically found at higher altitudes in the area.

2.4.2 Earthworm Diversity and Function

The Columbian Forest Region is also found only in British Columbia in Canada; this forest region has very few reported collection sites where the following earthworm species have been found: *Aporrectodea tuberculata*, *Lumbricus rubellus* *Aporrectodea turgida*, *Octolasion tyrtaeum*, *Dendrobaena octaedra*, *Dendrodrilus rubidus*, and *Lumbricus terrestris*.

There are no other earthworm species reported in the Columbian Forest.

2.5 Deciduous/Carolinian Forest Region

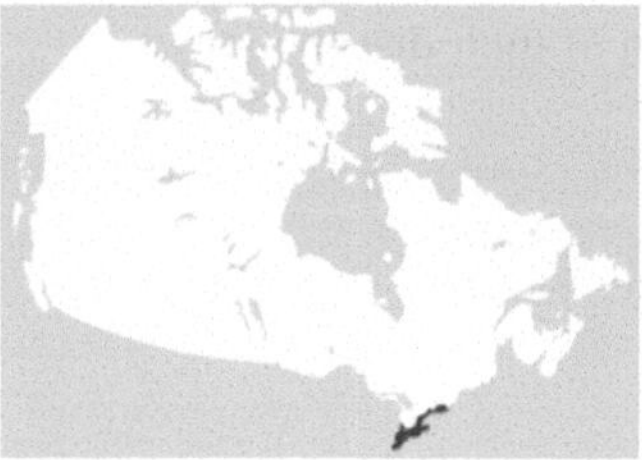

2.5.1 Forest Description

This region, which is the smallest forested area in Canada, is located along the northern shores of Lakes Erie and Ontario and the southeastern side of Lake Huron. Although it is modest in area, this region has the highest number of indigenous tree species compared to any other location. The Great Lakes-St Lawrence Forest Region is home to various broad-leaved trees, including the cucumber tree (*Magnolia acuminata* (Linnaeus)), tulip tree (*Liriodendron tulipifera* L.), black gum (*Nyssa sylvatica* Marshall), blue ash (*Fraxinus quadrangulata* Michaux), sassafras

(*Sassafras albidum* (Nuttall) Nees), walnut/butternut (*Juglans cinerea* L.). These trees are prevalent at the northern boundaries of their distribution range. Only a scattered distribution of eastern white pine (*Pinus strobus* L.), tamarack (*Larix laricina* (Du Roi) K. Koch), eastern red cedar (*Juniperus virginiana* L.) and eastern hemlock (*Tsuga canadensis* (L.) Carrière) are considered conifers.

2.5.2 Earthworm Diversity and Function

Ontario is the only Canadian province that has the Deciduous Forest Area with its Carolinian Remnant. There are 13 species which have been collected in this forest type: *Aporrectodea longa*, *Aporrectodea rosea*, *Aporrectodea trapezoides*, *Aporrectodea tuberculata*, *Aporrectodea turgida*, *Dendrobaena octaedra*, *Dendrodrilus rubidus*, *Lumbricus castaneus*, *Lumbricus festivus*, *Lumbricus rubellus*, *Lumbricus terrestris*, *Octolasion cyaneum*, and *Octolasion tyrtaeum*.

Their specific distributions can be identified in Reynolds (1976c, 2014a, 2014c). Ontario is home to five species, namely *Allolobophora chlorotica*, *Aporrectodea icterica* (Savigny 1826) *Eisenia fetida*, *Eiseniella tetraedra*, and *Sparganophilus tamesis*. However, these species are not found in this particular forest type.

2.6 *Great Lakes-St Lawrence Forest Region*

2.6.1 Forest Description

The Great Lakes-St Lawrence Forest Area is the second biggest forest area in Canada, while being less than tenth the size of the boreal forest. Except for a 322 km discontinuity where the boreal area meets the north coast of Lake Superior, this forest extends from southeastern Manitoba to the Gaspé Peninsula. It is a transitional forest between the broad-leaved and the coniferous zones, bounded to the south by the deciduous forest area. Eastern white pine (*Pinus strobus* L.), red pine (*Pinus resinosa* Aiton), eastern hemlock (*Tsuga canadensis* (L.) Carrière), and yellow birch (*Betula alleghaniensis* Britton) are illustrative species. The presence of Sugar and red maples (*Acer saccharum* Marshall and *A. rubrum* L., respectively), beech (*Carpinus caroliniana* Walter var. *virginiana* (Marshall) Fernald), red oak (*Quercus rubra* L.), basswood (*Tilia americana* L.), and white elm (*Ulmus americana* L.) may be discovered, along with several boreal plants.

2.6.2 Earthworm Diversity and Function

While southeastern Manitoba does feature a tiny piece of the Great Lakes-St Lawrence Forest Region, there have been no documented sightings of earthworms in this area of Manitoba. The following earthworm species were found in this forest type in both Ontario and Quebec: *Aporrectodea longa*, *Aporrectodea rosea*, *Aporrectodea trapezoides*, *Aporrectodea tuberculata*, *Aporrectodea turgida*, *Dendrobaena octaedra*, *Dendrodrilus rubidus*, *Lumbricus castaneus*, *Lumbricus festivus*, *Lumbricus rubellus*, *Lumbricus terrestris*, *Octolasion cyaneum*, and *Octolasion tyrtaeum*.

The following seven species are present in some jurisdictions, but not in forested areas: *Allolobophora chlorotica*, *Bimastos beddardi* (Michaelsen 1894), *Eisenia fetida*, *Eisenia hortensis* Michaelsen, 1890, *Eiseniella tetraedra*, *Satchellius mammalis,* and *Sparganophilus tamesis*. Their specific distributions can be observed in Reynolds (2014a, 2014c).

2.7 *Montane Forest Region*

2.7.1 Forest Description

This area encompasses the central part of British Columbia, as well as various valley pockets that are located close to the Alberta border. These regions have a predominant dry climate in common. The dominant tree species in this area is the blue Douglas fir (*Pseudotsuga menziesii* var. *glauca* (Mayr) Franco), which is a smaller variant of the coastal form. Lodgepole pine (*Pinus contorta* Douglas var. *latifolia* Englemann) and trembling aspen (*Populus tremuloides* Michaux) are often found, whereas white spruce is mostly seen in colder, shaded valleys. Ponderosa pine (*Pinus ponderosa* Lawson) is common. Engelmann spruce (*Picea engelmannii* Parry) and alpine fir (*Abies lasiocarpa* (Hooker) Lindley), together with western white birch (*Betula papyrifera* Marshall var. *commutata* (Regel) Fernald), are significant species found in the subalpine region's northern boundaries.

2.7.2 Earthworm Diversity and Function

This is another forest region which is only found in British Columbia, Canada. There are six earthworm species reported from the Montane Forest Region: *Aporrectodea trapezoides*, *Aporrectodea tuberculata*, *Dendrodrilus rubidus*, *Lumbricus rubellus*, *Lumbricus terrestris*, and *Octolasion tyrtaeum*.

The Montane Forest Zone is home to three earthworm species: *Allolobophora chlorotica*, *Aporrectodea turgida*, and *Eisenia fetida*. However, these species are not found inside the forests themselves. Their specific distributions can be seen in Reynolds (2015).

2.8 Subalpine Forest Region

2.8.1 Forest Description

There are two distinct zones of vegetation in the Subalpine Forest Area. The two regions exhibit clear differentiation, with one extending along the Pacific coastline and the other situated interior Alberta and British Columbia. Various plant populations, impacted by soil erosion, the hydrographic network, human activities, and factors such as climate, the environment, and composition, characterize the two vegetation areas.

Therefore, on the Pacific coast region, particularly in British Columbia, we may see the presence of yellow cypress (*Chamaecyparis nootkatensis* (D. Donn) Spach), mountain hemlock (*Tsuga mertensiana* (Bongard) Carrière), western hemlock (*Tsuga heterophylla* (Rafinesque) Sargent), and fir (*Abies lasiocarpa* (Hooker) Nuttall). The subalpine forest in the inland regions of British Columbia and Alberta consists mostly of Engelmann spruce (*Picea engelmannii* Parry) and white spruce (*Picea glauca* (Moench) Voss).

At high elevations, the subalpine fir species are the most prevalent. The other two species are western larch (*Larix occidentalis* Nuttall), white pine (*Pinus monticola* Douglas), and limber pine (*Pinus flexilis* James). The Pacific coast has significant snowfall, which serves as a protective barrier against frost for the soil. The primary tree species found in this area are mountain hemlock, Pacific silver fir (*Abies amabilis* Douglas ex J. Forbes), and yellow cypress. In inland regions, the land experiences freezing temperatures prior to the arrival of snowfall.

2.8.2 Earthworm Diversity and Function

There are very few earthworm species found in this forest region. The following seven species are reported from both Alberta and British Columbia sections of this forest type: *Aporrectodea tuberculata*, *Aporrectodea turgida*, *Dendrobaena octaedra*, *Dendrodrilus rubidus*, *Lumbricus rubellus*, *Lumbricus terrestris*, and *Octolasion tyrtaeum*.

There are no other earthworm species found in the Subalpine Forest Region. Their precise distributions could be seen in Reynolds (2015).

3 Earthworm Species in Canadian Forests

A complete discussion of each species was presented in detail in Reynolds (1977a, 1977b, 2022) and Reynolds et al. (1974) including synonomy, diagnosis, biology, line drawings, and illustrated glossary.

3.1 Lumbricidae

***Aporrectodea limicola* (Michaelsen, 1890)**—is of small to medium size, measuring between 42 and 90 mm in length and 3 to 4 mm in diameter. It consists of 103 to 127 segments. The prostomium is epilobic and ½ open, with the first dorsal pore 4/5. The clitellum is saddle-shaped and spans segments 1/3 28, 29 to 35, and 36, Tubercula pubertatis, two pairs on segments 33, 34, setae closely paired; *Ap. limicola* is a European introduction to North America. *Ap. limicola* is a genuine soil species that lives inside the soil. It is an acid-loving species that prefers humid conditions and may be found in meadows and forests. Color, unpigmented, live material appears flesh colored. Reproduction by *Ap. limicola* is obligatorily amphimictic (Reynolds 2022, 1975a). This species is widespread in Europe, but limited distribution in North America (Reynolds 2018).

***Aporrectodea longa* (Ude 1885)**—is a large species of earthworm measuring between 90 and 150 mm in length, with a diameter ranging from 6 to 9 mm. It typically has between 150 and 222 segments with a partially closed prostomium and epilobic ½ closed, first dorsal pore 12/13, clitellum saddle on segments 27, 28–34, 35. Tubercula pubertatis on segments 32–34, setae closely paired; *Ap. longa* is a European introduction to North America. *Ap. longa* is a genuine soil species that exhibits an anecic behavior. It feeds on the soil surface throughout the night, displaying selective eating habits. Occasionally, it drags leaves into its holes (Reynolds 2022). This species has been seen throughout Europe in many habitats, including cultivated soil, gardens, grasslands, and forests. It is particularly numerous in soils next to lakes and rivers. The information is not well covered in North America (Reynolds 2018). The color of the organism is mostly gray or brown, with a little iridescent quality on its back. *Aporrectodea longa* reproduces only by sexual reproduction, with copulation occurring underground (Reynolds 1975a).

***Aporrectodea rosea* (Savigny, 1826)**—is a tiny organism measuring between 25 and 85 mm in length, with a diameter of 3 to 5 mm and typically has 120 to 150 segments. The prostomium is partially closed, covering about half to two-thirds of its width. The first dorsal hole is located around four-fifths of the way up the worm's body. *Aporrectodea rosea* also has a clitellum, produces a saddle-like structure. This saddle is slightly flared on the ventral side of segments 25 and 26 to 32, Tubercula pubertatis on segments 34–36, setae closely paired; *Ap. rosea* is a European introduction to North America. *Ap. rosea*, a genuine soil species (endogeic), is often found in gardens, fields, grasslands, woodlands, as well as beneath debris and leaves. Despite being classified as unpigmented, this species often exhibits a pinkish hue when living (Reynolds 1974). *Ap. rosea* reproduces by amphimictic reproduction, which involves copulation beneath the soil surface. Nonetheless, there have been some reported cases of restricted parthenogenetic communities. These communities are probably a result of a high occurrence of male sterility within those communities (Gates 1970; Reynolds 1974, 1977a; Reynolds et al. 1974). The extensive dispersion of *Ap. rosea* in North America is shown in Reynolds and Wetzel (2008, 2012).

***Aporrectodea trapezoides* (Dugès, 1828)**—is a species of earthworm that typically measures between 80 and 140 mm in length, with a diameter ranging from 3 to 7 mm. It has a segment count of 93 to 169, usually exceeding 130 segments, prostomium epilobic ½–¾ closed, first dorsal pore 12/13, usually, clitellum saddle on segments 27, 28–33, 34, Tubercula pubertatis on segments 31–33, setae closely paired; *Ap. rosea* is a European species that lives in the soil (endogeic) and has been introduced to North America. *Aporrectodea trapezoides* has been found in several settings, such as gardens, soil around the roots of plants in pots, farming land, different types of forest soils, occasionally in sandy soils, along stream banks, and in caves (Reynolds 1977a). Coloration in *Ap. trapezoides* is variable—often lighter in segments posterior to the clitellum, then darkening to a richer hues (such as brown, slate, brownish, reddish brown, excluding purple) (Reynolds 1977a). *Ap. trapezoides* reproduces by parthenogenesis, often with pseudogamy. Sterility in males is also frequently observed (Gates 1970; Reynolds 1974). In North America, there are also some amphimictic populations where maturing sperm is present in seminal vesicles (Gates 1970). The extensive dispersion of *Ap. trapezoides* over North America was demonstrated in Reynolds and Wetzel (2008, 2012).

***Aporrectodea tuberculata* (Eisen, 1874)**—is a species of earthworm that is of medium size, measuring between 90 and 150 mm in length and having a diameter of 4 to 8 mm. It typically has a segment count ranging from 146 to 194. The prostomium is epilobic and the clitellum is saddle-shaped, located on segments 27 to 34. The worm's first dorsal pore is found on either the 11th and 12th segments or the 12th and 13th segments. Tubercula pubertatis on segments 30, 31–33, 34, setae closely paired; *Ap. tuberculata* is a European species that lives in the soil and was introduced to North America. *Aporrectodea tuberculata* is commonly collected from pastureland, from wet areas near streams and springs with elevated concentrations of organic matter, and from ditches, under logs, and woody debris (logs, brush), turf, compost, peat, bogs, but rarely from manure. The coloration of *Ap. tuberculata* is devoid of pigmentation, giving it a white or grayish appearance. Differentiating between *Ap. tuberculata* and *Ap. turgida* may be challenging in several instances. This is due in part as their clitellums and Tuberculata Pubertatae overlap the same segments. Genital Tumescences are on alternate segments in *Ap. tuberculata*, but not in *Ap. turgida*. Additionally, Genital Tumescences occur on segment 26 in *Ap. tuberculata* but on segment 27 in *Ap. turgida*. In the case of juveniles—where these sexual characters are not yet developed—*Ap. tuberculata* is the larger of the two species with a range of 146–191 segments (mean: 158 ± 15), vis-à-vis *Ap. turgida* with a range of 134–151 segments (mean: 143 ± 8) (see Gates 1970). *Ap. tuberculata* reproduction is invariably amphimictic, as it involves copulation taking place below the soil's surface (Reynolds 1974). *Ap. tuberculata* has been documented on each of the continents, with the exception of Antarctica. However, the majority of North American sightings have been concentrated across Canada and the northern territories of the United States (Reynolds and Wetzel 2008, 2012).

***Aporrectodea turgida* (Eisen, 1873)**—is a prevalent and extensively spread earthworm species. It has a medium size, ranging from 60 to 85 mm in length and 3.5 to 5.0 mm in diameter. It typically has 130 to 168 segments. The prostomium is located on the epilobic region, and the first dorsal pore is found on segments 12/13 or 13/14. The clitellum is saddle-shaped and spans segments 27, 28, 29–34, and 35, Tubercula pubertatis on segments 31–33, setae closely paired; *Ap. turgida* is a European species that dwells in the soil and was introduced to North America. *Aporrectodea turgida* is often found in several environments such as fields, grass, forest humus, gardens, compost, riparian sections of streams, landfills, wasteland, and caves (Reynolds 1977a). *Aporrectodea turgida* is unpigmented and appears white or pale gray in color, occasionally pink in the region anterior to the crop, and with light pigmentation on the dorsum. Identifying this species may sometimes be challenging because of its similarity to its congener, *Ap. tuberculata* (see to the comment in the species account for *Ap. tuberculata*). *Ap. turgida* reproduces by obligatory amphimictic reproduction, where copulation takes place below the surface (Reynolds 1974). *Ap. turgida* is mostly found in the central states of the United States and all regions in Canada (Reynolds and Wetzel 2008).

***Bimastos lawrenceae* Fender, 1994**—is an earthworm of modest to moderate size, measuring around 80 mm in length and 5 mm in diameter, segment number 103–134, and known from only five individuals, therefore, a very rare earthworm in Canada. Prostomium broad, epilobous, open, first dorsal pore 5/6, clitellum saddle on segments 25–34, Tuberculata pubertatis, absent, setae narrowly paired; Reynolds (2018) describes *Bimastos lawrenceae* as a native corticole-epigeic species. The paratypes were obtained from areas with damp soil and densely packed plants in wet conditions. The holotype was acquired from a sample of soil and plants, although extensive excavation in the region yielded just one paratype from an exceptionally damp location. The color is reddish, with a light reddish and translucent appearance in life (McKey-Fender et al. 1994: 1338). *Bimastos lawrenceae* reproduces by parthenogenesis, as stated by McKey-Fender et al. (1994). The occurrence of this species in Canada is extremely uncommon.

***Dendrobaena octaedra* (Savigny, 1826)**—is a tiny earthworm that measures between 17 and 60 mm in length, with a diameter of 3 to 5 mm. It typically has 60 to 100 segments. The prostomium is closed and the first dorsal pore is located between segments 4/5 and 6/7. The clitellum is saddle-shaped and found on segments 27, 28, 29 to 33, and 34. Tubercula pubertatis are usually found on segments 31 to 33. The setae have a wide distribution and separated from each other; *D. octaedra* is a surface living epigeic species and European introduction to North America (Reynolds 2022). *Dendrobaena octaedra* inhabits soil, sod, moss-covered stream banks, logs, leafy debris, as well as cold, damp ravines, highland seepage places, and soils of forests. The color is a shade of red, specifically ranging from dark red to purple, as described by Reynolds et al. (1974: 31) and Reynolds (1977a: 65). The reproduction of *D. octaedra* is characterized by parthenogenetic polymorphism in the majority of individuals, making it likely to be more prevalent than in other lumbricid species. A comprehensive analysis of the various morphs is available in a publication authored by Gates (1974). *Dendrobaena octaedra* has been documented on all continents, and the extensive presence of *D. octaedra* in North America has been demonstrated in Reynolds and Wetzel's publications in 2008 and 2012.

***Dendrodrilus rubidus* (Savigny, 1826)**—is a species of tiny earthworm with a length range of 20–90 mm and a diameter of 2–5 mm. It typically has 50–120 segments. The prostomium is open and located on the epilobic region. The first dorsal pore is located on the fifth or sixth segment. The clitellum is saddled on segments 26, 27–31, and 32. If available, the Tubercula pubertatis can be observed on segments 28, 29–30, setae widely paired; *D. rubidus* is a European introduction to North America. As an epigeic and corticole species, *Dendrodrilus rubidus* has been identified in a variety of environments, involving as fields of crops, gardens, banks of streams, moss in springs and water wells, compost, peat, and even manure. *Dd. rubidus* primarily inhabits the upper layers of soil. However, it has been observed that during moist nights, these worms exhibit behavior of venturing onto the soil surface or even scaling trees. Reports indicate that this organism is often found under the bark of aged trees, as well as below moss, leaf debris, or decaying wood in damp environments. The coloration is red and has a deeper shade on the back (according to Reynolds et al. 1974: 32–33; Reynolds 1977a: 71). *Dendrodrilus rubidus* exhibits facultative parthenogenesis, often accompanied by male sterility and the frequent lack of spermathecae (Gates 1972; Reynolds 1974). *Dendrodrilus rubidus* has been documented on every continent, and its extensive presence in North America was shown in Reynolds and Wetzel's publications in 2008 and 2012.

***Lumbricus castaneus* (Savigny, 1826)**—is a tiny earthworm regarding 30–50 mm in length, with a diameter of 3–5 mm. It typically has 70–100 segments, a tanylobic prostomium, and a saddle-shaped clitellum located on segments 28–33. The earthworm also has Tubercula pubertatis on segments 29–32, and its setae are firmly paired; *L. castaneus* is a species that was introduced from Europe to North America. According to Reynolds (2018), *Lumbricus castaneus* is an epigeic species. Gerard (1964) documented its habitat as mostly terrestrial, found predominantly in soil that is rich in organic materials. It can be observed in many locations such as parks, gardens, grasslands, woodlands, along river and marshlands, and below leaves, stones, and manure. The color is very saturated, with a deep red hue that leans toward chestnut and violet-brown. It also has a strong iridescent aspect. *Lumbricus castaneus* is a species that reproduces only by sexual reproduction (Reynolds 1974). This species, known as peregrine, is now found in both North America and Europe.

Lumbricus festivus **(Savigny, 1826)**—this medium to large earthworm has the following characteristics: length ranging from 48 to 105 mm, diameter from 4 to 5 mm, and segment number from 100 to 143. It has a tanylobic prostomium, a first dorsal pore 5/6, a clitellum saddle on segments 34 to 39, Tubercula pubertatis on segments 35 to 37 and 38, and setae that are closely paired. *L. festivus* was introduced to North America by Europeans. As noted by Reynolds (2018), the epigeic species *Lumbricus festivus* can be found in a variety of settings, including woodlands, river banks, grasslands, and even beneath stones, leaves, and dungs. The coloration is a reddish brown, with an iridescent appearance on the upper side and a lighter color on the underside. *Lumbricus festivus* is believed to engage only in sexual reproduction (Reynolds 1974), with copulation taking place below the surface of the soil. The peregrine species is now found in Europe, many Canadian provinces, and one US state that border Quebec.

Lumbricus rubellus **Hoffmeister, 1843**—is a species that typically measures between 50 and 150 mm in length, with a diameter of 4 to 6 mm. It has a segment count ranging from 70 to 120. The prostomium is tanylobic, and it has 5/6 to 8/9 first dorsal pores. The clitellum is located on segments 26, 27 to 32, and 33, and there are Tubercula pubertatis on segments 28 to 31 and 32, setae closely paired; *L. rubellus* is a surface-dwelling (epi-endogeic) and a European introduction to North America. *Lumbricus rubellus* is commonly collected from soils under debris, logs, dung pats in pastures, in gardens and parks, in stream banks, and from forests (Reynolds 1977a). Specimens are red in color. Differentiating between the juvenile forms of *Lumbricus rubellus* and *Lumbricus terrestris* poses a challenge: *L. rubellus* is an epi-endogeic species with a cylindrical body which may be slightly dorso-ventrally compressed posteriorly, vis-à-vis *L. terrestris*—an anecic species with a cylindrical body which is strongly dorso-ventrally compressed posteriorly. Reproduction by *Lumbricus rubellus* is obligatorily amphimictic (Reynolds 1974, 1977a). The species is found worldwide, and its specific distribution in North America is shown in Reynolds and Wetzel's publications from 2008 and 2012.

Lumbricus terrestris **Linnaeus, 1758**—is quite large, measuring between 90 and 300 mm in length and 6 to 10 mm in diameter. It has a total of 120 to 160 segments. The prostomium is tanylobic. The worm has a dorsal pore on the seventh and eighth segments. The citellum is located on segments 31, 32 to 37. Tubercula pubertatis can be found on segments 33 to 36. The setae are larger and more widely spaced in the tail and head regions of the earthworm. This species is a European introduction to North America. A deep-dwelling (anecic) casting species, *L. terrestris* surfaces to draw leaf litter into its burrow and often can be found on the surface during and after periods of moderate to heavy rains. The specimen is often gathered from many environments, such as gardens, cultivated and grazing fields, river banks, streams, muddy areas, peat with trees, under logs in watercourses, under cow dung, in compost, in greenhouses as well, and in caves. However, it is less frequently found in woodlands (Reynolds 1977a). Specimens of *L. terrestris* are red-brown in color; it is challenging to differentiate between the young individuals of *Lumbricus terrestris* and *L. rubellus* (see note in *L. rubellus*). Reproduction by *Lumbricus terrestris* is obligatorily amphimictic (Reynolds 1974, 1977a). *Lumbricus terrestris* is widely distributed worldwide and is well known as the most prevalent species of earthworm. This is mostly owing to its extensive usage as a standard example of megadriles in elementary and secondary education, as well as its frequent use as bait for fishing. In actuality, however, *L. terrestris* is less widespread in North America than many other species (e.g., *L. rubellus*, *Ap. tuberculata*, *Ap. turgida*, *Ap. trapezoides*, and *Octolasion tyrtaeum*) Reynolds and Wetzel (2008, 2011, 2012).

***Octolasion cyaneum* (Savigny, 1826)**—is a medium in size species measuring between 65 and 180 mm in length, with a diameter of 7–8 mm. It typically has 140–158 segments. The prostomium located at the anterior end is partially closed, covering about one-third to one-half of the opening. The first dorsal pore is found on segments 9/10, 11/12, or 12/13. The clitellum is saddle-shaped and spans segments 29–34, Tubercula pubertatis on segments 30–33, setae closely paired anteriorly, and widely paired posteriorly. The species under discussion is an introduced European species that has successfully established populations in various regions, including North America, Europe, South America, and Australasia. *Octolasion cyaneum*, as indicated by Reynolds (2018), is classified as an endogeic species. The presence of this species has been documented beneath submerged stones, as well as within moss, along stream banks, and other habitats associated with freshwater ecosystems. It is also observed in agricultural fields that have undergone ploughing, as well as in moist sandy areas and soil ecosystems within forests. The reproductive strategy of *O. cyaneum* is exclusively parthenogenetic, as documented by Reynolds (1974). There is no recorded evidence of copulation, and it is possible that such behavior has never been observed, as suggested by Gates (1973). The color is a shade of blue-gray or white, as described by Reynolds et al. (1974: 45) and Reynolds (1977a: 105). This species is uncommon in the United States and Canada.

***Octolasion tyrtaeum* (Savigny, 1826)** —is typically measures between 25 and 130 mm in length, with a diameter of 3 to 6 mm. It has a segment count ranging from 75 to 165. The prostomium is epilobic and 1/3–2/3 closed. The worm has a dorsal pore located between the ninth and tenth segments, or between the 13th and 14th segments, with the most common location being between the 11th and 12th segments. It also has a clitellum, and a saddle-like structure on segments 30 to 35, Tubercula pubertatis on segments 31–34, setal widely paired. This species is a European introduction to North America. This endogeic species is commonly observed in forested areas and woodlots. It frequently appears in environments containing decomposing logs and leaf mold, compost, peat, forest debris, gardens, agricultural and grass lands, swamps, springs, stream banks, caves, and among the roots of plants that are submerged (Reynolds 1977a). Coloration of *O. tyrtaeum* is variable, from unpigmented to milky white, to gray, blue or pink (Reynolds 1974, 1977a, 2022). Reproduction by *O. tyrtaeum* is obligatorily amphimictic (Reynolds 1974, 1977a); additionally, there have been recorded instances of parthenogenetic populations (Jaenike et al. 1980; Jaenike et al. 1980). The species has a worldwide range, and its documented distribution in North America is depicted in Reynolds and Wetzel's publications (2008, 2012, 2021).

3.2 Megascolecidae

***Amynthas agrestis* (Goto and Hatai 1899)**—is a species of earthworm that falls within the medium to large size range, measuring from 70 to 200 mm in length and 5 to 8 mm in diameter. It typically has 70 to 110 segments and has an epilobic prostomium with an exposed tongue. The first dorsal pore is often found on the 12th or 13th segment, whereas the clitellum is annular and spans segments 13/14 to 16/17. Tubercula pubertatis absent, setae completely encircle the segment. *Amynthas agrestis* is an indigenous species originally from Japan. Presently, it has been documented in Canada (Reynolds 2014a; Reynolds and McTavish 2021), Korea, Japan, and the United States (Reynolds 2018). As noted by Zhang et al. (2013), nutritional adaptability has been a contributing factor in the successful invading of US woodlands by the epi-endogeic species *Amynthas agrestis* (Reynolds 2018). *A. agrestis* has been found in numerous human-created places as well as wooded places. Color, in living samples, is reddish on preclitellar part, turning bright brownish in the postclitellar zone (following Tandy 1969; Reynolds 1978). The reproduction of *A. agrestis* in North America is exclusively parthenogenetic, as stated by Reynolds (1974). *A. agrestis* completes its life cycle within a year, survives the colder months merely in the form of cocoons. The adults of this species breed throughout the warmer months and perish by the conclusion of autumn (Reynolds 2022). Reynolds (2018, 2022) and Reynolds and McTavish (2021) provided a comprehensive inventory of species along with their first documentation in the United States and Canada.

***Amynthas hilgendorfi* (Michaelsen 1892)**—is a species of earthworm that falls within the medium to large size range, measuring between 110 and 170 mm in length and 6 to 8 mm in diameter. It typically has a total of 90 to 118 segments. The prostomium of this earthworm is open and located on the epilobic region. The first dorsal pore is generally found on the 11th or 12th segment. The clitellum is annular and spans segments 13/14 to 16/17, Tubercula pubertatis absent, setae completely encircle the segment. *Amynthas hilgendorfi* is an indigenous species originating from the Orient, namely Japan. Currently, it is only documented in Canada (Reynolds 2014a; Reynolds and McTavish 2021; McAlpine et al. 2022), Japan, Korea, Russia, United States (Reynolds 2018), and Vietnam. *Amynthas hilgendorfi* is an epi-endogeic species frequently within the leaf litter (Reynolds 1974). This species exhibits a widespread distribution in both urban and rural environments and has been observed to encroach upon deciduous woodland ecosystems as well. This species is known to exhibit obligate parthenogenesis, as documented by Reynolds

(1974). In natural habitats, *A. hilgendorfi*, a species with a 1-year life cycle, survives the winter solely in the form of cocoons. The adult individuals engage in reproduction during the summer season and subsequently perish by the conclusion of autumn (Chang et al. 2017). The color on the back is reddish or brownish, according to Tandy (1969) and Reynolds (1978). Reynolds (2018, 2022), Reynolds and McTavish (2021), and McAlpine et al. (2022) provided a comprehensive compilation of species and their first documentation in the United States and Canada.

***Amynthas tokioensis* (Beddard 1892)**—is a small earthworm with a length ranging from 30 to 125 mm and a diameter of 3 to 7 mm. It typically possesses a segment count ranging from 84 to 102. The prostomium of this species features an epilobic open tongue, while the first dorsal pore is located on the 12th or 13th segment. The clitellum appears as an annular structure spanning segments 13/14 to 16/17, Tubercula pubertatis absent, setae completely encircle the segment. *Amynthas tokioensis* is an indigenous species of the Orient, namely Japan. Presently, it is only documented in Canada (Reynolds 2014a; Reynolds and McTavish 2021; McAlpine et al. 2022), Japan, Korea, and the United States (Reynolds 2018). *Amynthas tokioensis* is an epi-endogeic species frequently within the leaf litter; it is prevalent in both urban and suburban areas and is also intruding into deciduous woods (Chang et al. 2017). Color of live specimens is red or brown. Reynolds (2018, 2022), Reynolds and McTavish (2021), and McAlpine et al. (2022) provided a comprehensive compilation of species and their first documentation in the United States and Canada. In the past, *A. tokioensis* has been called *Metaphire levis* or *Pheretima levis* in publications from North America.

***Arctiostrotus fontinalis* McKey-Fender, 1994**—is a species of earthworm that falls within the smaller size range (34–104 mm in length, with a diameter of 1–2 mm) and possesses a varying number of segments (ranging from 67 to 105). The prostomium is tanylobic and wide. In the caudal bulb, there are typically two or three dorsal pores, with three being the most common occurrence, clitellum annular on segments 13–18, Tuberculata pubertatis absent, setae separate. *Arctiostrotus fontinalis* is an indigenous species of North America, specifically found only in the northwestern part of the United States and the Yukon territory in Canada. The *Arctiostrotus* species in question is the smallest and most hygrophilous of those that are currently recognized. Only a small number of specimens of this species were collected from land-based locations, with the majority being discovered in regions where water seeps out, at the edges of springs or spring-fed streams, or in close proximity to springs, where saturated organic matter meets mineral soil. In the Yukon region, the specimen was obtained from a moist black organic soil known as Organic Cryosol. This soil was located at the bottom of a slope, where birch and willow trees were present (Reynolds et al. 2019). Additionally, the specimen was also found in a damp mossy area within the soil of an alpine tundra pasture. This pasture was situated in a depression and featured diamond-leaf willow (*Salix pulchra* Cham.) and glandular birch (*Betula glandulosa* Michx.) (Reynolds et al. 2019). Color is unpigmented (after McKey-Fender et al. 1994). *Arctiostrotus fontinalis* is a species that lives in the mud and on the surface of the soil (limicolous-epigeic) (Reynolds 2018). The method of reproduction is currently uncertain; however, it is likely to be amphimictic until observed otherwise. Reynolds (2018, 2022) provided a comprehensive inventory of species along with their first documentation in the United States and Canada. The occurrence of this species in North America is quite uncommon.

***Arctiostrotus perrieri* (Beddard, 1892)**—is an earthworm that falls within the size range of small to fairly large, with a length range of 50–204 mm, a diameter of 2.25–4 mm, and a segment count of 135–170. Its prostomium is tanylobic, with a hard prostomial lobe that is noticeably wider than the tongue. The sides of the tongue are parallel, and the clitellum is annular and located on segments 13–18, Tuberculata pubertatis absent, dorsal pores lacking, setae separate. *Ar. perrieri* is an indigenous species found only in the northwestern parts of the United States and British Columbia in Canada. The specimen was discovered in several habitats such as woodland soils, leaves mold, decaying stump, stream bank, and moss, specifically in moist soil in British Co1umbia (McKey-Fender et al. 1994). The coloration of this species ranges from pigmented purple-brown to relatively pale or sometimes unpigmented, potentially indicating bleaching in some cases (McKey-Fender et al. 1994). This species exhibits an epigeic nature, as documented by Reynolds (2018). Additionally, it displays parthenogenetic characteristics, as observed by Fender and McKey-Fender (1990). Reynolds (2018, 2022) provided a comprehensive inventory of species along with their initial documentation in Canada and the United States.

***Arctiostrotus vancouverensis* McKey-Fender, 1994**—is a worm of intermediate size, measuring between 167 and 381 mm in length, with a diameter ranging from 3.2 to 4.5 mm. It consists of 147 to 204 segments. The prostomium is tanylobic, and the clitellum is annular, located on segments 13 to 18. The clitellum extends dorsally to 9/12 but is absent ventrally in the male's field region. The Tuberculata pubertatis are not present, and the dorsal pores are often missing. The setae are distinct and not connected. *Ar. vancouverensis* is a native species of North America, found only in British Columbia, Canada, and the northwest region of the United States. The presence of this species has been observed within the logs and the organic matter layer of woodland ecosystems, as documented by McKey-Fender et al. (1994). The reproductive behavior of the organism remains uncertain presently, with the prevailing assumption being amphimictic in the absence of compelling evidence suggesting otherwise. The species *Arctiostrotus vancouverensis* is known to exhibit an epigeic behavior, as described by Reynolds (2018). The absence of pigmentation is observed in the coloration (McKey-Fender et al. 1994).

***Pithemera bicincta* (Perrier 1872)**—is a tiny earthworm measuring between 40 and 80 millimeters in length, with a diameter of 2 to 3 millimeters and consisting of 77 to 101 segments. Its prostomium is epilobic, and its tongue is exposed, clitellum annular on segments 14–16, Tuberculata pubertatis absent, first dorsal pore at 11/12 or 12/13, setae completely encircle the segment. *Pithemera bicincta* is a native of the Orient (probably the south Pacific), the exact origin is unknown and currently it is only known from the United States (Florida, Georgia, Hawai′i (Reynolds 2018), Canada (Reynolds and McTavish 2021), and various south Pacific and Caribbean Islands (Gates 1972). Reproduction of *Pithemera bicincta* is unknown at this time, but likely to be parthenogenetic. It is an epi-endogeic species living in the soil and litter (Reynolds 2018). Color of live specimens is red. Reynolds (2018) provided a comprehensive inventory of species and their first documentation in both the United States and Canada. The occurrence of this species in North America is quite uncommon.

***Toutellus oregonensis* (Smith 1937)**—is a big worm, measuring between 74 and 88 mm in length and 4.5 to 6 mm in diameter. It consists of 113 to 121 segments, with the clitellum located on segments 13 to 18. Tubercula pubertatis is absent, the prostomium is partially open (½ half), the sides of the tongue converge, dorsal pores are located behind the clitellum, and the setae are distinct and separated. *Toutellus oregonensis* is a species indigenous to North America, restricted to the coast in the northwest of the United States and the province of British Columbia (Canada). Based on Fender and McKey-Fender (1990), until contrary evidence is presented, it is hypothesized to be an amphibious species. As indicated by Reynolds (2018), *Toutellus oregonensis* is an endogeic species. It was obtained on MacMillan Bloedel property (Cameron Sector) on the island of Vancouver, in the soil beneath western red cedar (*Thuja plicata*). This species lacks pigmentation. Reynolds (2018) provided a comprehensive inventory of species and their initial documentation in the United States of America and Canada. This species is quite uncommon in North America.

4 Discussion

Although there are 26 earthworm species in the forest soils of the Canadian Forest Regions, they are not all performing the same function. Earthworm ecological types were first described by Bouché (1970, 1972). He had only three forms: anéciques (anecic), épigées, and endogées. Although more widely known as designations by Bouché, he patterned his forms from earlier work by Gisin (1943) who had only two forms: épigées and endogées. He also modelled his system after concepts presented earlier by Wicke (1953) and Lee (1959) although they did not use these terms. Bouché (1972: 475–476) constructed his system on several factors: (1) color or pigmentation, (2) musculature, (3) respiration, (4) regeneration, (5) nephropores, (6) diapause, (7) tail morphology, and (8) tail size. His system was qualitative, not quantitative. Recently, Bouché's classification has been quantified from his original dataset utilizing 125 species (Table 1) (Bottinelli et al. 2020). Based on my work, primarily in North America, I added two more types in 1977, e.g. **corticole** and **limicolous**. Since then others have added a number of other types: amphibious, aquatic, arboreal, epi-anecic, epi-endogeic, endo-anecic, geophagous, hydrophilic, marine, ripicolous, and saproxylic. These are primarily subdivisions of the five types I use and may be used in some cases for aquatic oligochaetes; it depends on whether you are a lumper or splitter!

Anecic (anécique) species are deep burrowers (up to 2 meters); once they have completed their burrow system they do not eat their way through the soil as endogeic species, instead returning to the topsoil to consume soils, plant litter, duff, and

Table 1 Summary of a mean values for morphoanatomical characteristics that define the most often researched ecological groups by Bouché (1972) (*according to* Bottinelli et al. 2020)

Morphoanatomical traits	Epigeic	Endogeic	Anecic	Epi-anecic
Length (mm)	62	79	349	143
Diameter (mm)	2	3	9	6
Weight (mg)	518	751	1235	3054
Pigmentation (%)	100	0	100	100
Skin color	red (86%)	Pink (61%)	Gray (100%)	Gray (55%)
Anterior-posterior gradient (%)	100	0	100	100
Dorso-ventral gradient (%)	95	0	84	100
Flattening index	No (43%) Tail (29%)	No (45%) Clitellum (29%)	Clitellum-Tail (47%) Tail (47%)	Clitellum-Tail (45%) Tail (23%)

to copulate (Fig. 2: **1**). These species are very large, measuring 90–300 mm in length as adults. Anecic species are strongly pigmented, commonly reddish brown in color and consume large amounts of litter annually. Examples of earthworm species in Canadian forest soils are: *Aporrectodea longa* and *Lumbricus terrestris.*

Epigeic (épigée) species are small to medium-sized earthworms, measuring usually 1–7 cm in length as adults. They also appear reddish brown on the dorsal surface and are litter and surface-dwelling species found in the litter or in the A horizon of the soil. Epigeic mostly consume plant debris and reside on the top of the soil or within layers of litter (Fig. 2: **3**). As they consume humus or soil, their castings neutralize the soil and—by aggregation of soil particles—improve water infiltration and percolation. Examples of earthworm species in Canadian forest soils are *Bimastos lawrenceae*, *Dendrobaena octaedra, Lumbricus castaneus, Lumbricus festivus*, *Arctiostrotus perrieri,* and *Arctiostrotus vancouverensis.*

Endogeic (endogée) species—medium-sized worms (usually measuring 20–120 mm in length as adults), living in horizontal burrows as soil feeders (Fig. 2: **2**). They are shallow, horizontal burrowers with no preferential orientation, often refilling burrows with casts. Endogeic species frequently are unpigmented, but some species may appear pinkish, whitish, green, yellowish-brown, or gray-blue. They live and feed in the mineral soil horizons getting their energy from minerals and carbohydrates (litter) found there. These species can tunnel through the soil to depths of almost 50 cm. Examples of earthworm species in Canadian forest soils are: *Aporrectodea limicola*, *Ap. rosea*, *Aporrectodea trapezoides*, *Ap. tuberculata*, *Ap. turgida*, *Octolasion cyaneum*, *O. tyrtaeum,* and *Toutellus oregonensis.*

Some endogeic species may be collected from habitats more commonly supporting epigeic species—frequenting the soil surface during favorable weather conditions (periods of increased precipitation and cooler temperatures) but burrowing back into the soil during periods of dry and warmer weather, or to aestivate or hibernate, e.g., *Lumbricus rubellus* (Reynolds 2018).

Additionally, I have utilized two other varieties for years: *corticole* and *limicolous.*

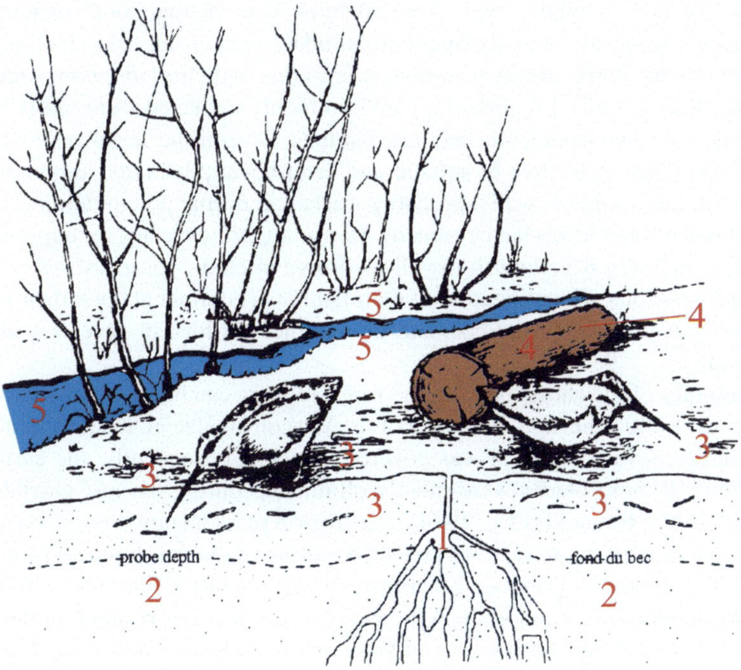

Fig. 2 The earthworm ecological categories are categorized as follows: 1 = anecic/anécique, 2 = endogeic/endogée, 3 = epigeic/épigée, 4 = corticole/corticol; and 5 = limicolous/limicole (*adapted from* Reynolds 1977b)

Corticole species—are typically small worms, with adult lengths generally reaching less than 100 mm. These worms reside beneath the bark of collapsed trees, namely in the frass, which is the sawdust-like substance found between the bark and the cambium layer of the tree's wood (Fig. 2**: 4**). Many of these species are found in the litter adjacent to the fallen trees. Habitat usage by corticole species (e.g., *Bimastos beddardi* [Michaelsen 1894], *B. heimburgeri* [Smith 1928], *B. longicinctus*, *B. parvus*, *B. tumidus* [Eisen 1874] and *B. zeteki* [Smith and Gittens 1915]) depends on their geographical location and can also be influenced by the existence of other species of earthworms (e.g., *Dendrodrilus rubidus*) (Černosviton and Evans 1947; Gerard 1964; Reynolds 1977b). For example, in areas where one or more *Bimastos* spp. are found, they usually are the dominant corticole species—perhaps influencing *Dendrodrilus rubidus* individuals to move into habitat more commonly occupied by epigeic species. *Bimastos lawrenceae* may be a corticole species, but so few individuals from a limited area are known that it is not possible to define fully its ecological role at this time.

Limicolous (limicole) species are mud dwellers, living in very wet habitats that include the saturated soils along the banks of streams, ponds, lakes, wetlands, flood plains, as well as saturated tree-wood (Fig. 2**: 5**), and moist litter covering the soil / muddy surfaces in these areas; species are also often collected from the sediments in truly aquatic habitats, and from depths approaching one meter. Most *Sparganophilus* species (Family Sparganophilidae) are considered to be limicolous (Reynolds 1980, 2001, 2007). *Eiseniella tetraedra* and *Allolobophora chlorotica* (Savigny 1826) (family Lumbricidae) are commonly collected from moist to saturated soils, mud, and from lentic and lotic habitats. Several species in yet other families are also discovered, live in aquatic and semi-aquatic habitats, although likely there only temporarily (as waifs, probably washed in during precipitation events, or stranded when water levels have receded). *Sparganophilus* species are unpigmented; *Eiseniella tetraedra* is yellowish to yellow-brown in color. These are not generally found in forest soils. The utilization of their habitat is contingent upon their particular geographical spot and the coexistence of other earthworm species within the ecosystem.

A summary of the earthworms in the forest regions can be found in Table 2.

Around 11,000 years ago, in the era of the Wisconsin glacial, the northern part of the United States and Canada was covered by ice. Consequently, the earthworm populations in these places were mostly eliminated during the last glacial epoch (Reynolds 1994; Reynolds et al. 1974). Earthworm recolonization has been delayed in these places after the retreat of glaciers, resulting in the development of northern woodlands without the presence of earthworm communities (Reynolds 2018). The rise in worldwide travel and trade in the past few decades has resulted in the arrival of European and Asian earthworms to sites that were historically devoid of earthworms, as well as to regions that already had indigenous earthworm communities (Reynolds 1976b, 2002, 2003, 2004; Addison 2009; Reynolds et al. 2019). The forest floor, referred to as the "horizon O" or litter layer, consists of decomposing material from plants. The primary source of carbon accumulation in forest soils is mostly derived from the plant's life cycles. Through the stoma, carbon dioxide enters leaves where it is used for photosynthesis and then stored within the plant's tissues until it dies. After a plant's roots perished the organic debris from the plants drop to the forest floor and are either integrated into the litter layer or deeper horizons (Janzen 2004). Carbon is only stable when it reaches the A or B horizons. Earthworms contribute to this process; however, their specific involvement varies depending on their habitat, the available food supplies, and their dietary preferences (Serniak 2017). The earthworms found in the forest areas of Canada are mostly imported European species. These worms typically reduce the forest floors overall quantity and thickness. In extreme instances, they can lead to the full disappearance of the litter layer (Wironen and Moore 2006).

There is a difference in the leaf litter in the forest, and earthworms have shown a preference for different leaf species. Initial studies in leaf litter breakdown were conducted by Edwards and Heath (1963) and Heath et al. (1966). They found the disappearance rates of leaf discs in fine mesh bags in woodland soil to be: elm>ash = lime>oak = beech sometimes the rates of oak and beech were similar, but oak always disappeared first. In subsequent studies (Reynolds and McBrayer, unpublished data), with additional leaf species we were able to expand the preference list as follows: *Robinia* (locust) > > *Alnus* (alder) = *Populus* (poplar) > *Prunus* (cherry) = *Salix* (willow) = *Ulmus* (elm) = *Acer* (maple) = *Liriodendron* (tulip

Table 2 Summary of earthworms in Canadian forests

Earthworm (Ecological type)	Forest region	Dominant tree Species	Soil type Canadian system	Soil type American system
Lumbricidae				
Aporrectodea limicola (endogeic)	Coast	*Tsuga heterophylla*	Humo-Ferric Podzol	Haplorhods
Aporrectodea longa (anecic)	Acadian	*Acer saccharum, Tsuga canadensis*	Humo-Ferric Podzol	Haplorhods
	Boreal	*Betula papyrifera, Populus tremuloides*	Melanic Brunisol	Hapludolls
	Coast	*Pseudotsuga menziesii, Populus tremuloides*	Humo-Ferric Podzol	Haplorhods
	Deciduous	*Plantanus occidentalis, Sassafras albidum*	Luvisols and Gleysol	Hapludalfs
	Deciduous	*Acer saccharum, Fagus grandifolia*	Gray Brown Luvisol	Hapludalfs
Aporrectodea rosea (endogeic)	Acadian	*Picea mariana, Betula papyrifera Ulmus americana*	Humo-Ferric Podzol	Haplorhods
	Boreal	*Betula papyrifera, Populus tremuloides*	Melanic Brunisol	Hapludolls
	Boreal	*Abies balsamea, Betula papyrifera*	Brunisol	Inceptisols
	Coast	*Pseudotsuga menziesii, Tsuga heterophylla*	Humo-Ferric Podzol	Haplorhods
	Deciduous	*Plantanus occidentalis, Sassafras albidum*	Luvisols and Gleysol	Hapludalfs
	Deciduous	*Liriodendron tulipifera, Magnolia acuminata*	Gleysol	Aquents
	Deciduous	*Acer saccharum, Fagus grandifolia*	Gray Brown Luvisol	Hapludalfs
	Great Lakes-St. Lawrence	*Alnus rugosa*	Brunisol	Inceptisols
	Great Lakes-St. Lawrence	*Fagus grandifolia, Acer rubrum*	Humic Gleysol	Aquaolls
	Great Lakes-St. Lawrence	*Acer rubrum, Fagus grandifolia*	Dystric Brunisol	Dystrochrepts
Aporrectodea trapezoides (endogeic)	Acadian	*Abies balsamea, Betula papyrifera*	Humo-Ferric Podzol	Haplorhods

(continued)

Table 2 (continued)

Earthworm (Ecological type)	Forest region	Dominant tree Species	Soil type Canadian system	Soil type American system
	Acadian	*Acer saccharum, Betula alleghaniensis*	Orthic Podzol	Cryorthods
	Boreal	*Abies balsamea, Betula papyrifera*	Brunisol	Inceptisols
	Boreal	*Populus tremuloides, Salix nigra*	Gray Brown Podzol	Udalfs
	Coast	*Alnus rubra, Pseudotsuga menziesii*	Dystric Brunisol	Dystrochrept
	Coast	*Pseudotsuga menziesii, Tsuga heterophylla*	Humo-Ferric Podzol	Haplorhods
	Coast	*Acer macrophyllum, Pseudotsuga menziesii*	Humo-Ferric Podzol	Haplorhods
	Columbia	*Pseudotsuga menziesii, Betula papyrifera*	Eutric Brunisol	Eutrochrepts
	Deciduous	*Plantanus occidentalis, Sassafras albidum*	Orthic Luvisol	Hapludalfs
	Deciduous	*Liriodendron tulipifera, Magnolia acuminata*	Gleysol	Aquents
	Deciduous	*Acer saccharum, Fagus grandifolia*	Gray Brown Podzol	Udalfs
	Great Lake-St. Lawrence	*Abies balsamea, Betula alleghaniensis*	Humo-Ferric Podzol	Haplorhods
Aporrectodea tuberculata (endogeic)	Acadian	*Pinus strobus, Picea rubens*	Humo-Ferric Podzol	Haplorhods
	Acadian	*Abies balsamea, Betula papyrifera*	Ferro-Humic Podzol	Cryorthods
	Acadian	*Abies balsamea, Betula papyrifera*	Humo-Ferric Podzol	Haplorhods
	Acadian	*Acer saccharum, Tsuga canadensis*	Humo-Ferric Podzol	Haplorhods
	Acadian	*Acer saccharum, Tsuga canadensis*	Gray Brown Podzol	Hapludalfs
	Boreal	*Abies balsamea, Betula papyrifera*	Brunisol	Inceptisols
	Boreal	*Picea glauca, Abies balsamea*	Orthic Regosol	Udorthents

(continued)

Table 2 (continued)

Earthworm (Ecological type)	Forest region	Dominant tree Species	Soil type Canadian system	Soil type American system
	Boreal	*Betula papyrifera, Populus tremuloides*	Humo-Ferric Podzol	Haplorhods
	Boreal	*Picea glauca, Picea mariana*	Turbic Cryosol	Turbels
	Coast	*Pseudotsuga menziesii, Tsuga heterophylla*	Humo-Ferric Podzol	Haplorhods
	Coast	*Alnus rubra*	Fibrisol	Fibrists
	Columbian	*Thuja plicata, Tsuga heterophylla*	Humo-Ferric Podzol	Haplorhods
	Columbian	*Pseudotsuga menziesii,*	Eutric Brunisol	Eutrochrepts
	Deciduous	*Plantanus occidentalis, Sassafras albidum*	Luvisols and Gleysol	Hapludalfs
	Deciduous	*Liriodendron tulipifera, Magnolia acuminata*	Gleysol	Aquents
	Deciduous	*Acer saccharum, Fagus grandifolia*	Gray Brown Podzol	Udalfs
	Great Lakes-St. Lawrence	*Fagus grandifolia, Acer rubrum*	Humic Gleysol	Aquaolls
	Great Lakes-St. Lawrence	*Acer rubrum, Fagus grandifolia*	Dystric Brunisol	Dystrochrepts
	Great Lake-St. Lawrence	*Picea glauca, Abies balsamea*	Humo-Ferric Podzol	Haplorhods
	Great Lake-St. Lawrence	*Picea glauca, Abies balsamea*	Dystric Brunisol	Dystrochrepts
	Montane	*Abies lasiocarpa, Pseudotsuga menziesii, Picea glauca*	Humo-Ferric Podzol	Haplorhods
	Subalpine	*Picea engelmannii, Abies lasiocarpa*	Ferro-Humic Podzol	Cryorthod
Aporrectodea turgida (endogeic)	Acadian	*Acer saccharum, Betula alleghaniensis*	Orthic Podzol	Cryorthods
	Acadian	*Acer saccharinum, Fraxinus pennsylvanica*	Orthic Podzol	Cryorthods
	Acadian	*Abies balsamea, Betula papyrifera*	Ferro-Humic Podzol	Cryorthods

(continued)

Table 2 (continued)

Earthworm (Ecological type)	Forest region	Dominant tree Species	Soil type Canadian system	Soil type American system
	Acadian	*Abies balsamea, Betula papyrifera*	Orthic Gleysol	Typic Cyaquepts
	Boreal	*Abies balsamea, Betula papyrifera*	Brunisol	Inceptisols
	Boreal	*Picea glauca, Abies balsamea*	Orthic Regosol	Udorthents
	Boreal	*Betula papyrifera, Populus tremuloides*	Humo-Ferric Podzol	Haplorhods
	Boreal	*Acer saccharum, Tilia americana*	Eutric Brunisol	Eutrochrepts
	Coast	*Alnus rubra*	Ferro-Humic Podzol	Cryorthods
	Coast	*Sphagnum* bog	Fibrisol	Fibrists
	Deciduous	*Acer saccharum, Fagus grandifolia*	Gray Brown Podzol	Udalfs
	Deciduous	*Plantanus occidentalis, Sassafras albidum*	Orthic Luvisol	Hapludalfs
	Deciduous	*Liriodendron tulipifera, Magnolia acuminata*	Gleysol	Aquents
	Great Lakes-St. Lawrence	*Alnus rugosa*	Brunisol	Inceptisols
	Great Lakes-St. Lawrence	*Acer saccharum, Betula alleghaniensis*	Humo-Ferric Podzol	Haplorhods
	Great Lakes-St. Lawrence	*Fagus grandifolia, Acer rubrum*	Dystric Brunisol	Dystrochrepts
	Great Lakes-St. Lawrence	*Picea glauca, Abies balsamea*	Humo-Ferric Podzol	Haplorhods
	Montane	*Pseudotsuga menziesii, Populus tremuloides*	Humo-Ferric Podzol	Haplorhods
	Subalpine	*Picea engelmannii, Abies lasiocarpa*	Ferro-Humic Podzol	Cryorthods
Bimastos lawrenceae (corticole-epigeic)	Coast	*Tsuga heterophylla*	Humo-Ferric Podzol	Haplorhods
Dendrobaena octaedra (epigeic)	Acadian	*Acer saccharum, Betula papyrifera*	Orthic Podzol	Cryorthods

(continued)

Table 2 (continued)

Earthworm (Ecological type)	Forest region	Dominant tree Species	Soil type Canadian system	Soil type American system
	Acadian	*Acer saccharum, Tsuga canadensis*	Humo-Ferric Podzol	Haplorhods
	Boreal	*Abies balsamea, Betula papyrifera*	Brunisol	Inceptisols
	Boreal	*Abies balsamea, Betula tremuloides*	Gray Luvisol	Boralfs
	Boreal	*Betula papyrifera, Populus tremuloides*	Humo-Ferric Podzol	Haplorhods
	Boreal	*Abies balsamea, Betula tremuloides*	Gray Luvisol	Boralfs
	Boreal	*Betula glandulosa, Populus balsamifera*	Melanic Brunisol	Hapludolls
	Coast	*Alnus rubra*	Ferro-Humic Podzol	Cryorthods
	Coast	*Tsuga heterophylla*	Ferro-Humic Podzol	Cryorthods
	Deciduous	*Acer saccharum, Fagus grandifolia*	Gray Brown Podzol	Udalfs
	Deciduous	*Plantanus occidentalis, Sassafras albidum*	Luvisols and Gleysol	Hapludalfs
	Deciduous	*Liriodendron tulipifera, Magnolia acuminata*	Gleysol	Aquents
	Great Lakes-St. Lawrence	*Acer saccharum, Betula alleghaniensis*	Humo-Ferric Podzol	Haplorhods
	Great Lakes-St. Lawrence	*Fagus grandifolia, Acer rubrum*	Dystric Brunisol	Dystrochrepts
	Great Lakes-St. Lawrence	*Picea glauca, Abies balsamea*	Dystric Brunisol	Dystrochrepts
	Montane	*Populus tremuloides, P. balsamifera*	Gray Luvisol	Boralfs
	Montane	*Pinus contorta, Abies glauca, Populus tremuloides*	Eutric Brunisol	Cryochrepts
	Montane	*Pinus contorta, Abies lasiocarpa, Picea glauca*	Gray Luvisol	Boralfs

(continued)

Table 2 (continued)

Earthworm (Ecological type)	Forest region	Dominant tree Species	Soil type Canadian system	Soil type American system
	Montane	*Abies lasiocarpa, Pseudotsuga menziesii, Picea glauca*	Humo-Ferric Podzol	Haplorhods
	Subalpine	*Picea engelmannii, Abies lasiocarpa*	Ferro-Humic Podzol	Cryorthods
Dendrodrilus rubidus (corticole-epigeic)	Acadian	*Acer saccharum, Betula alleghaniensis*	Orthic Podzol	Cryorthods
	Acadian	*Acer saccharinum, Fraxinus pennsylvanica*		
	Acadian	*Abies balsamea, Betula papyrifera*	Orthic Gleysol	Typic Cryaquepts
	Acadian	*Abies balsamea, Betula papyrifera*	Humo-Ferric Podzol	Haplorhods
	[3]Boreal	*Abies balsamea, Betula papyrifera*	Brunisol	Inceptisols
	Boreal	*Abies balsamea, Betula tremuloides*	Gray Luvisol	Boralfs
	Boreal	*Populus tremuloides, P. balsamifera, Picea glauca*	Gray Luvisol	Boralfs
	Boreal	*Abies balsamea, Betula tremuloides*	Gray Luvisol	Boralfs
	Coast	*Tsuga heterophylla, Pseudotsuga menziesii Thuja plicata*	Humo-Ferric Podzol	Haplorhods
	Coast	*Tsuga heterophylla*	Ferro-Humic Podzol	Humic Cryorthods
	Deciduous	*Acer saccharum, Fagus grandifolia*	Gray Brown Podzol	Udalfs
Dendrodrilus rubidus	Deciduous	*Plantanus occidentalis, Sassafras albidum*	Luvisols and Gleysol	Hapludalfs
	Deciduous	*Liriodendron tulipifera, Magnolia acuminata*	Gleysol	Aquents
	Great Lakes-St. Lawrence	*Acer saccharum, Betula alleghaniensis*	Humo-Ferric Podzol	Haplorhods
	Great Lakes-St. Lawrence	*Picea glauca, Abies balsamea*	Dystric Brunisol	Dystrochrepts

(continued)

Table 2 (continued)

Earthworm (Ecological type)	Forest region	Dominant tree Species	Soil type Canadian system	Soil type American system
	Montane	*Abies lasiocarpa, Pseudotsuga menziesii, Picea glauca*	Humo-Ferric Podzol	Haplorhods
	Subalpine	*Picea engelmannii, Abies lasiocarpa*	Ferro-Humic Podzol	Cryorthods
	Subalpine	*Picea engelmannii, Picea glauca*	Ferro-Humic Podzol	Cryorthods
Lumbricus castaneus (epigeic)	Acadian	*Abies balsamea, Betula papyrifera*	Orthic Gleysol	Typic Cryaquepts
	Acadian	*Acer saccharum, Tsuga canadensis*	Humo-Ferric Podzol	Haplorhods
	Boreal	*Acer saccharum, Tilia americana*	Eutric Brunisol	Eutrochrepts
	Boreal	*Acer saccharum, Betula alleghaniensis*	Humo-Ferric Podzol	Haplorhods
	Coast	*Alnus rubra*	Ferro-Humic Podzol	Cryorthods
	Deciduous	*Acer saccharum, Fagus grandifolia*	Gray Brown Podzol	Udalfs
	Deciduous	*Plantanus occidentalis, Sassafras albidum*	Luvisols and Gleysol	Hapludalfs
	Deciduous	*Liriodendron tulipifera, Magnolia acuminata*	Gleysol	Aquents
	Great Lakes-St. Lawrence	*Acer saccharum, Betula alleghaniensis*	Humo-Ferric Podzol	Haplorhods
	Great Lakes-St. Lawrence			
Lumbricus rubellus (epi-endogeic)	Acadian	*Acer saccharum, Fagus grandifolia*	Orthic Podzol	Cryorthods
	Acadian	*Acer saccharinum, Fraxinus pennsylvanica*	Gleyed Humic Podzol	Cryohumods
	Acadian	*Abies balsamea, Betula papyrifera*	Ferro-Humic Podzol	Cryorthods
	Acadian	*Abies balsamea, Betula papyrifera*	Orthic Gleysol	Typic Cryaquepts

(continued)

Table 2 (continued)

Earthworm (Ecological type)	Forest region	Dominant tree Species	Soil type Canadian system	Soil type American system
	Acadian	*Abies balsamea, Betula papyrifera*	Humo-Ferric Podzol	Haplorhods
	Boreal	*Abies balsamea, Betula papyrifera*	Brunisol	Inceptisols
	Boreal	*Abies balsamea, Betula tremuloides*	Gray Luvisol	Boralfs
	Coast	*Pseudotsuga menziesii, Tsuga heterophylla*	Humo-Ferric Podzol	Haplorhods
	Coast	*Tsuga heterophylla*	Ferro-Humic Podzol	Cryorthods
	Coast	*Alnus rubra*	Humo-Ferric Podzol	Haplorhods
	Columbia	*Thuja plicata, Tsuga heterophylla*	Humo-Ferric Podzol	Haplorhods
	Deciduous	*Plantanus occidentalis, Sassafras albidum*	Luvisols and Gleysol	Hapludalfs
	Deciduous	*Liriodendron tulipifera, Magnolia acuminata*	Gleysol	Aquents
	Deciduous	*Acer saccharum, Fagus grandifolia*	Gray Brown Podzol	Udalfs
	Great Lakes-St. Lawrence	*Alnus rugosa*	Brunisol	Inceptisols
	Great Lakes-St. Lawrence	*Acer rubrum, Fagus grandifolia*	Dystric Brunisol	Dystrochrepts
	Great Lakes-St. Lawrence	*Picea glauca, Abies balsamea*	Humo-Ferric Podzol	Haplorhods
	Montane	*Picea contorta, P. glauca, Populus tremuloides*	Gray Luvisol	Boralfs
	Subalpine	*Picea engelmannii, Picea glauca*	Ferro-Humic Podzol	Cryorthods
Lumbricus terrestris (anecic)	Acadian	*Acer saccharum, Tsuga canadensis*	Humo-Ferric Podzol	Haplorhods

(continued)

Table 2 (continued)

Earthworm (Ecological type)	Forest region	Dominant tree Species	Soil type Canadian system	Soil type American system
	Boreal	*Abies balsamea, Betula papyrifera*	Brunisol	Inceptisols
	Boreal	*Abies balsamea, Betula tremuloides*	Gray Luvisol	Boralfs
	Coast	*Pseudotsuga menziesii*	Humo-Ferric Podzol	Haplorhods
	Coast	*Tsuga heterophylla*	Ferro-Humic Podzol	Cryorthods
	Coast	*Alnus rubra*	Ferro-Humic Podzol	Cryorthods
	Columbia	*Thuja plicata, Tsuga heterophylla*	Humo-Ferric Podzol	Haplorhods
	Deciduous	*Plantanus occidentalis, Sassafras albidum*	Luvisols and Gleysol	Hapludalfs
	Deciduous	*Liriodendron tulipifera, Magnolia acuminata*	Gleysol	Aquents
	Deciduous	*Acer saccharum, Fagus grandifolia*	Gray Brown Podzol	Udalfs
	Great Lakes-St. Lawrence	*Acer rubrum, Fagus grandifolia*	Dystric Brunisols	Dystrochrepts
	Great Lakes-St. Lawrence	*Fagus grandifolia, Acer rubrum*	Humic Gleysol	Aquaolls
Lumbricus Terrestris	Great Lake-St. Lawrence	*Picea glauca, Abies balsamea*	Humo-Ferric Podzol	Haplorhods
	Montane	*Populus tremuloides, Populus balsamifera*	Gray Luvisol	Boralfs
Octolasion cyaneum (endogeic)	Boreal	*Abies balsamea, Betula papyrifera*	Brunisol	Inceptisols
	Coast	*Pseudotsuga menziesii*	Humo-Ferric Podzol	Haplorhods
	Great Lake-St. Lawrence	*Acer saccharum, Tilia americana*	Humo-Ferric Podzol	Haplorhods
	Great Lake-St. Lawrence	*Acer saccharum, Fagus grandifolia*	Gray Brown Podzol	Udalfs

(continued)

Table 2 (continued)

Earthworm (Ecological type)	Forest region	Dominant tree Species	Soil type Canadian system	Soil type American system
Octolasion tyrtaeum (endogeic)	Acadian	*Acer saccharinum, Fraxinus pennsylvanica*	Humo-Ferric Podzol	Haplorhods
	Acadian	*Acer saccharum, Tsuga canadensis*	Humo-Ferric Podzol	Haplorhods
	Boreal	*Abies balsamea, Betula papyrifera*	Brunisol	Inceptisols
	Boreal	*Abies balsamea, Betula tremuloides*	Gray Luvisol	Boralfs
	Coast	*Alnus rubra, Pseudotsuga menziesii*	Humo-Ferric Podzol	Haplorhods
	Columbian	*Thuja plicata, Tsuga heterophylla*	Ferro-Humic Podzol	Humic Haplorthods
	Deciduous	*Plantanus occidentalis, Sassafras albidum*	Luvisols and Gleysol	Hapludalfs
	Deciduous	*Liriodendron tulipifera, Magnolia acuminata*	Gleysol	Aquents
	Deciduous	*Acer saccharum, Fagus grandifolia*	Gray Brown Podzol	Udalfs
	Great Lakes-St. Lawrence	*Alnus rugosa*	Brunisol	Inceptisols
	Great Lakes-St. Lawrence	*Picea glauca, Abies balsamea*	Dystric Brunisol	Dystrochrepts
	Montane	*Populus tremuloides, Populus balsamifera*	Gray Luvisol	Boralfs
	Subalpine	*Picea engelmannii, Abies lasiocarpa*	Ferro-Humic Podzol	Cryorthods
Megascolecidae				
Amynthas agrestis (epi-endogeic)	Deciduous	*Acer saccharum, Fagus grandifolia*	Gray Brown Podzol	Udalfs
Amynthas hilgendorfi (epi-endogeic)	Acadian	*Acer saccharum, Fraxinus pennsylvanica*	Humo-Ferric Podzol	Haplorhods
	Deciduous	*Acer saccharum, Fagus grandifolia*	Gray Brown Podzol	Udalfs
Amynthas tokioensis (epi-endogeic)	Acadian	*Acer saccharum, Fraxinus pennsylvanica*	Humo-Ferric Podzol	Haplorhods

(continued)

Table 2 (continued)

Earthworm (Ecological type)	Forest region	Dominant tree Species	Soil type Canadian system	Soil type American system
	Deciduous	*Acer saccharum*, *Fagus grandifolia*	Gray Brown Podzol	Udalfs
Arctiostrotus fontinalis (epi-endogeic)	Boreal	*Salix pulchra*, *Betula glandulosa*	Organic Cryosol	Gleic Histosols
Arctiostrotus perrieri (epigeic)	Coast	*Pseudotsuga menziesii*, *Tsuga heterophylla*	Dystric Brunisol	Dystrochrepts
	Coast	*Thuja plicata*, *Tsuga heterophylla*	Ferro-Humic Podzol	Humic Haplorthods
	Coast	*Alnus rubra*, *Pseudotsuga menziesii*	Ferro-Humic Podzol	Humic Haplorthods
	Coast	*Acer macrophyllum*	Ferro-Humic Podzol	Humic Haplorthods
	Coast	*Thuja plicata*, *Tsuga heterophylla*	Fibrisol	Fibrists
Arctiostrotus vanouverensis (epigeic)	Coast	*Tsuga heterophylla*, *Abies amabilis*	Humo-Ferric Podzol	Haplorthods
	Coast	*Thuja plicata*, *Tsuga heterophylla*	Dystric Brunisol	Dystrochrepts
		Thuja plicata, *Tsuga heterophylla*	Ferro-Humic Podzol	Humic Haplorthods
Pithemera bicincta (epigeic)	Deciduous	*Acer saccharum*, *Fagus grandifolia*	Gray Brown Podzol	Udalfs
Toutellus oregonensis (endogeic)	Coast	*Pseudotsuga menziesii*	Humo-Ferric Podzol	Haplorthods
	Coast	*Thuja plicata*	Ferro-Humic Podzol	Humic Haplorthods

Sources: Addison (2009); Bottinelli et al. (2020); Cameron et al. (2007); Dymond et al. (1997); Eisenhauer et al. (2007); McAlpine and Reynolds (2018, 2019, 2020); McKey-Fender et al. (1994); McLean and Parkinson (1997); Moore and Reynolds (2003); Moore et al. (2009, 2018, 2022), Panesar et al.(2000); Reynolds (1975a, 1975b, 1975c, 1975d, 1976a, 1976b, 1977a), Reynolds and McTavish (2021), Reynolds and Reynolds (1992), Reynolds et al. (2014, 2015, 2019), Welke and Parkinson (2003); Whalen (2004), Wironen and Moore (2006)

poplar) > *Fraxinus* (ash) = *Tilia* (basswood) = *Gleditsia* (locust) = *Betula* (birch) > *Quercus* (oak) = *Fagus* (beech) > *Picea* (spruce) = *Pinus* (pine) = *Abies* (fir) = *Thuja* (cedar) = *Tsuga* (hemlock) = *Pseudotsuga* (Douglas fir) = *Larix* (larch/tamarack) (Reynolds 1972; Reynolds and Jordan 1975). I have termed this sequence the leaf palatability index since the earthworm preference is the same as the natural leaf disappearance order. This is not surprising since the greater amount of lignins and tannins in the oak and beech leaves and conifer needles are more difficult for earthworms to digest.

Soil temperature is important for earthworm development with the optimum being 12–25° C for most species. Soil reaction is also a consideration and earthworms have been reported in forest soils as low as pH 2.8 (KCl). Most reported soil pH for earthworm habitats is 4.0 to 8.1 (H_2O) (Reynolds and Jordan 1975). This was recently supported by Ruiz et al. (2021):

> The worldwide distribution of earthworms is mostly influenced by favorable soil moisture and mechanical circumstances. Earthworm activity is prevented in many areas of the globe due to other limitations, such as the presence of permafrost soil and very low mean annual temperatures below freezing. Although there is data indicating that soil acidity (with a pH below 4.5) may restrict the dispersal of earthworms, it was discovered that the worldwide distribution of earthworms was not significantly affected by low soil pH levels. The predominant factor influencing the distribution of earthworms seems to be soil physical (hydromechanical) conditions, which are influenced by soil water content and the physiological constraints of earthworms in soils that are unfrozen.

The distribution of soil orders in Canada is shown in Fig. 3.

The results in **Appendix 1** should come as no surprise, considering 15.8% of the land mass contains podzolic soils (= Spodosols). Therefore, it would be suspected

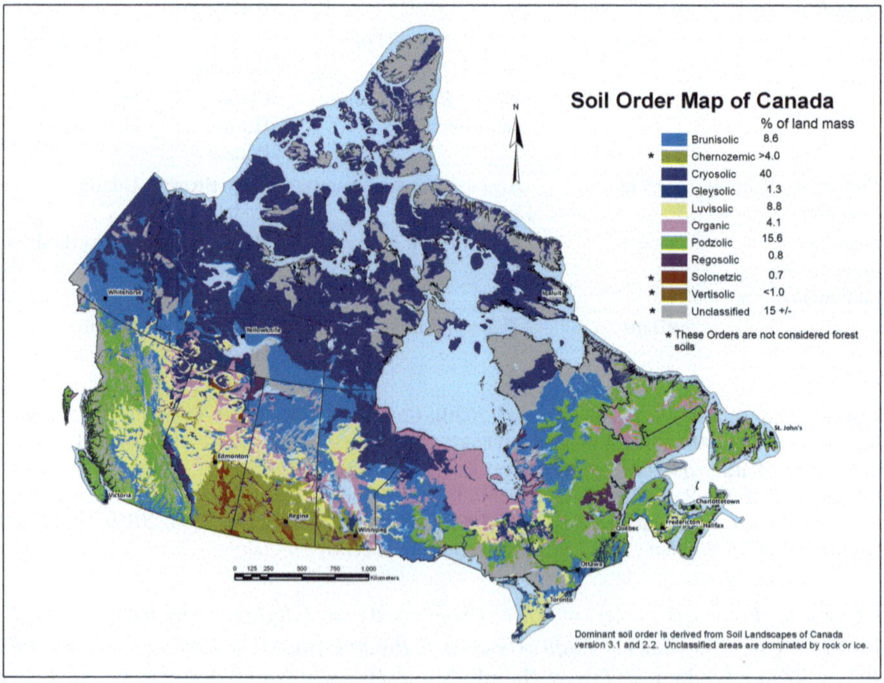

Fig. 3 Soil order map of Canada with percent of land mass (Canadian Society of Soil Science 2020)

that earthworms would be recorded primarily in these soils. It should be pointed out that 20.7% of the land mass is not considered forest or forest soils (marked * on Fig. 3). Also, cryosolic soils (= Gelisols) which represent 40% of the land mass are in areas which are for the most part devoid of earthworms. If you add these two groups together, it represents approximately 61% of the land mass, so that podzolic soils comprise about 62.5% of the forested land mass which could be inhabited by earthworms. The second largest area where earthworms have been recorded are the brunisolic order (= Inceptisols) which represents approximately 31% of the forested land mass. The remaining 6.5% of the forested land mass where earthworms are reported is divided between Gleysols and Organic (= Histosols), Luvisols (= Alfisols), and Regosols (= Entisols) soils.

A summary of earthworm species recorded from the various forest soils in Canada is listed in **Appendix 2.** A summary of the soil type profiles from which earthworms have been detected in Canada can be identified in **Appendix 3.**

At one time, I held the belief that earthworms in their entirety were beneficial, devoid of any detrimental consequences associated with their activities within the soil. Their proliferation as invasive species and the impact of climate change have compelled me to reconsider my initial stance in recent years. I frequently envisioned myself as the opposite of Johnny Appleseed, journeying around the globe to collect samples in a haphazard manner, rather than dispersing apple seeds indiscriminately wherever he went (Hillis 1917; Means 2011).

The question that has been on the minds of many scientists is—what effect is climate change going to have on earthworms in the Canadian Forest Regions?

The presence of invasive pheretimoid earthworms has caused changes in the physical characteristics of soils. These earthworms generate a casting layer that serves as a thermal shelter (Görres et al. 2019). In their study, Ali and Naaz (2013) used earthworms as biological markers to evaluate the environmental repercussions on soils. In terrestrial environments, organisms above as well as below the surface of the soil could be significantly impacted by climate change (Singh et al. 2019). Various ecological conditions could influence the population densities, biomass, and species composition of earthworms. Without any question, the prominent and often mentioned elements consist of soil water content, temperature of the soil, and pH of the soil, all of which have the ability to function as significant limitations. Additional variables, such as the depth and texture of the soil, the availability of nutrients and vegetation, the physical characteristics of the site (aspect, elevation location, slope), and even the color of the soil, can combine with the mentioned factors and affect conditions in the soil. An initial model that takes into account these combinations was released almost half a century ago (Reynolds and Jordan 1975). We proposed that the measurement of soil water stress (tension) has more significance compared to the amount of moisture present in the soil. However, it appears that this aspect has been overlooked in the majority of subsequent investigations, such as those conducted by Diehl and Williams (1992) and Bessolitsyna (2012). Earthworm populations fluctuate in response to variations in patterns of rainfall (Tondoh 2006; Eisenhauer et al. 2019). However, if earthworms have insufficient access to enough water, this fluctuation might not be viable in the long term. Thus, soil water stress provides a more accurate estimate of the earthworm's capability to breed and consume food in a certain environment (Reynolds 2021).

Biological invasions can have strong impacts on biodiversity as well as ecosystem functioning and introduced below ground organisms (e.g., earthworms) may have strong effects. The impacts of introduced or invading earthworms on plant diversity and community composition in Canadian forests can be significant (Craven et al. 2017).

Over the past five decades, I have seen an obvious variation in the variety of earthworm species in the United States and Canada. During the period from the middle to the late 1960s, the Oriental pheretimoid species, including *Amynthas*, *Metaphire*, *Pithemera*, and *Polypheretima* spp., were primarily discovered in the southern region of the United States, specifically below Tennessee (Reynolds 1978, 2018). Their parthenogenetic reproduction allows them to rapidly expand their territory in response to fluctuations in climate. Over the course of the last five decades, they have expanded their presence to include the soils of America's Midwest as well as Northeast (Reynolds 2010c, 2011b). Only recently did I document the first occurrences of these earthworms in Canada, as reported by Reynolds in (2014a), then subsequently in (2021) by Reynolds and McTavish, and in (2021) by McAlpine et al. Although I might not witness it, the warming of soils in Canada and the subsequent increase in rainfall and flora diversity may eventually lead to earthworms becoming a significant part of the soil fauna in the Canada's far north.

The boreal wooded areas of western Canada are experiencing tremendous expansion in leisure and financial transactions. These activities have been proposed to be promoting the importation and proliferation of non-indigenous species, namely exotic European earthworms, including *Dendrodrilus rubidus*, *Dendrobaena octaedra*, and *Lumbricus rubellus* (Cameron et al. 2007; Reynolds et al. 2019). Cameron et al. proposed four pathways for the introduction of earthworms due to their limited ability to travel independently: (1) discarded bait used for fishing, (2) transportation by automobiles, (3) dissemination by vertebrates, and (4) dispersal through river systems. According to their analysis, the first two pathways are the most probable. Based on my own personal observations, especially in the United States, I have seen that river-flooding is an effective method of dispersion. In the northernmost regions where the waterways converge and merge with the Arctic area, this is a very probable occurrence.

Dr. Erin K. Cameron, a researcher in environmental science at Saint Mary's University, discovered that almost all of the earthworms at a specific research site in Alberta are a species that consumes leaf debris but does not dig through the soil. She conducted a research in 2015 where she simulated the long-term impacts of the worms. The carbon content of the forest floor is decreased by 50 to 94%, mostly during the first four decades, and this carbon is subsequently emitted into the earth's atmosphere (Cameron et al. 2015). In an earlier study, Cameron et al. (2007) determined the worms' range would increase from 9% of northeastern Alberta to half of the region by 2049.

Currently, the Arctic region is characterized by a lack of trees and is mostly covered by sparsely distributed tundra flora, including sedges, grasses, and a limited number of flowering species (Reynolds 2020). Conversely, the sediment cores from Russia included pollen derived from tree species such as fir (*Abies* spp.), hemlock (*Tsuga* spp.), larch (*Larix* spp.), and spruce (*Picea* spp.). This observation indicates that the current distribution of boreal woodlands, which currently extends hundreds of kilometers west and south in the Russian Federation and up to the Arctic Circle in Alaska and Canada, was historically much more extensive. In the past, these forests stretched all the way to the Arctic Ocean, covering vast areas of Arctic North America and Arctic Russia. Modifications in Arctic flora may have significant consequences for the relationships between organisms at different trophic levels and the overall functioning of the ecosystem, which in turn can contribute to climate feedbacks. Arctic vegetation populations and species are often responsive to increases in temperature, however changes noticed across time are varied and intricate (Bjorkman et al. 2018).

Scientists have observed significant summer melting rates over the Greenland Ice Sheet, providing more proof of the alarming Arctic warming. Canada's last surviving ice shelf, located in Nunavut, disintegrated and fell into the ocean at the beginning of August 2020. In the summertime of 2020, several areas of Arctic Siberia and Svalbard, a cluster of Norwegian islands in the Arctic Ocean, experienced extremely high temperatures, reaching a record-breaking 38°C (Brigham-Grette and Petsch 2020).

Earthworms must adapt to the temperature conditions generally prevailing in the soil. Reynolds (1973) reported the optimum soil temperature for earthworm development for several species: *Aporrectodea rosea* 12°C, *Aporrectodea turgida* 12°C, *Dendrodrilus rubidus* 18–20°C, *Eisenia fetida* 25°C, *Lumbricus rubellus* 15–18°C, and *Octolasion cyaneum* 15°C.

Soil has a higher capacity to retain heat or cold compared to air. The insulation of the area is provided by the soil layer above it, as well as by the presence of flora and snow. As you descend farther, the level of insulation increases and the mean temperature either rises or falls, up to a certain extent. While air temperature can exhibit changes of 50°F (10°C) in 24 h, the change in soil temperature can take days to change even a couple of degrees.

According to research conducted in the Swedish Boreal Woodland by Jungqvist et al. (2014), the greatest variation was anticipated to happen in the top soil, with variations between the responses of the top and bottom layers being most noticeable in the end of summer and the beginning of the fall season. However, air temperature may also be altering more quickly than soil temperatures in this catchment on a yearly scale. This study demonstrated the intricate nature of the anticipated impacts of climate change on soil temperature in locations characterized by snow. It emphasizes that relying just on air temperature to predict future soil temperature responses to climate change is insufficient and should be ignored in boreal zones.

Consequently, climate change will mostly impact epigeics that inhabit the layer of organic matter and topsoil, e.g., *Arctiostrotus fontinalis*, *Ar. perrieri*, *Ar. vancouverensis*, *Bimastos lawrenceae*, *Dendrobaena octaedra*, *Dendrodrilus rubidus*, *Lumbricus castaneus*, and *L. festivus*. The endogeic species which live in the surface soil will be less affected by the early stages of climate change, e.g., *Aporrectodea limicola*, *Ap. rosea*, *Ap. trapezoides*, *Ap. tuberculata*, *Ap. turgida*, *Lumbricus rubellus*, *Octolasion cyaneum*, and *O. tyrtaeum*. The anecic species, *Aporrectodea longa* and *Lumbricus terrestris*, have a higher resilience to the impacts of climate change due to their ability to tunnel deeper into the soil profile and surface only for the purposes of feeding and reproduction.

One species in particular found in western Canada, *Arctiostrotus perrieri*, will definitely be affected by climate change, "This taxon is very sensitive to high temperatures and will rapidly degrade in an environment that is pleasantly comfortable for humans." Fender and McKey-Fender (1990: 364).

Grant (1955) using three different species determined their upper lethal temperatures to be 27.7°C (*Eisenia fetida*), 24.9 °C (*Amynthas hupeiensis*) and 24.7 °C (*Aporrectodea turgida*) following acclimatization at 22°C. Acclimatization results in a mean increase in resistance to heat of 0.3°C for every 1°C rise in the conditioning temperature (Tables 3 and 4).

The existence of parthenogenetic morphs, together with their extensive adaptability to various soil and habitat conditions, makes these species very proficient invaders, especially in tropical, subtropical, and even temperate areas (Chang et al. 2017; Brown et al. 2006).

5 Arctic Earthworm Migration and Climate Change

The inquiry pertains to the potential impact of earthworm invasion in the Northern region on the soil's carbon emission or the augmentation of atmospheric carbon dioxide sequestration in the soil. Currently, however, it is challenging to ascertain, and the findings may frequently be in opposition to one another. Zhang et al. (2013) and Coleman et al. (2017) provided evidence of the contribution of earthworms to the process of net carbon sequestration. This process involves the removal of carbon dioxide from the atmosphere and its storage in the soil in a solid or liquid form.

Earthworms have been perceived to accelerate frequently the release of CO_2, particularly in short-term investigations. However, in some over time trials, they have additionally been shown to improve the process of carbon stabilization in soil aggregates (Coleman et al. 2017). However, further empirical evidence corroborates the perspective that earthworms diminish carbon sequestration, since the detection of CO_2 emissions is more straightforward compared to the stability of carbon. Consequently, a new meta-analysis investigation discovered that the existence of earthworms would lead to a 33% rise in carbon dioxide emissions from the soil (Lubbers et al. 2013). Contrarily, Zhang et al. (2013) discovered that earthworms may enhance the process of carbon sequestration by unevenly increasing the stability of carbon in comparison to its mineralization. Zhang et al. suggested that the effects of earthworms on carbon sequestration cannot be solely attributed to changes in carbon dioxide emissions or stabilized carbon. They emphasized the need to investigate the simultaneous effects of earthworms on both carbon mineralization and carbon stabilization procedures. They discovered that earthworms enhance the

Table 3 Summary of the earthworm species in the various forest regions

Species	Acadian	Boreal	Coast	Columbian (Carolinian)	Deciduous	Great Lakes-St. Lawrence	Montane	Subalpine
Lumbricidae								
Aporrectodea limicola	–	–	●	–	–	–	–	–
Aporrectodea longa	●	–	●	●	–	●	–	–
Aporrectodea rosea	●	●	●	–	●	●	–	–
Aporrectodea trapezoides	●	●	●	●	●	–	–	–
Aporrectodea tuberculata	●	●	●	●	●	●	●	●
Aporrectodea turgida	●	●	●	–	●	●	●	●
Bimastos lawrenceae	–	–	●	–	–	–	–	–
Bimastos parvus	–	●	–	–	–	–	–	–
Dendrobaena lusitana	●	–	–	–	–	–	–	–
Dendrobaena octaedra	●	●	●	–	●	●	●	●
Dendrodrilus rubidus	●	●	●	–	●	●	●	●
Lumbricus castaneus	●	●	●	–	●	●	–	–
Lumbricus festivus	●	●	●	–	●	●	–	–
Lumbricus rubellus	●	●	●	●	●	●	●	●
Lumbricus terrestris	●	●	●	●	●	●	●	–
Octolasion cyaneum	–	●	●	–	–	●	–	–
Octolasion tyrtaeum	●	●	●	●	●	●	●	●
Megascolecidae								
Amynthas agrestis	–	–	–	–	●	–	–	–
Amynthas hilgendorfi	●	–	–	–	●	–	–	–
Amynthas minimus	●	–	–	–	–	–	–	–
Amynthas tokioensis	●	–	–	–	●	–	–	–
Arctiostrotus fontinalis	–	●	–	–	–	–	–	–

Species	Acadian	Boreal	Coast	Columbian (Carolinian)	Deciduous	Great Lakes-St. Lawrence	Montane	Subalpine
Arctiostrotus perrieri	–	–	●	–	–	–	–	–
Arctiostrotus vancouverensis	–	–	●	–	–	–	–	–
Pithemera bicincta	–	–	–	–	●	–	–	–
Toutellus oregonensis	–	–	●	–	–	–	–	–

Additional species not considered to be present in Canadian forest soils: Lumbricidae—*Allolobophora chlorotica, Aporrectodea bowcrowensis, Aporrectodea icterica, Bimastos beddardi, Dendrobaena attemsi, Eisenia fetida, Eiseniella tetraedra, Satchellius mammalis*, Megascolecidae—*Amynthas minimus*, and Sparganophilidae—*Sparganophilus tamesis*

Table 4 Summary of earthworm species in Canadian forest soils

Earthworm species	Forest soils										
	Humo-Ferric Podzols	Ferro-Humic Podzols	Gray Brown Podzols	Dystric Brunisols	Eutric Brunisols	Melanic Brunisols	Humic/Orthic Gleysols	Orthic/Humic Regosols	Turbic/Organic Cryosols	Gray/Brown Luvisols	Fibrisol
Lumbricidae											
Aporrectodea limicola	●	–	–	–	–	–	–	–	–	–	–
Aporrectodea longa	●	–	●	–	–	–	–	–	–	–	–
Aporrectodea rosea	●	●		●		●				●	–
Aporrectodea trapezoides	●	–	●	●	●	–	●			●	–
Aporrectodea tuberculata	●	●	●	●	●	●	●	●	●	●	●
Aporrectodea turgida	●	●	–	●	●	–	–	●	●	–	●
Bimastos lawrenceae	●	–	–	–	–	–	–	–	–	–	–
Bimastos parvus	–	–	–	–	–	–	–	–	–	●	–
Dendrobaena lusitana	●	–	–	–	–	–	–	–	–	–	–
Dendrobaena octaedra	●	●	–	●	●	●	●	–	–	●	–
Dendrodrilus rubidus	●	●	●	●	–	–	●	–	–	●	–

(continued)

Table 4 (continued)

Earthworm species	Forest soils										
	Humo-Ferric Podzols	Ferro-Humic Podzols	Gray Brown Podzols	Dystric Brunisols	Eutric Brunisols	Melanic Brunisols	Humic/Orthic Gleysols	Orthic/Humic Regosols	Turbic/Organic Cryosols	Gray/Brown Luvisols	Fibrisol
Lumbricus castaneus	●	●	●	–	●	–	●	–	–	–	–
Lumbricus festivus	●	–	●	–	–	●	–	●	–	–	●
Lumbricus rubellus	●	●	●	●	–	–	●	–	–	●	–
Lumbricus terrestris	●	●	●	●	–	–	●	–	–	●	–
Octolasion cyaneum	●	–	●	●	–	–	–	–	–	–	–
Octolasion tyrtaeum	●	●	●	●	–	–	●	–	–	●	–
Megascolecidae											
Amynthas agrestis	–	–	–	–	–	–	●	–	–	●	–
Amynthas hilgendorfi	–	–	–	–	–	–	●	–	–	●	–
Amynthas minimus	–	–	–	–	–	–	–	–	–	●	–
Amynthas tokioensis	–	–	–	–	–	–	●	–	–	●	–
Arctiostrotus fontinalis	–	–	–	–	–	–	–	–	●	–	–

Earthworm species	Forest soils										
	Humo-Ferric Podzols	Ferro-Humic Podzols	Gray Brown Podzols	Dystric Brunisols	Eutric Brunisols	Melanic Brunisols	Humic/Orthic Gleysols	Orthic/Humic Regosols	Turbic/Organic Cryosols	Gray/Brown Luvisols	Fibrisol
Arctiostrotus perrieri	●	●	–	●	–	–	–	–	–	–	–
Arctiostrotus vancouverensis	●	●	–	●	–	–	–	–	–	–	–
Pithemera bicincta	–	–	–	–	–	–	●	–	–	–	–
Toutellus oregonensis	●	●	–	–	–	–	–	–	–	–	–

process of carbon mineralization. However, the overall amount of carbon dioxide that could possibly be released from the soil with earthworms is not significantly different from soil without earthworms. This is because both earthworms and the soil's microbes have the same capacity to mineralize carbon. Prior studies frequently failed to acknowledge this fact, leading to the erroneous conclusion that earthworms reduce carbon sequestration only due to the frequent increase in CO_2 emissions caused by earthworms (Lavelle 1997; Liski et al. 2003). Furthermore, the net carbon sequestration is determined by the size of the activated carbon pool (C_{act}) and how it is allocated between carbon mineralization and carbon stabilization, which could naturally enhance due to a surge in carbon mineralization (C_{min}) and carbon stabilization (C_{sta}). Zhang et al. have made a significant contribution by introducing the innovative notion of the sequestration quotient (SQ, C_{sta}/C_{act}). This concept allows us to accurately measure the influence of earthworms on the delicate equilibrium between carbon stabilization and mineralization. The findings of the research indicate that the existence of earthworms plays a significant role in promoting carbon sequestration. This is because the carbon stabilized by earthworms surpasses the amount that is transformed into carbon dioxide during the process of carbon mineralization. Consequently, sequestration quotient values are notably higher in areas where earthworms are present.

Earthworms have a crucial role in regulating the greenhouse gas levels in soils globally, and their impact is projected to increase in the coming decades. It is believed that they promote the process of carbon sequestration in soil aggregates, but they also lead to a rise in the release of the primary greenhouse gases, namely CO_2 (carbon dioxide) and N_20 (nitrous oxide). Thereby, the question of whether earthworms primarily contribute to or reduce greenhouse gas emissions in soils remains a subject of intense discussion. Lubbers et al. (2013) conducted a quantitative analysis to assess the overall impact of earthworms on the rise of greenhouse gas emissions from soils.

A new study conducted by Blume-Werry et al. (2020) has shown that invasive earthworms have the ability to release nitrogen in Arctic plants. The development of plants in the arctic is often constrained by a scarcity of nitrogen, which is caused by the combination of cold temperatures and sluggish soil microbial activity. Their research demonstrated that the nitrogen cycle in arctic plant soil systems was limited by the absence of earthworms and their associated behaviors. The presence of earthworms led to an upsurge in both the height and quantity of floral shoots, as well as an improvement in fine root formation and plant sustainability (greenness). The implications of earthworms indicate that the introduction of these organisms by humans could potentially result in significant alterations to the composition and operation of arctic ecological systems. The authors only identified the earthworms as *Aporrectodea* sp. and *Lumbricus* sp. Given their citation of Reynolds (2000a), it might be concluded that they are referring to *Ap. tuberculata* and *L. rubellus* or *L. terrestris*.

Another factor which is receiving more attention due to global climate change is wildfires, which have become more frequent and intense worldwide. This is presenting an acute problem for forested countries, and northern Canada in particular. Forest ecosystems are decreasing rapidly, which causes an irreparable loss of biodiversity. This situation may cause a permanent loss of some species and animal communities. Even if recovery is possible, it may take a long time to recover. There are competing views that wildfires increase biodiversity (Geraskina et al. 2021). Currently, vegetation in the Canadian north is sparse compared to the south. If the widespread forest fires of recent years continue as the climate changes, the spread of earthworms into the northern forest regions may be delayed.

A recent report has shown that earthworms are bioindicators of soil health (Alves et al. 2022). The use of biological indicators, such as earthworms, to assess the health of soil has grown because they allow an integrated assessment of the terrestrial ecosystem's functioning.

The Arctic is now experiencing unprecedented warmth not seen in the last 3 million years. Geologists are very concerned on the implications of these changes in the Arctic and their potential global impact. In a little span of two centuries, humanity have managed to entirely undo the course set in motion 50 million years ago and bring the planet back to CO_2 levels that have not been seen for millions of years. According to scientific projections, it is expected that the Arctic will be entirely devoid of ice during the summer season over the next 20 years, namely by the year 2040 (Brigham-Grette and Petsch 2020).

Recent research reveals that the Canadian North is seeing growth. Canada has been seeing a growth rate of 5%, however, the northern regions are expanding at a quicker pace on average. Based on data gathered in May 2016, the three territories in the Canadian North are seeing a higher rate of growth compared to the rest of Canada, with Nunavut leading with a growth rate of 12%. From 2016–2021 the Yukon had a growth rate of 12.1% (Statistics Canada 2022).

As the north warms, there will be vegetational changes, the population will increase, and tourism will also increase. Given the limited natural migration abilities of earthworms, it is important to acknowledge that human activities play a significant role in the spread of earthworm species into northern woodlands. Based on my professional assessment, I anticipate a significant increase in both the abundance and variety of earthworms in Arctic and forest soils by the year 2050.

Acknowledgments I would like to express my gratitude to W.M. Reynolds from the Oligochaetology Laboratory in Kitchener, Dr. Frederick W. Kutz, formerly of the US EPA in Washington, and Mark J. Wetzel from the Illinois Natural History Survey, Prairie Research Institute at the University of Illinois Urbana-Champaign, for their valuable contributions in reviewing the manuscript and providing insightful comments and suggestions.

Appendices

Soil Type Profiles Where Earthworms Have Been Recorded

(Soil Classification Working Group 1998)

Cryosolic Order

The soils include permafrost, which refers to material that remains frozen indefinitely. This permafrost is located within one meter of the surface, or two meters if the soil has been significantly disturbed by frost activity, a process known as cryoturbation. Permafrost acts as an obstacle for roots and water, causing the layer above it, known as the active layer, to potentially become a saturated and semifluid substance during the spring thaw. Frequently, the layer of permafrost in close proximity to the soil's surface is rich in ice. The thawing of ice and frozen substances, triggered by the disruption of surface plants in boreal forests or tundra regions, can lead to soil slumping and the consequent disturbance of infrastructure such as roadways and pipelines, as well as building structures. Cryosolic soils, encompassing approximately 3,672,000 square kilometers (equivalent to roughly 40% of Canada's entire land area), exhibit prevalence across significant portions of the Yukon, Northwest Territories, and Nunavut. These soils are also present in northern regions of all provinces, with the exception of the Atlantic Provinces (with the exception of Labrador). Turbic Cryosols have a structured surface with distinct features such as stone networks and hummocks. They also display mixed horizons or other indications of cryoturbation. Static Cryosols exhibit little signs of cryoturbation and are often seen in conjunction with gravelly or sandy substances. Organic Cryosols consist mostly of organic substances, such as peat. Organic Cryosols are found in latitudes beyond the ongoing permafrost barrier due to the insulating properties of organic materials.

Earthworms present: *Ap. tuberculata*, *Ap. turgida*, *Arctiostrotus fontinalis*.

Organic Order

The composition of these soils is mostly organic matter, with a concentration of over 30% by weight in the top 1/2 meter. Additionally, there is no permafrost in close proximity to the soil surface. The soils found in peatlands, such as bogs, swamps, and fens, are the main types. Organic soils often form from the accumulation of vegetative materials from species that thrive in waterlogged environments. Certain organic soils consist mostly of vegetation materials that have been accumulated in lakes, while others are primarily formed of woodland leaf litter found on steep slopes in regions with frequent precipitation. Canada's geographical surface is covered by organic soils, which span around 374,000 km^2, accounting for 4.1% of the total area. These soils are mostly found in Ontario, Manitoba, and northern Alberta, with lesser patches in other territories and provinces.

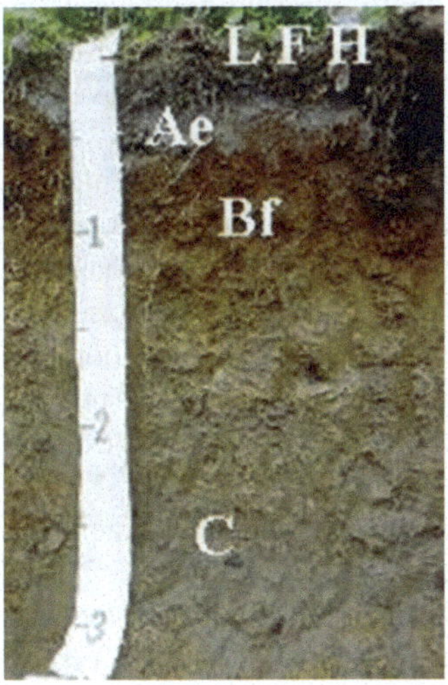

Organic soils are classified into four major categories. Fibrisols, which are often found in Canada, primarily include organic material that is largely undecomposed and contains easily identifiable plant pieces. Resistant fibers constitute more than

40% of the volume of these soils. Fibrisols are the most common kind of soil that originates from Sphagnum mosses. Mesisols exhibit a higher degree of decomposition and have a lower proportion of fibrous debris compared to Fibrisols (10–40% by mass). Humisols mostly consist of decomposed organic matter and may include a maximum of 10% fiber in terms of mass. Folisols mostly include substantial accumulations of forest detritus that overlay either bedrock, cracked bedrock, or loose debris. They are often found in moist highlands in the British Columbia coastal regions.

Earthworms present: *Ap. tuberculata, Ap. turgida, Lumbricus festivus.*

Podzolic Order

The acid soils possess a B horizon that contains deposits of shapeless substances made up of decomposed organic matter linked with iron and aluminum. They primarily form on sandy substrates within regions characterized by cold, humid climates and are often found under forests or shrubs. Water percolating downwards through the fairly porous substance extracts essential components (such as calcium), leading to the formation of an acidic environment. Decomposed forest litter produces soluble organic compounds that corrode soil minerals in the upper layers of the soil. A significant portion of the aluminum and iron that is freed from this process binds with organic matter. Once the ratio of iron and aluminum to organic matter surpasses a threshold, the organic complex undergoes a transformation, rendering it insoluble and subsequently leading to its deposition within the B horizon. Aluminum and iron that have dissolved might additionally migrate downwards in inorganic forms and settle as complexes of aluminum-silicon and oxides of iron. Ae horizon, which is typically light gray and highly leached, often lies above the Podzolic B horizon.

Podzolic soils cover about 1,429,000 square kilometers, which accounts for 15.6% of the total land area of Canada. These soils are prevalent in large portions of the humid Appalachian and Canadian Shield zones, as well as in the moist coastline zone of British Columbia. They are categorized into three major groupings based on the kind of Podzolic B horizon. Humic Podzols are characterized by a B horizon that is black in color and has a relatively low concentration of iron. They mostly occur in moist areas with high levels of humidity and are significantly less prevalent compared to other types of Podzolic soils.

Ferro-Humic Podzols feature a B horizon that is dark brownish-red or black and has a minimum of 5% organic carbon together with significant levels (usually 2% or higher) of both aluminum and iron in organic compounds. They frequently appear in the more wet places of the Podzolic soil region, such as British Columbia coasts and some areas of Newfoundland and southern Quebec. Approximately 36% of the soils of Labrador, namely in the Churchill River valley, consist of Ferro-Humic Podzols. The most prevalent podzolic soils in Canada are called Humo-Ferric Podzols. They feature a brownish-red B horizon with no more than 5% organic carbon linked to iron as well as aluminum compounds.

Earthworms present: *Aporrectodea limicola, Ap. longa, Ap. rosea, Ap. trapezoides, Ap. tuberculata, Ap. turgida, Bimastos lawrenceae, Dendrobaena lusitana, Db. octaedra, Dendrodrilus rubidus, Lumbricus castaneus, L. festivus, L. rubellus, L. terrestris, Octolasion cyaneum, O. tyrtaeum, Arctiostrotus perrieri, Ar. vancouverensis, Toutellus oregonensis.*

Gleysolic Order

These soils experience occasional or persistent water saturation and oxygen depletion. They often appear in shallow indentations and flat regions in moderately moist and wet climates, often with other types of soil on inclines and hillsides. Following the melting of snow or the occurrence of significant rainfall, lowlands in the terrain may get inundated. Flooding ensues while the soil temperature surpasses around 5 °C, leading to microbial activity that swiftly depletes oxygen during a couple of days. In such circumstances, soil components that have undergone oxidation (such as nitrate and ferric oxide) face reduction. The removal of ferric oxide results in the elimination of the brownish hue that is often seen in several soils, resulting in a gray appearance. As the soil becomes dry and oxygen returns, the decreased iron might undergo oxidation in certain locations, resulting in the formation of vivid yellow-brown blotches known as mottles. Gleysolic soils are often characterized by their inadequate drainage and dull gray hue, occasionally mixed with brown mottles. Canada's geographical area has around 117,000 square kilometer (1.3%) of Gleysolic soils.

Three distinguished classifications of Gleysolic soils are delineated. Humic Gleysols exhibit the characteristic of possessing a prominently dark A horizon that is notably enriched in organic substances. Gleysols are characterized by the absence of a distinct horizon. Luvic Gleysols exhibit the characteristic presence of a leached (A_e) horizon, which is situated beneath a B horizon where clay accumulation is observed. Additionally, it is worth noting that these soils may display a surface horizon with a darker hue.

Earthworms present: *Ap. tuberculata, Ap. turgida, Dendrobaena octaedra, Dendrodrilus rubidus, Lumbricus castaneus, Lumbricus rubellus, L. terrestris, Octolasion tyrtaeum, Amynthas agrestis, Am. hilgendorfi, Am. tokioensis, Pithemera bicincta.*

Luvisolic Order

The soils under consideration consist of eluvial horizons, which have experienced clay leaching due to snow melting or excessive rainfall, as well as illuvial horizons, which have witnessed clay deposition. These horizons are referred to as A_e and B_t, accordingly. In the context of saline or calcareous substances, it is essential to note that the process of clay transfer usually commences by the leaching of salts and carbonates. Luvisolic soils are commonly found in regions characterized by subhumid to moist climates, particularly in woodland regions. These soils develop when the parent materials possess a significant amount of clay. Luvisolic soils include around 809,000 square kilometers, which accounts for 8.8% of Canada's total geographical area. The middle to northern Interior Plains include broad expanses of Luvisolic soils, while smaller patches may be found in all locations situated south of the permafrost area.

The classification of Luvisolic soils is mostly based on soil temperature, resulting in the formation of two major groupings. A dark A_h horizon, where organic matter has been incorporated with the mineral material (often through the action of earthworms), an A_e horizon, which is brown and frequently platy, and a B_t horizon, which is illuvial and frequently contains blocky structures. Their annual average soil temperature is at least 8 degrees Celsius. The southern region of the Great Lakes-St. Lawrence Lowlands is home to the majority of Gray Brown Luvisols. Gray Luvisols consist of eluvial and illuvial horizons, and if the average yearly soil temperature is less than 8 °C, they might also contain an A_h horizon. The Boreal Forest Zone of the Interior Plains has extensive regions of Gray Luvisols, characterized by deep, light-gray eluvial layers in the forest litter. These horizons are followed by thick B_t horizons, which have a clay covering on the uppermost layer of aggregates.

Earthworms present: *Aporrectodea longa*, *Ap. rosea*, *Ap. trapezoides*, *Ap. tuberculata*, *Bimastos parvus*, *Dendrobaena octaedra*, *Dendrodrilus rubidus*, *Lumbricus castaneus*, *L. festivus*, *L. rubellus*, *L. terrestris*, *Octolasion tyrtaeum*, *Amynthas agrestis*, *Am. hilgendorfi*, *Am. minimus*, *Am. tokioenis*.

Brunisolic Order

This order encompasses all soils that possess B horizons yet lack the ability to fulfill the criteria of any of the prior orders. Numerous Brunisolic soils exhibit brownish B horizons devoid of substantial indications of clay deposition, as observed in Luvisolic soils, or amorphous substance accumulation, as in Podzolic soils. Over time and under consistent climatic circumstances, some Brunisolic soils will develop into Luvisolic soils, while others will transform into Podzolic soils. In conjunction with other soil types, Brunisolic soils cover nearly 790,000 square kilometers (8.6%) of the total land area of Canada. They are found in everywhere south of the permafrost region.

Based on acidity and organic material abundance in the A horizon, four major groupings are identified. Melanic Brunisols possess an Ah horizon that is a minimum of 10 centimeters in thickness and a pH level higher than 5.5. They are widespread throughout southern Ontario and Quebec. Eutric Brunisols share fundamental characteristics with Melanic Brunisols, with the exception that the Ah horizon, if present, has a thickness of 10 centimeters or less. Sombric Brunisols possess an A_h horizon with a minimum thickness of 10 centimeters. They exhibit acidity, with a pH level less than 5.5. Dystric Brunisols are characterized by their acidity and lack of a 10 centimeters thick A_h horizon.

Earthworms present: *Aporrectodea rosea, Ap. trapezoides, Ap. tuberculata, Ap. turgida, Dendrobaena octaedra, Dendrodrilus rubidus, Lumbricus castaneus, L. festivus, L. rubellus, L. terrestris, Octolasion cyaneum, O. tyrtaeum, Amynthas perrieri, Am. tokioensis.*

Regosolic Order

The development of these soils is insufficient to fulfill the requirements of any other order. Insufficient time for the development of genetic horizons or the unstable state of materials can both contribute to their feeble development or absence. Regosolic soils exhibit features that are mostly determined by the characteristics of the parent materials. Two exceptional groupings are delineated. Regosols are predominantly composed of C horizons. Humic regosols possess a minimum 10-centimeter-thick A_h horizon. Approximately 73,000 square kilometers (around 0.8%) of the land area of Canada is composed of regosolic soils.

Earthworms present: *Ap. tuberculata, Ap. turgida, Lumbricus festivus.*

Tree Species in Canadian Forest Regions

Predominant tree species in the Forest Regions of Canada

Forest region	Location	Predominant tree species
Acadian	Maritimes	Red spruce (*Picea rubens*), balsam fir (*Abies balsamea*), yellow birch (*Betula alleghaniensis*).
Boreal	Northern Canada	White spruce (*Picea glauca*), black spruce (*P. mariana*), balsam fir (*Abies balsamea*), jack pine (*Pinus banksiana*), white birch (*Betula papyrifera*), trembling aspen (*Populus tremuloides*), tamarack (*Larix laricina*), willow (*Salix*).
Carolinian (Deciduous)	Southwestern Ontario	Beech (*Fagus grandifolia*), maple (*Acer saccharum*), black walnut (*Juglans cinerea*), hickory, oak (*Quercus*).
Coast	British Columbia	Western redcedar (*Thuja plicata*), western hemlock (*Tsuga heterophylla*), Sitka spruce (*Picea sitchensis*), Douglas fir (*Pseudotsuga menziesii*).
Columbia	British Columbia	Western redcedar (*Thuja plicata*), western hemlock (*Tsuga heterophylla*), Douglas fir (*Pseudotsuga menziesii*).
Great Lakes–St Lawrence	Central Canada	Red pine (*Pinus resinosa*), eastern white pine (*Pinus strobus*), eastern hemlock (*Tsuga canadensis*), yellow birch (*Betula alleghaniensis*), maple (*Acer saccharum*), oak (*Quercus rubra*).
Montane	British Columbia and Alberta	Douglas fir (*Pseudotsuga menziesii*), lodgepole pine (*Pinus contorta*), ponderosa pine (*Pinus ponderosa*), trembling aspen (*Populus tremuloides*).
Subalpine	British Columbia and Alberta	Engelmann spruce (*Picea engelmannii*), subalpine fir (*Abies lasiocarpa*), lodgepole pine (*Pinus contorta*).

References[2]

Addison JA (2009) Distribution and impacts of invasive earthworms in Canadian forest ecosystems. Biol Invasions 11(1):59–79

Ali AS, Naaz I (2013) Earthworm biomarkers: the new tools of environmental impact assessment. Biosci Biotech Res Comm 6(2):163–169

Alves PRL, Bandeira FO, Hennig TB (2022) Ecological role of earthworms as bioindicators of soil health. In: Vig AP, Suthar SS, Singh J (eds) Earthworm engineering and applications, 1st edn. Nova Science Publishers, Hauppauge, NY, pp 51, 270–81. https://doi.org/10.52305/VLNM4957

Bessolitsyna EP (2012) Ecological and geographic distribution patterns of earthworms (Oligochaeta, Lumbricidae) in landscapes of southern middle Siberia. Russ J Ecol 43:82–85. https://doi.org/10.1134/S106741361106004X

[2] All Megadrilogica references are available for free download on ResearchGate.

Bjorkman AD, Myers-Smith IH, Thomas HJD, Elmendorf SC, Tundra, ITEX, and Tundra Trait Team (2018) Changes in plant functional traits across a warming tundra biome. Nature 562:57–62. https://doi.org/10.1038/s41586-018-0563-7

Blume-Werry G, Krab EJ, Olofsson J, Sundqvist MK, Väisänen M, Klaminder J (2020) Invasive earthworms unlock arctic plant nitrogen limitation. Nat Commun 11:1766. https://doi.org/10.1038/s41467-020-15568-3

Bohlen PJ, Scheu S, Hale CM, McLean MA, Migge S, Groffman PM, Parkinson D (2004) Non-native invasive earthworms as agents of change in northern temperate forests. Front Ecol Environ 2(8):427–435

Bottinelli N, Hedde M, Jouquet P, Capowiez Y (2020) An explicit definition of earthworm ecological categories—Marcel Bouché's triangle revisited. Geoderma 372(2020):1–7

Bouché MB (1970) Relations entre les structures spatiales et fonctionelles des écosystèmes, illustrées par le rôle pédobiologique es vers de terre. In: Pesson P (ed) La vie dans le sol. Aspects nouveaux, Études expérimentales. Gauthiers-Villars, Paris, pp 187, 471–209

Bouché MB (1972) Lombriciens de France, écologie et systématique. Inst Nat Rech Agron:671 pp

Brigham-Grette J, Petsch S (2020) The Arctic hasn't been this warm for 3 million years—and that foreshadows big changes for the rest of the planet. The conversation (September 30, 2020)

Brown GG, James SW, Pasini A, Nunes DH, Benito NP, Trigo Martins P, Sauttter KD (2006) Exotic, peregrine, and invasive earthworms in Brazil: diversity, distribution, and effects on soils and plants. Caribbean J Sci 42(3):339–358

Cameron EK, Bayne EM, Clapperton JM (2007) Humanfacilitated invasion of exotic earthworms into northern boreal forests. Ecoscience 14(4):482–490

Cameron EK, Shaw ECH, Bayne M, Kurz WA, Kull SJ (2015) Modelling interacting effects of invasive earthworms and wildfire on forest floor carbon storage in the boreal forest. Soil Biol Biochem 88:189–196

Canadian Society of Soil Science (2020) Soils of Canada. Online: Available: soilsofcanada.ca. Accessed 30 Jan 2022

Černosviton L, Evans AC (1947) Lumbricidae. Synopsis of the British fauna. 6. Linnaeus Soc. London 36

Chang C-H, Johnston M, Görres J, Dávalos A, McHugh D, Szlávecz K (2017) Co-invasion of three Asian earthworms, *Metaphire hilgendorfi, Amynthas agrestis*, and *Amynthas tokiensis* in the USA. Biol Invasions 20:843–848

Clapperton MJ, Miller JJ, Larney FJ, Lindwall CW (1997) Earthworm populations as affected by long-term tillage practices in southern, Alberta, Canada. Soil Biol Biochem 29(3–4):631–633

Coleman DC, Callaham MA Jr, Crossley DA Jr (2017) Fundamentals of soil ecology, 3rd edn. Academic Press, London, p 369

Craven D, Thanur MP, Cameron EK, Frelich LE, Beauséjour R, Blair RB, Blossey B, Burtis J, Choi A, Dávalos A, Fahey TJ, Fisichelli NA, Gibson K, Handa IT, Hopfensperger K, Loss SR, Nuzzo V, Maerz JC, Sackett T, Scharenbroch BC, Smith SM, Vellend M, Umek LG, Eisenhauer N (2017) The unseen invaders: introduced eart hworms as drivers of change in plant communities in north American forests (a meta-analysis). Glob Chang Biol 23:1065–1074

Diehl WJ, Williams DL (1992) Distribution of polymorphisms among pathways of carbohydrate metabolism in the earthworm *Eisenia fetida* (Oligochaeta). Comp Biochem Physiol 101B:83–90

Dymond P, Scheu S, Parkinson D (1997) Density and distribution of *Dendrobaena octaedra* (Lumbricidae) in aspen and pine forests in the Canadian Rocky Mountains (Alberta). Soil Biol Biochem 29(3/4):265–273

Edwards CA, Heath GW (1963) The role of soil animals on breakdown of leaf material. In: Doeksen J, van der Drift J (eds) Soil organisms. North Holland Publ. Co, Amsterdam, pp 76–84. viii + 453

Eisenhauer N, Partsch S, Parkinson D, Scheu S (2007) Invasion of a deciduous forest by earthworms: changes in soil chemistry, microflora, microarthropods and vegetation. Soil Biol Biochem 39:1099–1110

Eisenhauer N, Bonn A, Guerra CA (2019) Recognizing the quiet extinction of invertebrates. Nat Commun 10:50. https://doi.org/10.1038/s41467-018-07916-1

Fender WM, McKey-Fender D (1990) Oligochaeta: Megascolecidae and other earthworms from western North America. In: Dindal DL (ed) Soil biology guide. John Wiley & Sons, Inc, New York, pp 357–378. xviii + 1349

Gates GE (1970) Miscellanea Megadrilogica VII. Megadrilogica 1(2):1–6

Gates GE (1973) The earthworm genus Octolasion in America. Bull. Tall Timbers Res. Stn. No. 14:29–50

Gates GE (1974) Contributions to a revision of the family Lumbricidae. X. *Dendrobaena octaedra* (Savigny, 1826) with special references to the importance of its parthenogenetic polymorphism for the classification of earthworms. Bull. Tall Timbers Res. Stn. No. 15: 15–57

Gates GE (1972) Toward a revision of the earthworm family Lumbricidae. IV. The trapezoids species complex. Bull. Tall Timbers Res. Stn. No. 12:1–146

Gerard BM (1964) Lumbricidae No.6 (Annelida) with keys and descriptions, 2nd ed. The Linnean Society of London Synopses of the British Fauna

Geraskina AP, Tebenkova DN, Ershov DV, Ruchinskaya EV, Sibirtseva NV, Lukina NV (2021) Wildfires as a factor of the loss of biodiversity and functions of forest ecosystems. Forest Sci Issues 4(2):1–76

Gisin H (1943) Oekologie und Lebengemeinschaften der Collembolen im schweizerischen Excursions-gebiet Basel. Rev Suisse Zool 50:131–224

Görres JH, Martin C, Aiin MN, Bellitürk K (2019) Physical properties of soils altered by invasive pheretimoid earthworms: does their casting layer create thermal refuges? Soil Syst 3(3):52. https://doi.org/10.3390/soilsystems3030052

Grant WC Jr (1955) Temperature relationships in the megascolecid earthworm *Pheretima hupeiensis*. Ecology 36(3):412–417

Heath GW, Arnold MK, Edwards CA (1966) Studies in leaf litter breakdown. I. Breakdown of leaves of different species. Pedobiologia 6(1):1–12

Hillis ND (1917) The quest of John Chapman: the story of a forgotten hero. The Macmillan Co, New York, p 308

Jaenike J, Selander RK (1980) Evolution and Ecology of Parthenogenesis in Earthworms. American Zoologist 19(3):729–737

Jaenike J, Parker ED Jr, Selander RK (1980) Clonal niche structure in the parthenogenetic earthworm *Octolasion tyrtaeum*. American Naturtalist 116(2):196–205

Janzen HH (2004) Carbon cycling in earth systems—a soil science perspective. Agric Ecosyst Environ 104:399–417

Jungqvist G, Oni SK, Teutschbein C, Futter MN (2014) Effect of climate change on soil temperature in Swedish boreal forests. PLoS One 9(4):e93957. https://doi.org/10.1371/journal.pone.0093957

Lavelle P (1997) Faunal activities and soil processes: adaptive strategies that determine ecosystem function. Adv Ecol Res 27:93–132. https://doi.org/10.1016/S0065-2504(08)60007-0

Lee KE (1959) The earthworm fauna of New Zealand. Wellington: New Zealand Dept Sci Indust Res Bull 130:486

Liski J, Nissinen A, Eerhard M, Taskinen O (2003) Climatic effects on litter decomposition from arctic tundra to tropical rainforests. Glob Chang Biol 9:575–584

Lubbers IM, van Groenigen KS, Fonte SJ, Six J, Brussaard L, van Groenigen JW (2013) Greenhouse-gas emissions from soils increased by earthworms. Nat Clim Chang 3:187–194

McAlpine DF, Reynolds JW (2018) BiotaNB 2009–2010 earthworms (Oligochaeta: Lumbricidae) of the Jacquet river gorge protected natural area and region, New Brunswick, with additions to the province. Megadrilogica 23(7):103–107

McAlpine DF, Reynolds JW (2019) Earthworms (Oligochaeta: Lumbricidae) of the Nepisiquit protected natural area, New Brunswick, Canada. Megadrilogica 24(4):39–49

McAlpine DF, Reynolds JW (2020) Earthworms (Oligochaeta: Lumbricidae) of the Caledonia gorge protected natural area and region, New Brunswick, with additions to the province. Megadrilogica 25(4):53–60

McAlpine DF, Sullivan SA, McAlpine FJT, Lewis JH, Reynolds JW (2020) Earthworms (Lumbricidae and Sparganophilidae) of the grand Lake protected natural area, New Brunswick, Canada. Megadrilogica 25(6):89–98

McAlpine DF, Reynolds JW, Manzer L, Elton K (2022) First reports of invasive earthworms (Oligochaeta: Megascolecidae) of Asian origin in Atlantic Canada. Biol Invasions 11:9

McKey-Fender D, Fender WM (1982) *Arctiostrotus* (gen. Nov.). I. The identity of *Plutellus perrieri* Benham 1892 and its distribution in relation to glacial refugia. Megadrilogica 4(3):81–86

McKey-Fender D, Fender WM, Marshall VG (1994) North American earthworms native to Vancouver Island and the Olympic peninsula. Canadian J Zool 72:1325–1339

McLean MA, Parkinson D (1997) Changes in structure, organic matter and microbial activity in pine forest soil following the introduction of *Dendrobaena octaedra* (Oligochaeta, Lumbricidae). Soil Biol Biochem 29(3/4):537–540

McLean MA, Parkinson D (2000a) Field evidence of the effects of the epigeic earthworm *Dendrobaena octaedra* on the microfungal community in pine forest floor. Soil Biol Biochem 32(3):351–360

McLean MA, Parkinson D (2000b) Introduction of the epigeic earthworm *Dendrobaena octaedra* changes the oribatid community and microarthropod abundances in a pine forest. Soil Biol Biochem 32(11–12):1671–1681

Means H (2011) Johnny Appleseed: the man, the myth, the American story. Simon & Schuster, New York, p 19

Moore J-D, Reynolds JW (2003) Premières mentions de vers de terre (Oligochaeta: Lumbricidae) dans deux écosystèmes forestiers du Bouclier canadien, Québec, Canada. Megadrilogica 9(9):53–60

Moore J-D, Ouimet R, Reynolds JW (2009) Premières mentions de vers de terre dans trois écosystèmes forestiers du Bouclier canadien. Nat Canadien 133(1):31–37

Moore J-D, Görres JH, Reynolds JW (2018) Colonization front and potential effects of forest ecosystem invasions by exotic Asian pheretimoid earthworms (*Amynthas* spp., *Metaphire* spp.) in eastern Canada. Environ Rev 26(2):113–120

Moore J-D, Ouimet R, Reynolds JW (2022) A review of exotic earthworm observations in the Canadian boreal forest and taiga zones. Environ Rev 30:289–305. https://doi.org/10.1139/er-2021-0074

Natural Resources Canada (2007) Forest regions classifications. Retrieved July 23, 2021, from ecosys.cfl.scf.rncan.gc.ca/images/classif/forest-reg2_e.pdf

Panesar TS, Marshall VG, Barclay HG (2000) The impact of clearcutting and partial harvesting systems on population dynamics of soil nematodes in coastal Douglas-fir forests. Pedobiologia 44:641–665

Parkinson D, McLean MA, Scheu S (2004) Impacts of earthworms on other biota in forest soils, with some emphasis on cool temperate montane forests. In: Edwards CA (ed) Earthworm ecology, 2nd edn. CRC Press, Boca Raton, FL, pp 241, 441–259

Reynolds JW (1972) The relationship of earthworm (Oligochaeta: Acanthodrilidae and Lumbricidae) distribution and biomass in six heterogeneous woodlot sites in Tippecanoe County, Indiana. J Tennessee Acad Sci 47(2):63–67

Reynolds JW (1973) Earthworm (Annelida, Oligochaeta) ecology and systematics. In: Dindal DL (ed) Proc. First Microcommunities Conf. CONF 711076, U.S.A.E.C., Natn. Tech. Inform. Serv. U.S. Dept. Com., Springfield, VA, pp 95–120. 22151, 285

Reynolds JW (1974) Are oligochaetes really hermaphroditic amphimictic organisms? Biologist 56(2):90–99

Reynolds JW (1975a) Les lombricidés (Oligochaeta) des Îles-de-la-Madeleine. Megadrilogica 2(3):1–8

Reynolds JW (1975b) Les lombricidés (Oligochaeta) de la Gaspésie. Québec Megadrilogica 2(4):4–9
Reynolds JW (1975c) Boiteagan (Oligochaeta: Lumbricidae) cheap Breatunn. Megadrilogica 2(6):1–7
Reynolds JW (1975d) The earthworms (Oligochaeta: Lumbricidae) of Prince Edward Island. Megadrilogica 2(7):4–10
Reynolds JW (1976a) The distribution and ecology of the earthworms of Nova Scotia. Megadrilogica 2(8):1–8
Reynolds JW (1976b) Un aperçu vers de terre dans les forêts nord-américanes, leurs activitiés et leurs répartition. Megadrilogica 2(9):1–11
Reynolds JW (1976c) *Aporrectodea icterica* (Savigny, 1826) une espèce europeen de vers de terre récemment découverte en Amérique du Nord. Megadrilogica 2(12):3–4
Reynolds JW (1977a) The earthworms (Lumbricidae and Sparganophilidae) of Ontario. Life Sci Misc Publ, Roy Ont Mus *xi*:+141
Reynolds JW (1977b) Earthworms utilized by the American woodcock. Proc Woodcock Symp 6:161–169
Reynolds JW (1978) The earthworms of Tennessee (Oligochaeta). IV. Megascolecidae, with notes on distribution, biology and a key to the species in the state. Megadrilogica 3(7):117–129
Reynolds JW (1994) The distribution of the earthworms (Oligochaeta) of Indiana: a case for the post quaternary introduction theory for megadrile migration in North America. Megadrilogica 5(3):13–32
Reynolds JW (2000a) A contribution to our knowledge of the earthworm fauna of Newfoundland and Labrador, Canada (Oligochaeta, Lumbricidae). Megadrilogica 8(2):5–8
Reynolds JW (2000b) A contribution to our knowledge of the earthworm fauna of Manitoba, Canada (Oligochaeta, Lumbricidae). Megadrilogica 8(3):9–12
Reynolds JW (2001) The earthworms of New Brunswick (Oligochaeta: Lumbricidae). Megadrilogica 8(8):37–47
Reynolds JW (2002) First earthworm record from Nunavut, Canada and the second from the Northwest Territories. Megadrilogica 9(6):40
Reynolds JW (2003) A second earthworm species (Lumbricidae) from Nunavut, Canada. Megadrilogica 9(8):52
Reynolds JW (2004) More earthworm (Lumbricidae) collections from Nunavut, Canada. Megadrilogica 10(3):13–14
Reynolds JW (2007) First earthworm records from Wentworth township, Argenteuil County, Quebec. [Premiers signalements de lombrics dans le canton de Wentworth, comté d'Argenteuil (Québec).]. Megadrilogica 11(5):58–62
Reynolds JW (2010a) Earthworms (Oligochaeta: Lumbricidae, Sparganophilidae) of the Atlantic maritime Ecozone. Chapter 13. In: McAlpine DF, Smith I (eds) Assessment of species diversity in the atlantic maritime ecozone. NRC Press, Ottawa, pp 225–281. xii + 785
Reynolds JW (2010b) The earthworms (Oligochaeta: Lumbricidae) of Nova Scotia, revisited. Megadrilogica 14(6):77–100
Reynolds JW (2010c) The earthworms (Oligochaeta: Acanthodrilidae, Lumbricidae, Megascolecidae and Sparganophilidae) of northeastern United States, revisited. Megadrilogica 14(7):101–157
Reynolds JW (2011a) The earthworms (Oligochaeta: Lumbricidae) of Pelee Island, Ontario, Canada. Megadrilogica 15(3):23–33
Reynolds JW (2011b) The earthworms (Oligochaeta: Acanthodrilidae, Eudrilidae, Glossoscolecidae, Komarekionidae, Lumbricidae, Lutodrilidae, Ocnerodrilidae, Octochaetidae, Megascolecidae and Sparganophilidae) of southeastern United States. Megadrilogica 14(9–12):175–318
Reynolds JW (2014a) A checklist by counties of earthworms (Oligochaeta: Lumbricidae, Megascolecidae and Sparganophilidae) in Ontario, Canada. Megadrilogica 16(10):111–135

Reynolds JW (2014b) A checklist by counties of earthworms (Oligochaeta: Lumbricidae and Sparganophilidae) in the Canadian Maritime provinces. Megadrilogica 16(11):137–156

Reynolds JW (2014c) A checklist by counties of earthworms (Oligochaeta: Lumbricidae and Sparganophilidae) in Quebec, Canada. Megadrilogica 17(6):73–103

Reynolds JW (2015) A checklist of earthworms (Oligochaeta: Lumbricidae, Megascolecidae, and Sparganophilidae) in western and northern Canada. Megadrilogica 17(10):141–156

Reynolds JW (2018) First earthworm (Annelida: Oligochaeta) species' collections in Canada and the continental United States. Megadrilogica 23(1):1–50

Reynolds JW (2020) Earthworms (Annelida: Oligochaeta) in Canadian Ecozones. Megadrilogica 26(1):1–35

Reynolds JW (2021) The earthworms (Oligochaeta: Lumbricidae, Megascolecidae and Sparganophilidae) of Canada. Canada Food Inspection Agency, Ottawa, p ii + 185. (in press)

Reynolds JW (2022) The Earthworms (Lumbricidae, Megascolecidae and Sparganophilidae) in Canada. Ottawa: Canada Food Inspection Agency, i + 179 pp.

Reynolds JW, Clapperton MJ (1996) New earthworm records for Alberta (Oligochaeta: Lumbricidae) including the description of a new Canadian species. Megadrilogica 6(8):73–82

Reynolds JW, Jordan GA (1975) A preliminary conceptual model of megadrile activity and abundance in the Haliburton highlands. Megadrilogica 2(2):1–9

Reynolds JW, Khan MN (1999) A contribution to our knowledge of the earthworm fauna of Saskatchewan, Canada. Megadrilogica 7(12):81–82

Reynolds JW, Mayville PN (1994) New earthworm records from Rainy River district in north western Ontario (Oligochaeta: Lumbricidae). Megadrilogica 6(2):13–16

Reynolds JW, McTavish MJ (2021) New Asian pheretimoid "jumping earthworm" records (Oligochaeta: Megascolecidae) in Canada. Megadrilogica 26(7):83–89

Reynolds JW, Moore SM (1996) Note on the first earthworm record from the Northwest Territories, Canada. Megadrilogica 6(10):96

Reynolds JW, Reynolds KW (1992) Les vers de terre (Oligochaeta: Lumbricidae et Sparganophilidae) sur la rive nord du Saint-Laurant (Québec). Megadrilogica 4(9):145–161

Reynolds JW, Clebsch EEC, Reynolds WM (1974) The earthworms of Tennessee (Oligochaeta). I. Lumbricidae. Contr. N. Amer. Earthworms, No. 3. Bull. Tall Timbers Res. Stn., No. 17, *viii* + 133

Reynolds JW, McAlpine DF, Vanderwolf KJ, Huynh HM (2014) Earthworm (Oligochaeta: Lumbricidae) occurrences from Anticosti Island, Quebec, Canada. Megadrilogica 17(1):1–6

Reynolds JW, Hasson AJ, McAlpine DF (2015) Earthworms (Oligochaeta: Lumbricidae) of the Nepisiquit protected natural area and region, New Brunswick, with additions to the province. Megadrilogica 19(1):1–10

Reynolds JW, Cameron EK, Sweeney LM, Bennett B, Godin B (2019) The earthworms (Oligochaeta: Lumbricidae and Megascolecidae) in the Yukon and a first record for Skagway, Alaska, USA. Megadrilogica 25(1):1–17

Reynolds JW, Wetzel MJ (2008) Terrestrial Oligochaeta (Annelida: Clitellata) in North America, including Mexico, Puerto Rico, Hawaii, and Bermuda. Megadrilogica 12(12):157–208

Reynolds JW, Wetzel MJ (2011) The earthworms (Oligochaeta: Acanthodrilidae, Komarekionidae, Lumbricidae, Megascolecidae and Sparganophilidae.) of Illinois, USA, revisited. Megadrilogica 15(4):35–67

Reynolds JW, Wetzel MJ (2012) Terrestrial Oligochaeta (Annelida: Clitellata) in North America, including Mexico, Puerto Rico, Hawaii, and Bermuda. III. Megadrilogica 15(8):191–211

Rowe JS (1972) Forest regions of Canada, vol 1300. Department of Environment, Publ, Ottawa, Canada, p x + 172

Ruiz SA, Bickel S, Or D (2021) Global earthworm distribution and activity windows based on soil hydromechanical constraints. Commun Biol 4(612):1–9. https://doi.org/10.1038/s42003-021-02139-5

Sackett TE, Smith SM, Basiliko N (2013) Indirect and direct effects of exotic earthworms on soil nutrient and carbon pools in North American temperate forests. Soil Biol Biochem 57(2013):459–467

Scheu S, McLean MA (1993) The earthworm (Lumbricidae) distribution in Alberta (Canada). Megadrilogica 4(11):175–180

Scheu S, Parkinson D (1994) Effects of earthworms on nutrient dynamics carbon turnover and microorganisms from cool temperate forests of the Canadian Rocky Mountains—laboratory studies. Appl Soil Ecol 1(1994):113–125

Serniak LT (2017) The effects of earthworms on carbon dynamics in forest soils. Biol Invasions 12(022017):213–229

Singh J, Schädler M, Demetrio W, Brown GG, Eisenhauer N (2019) Climate change effects on earthworms – a review. Soil Org 91(3):114–138. https://doi.org/10.25674/so91iss3pp114

Soil Classification Working Group (1998) The Canadian system of soil classification, 3rd ed. Agriculture and Agri-Food Canada Publication 1646, p 187

Spiers GA, Gagnon D, Nason GE, Packee EC, Lousier JD (1986) Effects and importance of indigenous earthworms on decomposition and nutrient cycling in coastal forest ecosystems. Canadian J For Res 16:983–989

Stanton C, Bourchier R (2015) Forest regions. In: The Canadian encyclopedia. https://www.thecanadianencyclopedia.ca/en/article/forest-regions

Statistics Canada (2022) Census profile. 2021 Census Statistics Canada Catalogue number 98-316-X2021001. Ottawa. Released February 9, 2022. https://www12.statcan.gc.ca/census-recensement/2021/dp-pd/prof/index.cfm?Lang=E. Accessed 13 Feb 2022

Tandy RE (1969) The earthwqorm genus Pheretima Kinberg, 1866 in Louisiana. Unpublished Ph.D, Dissertation. Louisiana State University, Baton Rouge, LA, 155 pp

Tomlin AD, Fox CA (2003) Earthworms and agricultural systems: status of knowledge and research in Canada. Canadian J Soil Sci 83:265–278

Tondoh JE (2006) Seasonal changes in earthworm diversity and community structure in Central Côte d'Ivoire. European J Soil Biol 42:S334–S340. https://doi.org/10.1016/j.ejsobi.2006.09.003

Welke SE, Parkinson D (2003) Effects of *Aporrectodea trapezoides* activity on seedling growth of *Pseudotsuga menziesii*, nutrient dynamics and microbial activity in different forest soils. For Ecol Manag 173:169–187

Wetzel MJ, Reynolds JW (2021) A preliminary inventory of earthworms (Annelida, Clitellata) of the Nachusa Grasslands Area, Lee and Ogle Counties, Illinois, USA. Megadrilogica 26(8):91–124

Whalen JK (2004) Spatial and temporal distribution of earthworm patches in cornfield, hayfield and forest systems of southwestern Quebec, Canada. Appl Soil Ecol 27(2):143–151

Wicke DE (1953) Ueber die vertikale Verteilung der Lumbriciden im Boden. Zeitschr Morphol Ökol Tiere 41:372–385

Wironen M, Moore TR (2006) Exotic earthworm invasion increases soil carbon and nitrogen in an old-growth forest in southern Quebec. Canadian J For Res 36:845–854

Zhang W, Hendrix PF, Dame LE, Burke RA, Wu KJ, Neher DA, Li J, Shao Y, Fu S (2013) Earthworms facilitate carbon sequestration through. Unequal amplification of carbon stabilization compared with mineralization. Nat Commun 4:2576. https://doi.org/10.1038/ncomms3576

Earthworm Ecology in Northern European Forests

Frank Ashwood, Justine Lejoly, Aidan Keith, and Olaf Schmidt

1 Introduction

Forests cover roughly 35% of Europe's land area and are mostly managed, with around 75% of the total forested area available for wood supply, and 24% in protected areas (Forest Europe 2020). Forest composition varies greatly across the continent, including conifer-dominated boreal forests in Northern countries, mixed forests in Central-Western Europe and greater dominance of broadleaved trees in South-Western countries. Although in general they are less disturbed than agricultural land, European forests are subject to numerous natural disturbances or 'external' threats to biodiversity, such as windthrow, wildfire and snow damage, biological invasions and land-use change, many of which will likely increase with climate change (Forest Europe 2020; Machado Nunes Romeiro et al. 2022). Forest management practices may also pose 'internal' threats to forest biodiversity, and these may interact with external threats to increase the overall vulnerability of individual species (Muys et al. 2022). In general, managed forests possess lower biodiversity than natural forest systems, owing to reduced tree species diversity and similarity in

F. Ashwood (✉)
University of Canterbury, Private Bag, Christchurch, New Zealand
e-mail: francis.ashwood@canterbury.ac.nz

J. Lejoly
Netherlands Institute of Ecology (NIOO-KNAW), Wageningen, The Netherlands
e-mail: j.lejoly@nioo.knaw.nl

A. Keith
UK Centre for Ecology and Hydrology, Lancaster, UK
e-mail: ake@ceh.ac.uk

O. Schmidt
School of Agriculture and Food Science, University College Dublin, Dublin, Ireland
e-mail: olaf.schmidt@ucd.ie

Y. Kooch, Y. Kuzyakov (eds.), *Earthworms and Ecological Processes*,
https://doi.org/10.1007/978-3-031-64510-5_3

stand age, and lack of forest microhabitats such as deadwood to support invertebrate populations, including earthworms (Muys et al. 2022).

Earthworms play an important ecological and biogeochemical role in European forest ecosystems, driving litter decomposition and acting as key bioturbators, but also steering soil properties. Earthworm community composition and dynamics in forest ecosystems are unique, with a notably higher abundance of epigeic earthworms and lower abundances of endogeic earthworms compared to croplands and grasslands (Phillips et al. 2019). Earthworms are generally more abundant in broadleaf and mixed forests compared to coniferous forests (Phillips et al. 2019), but earthworm heterogeneity is immense within forest ecosystems. At higher latitudes in Europe, boreal forests have mostly developed in the absence of earthworms and are now being invaded by European/Lumbricid earthworm species which, paradoxically, are considered non-native, having originated in more temperate areas of Eurasia (Terhivuo 1988; Wackett et al. 2018). Within forest stands, earthworms can be found in various microhabitats, including deadwood and leaf litter, even in the canopy or under loose bark, revealing a higher earthworm diversity than typically found with standard soil sampling methods (Schmidt et al. 2015; Römbke et al. 2017; Ashwood et al. 2019b). In this chapter, we bring a new perspective to earthworm ecology in forest ecosystems, with a focus on European temperate and boreal forests. We explore large-scale (including non-native earthworms) and small-scale (microhabitat) heterogeneities, discuss how global threats (e.g. climate and land-use change) can reshape our forests and their earthworm communities and include important considerations for future research into earthworm ecology in forest systems.

2 Earthworms in Temperate Forests and Land-Use Change

Land use is a dominant factor controlling earthworm community composition and population structure, ultimately through its influence on food resources and soil physico-chemical characteristics. Land-use change from a non-wooded to a wooded habitat presents a significant shift in biotic and abiotic conditions and demonstrates the links between above ground and below ground systems (Wardle et al. 2004). It can include considerable changes in plant species richness, the heterogeneity of the understory layer and the structural complexity of vegetation (particularly as the wooded habitat matures). These changes affect the quality and quantity of above- and below ground litter entering the soil system. There are also important impacts on microclimate and soil hydrology, for example, through changes in rainfall interception, transpiration, rooting patterns and their influence on soil architecture through biologically driven processes. The magnitude and variation in the potential impacts of land-use change to wooded habitat on earthworm populations and species prevalence are dependent on a range of factors, the most important being previous land use, tree species and time. There is a legacy effect of the previous land use to consider when the impacts of land-use change are assessed over time. Considering the starting point of the recent land-use change (tens to hundreds of years), it is typical that change to a wooded or forest system has begun at either end of a land-use spectrum, often either as semi-natural heathland/moorland or as agricultural land. Clearly, there are vastly different trajectories of vegetation, soil biogeochemistry and structure represented by these two land-use changes; and there is a rich history in the study of both in the context of earthworms.

In highly organic soils, such as in heath and moorland, earthworms are present but typically at low densities and dominated by a few epigeic species (e.g. *Dendrobaena octaedra* and *Bimastos eiseni*). There has been interest in the effects of afforestation in these organic semi-natural habitats for at least 70 years, and some early work focused on earthworms and their potential role in accelerating soil development. In the UK, experimental plots were established to determine the role of birch as a 'soil improver' and whether it could transform the podzolic soil of heather moorland to a forest brown earth (Dimbleby 1952). Dimbleby noted that this regeneration of a mull-type soil from a podzolised mor-type soil was evidenced in earlier studies as taking as little as 12 to 15 years. Qualitative investigation through visual searches under groups of young birch found a greater diversity of soil macroinvertebrates than surrounding moorland, including earthworm species more typical of woodland mull soils (Dimbleby 1952). While the focus of Dimbleby's study was on changing soil properties in these plots, he stated that 'there can be little doubt that the soil fauna plays a vital role'. This work was followed up over the subsequent 30 years at Silpho with a specific investigation of the earthworm communities, but only limited change was evidenced (Satchell 1980). Similar research using a Scottish chronosequence found earthworms to increase in abundance and diversity with time since birch establishment on moorland (Miles 1981; Miles and Young 1980). It was therefore suggested that particularly nutrient-poor sites, such as Silpho in the North York Moors, UK, may take centuries rather than decades to regenerate a brown earth (Miles 1981) and, in relation, that the rate of change in soils and their biological communities is linked to the growth rate of birch (Mitchell et al. 2007). However, a study of earthworm communities in Iceland (Sigurdsson and Gudleifsson 2013) found little difference in earthworm biomass between heathland and native mountain birch woodland, and this was linked to the favourable pH across these soils, which are basaltic. The context-dependent mechanisms that drive changes in earthworm communities during afforestation of organic soils are not well quantified in space or time, and further research is needed to pin these down.

Agricultural soils themselves represent relatively diverse land-use starting conditions for afforestation. Arable soils tend to have lower concentrations of organic matter, typically involve crop rotations and may be subjected to disturbance through tillage operations. Grasslands and pastures generally have higher concentrations of organic matter, vegetative cover all year round and limited physical disturbance; grazed systems will also have manure inputs from livestock, which promotes earthworm abundance. The abundance and composition of earthworm communities in agricultural soils is greatly influenced by these land management practices, with cultivated systems tending to contain a relatively low abundance of earthworms (typically ~50–200 individuals m^{-2}), somewhat depending on time since cultivation and crop type and 3–5 species per sampling unit, whereas grassland or pasture systems generally contain much greater abundances (typically ~300–600 individuals m^{-2}) and 4–7 species per sampling unit (Rutgers et al. 2009; Keith et al. 2012); earthworm abundance and composition is also known to increase quickly when arable is put under grass ley. How afforestation may impact residual earthworm communities is, therefore, contextual to previous land management. For example, in a study of earthworm communities under birch in Finland, Räty and Huhta (2004) found that birch stands which had been arable land (30 years previously) retained soil properties and earthworm species composition similar to that in agricultural soils. Ashwood et al. (2019a) measured earthworm abundance and composition across a regional network of sites in the UK, designed to study the long-term effects of woodland creation. This site network included arable and pasture land use, paired with secondary and ancient woodland sites. While total earthworm abundance was not significantly different between arable and any woodland age (young, mature, ancient), woodland sites hosted greater earthworm diversity including more epigeic and endogeic species; young woodland also had significantly greater diversity than arable land (Ashwood et al. 2019a). Large-scale tree planting for climate mitigation is increasing, and (with large gains in carbon stocks foreseen) agricultural land is likely to contribute significant land area to this campaign; Lee et al. (2023) estimated forest expansion in the EU to be 2.44 Mha since 2010 and the area of abandoned agricultural land at 2.6 Mha since 2009, with the latter able to serve future forest expansion. Developing a better understanding of how earthworm abundances and species composition may change in this land-use shift is both topical and important, as we know earthworms have manifested impacts on soil biophysical processes.

Land-use change to a forest or woodland, as previously mentioned, involves a clear shift in habitat structure and associated conditions. While this shift has a great influence on soil biogeochemistry and ecology, the tree species or mixture of tree species present also plays a significant role in determining the resulting earthworm community. There is a broad spectrum of traits across tree species in northern European forests. Hence, broad classification of tree types and considering the range of functional effect traits of tree species are useful. The broad differences between effects of coniferous and deciduous tree species on soil biogeochemical processes are well evidenced (see review by Augusto et al. 2015), and a body of literature has been developed, highlighting relatively consistent differences in earthworm abundance and diversity under coniferous and deciduous tree species (e.g. Scheu et al. 2003; Ammer et al. 2006; Olsson et al. 2012; Schelfhout et al. 2017; Desie et al. 2019). Scheu et al. (2003) compared soil invertebrates under pure and mixed stands of contrasting tree species and found exceptionally low earthworm biomass under spruce compared to beech. In a replicated forestry experiment in south-west Sweden, birch supported approximately six times the earthworm abundance as pine and spruce, with higher numbers of *Dendrobaena octaedra* and *Lumbricus rubellus*, the two species present (Olsson et al. 2012). Though the split between conifer and broadleaf is useful, more detailed evidence and further mechanistic insights into the impacts of afforestation on earthworm communities can be gained by comparing tree species.

The use of in situ 'common garden' experiments is a powerful approach to investigate the effects of different tree species on soil biogeochemistry and ecology, particularly where these are long term and have been regularly monitored (e.g. Reich et al. 2005; Vesterdal et al. 2008). Reich et al. (2005) investigated leaf litter properties, soil properties and earthworms in a common garden experiment in Poland that included 14 tree species. The variation presented across these tree species facilitated the inference of a positive relationship between litter calcium concentration and both soil pH and earthworm biomass (Reich et al. 2005). While earthworms require calcium and therefore may be impacted by litter quality, they can also make calcium more available (Piearce 1972; Lambkin et al. 2011); it is difficult to tease these apart but likely that these processes are reinforcing over time. Schelfhout et al. (2017) also studied a common garden experiment, established for 36 years at time of sampling, with six tree species replicated at six sites across Denmark. This research provided further evidence for greatest positive effects on earthworm density and biomass for *Acer pseudoplatanus*, *Tilia cordata* and *Fraxinus excelsior*, particularly for endogeic and anecic earthworm species, and intermediate effects of *Fagus sylvatica* and *Quercus robur*. Similarly, Neirynck et al. (2000) studied 60–65-year-old stands in the Forest of Halle in Belgium, recording greater earthworm biomass under mull-forming tree species (*Tilia*, *Acer*, *Fraxinus*), compared to moder-forming tree species (*Quercus* and *Fagus*), particularly the epigeics *Lumbricus rubellus* and *Dendrobaena octaedra* and the endogeic *Aporrectodea limicola*. There were also significant differences between the effects of tree species on epigeic, endogeic, anecic and total earthworm biomass, in a study exploring grassland afforestation using experimental forest in Belgium (Muys et al. 1992). Finally, Frouz et al. (2013) investigated the effects of six tree species planted on mining spoil which have been established for 22–32 years at the time of sampling. Soil formation differed between tree species with Alder and Tilia having the greatest earthworm densities, A-horizon thickness and carbon accumulation.

Several studies have tested the relative importance of tree species diversity versus tree species identity. Scheu et al. (2003) found that earthworm biomass in mixed stands of beech and spruce were more similar to spruce monoculture, and this was more pronounced in the oldest stand (120 years). Using data from across six European regions, De Wandeler et al. (2018) found that while taxonomic tree diversity did not have an effect on earthworm abundance or species richness, tree functional diversity had a positive effect on earthworm biomass, and the proportion of evergreen leaf litter had a negative relationship with earthworm biomass and species richness. It is likely that the richness of tree species may be important at larger spatial scales and considering niches other than soil such as deadwood, moss and tree holes (e.g. Burton and Eggleton 2016; Ashwood et al. 2019a).

We have discussed how previous land use, tree species and time since change are all key determinants of the earthworm community and its species richness, but it is also key to consider the critical functional implications of these changes. There is robust and consistent evidence that the litter inputs of different tree species play a key role in determining change in the soil ecosystem. Desie et al. (2019) demonstrated that the impact of differences in tree litter types on the soil system (comparing trajectories under spruce and mixed deciduous forest) was driven by shifts along the soil-buffering gradient with non-linear shifts from one domain to the other (aluminium and exchange domains). Such shifts in pH feedback to alter carbon stability, litter and humus layers may positively reinforce earthworm abundance (Desie et al. 2019, 2020). Hansson et al. (2013) posited that the shorter mean residence time (MRT) of soil carbon under birch, compared to spruce and pine, was related to much higher earthworm populations (as reported in Olsson et al. 2012), favoured due to higher pH and more palatable litter. Positive relationships between tree species, earthworms and bioturbation were also noted by Frouz et al. (2013), with bioturbation activity being a strong predictor of soil formation (thickness of A-horizon) and carbon accumulation in this layer. Frouz et al. (2013) concluded that 'the greatest effect of trees on soil may be indirect, i.e., may result from the effect of tree litter on the bioturbation activity of soil fauna'. Differences between tree species and their impacts on the soil ecosystem can also have an impact on water infiltration, with implications for surface run-off generation and flooding (Chandler et al. 2018), though the role of earthworms in this requires further study.

Our understanding of the impacts of different land-use changes on earthworm communities and their feedbacks to soil functions is very good. There does remain, however, a number of knowledge gaps that would benefit from being filled. Studies that have explored effects of afforestation on earthworm communities have used few timepoints, usually comparing young and very old stands (e.g. Scheu et al. 2003; Ashwood et al. 2019a). This coarse temporal assessment is often necessary because 'space-for-time' approaches are required and the focus may be on end-points; but monitoring changes in earthworm communities at greater temporal frequency could help provide detailed evidence on non-linear changes and feedbacks to soil functioning, the time taken for species composition to reach new dynamic equilibria and the role of species immigration and its link to the surrounding landscape configuration. There is also the question of whether community trajectories are different between afforestation through planting and natural regeneration. Future work using DNA-based and eco-acoustic approaches (see Sect. 3.5 of this chapter) could provide efficient methods to realise more intensive monitoring.

3 Earthworms in Boreal Forests

We have so far discussed earthworms as natural components of European forests; however, it is important to note that this is not always the case. During the last glaciation, the northern European high latitudes were covered by a thick layer of ice. As a result, all earthworm species were extirpated and started recolonising previously glaciated areas only recently (Terhivuo 1988; Cameron and Bayne 2009). Boreal forests have mostly developed in the absence of earthworms, which, together with low temperatures limiting decomposition, resulted in the accumulation of thick layers of partly decomposed organic matter—the forest floor. By accessing this previously unused food resource, non-native earthworms can drastically change the ecosystem (e.g. Bohlen et al. 2004; Frelich et al. 2019). While all earthworm species currently found in European boreal forests are Lumbricidae originating from Europe, most of them are paradoxically considered non-native. Only two species—both litter-dwelling—are considered native to European boreal forests: *Dendrobaena octaedra* (by far the most common) and *Bimastos rubidus* (Wackett et al. 2018). The parthenogenetic reproduction and spontaneous dispersal of *D. octaedra* can explain why this species has the widest distribution at high latitudes (Terhivuo and Saura 2006). Additionally, cocoons of *D. octaedra* can tolerate temperatures below −35 °C through dehydration mechanisms (Holmstrup and Westh 1994; Leirikh et al. 2004). In contrast, soil-burrowing or *geoengineering* species are often dispersed through human-mediated mechanisms and are considered invasive, both in European and North American boreal forests (Cameron et al. 2007; Terhivuo 1988; Wackett et al. 2018).

Compiling available records of earthworms in Scandinavia, 11 species have been observed in European boreal forests above a latitude of 60°N (Table 1). Terhivuo (1988) synthesised the records of Lumbricidae in Finland and neighbouring areas of Norway, Sweden, Estonia and Russia. More recent sampling efforts include birch (Räty and Huhta 2004; Thørring et al. 2020; Wackett et al. 2018), mixed (Desie et al. 2020) and coniferous forests (Sveistrup et al. 2005). Native earthworms (*D. octaedra and B. rubidus*) are found in 60 to 70% of the surveyed areas in Sweden and Finland, including sites with no apparent human impact (Desie et al. 2020; Wackett et al. 2018). Geoengineering, non-native, species are found less often (<15% in Desie et al. (2020) and include multiple species of the genera *Lumbricus*, *Aporrectodea* and *Octolasion*, as well as *Allolobophora chlorotica*, most of which are also found in boreal forests of Russia, Canada and the USA (Addison 2009). While it is commonly believed that low pH and recalcitrant litter limit the establishment of viable earthworm populations and they are more often found in deciduous forests, they are also observed in coniferous forests, including soils with pH values below 3 (Addison 2009).

Although their relative importance in soil food webs is smaller in boreal forests compared to temperate forests (Gongalsky et al. 2021), earthworms have the potential to tremendously alter ecosystem functioning and composition, especially as invasive species (Ferlian et al. 2018; Frelich et al. 2019). The presence of non-native geoengineering earthworms in northern forests is associated with similar changes in soil morphology in Canadian boreal forests (Lejoly et al. 2021), northern North American temperate forests (Alban and Berry 1994; Resner et al. 2015) and Swedish subalpine arctic birch forests (Wackett et al. 2018). The thickness of the organic surface horizon (forest floor) is greatly reduced, with humus forms evolving from mor/moder to (vermi)mull, and is well-mixed with the surface mineral horizon (A), where earthworm casts are abundant. These changes are typically associated with a net reduction of soil C stocks (Lejoly et al. 2021). This is highly concerning, considering that boreal forests represent a major C terrestrial reservoir, most of which is below ground (Bradshaw and Warkentin 2015). However, there is some evidence that the remaining soil C may be more protected against microbial decomposition and thus more stable (Knowles et al. 2016; Angst et al. 2019). Beyond

Table 1 Records of Lumbricidae in European boreal forests

	Origin	Reproduction[a]	Records	Forest types and region	Northern limit
Dendrobaena octaedra	N	P	Terhivuo (1988)	Coniferous and deciduous forests (Finland)	Arctic Ocean
			Räty and Huhta (2004)	Birch forests (Finland—Jyväskylä)	62°N
			Wackett et al. (2018)	Birch forests (Sweden—Abisko and Padjelanta National Park)	68°N
			Desie et al. (2020)	Mixed (pine-birch) forests (Finland North Karelia region)	62°N
Bimastos rubidus[b]	N	PB	Terhivuo (1988)	Coniferous and deciduous forests (Finland)	Arctic Ocean
			Räty and Huhta (2004)	Birch forests (Finland—Jyväskylä)	62°N
			Sveistrup et al. (2005)	Coniferous (spruce-pine) forest (Norway—Vansemb)	60°N
			Wackett et al. (2018)	Birch forests (Sweden—Abisko and Padjelanta National Park)	68°N
			Desie et al. (2020)	Mixed (pine-birch) forests (Finland—North Karelia)	62°N
Lumbricus terrestris	E	B	Terhivuo (1988)	Deciduous forests (Finland)	Arctic Ocean
			Räty and Huhta (2004)	Birch forests (Finland—Jyväskylä)	62°N
			Desie et al. (2020)	Mixed (pine-birch) forests (Finland—North Karelia)	62°N
Lumbricus rubellus	E	B	Terhivuo (1988)	Coniferous and deciduous forests (Finland)	Arctic Ocean
			Räty and Huhta (2004)	Birch forests (Finland—Jyväskylä)	62°N
			Sveistrup et al. (2005)	Planted spruce forest (Norway—Vansemb)	60°N
			Wackett et al. (2018)	Birch forests (Sweden—Abisko and Padjelanta National Park)	68°N
Lumbricus castaneus	E	B	Terhivuo (1988)	Deciduous forests (Finland)	63°N

(continued)

Table 1 (continued)

	Origin	Reproduction[a]	Records	Forest types and region	Northern limit
			Räty and Huhta (2004)	Birch forests (Finland—Jyväskylä)	62°N
Aporrectodea sp.	E	–	Wackett et al. (2018)	Birch forests (Sweden—Abisko)	68°N
Aporrectodea caliginosa	E	B(P)	Terhivuo (1988)	Coniferous and deciduous forests (Finland)	Arctic Ocean
			Räty and Huhta (2004)	Birch forests (Finland—Jyväskylä)	62°N
			Sveistrup et al. (2005)	Planted spruce forest (Norway—Vansemb)	60°N
			Desie et al. (2020)	Mixed (pine-birch) forests (Finland—North Karelia)	62°N
			Thørring et al. (2020)	Birch forests (Norway—Orrefjell)	69°N
Aporrectodea rosea	E	P(B)	Terhivuo (1988)	Coniferous and deciduous forests (Finland)	63°N
			Thørring et al. (2020)	Birch forests (Norway—Orrefjell)	69°N
Aporrectodea longa	E	B	Terhivuo (1988)	Deciduous forests (Finland)	66°N
			Thørring et al. (2020)	Birch forests (Norway—Orrefjell)	69°N
Octolasion tyrtaneum	E	P(B)	Terhivuo (1988)	Coniferous and deciduous forests (Finland)	65°N
			Räty and Huhta (2004)	Birch forests (Finland—Jyväskylä)	62°N
Octolasion cyaneum	E	P	Terhivuo (1988)	Deciduous forests (Finland)	61°N
Allolobophora chlorotica	E	B	Terhivuo (1988)	Deciduous forests (Finland)	62°N

Origin: N = native; E = exotic. Reproduction: P = parthenogenesis; B = biparental; if both reproduction types have been observed but one is dominant, the minor one is in parentheses. For studies where multiple land uses are presented, only data from forests are presented. The latitude of the most northern location is mentioned for each study as the northern limit

[a]According to Terhivuo (1988)

[b]Previously considered *Dendrodrilus rubidus* (Csuzdi et al. 2017)

morphological changes, research on the effects of invasive earthworms on European boreal forests is very limited, but we can expect the consequences of earthworm invasion to be similar to those observed in North America. The introduction of earthworms can have negative effects on the abundance of endemic soil fauna, including nematodes (Jochum et al. 2021a), enchytraeids (Räty 2004), microarthropods (Cameron et al. 2013; McAdams 2017; McLean and Parkinson 1998) and above ground arthropods (Jochum et al. 2021b). Non-native earthworms were also shown to alter plant communities, decreasing plant species richness and increasing non-native plant cover (Craven et al. 2017) while increasing tree drought sensitivity (Larson et al. 2009) and herbivory (Thakur et al. 2020). The effects can be so profound that it is possible to identify a tree ring signature associated with the time of earthworm invasion (Larson et al. 2009). Geoengineering earthworms can even survive in the arctic region, beyond the tree line (Blume-Werry et al. 2020). Survival may happen in the form of cocoons, although frost tolerance of deep-burrowing Lumbricid species is lower than that of litter-dwelling species such as *D. octaedra* (Holmstrup 1994). In tundra soils of Europe, their introduction was shown to affect soil structure, including increased macroporosity and decreased soil moisture (Klaminder et al. 2023), as well as vegetation performance, namely, by increasing plant N uptake (Blume-Werry et al. 2020). Considering the changes observed both in the European arctic and in North American boreal forests, it is very likely that geoengineering earthworms are also profoundly reshaping European boreal forests.

While current research efforts are advancing our understanding of the changes in soil biodiversity and functioning associated with non-native earthworms, their invasion is happening in a rapidly changing environment. Notably, climate change predictions forecast increased wildfire frequency in most parts of the boreal biome (de Groot et al. 2013; Kilpeläinen et al. 2010). The effects of fire on earthworms are not well understood, but available literature suggests limited long-term effects for native earthworms (Certini et al. 2021). In the case of exotic earthworms, Ikeda et al. (2015) successfully decreased cocoon survival and the abundance of a non-native Megascolecidae earthworm (*Amynthas agrestis,* an Asian earthworm species invasive in North America) with prescribed fires. Heat waves can also negatively affect earthworm survival, as a study found that the cocoons of *Amynthas* spp. became non-viable after exposure to temperatures above 38 °C (Johnston and Herrick 2019). However, major life trait differences exist between Megascolecidae and Lumbricidae, the former having an annual life cycle, and this finding should therefore not be directly applied to all European earthworm species. While experimental approaches testing the interactive effects of invasive earthworms and wildfires are still lacking, a model-based study found that the forest floor carbon losses were 2.5 times higher in the presence of both fire and earthworms (3.17 Mg C ha^{-1}) than in the presence of fire only (0.91 Mg C ha^{-1}; Cameron et al. 2015). Additionally, high latitudes are warming up at a faster rate, shifting the geographical range of deciduous tree species northwards (Boilard et al. 2023; Searle and Chen 2017). Considering that earthworms are generally more abundant in deciduous forests than coniferous forests, a compositional shift towards more deciduous broadleaves may further hasten their establishment.

4 Earthworms in Forest Microhabitats

Earthworm ecology continues to be a topic of great research interest globally, yet earthworm community dynamics in ex situ forest microhabitats (e.g. leaf litter, deadwood and trees with 'suspended soils') is still poorly investigated (Lavelle et al. 1997; Blouin et al. 2013; White et al. 2020; Zuo et al. 2023). Earthworms are commonly classified into three main ecological categories, anecic (deep-burrowing), endogeic (shallow-burrowing) and epigeic (surface-living) (Bouché 1972; Lee 1985), although in reality such grouping is far more variable and nuanced (Bottinelli et al. 2020). Besides the three main traditional classifications, an additional group of earthworms was named 'arboreal' in tropical forests by Lee (1985) and 'corticolous' in temperate systems by Bouche (1972), to recognise those earthworm species that live in close association with trees, such as in the canopy or under the bark of trees, or within deadwood. Recently, novel studies have revealed that where microhabitats such as deadwood are not included in forest earthworm surveys, valuable information on earthworm species ecologies may be missed, and even whole species excluded from those surveys (Geraskina 2016; Vorobeichik et al. 1998; Schmidt et al. 2015; Römbke et al. 2017; Ashwood et al. 2019b). This subchapter aims to highlight the latest findings regarding earthworm ecology in northern European temperate and boreal forest microhabitats, including novel methodologies, and point to where future work could be usefully directed in this emerging research area.

It is known that earthworms colonise deadwood following the early stages of decay by fungi and invertebrate-led wood channelisation, and they further advance decomposition processes through the transport and regulation of nutrients, water and microbes (Ausmus 1977; Caldwell 1993; Hendrix 1996). There is an interactive relationship between earthworm populations and deadwood breakdown; earthworm populations in deadwood are influenced by factors such as tree species, decay stage and the biotic and abiotic conditions of the surrounding forest—but equally, the diversity of earthworms in a forest may affect the rate of deadwood decay (Geraskina 2016; Zuo et al. 2023). Since the colonisation process of deadwood by earthworms has been covered in previous literature, particularly for North American forests (e.g. Hendrix 1996), this subchapter focuses instead on recent ecological findings pertinent to earthworm populations in Northern European forest microhabitats.

Earthworm abundance and community composition in deadwood is driven by a range of factors, including forest age, tree species identity, wood decay stage and the surrounding environmental conditions (Muys et al. 1992; Zuo et al. 2016). In two contrasting temperate forests in the Netherlands, Zuo et al. (2023) found that deadwood decay stage and tree species influenced earthworm abundances, with higher earthworm populations in logs of *Picea abies* and *Populus* spp. after 1 and 2 years of decay, compared to a range of other native coniferous and broadleaf European tree species. After 4 years, however, this pattern reversed, although overall, *Populus* spp. deadwood supported the highest earthworm abundance, followed by *P. abies* and *Quercus robur* (Zuo et al. 2023). The authors suggest that the dynamic changes in earthworm populations across decay years indicates a potential succession process of earthworm species turnover throughout the deadwood decomposition process, attributed to changes in resource quantity and quality (Zuo et al. 2023). Ashwood et al. (2019b) investigated mid- and late decay-stage deadwood in mixed oak-birch forests in southeast England and found little effect of decay stage on earthworm population, except in the abundance of *Bimastos eiseni*. They did, however, observe an increase in the proportion of endogeic earthworms within the deadwood earthworm community in the late decay stage—a trend hypothesised by Hendrix (1996) and confirmed by Kooch (2012) in mixed broadleaf deadwood in northern Iran. Likewise, Geraskina and Shevchenko (2019) observed mid- to late decay stage deadwood was an important habitat for epigeic species, particularly *D. attemsi*, in a range of forest types in the northwestern Caucasus.

The high abundance and species diversity observed in early-mid-decay stage deadwood by Zuo et al. (2023) correspond with changes in the integrity of the wood structure from intact with tight bark to softening wood with loosening bark. Using a bark exchange methodology on mid-decay deadwood in the Netherlands, Chang et al. (2023) found that invertebrate abundance, including Clitellata, was positively correlated with bark pH and thickness. Conversely, tannin content showed a negative correlation with Clitellata abundance, and the authors link these correlations to the role of tree species biochemistry affecting resource availability (Chang et al. 2023). Previously at the same forest site, Zuo et al. (2016) found that after 1 year of decomposition, deadwood bark looseness and thickness were significantly correlated with annelid abundance, and the ratio of inner-to-outer bark significantly negatively correlated, across 11 native European angiosperm tree species. Chang et al. (2023) extended this to demonstrate that bark trait differences in gymnosperm tree species can also explain invertebrate (including Clitellata) community composition of deadwood. Similarly, Ashwood et al. (2019b) found that, in a mixed oak-birch forest, the looseness of bark influenced earthworm abundance, and very few earthworms were recovered from deadwood lacking a bark cover.

Recent studies in temperate and boreal forests have shown that all life stages of earthworms may be present in decaying wood (Geraskina 2016; Ashwood et al. 2019b; Kostecka et al. 2020; Zuo et al. 2023). In common oak (*Q. robur*) and silver birch (*Betula pendula*) forest stands in southeast England, deadwood was found to support proportionately higher populations of juvenile earthworms than the surrounding soil habitat, where microenvironmental conditions (temperature and moisture) were more favourable (Ashwood et al. 2019b). This suggests that decaying wood may be an important habitat for earthworm reproduction, as well as enabling overwintering of cocoons and juvenile earthworms (Ashwood et al. 2019b; Kostecka et al. 2020; Zuo et al. 2023). Earthworms (particularly juveniles) are highly sensitive to the moisture and temperature conditions of their surrounding environment, and bark and inner deadwood provide a buffering effect against external moisture and temperature extremes. Thus, deadwood potentially extends periods of earthworm activity in northern European forests throughout summer (when the soil and litter layers are usually very dry) and during winter freezing conditions (Hendrix 1996; Geraskina 2016; Zuo et al. 2016); Ashwood et al. 2019b). The same may be true for ‘suspended soils’ in tree holes, bark pockets and epiphytes such as mosses, affording protection against freezing and desiccation to arboreal earthworm species (Römbke et al. 2017).

The study of arboreal earthworm species is scarce in European forests; however, there are many examples from the tropics, where the preference of some earthworms for above ground habitats is well-known (Lee 1985, Fragoso and Rojas-Fernandez 1996, James and Brown 2006). In wet tropical forests, earthworms may be regularly found living in the forest canopy, for example, in debris-filled bromeliad rosettes (Richardson et al. 2006). Since earthworms are especially vulnerable to desiccation above ground, these behaviours are mostly found in forests with high humidity (James 2005). It is not yet clear why arboreal earthworm species are present in temperate and boreal European forests, which do not seemingly provide such permanently moist above ground conditions. Römbke et al. (2017) suggest resource availability and protection from predators, as well as lack of competition with other native earthworm species, may all be attractive factors to tree-dwelling earthworm species. With limited available studies and methodologies concerning tree-inhabiting earthworms in European forests, we know very little of the life history and ecological roles of these unique earthworms.

What is clear, however, is that where alternative habitats such as deadwood are not included in earthworm surveys of European forests, arboreal species may be missed entirely (Butt and Lowe 2004; Schmidt et al. 2015; Römbke et al. 2017). For example, in oak forest stands in south-east England, Ashwood et al. (2019b) found the nationally rare *Dendrobaena pygmaea* only underneath deadwood, not in the surrounding open soil habitat. On the Scottish Isle of Rum, Butt and Lowe (2004) recovered the species *B. rubidus* from beneath the bark of a dead tree. In boreal old-growth coniferous (fir-spruce) forests of the Pechora-Ilych Nature Reserve in Russia, Geraskina (2016) found two species, *Dendrobaena octaedra* and *B. rubidus*, only in deadwood and not in the soil. Likewise, in the Northwestern Caucasus, Geraskina and Shevchenko (2021) found the species *B. rubidus* and *Dendrobaena attemsi* only in deadwood and not in soil pits during spring and summer sampling of beech-fir forests. They report that the species *B. rubidus*, *D. attemsi* and *D. octaedra* are closely associated with trees and thus easily missed from soil surveys. Interestingly, Geraskina (2016) also only found specimens of the typically soil-living *Aporrectodea caliginosa* in the moss cover of a fallen spruce (*Picea obovata*), rather than during soil pit sampling in a boreal forest.

Overcoming the current under-sampling of forest microhabitats would not only fill gaps in our understanding of tree-associated earthworm species ecologies but may also generate more accurate species distribution and conservation classifications (Ashwood et al. 2019b). For example, Schmidt et al. (2015) were able to confirm the presence of *D. attemsi* (Fig. 1) in Ireland by surveying organic matter residues under deadwood and in moss covering the base of large sessile oak (*Quercus petraea*) trees. The same species was also recently reclassified in Germany from 'rare' to 'moderately common', following forest microhabitat surveys (Lehmitz et al. 2016). Woodland and deadwood earthworm sampling in the UK has also increasingly revealed *D. attemsi* in recent years (Butt and Lowe 2004; Eggleton et al. 2009), with Ashwood et al. (2019b) finding it alongside specimens of *D.*

Fig. 1 *Dendrobaena attemsi*, an 'arboreal' European earthworm species (© Frank Ashwood)

octaedra in the saddle of a silver birch (*B. pendula*) approximately 1 m above ground in a mixed oak forest in southeast England. The acid-tolerant, stenotopic nature of this species may make it locally abundant in forest stands where other species are excluded; for example, in the Netherlands, Zuo et al. (2023) found *D. attemsi* was present only in deadwood on an acidic and infertile sandy podzolic soil, and not in their twinned fertile calcareous reclaimed marine clay site. Similarly, *D. attemsi* showed very localised but abundant populations in acidic and organic matter-rich mixed oak-birch woodland plots in southeast England (Ashwood et al. 2019b).

Another seemingly acid-tolerant European arboreal earthworm species of interest is *Bimastos eiseni* (previously *Allolobophoridella eiseni*—see Csuzdi et al. 2017), which was reported to live under the bark of trees in Austria by Zicsi (1965) even before Bouché (1972) presented the 'corticole' ecological category in France. Similar description of this species' ecology was given for Hungary by Csuzdi and Zicsi (2003), although tree bark was not included in the habitat description for *B. eiseni* in the UK by Sims and Gerard (1999). Butt and Lowe (2004) found *B. eiseni* alongside *B. rubidus* under rocks in a crag on the Scottish Isle of Rum, while Ashwood et al. (2019b) found *B. eiseni* individuals beneath the bark of a standing dead tree and within a bleeding bark fissure on a diseased common oak (*Q. robur*) in mixed oak-birch forests in southeast England. Ashwood et al. (2019b) also found a significantly greater abundance of *B. eiseni* in deadwood than in the surrounding forest soil. In the LOGLIFE common garden/tree cemetery experiment in the Netherlands, where logs of ten native tree species were reciprocally transplanted between two sites, Zuo et al. (2023) reported high abundance of *B. eiseni* within deadwood at their acidic and infertile sandy podzolic forest site, compared with only one specimen retrieved from deadwood in their fertile calcareous reclaimed marine clay site. Systematic forest microhabitat experiments such as LOGLIFE, as well as more opportunistic and novel forest microhabitat surveying, may go a long way towards identifying the specific ecological niches required by arboreal European earthworm species such as *B. eiseni* and *B. attemsi*.

A particularly novel and insightful European arboreal earthworm assessment in recent years was a long-term survey of six Strict Forest Reserves along with five managed comparison forests in Hesse (Central Germany) by Römbke et al. (2017). Recognising that traditional standardised sampling methods (i.e. soil pit surveys) typically miss surface-living and arboreal earthworm species, these researchers deployed ground-level pitfall traps and eclector traps on standing tree trunks (at 1.8–2.0 m in height) and decaying logs laying on the ground in beech (*Fagus sylvatica*) and oak (*Q. robur*) forests, over a period of 2 years. The results were quite surprising, with nine earthworm species returned in eclector traps on the trunks of standing trees (both alive and dead), versus 16 earthworm species in pitfall traps. There was a difference in the dominant species returned between the two trap types: *Lumbricus rubellus* (22–43% of adults)- and *D. octaedra* (10–16%)-dominated pitfall traps, while *B. eiseni* was by far the most common species in eclector traps on standing trees (71–83%), followed by *B. rubidus* (3–21%) and *D. octaedra* (6–8%) (Römbke et al. 2017). This study conclusively demonstrated that *B. eiseni* is a predominantly arboreal European earthworm species that routinely climbs up the bark of living and dead trees, where it is far more commonly found than in forest floor habitats such as litter or soil. Whereas Graff (1953) and Bouché (1972) classified *B. eiseni* as a rare species in Germany and France, Römbke et al. (2017)'s unique study provided data that helped classify the species as 'moderately common' and 'not endangered' in Germany (Lehmitz et al. 2016). As seen here, and in the previously described finding of *D. attemsi* in Ireland by Schmidt et al. (2015), there is much yet to be learned about arboreal earthworm ecologies and distributions in European forests, if researchers are willing to adopt non-soil-based earthworm sampling methodologies.

A recent citizen science project in the UK (Petrovan et al. 2022) made use of the survey and inspection of over 6000 tree cavities as part of the Bat Tree Habitat Key (BTHK) project. During the inspection, the surveyors (trained volunteers and professionals) identified and surveyed trees across the UK, searching for potential bat roost features and recognisable (vertebrate and invertebrate) animals inside cavities using camera endoscopes. Of the 7875 observations, 78 in England reported 'worms' as being present and visible in the inspected tree holes. All trees were broadleaved species, in the woodland/broadleaved/semi-natural habitat type, mostly alive but some dead, and the tree holes were on average at a height of 1.71 m (range 0.68–6.49 m). Due to the citizen science nature of this project, no further identification apart from 'worms' is available (Henry Andrews, pers. comm.); however, it is remarkable that about 1% of all tree cavities may have harboured earthworms.

Perhaps the biggest barrier to the inclusion of microhabitats such as deadwood into more formal forest earthworm surveys is a lack of standardised systematic and quantitative methodologies for doing so (Ashwood et al. 2019b). Since the quantity and quality of such microhabitats can vary widely between woodland types and ages, standardisation of sampling and data analysis would appear problematic (Burton and Eggleton 2016). However, decaying wood and leaf litter are discrete units that can be easily experimentally manipulated, and are available in a wide range of sizes and qualities, making them useful subjects for ecological research (Carroll 1996; Cornelissen et al. 2012; Zuo et al. 2018; Ashwood et al. 2019b). Besides the study of Römbke et al. (2017), two other notable recent pieces of research in northern European forests have embraced the challenge of systematically investigating earthworm-deadwood relationships further.

In an attempt to develop a standardised approach to deadwood earthworm assessment in natural forest settings, Ashwood et al. (2019b) systematically surveyed decaying wood and the underlying soil, as well as the surrounding forest soil within 12 10 m^2 plots in a chronosequence of *Q. robur* and *B. pendula* forest stands in Surrey, England. Alongside the results from this study already reported, it was found that the surrounding soil contained significantly higher total earthworm abundance and biomass than deadwood. For the first time, soil and deadwood habitats were directly comparable, because the authors calculated earthworm abundance per m^2 surface area of deadwood, unlike previous studies that instead reported earthworm populations per volume of deadwood (e.g. Kooch 2012; Geraskina 2016). This approach was subsequently adopted by Geraskina and Shevchenko (2021) to report earthworm community dynamics within deadwood in a variety of forests in Northwest Caucasus. While microhabitat surveying yielded additional species abundance and biomass data in their plots, Ashwood et al. (2019b) concluded that systematic deadwood surveys should be considered as additional and complementary to traditional soil sampling protocols, rather than a direct replacement.

In the Netherlands, researchers at the LOGLIFE project have embarked on a decade-long (to date) and highly systematic body of deadwood life cycle research (Cornelissen et al. 2012). This multispecies 'tree cemetery' or 'common garden' experiment was established to disentangle the interactive effects of wood traits, tree species, forest site effects and other drivers of deadwood decay and has subsequently included studies into a range of deadwood-associated invertebrate community dynamics, including earthworms (Zuo et al. 2018, 2023; Chang et al. 2023). Logs of uniform age and dimensions were incubated in two strongly contrasting (in soil and microclimate) temperate forest sites in central Netherlands (Cornelissen et al. 2012). Tree species were selected to represent native northern European forestry species: *Abies grandis, Larix kaempferi, P. abies, Pseudotsuga menziesii* (coniferous), *B. pendula, F. sylvatica, Fraxinus excelsior, Populus x canadensis, Populus tremula* and *Q. robur* (broadleaf). Over the course of 4 years, Zuo et al. (2023) surveyed deadwood earthworm community dynamics within both experimental stands and successfully identified how abiotic and biotic factors such as tree species, duration of wood decay and decay environment all affected earthworm abundance and community composition in the decaying logs. Their findings show that the range of deadwood of different tree species and decay stages typically found in old-growth and unmanaged temperate mixed forests will be likely to support greater earthworm diversity and abundances at the stand scale, compared with managed monoculture plantations (Zuo et al. 2023).

Having summarised the recent advances in northern European forest microhabitat earthworm research, it is appropriate to conclude with suggestions for future work in this area. Firstly, more studies are urgently needed to refine systematic methodologies for investigating earthworm communities in forest microhabitats. To gain a full understanding of deadwood earthworm community dynamics, studies are also required which consider the full decay process, which may range from a decade to well over a century (Shorohova and Kapitsa 2014; Zuo et al. 2023). Future research could also usefully investigate further the feedback loop between deadwood decay processes and earthworm communities (Zuo et al. 2023). Additionally, the role of forest soil as a source or co-habitat for European corticolous earthworm species requires further investigation, with limited studies available which investigate the interaction between the two habitats (Ashwood et al. 2019b). Greater consideration of above-soil habitats as multidimensional 3D systems for earthworm macro-ecological modelling, such as species distribution models (SDMs), may help to improve our understanding of arboreal earthworm ecologies (White et al. 2020). Finally, since deadwood in European forests appears to possess high populations of juvenile earthworms, which cannot be morphologically identified, the application of molecular techniques would be a sensible approach to identifying the full range of species present in these and similar forest microhabitats (Ashwood et al. 2019b; Zuo et al. 2023).

5 Future Research Perspectives

To predict or model future earthworm distributions in forest ecosystems such as we have discussed throughout this chapter, it is first essential to predict the nature of future forests. Forest landscape models can simulate the dynamics of forest ecosystems at the landscape level over decades or centuries. For example, a comparison of four such models applied to alpine landscapes (Petter et al. 2020) showed that projections using past climate conditions agreed well with observations, yet uncertainty under future climate conditions was high. Such models can incorporate various factors including temperature and CO_2 concentrations, N deposition, disturbance (e.g. fires and insect attack) and water balance. Earthworms are likely to be affected by many of these factors directly, but also indirectly, for example, via changed litter inputs. Based on findings from a long-term (8-year) field study on Enchytraeidae (Holmstrup et al. 2015), it can be assumed that the most important direct factor for earthworm populations under projected future climate conditions is soil water rather than temperature or atmospheric CO_2 concentrations. Analysis of the drought years of 2018 and 2019 in northern Central Europe suggests that one can expect not only severe growth reductions for deciduous tree species under the predicted dryer and warmer future conditions but also impacts on the whole structure and species composition of these forests (Scharnweber et al. 2020). Future research should therefore focus intensely on the direct and indirect effects of water dynamics on earthworm populations in forests.

It has been argued recently (Buřivalová et al. 2023) that the heightened attention paid currently to the role of forests as C stores in mitigating climate change also creates opportunities to increase attention to other 'co-benefits' of forest conservation, such as biodiversity conservation. This is likely to focus on above ground aspects of biodiversity, and in particular biodiversity that is easy to monitor, for example, by remote sensing (Buřivalová et al. 2023). However, there are opportunities to include research on below ground biodiversity, and in particular on earthworms, in future projects that primarily focus on carbon and climate mitigation. This should make use of novel promising methods that have emerged in the last decade. The use of eDNA to detect earthworm presence has proven successful for fresh soil samples (Bienert et al. 2012) as well as 30-year-old soil samples (Jackson et al. 2017). For instance, this molecular approach would be highly suitable to reconstruct the invasion timeline of a given site on boreal forests, when archival samples are available. Recent advances are promising to identify earthworms to the species level (J. Lejoly, pers. comm.).

Soil acoustic emissions have been successfully used to quantify earthworm activity in the laboratory (Lacoste et al. 2018). In the field, the presence of non-native earthworms can alter soundscape characteristics through changes in soil structure (Keen et al. 2022). The same method also showed great potential to monitor below ground biodiversity during restoration (Robinson et al. 2023). This non-invasive innovative method represents an opportunity to jointly study earthworm activity and their effects on soil structure. A third example of research methods is natural abundance stable isotope ratio measurements (mainly $\delta^{15}N$ and $\delta^{13}C$) that have been used in the last two decades to decipher feeding relations of soil fauna (Potapov et al. 2019). Most recently, Hsu et al. (2023) were able to identify nine earthworm isotopic niches and showed that the niches form a continuum, rather than distinct clusters. Methods such as this will allow us to not just to monitor earthworm populations, but also to study the functions of earthworms and their roles in nutrient and carbon dynamics, soil structure dynamics and overall stability of forest ecosystems (Powell 2007).

Coordinated efforts and distributed environmental research infrastructures are needed to monitor forests and other terrestrial ecosystems under global change (Musche et al. 2019). Hopefully this chapter has made it clear that any such endeavours studying forest ecosystem structures and processes, the impacts of anthropogenic drivers and the provisioning of ecosystem services in forests should consider including earthworms as key representatives of the below ground biodiversity and important modulators of carbon and nutrient cycling. Indeed, since earthworms are crucial for the regulation and cycling of forest carbon and nutrients in most northern European forest systems (excluding boreal), supporting earthworm populations through reduced forest management, increased tree species diversity and the retention of forest microhabitats may greatly enhance wider forest biodiversity and ecosystem functioning.

References

Addison JA (2009) Distribution and impacts of invasive earthworms in Canadian forest ecosystems. Biol Invasions 11:59–79

Alban DH, Berry EC (1994) Effects of earthworm invasion on morphology, carbon, and nitrogen of a forest soil. Appl Soil Ecol 1:243–249

Ammer S, Weber K, Abs C, Ammer C, Prietzel J (2006) Factors influencing the distribution and abundance of earthworm communities in pure and converted Scots pine stands. Appl Soil Ecol 33(1):10–21

Angst G, Mueller C, Prater I, Angst Š, Peterse F, Nierop K (2019) Earthworms act as biochemical reactors to convert labile plant compounds into stabilized soil microbial necromass. Commun Biol 2(441):1–7

Ashwood F, Watts K, Park K, Fuentes-Montemayor E, Benham S, Vanguelova EI (2019a) Woodland restoration on agricultural land: long-term impacts on soil quality. Restor Ecol 27(6):1381–1392

Ashwood F, Vanguelova EI, Benham S, Butt KR (2019b) Developing a systematic sampling method for earthworms in and around deadwood. For Ecosyst 6(1):1–12

Augusto L, De Schrijver A, Vesterdal L, Smolander A, Prescott C, Ranger J (2015) Influences of evergreen gymnosperm and deciduous angiosperm tree species on the functioning of temperate and boreal forests. Biol Rev 90(2):444–466

Ausmus BS (1977) Regulation of wood decomposition rates by arthropod and annelid populations. Ecol Bull 25:180–192

Bienert F, De Danieli S, Miquel C, Coissac E, Poillot C, Brun JJ, Taberlet P (2012) Tracking earthworm communities from soil DNA. Mol Ecol 21:2017–2030

Blouin M, Hodson ME, Delgado EA, Baker G, Brussaard L, Butt KR, Dai J, Dendooven L, Pérès G, Tondoh JE, Cluzeau D (2013) A review of earthworm impact on soil function and ecosystem services. Eur J Soil Sci 64(2):161–182

Blume-Werry G, Krab EJ, Olofsson J, Sundqvist MK, Väisänen M, Klaminder J (2020) Invasive earthworms unlock arctic plant nitrogen limitation. Nat Commun 11:1–10

Bohlen PJ, Scheu S, Hale CM, McLean MA, Migge S, Groffman PM, Parkinson D (2004) Non-native invasive earthworms as agents of change in northern temperate forests. Front Ecol Environ 2(8):427–435

Boilard G, Bradley RL, Houle D (2023) A northward range shift of sugar maple (*Acer saccharum*) in Eastern Canada should reduce soil carbon storage, with no effect on carbon stability. Geoderma 432:116403

Bottinelli N, Hedde M, Jouquet P, Capowiez Y (2020) An explicit definition of earthworm ecological categories–Marcel Bouché's triangle revisited. Geoderma 372, p 114361

Bouché MB (1972) Lombriciens de France: écologie et systématique. Institut national de la recherche agronomique, Paris

Bradshaw CJA, Warkentin IG (2015) Global estimates of boreal forest carbon stocks and flux. Glob Planet Change 128:24–30

Buřivalová Z, Yoh N, Butler RH, Sagar HSSC, Game ET (2023) Broadening the focus of forest conservation beyond carbon. Curr Biol 33:R621–R635

Burton VJ, Eggleton P (2016) Microhabitat heterogeneity enhances soil macrofauna and plant species diversity in an Ash–Field Maple woodland. Eur J Soil Biol 75:97–106

Butt KR, Lowe CN (2004) Anthropic influences on earthworm distribution, Isle of Rum National Nature Reserve, Scotland. Eur J Soil Biol 40(2):63–72

Caldwell RS (1993) Macroinvertebrates and their relationship to coarse woody debris: with special reference to land snails. In: JW MW, Crossley DA Jr (eds) Biodiversity and coarse woody debris in southern forests, proceedings of the workshop on coarse woody debris in southern forests: effects on biodiversity. General technical report SE-94. USDA Forest Service, Southern Research Station

Cameron EK, Bayne EM (2009) Road age and its importance in earthworm invasion of northern boreal forests. J Appl Ecol 46(1):28–36

Cameron EK, Bayne EM, Jill M (2007) Human-facilitated invasion of exotic earthworms into northern boreal forests. Ecoscience 14:482–490

Cameron EK, Knysh KM, Proctor HC, Bayne EM (2013) Influence of two exotic earthworm species with different foraging strategies on abundance and composition of boreal microarthropods. Soil Biol Biochem 57:334–340

Cameron EK, Shaw CH, Bayne EM, Kurz WA, Kull SJ (2015) Modeling interacting effects of invasive earthworms and wildfire on forest floor carbon storage in the boreal forest. Soil Biol Biochem 88:189–196

Carroll CR (1996) Coarse woody debris in forest ecosystems: an overview of biodiversity issues and concepts. In: JW MW, Crossley DA Jr (eds) Biodiversity and coarse woody debris in southern forests, proceedings of the workshop on coarse woody debris in southern forests: effects on biodiversity. General technical report SE-94. USDA Forest Service, Southern Research Station

Certini G, Moya D, Lucas-Borja ME, Mastrolonardo G (2021) The impact of fire on soil-dwelling biota: A review. For Ecol Manag 488:118989

Chandler KR, Stevens CJ, Binley A, Keith AM (2018) Influence of tree species and forest land use on soil hydraulic conductivity and implications for surface runoff generation. Geoderma 310:120–127

Chang C, Berg MP, van Logtestijn RS, Zuo J, Lin L, Bom C, Wolters J, Biesbroeck M, de Ruijter P, Hefting MM, Sass-Klaassen U (2023) Reciprocal bark exchange helps to disentangle tree species-dependent bark and wood trait effects on invertebrate diversity. J Ecol 111(1):125–138

Cornelissen JH, Sass-Klaassen U, Poorter L, van Geffen K, van Logtestijn RS, van Hal J, Goudzwaard L, Sterck FJ, Klaassen RK, Freschet GT, van der Wal A (2012) Controls on coarse wood decay in temperate tree species: birth of the LOGLIFE experiment. Ambio 41:231–245

Craven D, Thakur MP, Cameron EK, Frelich LE, Beauséjour R, Blair RB, Blossey B, Burtis J, Choi A, Dávalos A, Fahey TJ, Fisichelli NA, Gibson K, Handa IT, Hopfensperger K, Loss SR, Nuzzo V, Maerz JC, Sackett T, Scharenbroch BC, Smith SM, Vellend M, Umek LG, Eisenhauer N (2017) The unseen invaders: introduced earthworms as drivers of change in plant communities in North American forests (a meta-analysis). Glob Change Biol 23:1065–1074

Csuzdi C, Zicsi A (2003) Earthworms of Hungary (Annelida: Oligochaeta, Lumbricidae). Hungarian Natural History Museum, Budapest, p 271

Csuzdi C, Chang CH, Pavlícek T, Szederjesi T, Esopi D, Szlávecz K (2017) Molecular phylogeny and systematics of native North American lumbricid earthworms (Clitellata: Megadrili). PLoS One 12(8):e0181504

de Groot WJ, Flannigan MD, Cantin AS (2013) Climate change impacts on future boreal fire regimes. For Ecol Manag 294:35–44

De Wandeler H, Bruelheide H, Dawud SM, Dănilă G, Domisch T, Finér L, Hermy M, Jaroszewicz B, Joly FX, Müller S, Ratcliffe S (2018) Tree identity rather than tree diversity drives earthworm communities in European forests. Pedobiologia 67:16–25

Desie E, Vancampenhout K, Heyens K, Hlava J, Verheyen K, Muys B (2019) Forest conversion to conifers induces a regime shift in soil process domain affecting carbon stability. Soil Biol Biochem 136:107540

Desie E, Van Meerbeek K, De Wandeler H, Bruelheide H, Domisch T, Jaroszewicz B, Joly F-X, Vancampenhout K, Vesterdal L, Muys B (2020) Positive feedback loop between earthworms, humus form and soil pH reinforces earthworm abundance in European forests. Funct Ecol 34:2598–2610

Dimbleby GW (1952) Soil regeneration on the north-east Yorkshire moors. J Ecol 40(2):331–341

Eggleton P, Inward K, Smith J, Jones DT Sherlock E (2009) A six year study of earthworm (Lumbricidae) populations in pasture woodland in southern England shows their responses to soil temperature and soil moisture. Soil Biol Biochem 41(9):1857–1865

Ferlian O, Eisenhauer N, Aguirrebengoa M, Camara M, Ramirez-Rojas I, Santos F, Tanalgo K, Thakur MP (2018) Invasive earthworms erode soil biodiversity: a meta-analysis. J Anim Ecol 87:162–172

Forest Europe (2020) State of Europe's forests 2020. https://foresteurope.org/wp-content/uploads/2016/08/SoEF_2020.pdf

Fragoso C, Rojas-Fernandez P (1996) Earthworms inhabiting bromeliads in Mexican tropical rainforests: ecological and historical determinants. J Trop Ecol 12(5):729–734

Frelich LE, Blossey B, Cameron EK, Dávalos A, Eisenhauer N, Fahey T, Ferlian O, Groffman PM, Larson E, Loss SR, Maerz JC, Nuzzo V, Yoo K, Reich PB (2019) Side-swiped: ecological cascades emanating from earthworm invasions. Front Ecol Environ 17:502–510

Frouz J, Livečková M, Albrechtová J, Chroňáková A, Cajthaml T, Pižl V, Háněl L, Starý J, Baldrian P, Lhotáková Z, Šimáčková H (2013) Is the effect of trees on soil properties mediated by soil fauna? A case study from post-mining sites. For Ecol Manag 309:87–95

Geraskina AP (2016) The population of earthworms (Lumbricidae) in the main types of dark coniferous forests in Pechora-Ilych Nature Reserve. Biol Bull 43(8):819–830

Geraskina AP, Shevchenko NE (2019) Spatial distribution of the epigeic species of earthworms *Dendrobaena octaedra* and *D. attemsi* (Oligochaeta: Lumbricidae) in the forest belt of the northwestern Caucasus. Turk J Zool 43(5):480–489

Geraskina AP, Shevchenko NE (2021) Biotopic allocation of morpho-ecological groups of earthworms (Oligochaeta, Lumbricidae) to the main forest types in the Bol'shaya Laba river basin (Northwestern Caucasus). Biol Bull 48(8):1176–1188

Gongalsky KB, Zaitsev AS, Korobushkin DI, Saifutdinov RA, Butenko KO, de Vries FT, Ekschmitt K, Degtyarev MI, Gorbunova AY, Kostina NV, Rakhleeva AA, Shakhab SV, Yazrikova TE, Wolters V, Bardgett RD (2021) Forest fire induces short-term shifts in soil food webs with consequences for carbon cycling. Ecol Lett 24:438–450

Graff O (1953) Die Regenwürmer Deutschlands.–Schriftenreihe der Forschungsanstalt für Landwirtschaft Braunschweig-Völkenrode 7. Schaper, Germany

Hansson K, Fröberg M, Helmisaari HS, Kleja DB, Olsson BA, Olsson M, Persson T (2013) Carbon and nitrogen pools and fluxes above and below ground in spruce, pine and birch stands in southern Sweden. For Ecol Manag 309:28–35

Hendrix PF (1996) Earthworms, biodiversity, and coarse woody debris in forest ecosystems of the southeastern U.S.A. In: JW MW, Crossley DA Jr (eds) Biodiversity and coarse woody debris in southern forests, proceedings of the workshop on coarse woody debris in southern forests: effects on biodiversity. General technical report SE-94. USDA Forest Service, Southern Research Station, pp 43–48

Holmstrup M (1994) Physiology of cold hardiness in cocoons of five earthworm taxa (Lumbricidae: Oligochaeta). J Comp Physiol B 164(3):222–228

Holmstrup M, Westh P (1994) Dehydration of earthworm cocoons exposed to cold: a novel cold hardiness mechanism. J Comp Phys B 164:312–315

Holmstrup M, Schmelz RM, Carrera N, Dyrnum K, Larsen KS, Mikkelsen TN, Beier C (2015) Responses of enchytraeids to increased temperature, drought and atmospheric CO2: results of an eight-year field experiment in dry heathland. Eur J Soil Biol 70:15–22

Hsu GC, Szlavecz K, Csuzdi C, Bernard M, Chang CH (2023) Ecological groups and isotopic niches of earthworms. Appl Soil Ecol 181:104655–104655

Ikeda H, Callaham MA Jr, O'Brien JJ, Hornsby BS, Wenk ES (2015) Can the invasive earthworm, *Amynthas agrestis,* be controlled with prescribed fire? *Soil Biol.* Biochemist 82:21–27

Jackson M, Myrholm C, Shaw C, Ramsfield T (2017) Using nested PCR to improve detection of earthworm eDNA in Canada. Soil Biol Biochem 113:215–218

James SW (2005) New genera and species of pheretimoid earthworms (Clitellata: Megascolecidae) from southern Luzon, Philippines. Syst Biodivers 2(3):271–279

James SW, Brown GG (2006) Earthworm ecology and diversity in Brazil. In: Soil biodiversity in Amazonian and other Brazilian ecosystems, pp 56–116

Jochum M, Ferlian O, Thakur MP, Ciobanu M, Klarner B, Salamon JA, Frelich LE, Johnson EA, Eisenhauer N (2021a) Earthworm invasion causes declines across soil fauna size classes and biodiversity facets in northern North American forests. Oikos 130:766–780

Jochum M, Thouvenot L, Ferlian O, Zeiss R, Klarner B, Pruschitzki U, Johnson EA, Eisenhauer N (2021b) Aboveground impacts of a belowground invader: how invasive earthworms alter aboveground arthropod communities in a northern North American forest. Biol Lett 18:1–7

Johnston M, Herrick B (2019) Cocoon heat tolerance of Pheretimoid earthworms *Amynthas tokioensis* and *Amynthas agrestis*. Am Midl Nat 181(2):299–309

Keen SC, Wackett AA, Willenbring JK, Yoo K, Jonsson H, Clow T, Klaminder J (2022) Non-native species change the tune of tundra soils: Novel access to soundscapes of the Arctic earthworm invasion. Sci Total Environ 838:155976

Keith AM, Boots B, Hazard C, Niechoj R, Arroyo J, Bending GD, Bolger T, Breen J, Clipson N, Doohan FM, Griffin CT (2012) Cross-taxa congruence, indicators and environmental gradients in soils under agricultural and extensive land management. Eur J Soil Biol 49:55–62

Kilpeläinen A, Kellomäki S, Strandman H, Venäläinen A (2010) Climate change impacts on forest fire potential in boreal conditions in Finland. Clim Chang 103:383–398

Klaminder J, Krab EJ, Larsbo M, Jonsson H, Fransson J, Koestel J (2023) Holes in the tundra: Invasive earthworms alter soil structure and moisture in tundra soils. Sci Total Environ 859:160125

Knowles ME, Ross DS, Görres JH (2016) Effect of the endogeic earthworm Aporrectodea tuberculata on aggregation and carbon redistribution in uninvaded forest soil columns. Soil Biol Biochem 100:192–200

Kooch Y (2012) Response of earthworms' ecological groups to decay degree of dead trees (case study: Sardabrood forest of Chalous, Iran). Eur J Exp Biol 2:532–538

Kostecka J, Butt KR, Mazur-Pączka A, Pączka G, Garczyńska M, Podolak A (2020) Aspects of the ecology of the earthworm *Eisenia lucens* (Waga 1857) studied in the field and in laboratory culture. Environ Sci Pollut Res 27(27):33486–33492

Lacoste M, Ruiz S, Or D (2018) Listening to earthworms burrowing and roots growing – acoustic signatures of soil biological activity. Sci Rep 8:10236

Lambkin DC, Gwilliam KH, Layton C, Canti MG, Piearce TG, Hodson ME (2011) Production and dissolution rates of earthworm-secreted calcium carbonate. Pedobiologia 54:S119–S129

Larson ER, Kipfmueller KF, Hale CM, Lee A, Frelich E, Reich PB (2009) Tree rings detect earthworm invasions and their effects in northern hardwood forests. Biol Invasions 12:1053–1066

Lavelle P, Bignell D, Lepage M, Wolters V, Roger P, Ineson P, Heal OW, Dhillion S (1997) Soil function in a changing world: the role of invertebrate ecosystem engineers. Eur J Soil Biol 33:159–193

Lee KE (1985) Earthworms: their ecology and relationships with soils and land use. Academic Press Inc, Sydney

Lee H, Pugh TAM, Patacca M, Seo B, Winkler K, Rounsevell M (2023) Three billion new trees in the EU's biodiversity strategy: low ambition, but better environmental outcomes? Environ Res Lett 18:034020

Lehmitz R, Römbke J, Graefe U, Beylich A, Krück S (2016) Rote Liste und Gesamtartenliste der Regenwürmer (Lumbricidae et Criodrillidae) Deutschlands. Naturschutz und Biologische Vielfalt 70:565–590

Leirikh AN, Meshcheryakova EN, Berman DI (2004) The mechanism of cold hardiness of egg cocoons of the earthworm *Dendrobaena octaedra* (Sav.) (Lumbricidae: Oligochaeta). Doklady Biol Sci 398:385–387

Lejoly J, Quideau S, Laganière J (2021) Invasive earthworms affect soil morphological features and carbon stocks in boreal forests. Geoderma 404:1–13

Machado Nunes Romeiro J, Eid T, Antón-Fernández C, Kangas A, Trømborg E (2022) Natural disturbances risks in European Boreal and Temperate forests and their links to climate change—A review of modeling approaches. For Ecol Manag 509:120071

McAdams BN (2017) Oribatid mite communities after ecosystem disturbance in Alberta. MSc thesis,. University of Alberta, Edmonton

McLean MA, Parkinson D (1998) Impacts of the epigeic earthworm *Dendrobaena octaedra* on oribatid mite community diversity and microarthropod abundances in pine forest floor: a mesocosm study. Appl Soil Ecol 7:125–136

Miles J (1981) Effects of birch on moorlands. ITE Report. Cambrian News, Aberystwyth

Miles J, Young WF (1980) The effects on heathland and moorland soils in Scotland and northern England following colonisation by birch. Bull Ecol 11:233–242

Mitchell RJ, Campbell CD, Chapman SJ, Osler GH, Vanbergen AJ, Ross LC, Cameron CM, Cole L (2007) The cascading effects of birch on heather moorland: a test for the top-down control of an ecosystem engineer. J Ecol 95(3):540–554

Musche M, Adamescu M, Angelstam P, Bacher S, Bäck J, Buss HL, Duffy C, Flaim G, Gaillardet J, Giannakis GV, Haase P (2019) Research questions to facilitate the future development of European long-term ecosystem research infrastructures: a horizon scanning exercise. J Environ Manag 250:109479

Muys B, Lust N, Granval PH (1992) Effects of grassland afforestation with different tree species on earthworm communities, litter decomposition and nutrient status. Soil Biol Biochem 24(12):1459–1466

Muys B, Angelstam P, Bauhus J, Bouriaud L, Jactel H, Kraigher H, Müller J, Pettorelli N, Pötzelsberger E, Primmer E, Svoboda M (2022) Forest biodiversity in Europe. From Science to Policy 13. European Forest Institute

Neirynck J, Mirtcheva S, Sioen G, Lust N (2000) Impact of *Tilia platyphyllos* Scop., *Fraxinus excelsior* L., *Acer pseudoplatanus* L., *Quercus robur* L. and *Fagus sylvatica* L. on earthworm biomass and physico-chemical properties of a loamy topsoil. For Ecol Manag 133(3):275–286

Olsson BA, Hansson K, Persson T, Beuker E, Helmisaari HS (2012) Heterotrophic respiration and nitrogen mineralisation in soils of Norway spruce, scots pine and silver birch stands in contrasting climates. For Ecol Manag 269:197–205

Petrovan SO, Al-Fulaij N, Christie A, Andrews H (2022) Why link diverse citizen science surveys? Widespread arboreal habits of a terrestrial amphibian revealed by mammalian tree surveys in Britain. PLoS One 17(7):e0265156

Petter G, Mairota P, Albrich K, Bebi P, Brůna J, Bugmann H, Haffenden A, Scheller RM, Schmatz DR, Seidl R, Speich M (2020) How robust are future projections of forest landscape dynamics? Insights from a systematic comparison of four forest landscape models. Environ Model Softw 134:104844

Phillips HRP, Guerra CA, Bartz MLC, Briones MJI, Brown G, Crowther TW, Ferlian O, Gongalsky KB, Van den Hoogen J, Krebs J, Orgiazzi A, Routh D, Schwarz B, Bach EM, Bennett J, Brose U, Decaëns T, König-Ries B, Loreau M, Mathieu J, Mulder C, Van der Putten WH, Ramirez KS, Rillig MC, Russell D, Rutgers M, Thakur MP, De Vries FT, Wall DH, Wardle DA, Arai M, Ayuke FO, Baker GH, Beauséjour R, Bedano JC, Birkhofer K, Blanchart E, Blossey B, Bolger T, Bradley RL, Callaham MA, Capowiez Y, Caulfield ME, Choi A, Crotty FV, Dávalos A, Cosin DJD, Dominguez A, Duhour AE, van Eekeren N, Emmerling C, Falco LB, Fernández R, Fonte SJ, Fragoso C, Franco ALC, Fugère M, Fusilero AT, Gholami S, Gundale MJ, López MG, Hackenberger DK, Hernández LM, Hishi T, Holdsworth AR, Holmstrup M, Hopfensperger KN, Lwanga EH, Huhta V, Hurisso TT, Iannone BV, Iordache M, Joschko M, Kaneko N, Kanianska R, Keith AM, Kelly CA, Kernecker ML, Klaminder J, Koné AW, Kooch Y, Kukkonen ST, Lalthanzara H, Lammel DR, Lebedev IM, Li Y, Lidon JBJ, Lincoln NK, Loss SR, Marichal R, Matula R, Moos JH, Moreno G, Morón-Ríos A, Muys B, Neirynck J, Norgrove L, Novo M, Nuutinen V, Nuzzo V, Rahman PM, Pansu J, Paudel S, Pérès G, Pérez-Camacho L, Piñeiro R, Ponge J-F, Rashid MI, Rebollo S, Rodeiro-Iglesias J, Rodríguez MÁ, Roth AM, Rousseau GX, Rozen A, Sayad E, van Schaik L, Scharenbroch BC, Schirrmann M, Schmidt O, Schröder B, Seeber J, Shashkov MP, Singh J, Smith SM, Steinwandter M, Talavera JA, Trigo D, Tsukamoto J, de Valença AW, Vanek SJ, Virto I, Wackett AA, Warren MW, Wehr

NH, Whalen JK, Wironen MB, Wolters V, Zenkova IV, Zhang W, Cameron EK, Eisenhauer N (2019) Global distribution of earthworm diversity. Science 366:480–485

Piearce T (1972) The calcium relations of selected Lumbricidae. J Anim Ecol 41:167–188

Potapov AM, Tiunov AV, Scheu S (2019) Uncovering trophic positions and food resources of soil animals using bulk natural stable isotope composition. Biol Rev 94:37–59

Powell JR (2007) Linking soil organisms within food webs to ecosystem functioning and environmental change. Adv Agron 96:307–350

Räty M (2004) Growth of *Lumbricus terrestris* and *Aporrectodea caliginosa* in an acid forest soil, and their effects on enchytraeid populations and soil properties. Pedobiologia 48:321–328

Räty M, Huhta V (2004) Earthworm communities in birch stands with different origin in central Finland. Pedobiologia 48:283–291

Reich PB, Oleksyn J, Modrzynski J, Mrozinski P, Hobbie SE, Eissenstat DM, Chorover J, Chadwick OA, Hale CM, Tjoelker MG (2005) Linking litter calcium, earthworms and soil properties: a common garden test with 14 tree species. Ecol Lett 8(8):811–818

Resner K, Yoo K, Sebestyen SD, Aufdenkampe A, Hale C, Lyttle A, Blum A (2015) Invasive earthworms deplete key soil inorganic nutrients (Ca, Mg, K, and P) in a northern hardwood forest. Ecosystems 18:89–102

Richardson BA, Borges S, Richardson MJ (2006) Differences between epigeic earthworm populations in tank bromeliads from Puerto Rico and Dominica. Caribbean J Sci 42(3):380–385

Robinson JM, Breed M, Abrahams C (2023) The sound of restored soil: Measuring soil biodiversity in a forest restoration chronosequence with ecoacoustics. Restor Ecol:e13934

Römbke J, Blick T, Dorow WH (2017) *Allolobophoridella eiseni* (Lumbricidae), a truly arboreal earthworm in the temperate region of Central Europe. Soil Org 89(2):75–84

Rutgers M, Schouten AJ, Bloem J, Van Eekeren N, De Goede RGM, Jagersop Akkerhuis GAJM, Van der Wal A, Mulder C, Brussaard L, Breure AM (2009) Biological measurements in a nationwide soil monitoring network. Eur J Soil Sci 60(5):820–832

Satchell JE (1980) Earthworm populations of experimental birch plots on a *Calluna* podzol. Soil Biol Biochem 12(4):311–316

Scharnweber T, Smiljanic M, Cruz-García R, Manthey M, Wilmking M (2020) Tree growth at the end of the 21st century-the extreme years 2018/19 as template for future growth conditions. Environ Res Lett 15(7):074022

Schelfhout S, Mertens J, Verheyen K, Vesterdal L, Baeten L, Muys B, De Schrijver A (2017) Tree species identity shapes earthworm communities. Forests 8(3):85

Scheu S, Albers D, Alphei J, Buryn R, Klages U, Migge S, Platner C, Salamon JA (2003) The soil fauna community in pure and mixed stands of beech and spruce of different age: trophic structure and structuring forces. Oikos 101(2):225–238

Schmidt O, Shutenko GS, Keith AM (2015) Multiple records confirm presence of *Dendrobaena attemsi* (Oligochaeta: Lumbricidae) in Ireland. Ir Nat J 34(2):110–112

Searle EB, Chen HYH (2017) Persistent and pervasive compositional shifts of western boreal forest plots in Canada. Glob Change Biol 23:857–866

Shorohova E, Kapitsa E (2014) Influence of the substrate and ecosystem attributes on the decomposition rates of coarse woody debris in European boreal forests. For Ecol Manag 315:173–184

Sigurðsson BD, Guðleifsson BE (2013) Impact of afforestation on earthworm populations in Iceland. Icel Agric Sci 26:21–36

Sims RW, Gerard BM (1999) Earthworms: notes for the identification of British species. Linnean Society of London and the Estuarine and Coastal Sciences Association

Sveistrup TE, Haraldsen TK, Langohr R, Marcelino V, Kværner J (2005) Impact of land use and seasonal freezing on morphological and physical properties of silty Norwegian soils. Soil Tillage Res 81:39–56

Terhivuo J (1988) The Finnish Lumbricidae (Oligochaeta) fauna and its formation. Ann Zool Fenn 25:229–247

Terhivuo J, Saura A (2006) Dispersal and clonal diversity of north-European parthenogenetic earthworms. Biol Invasions 8:1205–1218

Thakur MP, Kunne T, Unsicker SB, Biere A, Ferlian O, Pruschitzki U, Thouvenot L, Turke M, Eisenhauer N, Künne T, Unsicker SB, Biere A, Ferlian O, Pruschitzki U, Thouvenot L, Türke M, Eisenhauer N (2020) Invasive earthworms reduce chemical defense and increase herbivory and pathogen infection in native trees. J Ecol 109:763–775

Thørring H, Wærsted FM, Raaness A, Skipperud L, Jensen LK (2020) Elevated natural radioactivity in undisturbed forest and mountain areas of arctic Norway—local geology, soil characteristics, and transfer to biota. J Environ Radioact 222:106291

Vesterdal L, Schmidt IK, Callesen I, Nilsson LO, Gundersen P (2008) Carbon and nitrogen in forest floor and mineral soil under six common European tree species. For Ecol Manag 255(1):35–48

Vorobeichik EL (1998) Populations of earthworms (Lumbricidae) in forests of the Middle Urals in conditions of pollution by discharge from copper works. Russ J Ecol 29(2):85–91

Wackett AA, Yoo K, Olofsson J, Klaminder J (2018) Human-mediated introduction of geoengineering earthworms in the Fennoscandian arctic. Biol Invasions 20:1377–1386

Wardle DA, Bardgett RD, Klironomos JN, Setala H, Van Der Putten WH, Wall DH (2004) Ecological linkages between aboveground and belowground biota. Science 304(5677):1629–1633

White HJ, León-Sánchez L, Burton VJ, Cameron EK, Caruso T, Cunha L, Dirilgen T, Jurburg SD, Kelly R, Kumaresan D, Ochoa-Hueso R (2020) Methods and approaches to advance soil macroecology. Glob Ecol Biogeogr 29(10):1674–1690

Zicsi A (1965) Beiträge zur Kenntnis der Lumbricidenfauna Österreichs. – Opuscula Zoologica Budapest 5:247–265

Zuo J, Berg MP, Klein R, Nusselder J, Neurink G, Decker O, Hefting MM, Sass-Klaassen U, van Logtestijn RS, Goudzwaard L, van Hal J (2016) Faunal community consequence of interspecific bark trait dissimilarity in early-stage decomposing logs. Funct Ecol 30(12):1957–1966

Zuo J, Cornellissen JHC, Berg MP Muys B (2018) Earthworms above the soil, their community dynamics in deadwood. Oral contribution at ISEE11

Zuo J, Muys B, Berg MP, Hefting MM, van Logtestijn RS, van Hal J, Cornelissen JH (2023) Earthworms are not just "earth" worms: Multiple drivers to large diversity in deadwood. For Ecol Manag 530:120746

Earthworms, Ecological Processes, and Climate Change

Yahya Kooch

Abstract Gase emissions from soil are one of the most important sources of greenhouse gases (GHGs). As well as one-fifth of all global CO_2 emissions, soil contributes CH_4 emissions (approximately one-third) and N_2O emission (roughly two-thirds). GHGs are produced in soil during various biological processes. Earthworms play a fundamental role in the balance of soil GHG. However, the general influence of earthworms on the balance of GHG in the soil may seem uncomplicated; thus, a deeper look into earthworm activities, soil physicochemical processes, and soil GHG emissions is required to further understand the general impact of earthworms on GHG balance in soils. In a study, earthworm impact on CO_2 and N_2O emissions may be opposite in that they might boost emissions of CO_2 while reducing N_2O emissions or vice versa at the same time. This illustrates the intricacy of interactions between earthworms, other soil biota, and environmental factors. Studies have concluded that earthworms increase the amount of net greenhouse gas emissions from soil. Earthworms increased the emission of N_2O and CO_2 by 42 and 33%, respectively. There was a 16% boost in the soil net global warming potential caused by earthworms when both CO_2 and N_2O emissions were measured. According to my research, I believe that the anticipated changes in communities of earthworms in the following decades would have a major impact on (and likely increase) soil GHG emissions. As of yet, it is unclear how earthworms can negate increases in GHG emissions caused by their stimulating impacts on net primary production. It is overall necessary to conduct (i) research on undisturbed soils without earthworm activity legacy, (ii) long-run research, (iii) field research (particularly in natural ecosystems), and (iv) research of plant growing systems.

Y. Kooch (✉)
Faculty of Natural Resources & Marine Sciences, Tarbiat Modares University, Noor, Mazandaran, Iran
e-mail: yahya.kooch@modares.ac.ir

Y. Kooch, Y. Kuzyakov (eds.), *Earthworms and Ecological Processes*,
https://doi.org/10.1007/978-3-031-64510-5_4

1 Introduction

In addition to increasing the earth's atmosphere's radiative forcing, anthropogenic emissions of greenhouse gases (GHGs) are also widely credited with contributing to global warming, as one of the most important threats for our environment. The three main greenhouse gases—carbon dioxide (CO_2), methane (CH_4), and nitrous oxide (N_2O)—have increased by 4 1, 160, and 20%, respectively, since global industrialization began (IPCC 2013). Carbon dioxide is the vitally significant greenhouse gas, which is released in the atmosphere due to human activities and includes 64% of the total radiation force. From 1750, the global mean of carbon dioxide increased from 278 to 393 ppm. Burning fossil fuels and changing land use, especially deforestation, are the most vital human sources of CO_2 emissions. According to Al-Maliki et al. (2021), methane, as the second radiative forcing, accounts for 18% of the total radiative force. Its concentrations in the atmosphere enhanced from 700 ppb to 1819 parts per billion (ppb) in 1750. Methane is mainly emitted by natural sources such as natural wetlands. In addition, rice cultivation, the combustion of biomass, and livestock farming (enteric fermentation in ruminant animals) are other natural sources of methane production. Fossil fuels are among the most important industrial sources of methane production (IPCC 2007). Nitrous oxide (N_2O) is the third most fundamental greenhouse gas from human sources, which accounts for 6% of the total radiative forcing. Its global mean concentration increased from 270 ppm to 325 ppm in the postindustrial level. The impact of N_2O on warming the atmosphere is almost 300 times greater than carbon dioxide on a molar basis (Ramaswamy et al. 2001). Human activities that emit N_2O include industry, land use, transportation, fossil fuels, and particularly agriculture, which alone accounts for about 70% of anthropogenic N_2O emissions globally. Soil is roughly the source of one-fifth of global CO_2 emissions. Also, two-thirds of N_2O and one-third of global CH_4 are emitted from soils (Smith et al. 2003). GHGs are produced in soil during various biological processes. Soil fauna, plants roots, and respiration of soil microorganisms produce carbon dioxide (Rastogi et al. 2002). A group of archaea known as methanogens produce methane gas under anaerobic conditions. Based upon Wrage et al. (2005), N_2O is produced as a byproduct through denitrification, nitrification, and nitrifier denitrification processes. Substrate accessibility (for instance, labile carbon and mineral nitrogen for denitrification) together with soil physiochemical variables governs all (micro)biological processes that emit GHGs (such as soil moisture, diffusivity of gas, pH, and temperature). Because of the significant N input from fertilizer and the tillage-induced soil disturbances, agricultural soils might create favorable circumstances for the generation of GHGs. Therefore, the highest amounts of global soil CO_2 and N_2O are emitted from agricultural soils, but the amount of methane emission from these soils is low (Abail and Whalen 2021).

Human activities, mainly agricultural activities, reduce 42 to 78 Pg of soil organic C (SOC) of the total stock, which is about 2400 Pg (to a 2 m depth). According to estimates, the use of adaptive management practices in agriculture, including changing agriculture from conventional tillage to no tillage or reduced tillage, can reverse

SOC decline by 50 to 66% (Lal 2004). Although these agricultural systems do not affect CO_2 emissions, several studies have shown that N_2O emissions are increased in no-tilled soils compared to conventional tillage (Bottinelli et al. 2020a, 2020b; Butt and Quigg 2021; Gong et al. 2022). The degree to which higher N_2O emissions from soils exposed to reduced/no tillage may undermine strategies of carbon sequestration is currently unknown. In the conducted studies, the effects of management methods such as the application of fertilizer and soil residues and tillage on the emission of greenhouse gases from agricultural soils have been discussed, but the impacts of soil organisms or biota have not been investigated (Li et al. 2005; Kooch et al. 2022a). Earthworms are one of the most vital organisms in agricultural ecosystems which can act as a significant part in the of GHG balance emissions from the soil, in terms of total biomass and individual sizes (Rizhiya et al. 2007; Bottinelli et al. 2020a, 2020b; Zhong et al. 2023). Since the change of agricultural management from conventional tillage to reduced/no tillage leads to an increase in the abundance and diversity of earthworms, their effect on the soil GHGs emission will increase in the future decades (Kooch and Hosseini 2010). It seems that earthworms are beneficial to soil quality to a great extent by affecting the biotic and abiotic communities of the soil. Charles Darwin (1809 to 1882) was one of the first scientists who reported the earthworms' importance and recognized them as the cause of physicochemical decompositions, increasing the formation of humus and enhancing the structure of soil. Darwin discussed the importance of earthworms for the first time on November 1, 1837, in a lecture titled "On the formation of mould" at the "Geological Society of London". However, at that time, it was not much noticed by others (Desmond and Moore 1992). His final book only, published in 1881, attracted a sizable readership. Darwin made a brief statement on his most recent book in his autobiography: "I have now decided to submit a small book titled *The Formation of Vegetable Mould Through the Action of Worms*" to the manuscript printers (May 1, 1881). This is a relatively minor subject, and I have no idea if any people would find it interesting, but it has intrigued me. According to Barlow (1858), it is a brief article's conclusion that was presented to the Society of Geology more than 40 years ago, and it has brought back some old geological concepts.

Contrary to Darwin's expectations, his study of earthworms were not "curious little book of small vital", but became an important and influential work (Kutschera and Elliott 2010). His book sold very well that just 1 month after its publication, John Murray, a British publisher clerk (London), corresponded to Darwin: we have sold now 3500 worms (Feller et al. 2003). Due to the success of the book, Charles Darwin was portrayed as an earthworm scientist in the 1882 issue of the magazine *Punch* by a well-known cartoonist. Darwin's research changed society's understanding of earthworms and provided a positive and useful view of them. Before that time, earthworms were known as pests in soils that with their casts were able to deform well-manicured Victorian lawns. An important result was that Darwin's studies on earthworms created the research field of biology and introduced the idea of bioturbation; overall, he started an earthworm's study program that continues to this day (Kutschera and Elliott 2010). Based on many reasons, including soil amendment and interaction with microbes through burrowing, feeding, and casting activities, it can be proved that earthworms play a significant role in soil GHG emissions (Brown et al. 2000; Kooch and Jalilvand 2008; Kooch et al. 2022b). Through these activities, they affect the CO_2 and N_2O emissions and productions. The earthworm intestine is an anaerobic micro-site wherein local accumulation of accessible carbon and mineral nitrogen, a proper pH, and favorable moisture conditions all encourage denitrifier activities. These circumstances are perfect for denitrifying bacteria (Bottinelli and Capowiez 2021). Earthworm castings and burrow walls momentarily retain these "priming" impacts on denitrification (Brown et al. 2000). As a result, the emission of N_2O from them may be at most threefold than bulk soil (Kooch et al. 2007; Cardinael et al. 2019). Also, earthworms indirectly by ingestion, fragmentation, decomposing, and transferring new plant materials to the soil, by boosting soil porosity and density, and by altering dynamics of soil moisture and diffusivity of gas affect the CO_2 and N_2O production and emission (Giannopoulos et al. 2010) (Fig. 1).

It is possible that various earthworm species have different effects on soil GHG emissions. Darwin (1881) in the *Formation of Vegetable Mould Through the Action of Worms* examined earthworms' behaviors in terms of their sensory capacities and feeding behavior, their burrow construction, and their intelligence in leaf burying. Darwin may have previously noticed distinctions between earthworm species that fall into what are now known as several ecological classifications (Chen et al. 2021). These days, earthworm scientists usually split earthworms into three ecological groups depending on their burrowing and feeding habits (Fig. 2): (a) epigeic species that feed on non-decomposed litter and create temporary burrows (the activity of this category is restricted to some centimeters under the litter surface), (b) anecic species that bring their food from the surface litter in the soil and constant tunnels, and (c) endogeic species that live in temporary branching cavities below the surface and feed on soil and organic matters (Edwards 2004; Ferreira et al. 2018). These variations in feeding and burrowing behaviors among ecological groups have an

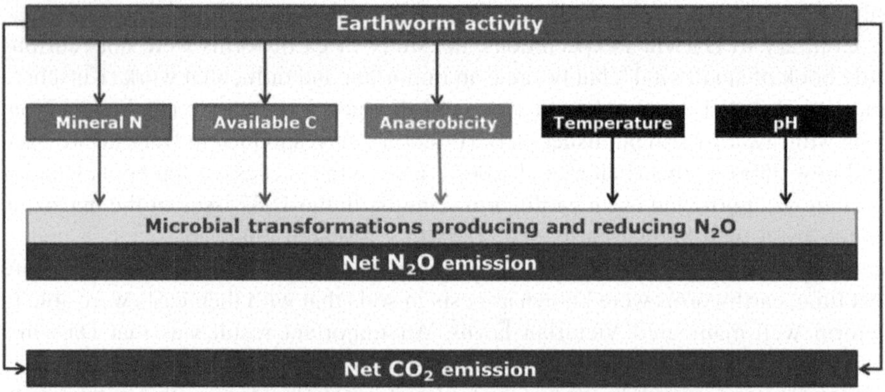

Fig. 1 Diagrammatic representation of how activity of earthworms affects abiotic soil elements, which in turn impact microbial activities that alter net emissions of CO_2 and N_2O

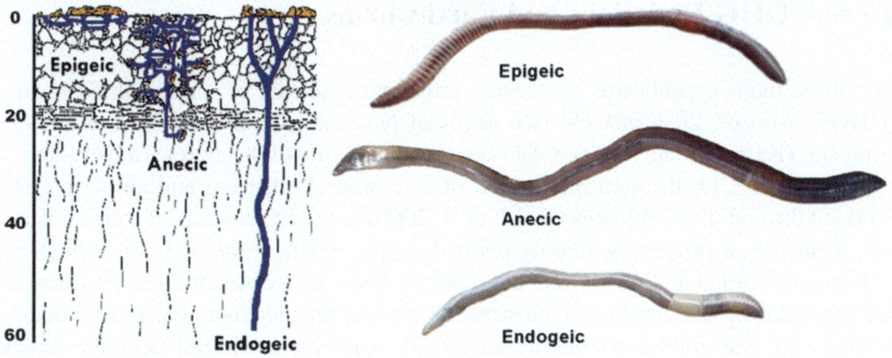

Fig. 2 Three earthworm species ecological behaviors

impact on the variables that regulate emissions of GHG. For instance, anecic species mineralize nitrogen from new residues of crops, but endogeic species from soil organic matter (Fierer 2019; Honvault et al. 2023). Additionally, the depth of residue assimilation differs across earthworm species belonging to various ecological groups, with anecic earthworms incorporating new residues at a greater depth than epigeics. This could alter the circumstances in which decomposition occurs (such as anaerobicity), which would then alter the N_2O generation and consumption. (Ferreira et al. 2018).

The potential of earthworms to accelerate the breakdown of plant matter and to enhance the plant nutrient availability is well established (Lavelle et al. 2004; Kooch et al. 2023). Additionally, by preserving soil carbon in the macro- and microaggregates created in their casts, earthworms can improve the stability of soil carbon (Kooch et al. 2020a, b). In particular, persistent microaggregate production inside C-enriched biogenic macroaggregates may be qualitatively significant for the long-term preservation of organic soil C (Golovanova et al. 2021). This has prompted several recommendations that earthworms improve carbon storage and hence lower emissions of CO_2. Nevertheless, it appears to be in contradiction to the immediate release of CO_2 and N_2O carried on by the behavior of earthworms in the soil (Giannopoulos et al. 2010). By affecting soil fertility, earthworms have a positive effect on the balance of GHGs in the soil. They prevent carbon loss by increasing plant growth and increasing soil carbon inputs by enhanced decomposition. Although many studies have been conducted on the impacts of earthworms on plant growth (Scheu 2003; Kooch et al. 2022c), the extent of this positive effect and its controlling factors is still unclear. Therefore, to what extent this issue balances GHGs in the soil should be further investigated.

2 Soil GHG Emissions and Earthworms

The three main greenhouse gases may either be a source of, or a sink for, soils (GHGs). Almost, 20% of CO_2, two-thirds of N_2O and one-third of CH_4 originate from soil (Rastogi et al. 2002). GHGs are produced in soil through various biological processes: CO_2 through respiration of soil (microbial, root, and animal), CH_4 released through methanogens (Smith et al. 2003), and N_2O released through a variety of microbial processes, mainly denitrification, nitrification, and denitrification of nitrifier (Kool et al. 2010; Gong et al. 2019). These processes of GHG production are regulated by availability of substrate as well as physicochemical factors determining soil microbial activities. Even though earthworms do not generate much greenhouse gas on their own, they can have a considerable impact on substrate availability and soil physiochemical properties, which can have an implicit impact on emissions. Since earthworms interact with microbes by their casting, feeding, and burrowing activities and also modify the structure of soil, they can be known as soil engineers (Brown et al. 2000). Because the earthworm gut is basically an anaerobic micro-site wherein mineral N local enrichment, accessible C, and favorable moisture conditions all encourage activity of denitrifier, the circumstances are optimal for denitrifying bacteria (Hodson et al. 2021). Casts, mucus, and burrow walls—the soil components immediately impacted by earthworm activity—are included in this ideal N_2O-producing environment. Therefore, compared to bulk soil, emissions of N_2O from burrow walls and casts might be a maximum threefold higher. By absorbing plant residues and blending the soil, encouraging soil aggregation, altering dynamics of soil moisture, and altering diffusivity of gas, earthworms also have an indirect impact on the generation of CO_2 and N_2O emissions (Kooch et al. 2010; Lubbers et al. 2011). Earthworms can boost the nutrient availability of plants by promoting the decomposition of plant residues (Zhong et al. 2023). Along with their well-known contribution to soil fertility, earthworms are also thought to contribute to the long-run stability of soil carbon by preserving it in the microaggregates that develop inside of big macroaggregates. Due to this, it has been suggested repeatedly that earthworms encourage soil carbon storage (Fig. 3) and hence lower net carbon dioxide emissions (Hoeffner et al. 2018). However, it appears to be in opposition to the rapid emission of CO_2 and N_2O caused by earthworm behaviors in the soil. The abundance of earthworms in ecosystems around the world will increase in the upcoming decades. For instance, for the first time since the last Ice Age, earthworms have invaded large parts of forest soils in North America (Hendrix and Bohlen 2002). The growth of the world's population and the use of more organic fertilizers in agricultural ecosystems to provide food for that populations provide the food needed by earthworms in the soil. Therefore, the importance and role of earthworms and their abundance in agricultural lands are constantly increasing in the upcoming decades. Changing from conventional land management methods to no tillage or protection tillage, by reducing soil disturbance, increases the activities of earthworms (Kooch et al. 2011). For instance, in no tillage soils, the density of earthworms increased by two to nine times and also the species diversity of

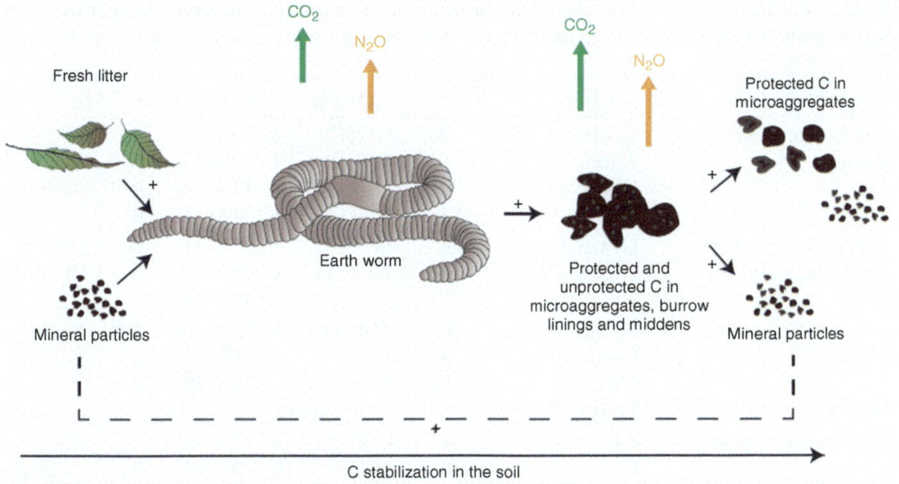

Fig. 3 An illustration of how soil is considered to be permanently stabilized by earthworms

earthworms changed (relative increase in the number of earthworms of the anecic species). Moreover, as more land is farmed, it may lead to decreases in the variation of earthworms, increases in biomass of earthworms under managed grasslands, and impacts on arable lands that are still unknown (Decaëns and Jiménez 2002; Hoeffner et al. 2019).

As shown by the + symbols, they continuously consume vast organic matter amounts, blend it with particles of mineral soil, transport this combination in their intestines, and expel it as casts. The earthworm's intestine totally destroys the soil's microstructure, and fresh microaggregates develop during intestine passage (Kooch et al. 2008). Worm casts and possibly burrow linings and middens are where organic material in an initial decomposition process is incorporated into the new microaggregates. Protecting labile soil organic matter requires the formation of these stable microaggregates within biogenic (worm-made) macroaggregates (Hoeffner et al. 2021). The earthworm-induced CO_2 and N_2O emissions (Table 1), which are frequently reported from laboratory trials, appear to be in strong contrast to the potential contribution of earthworms to long-term carbon stability (Lubbers et al. 2011).

Reviews of the research overwhelmingly support the idea that earthworms increase net soil greenhouse gases. The presence of earthworms raised soil emissions of CO_2 by 33% and N_2O by 42%. Earthworms' existence had little impact on SOC (Fig. 4). Investigations on earthworms indicated that their presence increased soils' net global warming potential (GWP) by 16% when CO_2 and N_2O emissions were recorded (Fig. 5).

Despite the apparent simplicity of the overall impact of earthworms on soil GHG balance, there are complex relationships between earthworm activities, biophysiochemical soil processes, and soil GHG emissions that require closer examination. For example, earthworms might have opposite impacts on CO_2 and N_2O emissions, allowing them to increase CO_2 emissions while reducing N_2O emissions

Table 1 For all subgroups of the regulating variables of CO_2 and N_2O emissions together with soil organic carbon, percentage change and effect size metrics on earthworm presence are available

Controlling factors	Subgroups		N_2O (44)		CO_2 (126)		SOC (67)
Earthworm functional group	Epigeic	0	27%(12)	+	26%(32)	0	6%(9)
	Anecic	+	46%(10)	+	50%(21)	N/A	0%(49)
	Endogeic	0	14%(10)	+	32%(60)	0	9%(9)
	mixture	+	75%(12)	+	34%(13)	0	
Earthworm numbers (individuals per m^2)	Low (<150)	+	48%(13)	+	13%(31)	0	2%(31)
	High (>150)	+	38%(31)	+	b*	0	1%(33)
					41%(95)a		
Experimental period (days)	Short (<30)	0	−10%(13)	+	73%(38)a	N/A	2%(28)
	Intermediate	+	b	+	21%(67)b	0	2%(31)
	(30–200)	N/A	57%(31)a		12%(19)b	0	
	Long (>200)						
Type of experiment	Laboratory	+	41%(41)	+	35%(112)	0	4%(27)
	field	0	52%(3)	0	20%(14)	0	−2%(38)
Nutrient inputs	Organic sources	+	69%(23)	+	26%(70	0	3%(32)
	Inorganic fertilizer	0	23%(8)	+	61%(10)	0	5%(11)
	None	0	18%(12)	+	40%(42)	0	−2%(21)
SOC	<2% C	+	27%(27)	+	53%(47)a	0	−2%(16)
	2–5% C	+	84%(17)	+	28%(22)	0	6%(6)
	5–30% C	N/A		0	ab	0	3%(15)
					10%(28)b		
C/N ratio of soil	Low (<12.5)	0	28%(14)	+	53%(51)a	0	3%(19)
	Intermediate	+	46%(30)	+	23%(56)b	0	5%(19)
	(12.5–30)	N/A		0	15%(8)ab	N/A	
	High (>30)						
Ecosystem (simulated)	Agroecosystem	+	41%(42)	+	45%(76)a	0	3%(40)
	Natural ecosystem	0	49%(2)	+	18%(50)b	0	1%(23)

95% confidence interval (CI) > zero; + denotes an effect magnitude larger than zero. 95% CI overlapped zero, meaning that the impact size is not substantial. Parentheses indicate the number of observations included in the effect size analysis. Substantial variations across categories are shown by distinct letters; categories are declared to be substantially different when their 95% (CI) do not overlap

at the same time, or vice versa, in the same research. (Contreras Ramos et al. 2009). It illustrates the interaction intricacy between earthworms, environmental exposures, and other soil biota. According to Fragoso et al. (1999), earthworms might play opposing functions at different time (and geographical) scales. They claimed that earthworms comminute, ingest, and degrade C over the course of hours, days, and weeks. However, earthworms have been demonstrated to slow down the decomposition of carbon by physically preserving the element in growing casts over a period of months, few years, and also few decades (Kooch et al. 2015; Kooch and Ghaderi 2023). This indicates that no one of these carbon transformations induced by earthworms occurs independently, yet they constantly operate at multiple spatial and temporal scales at the same time, with an overall effect on soil carbon balance that is still unclear. N_2O emissions is mostly produced through denitrification, which

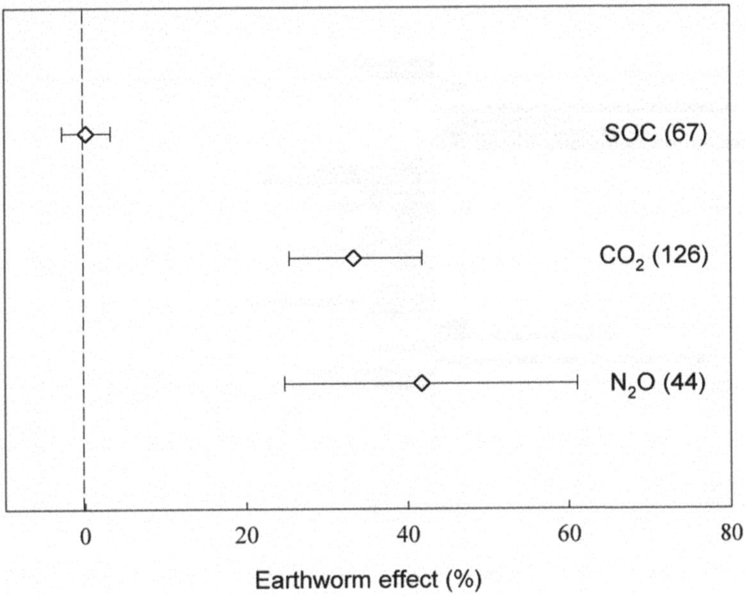

Fig. 4 Effect of earthworms in terms of percentage on soil and SOC emissions of CO_2 and N_2O. The reverse of the pooled variation was used to weight effect sizes across all meta-analyses. The 95% confidence intervals are indicated by error bars. Observational count numbers are in parenthesis (Lubbers et al. 2013b)

requires anaerobic conditions, but CO_2 emissions are directly related to the general decomposition rate by microbes. The nitrifier-denitrifier process is also an anaerobic process, requiring partly anaerobic conditions, which can produce N_2O with ammonia-oxidizing bacteria (Kool et al. 2010; Kooch 2012a, 2012b). Curry and Schmidt (2007) stated that earthworms play vital and beneficial roles in regulating ecosystem processes and soil properties, including the structure and organic matter of the soil, the food chain, and the abundance and activities of microbes. Plant nitrogen availability and production of grass (crop) are greatly enhanced by mineralization of organic matter in agroecosystems. By having both direct and indirect impacts on the microbial population, earthworms are considered to promote the nitrogen mineralization from organic materials. According to Cortez et al. (2000), earthworms significantly enhanced the amount of inorganic N (mostly NH_4^+) inside the soil. It was brought about by increased nitrogen mineralization from the $_{15}N$-labeled soil organic matter and residue.

Grasslands fertilized with cattle manure, slurry, or organic N fertilizers are significantly mineralized by earthworms (Van Vliet et al. 2007). The presence of earthworms might have a negative effect on greenhouse gas emissions, such as nitrous oxide (N_2O). During crop residue applications, earthworms increased N_2O fluxes; however, they did not increase them without residue (Speratti and Whalen 2008). Microbial processes determine the amount of N_2O produced in soils. Denitrification, nitrification, and nitrifier denitrification are the major microbial processes that

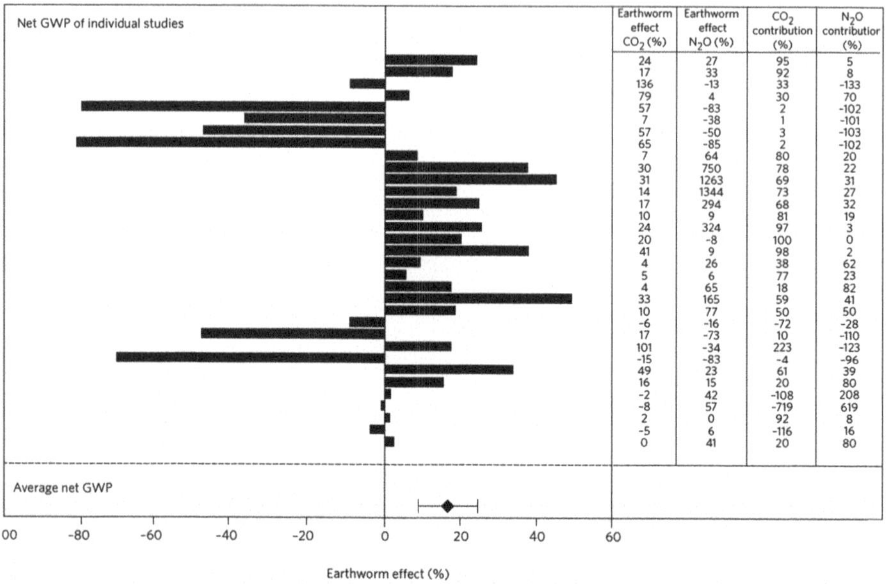

Fig. 5 Average of all observations and the percentage influence of presence of earthworms on the soil net GWP for every observation that contained measurements of the fluxes of CO_2 and N_2O. The pooled variance's inverse was used to weight the effect size. The 95% confidence interval is shown by the error bar. The four columns on the right for each observation reflect the impact of earthworms on emissions of CO_2 and N_2O. The first and second columns show how earthworm presence affects each of the two component gases; the third and fourth columns show how CO_2 and N_2O emissions caused by earthworms respectively contribute to the net GWP (see Lubbers et al. 2013b)

produce N_2O (Wrage et al. 2001). N_2O production and emission are affected in complex ways by earthworm interactions with soil microbes. N_2O-producing microorganisms thrive in the earthworm's intestine due to the perfect situations such as abundance of substrate, anaerobicity, high moisture content, and favorable pH (Kooch et al. 2017). Casts and burrow walls are primed by microbial activities in conjunction with mucus excretion by earthworms. In a more indirect path, earthworm activities can also alter emissions of N2O. Casting and burrowing by earthworms alter soil structure, forming aggregates of biogenic material and affecting porosity and pore size distribution (Jongmans et al. 2003). Thus, earthworms play a crucial role in soil water and air movement, which influences the redox potential, which determines N_2O production. It has been demonstrated that earthworms have different effects on N mineralization, availability of plant N, and emissions of N2O based on their functional groups (Rizhiya et al. 2007). Nevertheless, species interconnections can also be crucial, for instance, by interfering with each other's burrow systems (Felten and Emmerling 2009, Kooch et al. 2017). In the combinations of anecic, epigeic, and endogeic species, Postma-Blaauw et al. (2006) discovered impacts of species interactions on soil organic matter nitrogen mineralization, bacterial biomass, and growth rates. Other research found that species interactions between endogeic and epigeic species, as well as between anecic and epigeic species, had an impact on N_2O emissions (Giannopoulos et al. 2010). There is a strong relationship between land use and soil management and earthworm abundance and activity in soils (Rizhiya et al. 2007). The biggest earthworm populations are found in fertilized grasslands with a high level of organic matter in the rhizosphere and a steady supply of food (Van Vliet et al. 2007; Kooch et al. 2018).

3 The Role of Plants and Soil Characters

Based on literature reviews, the CO2 measurements become significantly more complicated when the balance of respiration and root growth is added to CO2 emissions from the decomposer system. However, numerous complications can be found due to plant removal. Firstly, plants and earthworms could interact in a complex manner in the soil root zone, influencing the emissions of N_2O. By rhizodepositing more substrate, plant roots can enhance denitrification, or by uptaking mineral nitrogen to promote plant growth, they can decrease substrate (Fonte and Six 2010). Additionally, through burrowing, earthworms are able to graze on roots of plants, consequently affecting the functional development of roots in the plants. Earthworms may also enhance plant growth by altering soil fertility and structure, contributing to a larger input of fresh carbon into the soil and increased rates of root respiration (Scheu 2003; Kooch et al. 2020a, 2020b). Therefore, earthworms may indirectly counteract some of the decomposition that they also cause. According to Scheu (2003), even though there has been no quantitative investigation of earthworm impacts on primary production, in 79% of all experiments, earthworms had notably enhanced the plant shoot biomass. According to Brown et al. (1999), 75% of the 246 cases examined in the tropical region experienced plant growth increases caused by the presence of earthworms, with a mean growth of 57% in biomass plant shoot. It has also been found by Brown et al. (1999) that earthworms have less of an effect on root production. There has been some evidence to support the hypothesis that the average value of shoot biomass in plant varies between 29 and 104% (Laossi et al. 2011). There are however few studies looking at grass and crop species (Scheu 2003; Kumar et al. 2021). Because plant biomass may be grazed or harvested, additional potential of SOC may be eliminated by grazing or harvesting from the system. Although earthworms enhance primary production, it remains unclear whether their increased emission of net soil GHGs would offset the increase. Until now, it is not completely clear whether production of plants in seminatural communities is affected by earthworms. Moreover, earthworm-induced GHG emissions might cancel out earthworm-induced plant growth in plant-earthworm studies because of a publication bias. Based on Barot et al. (2007), earthworms can enhance plant growth and mineralization in the short run but lessening the primary production in the long run because of increased nutrient losses. Based upon Laossi et al. (2011), earthworms may affect the nutrient availability immediately, but not in the long term. It is important not to use short-term effects of earthworms on plant growth as a prediction of long-term effects of earthworms on plant communities. In addition, in spite of the confounding factors, a research using plants (fertilized grass) in mesocosms observed an already high increase in N_2O emissions of 50.8%, along with an increase in grass biomass of 5.4% (Lubbers et al. 2011). Taking the evidence into account, earthworm-induced increases in soil N2O and CO2 emissions are unlikely to be offset by the stimulating effect of earthworms on plant growth.

When the soil organic carbon content and ratio of C/N in the soil grow, the impacts of earthworms on CO2 emissions decline. Furthermore, compared to agricultural systems, natural systems had fewer earthworm impacts. The impacts may reflect the same basic relationship, low soil organic carbon contents correlate with lower C/N ratios and agricultural systems, and greater amounts of organic carbon contents correlate with higher C/N ratios and natural systems. Due to decreased residue input into agricultural soils, carbon pools are smaller and older. It is known that earthworms can mobilize more resistant or conserved types of soil organic materials (Marhan et al. 2007; Law and Lai 2021). They are probably able to feed on organic materials that would have been problematic for other soil biota to decompose in soils with a poor quality source of food (low soil organic carbon content, lower ratio of soil C/N, indicating further organic compounds (decomposed) that are less accessible). Additionally, they could be able to graze on relatively degradable carbon fractions in the soil by selective feeding (Curry and Schmidt 2007). Therefore, earthworms might be capable of accelerating the carbon decomposition in certain soils and thereby boost emissions of carbon dioxide. The earthworm impact may be overshadowed by overall greater decomposition rates in soils with bigger and more accessible pools of carbon. Considering emissions of N2O, in soils with more organic carbon, in contrast, mean emissions induced by earthworms seemed to be significantly higher than soils with less organic carbon. The impacts of earthworms on emissions of N2O did not vary among the subgroups of soils with different C/N ratios, although they were only statistically significant at rather high C/N ratios. As previously mentioned, this may be connected to the fact that N2O emissions induced by earthworms usually happen after extended time periods, while the rates of decomposition have decreased, and when soils with a greater amount of organic carbon may provide a consistent (though relatively small) carbon source for denitrification. Either organic fertilizer was provided to the soil in place of no fertilizer or inorganic fertilizer generally, only then did earthworms considerably increase N2O emissions. Also this might highlight the necessity of a consistent carbon supply for the operations that produce N2O.

The most lab research that evaluate GHG emissions use simple and heavily controlled micro- and mesocosm experiments, which may not accurately reflect the actual world. The majority of experiments employed repacked and homogenized soils, which required earthworms to manipulate the soil before having any real impact on its structure. In contrast, it would be similarly challenging to research how earthworms have an impact on GHG emissions from soils that are not repacked since their effect on the structure of soil could already be observable as a result of activities of earlier earthworms, making it impossible to build any useful control treatment. According to Frelich et al. (2006), therefore, it is ideal to investigate the earthworms' effects on well-established soils that have never before hosted earthworms, such as the ecosystems in North America's cold temperate of mixed deciduous conifer forests, which cover millions of square kilometers. Earthworms are presently traveling the world at rates, routes, and in ways that have never been seen before, and their effects on ecosystems are not yet fully understood (Crooks 2002). For instance, nothing is known about the extent, nature, and effects of earthworms' invasion on North American forest ecosystems. Nonetheless, based upon Mazur-Paczka et al. (2021), investigations evaluating already occupied and impedingly occupied regions by worms might offer invaluable insights into the possible effects of earthworm invasion. Information regarding how earthworms affect the balance of soil GHG may be usefully obtained from these kinds of investigations. The main drivers of global environmental change—atmospheric CO2 increase, climatic changes, anthropogenically fixed nitrogen, habitat fragmentation and loss, and biotic invasion—can also affect the quantity and quality of resources plants return to the soil. Consequently, the decompositioner food web, including earthworms, can be affected multitrophically (Tylianakis et al. 2008). The most crucial ones are accessible carbon, mineral nitrogen, temperature, pH, and anaerobicity that together govern the ideal circumstances for microbial N2O generation in the soil (Cardinael et al. 2019). At the microscales, these governing elements can vary significantly across time and location. Biological and abiotic processes interact, affecting N2O production in the soil and possibly reducing it through the denitrification process in the final stage. By feeding and burrowing, for instance, earthworms can directly alter these regulating elements and, as a result, affect N2O production indirectly. It has been reported that they are able to increase the concentration of the availability of carbon and mineral nitrogen in the soil once it has been mixed with organic residues (Giannopoulos et al. 2010). They can also reduce anaerobicity by altering the structure of soil by casting and burrowing (McCay and Scull 2019). Based upon CO2 equivalents, it is estimated that 7.9% of all anthropogenic emissions of GHG (in N2O form) come from agricultural and related land-use changes (IPCC 2007). Depending on their ecological approach, earthworms can have a variety of effects on the variables regulating microbial N2O emissions (Edwards 2004; Golovanova et al. 2021).

4 Earthworms: Good or Bad?

In order to feed a fast growing population when decreasing our global footprint in the environment, our global food production system faces an unprecedented challenge (Godfray et al. 2010). Whether such "sustainable intensification" is achievable remains far from certain. A particular controversy relates to what factors determine the yield gap among more tolerable and conventional agricultural forms (Seufert et al. 2012). For agricultural systems to be sustained, earthworms are typically considered to be crucial. It is considered one of the most vital soil fauna, and they play a significant role in drainage, nutrient cycling, and regulating greenhouse gas emissions as ecosystem engineers (Lubbers et al. 2013a). Nevertheless, their alleged capacity to promote crop growth may be of greater importance to agriculture. Several years before artificial fertilizers and mechanization enabled greater crop production, this ability had already been proposed (Darwin 1881; Medina-Sauza et al. 2019). While earthworms have been described as having a beneficial effect on plant growth on numerous occasions (Brown et al. 2004; Hodson et al. 2021; Honvault et al. 2023), the mechanisms by which these effects might occur remain unclear. To determine if earthworms may help fill the yield gap among conventional and sustainable agricultures, however, this knowledge is crucial. The absence of a comprehensive study on earthworm effects has been hindered by the diverse conditions (soil fertility, climate, earthworm species, types of crops, and farming methods) in which they have been reported previously. Crop yield was dramatically boosted by 26% and aboveground biomass by 24% when earthworms were present (Fig. 6). According to Poorter and Nagel (2000), there was no appreciable rise in the ratio of shoot/root, indicating no relative shift in the allocation of carbon toward aboveground plant components. The presence of earthworms did not alter the N content in aboveground biomass either (Fig. 6), showing that the quality of crop was preserved.

The impact of earthworm presence on the aboveground biomass of the main grain crops and ryegrass was examined by Brown et al. (1999). All crops saw a large increase in aboveground biomass (Fig. 7a), with grain crop average increases of 33% and pasture grass average increases of 26% (Fig. 7b). How can earthworms encourage the growth of plants? According to Brown et al. (2004), earthworms can influence plant growth through six different mechanisms: dispersal of either

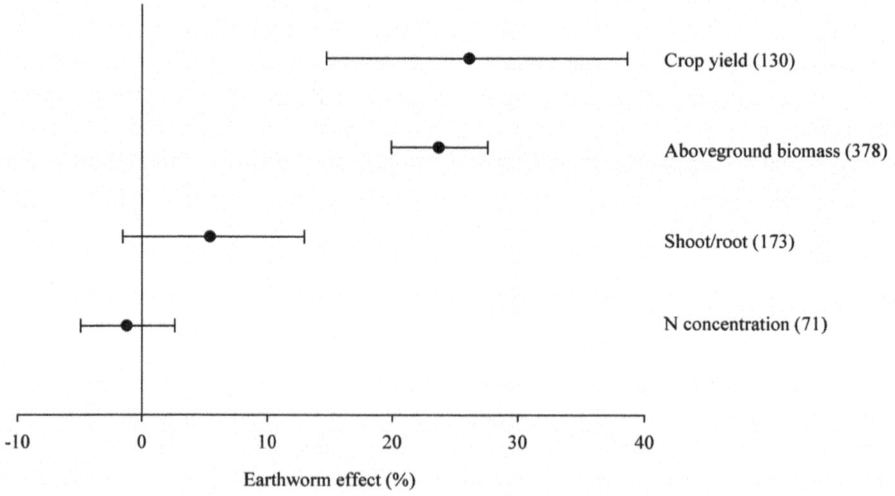

Fig. 6 How the presence of earthworms affected the key response variables of yield, aboveground biomasses, ratio of shoot/root, and concentration of nitrogen in aboveground biomasses (% increase or reduction). Between parentheses, for each class, the number of observations is displayed; error bars indicate the confidence range of 95%

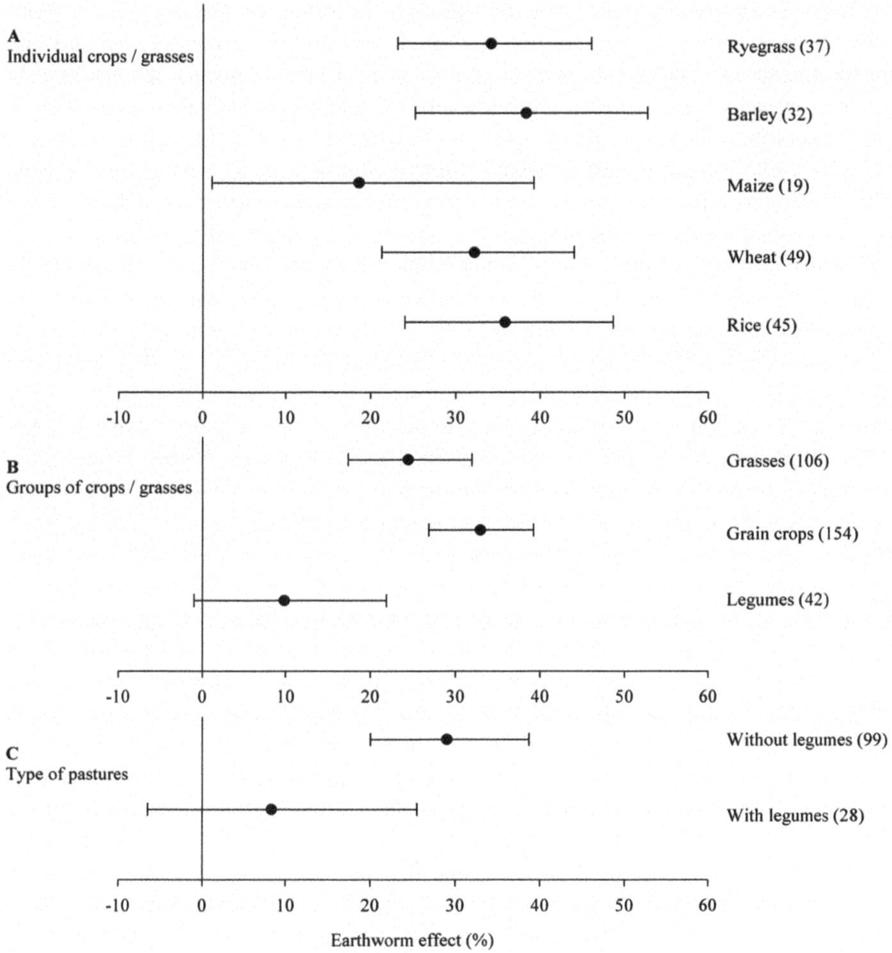

Fig. 7 The impact of earthworm abundance on aboveground biomass for (**a**) individual crops/grasses, (**b**) groups of crops/grasses, and (**c**) different pasture types (% increase or reduction). Between parentheses, for each class, the number of observations is displayed; error bars indicate the confidence range of 95%

microorganisms (beneficial) or microorganisms (detrimental), production of plant growth-controlling substances, root feeding, interaction with seeds, soil structure changes, and availability of nutrients. In early literature, the final two processes were the ones that were most frequently cited (White 1777). Enhanced plant tolerance and altered stress-related gene expression have been proposed as additional routes in more recent investigations (Jana et al. 2010; Moghimian et al. 2020).

The results imply that the predominant mechanism is enhanced nitrogen availability. In order to test this, Van Groenigen et al. (2006) divided my data into application rates for nitrogen fertilizer (Fig. 8a). Increasing N mineralization by earthworms stimulates plant growth, although the effect of earthworms on plant growth is insignificant when applied at a rate above 30 kg nitrogen per hectare, the mean atmospheric deposition in temperate regions. Earthworms had a considerably greater impact on plant growth in experiments using organic nitrogen fertilizers (+37%) compared to experiments using no fertilizer (+19%) or inorganic fertilizers (+10%), further demonstrating enhanced nitrogen mineralization as a key pathway. Since nitrogen mineralization is the primary mechanism, earthworms should have less of a positive impact on plants that can fix N2 symbiotically. Admittedly, the earthworm impact stopped being substantial for the legume crops in the study (Fig. 7b). Furthermore, the impact of earthworms on productivity of pasture was completely eliminated when legumes were available in the pastures (Fig. 7c). If earthworms have any impact on nutrients other than nitrogen, it is still vague. Even though it has been proposed that earthworms improve the availability of P in their castings (LeBayon and Milleret 2009), experimental investigations have not yet demonstrated how this affects plant growth. Despite having a higher P need than grasses, legumes did not benefit from the presence of earthworms (Fig. 7), which is consistent with an earthworm's little contribution to mobilization of P. Both plant residues and soil organic matter can be used by earthworms to facilitate nitrogen mineralization (Hoeffner et al. 2019). Van Groenigen et al. (2006) subdivided the dataset in various rate classes of residue applications to differentiate among the two (Fig. 8b). When residue application rates were the highest, the earthworm effect peaked at +50%, but stabilized at +20% at no residue and extremely low residue applications. This suggests that sources for earthworm-induced mineralization include both plant residue and soil organic matter. When earthworm survival rates were lower than 50%, crop yields were not increased by earthworm presence; therefore, the nitrogen effect must not originate from decomposing earthworms. According to previous experimental studies (Lubbers et al. 2011) and studies that used controls receiving dead earthworms, this corresponds to calculations on N released from decaying earthworm tissues. Only the greatest densities (individuals 400 m^{-2}) varied substantially from lower densities, despite the fact that density of earthworms had a greatly notable impact on aboveground biomass (Fig. 7b). The impact under pragmatic earthworms' densities differed between +12 and +22%. In all three ecological categories traditionally distinguished, the positive impact was observable. The results of my analysis do not indicate a remarkable difference between categories based on dynamics of nitrogen, as have been reported in several studies (Postma Blaauw et al. 2006; Rizhiya et al. 2007; Kooch et al. 2023). Nevertheless, this can be because of the research scarcity on species of anecic and epigeic (in particular). When crop residue is applied, anecics and epigeics may have a greater positive impact since they feed on fresh organic matter; endogeic species (which consume further decomposed SOM) may have less a reliant relationship with residue application (Giannopoulos et al. 2010). Compared to undisturbed soils, earthworm impact on aboveground biomass is almost twice as high in experiments in which soil has been disturbed (homogenized and repacked). These findings probably reflect a useful earthworm impacts on demolished soil structure restoration. It is likely that the impact of earthworms on plant's growth is temporary after soil tillage by altering soil structure. It was typically tough to determine this impact from enhanced nutrient availability; however, some investigations showed an additional earthworm impact on plant growth by boosting structure of soil in undisturbed soils (Edwards and Lofty 1980; Golovanova et al. 2021).

Significantly positive earthworms' impacts are seen over a wide variety of climatic zones, soil textures, soil C/N rates, and concentrations of soil organic matter (Fig. 9). The earthworms' impact is substantially weaker in soils with higher pH than in soils with lower pH. This might be a complicating impact because soils with higher pH were frequently associated with systems where no organic manure or residues were provided. In clay soils, texture impacts on earthworms were highest in comparison to sandy soils. This is in complete contradiction to a previous assessment (Brown et al. 1999), which found that the most positive earthworm impacts were obtained on sandy soils. Nevertheless, most clayey soil studies (75%) used disturbed soil, while sandy and loamy soil experiments (respectively 4 and 32%) did not. Earthworms are anticipated to be highly productive in infertile soils since increasing nitrogen availability in N-limited environments is the significant way by which earthworms enhance plant growth. Nonetheless, this causes a paradox since earthworms flourish in fertile soils in which higher levels of organic matter are available (Edwards 2004; Gong et al. 2022). According to Chadwick and Bradley (1948), their study revealed that "earthworms wouldn't survive in soil except if a high level of organic matter is preserved"; however, "there is no need to introduce earthworms when a highest concentration of organic matter is retained". In the case

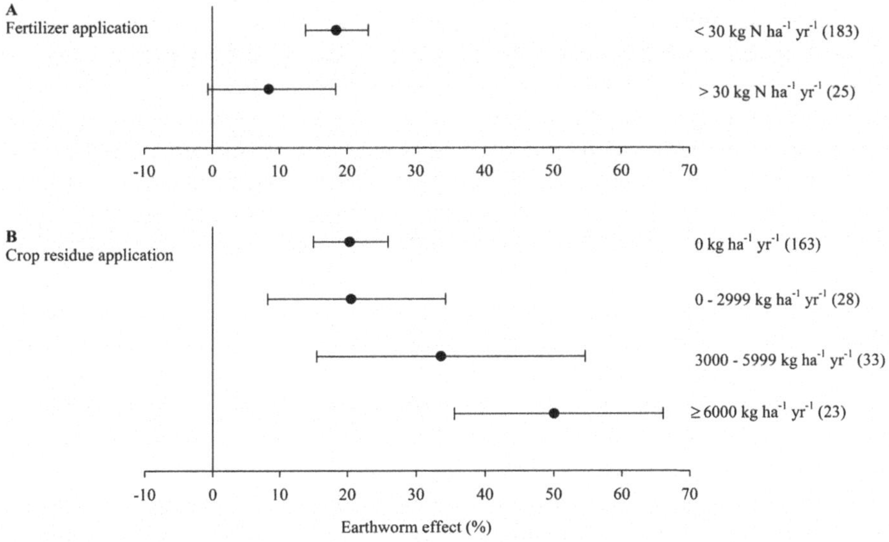

Fig. 8 How the presence of earthworms affects the aboveground biomass at various rates of (**a**) N fertilization and (**b**) crop residue application. Chemical and organic (manure) fertilizers are both included in N fertilization rates. Between parenthesis, for each class, the number of observations is displayed, and error bars represent the confidence range of 95%

of relatively low-quality soils that rely on crop residue treatments to preserve levels of soil fertility, this paradox fades. Residues of crops in those soils could provide food for earthworms' communities, and earthworms, by enhancing nitrogen mineralization, could boost productivity of crops. This mixture of low-quality soils and dependence on agricultural residues is most common in tropical lower input agricultural ecosystems and to a lesser degree in organic agriculture ecosystems in the rest of the developed countries (Feller et al. 2012). But in terms of earthworms' habitat quality, tropical systems with low inputs and organic agricultural systems differ substantially. Large amounts of high-quality crop residues or organic manures are applied in organic farming systems, which provides supreme situations for activity of earthworms. Earthworm activities may thus be critical in bridging the yield gap with traditional farming in those situations (Seufert et al. 2012; Kooch et al. 2021). In order to enhance earthworm populations, upcoming studies should thus concentrate on management approaches (Lavelle et al. 1989). Low residue quantity and quality are more prone to be the limiting factors in achieving the full value of earthworm activities in systems with low input in the tropical regions (Lavelle et al. 2001). As a consequence, studies in these systems should concentrate on maximizing the most use of the restricted residual resources that are accessible (Butt and Quigg 2021; Gong et al. 2022). A study I conducted demonstrates that earthworms increase productivity of crops in a wide range of agricultural systems and that the main mechanism involved is enhanced nitrogen mineralization. To conclude, in order to intensify agriculture sustainably and close the yield gap with traditional agriculture, earthworms are particularly beneficial for those farmers who cannot or do not wish to use N fertilizer.

Earthworms were considered soil pests prior to *The Formation of Vegetable Mould Through the Action of Worms* because their surface casts spoiled the smooth grasses meticulously maintained by Victorian gardeners. Furthermore, earthworms are not the primary cause of soil GHG emissions, even though they provide other ecosystem services such as drainage and nutrient cycling. Earthworms, on the other hand, are a proximal cause, one of the agents through whom humans contribute GHG emissions from agricultural systems: by providing vast volumes of nitrogen fertilizers, which could be transformed to N2O. In agriculture, earthworm-induced emissions, especially N2O, are largely dependent on humans using fertilizer to agricultural soils and are the unfortunate side effects of soil fertility. Are certain earthworm densities or ecological groups that result in a more favorable balance of GHG emissions and growth of plants than others? The findings of the two meta-analysis are merged and classified in Table 2 by earthworms' ecological density and category. As a result of this study, it seems that earthworms are capable of affecting plant growth, CO2 emissions, and N2O emissions universally rather than limited to certain ecological types of earthworms. The effect of N2O emissions fluctuated the highest between subcategories, while changes across categories had not been substantial. In contrast to low-density earthworms, high-density earthworms greatly increased growth of plants and emissions of carbon dioxide. Increased decomposition of soil residues at either low- or high-density earthworms negated any gains in soil carbon resultant from increased residue deposition. Is there a specific soil/

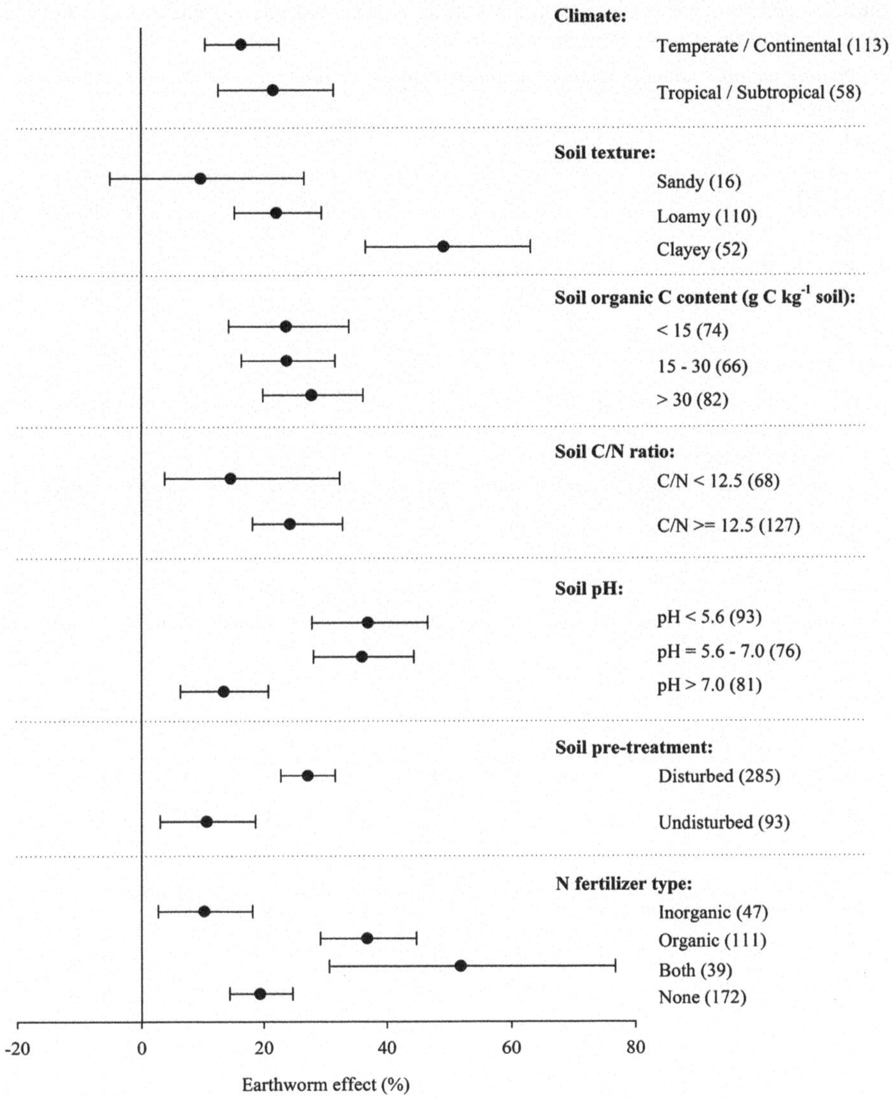

Fig. 9 How the presence of earthworms affects the amount of aboveground biomass as a function of the experimental circumstances (% increase or reduction). Between parenthesis, for each class, the number of observations is displayed; error bars indicate the confidence range of 95%

management feature in which the earthworm impact is actually useful? Table 3 represents the meta-analysis results by grouping them according to different experimental parameters connected with soil properties of distinct categories of agricultural systems. The earthworms' positive impact on plant growth seemed to be unaltered by soil C/N ratio or organic carbon level. Earthworms had a favorable effect on plant growth in both high and low SOC soils, regardless of the C/N ratio of organic

Table 2 Earthworm impacts on soil organic carbon, growth of plants, and emissions of CO_2/N_2O (%) for 'earthworm factors' (Lubbers et al. 2013a)

Earthworm factors—subgroups	Plant growth	SOC	CO_2	N_2O
Ecological category				
Epigeic	18	6	26	27
Endogeic	27	0	32	14
Anecic	38	ND	50	46
Mixture	15	9	34	75
Density (m^{-2})				
>150	11	2	13	48
=<150	27	1	41	38

ND = not determined

Table 3 Earthworms' impacts (%) on growth of plants, SOC, and CO2 and N2O emissions (Lubbers et al. 2013a)

Experimental factors—subgroups	Plant growth	SOC	CO_2	N_2O
Soil organic carbon content (g C kg^{-1} soil)				
<20	25	−2	55	27
20–50	27	6	25	84
50–300	25	3	10	ND
Soil C/N ratio				
<12.5	14.47	3	53	28
=>12.5	24.25	5	23	46
N fertilizer type				
Inorganic	10.27	5	61	23
Organic	36.90	3	26	69
Both	73.79	ND	ND	ND
None	19.39	−2	40	18
Residue application rate (kg C ha^{-1})				
0	22.72	ND	52	11
0–2999	19.98	ND	7	88
3000–59,999	33.43	ND	ND	ND
> 6000	51.11	ND	ND	ND

ND = not determined

material. Earthworm-induced plant growth was affected by fertilizer type and rate. When comparing to inorganic fertilizers or no fertilizers generally, the application of organic fertilizer greatly boosted the earthworms' impact on plant growth. Residue application in higher rates promoted far more earthworm-induced crop growth. All of this suggests earthworm-induced nitrogen mineralization from organic fertilizers and residues playing a significant role.

As only two subgroups of residue application rate were examined, we did not find a clear impact of fertilizer type on earthworm-induced CO2 emissions. Even though the kind of fertilizers had no influence on earthworm-induced carbon dioxide emission, its impact on emissions was greater, while the utilized organic fertilizer was lesser. When soil C/N ratio and SOC content are low, impacts of earthworms on CO2 emissions are quite strong. In addition, earthworms have been found to mobilize more resistant organic matter (Marhan et al. 2007) and to organic matter consumption that is hard to be decomposed by the microbes in soil (Curry and Schmidt 2007; Al-Maliki et al. 2021; Kooch et al. 2022a, 2022b). As a result, they might be able to speed up carbon decomposition in soils with lower SOC contents. N2O emissions from earthworms were unaffected by subgroups of C/N ratio and SOC, types of fertilizers, or rates of residue application. The results, however, were only statistically significant when ratios of C/N were comparably higher and organic fertilizers were used instead of inorganic fertilizers or no-fertilizer was applied (marginally significant, CI at 90%). According to Bottinelli et al. (2020a, 2020b), these findings point to the existence of stable carbon sources for heterotrophic N2O production. Additionally, it appears that these circumstances are ideal for earthworm-induced plant growth. In relation to the findings, which agroecosystems would benefit most from earthworm presence in terms of crops yields and regulation of greenhouse gas? And which ones are the worst? Figure 10 depicts a hypothetical graph that separates the fundamental agroecosystem types based on the rates at which N fertilizer is applied (x-axes) and the rates at which residue is applied (y-axes). Three variables were evaluated for these systems by Lubbers et al. (2013a): (i) habitat quality (HQ), which indicates whether the system is naturally suited to support earthworm populations; (ii) yield effect (YE), which indicates whether earthworms have an impact on crops yields when present; and (iii) GHG regulation

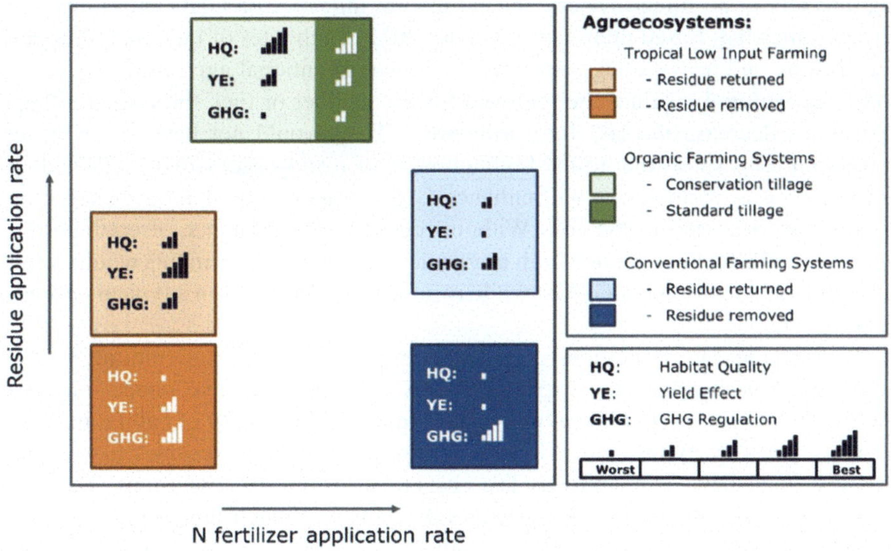

Fig. 10 Conceptual graph showing how the effects of earthworms on crop yields and GHG control vary depending on the kind of global agroecosystem and the rate at which nitrogen fertilizer and residue are applied

(GHG), which indicates whether earthworms have an impact on GHG emissions when present.

Low input tropical farming systems typically lack nitrogen fertilization inputs with crop residues often being burned for fuel or fed to animals, leaving soils deficient in nutrients (Feller et al. 2012). Due to a shortage of food, the HQ for the earthworms are typically not excellent in such a system. In contrast, when populations of earthworms are established, the yields will be significantly improved since these are the types of soils where earthworms have the greatest impact (no fertilization and low fertility). Therefore, an essential management strategy in these systems should focus on improving the quality of habitat by an effective residue management to maximize the benefits of earthworms' action. Since there is little possibility for N_2O emissions across residue systems, earthworms' impacts on emissions of GHG would be negligible. Nitrogen fertilizer supply rates are typically modest and applied organically in organic farming systems, whereas residue application rates are maximum (Skinner et al. 2014; Kooch et al. 2022c). Under conservation tillage management, the HQ will be ideal for earthworms' activities if located in parts of the world with favorable climates (humid tropics and temperate regions). When earthworms are present, crop yields increase because they may mineralize nitrogen, which plants can then absorb. The earthworm influence on crop yields is anticipated to diminish during tillage because plowing will speed up the integration of agricultural residues once the earthworm communities do initially. This will overcome the slow earthworm-induced nitrogen mineralization. In contrast, if tillage operations induce soil disturbance, earthworms may have a positive impact by rebuilding the structure of soil. However, since earthworms raise emissions of GHGs much more significantly in no-tillage system, the earthworm influence on GHG emissions will lessen somewhat. Based upon Seufert et al. (2012), high rates of inorganic nitrogen fertilization and heavy tillage operations define conventional agricultural systems. When agricultural residues are removed for use as fiber or fuel, the intense tillage operations decrease the HQ for earthworms. They would not have much of an impact on crop yields due to the high amounts of readily accessible N. Providing organic food to earthworms will enhance GHG emissions moderately when crop residues are returned to the soil. Without regard to the existence of earthworms, aggressive tillage practices and high inorganic nitrogen fertilizer inputs would result in significant emissions of GHGs, while residues are cleared from a typical agricultural system.

To conclude, Fig. 10 demonstrates that in tropical low input agricultural system, activities of earthworms are expected to have the most positive impact on crop yields. Therefore, it should be a top focus to improve the quality of habitat in those systems, particularly by optimizing the return of agricultural residues to the soil. These systems have comparatively low costs in terms of GHG emissions. The possibility for a significant rise in agricultural yields and much higher emissions of greenhouse gases present significant challenges for organic agricultural systems. Consequently, earthworms play a relatively small influence in GHG emissions and YE in traditional agricultural systems.

5 Future Research Directions

There is still more to learn about how earthworm activity affects GHG control and plant growth, and the proposed research for this field needs to include a number of new study objectives. As an illustration, the evaluation of YE, HQ, and GHG control for the major agroecosystems shown in Fig. 10 is based on experimental investigations that have been reported in peer-reviewed magazines; my results have not yet been confirmed in a single comprehensive study. To achieve this, I require field research conducted in low input agricultural system tropical areas to examine the possibility that earthworms might truly promote plant growth by optimizing the returning crop residues into the soils, in addition to the fertilizing benefits of the residues directly. The earthworm impact on GHG regulation is also crucial to study in such an investigation, as barely any research exists on that. The influence of earthworms on regulation of greenhouse gases has largely been researched in controlled circumstances of organic agricultural systems. The GHG emissions from organic agricultural systems, caused by earthworms, have not been reported in any fieldwork research so far. As comparison to fieldwork, the research on earthworm-induced GHG emissions has a significant bias in favor of lab experiments. There are several million square kilometers of earthworms in natural habitats in North American temperate and cold temperate woodlands, covering many millions of square miles (Frelich et al. 2006), must be investigated for earthworm effects in well-established soils where earthworms have never lived before. These types of soils enable for the establishment of control treatments that are completely deplete of earthworms and their leftovers (earthworms' impacts on structures of soil that last long after they are eliminated). Additionally, soils of tundra might become more and more fascinating for upcoming earthworm studies because climate change is probably to make them perfect habitats for communities of earthworms (IPCC 2001). A need for longer (er) term research to highlight earthworm impacts over the long run and seasonal variations accompanied the need for more realistic research. For earthworm impacts on N_2O emissions, there have been very few research lasting more than 200 days. Earthworm-induced N_2O emissions may be estimated over the long term in various seasons and throughout the year, and the turnover rate of earthworm-induced stabilized carbon might allow research on the impacts of earthworms on carbon. It is critical to conduct a full factorial long-run research in which SOC pools and GHG emissions are monitored intermittently on field plots with and without earthworms' presence. In addition to the impacts of earthworms on plant growth and greenhouse gas regulation, several interactions between earthworms and the rhizospheres should be taken into account. The time of year when roots of plants have colonized the soil is probably when earthworms are quite active; however, this depends on geographic area and temperature. There have been very few research conducted on earthworms feeding on roots of plants (Gong et al. 2022), or on how earthworms and plants interact with one another to stabilize carbon (Fonte et al. 2012). However, no studies have identified mechanisms by which earthworms interact with the rhizosphere that can explain how (detrimental or beneficial) microorganisms are dispersed or how plant growth is regulated. Lastly, investigations including more soil invertebrates should also be added to earthworms' GHG investigations (Kuiper et al. 2013). It is also important for upcoming studies to examine how invertebrates interact with soil to determine soil greenhouse gas emissions.

6 Conclusions

Earthworms stimulate plant growth, increasing the amount of carbon that permeates into the soil, however at the cost of additional CO2 and N2O emissions. Earthworm-induced plant growth and greenhouse gas emissions were also identified and studied. Enhanced nitrogen mineralization from SOM and residues is the key pathway for earthworm-induced plant growth. The earthworms' impact patterns for emissions of GHG are different. Generally, earthworm-induced N2O emissions are associated with relatively higher C/N ratios and SOC, proposing a causal relationship between increased carbon inputs and (earthworm induced) carbon sequestration. In soils with lower SOC contents, CO2 emissions induced from earthworms are particularly enhanced. Through mobilizing recalcitrant SOC pools and feeding on carbon fractions in the soil, earthworms can increase decomposition of carbon in these soils. As I have shown, earthworms contribute to soil C balance in a number of ways, including increasing fractions of SOC associated with C stabilization and enhanced CO2 emissions. As a final assessment of earthworm performance in various types of agroecosystems, I incorporated my research on the impacts of earthworms on crops yields and regulation of GHGs. Optimal residue management is crucial to maximizing earthworm benefits in tropical low input agricultural systems. In these systems, the costs are relatively lower in terms of greenhouse gas emissions. Earthworms have little effect on greenhouse gases in intensively controlled agroecosystems (which often offer inferior environment for earthworms), and their ability to increase yield is minimal. In organic agricultural systems, when favorable yield impacts are coupled with much higher GHG emissions, the interplay between how earthworms affect yields and greenhouse gas emissions is particularly noticeable. Based on my review, emissions of N2O and CO2 enhanced by 42% and 33%, respectively, via earthworms. I could find no evidence that earthworms have an impact on SOC stocks. For N2O, the earthworm-induced increase in emissions grew more noticeable over long periods (>1 month), but it decreased for carbon dioxide. High C/N ratios and SOC content were often associated with large earthworm impacts on N2O emissions. This makes one wonder if emissions of N2O induced by earthworms are also an unavoidable result of earthworm-induced sequestration of carbon and heavily implies that N2O emissions induced by earthworms are an inevitable result of enhanced soil carbon inputs. I come to the conclusion that anticipated changes in communities of earthworms over the following years will have a considerable impact on (and likely increase) fluxes of GHG. It is yet unknown how much earthworm-induced boosts in GHG emissions can be countered by the earthworms' stimulating impacts on net primary output. In general, more is required: (1) long-term research, (2) fieldwork investigations (particularly in natural habitats), (3) field research on undamaged soils without a background of earthworm activities, and (4) studies on systems with growing plants.

Acknowledgments The valuable and constructive comments/suggestions provided by the Dr. John Warren Reynolds (Oligochaetology Laboratory, 9-1250 Weber Street East, Kitchener, ON Canada N2A 4E1, and Research Associate, New Brunswick Museum, Saint John, NB Canada E2K 1E5) is highly appreciated.

References

Abail Z, Whalen JK (2021) Earthworm contributions to soil nitrogen supply in corn-soybean agroecosystems in Quebec, Canada. Pedosphere 31:405–412

Al-Maliki S, Al-Taey DK, Al-Mammori HZ (2021) Earthworms and eco-consequences: Considerations to soil biological indicators and plant function: A review. Acta Ecol Sin 41:512–523

Barlow N (1858) The Autobiography of Charles Darwin. Collins St. James's Place, London, UK

Barot S, Ugolini A, Brikci FB (2007) Nutrient cycling efficiency explains the long-term effect of ecosystem engineers on primary production. Functional ecology:1–10

Bottinelli N, Capowiez Y (2021) Earthworm ecological categories are not functional groups. Biol Fertil Soils 57:329–331

Bottinelli N, Hedde M, Jouquet P, Capowiez Y (2020a) An explicit definition of earthworm ecological categories—Marcel Bouché's triangle revisited. Geoderma 372:114361

Bottinelli N, Kaupenjohann M, Märten M, Jouquet P, Soucémarianadin L, Baudin F, Tran TM, Rumpel C (2020b) Age matters: Fate of soil organic matter during ageing of earthworm casts produced by the anecic earthworm Amynthas khami. Soil Biol Biochem 148:107906

Brown GG, Pashanasi B, Villenave C, Patrón JC, Senapati BK, Giri S, Boyer J (1999) Effects of earthworms on plant production in the tropics. In: Lavelle P, Brussaard L, Hendrix PF (eds) Earthworm management in tropical agroecosystems. CABI Publishing, Oxford, pp 87–147

Brown GG, Barois I, Lavelle P (2000) Regulation of soil organic matter dynamics and microbial activity in the drilosphere and the role of interactions with other edaphic functional domains. Eur J Soil Biol 36:177–198

Brown GG, Edwards CA, Brussaard L (2004) How earthworms affect plant growth: burrowing into the mechanisms. In: Edwards CA (ed) Earthworm ecology, 2nd edn. CRC Press, Boca Raton, Florida, p 441

Butt KR, Quigg SM (2021) Earthworm community development in soils of a reclaimed steelworks. Pedosphere 31:384–390

Cardinael R, Hoeffner K, Chenu C, Chevallier T, B'eral C, Dewisme A, Clzeau D (2019) Spatial variation of earthworm communities and soil organic carbon in temperate agroforestry. Biol Fertil Soils 55:171–183

Chadwick LC, Bradley K (1948) An experimental study of the effects of certain earthworms on crop production

Chen C, Ejack L, Chénier MR, Whalen JK (2021) Earthworm functional groups are related to denitrifier activity in riparian soils. Pedosphere 31:413–422

Contreras-Ramos SM, Alvarez-Bernal D, Montes-Molina JA, Van Cleemput O, Dendooven L (2009) Emission of nitrous oxide from hydrocarbon contaminated soil amended with waste water sludge and earthworms. Appl Soil Ecol 41:69–76

Cortez J, Billes G, Bouché MB (2000) Effect of climate, soil type and earthworm activity on nitrogen transfer from a nitrogen-15-labelled decomposing material under field conditions. Biol Fertil Soils 30:318–327

Crooks JA (2002) Characterizing ecosystem-level consequences of biological invasions: the role of ecosystem engineers. Oikos 97:153–166

Curry JP, Schmidt O (2007) The feeding ecology of earthworms – A review. Pedobiologia 50:463–477

Darwin C (1881) The formation of vegetable mould, through the action of worms, with observations on their habits. John Murray, London, UK

Decaëns T, Jiménez JJ (2002) Earthworm communities under an agricultural intensification gradient in Colombia. Plant Soil 240:133–143

Desmond A, Moore J (1992) Darwin. Penguin Books, London, UK

Edwards CA (2004) Earthworm ecology, 2nd edn. CRC Press LLC, Florida

Edwards CA, Lofty JR (1980) Effects of earthworm inoculation upon the root-growth of direct drilled cereals. J Appl Ecol 17:533–543

Feller C, Brown GG, Blanchart E, Deleporte P, Chernyanskii SS (2003) Charles Darwin, earthworms and the natural sciences: various lessons from past to future. Agric Ecosyst Environ 99:29–49

Feller C, Blanchart E, Bernoux M, Lal R, Manlay R (2012) Soil fertility concepts over the past two centuries: the importance attributed to soil organic matter in developed and developing countries. Arch Agron Soil Sci 58:S3–S21

Felten D, Emmerling C (2009) Earthworm burrowing behaviour in 2D terraria with single- and multi-species assemblages. Biol Fertil Soils 45:789–797

Ferreira T, Santos A, Demetrio WC, Cardoso GB, Moraes RA, Assis O, Niva C, Smokanit MA, Knópik JU, Sautter KD, Brown GG (2018) Earthworm species in public parks in Curitiba, Paraná, Brazil. Zootaxa 4496:535–547. https://doi.org/10.11646/zootaxa.4496.1.41

Fierer N (2019) Earthworms' place on Earth. Science 366(6464):425–426. https://doi.org/10.1126/science.aaz5670

Fonte SJ, Six J (2010) Earthworms and litter management contributions to ecosystem services in a tropical agroforestry system. Ecol Appl 20:1061–1073

Fonte SJ, Quintero DC, Velásquez E, Lavelle P (2012) Interactive effects of plants and earthworms on the physical stabilization of soil organic matter in aggregates. Plant Soil 359:205–214

Fragoso C, Lavelle P, Blanchart E, Senapati B, Jiménez JJ, Martínez MA, Tondoh J (1999) Earthworm communities of tropical agroecosystems: origin, structure and influence of management practices. In: Lavelle P, Brussaard L, Hendrix PF (eds) Earthworm management in tropical agroecosystems. CABI, Wallingford, pp 27–55

Frelich LE, Hale CM, Scheu S, Holdsworth AR, Heneghan L, Bohlen PJ, Reich PB (2006) Earthworm invasion into previously earthworm-free temperate and boreal forests. Biol Invasions 8:1235–1245

Giannopoulos G, Pulleman MM, Van Groenigen JW (2010) Interactions between residue placement and earthworm ecological strategy affect aggregate turnover and N_2O dynamics in agricultural soil. Soil Biol Biochem 42:618–625

Godfray HCJ, Beddington JR, Crute IR, Haddad L, Lawrence D, Muir JF, Toulmin C (2010) Food security: the challenge of feeding 9 billion people. Science 327:812–818

Golovanova E, Kniazev SY, Babiy KA, Tsvirko EI, Karaban K, Solomatin DV (2021) Dispersal of earthworms from the Rudny Altai (Kazakhstan) into Western Siberia. Ecologica Montenegrina 45:48–61

Gong X, Wang S, Wang Z, Jiang Y, Hu Z, Zheng Y, Chen X, Li H, Hu F, Liu M, Scheu S (2019) Earthworms modify soil bacterial and fungal communities through enhancing aggregation and buffering pH. Geoderma 347:59–69

Gong X, Li J, Chang SX, Wu Q, An Z, Huang C, Sun X, Li S, Wang H (2022) Cattle manure biochar and earthworm interactively affected CO_2 and N2O emissions in agricultural and

forest soils: Observation of a distinct difference. Front Environ Sci Eng 16:39. https://doi.org/10.1007/s11783-021-1473-8
Hendrix PF, Bohlen PJ (2002) Exotic earthworm invasions in North America: ecological and policy implications. Bioscience 52:801–811
Hodson ME, Corstanjeb R, Jones DT, Witton J, Burton VJ, Sloan T, Eggleton P (2021) Earthworm distributions are not driven by measurable soil properties. Do they really indicate soil quality? PLoS ONE 16(8):e0241945
Hoeffner K, Monard C, Santonja M, Cluzeau D (2018) Feeding behavior of epi-anecic earthworm species and their impacts on soil microbial communities. Soil Biol Biochem 125:1–9
Hoeffner K, Santonja M, Cluzeau D, Monard C (2019) Epi-anecic rather than strict-anecic earthworms enhance soil enzymatic activities. Soil Biol Biochem 132:93–100
Hoeffner K, Santonja M, Monard C, Barbe L, Le Moing M, Cluzeau D (2021) Soil properties, grassland management, and landscape diversity drive the assembly of earthworm communities in temperate grasslands. Pedosphere 31:375–383
Honvault N, Houben D, Lebrun M, Vedere C, Nobile C, Guidet J, Kervroëdan L, Aubertin ML, Rumpel C, Faucon MP, Dulaurent AM (2023) Positive or neutral effects of biochar-compost mixtures on earthworm communities in a temperate cropping system. Appl Soil Ecol 182:104684. https://doi.org/10.1016/j.apsoil.2022.104684
IPCC (2001) Climate change 2001: the scientific basis. Cambridge University Press, Cambridge, UK, p 881
IPCC (2007) Climate change 2007: synthesis report. In: Pachauri RK, Reisinger A (eds) Contribution of Working Group I, II and III to the Fourth Assessment Report of the Intergovernmental Panel on Climate Change. IPCC, Geneva, Switzerland, pp 13–42
IPCC (2013) Climate change 2013: the physical science basis. In: Working group I contribution to the IPCC fifth assessment report, AR5 edn, Stockholm, Sweden
Jana U, Barot S, Blouin M, Lavelle P, Laffray D, Repellin A (2010) Earthworms influence the production of above- and belowground biomass and the expression of genes involved in cell proliferation and stress responses in Arabidopsis thaliana. Soil Biol Biochem 42(2):244–252
Jongmans AG, Pulleman MM, Balabane M, van Oort F, Marinissen JCY (2003) Soil structure and characteristics of organic matter in two orchards differing in earthworm activity. Appl Soil Ecol 24:219–232
Kooch Y (2012a) Response of earthworms' ecological groups to decay degree of dead trees (case study: Sardabrood forest of Chalus, Iran). Eur J Exp Biol 3:532–538
Kooch Y (2012b) Soil variability related to pit and mound, canopy cover and individual trees in a Hyrcanian Oriental Beech stand. PhD Thesis of Forestry, Tarbiat Modares University, p 203 (In Persian)
Kooch Y, Ghaderi E (2023) The effect of Crataegus and Berberis canopy types on bioindicators of soil quality in a semi-arid climate. J Arid Environ 208:104862. https://doi.org/10.1016/j.jaridenv.2022.104862
Kooch Y, Hosseini SM (2010) Response of earthworm's biomass and diversity to wind throw events and soil properties in Hyrcanian forests of Iran. Folia Oecologica 37:181–190
Kooch Y, Jalilvand H (2008) Earthworms as ecosystem engineers and the most important detritivors in forest soils. Pak J Biol Sci 11:819–825
Kooch Y Jalilvand H, Bahmanyar MA, Pormajidian MR, Shahbeiki Gilkalayee M (2007) The effective soil factors on distribution of earthworms in forest ecosystem units. 10th congress of soil sciences in Iran, pp 221–223 (In Persian)
Kooch Y, Jalilvand H, Bahmanyar MA, Pormajidian MR (2008) Abundance, biomass and vertical distribution of earthworms in ecosystem units of Hornbeam forest. J Biol Sci 8:1033–1038
Kooch Y, Hosseini SM, Mohammadi J, Hojjati SM (2010) The effects of gap disturbance on soil chemical and biochemical properties in a mixed beech-hornbeam forest of Iran. Ecologia Balkanica 2:39–56

Kooch Y, Hosseini SM, Mohammadi J, Hojjati SM (2011) Analysis of earthworm's patchy distribution and variability of soil biochemical properties under single-tree influences. Int J Environ Sci 1:1813–1829

Kooch Y, Mollaei Darabi S, Hosseini SM (2015) The effects of pits and mounds following wind throw events on soil features and greenhouse gas fluxes in a temperate forest. Pedosphere 25:1–13

Kooch Y, Tarighat FS, Hosseini SM (2017) Tree species effects on soil chemical, biochemical and biological features in mixed Caspian lowland forests. Trees 31:863–872

Kooch Y, Tavakoli M, Akbarinia M (2018) Tree species could have substantial consequences on topsoil fauna: a feedback of land degradation/restoration. Eur J For Res 137:793–805

Kooch Y, Azizi Mehr M, Hosseini SM (2020a) The effect of forest degradation intensity on soil function indicators in northern Iran. Ecol Indic 114:1–9

Kooch Y, Ehsani S, Akbarinia M (2020b) Stratification of soil organic matter and biota dynamics in natural and anthropogenic ecosystems. Soil Tillage Res 200:1–11

Kooch Y, Ghorbanzadeh N, Wirth S, Novara A, Shah Piri A (2021) Soil functional indicators in a mountain forest-rangeland mosaic of northern Iran. Ecol Indic 126:107672

Kooch Y, Amani M, Abedi M (2022a) Vegetation degradation threatens soil health in a mountainous semi-arid region. Sci Total Environ 830:154827

Kooch Y, Ghorbanzadeh N, Hajimirzaaghaee S, Egli M (2022b) Soil functional indicators in mixed beech forests are clearly species-specific. J For Res 34(4):1033–1049. Accepted to publish

Kooch Y, Ghorbanzadeh N, Kuzyakov Y, Praeg N, Ghaderi E (2022c) Investigation of the effects of the conversion of forests and rangeland to cropland on fertility and soil functions in mountainous semi-arid landscape. Catena 210:105951

Kooch Y, Ghorbanzadeh N, Haghverdi K, Francaviglia R (2023) Soil quality cannot be improved after thirty years of land use change from forest to rangeland. Sci Total Environ 856:159132. https://doi.org/10.1016/j.scitotenv.2022.159132

Kool DM, Wrage N, Zechmeister-Boltenstern S, Pfeffer M, Brus D, Oenema O, Van Groenigen JW (2010) Nitrifier denitrification can be a source of N_2O from soil: a revised approach to the dual-isotope labelling method. Eur J Soil Sci 61:759–772

Kuiper I, De Deyn GB, Thakur MP, Van Groenigen JW (2013) Soil invertebrate fauna affect N_2O emissions from soil. Glob Chang Biol 19:2814–2825

Kumar S, Tripathi G, Mishra GV (2021) A comparative study on earthworm biodiversity and species habitat-relationship of hilly and plain areas of Sirohi District of Rajasthan, India. Appl Ecol Environ Sci 9:419–439

Kutschera U, Elliott JM (2010) Charles Darwin's observations on the behaviour of earthworms and the evolutionary history of a giant endemic species from Germany, *Lumbricus badensis* (Oligochaeta: Lumbricidae). Appl Environ Soil Sci 2010:11

Lal R (2004) Soil carbon sequestration impacts on global climate change and food security. Science 304:1623–1627

Laossi KR, Noguera DC, Decäens T, Barot S (2011) The effects of earthworms on the demography of annual plant assemblages in a long-term mesocosm experiment. Pedobiologia 54:127–132

Lavelle P, Barois I, Martin A, Zaidi Z, Schaefer R (1989) Management of earthworm populations in agro-ecosystems: a possible way to maintain soil quality? In: Clarholm M, Bergstrom L (eds) Ecology of arable land: perspectives and challenges. Kluwer Academic Press, London, pp 109–122

Lavelle P, Barros E, Blanchart E, Brown G, Desjardins T, Marian L, Rossi JP (2001) SOM management in the tropics: why feeding the soil macro fauna? Nutr Cycl Agroecosyst 61:53–61

Lavelle P, Charpentier F, Villenave C, Rossi J, Derouard L, Pashanasi B, Bernier N (2004) Effects of earthworms on soil organic matter and nutrient dynamics at a landscape scale over decades. In: Edwards C (ed) Earthworm ecology. CRC Press LLC, Boca Raton, FL, pp 145–160

Law MMS, Lai DYF (2021) Impacts of wetting-drying cycles on short-term carbon and nitrogen dynamics in Amynthas earthworm casts. Pedosphere 31:423–432

Le Bayon RC, Milleret R (2009) Effects of earthworms on Phosphorus dynamics – a review. Dyn Soil Dyn Plant 3:21–27

Li C, Frolking S, Butterbach-Bahl K (2005) Carbon sequestration in arable soils is likely toincrease nitrous oxide emissions, offsetting reductions in climate radiative forcing. Clim Chang 72(3):321–338

Lubbers IM, Brussaard L, Otten W, Van Groenigen JW (2011) Earthworm-induced N mineralization in fertilized grassland increases both N_2O emission and crop-N uptake. Eur J Soil Sci 62:152–161

Lubbers IM, López González E, Hummelink EWJ, van Groenigen JW (2013a) Earthworms can increase nitrous oxide emissions from managed grassland: a field study. Agric Ecosyst Environ 174:40–48

Lubbers IM, van Groenigen KJ, Fonte SJ, Six J, Brussaard L, van Groenigen JW (2013b) Greenhouse-gas emissions from soils increased by earthworms. Nat Clim Change 3, 187–194. www.nature.com/natureclimatechange, 8p

Marhan S, Langel R, Kandeler E, Scheu S (2007) Use of stable isotopes (13C) for studying the mobilization of old soil organic carbon by endogeic earthworms (Lumbricidae). Eur J Soil Biol 43:201–208

Mazur-Paczka A, Paczka G, Kostecka J, Butt KR, Jaromin M, Garczynska M, Podolak A (2021) Community structure of Lumbricidae in beech woodland of the Bieszczady National Park, Southeast Poland. Pedosphere 31:391–397

McCay TS, Scull T (2019) Invasive lumbricid earthworms in northeastern North American forests and consequences for leaf-litter fauna. Biol Invasions 21:2081–2093

Medina-Sauza RM, Álvarez-Jiménez M, Delhal A, Reverchon F, Blouin M, Guerrero-Analco JA, Cerdán CR, Guevara R, Villain L, Barois I (2019) Earthworms building up soil micro biota, a Review. Front Environ Sci 7:81. https://doi.org/10.3389/fenvs.2019.00081

Moghimian N, Jalali SG, Kooch Y, Rey A (2020) Downed logs improve soil properties in old-growth temperate forests of Northern Iran. Pedosphere 30:378–389

Poorter H, Nagel O (2000) The role of biomass allocation in the growth response of plants to different levels of light, CO_2, nutrients and water: a quantitative review. Aust J Plant Physiol 27:595–607

Postma-Blaauw MB, Bloem J, Faber JH, Van Groenigen JW, De Goede RGM, Brussaard L (2006) Earthworm species composition affects the soil bacterial community and net nitrogen mineralization. Pedobiologia 50:243–256

Ramaswamy V, Boucher O, Haigh J, Hauglustaine D, Haywood J, Myhre G, Solomon S (2001) Radiative forcing of climate change. Contribution if Working Group I to the Third Assessment Report of the Intergovernmental Panel on Climate Change. In: Houghton JT, Ding Y, Griggs DJ, Noguer M, Van der Linden PJ, Dai X, Maskell K, Johnson CA (eds) Climate change 2001: The scientific basis. Cambridge University Press, Cambridge, UK, p 881

Rastogi M, Singh S, Pathak H (2002) Emission of carbon dioxide from soil. Curr Sci 82:510–518

Rizhiya E, Bertora C, Vliet PCJ, Kuikman PJ, Faber JH, Groenigen JW (2007) Earthworm activity as a determinant for N_2O emission from crop residue. Soil Biol Biochem 39:2058–2069

Scheu S (2003) Effects of earthworms on plant growth: patterns and perspectives: The 7th international symposium on earthworm ecology, Cardiff, Wales, 2002. Pedobiologia 47:846–856

Seufert V, Ramankutty N, Foley JA (2012) Comparing the yields of organic and conventional agriculture. Nature 485(7397):229–U113

Skinner C, Gattinger A, Muller A, Mader P, Fliebbach A, Stolze M, Niggli U (2014) Greenhouse gas fluxes from agricultural soils under organic and non-organic management – A global meta-analysis. Sci Total Environ 468:553–563

Smith KA, Ball T, Conen F, Dobbie KE, Massheder J, Rey A (2003) Exchange of greenhouse gases between soil and atmosphere: interactions of soil physical factors and biological processes. Eur J Soil Sci 54:779–791

Speratti AB, Whalen JK (2008) Carbon dioxide and nitrous oxide fluxes from soil as influenced by anecic and endogeic earthworms. Appl Soil Ecol 38:27–33

Tylianakis JM, Didham RK, Bascompte J, Wardle DA (2008) Global change and species interactions in terrestrial ecosystems. Ecol Lett 11:1351–1363 van Groenigen KJ, Six J, Hungate BA, de Graaff MA, van Breemen N, van Kessel C (2006) Element interactions limit soil carbon storage. Proc Natl Acad Sci USA 103:6571–6574

Van Vliet PCJ, Van der Stelt B, Rietberg PI, De Goede RGM (2007) Effects of organic matter content on earthworms and nitrogen mineralization in grassland soils. Eur J Soil Biol 43(Suppl. 1):222–229

White G (1777) Letter LXXVII, Selborne

Wrage N, Velthof GL, van Beusichem ML, Oenema O (2001) Role of nitrifier denitrification in the production of nitrous oxide. Soil Biol Biochem 33:1723–1732

Wrage N, van Groenigen JW, Oenema O, Baggs EM (2005) A novel dual-isotope labelling method for distinguishing between soil sources of N_2O. Rapid Commun Mass Spectrom 19:3298–3306

Zhong L, Wu T, Ding J, Xu W, Yuan F, Liu BF, Zhao L, Li Y, Ren NQ, Yang SS (2023) Co-composting of faecal sludge and carbon-rich wastes in the earthworm's synergistic cooperation system: Performance, global warming potential and key microbiome. Sci Total Environ 857:159311. https://doi.org/10.1016/j.scitotenv.2022.159311

Part II
Earthworm Communities in Managed Ecosystems

Earthworms: Functional Traits and Soil Properties

Yahya Kooch, Mohammad Kazem Parsapour, and Donghui Wu

Abstract Earthworm populations are often vulnerable to physicochemical qualities, soil drying and wetting in different ecosystems, i.e., cultivated or noncultivated soils, that influence directly or indirectly their survival. The creation of various population patches for earthworms' species is supported by differences in soil qualities at diverse sites (i.e., croplands, forests, rangelands, and urban soils) or locations (e.g., catena landscape, pit and mound microtopography, gilgai position, single tree, deadwood, and canopy gap). The extent of the observed changes in earthworm populations is influenced by the ecosystem's characters, as well as by topographically and climatically controlled processes like redistribution of litter and creep of soil. Anyway, the soil environment can be deemed as a mosaic of soil earthworm activity in different locations that are influenced by site factors. Soil physicochemical features like texture, moisture, temperature, pH, and organic matter (OM) affect the distribution of earthworms in the soil. Soils with a neutral pH are usually preferred

Y. Kooch (✉)
Faculty of Natural Resources & Marine Sciences, Tarbiat Modares University, Noor, Mazandaran, Iran
e-mail: yahya.kooch@modares.ac.ir

M. K. Parsapour
Research Division of Natural Resources, Chaharmahal and Bakhtiari Agricultural and Natural Resources Research and Education Center, AREEO, Shahrekord, Iran

D. Wu
Key Laboratory of Wetland Ecology and Environment, State Key Laboratory of Black Soils Conservation and Utilization, Northeast Institute of Geography and Agroecology, Chinese Academy of Sciences, Changchun, China

State Environmental Protection Key Laboratory of Wetland Ecology and Vegetation Restoration, School of Environment, Northeast Normal University, Changchun, China

Key Laboratory of Vegetation Ecology, Ministry of Education, Northeast Normal University, Changchun, China

Jilin Provincial Key Laboratory of Animal Resource Conservation and Utilization, Northeast Normal University, Changchun, China
e-mail: wudonghui@iga.ac.cn

Y. Kooch, Y. Kuzyakov (eds.), *Earthworms and Ecological Processes*,
https://doi.org/10.1007/978-3-031-64510-5_5

by most species of earthworms. Soil moisture can play a vital role in the survival of earthworms, as their bodies have 75–90% of water by weight. There is a strong positive correlation between the soil organic matter and earthworm numbers and biomass. Earthworms are also recognized as soil managers due to their activities and alter the soil's physicochemical and biological properties.

1 Introduction

Earthworms are terrestrial invertebrates belonging to the Phylum Annelida, Class Chaetopoda, and Order Oligochaeta. They are believed to be originated about 600 million years ago, during the pre-Cambrian period. Aristotle referred to earthworms as "the earth guts and the soil fertility restoring agents" (Gonzalez et al. 2003). Earthworms with diverse habitats improve soil fertility by changing the physical and chemical composition of the soil, and many scientists have discussed the importance of earthworms in improving the soil fertility (Hale and Host 2005; Rutgers et al. 2016; Hoeffner et al. 2021). Earthworms convert organic matter (OM) into soil humus by ingesting soil and various mixtures of soil components that are then released as subsurface and surface castings. According to Szlavecz et al. (2018), earthworms play a major role in the decomposition of organic matter through feeding, aeration, fragmentation, turnover, and dispersion. The presence of optimal earthworms' population in soil induces a huge population of insects, spiders, bacteria, fungi, viruses, and other organisms, and thus a healthy soil. Thus, earthworms are biological indicators of soil quality (Thouvenot et al. 2021). The Annelida worms, known as earthworms, penetrate into the soil and decompose the organic matter, which turns into casts. In addition to using earthworms as bait for fishing, humans enrich soil with the nutrients that worms provide. In addition to turning over leaf matters, children enjoy digging in the soil with wriggling tubes. There is, however, a common misconception that each worm is the same. In addition to their diverse terrestrial, aquatic, and subaquatic habitats, worms are composed of a wide variety of fauna. The body of terrestrial worms are segmented cylindrical, and without limbs. Soil texture, moisture, and light are all factors that affect the skin of earthworms. However, earthworms are not native to any particular soil in spite of their widespread presence in the soil environment. Generally, gardeners hope that earthworms will colonize their cultivated lands, regardless of whether they are indigenous to the area where they are located. Climate-related variables, like soil water content, frequently cause obstacles, which earthworms cannot overcome. Nonetheless, when other parts of the globe were settled, for example, following Columbus' voyage to America, this led to the passive spread of earthworms to these new locations (Edwards and Bohlen 1996; Kooch et al. 2021).

The long-term sustainability of tallgrass prairies through conservation and restoration might have been affected positively or negatively by pedogenic mechanisms induced by earthworms through the bioturbation (Lukkari et al. 2004). Casts are formed after earthworms consume litter and soil organic matter (SOM) while moving through the layers of surface organic and mineral soil. There are different types of earthworms, which determine the shape and size of their casts (Neirynck et al. 2000). Earthworm casts have an impact on the formation of soil aggregates and soil quality (Kooch et al. 2023a). It is important to understand that soil functions and invertebrates are intricately related, both at micro and macro levels. According to some research, earthworms are the ideal bioindicators of the quality and health of a habitat (Phillips et al. 2008). Based upon Sautter et al. (2006), earthworms influence the soil's physicochemical, biological, and hydraulic features through their ingesting, moving, shredding, and secreting mucus. In spite of this, every earthworm species differently influences the soil properties and might or might not alter soil features and aboveground habitat in a beneficial way. Recent studies have attributed the surales landscape patterns of South America to earthworm foraging activities. Surales are tropical marshes with mounded grasslands that are periodically inundated. These mounds are formed, long-lasting, and continue to generate due to the *Andiorrhinus* species activity. Several species of *Andiorrhinus* spp. were prevailing in water-logged lands, while they did not occur in dry regions, and were thought to be responsible for the obvious self-organization and landscape patterns of the mounds of surales. The feeding habits and cast secretion of *Andiorrhinus* species led to the development of the different methods through which the mounds emerge. By creating permanent vertical burrows, the earthworms increase the mounds height and also create lowlands where water can accumulate (Smetak et al. 2007). Ecosystem services were assessed in 1991 near Manhattan, Kansas, in "the Konza Natural Prairie Area" (Szlavecz et al. 2006). In detail, native grassland prairies, communities of earthworms, played a key role in cycle of nutrients and decompositions. A prairie system with native earthworms consumed and ingested more soil than a prairie system with exotic earthworms. In addition, indigenous earthworms were more active for a longer period of time during the entire year than the nonnative earthworms that were found together with the indigenous earthworms, and this was ascribed to higher nutrient cycling and rates of decompositions (Xie et al. 2018).

The diversity and abundance of subterranean environments heavily influence the quality of soil, since aboveground diversity and biomass are linked to soil ecosystem services and functioning. In addition, the ratio of indigenous earthworms to nonnative earthworms affects the abundance and diversity of aboveground fauna, flora, and microreliefs (Ziemer 1968). There is a common perception that earthworms are useful to an ecosystem, but their population dynamics could greatly affect this perception (Tresch et al. 2018). For example, some casts of earthworms are humid and spherical. If the populations of these casts were unevenly distributed in an ecosystem or introduced to an agricultural system, their physical features would probably enhance soil bulk density. Both the soil medium's hydraulic conductivity and roots foraging capacity would be negatively affected by an increase in soil bulk density (Rutgers et al. 2016). Thus, negative effects would eventually impede the growth of aboveground plants, subsequently influencing nesting and forage materials for indigenous aboveground biota (Lukkari et al. 2004). Additionally, earthworm communities might have an impact on how seeds move along an environment and in the surface soil. It has been noted that *Lumbricus terrestris* is essential for plant invasion, recruitment, and the composition of plant communities (Mabuhay et al. 2003). Plant communities with diverse earthworm communities and *Lumbricus terrestris* may help prevent invasion by invasive plants, while plant communities without native earthworms or a lack of diversity may experience the opposite outcome (Kumar et al. 2021). Because of their strong physical and chemical impact on the soil environment, earthworms are considered ecosystem engineers (Lubbers et al. 2013; Kooch et al. 2022, 2023b). Their most prominent impacts on the soil ecosystem are bioturbation and nutrient cycling, which are performed by ingesting large amounts of litter and soil. The mostly positive services of earthworms for plant growth can benefit the sustainability of agriculture (Pfiffner 2014).

Since earthworm activity depends on soil moisture and soil temperature, it is suggested that earthworms are synchronizing the release of nutrients with plant uptake demands (Mabuhay et al. 2003). Earthworms may also enhance plant growth through altered water regulations, the stimulation of bacteria that can activate hormone signal pathways in plants, the increase of symbionts, and changes in soil structure (Scheyer and Hipple 2005). Earthworm populations vary greatly among land-use types and are subject to seasonal fluctuations (Kooch et al. 2023a). In a global meta-analysis, Wang et al. (2022) found the highest earthworm abundances under temperate climates across all ecosystem types, except for natural forests. Abundances in temperate climates were reported to range from an average of 123 individuals m^{-2} in arable fields to 425 individuals m^{-2} in managed pastures. In Mediterranean climates, where drought is a common phenomenon, abundances were generally much lower. Here arable fields counted only 50 individuals m^{-2} (Wardle 2002). Site-specific and seasonal changes are pronounced; therefore, single observations under favorable conditions can easily exceed 1000 individuals m^{-2}, even in arable fields (Xie et al. 2018). Management practices that affect earthworm abundance are tillage, crop sequence, organic inputs, and pesticide use (Zhou et al. 2007). Organic materials rich in protein and readily available carbohydrates are preferred by earthworms, even though some fresh litter is avoided due to phenolic compounds and other anti-nutritional factors; however, fresh litter becomes palatable after the initial degradation. Earthworms also feed on dead and living roots, but there is no evidence that roots are a preferred food source, or that earthworms could pose a threat to plants (Vilisics et al. 2012). Furthermore, there is growing evidence that earthworms can play a vital role in controlling plant diseases. For example, earthworms might play a role in keeping plant pathogenic nematodes below a tolerable level (Sautter et al. 2006). It was shown that earthworms assimilate fungi into their biomass and preferentially forage on litter infested with fungi (Pulleman et al. 2005). In this way, inoculum sources of fusarium head blight were reduced by earthworms (Mabuhay et al. 2003). Other examples in which earthworms reduced crop susceptibility to fungi were eyespot disease on wheat and Rhizoctonia solani Kühn on various crops (Lassauce et al. 2011). The enhancement of nutrient uptake by plants facilitated through earthworms was suggested to help plants to resist pathogen attacks (Kaiser et al. 2015). Recently it was shown that earthworms may reduce the number of sclerotia in the soil from the pathogen *Sclerotinia sclerotiorum* (Lib.) De Bary (Joimel et al. 2019).

2 Importance of Earthworms

- *As a growth enhancer of plants*: For the growth of plants, earthworms do the fantastic job of preparing the soil. Many research demonstrated that communities of earthworms have useful impacts on soils, as these effects increase plant growth as well as crop yields (Kim et al. 2012; Kooch et al. 2018). Auxins and cytokinins, which are beneficial substances for plant growth, are released by earthworms (Kooijman et al. 2008).
- *Vermicomposting*: Vermicomposting is a process through which earthworms and microorganisms biologically degrade and stabilize the organic waste into the so-called vermicompost. Earthworms stimulate microbial activity, fragment the organic waste, and increase the rate of mineralization (Nachtergale et al. 2002). All these processes convert wastes to humus-like substances with fine structures. Soil loosening takes place by adding vermicompost to clay soils, and finally enhancing the air entry into the soil and hence improving the soil porosity (Niemelä and Kotze 2009). Vermicast improves soil water holding capacity, since mucus associated with the cast is hygroscopic in nature. This also prevents water logging of soil due to its better-absorbing capacity. Nutrients, beneficial for plants, are released slowly and steadily from the organic carbon present in vermicompost.
- *In organic farming:* Various organic wastes (agricultural, domestic, and industrial) are burned, causing serious environmental problems and economic deficits. This problem can be overcome by earthworms, as they can process urban wastes, household garbage, sewage, and waste from food, paper, and wood industries. This composting process reduces the time needed for waste stabilization and produces vermicompost, an effective bioproduct. Vermiculture and vermicompost offer an approach to recycle and reuse organic agricultural wastes in an eco-friendly manner and are found to be economically very productive, compared with other biological inputs. By organic farming, we are able to eliminate the need for pesticides and chemical fertilizers by converting wastes into useful compost. Based upon Lubbers et al. (2013), this improves the health of humans and animals, plants and soils, producing a sustainable, eco-friendly, and economical biosystems models.
- *As a biomarker*: Biomarkers are physiological, biochemical, cellular, or behavioral differences that could be observed in samples of body fluids or tissues, or at the whole organisms' level to determine the exposure or/and the impacts of pollutants (Pfiffner 2014). Contaminant's impacts at lower levels, such as biochemical, cellular, or physiological level, are more rapid as compared to higher levels (for instance, the impacts at ecological level). Thus, it is possible to identify the toxicological impacts in populations of earthworms at an earlier stage. Rutgers

et al. (2016) have reported that earthworm biomarkers serve as practical tools for soil assessments and monitoring as a precursor to negative ecological consequences. Four distinct types of strategies can be utilized in controlling soil contamination according to Sarlo (2006): (1) Utilizing transplanted organisms in situ exposure bioassays, (2) analyzing biomarkers on indigenous earthworm communities, (3) exposing a chosen population of earthworms to the soil ambient medium, and (4) simulating field research. A growing number of contaminants have been observed in soil organisms that are affected by earthworms as biomarkers.

Depending on how many types and how many earthworms are present in an ecosystem, earthworms have various impacts on ecosystems (Zhong et al. 2023). Earthworms can be grouped according to morphological, behavioral, or physiological modifications allowing to share the available soil resources. Epigeic, anecic, and endogenic are the three basic ecophysiological categories (see Table 1). Epigeic worms are small- to medium-sized that feed on litter and live on the soil surface or inside the litter layer. They have pigmented skins and feed on litter. Anecic worms are huge, dorsally pigmented, and feed on soil and litter. They dwell in nearly vertical constant burrows and feed on soil and litter. Endogeic species are soil feeders

Table 1 The three ecophysiological categories of earthworms (Pfiffner 2014; Bottinelli et al. 2020; Bottinelli and Capowiez 2021)

Groups	Leaf litter dwellers	Shallow burrowers	Deep burrowers
	Surface dwellers, epigeic species	Shallow burrowing, endogeic species, horizontal burrowers	Vertically burrowing, anectic species
Representatives			
Examples	› Redworm (*Eisenia fetida*) › European earthworm (*Lumbricus rubellus*)	› *Octolasion lacteum* › Common field worm (*Allolobophora caliginosa*)	› Nightcrawler (*Lumbricus terrestis*) › Black-headed worm (*Allolobophora longa*)
Colour	Brownish-red overall	Pale	Reddish-brown, head darker
Habitat	› In litter layers, especially in grasslands, forests, and compost › Rarely found in cropland due to lack of permanent litter layers	› Topsoil (5–40 cm), humic mineral soil › Mostly horizontal, unstable burrows › Juveniles are generally found in the upper layers in the tangle of roots	› All soil layers, 3–4 m deep › Spend their entire lives in vertical, stable dwelling tubes (Ø 8–11 mm) › Important in agricultural soils
Size	Small, generally 2–6 cm long	Small to up to 18 cm long	Generally large, 15–45 cm long
Feeding behavior	› Feed on small plant parts on the surface of the soil	› Feed on plant parts incorporated in the topsoil	› Pull large plant parts into their dwelling tubes
Reproduction	› Vigorous › 100 cocoons per year	› Limited › 8–12 cocoons per year	› Limited › 8–12 cocoons per year
Lifespan	Short, 1–2 years	Medium, 3–5 years	Long, 4–8 years

with little pigmentation, broad horizontal burrow systems, and a wide range of sizes. Endogeic worms are split into three groups: polyhumics, mesohumics, and oligohumics, which are distinguished by the increasing size and the significance of organic mineral soil in their life. Earthworm species may not always fit strictly into the mentioned three classes, and they may show features from different groups at various periods of their lives or in different environments (Neirynck et al. 2000; Kooch et al. 2022, 2023b). These ecological groupings of earthworms play a vital role in soil ecosystem functioning. However, the impact of earthworms on plant growth and nutrient cycling is not only direct path but also indirect, as a result of changes in the microbial population. One to six species of earthworms make up most earthworm communities. Earthworm communities' relative abundance and species diversity are impacted by soil type, topography, and vegetation, as well as the biogeography of earthworms and the history of land use (van Groenigen et al. 2014). In a wide variety of fertile lands, from pastures, grasslands, and forests to rainforests and savannas, as well as many agro-environments, earthworms constitute the majority of soil biomass. By the way, earthworms are often lacking in arid grasslands along with deserts, as well as boreal extreme alpine ecosystems, and highly acidic forest soils covered with litter of poor quality (Vickery et al. 2009; Wang et al. 2022; Kooch and Ghaderi 2023). In North America, where about 100 endemic earthworm species have been found, many earthworm populations are made up of invasive foreign species. Many North American forest ecosystems are dominated by European-derived *Lumbricid* earthworms. According to a worldwide assessment of tropical earthworms, tropical forest ecosystems are home to 51 foreign and 151 native species of earthworms (Hale and Host 2005; Sautter et al. 2006).

3 Earthworms and Land Use/Cover

3.1 Croplands

Today, population growth has increased the area under cultivation. The physical destruction of burrows, cocoons, and the earthworm bodies themselves increases as the quantity and severity of tillage operations increases. Earthworms proliferate under reduced tillage techniques that leave crop residues on the surface during the year. These residues supply food, protect from the atmospheric agents, provide a cover to keep earthworms safe from predators such as birds on the surface, and shelter their burrows. Night crawlers (*Lumbricus terrestris*) benefit from reduced tillage disturbances, since they migrate in the same burrow between subsoil and topsoil searching for food. Earthworms will be unable to construct a different burrow to reach their source of food if ploughing destroys the burrow. Shallow-dwelling earthworms (Endogeic) can withstand yearly tillage since they keep digging new burrows to get much more food from the soil than from litter on the topsoil. Both types of earthworms can benefit from no-till and other conservation tillage approaches including ridge tillage and chisel plowing (Edwards and Bohlen 1996). Despite the fact that a single tillage operation may not significantly affect the population of earthworms, continuous tillage will result in a reduction in earthworm numbers over time. Tillage impacts surface residue breakdown and availability, whereas crop choice determines the residue quantity and quality available to earthworms as a source of food. In intensive ploughed cropping systems, removing crop residues and the absence of organic inputs like manure will drop earthworm populations to extremely low levels. The number of earthworms and the presence of crop residues (as amount and quality) in the soil are highly correlated. Crops covering the soil surface serve to boost earthworm populations by increasing their source of food (organic residue) and giving earthworms more time to feed and grow. Cover crops protect earthworms from the cold weather of the fall and the hot sun of the summer in southern regions. The higher food and soil cover brought by cover crops is extremely crucial since earthworms are taking up a large portion of crop residues (Abail and Whalen 2021; Zhong et al. 2023).

Earthworms benefit from nearly all organic fertilizers. Animal feces, wastewater, and leftover malt from brewery plants, paper pulp, and wastes from potato processing were all found to increase earthworm counts (Hodson et al. 2021). In a single year, organic material additions can double or triple the quantity of earthworms. Some liquid manure contains ammonia and salts, which can harm earthworms, but populations normally recover rapidly and continue to grow. In most cases, using inorganic fertilizers increases the quantity of earthworms. This is most likely an unintended consequence of increased crop biomass production and, as a result, of the higher organic residue levels (Law and Lai 2021). In a single year, organic material additions can double or triple the quantity of earthworms. Some liquid manure contains ammonia and salt, which can harm earthworms, but populations normally recover rapidly and continue to grow. Started in 1856, research at Park Grass (Rothamsted, UK) has shown that the earthworm populations were inversely related to the dose of nitrogen used after long exposure to various quantities of ammonium sulfate (0, 48, 97, and 145 kg ha^{-1}); the number of earthworms in a given area was inversely corresponding to the amount of nitrogen supplied based on Edwards and Lofty (1977), that is most likely related to the fact that the mentioned fertilizers lower soil pH. Up to 10% of the population will die if they are exposed to anhydrous ammonia during application. Farmers, on the other hand, report greater earthworms numbers in the long term as a result of higher crops and more food available for earthworms. To avoid killing earthworms during nitrogen application, a number of farmers have converted anhydrous ammonia to liquid nitrogen (28%). Others have switched to manures to protect and promote earthworm populations (Ernst 1995). Since most earthworm species prefer neutral pH values and require calcium for growing, lime appears to help earthworm populations in otherwise acid soils. Lime may assist earthworms indirectly by improving the growth of plants and thus the production of residues. In New Zealand, Edwards et al. (1995) found that adding 1 ton of lime per acre increased the number of surface-feeding earthworm species by 50%.

Generally, nearly all herbicides are safe for earthworms. Herbicides of the triazine class have a moderate impact on earthworm populations. Earthworms are mildly hazardous to the herbicides employed before the Second World War, such as copper sulfate and lead arsenate. The long-term development of these chemicals in the soil is the principal source of toxicity for earthworms (Kumar et al. 2021). The majority of pesticides in the carbamate class are poisonous to earthworms. Carbofuran (Furadan) toxicity has been considerably investigated. Aminocarb, aldicarb (Temik), bufencarb, carbaryl (Sevin), methomyl (Lannate), methiocarb (Measural), oxamyl (Vydate), propoxur (Baygon), promecarb, and thiofanox are other carbamate insecticides that have been found to be very poisonous to earthworms. In general, organophosphate pesticides are less hazardous to earthworms. Organophosphate insecticides such as ethyl-parathion, ethoprophos (Mocap), phorate (Thimet), chlorpyrifos (Equity, Dursban, Tenure, and others), and isazophos, on the other hand, are exceedingly or extremely poisonous. Aromatic organochlorine insecticides (often employed from the 1950s through the 1970s) are not particularly harmful. Chlordane, endrin, heptachlor, and isobenzan are exceptions. Earthworms have been found to be affected by carbamate fungicides (carbendazim and benomyl) and are extremely affected by other broad-spectrum fumigants (fungicides and nematicides) (Ernst 1995). Where soil moisture levels would normally be too low, a higher activity of earthworms can be supported by irrigating soils. Moreover, irrigation boosts crop yields and the availability of food for earthworms and their populations. Streams containing earthworms along with their cocoons could provide inoculum for some species (Edwards et al. 1995). By oxygenating the soil, draining imperfectly drained soils may give earthworms a more suitable medium to be more active. In general, earthworms improve soil quality by shredding wastes and enhancing microbial decomposition, as well as improving soil fertility and physical qualities including soil infiltration and aggregation. Food availability is the main factor restricting earthworm populations. Earthworm numbers can be increased by producing food from cover crop residues on the soil during conservation tillage operations. Fertilizers, in general, boost earthworm populations by providing more crop wastes, particularly under neutral pH values. Earthworms are largely unaffected by herbicides. Some pesticides, nematicides, and fungicides, on the other hand, are extremely poisonous to earthworms. Earthworm inoculation may be necessary in some cases to bring specific species to a region where favorable earthworm activities have been previously implemented (Honvault et al. 2023).

3.2 Forests and Rangelands

Forests and rangelands occupy a large area of the terrestrial ecosystems, affect the organic and mineral layers of the soil, and can influence earthworm populations (Kooch et al. 2020a, b; Kooch and Ghaderi 2023). Kooch et al. (2021) pointed out that earthworm abundance and biomass (in general, including epigeic, anecic, and endogeic) were highest in 10–20 cm soil depth, especially in forest covers (with dominant woody plants of *Fagus orientalis* Lipsky, *Carpinus betulus* L., *Fraxinus excelsior* L., and *Acer velutinum* Boiss.) compared to forests–rangelands (with dominant woody plants of *Fagus orientalis* Lipsky, *Carpinus betulus* L., *Crataegus microphylla* C. Koch., and *Berberis integerrima* Bunge.) and rangelands (with dominant woody plants of *Crataegus microphylla* C. Koch., *Berberis integerrima* Bunge., *Ribes Uva-crispa* L., and *Prunus spinosa* L.). Epigeic earthworms exhibited higher activities in the surface soils (0–10 cm), whereas endogeic and anecic earthworms had larger contributions in the 10–20 cm soil layer (see Fig. 1).

Earthworms affect forest floor structure acting as ecosystem engineers, by destroying the Oe and Oi layers, which are characterized by a Mor-type humus in soils with few earthworms (Ponge 2013). These changes can produce an A-horizon,

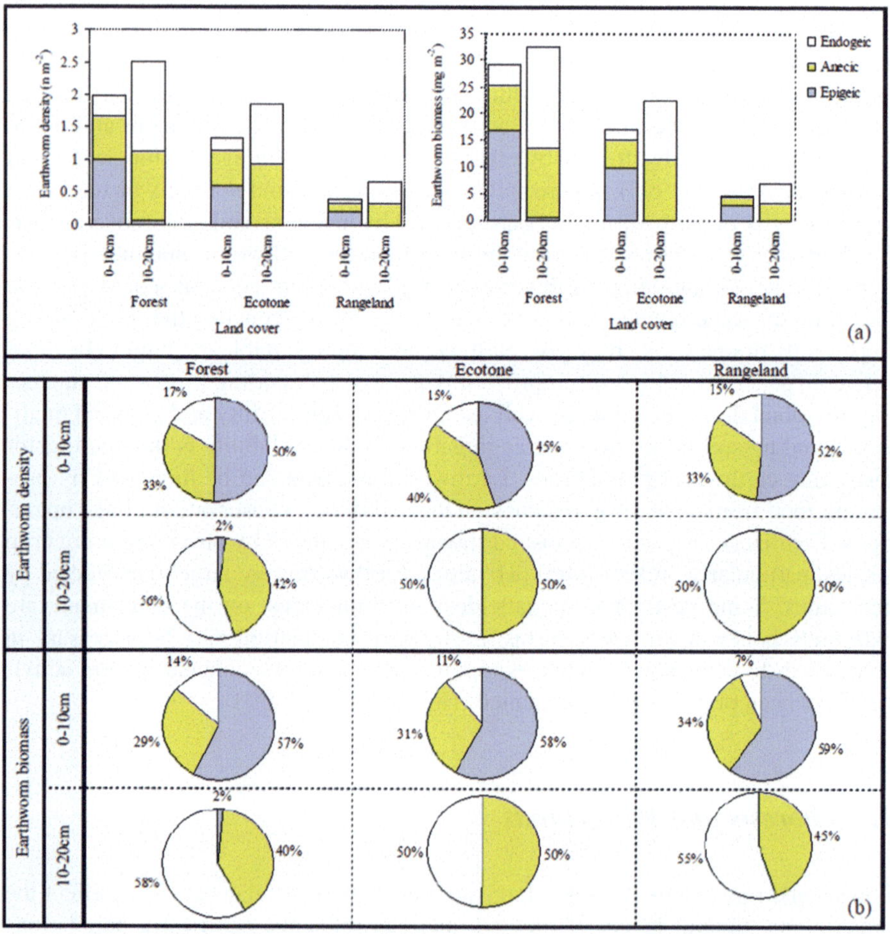

Fig. 1 Epigeic, anecic, and endogeic earthworm density and biomass (**a**) and their contribution (**b**) under different land covers in the Vaz watershed located close to Noor City, Mazandaran province, north of Iran (Kooch et al. 2021)

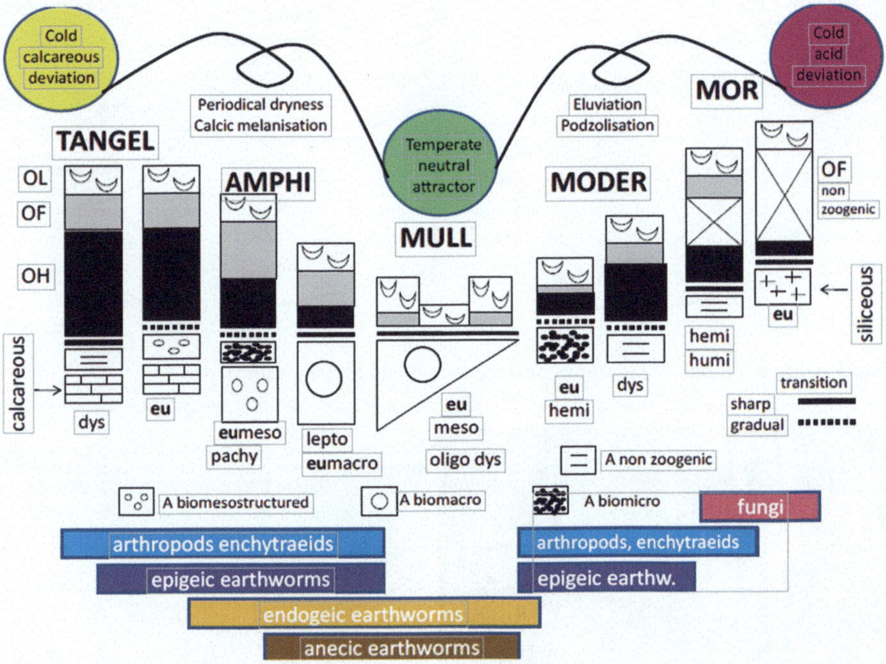

Fig. 2 Humus form transformations framework illustrating the role of anecic, epigeic, and endogeic earthworms in the mull type humus formation in temperate forests (Ponge et al. 2010)

which is characterized by a mull (more degraded) humus type, mixed with mineral debris. Mull is thought to act as an "attractor" for humus-forming organisms in forest ecosystems (see Fig. 2).

Natural ecosystems (i.e., forests and rangelands) have a variety of characteristics, as listed below that can lead to changes in soil properties and heterogeneity in earthworm activities.

Catena (as a Macro-landform) The character and distribution of soils are influenced by topography. The combined impacts of slope aspect, water dynamics, and/or erosion and deposition may cause topography effects (Gómez-Plaza et al. 2001). Soil activities at higher elevations on a slope often have an impact on soils at lower elevations within the same hillslope system. The vertical and horizontal components of slope curvature are separate. Vertical curvatures are caused by changes in slope gradient; convex vertical curvature develops when the slope gradient increases with length. Concave vertical curvatures take place when the opposite is true, i.e., when the gradient decreases toward the bottom part of the slope (see Fig. 3).

When the vertical curvature of the soil changes from convex to concave (V and Λ shapes), there is frequently a shift in soil type. Soils that develop on similar parent materials and under similar climatic conditions, but have different characteristics because of differences in relief and drainage (topographic position), would respond

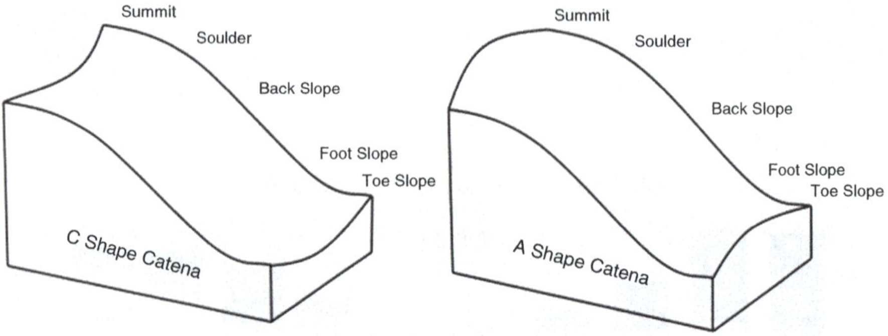

Fig. 3 Different forms and positions of slope (Fazlollahi Mohammadi 2016)

Fig. 4 Catena landform in forest ecosystem (**a**) and rangeland site (**b**), northern Iran (photo by Y. Kooch)

differently to the effects of topography. These soils are called catena (Rowell 2014) and can be visible in different ecosystems (e.g., forests and rangelands; see Fig. 4).

Fazlollahi Mohammadi (2016) investigated the quantity and biomass of earthworms at different soil depths (0–15, 15–30, and 30–45 cm) in northern Iran to define and assess the changes in soil attributes caused by slope location and catena forms. According to the findings, the number and biomass of earthworms were significantly higher in V-shaped catena, lower slope positions, and at a soil depth of 0–15 cm. The density and biomass of earthworms (epigeic, anecic, and endogeic) were greater in the V-shaped catena compared to the C-shaped one. There was substantial variation in the epigeic number and biomass in relation to catena form, soil depth, and slope position, with larger values observed in V-shaped catena, lower slope positions, and at a depth of 0–15 cm. Changes in soil moisture, pH, and fertility, according to Fazlollahi Mohammadi (2016), are the driving forces of earthworm activity in various forms and positions of the catena landscape.

Pit and Mound (as a Micro-landform) Tree uprooting is a common cause of bioturbation in forests, having considerable direct and indirect effects on soils, as well as pedogenesis and sediment transport. Soil profile reversal, soil mixing, sediment bulk local redistribution, and the formation of distinctive pit and mound topography are all direct results (see Fig. 5). Unprotected sediment is exposed to mass wasting

Fig. 5 Pit and mound structure in a windthrow site of beech forest, northern Iran (photo by Y. Kooch)

and erosion, and micro-scale weathering variations, moisture fluctuation, dynamic of organic matter, and microclimate conditions are created in the topography of pit and mound (Phillips et al. 2008).

Kooch (2012b) investigated the pit and mound landscape effects on soil engineers at local levels in the Hyrcanian forest. Data analysis revealed that the frequency of earthworms' ecological groupings differed significantly among microsites. The earthworms' density and biomass (epigeic, anecic, and endogeic) were higher in the pit bottom, whereas no earthworms were observed in the mound top or wall microsites. He underlined that earthworms require a certain amount of moisture in their surroundings and, as a result, are rarely found in dry areas. Since water makes up over 80% of fresh weight in earthworms' bodies, soil moisture is critical for their survival, and soil drying will kill them (Butt and Quigg 2021). Compared to other surfaces, mounds generate mountainous surfaces on topsoil and present higher volume and soil temperature (Londo 2001). However, soil temperature has an impact on the earthworm's number, biomass, and distribution (Brady 1990). As a result, earthworms in mound microsites die from a combination of low moisture and high temperature (Nachtergale et al. 2002). Thus, earthworm populations are typically restricted to mound microsites. The earthworm's hygrophilous nature is linked to the high assembly of epigeic biomass within the hole. Since some earthworms are semi-aquatic, they prefer damp environments with deep water. The rise in epigeic biomass in pits is a result of more tree litter being collected within this microsite, according to Nachtergale et al. (2002). Pits in the forest floor create a unique condition due to the increased litter thickness and water content, resulting in an increase in epigeic biomass. Therefore, the earthworm's presence in a pit can have had an indirect but positive impact on the nitrogen mineralization rate in the soil, by breaking down animal and plant residues and speeding up the mineralization process (Neirynck et al. 2000; Deleporte 2001). The abundance and biomass of earthworms decreased from closed canopy (undisturbed) to canopy gap, pit, and mound (disturbed) sites (Kooch and Hosseini 2010), and no earthworms were observed in mound microsites. Soil factors such as soil moisture, nitrogen, and available calcium support the dispersal of earthworms at the pit microsite (Kooch 2012b). Earthworms consume and digest a large amount of microfauna as part of their activities in the soil, which increases microbial activity. As a result of this process, soil microbial respiration improves, resulting in an increase in earthworm populations (Mosavi and Raisi 2009). Earthworms increase CO_2 by 33%, which means they can play a critical role in governing the global greenhouse gas balance of soils, and their effect is projected to grow in the coming decades (Lubbers et al. 2013). Research reviews (e.g., Kooch and Hosseini 2010; Kooch 2012b; Kooch et al. 2015; Thouvenot et al. 2021) demonstrate that earthworms can be used as a bioindicator to assess the changes in forest stands following disturbing events. The earthworms' activity and quantity were both reduced as a result of the windthrow, and findings showed that windthrow must be viewed as a significant component in forest ecology-related biodiversity. This is important for evaluating the effects of forest management policies and practices on soil, as well as using soils as forest ecosystem indicators.

Gilgai (as a Micro-landform) Gilgai is a series of depressions and mounds generated by shrink-swelling and cracking vertisols (clay soils); water is collected periodically in the depressions to make gilgai wetlands (see Fig. 6). This landform was historically abundant in Queensland's Brigalow Belt (Australia). Gilgai is an Aboriginal word that means "small waterhole" in the Kamilaroi, Wiradhuri, and kindred languages. Crabhole, melon hole, and gilgai were among the terms used by early Australian immigrants to describe these tiny relief characteristics. During periodic wetting and drying cycles, the clayey soil layers contract and expand,

Fig. 6 Gilgai landform in plain forest of northern Iran in the winter (**a**) and summer (**b**) seasons (photo by Y. Kooch)

resulting in gilgai microrelief. As a result, "blocks" of subterranean material are gradually pushed upward, forming mounds. Gilgai grows on black and grey vertisols, as well as brown and red vertisols to a lesser extent. Alluvial clayey sediments, sedimentary rocks, like mudstones, shales, and impure limestones, and basic igneous rocks (alkaline), specifically basalt, are the most prevalent soil parent materials of vertisols. The gilgai depressions fill with water during and after the rainy periods, and the terrain becomes scattered with a variety of shallow wetlands (Kabala et al. 2015).

Soil aggregation, soil organic matter breakdown, biota activities, and nutrient cycling all benefit from drying and rewetting (Golovanova et al. 2021). Organic substrates formed from organisms destroyed by the drying process, as well as other non-living soil organic materials, are the sources used in the degradation process following drying and rewetting. After soil drying and rewetting, the non-biomass soil organic carbon contribution to the mineralization flush is largely determined by the location of organisms in the soil structure relative to their substrates. The physical and chemical features of soil are predicted to be affected by air-drying and rewetting processes (Kaiser et al. 2015; Hoeffner et al. 2021); however, systematic assessments of the effects on biological traits such as earthworm populations are still missing. When the soil pores are completely or nearly completely filled with water, oxygen, the primary oxidizing agent, becomes scarce (Sajedi 2010). Since soil moisture has a major influence on the rates of biogeochemical processes in all terrestrial ecosystems, responses to moisture pulses caused by dynamic precipitation patterns are particularly complicated and difficult to predict (Evans and Wallenstein 2012). Soil rewetting and thawing have similar impacts on organisms, because both processes enhance water availability, boost microbial metabolism, rehydrate cells, and activate nutrients in soil (Kim et al. 2012). Sajedi (2010) found that excessive wetness resulted in poorer soil fauna such as quantity and biomass, as well as a higher frequency of anaerobic situations. Kooch (2012b) investigated the impact of seasons and excessive precipitation on soil earthworm activity in Iran's coastal woods and noted that throughout the summer, earthworm activity was substantially higher than during the winter. He also suggested that drainage, as a valuable silvicultural method, will promote the activities of soil earthworms and the related biological processes (such as carbon and nitrogen cycles) in the gilgai landscape.

Single Tree Zinke (1962) was the first to propose the concept of "single-tree influence circles." Generally, there are two basic aspects in the concept: first, within the circle of influence of each tree, soil properties vary predictably in relation to the distance from the bole and edge of the crown and second, at any location relative to the bole and crown, the degree of development of soil properties varies among tree species. Juxtaposition of individuals of contrasting species may thus result in abrupt changes in soil properties across the boundaries separating the influence zones (Boettcher and Kalisz 1990). In investigations of soil properties including pH and exchangeable cations, the effect of single trees has been well documented (e.g., Amiotti et al. 2000; Graham et al. 2004). Moreover, there has been research related to moisture loss in soil (Ziemer 1968), interception of rainfall (David et al. 2006), and abundance of earthworms (Boettcher and Kalisz 1990). Nutrient cycle and biomass distribution, mineral soil and forest floor can be influenced by each tree species (Kooch et al. 2021). Positive biogeochemical feedbacks are created by individual species (Richon and Tagliabue 2021). "Fertility islands," "regeneration microsites," "single-tree influence circles," and "theory of pipe model" have all been used to describe the influence of single trees on soils (Boettcher and Kalisz 1990). The soil attains qualities that vary geographically in response to the placement of the trees when it is influenced by a forest. The distribution of distinct ground flora species typically reflects this diversity in soil conditions. Each tree acts as a center of influence in the improvement or degradation of the forest soil (Scahrenbroch and Bockheim 2007). In the Hyrcanian forest (Alborz Mountain), the viability of the idea of "single-tree influence circles" was investigated in a forest area dominated by *Fagus orientalis* Lipsky (beech) and *Carpinus betulus* L. (hornbeam) on sheer slopes (Kooch et al. 2011). Also, three single trees (canopy cover of trees separated from others and covering distinctive area) were used as three replications for soil samples from each tree species and diameter class (45–55, 55–65, 65–75, 75–85, 85–95, and 95–105 cm). In terms of earthworm number and biomass, there was a considerable variation. The diameter classes of beech (65–75 cm; $P < 0.05$) and hornbeam (75–85 cm; $P < 0.01$) both presented higher values. When compared to beech single trees, the hornbeam site was better suitable for the assemblage of earthworms. The direct effects of tree species on soil biota and litter are induced by the organic matter inputs from above- and below-ground biomass of plants, while indirect effects include shading, soil conservation, and nutrient and water uptake through roots (Neher 1999). Generally speaking, pH in soil was found to be the most important determinant of earthworm population (Neirynck et al. 2000). A number of research studies (Wardle 2002; Hoeffner et al. 2021) reveal that plant chemical composition, particularly nitrogen, lignin concentration, and phenolic compounds, influences the abundance of soil and litter fauna by affecting edibility and decomposability. Soil organic content increased from acid loam to limestone in the mineral topsoil, which was connected to the increased activity of earthworms (Pulleman et al. 2005), particularly the shade of the hornbeam. Previous research has revealed that earthworms prefer some plant species residues above others (Lassauce et al. 2011; Kooch et al. 2018, 2023a). Furthermore, earthworm communities may have patchy distributions in situations where tree species producing litter of varying quality coexist. Since earthworms preferentially consume hornbeam litter, they may have decayed quickly (Cammeraat 2002; Aubert et al. 2003). Beech litter vanished after 1 year in monospecific stands on calcareous soil, where earthworm communities had no alternative (Kooijman et al. 2008). Earthworm biomass was shown to be substantially connected to canopy cover, which can show that earthworm aggregation is most likely to occur in shaded situations, owing to the improved microclimate generated by the tree shade (Sarlo 2006). Diverse tree species have different effects on the soil fauna abundance and composition, notably in forest floors. Soil structure is influenced by the composition of the overstory. Soil stability and structure differed by species of trees, most likely due to differences in earthworm activities (Graham et al. 1995). Kooch (2012b) indicated that single trees with distinct classes of diameter may be found in forest floors and soils with higher mineral layers, even in combined stands in steeply sloping terrains. The size of the changes is affected by the understory vegetation along with forest stand, as well as the processes controlled by topography and climate, including soil creep and redistribution of litter. Nevertheless, the soil landscape can be considered as a mosaic, with individual pedon qualities that make up the mosaic reflecting the present and biochemical and biological features of tree species. In some circumstances, the variation in the quality of substrate under the crowns of juxtaposed species might be big enough to cause temporary changes in plant development and soil properties. However, forest management has a big influence on the effect of species efficiency and effectiveness (e.g., mixed stands or low-density stands might accelerate the decomposition rate of litter), and more research must be done on this topic.

Deadwood Various ages of trees that are characterized by a large amount of deadwood in various stages of decay make up natural forests (see Fig. 7), along with the presence of old trees (Lassauce et al. 2011).

Deadwood decomposes over time and is incorporated into surface soil horizons, where it leads to alterations in the topsoil's physical, chemical, and biological characteristics (Moghimian et al. 2020). In temperate deciduous woods, however, just a few studies have been carried out (Brais and Drouin 2012). Similar to the controlled-release chemical fertilizers, nutrients, carbon, nitrogen, phosphorus, etc. can be released by coarse woody debris (CWD) during the decomposition. As a result, CWD can improve the fertility and productivity of surface forest soils (Zhou et al. 2007). As a result, it's critical to figure out how soil function and fertility are affected by wood decay, as well as to assess its impact in routinely managed woodlands. It's unclear if populations of earthworms are primarily influenced by the availability and quality of food, or their surroundings' chemical qualities (Aubert et al. 2003; Gonzalez et al. 2003). As a result, assessing the relationship between earthworm biomass and variety, the perturbations of the felled trees (dead trees), and the edaphic conditions are critical for forest ecosystem management. The decay degree of felled trees can be categorized into four classes, according to Marvi Mohadjer (2007): bork sloughing, generally fine longitudinal shakes in the wood (DC1); twigs

Fig. 7 Stages of the decay of deadwoods (Kooch 2012a)

sloping (DC2); the decay is progressed, longitudinal shakes spreading to furrows (DC3); the wood is totally deteriorated, log collapsing, wood friable, crown completely disintegrated (DC4). In northern Iran, research looked at the effects of felled beech and hornbeam trees on soil features and earthworm ecological groups. According to data analysis, the highest assembly of earthworms was found around dead trees with a high decay degree (DC4), primarily hornbeam dead trees. The activities of distinct ecological groups of earthworms surrounding the hornbeam dead trees with DC4 were promoted by higher soil moisture, pH, nitrogen, phosphorous, potassium, calcium, and magnesium, as well as lower soil carbon and the ratio of carbon to nitrogen (Kooch et al. 2008; Kooch 2012a). In the same way, Moghimian et al. (2020) found that felled logs boost soil characteristics and earthworm activity in old-growth temperate forests. Overall, it is worth noting that felled or dead trees can provide a unique habitat for a variety of living beings and play a key role in the biogeochemical cycles (Santiago and Roclewald 2005). Identifying the ecological consequences of dead trees on soil biota can provide a more accurate understanding of forest soil ecology that can be applied in managed forests.

Canopy Gap Wind-created gaps are a common disturbance event in temperate deciduous woods. The extent of the gaps varies widely, from a single crown to enormous open fields with diameters spanning many tree lengths (see Fig. 8).

The effects of canopy gaps on below-ground soil qualities, particularly on soil fauna such as earthworms, have been addressed in a few studies. The effects of small, medium, large, and very large gaps, respectively (85.12, 325.21, 512.11, and 723.85 m^2) on soil earthworm activity in the north of Iran in a mixed beech–hornbeam stand have been investigated (Kooch et al. 2010). According to the statistics, earthworm density and biomass decreased from a small gap to a very large gap. They claimed that the most essential effective components in small gaps could be higher soil moisture and lower temperature. The amount of soil moisture in the soil decreases as the canopy gap widens, yet the temperature of the soil rises (Scahrenbroch and Bockheim 2007). Due to the reduction in moisture and the increase in soil temperature, earthworm population and biomass decrease as the open areas in the canopy cover increase. Forest gaps have an irregular impact on the availability of materials and micro region resources, as well as the soil and the site. The presence of the aforementioned characteristics varies with time and location. It is obvious that using modest intervals to maintain the balance between the food

Fig. 8 Canopy gaps with different sizes in forest ecosystems (Kooch 2012b)

cycle and the climatic parameters of temperature, moisture, and light transmittance is a good guideline, especially in temperate habitats.

Fire Fire is an important disturbance agent in many terrestrial ecosystems that can be effective on soil characteristics along with population of earthworms. Commonly, the direct impacts of fire on soil-borne invertebrates are less evident than those on microorganisms, owing to the invertebrates' better ability to move, which allows them to avoid heat by burrowing deeper into the soil. Nevertheless, the indirect influences, notably a reduction in litter mass, can significantly reduce both the number of soil-borne invertebrate's species and the total mass. There is not much research on the effects on earthworms and is very scarce (Callaham et al. 2003; Dawes-Gromadzki 2007). Collett et al. (1993) indicated that planned fires with low severity generate a significant reduction in earthworm populations in Eucalyptus trees, and several years are required for their recovery to control levels. By the way, the time required to recover is unpredictably long, since it depends entirely hinges on water content in the soil after the burning period. Bhadauria et al. (2000) found that the earthworms' populations of *Eutyphoeus nanianus* and *Octolasion tyrtaeum* were significantly decreased by a wildfire in an oak–pine mixed woodland, but multiplied within a couple of months, possibly thanks to the better availability of nutrients and moisture in the soil that favored individuals from neighboring unburned regions. The fire effects on earthworms in deciduous forests are even more contradictory in the literature. In wooded ecosystems, fire has traditionally been utilized to manage fuel loads and understory growth (Keeley et al. 2006). Fire is also regarded as a viable strategy for soil restoration and nutrient recycling restoration. By either direct or indirect mortality, fire could play a vital role in limiting the spread of invasive earthworm species. Earthworms that reside and travel through the upper organic layers of soil are epigeic (litter dwelling) (Bossyut et al. 2006). As a consequence, removing litter or feeding materials could be a technique for mitigating the negative effects of fire on earthworms. Increased pH, nitrogen loss through combustion, and micronutrient release are just a few of the numerous changes that fire induces in soils (Mabuhay et al. 2003). All of these factors could have an indirect impact on earthworm survival, and it is unclear what reaction to fire, if any, earthworms will have in an Eastern deciduous forest.

3.3 Urban Soils

Earthworms in urban areas have been studied in some places (e.g., Ferreira et al. 2018; Xie et al. 2018); however, the majority of research was conducted in a single town and focused on just one or a few distinct environments, including vestige forests (Szlavecz et al. 2006), central strips (Tiho and Josens 2007), domestic yards (Smetak et al. 2007), parks (Schlaghamerský and Pižl 2009), and downtown gardens (Tresch et al. 2018). Almost no arthropod taxa such as springtails have been studied yet in multicity studies of ground beetles (Carabidae; Niemelä and Kotze

2009), millipedes (Diplopoda; Vilisics et al. 2012), woodlice (Isopoda; Szlavecz et al. 2018), and soil invertebrates (Collembola; Joimel et al. 2019). Although attempts are underway to map the diversity and quantity of earthworms on a continental scale (Rutgers et al. 2016), owing to the variability of ecosystems inside the urban lands these areas are typically excluded from such large-scale assessments. In metropolitan locations, poor soil conditions for earthworm survival are widespread. Urban soils are compacted, have a variety of textures and parent materials, are frequently alkaline, and are poorly aggregated. Furthermore, particularly in car park islands and roadway cutouts of trees, canopy covers are typically not adequately linked to shelter species that are surface-dwelling from intense light exposition. The bottoms of these islands and cutouts are often heavily compacted and separated from soil larger bodies, potentially impeding the migration of earthworms. Additionally, earthworms are known to be consumers and retainers of heavy metals and other contaminants that are thus transferred throughout the ecosystem (Kennette et al. 2002; Lukkari et al. 2004). Lastly, since people commonly remove leaves and other yard residues annually, these systems tend to contain varying levels of leaf litter and organic matter (Scheyer and Hipple 2005) and, eventually, the food sources for earthworms will be removed.

The dynamic of communities and populations of earthworms in urban settings, together with using earthworms for urban soils remediation, have received little attention in the literature. Cities are rapidly expanding, at the expense of forests and agricultural lands. As previously stated, urban soil functioning is frequently hindered, so that remediation is required to restore soil health and vegetation maintenance. Considering the comprehensive research on earthworms in agricultural and natural settings, as well as their widespread use in farming management, quite similar approaches may be used in the urban context. Likewise, studies of urban earthworms compare population of earthworms through a gradient from urban to rural soils (Szlavecz et al. 2006) or through an urbanization gradient in the urban region (Smetak et al. 2007). Various methods have been used to determine the urbanization gradient: location in relation to the city center (Steinberg et al. 1997), development type and distance around the stands (Szlavecz et al. 2006), or age of the site (Smetak et al. 2007). Szlavecz et al. (2006) found that most urban sites along the gradient had a higher quantity of earthworms and biomass. Earthworm density was higher in metropolitan regions, since the newly discovered foreign species possibly came from harbors in New York City (Steinberg et al. 1997). Contrarily, a reverse pattern was indicated by Smetak et al. (2007), with higher amounts and biomass of earthworms in lands located further away from the city center. This finding supports the assumption that urban soils are associated to a decrease in the organism's numbers and biomass, along with a decrease in diversity of species (Harris 1991).

4 Earthworm and Soil Properties

In various environments, numerous factors have been linked to the dynamics of earthworm communities. Since earthworms, due to their sensitive skin and high water content, are very sensitive to alternating environmental conditions, it is commonly recognized that earthworm communities are affected by soil texture, humidity, organic substances, pH, and sources of nutrition. The effect of a specific environmental condition can be evaluated to some extent; however, the synergistic impacts of changing environmental conditions are more difficult to be observed. Earthworms are influenced by soil pH, and it is generally accepted that earthworms tolerate a wide range of pH levels; however, every species has its own threshold for tolerating acidity and alkalinity. Earthworms' community diversity, number, and dispersion are constrained by these threshold levels. While the majority of earthworms prefer balanced pH levels, few, like *Lumbricus rubellus*, can withstand more acid levels. There is an important connection between soil moisture and all the organisms living in the soil. Since earthworms have ideal soil moisture levels, soil moisture content is a crucial factor. Earthworms have a slim cuticle covering the mucus-secreting surface epithelium, which allows the loss of water from the body. Based on Edwards and Bohlen (1996), besides organic matter, earthworms use the microbial biomass of the organic matter as a food source. Earthworms, like all other living beings, benefit from abundant food sources. Since earthworms consume organic matters, fungi and bacteria are able to further break down organic matters by creating an additional surface for decomposition (Fazlollahi Mohammadi 2016). Organic and minerals components are needed in the process of soil formation as well as in microbial activity. Therefore, there might be a chance to accelerate soil formation on reclaimed lands by adding organic matters and introducing soil organisms (Ferreira et al. 2018). Earthworms play a crucial role in soil formation by consuming, comminuting, and redistributing organic matter in the soil profile. As humic organic matters break down and stabilize, earthworms promote microbial activities (Evans and Wallenstein 2012). Hale and Host (2005) used earthworm activities to improve the soil maturation process in the Netherlands reclamation polders. Earthworm communities are established substantially more quickly in polders that had been converted to grasslands and farmlands, possibly because of the pesticides used in the gardens (Hodson et al. 2021). According to Keeley et al. (2006), in contrast to the Mor soils in adjoining polders with no earthworms, polders with earthworms formed mull soils. The beneficial impacts of earthworms' activities on the formation of polder soil include an improved aeration and aggregate formation, and an increased incorporation of leaf litter (Kennette et al. 2002).

Earthworm communities could modify soil carbon and phosphorus contents, alter C/N ratio, consolidate and combine soil horizons, and reduce fine roots biomass (Kooch 2012b). Earthworms' invasions eventually cause the movement of superficial organic matters into several centimeters of the mineral topsoil (Kooch et al. 2011). By creating burrows and middens, earthworms could change soil structure and enhance microhabitat heterogeneity. Invasive earthworms have the potential to significantly alter the occurrence and depth of soil horizons in the Great Lakes area. According to Lassauce et al. (2011), the moder humus in Minnesota hardwood forests soils is transformed into mull humus when burrowing earthworms are present. Moder soils have an upper surface of organic matter, which, when earthworms are absent, is being slowly degraded and integrated into the deeper mineral soil. In moder humus soils, the O horizon is unstructured toward the bottom and fibrous close to the horizon surface. Based on Lassauce et al. (2011), invasive burrowing earthworms are present if earthworms are found in mull soils, which differ from moder humus soils. Earthworms are the main organic material decomposers if earthworms are found in mull woodlands. Such soils have a surface "horizon A" that is dark, rich in organic materials, and composed of a combination of mineral soil, humus, and leaf litter, mixed by borrowings earthworms. Plant populations may be significantly impacted by the changes in structure of soil and decomposition rates caused by soils turned into mull humus. An increase in the horizon A and a decrease in the horizon O thickness are associated with invasive earthworms (Niemelä and Kotze 2009). Uninvaded hardwood soils and soils with only *Dendrobaena octaedra* (epigeic) earthworms present in the Chippewa Forest National Park (USA), according to Hale and Host (2005) were characterized by a dense organic horizon O and a light brown horizon A. In extensively invaded regions that included *Aporrectodea* spp., *Lumbricus terrestris*, and *Lumbricus rubellus*, an A horizon, black and thick, was present, but yet no organic O horizon was formed. An organic horizon of a moderate biomass and average thickness of *Octolasion tyrtaeum*, *Lumbricus rubellus*, and *Aporrectodea* species were present near the front edges of the earthworms' invasion. In the Chippewa Forest National Park, with sugar maple-basswood woodlands, Joimel et al. (2019) investigated the soil horizons A and O. Two layers were identified inside the O horizon: the O_1 horizon, which included recognizable leaf litter, and the O_2 horizon, which contained organic material that had been decomposed beyond identification. The A horizon is the dark layer of mineral-organic soil that lies underneath the O horizon and is defined as having a black color. The thickness of the A horizon was not significantly correlated with the occurrence of the epigeic earthworm *Dendrobaena octaedra*, as determined by Law and Lai (2021). Furthermore, the epi endogeic earthworm species *Lumbricus rubellus* caused the O_1 and O_2 horizons to become much thinner and the A horizon to become considerably thicker. Even though earthworms are commonly known as lowering soil bulk density via the burrowing process, soils of the uninvaded woodlands in the Great Lakes region tend to have a lower bulk density because of the appearance of a thick, porous O horizon and indigenous below-ground invertebrates including beetles, spiders, and centipedes that encourage soil aeration and bioturbation. Because of a reduced thickness of the O horizon, less native below-ground invertebrates, and soil consumption and intestine compression by earthworms, the soil bulk density has usually risen in forest areas of the region's Great Lakes where earthworms have invaded (Law and Lai 2021). By consuming soil that is transferred through their digestive system, physically and chemically changing the soil, and excreting castings, invasive earthworms may enhance bulk density in hardwood forests. In deciduous woodlands, earthworm castings are thick and usually contains more micro and macro pores than soils not disturbed by without earthworms (Lukkari et al. 2004). Whenever the bulk density of the soil increases, less water is able to penetrate into the forest floor, and there is also more surface erosion.

Significant alterations in the soil's nutrient cycle may be caused by invasive earthworms. Based on Mabuhay et al. (2003), greater levels of carbon and nitrogen have been detected in the mineral soils of Ottawa National Forest in the Great Lakes region (Michigan, USA) hardwoods, with a larger abundance of invasive earthworm communities. Carbon pools in soil could even be influenced by invasive earthworms (Nachtergale et al. 2002). Soil organic matters and C/N ratios were dramatically diminished by the presence of *Lumbricus rubellus* and *Lumbricus terrestris*, according to study by Phillips et al. (2008). Furthermore, invasive earthworms reduced northern hardwood woodlands soil carbon store by about 37% in the top 20 cm. Earthworms devour leaf litter, move debris throughout their intestinal tracts, and excrete waste as casts, and plants may easily absorb the easily available nutrients in earthworms' casts. Furthermore, according to Santiago and Roclewald (2005), invasive earthworms may induce a fast outflow of soil nutrients, including carbon, nitrogen, magnesium, calcium, potassium, and phosphorus, which is followed by depletion and leaching of nutrients. Consequently, woodlands with invasive earthworms have a faster nutrients cycle than non-invaded Great Lakes woodlands. The impact of invasive earthworms on net C sequestration has been the subject of much discussion. Several short-term investigations indicate that earthworms enhance CO_2 emissions (Scheyer and Hipple 2005), although some research suggests that soil aggregates are better stabilized when earthworms are present (Schlaghamerský and Pižl 2009; Thouvenot et al. 2021). More research seems to endorse the idea that invasive earthworms boost carbon emission, and a recent Meta analyses revealed that invasive earthworms might raise initial CO_2 emissions by about 33% (Tiho and Josens 2007). Nevertheless, according to Tresch et al. (2018), much research tends to concentrate on the impacts of earthworms on carbon dioxide generation, since its emission is an easier and more prevalent measurement than carbon stabilization. To evaluate the net impacts of invasive earthworms on terrestrial carbon storages, Zhang created a technique, known as the sequestration quotient, that measures both carbon stabilization and carbon mineralization. Apparently, earthworms are responsible for a minimum 25% net gain in early CO_2 emissions, as determined by a reexamination of both carbon stabilization and mineralization by Vilisics et al. (2012). Based upon Zhou et al. (2007), regardless of whether carbon emissions and stabilization are taken into account, it is challenging to estimate earthworms' impacts on carbon since earthworms may have indirect impacts on other organisms that affect carbon, such as changes in plants growth.

Soils in hardwood forests with earthworms would have a higher pH than those in hardwood forests without earthworms (Tresch et al. 2018). This was the result of earthworms that removed the litter layer and combined base-rich mineral soils from deeper depths with topsoils, as suggested by Tresch et al. (2018). The presence of earthworms has also been connected to higher soil pH, as discovered by Kumar et al. (2021); however, this could be due to earthworms' affinity to soils rich in calcium. A key component of soil's buffering ability, calcium is also an essential nutrient. It is possible that forest soils might lose their ability to function as a buffer as a consequence of the fast release and consequent calcium, potassium, and magnesium nutritional depletion mediated by earthworms (Lassauce et al. 2011). The hydrological cycle in forested habitats may be affected by earthworms due to their impact on the soil's physical properties. Hydrological cycles are affected as soil bulk density increases and the O horizon decreases, leading to more run-off and less infiltration. Due to these factors, xeric site conditions might become more common in forested regions that are subsequently occupied, potentially leading to a predominance change from mesic to dry-mesic plants communities (Kaiser et al. 2015). Earthworms have the ability to improve the variety of their environment, by causing changes in the soil's physical structure. This, in turn, may have a variety of consequences on the native fauna of the soil as well as the living creatures that live above ground. There is some evidence that earthworms enhance microhabitat complication in the short term, however, reduce it in the long term, according to a review by Graham et al. (2004). They also anticipate that invading earthworms would firstly boost faunal variety but will reduce it in the long term. However, the result in the long term would be somewhat dependent on the number of earthworms and the ecological groups that are already there. It appears that earthworms have mixed effects on micro-arthropods in the short term (both positively and negatively). However, according to Evans and Wallenstein (2012), in the long term, invasive earthworms are shown to be largely detrimental to micro-arthropods, possibly due to enhanced mechanical disturbances and decreased habitat complications. An extensive literature review conducted by David et al. (2006) indicated that endogeic earthworms had mostly negative impacts on soil micro-arthropods due to their food competition with them. On the other hand, the influence of anecic earthworms on soil micro-arthropods was indifferent at the habitat level and ideal at the microhabitat level. The review revealed only four papers that examined the impacts of epigeic species on soil micro-arthropods and indicated that the consequences of this ecological group deserved more research. Whereas enchytraeids and earthworms have similar functions, enchytraeids are classified under the Clittelata class of annelids, though earthworms are Oligochaeta. In contrast to their terrestrial counterparts, enchytraeids may be found across the Great Lakes region. According to a survey of hardwood forests in the Chippewa and Chequamegon Nicolet National Forests (USA) by Brais and Drouin (2012), invading European earthworms changed Enchytraeid community structure and had a detrimental impact on the richness of Enchytraeid species. According to the findings of a research conducted in the state of New York's hardwood forests by Bossyut et al. (2006), the presence of invading earthworms caused a decrease in the number of deer ticks (*Ixodes scapularis*), most probably by limiting the amount of appropriate habitat available on the forest floors. The research indicated that invading earthworms decreased the number of nymph *Ixodes scapularis* ticks by 46% and the number of larvae by 29%. Even though invading earthworms might have a detrimental impact on deer tick numbers, other variables, such as climate change and rising deer populations, might lead to future increases in tick populations.

The burrows operated by earthworms come from a variety of ecological communities, and burrows are placed at a variety of depths and orientations inside the soil profile. This improves water infiltration and drainage, in addition to water retention (Abail and Whalen 2021). While greater water retention and infiltration are typically positive soil qualities, they might also be detrimental in rare circumstances if the site is not adequately managed. Botticelli et al. (2020), for instance, found that inoculating earthworms into a rehabilitated mining area enhanced soil porosity in the top 20 cm. Nevertheless, this resulted into higher water logging in the soil, and artificial draining did not allow to adequately remove the extra water from the soil profile as a consequence of the greater porosity. Similarly, Butt and Quigg (2021) noticed a decrease in surface-bearing strength due to higher water infiltration in earthworm burrows, especially during the colonization stage of earthworms' settlement (Cammeraat 2002). Under flooded situations, stepping damages (e.g., animal trampling) have the ability to cancel the other soil structural advantages provided by earthworms' activities (David et al. 2006). Additionally, it has been demonstrated by Golovanova et al. (2021) that a tropical species of earthworms known as *Millsonia anomala* may enhance the level of drought stress experienced by paddy plants (*Oryza sativa*). A major environmental element that influences earthworms' activities and reproductive capacities is soil water content (Fazlollahi Mohammadi 2016). Under field conditions, the majority of earthworms are productive at water tension around the field capacity level (–10 kPa), yet their activities decrease over (–100 kPa) (Fazlollahi Mohammadi 2016). According to the majority of experimental studies, for a variety of tropical Lumbricid species, a soil moisture level of 25–30% is optimal (Evans and Wallenstein 2012). Numerous species have behavioral adaptations to resist droughts, such as *A. longa*, which attempts to travel to deeper and more humid soil layers and, if unsuccessful, enters a dormancy phase (diapause) (Gómez-Plaza et al. 2001). During the diapause earthworms prevent the loss of moisture by rolling into knots inside a mucus lined chamber in the soil, as long as environmental circumstances change. Graham et al. (2004) found that endogeic species such as *Allolobophora chlorotica* and *Aporrectodea caliginosa* exhibited a quite identical behavior. According to Hodson et al. (2021), it is possible that younger earthworms might not exhibit these behaviors, which indicates that they might be more susceptible to the effects of droughts. In an ideal soil condition, 1 square meter of land will have about one liter of soil held inside the intestine of earthworms' populations. Each year, earthworms will affect over 4–10% of the land's entire surface area (Gonzalez et al. 2003). This suggests that the egested materials definitely constitute the principal component of soil, and that after 10 years, almost half of the soil will have passed thru an earthworm, and that over 40 years, around 90% of the soil will be carried thru an earthworm (Golovanova et al. 2021). The effect that earthworms have on the ecology of the soil is significant at a worldwide scale, and these organisms play an essential role in the microbiological structure and physical–chemical qualities, which terrestrial animals and plants depend on. The burrowing of earthworms enhances drainage and aeration and water permeability, and they recycle organic materials by consuming leaf debris and coating their tunnels with nutrient-rich mucus (Deleporte 2001). The environment of the digestive tract is quite different from that of the surrounding soils; for example, the contents are subjected to anoxia, and the pH is neutralized (Hodson et al. 2021). Huge communities of earthworms occupy productive agricultural areas, and their interactions with sprayed herbicides are a continual and evolving process. This has enabled for the investigation of the earthworms' impact on soil microorganisms under such anthropogenic pressures. Nonetheless, it is sometimes impossible to recognize the host functioning from the roles played by bacteria inside and around earthworms. Based on Hoeffner et al. (2021), it has been widely accepted that earthworms impact the soil nitrification capability, and more recently, earthworms have also been considered as the source of 60% of worldwide N_2O emissions. In addition, their presence boosts the amount of nitrate in the soil and the amount of nitrate that plants can absorb owing to the enhanced bacterial activity. The fact that they are effective as a source for anoxic denitrifiers, on the other hand, points out the challenge of disentangling the hosts from the communities (Joimel et al. 2019).

In the scientific literature, research findings addressing the influence of earthworms' species on the soil micro-biome are diverse, and frequently conflicting due to the great variety of soil conditions and species of earthworms. In the topsoils of hardwood forests, invading earthworms enhanced the overall microbial biomass of the surrounding ecosystem (Kabala et al. 2015). Nevertheless, *A. caliginosa* and *Lumbricus terrestris* decreased the number of bacteria recognizable from leaf litter and soil while consumed, especially in the gizzard/crop organs, although the quantity of *Cytophaga flavobacterium* (Bacteroidetes) increased in the cast materials and hind intestine system (Kaiser et al. 2015). Earthworms have been demonstrated to alter the natural microbial population, resulting in higher oxidation of methane (CH_4) at landfills, which are a significant source of human-induced CH_4 (Hodson et al. 2021). Notwithstanding the varying amounts of organic matter on which the hosts fed, the earthworm species *Eudrilus eugeniae* displayed significant amounts of methane generation, which is hypothesized to be produced by related Methanobacteriaceae and Methanosarcinaceae (Gonzalez et al. 2003). This impact was not observed in other species, including Lumbricids, which may be the result of different redox potentials inside their nutritive canals. On the other hand, based upon Evans and Wallenstein (2012), earthworms have long been considered a stimulus to study the concept of generalized antimicrobial functionality. Liquid isolations from entire earthworms indicate antifungal and antibacterial characteristics that influence the surrounding soils, which pass into the digestive tract (Ferreira et al. 2018). In particular, the coelomic liquid of *Eisenia fetida* has considerable antibacterial impacts (Gómez-Plaza et al. 2001). Vermicomposting (preserving a community of earthworms, typically *E. fetida*, to transform organic wastes to fertilizers) is a significant human use of earthworms owing to the increases in polysaccharides and other nutrients, which encourage the growth of plants (Graham et al. 2004). In addition to the generalized antibacterial impacts discussed above, vermicomposts display antifungal capabilities (Hodson et al. 2021). These antifungal qualities are mediated by *E. fetida's* promotion of bacterial communities that show chitinase activities in vermicomposts.

Earthworms' species almost always host a specific *Acidovorax*-like extracellular symbiotic species in the ampulla of the osmoregulatory nephridia. This bacterium, known as *Verminephrobacter*, has been reported by Keeley et al. (2006). The nephridial symbiont is recognized as an essential component of earthworm physiology since it is present in the majority of earthworm species (Kooch et al. 2020a). According to Kooch et al. (2011), it is a vertically transmitting symbiont that has differentiated for a long time with the host during the evolutionary history (estimates presume that *Verminephrobacter* has been linked to Lumbricidae because the earthworm's family formed 62–132 million years ago) and exhibits the fast rate of mutation typical of other symbionts. Based on its historical connection, numerous hypothetical functions for Verminephrobacter have been proposed (Kooijman et al. 2008). A primary function in protein and nitrogen restoration was initially postulated due to anatomical position and nephridial functionality (Lukkari et al. 2004). However, this has been questioned because of the lack of extracellular proteases in the genome of *Verminephrobacter eisenia* and analyzation of aposymbiotic-reared individuals (Nachtergale et al. 2002). Based on Kumar et al. (2021), the lack of the symbiont is deleterious to host health and fertility in poor nutritional environments for the earthworm species *Aporrectodea tuberculata*, emphasizing the fundamental role that it should fulfill for the host. In their intestines, earthworms establish a micro-habitat for transitory soil microorganisms (Fig. 9). The initial release of calcium carbonate from the calciferous glands in the earthworms' anterior section, which quickly lowers the acidity to a more neutral level, and the anoxia imposition throughout the intestinal lumen, which creates a special reservoir from the outer soil, are two of the most prominent features of earthworms (Mabuhay et al. 2003). Furthermore, the carbon and nitrogen compounds secreted into the hind intestine lumen supply diverse soil-derived energy sources (Neirynck et al. 2000; Zhong et al. 2023). Studies that had been conducted in the past on the microbiota linked with earthworms relied heavily on culture and molecular methods, and there has not been any research using the High Throughput Sequencing (HTS) techniques that are currently available and widespread. The results of the earliest investigations revealed that the number of microorganisms increased by approximately 1000-fold and produced a rise in the concentration of soluble organic carbon while passing thru the digestive tract of an earthworm (Phillips et al. 2008). Rutgers et al. (2016) hypothesized that earthworm microbiome would simply reflect the digested plant and soil microbiota, yet scanning electron microscopy detected the existence of bacterial species in *L. terrestris*' hind intestine (Sarlo 2006). Scheyer and Hipple (2005) determined that Pseudomonas and Firmicute species become more prevalent throughout the *L. rubellus* digestive tract, but more recently, the examination of *L. terrestris* using the T-RFLP method has revealed very identical microbial profiles within the different compartments (soil, transient intestinal contents, and casts), suggesting the existence of a microbiome derived from soil (Szlavecz et al. 2018).

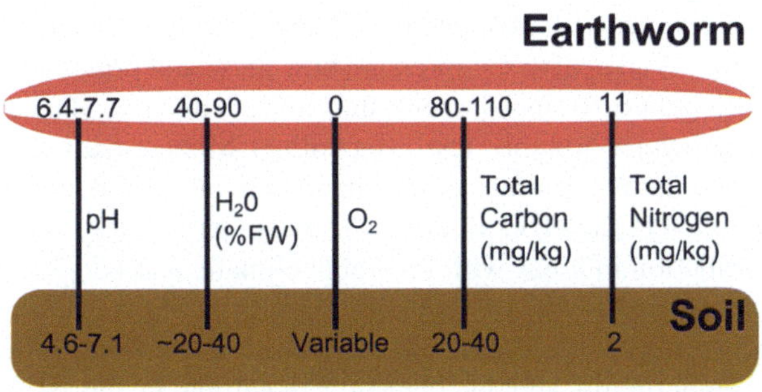

Fig. 9 Different conditions exist between the microhabitat of the earthworm intestine and the surrounding soils; FW refers fresh weight. Based on study by Drake and Horn (2007)

The most of the earthworm's microbial activity is probably due to the transitory community being "stimulated" by the peculiar habitat it experiences during movement, when the intestine functions as a "fermenter" (Tiho and Josens 2007). This resulted in the formation of nitrogenous gas and stimulated the growth of Enterobacteriaceae and Clostridia populations through the metabolism of plant and mucus-derived saccharides. In the castings of *A. caliginosa* and *L. terrestris*, several taxonomy groupings have been detected at high concentrations, most importantly Bacteroidetes species, which may play a role in the breakdown of organic materials (Szlavecz et al. 2018). Schlaghamerský and Pižl (2009) reported the presence of an Actinobacterium, Proteobacteria, and Firmicutes species on the intestine walls of various species of earthworms. The presence of bacteria that are not found in the soil around *A. caliginosa*, *E. fetida*, and *L. terrestris* has led researchers to hypothesize that these organisms have symbionts. These bacteria, according to Santiago and Roclewald (2005), include members of the families Enterobacteriaceae, Flavobacteriaceae, Sphingobacteriaceae (Bacteroidetes), Pseudomonadaceae, and numerous Actinobacteria. According to Gómez Plaza et al. (2001), the impacts of *Eisenia andrei* on the transitory community lead to a decrease in the overall diversity of the soil while simultaneously leading to an enhancement in the microbial activities. An application of isolated gut liquid has directly eliminated some bacterial species in vivo, while others persisted or increased as a result of the complicated antimicrobial impact that the intestine of *A. caliginosa* exerts on transitory species (Lukkari et al. 2004). The suppression of several taxonomic targets in the anterior and posterior portions differed depending on the body posture. *Alcaligenes faecalis* and *Bacillus megaterium* populations were reported to be declining in the anterior parts, but no microbial activity was discovered in the posterior area, where instead an increase was found in the ability of several *Pseudomonas* species to form colonies. This is consistent with the increases in microbial diversity and load that have been documented in the hind intestine (Lassauce et al. 2011). It should not be surprising that the very presence of earthworm-associated bacterial species and populations is shrouded in a great deal of skepticism and debate. By making use of the most recent developments in High Throughput Sequencings technology, it will be possible to elucidate the microbiome at an unprecedented level of detail and to gain a broader knowledge of this biotic system. Since earthworms are soil-living beings, they are in close contact with changes in soil chemistry and bioavailable contaminants. As a result, it is critical to comprehend how these soil organisms adjust genetically and physiologically to environments that are very diverse. Since the conception of the "acute earthworm toxicity test" (Fazlollahi Mohammadi 2016), extensive research has led to the recognition of earthworms as model species for evaluating environmental toxicity (Kooch et al. 2015). They are particularly useful due to their widespread distribution in areas with favorable climatic and edaphic conditions. In addition, the ubiquity of earthworm species extends to very hazardous conditions. Individual earthworms have a significant number of interactions with the soil environment, both epidermally and by ingestion, resulting in the exposure to large amounts of pollutants. Notwithstanding this, earthworm species are frequently found in large numbers, indicating their tolerance for habitats that many other species may find harmful. Earthworm field communities have been found to exist in soils with metal/metalloid concentrations that often exceed the LC50 limits for acute toxicity in laboratory conditions by many orders of magnitudes (Cammeraat 2002). Due to the diversity of environmental stresses, the processes by which invertebrates, and especially earthworms, are able to tolerate such environments have been extensively explored; however, these processes have yet to be firmly proven. Diverse invertebrate species have genetically encoded processes related to their resistance to metal toxins, such as cadmium and copper (Butt and Quigg 2021). The control of current tolerance mechanisms, such as the activity of metallothionein, a metal ion-binding peptide, may be used to increase an organism's ability to withstand higher toxicity levels. This can be accomplished by the use of metallothionein, a known strategy used by several earthworm species, that provides a variety of purposes, including the sequestration of metal into tissues, which, in turn, leads to a reduction in toxicity (Bossyut et al. 2006).

5 Earthworm and Ecosystem Functions

As ecosystems provide advantages to humans as well as to other species, these services or functions have been classified. Services have been categorized, worldwide supply and demand identified, and financial benefits and mechanisms of valuing ecosystem services identified (Golovanova et al. 2021). There are four distinct types of ecosystem services: regulating services, supporting services, provisioning services and cultural services. Disease control, flood control, and climate regulations are all covered by the regulatory services. Nutrients cycling, primary production, soil structure, pollination, and other crucial activities that support ecosystems and biodiversity are instances of supporting services. Services that supply tangible products such as food, clean water, wood, and fuel are referred to as provisioning services. Cultural services offer educational, entertaining, and visual appeal benefits to people (Hodson et al. 2021). The structure of ecosystems and biodiversity are the sources of most of mentioned ecosystem functions. Assessing how biodiversity correlates to environmental services supply in a direction that can influence management and conservation aims, however, is currently missing for many types of functions (Kaiser et al. 2015). Identification of the connections among ecological elements and ecosystem services or functions and the development of managing strategies are necessary to achieve organization targets (Kim et al. 2012). Creating and managing a regenerated natural environment so that ecosystem services are provided is key to its effectiveness. The development and expansion of flora is a crucial measure of the success of creating green spaces (Lukkari et al. 2004). Soil quality, which is a result of several interconnected physicochemical and biological characteristics, has a significant impact on the health of the flora. Outcomes of biological processes in the soil, including the carbon transformation in connection with organic matter in the soil, the indispensable plant nutrients cycling, soil structural maintenance, and soil biodiversity regulation, are frequently the basis for the provision of soil ecosystem services and functions. Such activities, which are regulated by sophisticated biological interactions with the soil's physicochemical characteristics, are crucial to the soil's ability to perform critical functions (Mosavi and Raisi 2009). Ecologists have not paid much attention to the biodiversity of soils until lately, although it is now widely acknowledged that soil fauna and flora are important regulators of many ecosystem functions (Niemelä and Kotze 2009). The principal outcomes of bioactivities inside the soil are frequently those used for primary production (for instance, tree growing in forests), the supply of soil function, as well as other associated ecosystem services. The services related to soils have been outlined by several research studies (Rutgers et al. 2016; Szlavecz et al. 2018; Honvault et al. 2023). Since distinct soil ecosystem functions might perform various services, soil biota is frequently categorized into "functional groupings" based on how they relate to such services (Fig. 10) (Wall et al. 2012).

It is valuable for ecosystem managers to understand if all organism species participate equally in providing ecosystem services or if certain organisms are more important than others (van Groenigen et al. 2014). Managers of ecosystems might therefore concentrate their attention more effectively on those specific ecological components that have been shown to be highly functionally relevant. Ecosystem engineers are those organisms that have an incommensurate influence on the structure of their habitats, adjacent biota, and ecosystem services (Wall et al. 2012; Thouvenot et al. 2021; Kooch and Ghaderi 2023). Based on Wall et al. (2012),

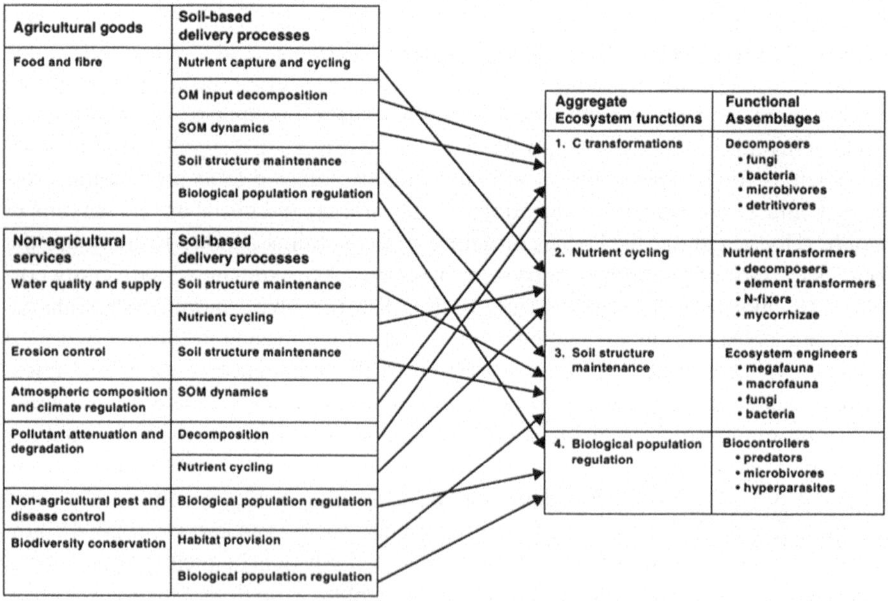

Fig. 10 Relations among the soil biota actions and a variety of ecosystem services and products goods that civilization may anticipate from soils. Organic matter (OM) is also known as soil organic matter (SOM)

termites, ants, and earthworms are the soil's primary ecosystem engineers. Despite being very few, earthworms frequently produce more biomasses than all other soil organisms in temperate and fertile soils, due to their special (larger) physical appearance. Earthworms are recognized ecosystem engineers, which constantly change their surrounding environments by digging, combining soil with organic materials, and other processes. These actions have a profound impact on the local plants and soil biota (Schlaghamerský and Pižl 2009). Since Darwin (1881) first highlighted the benefits of earthworms, a great deal of studies have been carried out on their ecology and their function in the provision of soil ecosystem services (Szlavecz et al. 2018). According to van Groenigen et al. (2014), earthworms are the catalysts for the two "supporting services" of nutrient cycling and soil forming. Furthermore, a variety of additional services depend on these soil functions (Tresch et al. 2018).

6 Conclusions

Soils are the most diverse components in the biosphere, with dramatic differences in characteristics and processes at nano- to macroscales. The population of earthworms is one of the most vital indicators of soil health and quality, with implications also for nitrogen cycle in terrestrial ecosystem. Different literature reviews have shown, even at the small scales, that the effects of various sites (i.e., croplands, rangelands, forests, and urban soils) on earthworm activities can be detected in the soil. The extent of the observed changes in earthworm communities is influenced by the ecosystem's characters, as well as by topographically and climatically controlled processes like redistribution of litter and creep of soil. Anyway, the soil environment can be understood as a mosaic of earthworm's activity in different places that are influenced by site factors. Soil physicochemical features like texture, moisture, temperature, pH, and organic matter affect the distribution of earthworms in the soil. As earthworms are sensitive to the concentration of H^+ ions, their distribution and abundance are very much affected by pH and pH-associated factors. Soils, having neutral pH, are usually preferred by most species of earthworms. Soil moisture plays a significant role in the survival of earthworms, as their bodies have 75–90% of water by weight. By moving to a moister area or by aestivating, they can survive even in unfavorable moisture situations. There is a steady positive connection among soil organic matters and earthworms' numbers and biomass. Earthworms are also recognized as soil managers because of their activities that alter the soil's physicochemical and biological properties.

Acknowledgments The valuable and constructive comments/suggestions provided by Dr. Rosa Francaviglia (Council for Agricultural Research and Economics, Research Centre for Agriculture and Environment, Rome, Italy) are highly appreciated.

References

Abail Z, Whalen JK (2021) Earthworm contributions to soil nitrogen supply in corn-soybean agro-ecosystems in Quebec, Canada. Pedosphere 31:405–412

Amiotti NM, Zalba P, Sanchez LF, Peinemann N (2000) The impact of single trees on properties of loess-derived grassland soils in Argentina. Ecology 81:3283–3290

Aubert M, Hedde M, Decaens T, Bureau F, Margerie P, Alard D (2003) Effects of tree canopy composition on earthworms and other macro—invertebrates in beech forests of Upper Normandy (France). Pedobiologia 47:904–912

Bhadauria T, Ramakrishnan PS, Srivastava KN (2000) Diversity and distribution of endemic and exotic earthworms in natural and regenerating ecosystems in the central Himalayas, India. Soil Biol Biochem 32:2045–2054

Boettcher SE, Kalisz PJ (1990) Single-tree influence on soil properties in the mountains of Eastern Kentucky. Ecology 71:1365–1372

Bossyut H, Six J, Hendrix PF (2006) Interactive effects of functionally different earthworm species on aggregation and incorporation and decomposition of newly added residue carbon. Geoderma 130:14–25

Bottinelli N, Capowiez Y (2021) Earthworm ecological categories are not functional groups. Biol Fertil Soils 57:329–331

Bottinelli N, Kaupenjohann M, Märten M, Jouquet P, Soucémarianadin L, Baudin F, Tran TM, Rumpel C (2020) Age matters: Fate of soil organic matter during ageing of earthworm casts produced by the anecic earthworm Amynthas khami. Soil Biol Biochem 148:107906

Brady NC (1990) The nature and properties of soils, 10th edn. Macmillan, New York. 621p

Brais S, Drouin P (2012) Interactions between deadwood and soil characteristics in a natural boreal trembling aspen-jack pine stand. Can J For Res 42:1456–1466

Butt KR, Quigg SM (2021) Earthworm community development in soils of a reclaimed steel-works. Pedosphere 31:384–390

Callaham MA, Blair JM, Todd TC, Kitchen DJ, Whiles MR (2003) Macro invertebrates in North American tallgrass prairie soils: effects of fire, mowing and fertilization on density and biomass. Soil Biol Biochem 35:1079–1093

Cammeraat LH (2002) A review of two strongly contrasting geomorphologic systems within the context of scale. Earth Surf Process Landf 27:1201–1222

Collett NG, Neumann FG, Tolhurst KG (1993) Effects of two short rotation prescribed fires in spring on surface-active arthropods and earthworms in dry sclerophyll eucalypt forest of west-central Victoria. Aust For 56:49–60

Darwin C (1881) Movements of plants. Nature 23(592):409–409

David TS, Gash JHC, Valente F, Pereira JS, Ferreira MI, David JS (2006) Rainfall interception by an isolated evergreen oak tree in a Mediterranean savannah. Hydrol Process 20:2713–2726

Dawes-Gromadzki TZ (2007) Short-term effects of low intensity fire on soil macro invertebrate assemblages in different vegetation patch types in an Australian tropical savanna. Austral Ecol 32:663–668. Blackwell

Deleporte S (2001) Changes in the earthworm community of an acidophilus lowland beech forest during a stand rotation. Eur J Soil Biol 37:1–7

Drake HL, Horn MA (2007) As the worm turns: the earthworm gut as a transient habitat for soil microbial biomes. Ann Rev Microbiol 61:169–189

Edwards CA, Bohlen PJ (1996) Biology and ecology of earthworms, 3rd edn. Chapman and Hall, London. 426p

Edwards CA, Lofty R (1977) The biology of earthworms, 2nd edn. Chapman & Hall, London, UK

Edwards CA, Bohlen PJ, Linden DR, Subler S (1995) Earthworms in agroecosystems. In: Hendrix PF (ed) Earthworm ecology and biogeography. Lewis, Boca Raton, FL, pp 185–206

Ernst D (1995) The farmer's earthworm handbook: Managing your underground moneymakers. Lessiter, Brookfield, WI

Evans SE, Wallenstein MD (2012) Soil microbial community response to drying and rewetting stress: does historical precipitation regime matter? Biogeochemistry 109:101–116

Fazlollahi Mohammadi M (2016) The effect of catena landscape on soil Eco physiological indices and some features of Oriental Beech (*Fagus orientalis* Lipsky). PhD thesis of forestry, Tarbiat Modares University, 148p

Ferreira T, Santos A, Demetrio WC, Cardoso GB, Moraes RA, Assis O, Niva C, Smokanit MA, Knópik JU, Sautter KD, Brown GG (2018) Earthworm species in public parks in Curitiba, Paraná, Brazil. Zootaxa 4496:535–547. https://doi.org/10.11646/zootaxa.4496.1.41

Golovanova E, Kniazev SY, Babiy KA, Tsvirko EI, Karaban K, Solomatin DV (2021) Dispersal of earthworms from the Rudny Altai (Kazakhstan) into Western Siberia. Ecologica Montenegrina 45:48–61

Gómez-Plaza A, Martınez-Mena M, Albaladejo J, Castillo VM (2001) Factors regulating spatial distribution of soil water content in small semiarid catchments. J Hydrol 253:211–226

Gonzalez G, Seastedt TR, Donato Z (2003) Earthworms, arthropods and plant litter decomposition in aspen and lodge pole pine forests in Colorado, USA. Pedobiologia 47:863–869

Graham RC, Ervin JO, Wood HB (1995) Aggregate stability under oak and pine after four decades of soil development. Soil Sci Soc Am J 59:1740–1744

Graham S, Wilson BR, Reid N (2004) Scattered paddock trees, litter chemistry and surface soil properties in pastures of the New England Tablelands, NSW. Aust J Soil Res 42:905–912

Hale C, Host E (2005) Assessing the impacts of European earthworm invasions in beech- maple hardwood and aspen- fir boreal forests of the western Great Lakes region. National park service Great Lakes inventory and monitoring network report GLKN/2005/11

Harris JA (1991) Chapter 8: The biology of soils in urban areas. In: Bullock P, Gregory PJ (eds) Soils in the urban environment. Blackwell Scientific, Oxford

Hodson ME, Corstanjeb R, Jones DT, Witton J, Burton VJ, Sloan T, Eggleton P (2021) Earthworm distributions are not driven by measurable soil properties. Do they really indicate soil quality? PLoS One 16(8):e0241945

Hoeffner K, Santonja M, Monard C, Barbe L, Le Moing M, Cluzeau D (2021) Soil properties, grassland management, and landscape diversity drive the assembly of earthworm communities in temperate grasslands. Pedosphere 31:375–383

Honvault N, Houben D, Lebrun M, Vedere C, Nobile C, Guidet J, Kervroëdan L, Aubertin ML, Rumpel C, Faucon MP, Dulaurent AM (2023) Positive or neutral effects of biochar-compost mixtures on earthworm communities in a temperate cropping system. Appl Soil Ecol 182:104684. https://doi.org/10.1016/j.apsoil.2022.104684

Joimel S, Schwartz C, Maurel N, Magnus B, Machon N, Bel J, Cortet J (2019) Contrasting homogenization patterns of plant and collembolan communities in urban vegetable gardens. Urban Ecosyst 22:553–566

Kabala C, Plonka T, Przekora A (2015) Vertic properties and gilgai-related subsurface features in soils of south-western Poland. Catena 128:95–107

Kaiser M, Kleber M, Berhe AA (2015) How air-drying and rewetting modify soil organic matter characteristics: an assessment to improve data interpretation and inference. Soil Biol Biochem 80:324–340

Keeley JE, Fotheringham CJ, Baer-Keeley M (2006) Demographic patterns of postfire regeneration in Mediterranean-climate shrublands of California. Ecol Monogr 76:235–255

Kennette D, Hendershot W, Tomlin A, Sauvé S (2002) Uptake of trace metals by the earthworm *Lumbricus terrestris* L. in urban contaminated soils. Appl Soil Ecol 19:191–198

Kim DG, Vargas R, Bond-Lamberty B, Turetsky MR (2012) Effects of soil rewetting and thawing on soil gas fluxes: a review of current literature and suggestions for future research. Bio Geosci 9:2459–2483

Kooch Y (2012a) Response of earthworms' ecological groups to decay degree of dead trees (case study: Sardabrood forest of Chalous, Iran). Eur J Exp Biol 3:532–538

Kooch Y (2012b) Soil variability related to pit and mound, canopy cover and individual trees in a Hyrcanian Oriental Beech stand. PhD thesis of forestry, Tarbiat Modares University, 203p (In Persian)

Kooch Y, Ghaderi E (2023) The effect of Crataegus and Berberis canopy types on bioindicators of soil quality in a semi-arid climate. J Arid Environ 208:104862. https://doi.org/10.1016/j.jaridenv.2022.104862

Kooch Y, Hosseini SM (2010) Response of earthworm's biomass and diversity to wind throw events and soil properties in Hyrcanian forests of Iran. Folia Oecologica 37:181–190

Kooch Y, Jalilvand H, Bahmanyar MA, Pormajidian MR (2008) Abundance, biomass and vertical distribution of earthworms in ecosystem units of Hornbeam forest. J Biol Sci 8:1033–1038

Kooch Y, Hosseini SM, Mohammadi J, Hojjati SM (2010) The effects of gap disturbance on soil chemical and biochemical properties in a mixed beech-hornbeam forest of Iran. Ecologia Balkanica 2:39–56

Kooch Y, Hosseini SM, Mohammadi J, Hojjati SM (2011) Analysis of earthworm's patchy distribution and variability of soil biochemical properties under single-tree influences. Int J Environ Sci 1:1813–1829

Kooch Y, Mollaei Darabi S, Hosseini SM (2015) The effects of pits and mounds following wind throw events on soil features and greenhouse gas fluxes in a temperate forest. Pedosphere 25:1–13

Kooch Y, Tavakoli M, Akbarinia M (2018) Tree species could have substantial consequences on topsoil fauna: a feedback of land degradation/restoration. Eur J For Res 137:793–805

Kooch Y, Azizi Mehr M, Hosseini SM (2020a) The effect of forest degradation intensity on soil function indicators in northern Iran. Ecol Indic 114:1–9

Kooch Y, Ehsani S, Akbarinia M (2020b) Stratification of soil organic matter and biota dynamics in natural and anthropogenic ecosystems. Soil Tillage Res 200:1–11

Kooch Y, Ghorbanzadeh N, Wirth S, Novara A, Shah Piri A (2021) Soil functional indicators in a mountain forest-rangeland mosaic of northern Iran. Ecol Indic 126:107672

Kooch Y, Ghorbanzadeh N, Kuzyakov Y, Praeg N, Ghaderi E (2022) Investigation of the effects of the conversion of forests and rangeland to cropland on fertility and soil functions in mountainous semi-arid landscape. Catena 210:105951

Kooch Y, Ghorbanzadeh N, Haghverdi K, Francaviglia R (2023a) Soil quality cannot be improved after thirty years of land use change from forest to rangeland. Sci Total Environ 856:159132. https://doi.org/10.1016/j.scitotenv.2022.159132

Kooch Y, Ghorbanzadeh N, Hajimirzaaghaee S, Egli M (2023b) Soil functional indicators in mixed beech forests are clearly species-specific. J For Res 34(4):1033–1049

Kooijman AM, Kooijman-Schouten MM, Martinez-Hernandez GB (2008) Alternative strategies to sustain N-fertility in acid and calcaric beech forests: low microbial N-demand versus high biological activity. Basic Appl Ecol 9:410–421

Kumar S, Tripathi G, Mishra GV (2021) A comparative study on earthworm biodiversity and species habitat-relationship of hilly and plain areas of Sirohi District of Rajasthan, India. Appl Ecol Environ Sci 9:419–439

Lassauce A, Paillet Y, Jactel H, Bouget C (2011) Deadwood as a surrogate for forest biodiversity: Meta-analysis of correlations between deadwood volume and species richness of saproxylic organisms. Ecol Indic 11:1027–1039

Law MMS, Lai DYF (2021) Impacts of wetting-drying cycles on short-term carbon and nitrogen dynamics in Amynthas earthworm casts. Pedosphere 31:423–432

Londo AJ (2001) Bucket mounding as a mechanical site preparation technique in wetlands. NIAF 18:7–13

Lubbers IM, Van Groenigen KJ, Fonte SJ, Six J, Brussard L, Groengen WV (2013) Greenhouse-gas emissions from soils increased by earthworms. Nat Clim Chang 3:187–194

Lukkari T, Taavitsainen M, Vaisanen A, Haimi J (2004) Effects of heavy metals on earthworms along contamination gradients in organic rich soils. Ecotoxicol Environ Saf 59:340–348

Mabuhay JA, Nakagoshi N, Horikoshi T (2003) Microbial biomass and abundance after forest fire in pine forests in Japan. Ecol Res 18:431–441. Springer

Marvi Mohadjer MR (2007) Silviculture. Tehran University Publication, 387p (in Persian)

Moghimian N, Jalali SG, Kooch Y, Rey A (2020) Downed logs improve soil properties in old-growth temperate forests of Northern Iran. Pedosphere 30:378–389

Mosavi F, Raisi F (2009) The effect of earthworm and organic matter on carbon mineralization and microbial biomass. In: Articles collection of 11th soil sciences congress in Iran, pp 112–113 (in Persian)

Nachtergale L, Ghekiere K, De Schrijver A, Muys B, Luyssaert S, Lust N (2002) Earthworm biomass and species diversity in wind throw sites of a temperate lowland forest. Pedobiologia 46:440–451

Neher DA (1999) Soil community composition and ecosystem processes: Comparing agricultural ecosystems with natural ecosystems. Agrofor Syst 45:159–164

Neirynck J, Mirtcheva S, Sioen G, Lust N (2000) Impact of *Tilia platyphyllos* Scop., *Fraxinus excelsior* L., *Acer pseudoplatanus* L., *Quercus robur* L., and *Fagus sylvatica* L. on earthworm biomass and physico-chemical properties of loamy topsoil. For Ecol Manag 133:275–286

Niemelä J, Kotze DJ (2009) Carabid beetle assemblages along urban to rural gradients: a review. Landsc Urban Plan 92:65–71

Pfiffner L (2014) Earthworms-architects of fertile soils: their significance and recommendations for their promotion in agriculture. Technical guide on earthworms, Order no. 1629, International edition, FiBL 2014, www.fibl.org

Phillips JD, Marion DA, Turkington AV (2008) Pedologic and geomorphic impacts of a tornado blow down event in mixed pine–hardwood forest. Catena 75:278–287

Ponge JF (2013) Plant–soil feedbacks mediated by humus forms: a review. Soil Biol Biogeochem 57:1048–1060

Ponge JF, Zanella A, Sartori G, Jabiol B (2010) Terrestrial humus forms: ecological relevance and classification <Hal-00521337>

Pulleman MM, Six J, Breemen N, Van Jongmans AG (2005) Soil organic matter distribution and micro aggregate characteristics as affected by agricultural management and earthworm activity. Eur J Soil Sci 56:453–467

Richon C, Tagliabue A (2021) Biogeochemical feedbacks associated with the response of micronutrient recycling by zooplankton to climate change. Global change biology 27(19):4758–4770

Rowell DL (2014) Soil science: Methods & applications. Routledge, London

Rutgers M, Orgiazzi A, Gardi C, Römbke J, Jänsch S, Keith AM, Neilson R, Boag B, Schmidt O, Murchie AK, Blackshaw RP, Pérès G, Cluzeau D, Guernion M, Briones MJI, Rodeiro J, Piñeiro R, Cosín DJD, Sousa JP, Suhadolc M, Kos I, Krogh PH, Faber JH, Mulder C, Bogte JJ, va Wijnen HJ, Schouten AJ, de Zwart D (2016) Mapping earthworm communities in Europe. Appl Soil Ecol 97:98–111

Sajedi T (2010) The effects of excessive moisture on soil carbon and nitrogen mineralization and forest productivity. PhD thesis, University of British Colombia, 160p

Santiago JM, Roclewald AD (2005) Dead trees are sources for forest wildlife, Extension fact sheet. Ohio State University Express, 5p

Sarlo M (2006) Individual tree species effects on earthworm biomass in a tropical plantation in Panama. Caribb J Sci 42:419–427

Sautter KD, Brown CG, James SW, Pasini DH, Nunesand N, Benito P (2006) Present knowledge on earthworm biodiversity in the state of Parana, Brazil. Eur J Soil Biol 42:321–332

Scahrenbroch BC, Bockheim JG (2007) Pedodiversity in an old-growth northern hardwood forest in the Huron Mountains, Upper Peninsula, Michigan. Can J For Res 37:1106–1117

Scheyer JM, Hipple KW (2005) Urban soil primer. United States Department of Agriculture, Natural Resources Conservation Service, National Soil Survey Center, Lincoln, NE. http://soils.usda.gov/use

Schlaghamerský J, Pižl V (2009) Enchytraeids and earthworms (Annelida: Clitellata: Enchytraeidae, Lumbricidae) of parks in the city of Brno, Czech Republic. Soil Organ 81:145–173

Smetak KM, Johnson-Maynard JL, Lloyd JE (2007) Earthworm population density and diversity in different-aged urban systems. Appl Soil Ecol 37:161–168

Steinberg DA, Pouyat RV, Parmelee RW, Groffman PM (1997) Earthworm abundance and nitrogen mineralization rates along an urban-rural land use gradient. Soil Biol Biochem 29:427–430

Szlavecz K, Placella SA, Pouyat RV, Groffman PM, Csuzdi C, Yesilonis I (2006) Invasive earthworm species and nitrogen cycling in remnant forest patches. Appl Soil Ecol 32:54–62

Szlavecz K, Chang CH, Bernard MJ, Pitz SL, Xia L, Ma Y, McCormick MK, Filley T, Yarwood SA, Yesilonis ID, Csuzdi C (2018) Litter quality, dispersal and invasion drive earthworm community dynamics and forest soil development. Oecologia 188:237–250

Thouvenot L, Ferlian O, Beugnon R, Künne T, Lochner A, Thakur MP, Türke M, Eisenhauer N (2021) Do invasive earthworms affect the functional traits of native plants? Front Plant Sci 12:627573. https://doi.org/10.3389/fpls.2021.627573

Tiho S, Josens G (2007) Co-occurrence of earthworms in urban surroundings: a null model analysis of community structure. Eur J Soil Biol 43:84–90

Tresch S, Moretti M, Le Bayon RC, Mäder P, Zanetta A, Frey D, Fliessbach A (2018) A gardener's influence on urban soil quality. Front Environ Sci 6:25. https://doi.org/10.3389/fenvs.2018.00025

van Groenigen JW, Lubbers IM, Vos HMJ, Brown GG, De Deyn GB, van Groenigen KJ (2014) Earthworms increase plant production: a meta-analysis. Sci Rep 4:6365

Vickery JA, Feber RE, Fuller RJ (2009) Arable field margins managed for biodiversity conservation: A review of food resource provision for farmland birds. Agric Ecosyst Environ 133:1–13

Vilisics F, Bogyó D, Sattler T, Moretti M (2012) Occurrence and assemblage composition of millipedes (Myriapoda, Diplopoda) and terrestrial isopods (Crustacea, Isopoda, Oniscidea) in urban areas of Switzerland. Zookeys 176:199–214. https://doi.org/10.3897/zookeys.176.2153

Wall DH, Bardgett RD, Behan-Pelletier V, Herrick JE, Jones TH, Ritz K, Six J (2012) Soil ecology and ecosystem services. Oxford University Press, Oxford

Wang H, Wang L, Liu J, Nie Y, Wang D (2022) Biological engineering analysis of vermicompost based on image features and machine learning. Hindawi Mob Inf Syst 2022:7347142. 14 pages. https://doi.org/10.1155/2022/7347142

Wardle DA (2002) Communities and ecosystems: linking the above ground and below ground components. Princeton University Press, Princeton. 735p

Xie T, Wang M, Chen W, Uwizeyimana H (2018) Impacts of urbanization and landscape patterns on the earthworm communities in residential areas in Beijing. Sci Total Environ 626:1261–1269

Zhong L, Wu T, Ding J, Xu W, Yuan F, Liu BF, Zhao L, Li Y, Ren NQ, Yang SS (2023) Co-composting of faecal sludge and carbon-rich wastes in the earthworm's synergistic cooperation system: Performance, global warming potential and key microbiome. Sci Total Environ 857:159311. https://doi.org/10.1016/j.scitotenv.2022.159311

Zhou L, Dai LM, Gu HY, Zhong L (2007) Review on the decomposition and influence factors of coarse woody debris in forest ecosystem. J For Res 18:48–54

Ziemer RR (1968) Soil moisture depletion patterns around scattered trees. Research Note PSW-166, US Forest Service, Berkeley, 75p

Zinke PJ (1962) The pattern of influence of individual forest trees on soil properties. Ecology 43:130–133

Earthworms and Humus Forms

Renée-Claire Le Bayon, Jean-François Ponge, and Augusto Zanella

Abstract Soils play a central role in global biogeochemical cycles and host plants and invertebrates whose engineering activities mainly occur in the upper soil layers, the so-called humipedon. This latter is the critical zone where most chemical, physical, and biological ecosystem processes arise. This chapter overviews the main definitions and concepts (diagnostic horizon, humus system, humus form) and functional approaches to humipedon intimately linked to ecosystem engineers, mainly plant roots and earthworms. Biological activities are decisive in integrating organic matter and forming soil aggregates and galleries. We highlight the relevance of studying humus forms based on field observations in various environments (grasslands, forests, mountains, floodplains) and pioneer ecosystems and underline that more research is needed on the role played by earthworms in the evolution of terrestrial biomes.

1 Exploring the Underground

The world's population reached a record eight billion people on 15 November 2022, according to the Department of Economic and Social Affairs of the United Nations (2022). Human population growth impacts the Earth's system in various ways, including the extraction of natural resources such as fossil fuels, minerals, trees, water and wildlife. Moreover, more than half of the world's population now lives in urban areas (Ritchie et al. 2024), and urbanisation is still in expansion. Such increased pressures endanger maintaining and expanding essential ecosystem

R.-C. Le Bayon (✉)
University of Neuchâtel, Neuchâtel, Switzerland
e-mail: claire.lebayon@unine.ch

J.-F. Ponge
Muséum National d'Histoire Naturelle, Paris, France
e-mail: ponge@mnhn.fr

A. Zanella
University of Padua, Padua, Italy
e-mail: augusto.zanella@unipd.it

Y. Kooch, Y. Kuzyakov (eds.), *Earthworms and Ecological Processes*,
https://doi.org/10.1007/978-3-031-64510-5_6

benefits and services required for human livelihood (Larondelle et al. 2014). These latter are commonly defined as the "benefits people obtain from the ecosphere and its ecosystems" (Millennium Ecosystem Assessment 2005) and listed into four broad categories: provisioning, regulating, cultural and supporting services (Dominati et al. 2010). To preserve as best as possible ecosystems' services, the conservation and management of ecosystems appear crucial because this approach considers the entire communities of species and their interactions with the physical environment.

Usually, the most common and easiest way to understand terrestrial ecosystems is to study vegetation… provided you know botany and phytosociology! Vegetation relevés allow the establishment and monitoring of plant communities' structure and composition using a percentage cover of species in each layer of phytocenosis (Braun-Blanquet's scale). This approach is beneficial for preserving and predicting the ecosystems' evolution due, for instance, to anthropic impacts such as climate change. To complement this descriptive vision, energy, water and nutrient fluxes provide a functional approach to ecosystems. However, more than staying at the level of the upper part of plants is required: we need to follow the plants' dark side underground, which means their roots. This guides us to the most concealed part of the ecosystem, the underground black box and perhaps one of the last final frontiers of the research and knowledge: the soil (see the review in Science, Sugden et al. 2004).

As a three-dimensional effective, almost non-renewable system, soils represent the primary living matrix for many ecosystems and human activities (Dominati et al. 2010; Adhikari and Hartemink 2016; Blanchart et al. 2018; Sofo et al. 2022). They play a central role in global biogeochemical cycles and host the most extensive diversity of organisms (Smith et al. 2015). Among these organisms, plants and invertebrates are essential components of soil genesis, functions and properties. Their engineering activities mainly occur in the so-called humipedon, corresponding to the upper soil layers particularly enriched in organic matter, which are the core of the interactions between vegetation and soil (Zanella et al. 2018a, b). The activity of soil organisms is exceptionally high in humipedons and intensely involved in organic matter recycling and provisioning, as well as biogenic structures such as burrows, pellets and aggregates. Earthworms and plant roots mainly govern these latter: as allogenic engineers, they are responsible for physical modifications but may also be involved in biological and biochemical processes. In temperate ecosystems, they are thus the leading soil engineers creating habitats for other organisms, such as small arthropods such as springtails or mites (Eisenhauer 2010; Cameron et al. 2013; Liu et al. 2013). Both plants and earthworms also produce chemical substances, exudates (Bais et al. 2006) and mucus (Salmon 2001), which are enriched in water, carbon (C) and nitrogen (N). Plant roots interact with earthworms, and their concomitant activity increases aggregate stability (Fonte et al. 2012; Schomburg et al. 2018a, b). The critical zone where most chemical, physical and biological processes occur is the humipedon (Fig. 1).

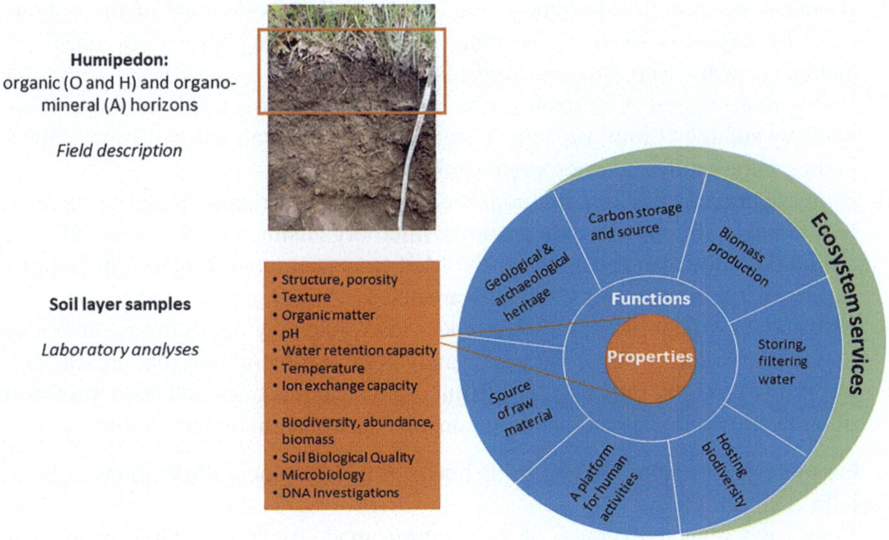

Fig. 1 The humipedon is crucial in ecosystem services as it interfaces between vegetation and mineral soil layers. The humipedon is involved in the carbon cycle, the hosting of biodiversity and the regulation of water and biomass production at the ecosystem level. Moreover, the humipedon has its own spatiotemporal functioning scale. Adapted from Adhikari and Hartemink (2016)

2 Humipedon, Humus Systems and Humus Forms: What Are We Talking About

2.1 Vocabulary

Since Zanella et al. (2018e), the "humipedon" represents the surface organic (O, H) and organo-mineral (A) soil layers, thus avoiding confusion with previous "topsoil", "humus" or "forest humus" definitions. In terrestrial ecosystems, the humipedon organises into five *humus systems*: Mull, Moder, Mor, Amphi and Tangel, which can be eventually submerged a few days per year.

Humus systems differ by morpho-functional humus *diagnostic horizons* and are generated by specific groups of organisms. In each humus system, the thickness of diagnostic horizons defines some (3–4) similar *humus forms*.

The humipedon is made up of organic horizons (O, organic; OL, litter; OF, fragmented/fermented litter; OH, humic, i.e. transformed in rather organic minute droppings, small particle litter) and of organo-mineral A horizon (organic derivate integrated within a mineral layer). These horizons are placed on top of each other: OL at the surface, then OF, and OH, on A at the bottom of the humipedon. The OL layer is composed of recognisable leaves at different decomposition stages. The OF layer consists mainly of fragmented debris, and the OH layer contains a majority of holorganic droppings.

It is relatively easy to distinguish three zoogenic and one non-zoogenic type of organo-mineral A horizons:

1. Biomacrostructured A horizon (code: maA): at least one-third of the volume made by large (>4 mm) organo-mineral biogenic aggregates; origin: anecic or endogeic earthworms and microorganisms.
2. Biomicrostructured A horizon (code: miA): at least two-thirds of the volume made by small (≤1 mm) organic or organo-mineral faecal pellets; origin: arthropods, enchytraeids and microorganisms.
3. Biomesostructured A horizon (code meA): zoogenic horizon, not maA, not miA; origin: anecic or endogeic earthworms, microorganisms, insect larvae and large arthropods may also intervene in the Mediterranean environment. The appearance is that of a finer and very amalgamated maA aggregates.
4. Non-zoogenic A horizon: massive (code: msA, like clay organo-mineral blocky horizon) or single grain (code: sgA, sandy organo-mineral horizon, organic pellets ≤10%); origin: rather due to chemico-physical reactions and water transport of molecules and minerals, microorganisms may play an unknown role too.

For more information on diagnostic horizons and graphical illustrations, refer to Zanella et al. (2018f).

Table 1 is a revisited version of the synthesis made by Ponge (2003), which has been completed with Tangel and Amphi systems. From Tangel to Mor systems, a gradient of variations occurs, especially regarding the horizon sequences relative to the activity of soil organisms. The ecosystems and soil types differ, as do the recycling of organic matter and nutrient availability. At a finer level, in each humus system, diagnostic horizons' presence and thickness define several humus forms, differing slightly in between (i.e. Eumull, Mesomull, Oligomull, Dysmull; Zanella et al. 2018g).

In addition to plant roots, soil fauna and microorganisms (bacteria, fungi) are intensely involved and often used to discriminate the humus system at the inter- and intra-level. The same occurs for arthropods: while springtails and mites are the most common arthropods in the humipedon, enchytraeids and earthworms are the most efficient in organic matter recycling bioturbation processes (Table 1). In addition, arthropods and enchytraeids mainly live in O layers, while earthworms can live in soil surface layers or more profoundly. Figure 2 is an updated version of Zanella et al. (2018c), an extension of the presence of earthworms to a broader range of humus forms. New in situ observations revealed that epigeics are mostly ubiquitous in humus systems, while anecics range from Eumesoamphi to Eumoder.

2.2 *The Amphi System: A Historical Overview*

The introduction of Amphi forms in the classification of humus forms results from field observations and discussions within a consortium of researchers over the last two decades, as briefly described below.

In 1882, Peter Erasmus Müller (1887) described two humus forms in Germany: Mull (made by earthworms) and Torf (made by arthropods). In 1970, in the Austrian and Italian Alps, Franz Hartmann (1970) discovered that these two humus forms might overlap in what he called "Humus gemellare", literally "twin humus". This system was composed of a grey-brown organic-mineral layer made of large

Table 1 Main biological features of the five humus systems at the ecosystem level

	Tangel	Amphi	Mull	Moder	Mor
Ecosystems	Grasslands, coniferous woodlands, alpine meadows, cold and humid conditions	Deciduous and coniferous woodlands, transitional ecosystems such as abandoned pastures	Grasslands, deciduous woodlands with rich herb layer, Mediterranean scrublands	Deciduous and coniferous woodlands with poor herb layer	Heathlands, coniferous woodlands, sphagnum bogs, alpine meadows
Biodiversity	Medium	High in the A, low in O layers	High	Medium	Low
Productivity	Medium	Medium to high	High	Medium	Medium to low, adapted to acidophilic species
Litter horizons	OL, OF, OH	OL, OF, OH	OL, OF	OL, OF, OH	OL, OF, OH
Organo-mineral horizon	nozA, meA	miA, meA, maA	meA, maA	meA, miA	When present, nozA
Soil type	Holorganic soils, humo-calcic soils	Brown calcic soils	Brown soils	Grey-brown podzolic soils	Leached brown soils to podzols
Phenolic content of litter	Medium to high	Medium	Poor	Medium	High
Humification	Slow	Slow to medium	Rapid	Slow	Very slow
Humified organic matter	Slow but continuous degradation of plant debris	Holorganic faecal pellets in the O horizons, organo-mineral aggregates in the A horizon	Organo-mineral aggregates with clay-humus complexes	Holorganic faecal pellets	Slow oxidation of plant debris

(continued)

Table 1 (continued)

	Tangel	Amphi	Mull	Moder	Mor
Exchange sites	Organic (rich near the rock)	Organic (rich) to mineral	Mineral	Organic (rich)	Organic (poor)
Mineral weathering	Poor, except in the vicinity of the rock	Medium to high	High	Medium	Poor
Mineral buffer type	Carbonate range	Carbonate to silicate	Carbonate to silicate range	Silicate range	Iron/aluminium range
Impact of fire	High	Medium to high	Low (except in Mediterranean ecosystems)	Medium	High
Regeneration of trees	Poor	Medium, depending on A type (maA to meA)	Easy (permanent)	Poor (cyclic processes)	Young trees may develop on local degrading stumps or surface-stripped soils
Dominant mycorrhizal types	Ectomycorrhizae	VA-mycorrhizae to ectomycorrhizae	VA-mycorrhizae	Ectomycorrhizae	Ericoid and arbutoid mycorrhizae
Mycorrhizal partners	Unknown	Basidiomycetes	Zygomycetes	Basidiomycetes	Ascomycetes
Nitrogen forms	Protein	Protein, ammonium, nitrate	Protein, ammonium, nitrate	Protein, ammonium	Protein
Nutrient availability to plants	Direct from the OH horizon	Direct to indirect	Direct (through absorbing hairs)	Indirect (through extramatrical mycelium)	Poor
Nutrient use efficiency	Low to medium	Medium to high	Low	Medium	High
Fauna	Mesofauna (high), microfauna (high)	Macrofauna in the A horizon, mesofauna and microfauna in O horizons	Megafauna, macrofauna, mesofauna, microfauna	Macrofauna (poor), mesofauna (rich), microfauna	Mesofauna (poor), microfauna (poor)

	Tangel	Amphi	Mull	Moder	Mor
Earthworms: main ecological categories	Epigeics, some endogeics	Epigeics, endogeics, anecics	Endogeics and anecics	Epigeic, some endogeics	Some epigeics in thin mors
Faunal group dominant in biomass	Mites, springtails, a few earthworms, enchytraeids	Endogeic earthworms in the A horizon, macro- and microarthropods, epigeic earthworms and enchytraeids in O horizons	Earthworms	Enchytraeids, a few earthworms	Mites, springtails
Microbial group dominant in biomass	Fungi	Bacteria and fungi	Bacteria	Fungi and bacteria	Fungi
Affinities with polluted condition	Low	Low	Low	Medium	High

Adapted and revised from Ponge (2003)

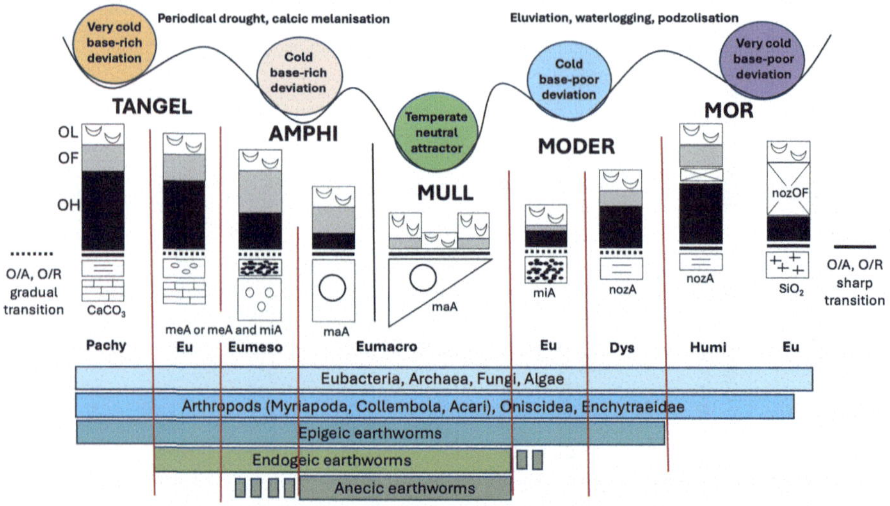

Fig. 2 An updated version of the simplified scheme of the terrestrial humus system classification (updated, from Zanella et al. 2018c). The scheme shows the main climatic and parent material determinants (top of the picture), diagnostic horizons (middle) and biological actors of organic matter transformation (bottom). OL, OF, nozOF, OH, maA, miA and nozA: diagnostic horizons described in Zanella et al. (2018f) transition between OH and A horizons: dashed line, gradual; continuous line, sharp; A horizon aggregates: two lines, non-zoogenic; small black circles, biomicrostructured; white small circles, biomesostructured; large white circles, biomacrostructured; lithopedon: bricks, base-rich substrate; +, base-poor substrate; Pachy, thick; Eu, typical; Dys, acid; Humi, rich in undecayed organic matter; Eumacro and Eumeso, large or medium biogenic structures in the A horizon. Humps and troughs of the continuous blue line refer to the hypothesis of humus systems as ecological attractors, with Mull as the "final" attractor

earthworms' dejections, overlapped by an organic black layer made of small aggregates generated by arthropods.

In 1995, Brêthes et al. (1995) set a similar humus form in their classification system; they called it "Amphimull" ("double Mull") and placed it in the Mull system. In 1996, Calabrese et al. (1996) revealed that such "twin humus" was relatively common in beech forests and Mediterranean ecosystems. This "Amphimull" evolved into "Amphimus" in 2009 (Jabiol et al., in Baize and Girard 2009), showing the importance of discriminating this system from the Mull one and the specificity of this particular humus form.

In 2003, 26 soil specialists met in Trento (Italy) and founded a Humus Group. After 3 years of exchanges, the Humus Group presented a poster at the World Congress of Soil Science in Philadelphia (USA, 2006) on which the Amphi reference was displayed aside from the well-known ones of Mull, Moder and Mor (Fig. 3). To provide a complete and updated framework in this chapter dedicated to earthworms, in Fig. 3, we added the Tangel system (bottom, in orange), which was not present on the drawing presented in Philadelphia because it was considered only in 2018.

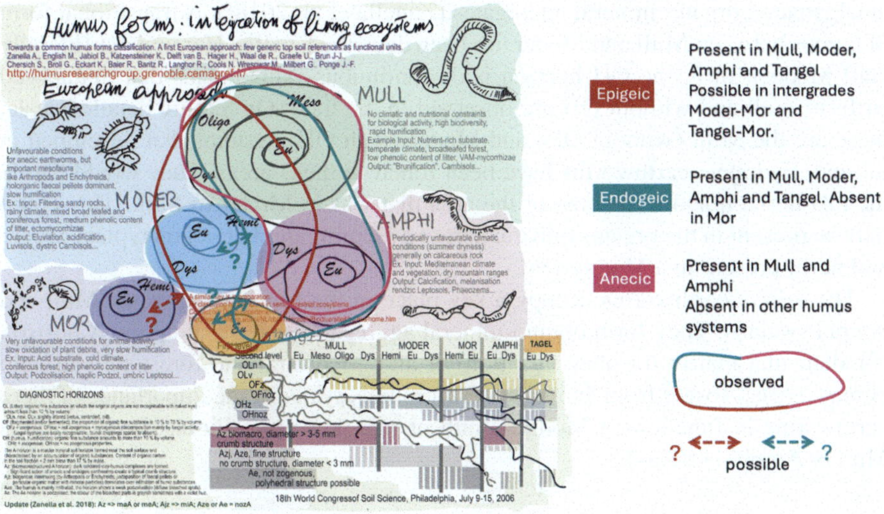

Fig. 3 Humus forms: integration of living ecosystems. Poster presented at the 18th World Congress of Soil Science at Philadelphia in July 2006, modified. Mull, Moder, Mor, Amphi and Tangel "humus form references" (corresponding to modern "humus systems") were represented as evolutionary holes in which the humipedon ecosystems fall driven by climatic, geological and vegetational relationships and constraints. In 2006, the Amphi "hole" was filled with two humus forms: Euamphi and Dysamphi. Since 2018, they became four: Leptoamphi, Eumacroamphi, Eumesoamphi and Pachyamphi. Moder did not change; Mor got a third Humimor form in 2018. In 2018, a Tangel fifth hole was added at the bottom of the picture, touching Amphi, Moder and Mor. This supplementary humus system, mostly generated by arthropods in calcareous or dolomitic high mountain environments, was first described by Walter Kubiëna in 1953 (Kubiëna 1953)

The group of *epigeic* earthworms has a habitat relegated to organic horizons (vOL, zoOF, zoOH; v, verbleit, bleached, old and packed litter; zo, zoogenic, with evident signs of the presence of animals that feed on litter, such as holes, bites, droppings, etc.). The epigeic earthworms produce small organic droppings (organic carbon > 20% by weight, measured with the elemental analyser) that remain in the organic part of the soil profile (Galvan et al. 2008). These earthworms initiate the biotransformation of litter but only indirectly participate in the genesis of organo-mineral aggregates, which are due only to endogeic and anecic groups; they are active in all humus systems presenting zoogenic organic diagnostic horizons, as in Mull, Moder, Amphi and Tangel (absent on Mor).

The *endogeic* group does not inhabit the organic horizons of the humipedon but prefers the superficial part of the organo-mineral horizon, feeding on organic horizons deposited above their habitat, but creating an organo-mineral horizon composed of their excrements. This also happens in environments of very superficial soils, as long as they are also covered by organic horizons, both in acidic and calcareous environments. The A horizon formed by endogeic earthworms is slightly smaller in size but very similar to that formed by anecics, made up of uniformly amalgamated organo-mineral aggregates (Zanella et al. 2018f). A transitional form of humus between Mull and Moder has been described, called Hemimoder (literally Half Moder). In it, the OH horizon is discontinuous. This means that juxtaposed surfaces with and without OH are observed in the field. On a more detailed scale, these are the Mull (without OH) and Moder (with OH) systems that developed in mosaic. Endogeic earthworms have been found in these areas. They would eat up the OH horizon and turn some of the area that was in Moder to Mull. This group may be present in the organo-mineral part of a Hemimoder but not on typical Moder and never present in a Mor system.

The *anecic* earthworms occupy a key place in the functioning of the Mull and Amphi systems. They form organo-mineral aggregates that make up the A horizon (fundamental matrix for plant nutrition) of all agricultural and forest soils in mild climate areas. Absent from Tangel and Mor humus systems, the importance of these earthworms in functioning Moder is null or marginal (transitional forms towards Mull or Amphi systems).

3 Earthworm Engineering: Opportunity or Threat?

Earthworms are known for a long time for the ecosystem services they provide through their engineering activity: litter decomposition (Bocock 1964), soil structure (Oades 1993), and nutrient cycling (Bohlen and Edwards 1995; Edwards and Arancon 2022), among many others, are improved under the influence of their multiple interactions with plants, microbes and other soil fauna (Brown et al. 2000), stemming in the formation of earthworm Mull (Zanella et al. 2018b). Less well-known despite of far-reaching consequences for global warming mitigation is the role played by earthworms in carbon sequestration in Mull humus systems in which organic matter is buried and mixed with mineral particles in physically and chemically stable aggregates (Zhang and Schrader 1993). Although more research on the subject is urgently needed, earthworm (macroaggregated) Mull and Amphi thereby might contain more organic carbon (although invisible to the naked eye) over their whole thickness and thus might sequester more carbon than Moder or Mor humus systems that visibly accumulate more organic matter at the soil surface (Andreetta et al. 2011; de Nicola et al. 2014). Another example of still poorly known ecosystem services provided by earthworms is the stimulation of forest regeneration through the building of humus forms favourable to tree seedling establishment(Ponge et al. 1998). As a consequence of benefits provided by earthworm activity, the colonisation of earthworm-free land by earthworms has been shown to favour crop production. Widely reported examples are the colonisation of Dutch polders by indigenous earthworm species and New Zealand pastures by exotic (European) earthworm species (Curry and Cotton 1983). In both cases, earthworms were considered as an essential tool for the successful development of agriculture.

However, and quite surprisingly, earthworm processes, e.g. casting deposition, may favour biodiversity by increasing heterogeneity at the soil surface and creating niches for plants (Milcu et al. 2006) and animals (Hamilton and Sillman 1989), but these same processes may also degrade the soil by favouring erosion along slopes (Le Bayon and Binet 1999; Le Bayon et al. 2002) or creating impervious crusts on the ground (Chauvel et al. 1999). Many reports have been done on detrimental aspects of earthworm activity. The most threatening is the still the incoming invasion of North American biomes by European (several lumbricid species) and Asian (several pheretimoid species) earthworms introduced in the past by European settlers (Nuzzo et al. 2009). In US northeastern forests, indigenous earthworm faunas, restricted to a few species living in their typical thick forest floors, are now outcompeted by invasive earthworms through changes in humus forms mediated by their mixing and digging activities, even if some cases of resistance have been observed (Hendrix et al. 2006). Dramatic changes caused by the establishment of exotic earthworm populations are also visible at the scale of forest vegetation (Hale et al. 2006). It might even be postulated that in a more or less near future tree populations will be affected, too, in particular if tree species favoured by Mull for their regeneration outcompete tree species favoured by Moder or even Mor.

In China, the introduction of the European manure worm *Eisenia fetida*, as yet commonly used for composting organic wastes, has been suggested as a mean to reclaim soils improper for agriculture (Zhang et al. 2015). This voluntary introduction, besides the expected development of agriculture, might have far-reaching detrimental consequences for the conservation of indigenous ecosystems, even if such deleterious effects are not purported at the beginning.

The main reason for these contrasted aspects of earthworm activity, whether opportunities of threats according to circumstances, is that earthworms, in particular those species capable of digging the soil and mixing it with organic matter (i.e. mull-forming species) create disturbances at a scale ranging from the drilosphere to whole forest or grassland ecosystems. These disturbances propagate themselves through scales by self-entertained chains of action and reaction, also called positive feedback loops until some equilibrium is reached and stabilised through negative feedbacks (Jouquet et al. 2006). Here, the Mull humus system can be considered as the climax, generating by itself the conditions of its own development, earthworm activity favouring the nutrient-rich vegetation favouring in turn the development of nutrient-eager earthworm populations (Ponge 2003; Arim et al. 2006). Detrimental effects of earthworm invasion can be alleviated only by changing the humus form, which is an unattainable task, or waiting enough time for a new equilibrium once the spread of invaders arrives to completion (Arim et al. 2006).

4 Let Us Go to the Field

Depending on the environment in which they form, there are different types of O and A horizons. In a mesophilic environment (plains, low mountains, rainy tropical), the O horizons last very little (animals eat them quickly); the OH horizon is rarely present. An earthworm biomacrostructured A horizon develops. This humipedon is said to belong to a Mull system. Anecic and endogeic earthworms make the whole A horizon of Mull and Amphi humipedons. When all organic horizons have been integrated in the A horizon, these animals eat their own matured in the soil droppings, living on the stock of organic molecules they previously set into the soil. In doing so, they function as natural refills of the soil, preventing it from being stripped and returning to an inert earth phase (Bouché 2014). It is known that the intense exploitation of agricultural soils produces the joint loss of organic carbon and earthworm populations (Pelosi et al. 2014).

4.1 Earthworms' Horizons in Grassland and Forest Ecosystems

Pedoclimatic conditions (temperature, humidity, acidity) regulate biological activities, especially for earthworms that are active most of the time in Spring and Autumn. Depending on the environment they form, there are thus different O and A horizons.

In a mesophilic environment (plains, hills, rainy tropical environments), the O horizon disappears quickly, the organic matter being swiftly decomposed by organisms. Moreover, the OH horizon is rare, and an earthworm biomacrostructured A horizon usually develops. The resulting humipedon ranges in the Mull system with anecic and endogeic earthworms as being the main ones responsible for forming the A layer. Humus forms, following a transition from mesophilic to aridic or colder conditions, may progressively be a Eumull (OL/maA), Mesomull (OL/(vOL)/maA; vOL being an old, whitish litter; brackets mean a discontinuous layer), Oligomull (OL/(OF)/meA) and Dysmull (OL/OF/meA) (Fig. 4).

Going from mesophilic (warm and humid climate) towards conditions less favourable to anecic and endogeic earthworms (drought, low temperature, acidity), we progressively encounter Eumull: OL, maA; Mesomull: OL, (vOL), maA; vOL is

Fig. 4 Pictures from two permanent grasslands (the two photos on the left) and beech forests in Switzerland (the two photos on the right). A sequence in a Mull system, from the best integration of organic matter to the lowest one: Eumull, Mesomull, Oligomull and Dysmull, respectively from left to right

Fig. 5 Mull (left) and Amphi (right) on calcareous substrate. The OH horizon is present only in the Amphi humipedon. In the Amphi profile, earthworms are less active at the top, but can refuge in the cracks of the rock during dry periods

an old, whitish litter: brackets means that it is a discontinuous layer; Oligomull: OL, (OF), meA; Dysmull: OL, OF continuous, and meA.

If the climatic conditions become even more difficult, the genesis of humipedon knows a bifurcation:

1. On a basic substrate, the evolution goes towards an Amphi system (Fig. 5, right). Compared to a Mull (Fig. 5, left), anecic and endogeic earthworms are living deeper into the soil and cracks of carbonate rock. Progressively, less favourable conditions for anecic and endogeic earthworms lead from Leptoamphi (OH < 1 cm or discontinuous, maA) to Eumacroamphi (OH ≥ 1 cm, maA), Eumesoamphi (OH < 3 cm, meA) and finally to Eupachy amphi (OH ≥ 3 cm, meA).
2. On an acidic substrate, the evolution leads to a Moder system. An OH horizon appears above an A horizon no longer made by earthworms (Fig. 6, left). The sequence of horizons of a Moder system is as follows: OL, OF, OH and miA from arthropods or enchytraeids (presence of small lumps, mostly <1 mm) or non-zoogenic sgA or msA. Intergrades towards an Amphi system may show the presence of aggregates made by endogeic earthworms. A Mediterranean Amphi, with its typical maA horizon, is displayed (Fig. 6, right) as a comparison.

If the situation becomes even more difficult (high cold mountains, coniferous forests), the system moves towards Mor (acidic environment) or Tangel (limestone) systems:

1. On an acidic substrate, the sequence is OL, OF, OH and absence of A [the organic horizon lies directly on mineral horizons, generally an E horizon] (Fig. 7, left). A true Mor system (with a clear break between organic and mineral horizons,

Fig. 6 Moder (left) and Amphi (right) in Mediterranean ecosystems. The Moder generates on an acidic substrate and the Amphi on a calcareous one. Under a black organic OH horizon, a thin (2–4 cm) grey arthropod' A horizon characterises the Moder system; under a brown OH horizon, a thick (>10 cm) red-brown earthworm' A horizon takes place in the Amphi system

Fig. 7 Mor (left) and Tangel (right) systems. The first lies on the acidic sand of Fontainebleau (Parisian region, France); consisting of a pack of dry moss and invaded by yellow fungal hyphae, this environment is not suitable for the development of earthworms. Although at 2000 m, on the slope facing east of the basin of Lake Misurina (Belluno, Alps, Italy), earthworms are instead numerous in the Tangel, which develops in the Swiss stone pine forest, and the water flows to the bottom and moistens a very organic soil, in the coldest and most unfavourable periods, earthworms are probably able to overwinter in deep shelters among the cracks in the limestone rock

with degradation of the litter due to fungi) is always devoid of any category of earthworms (or other pedofauna).

2. On a limestone or dolomitic rock, OL, OF, OH and sometimes a thin clay-humic A is in contact with the rock, due to endogeic earthworms' activity or by decarbonatation of the carbonatic rock. Superficial Tangel discovered at high altitudse in stone pine forests (>2000 m) were inhabited by numerous endogeic earthworms (Fig. 7, right).

4.2 *Pioneer Ecosystems: Bryo and Rhizo Biological Connections*

Biotic and abiotic factors condition the formation of the humipedon, as reviewed in Zanella et al. (2018a). At the same altitude, a southern exposure is hotter and drier than a northern one. The micromorphology plays also an important role and greatly influences the amount of water present in the soil. Earthworms generally love aerated soils that are well-fed with water. It is very easy to find earthworms in dried up anmoors, for example. Everyone knows that earthworms are plentiful in ripening manure. Anecic earthworms do not like high C/N litter. Many conifers produce such type of litter.

Earthworms are numerous in pioneer environments such as Bryo and Rhizo topsoils (Fig. 8). These humipedons exist as a single system in the high mountains, forming thin layers in contact with the rock; in forest and grasslands at a lower altitude, they may overlap other systems. They can also be found at low altitudes, in environments of initial soils, including anthropised ones such as the boundary walls, sidewalks edges and house roofs. A Bryo system typically develops under a carpet

Fig. 8 Top left and right: a Rhizo system at 2500 m in calcareous alpine mountains. The clod appears as it is just detached from the rock and pours out. In it, you can see the roots of the plants immersed in a brownish-grey mass (bryAOH) created by the young earthworm captured and set on the hand (do not worry, we released it immediately and put it back under the clod). Down left and right: a Bryo system, on Cambrian acid red schists. Beneath its surface living layer, this system typically shows an organic horizon of decomposing mosses and/or lichens (bryOF and thin bryOH), which rests on this more or less fragmented rock. In this fractured rock took refuge two epigeic earthworms

of mosses or lichens and shows organic horizons due the degradation of the low part of the thin-leaved steams of these plants; a Rhizo system takes place under a carpet of grasses and is characterised by the dense intertwining of secondary roots of these plants. In the high mountains, these systems form thin humipedons in contact with the rock; at lower altitudes, they lie son on other systems (often Moder systems for Bryo, Mull systems for Rhizo), in forest and grasslands as well. Endogeic earthworms are common in Bryo systems under mosses and/or lichens between the organic rests and the fragmented rock (that may be acidic or basic). Epigeic earthworms live in thin Rhizo systems at high altitude (2000–2500 m) in dolomitic and calcareous mountains. At a lower altitude, numerous endogeic and anecic earthworms are found in more or less humid Rhizo Mull systems all over the world, as explained in Domínguez et al. (2018).

4.3 Earthworms and Humus Forms in Alluvial Ecosystems

In near-natural floodplains, the high turnover of habitats and ecosystems due to the alluvial dynamics leads to rapid succession and community-assembly changes compared to other ecosystems (Corenblit et al. 2009; Milner and Tockner 2010). The discharge fluctuation and the frequency and intensity of flooding events cause erosion and/or deposition of sediments with a wide texture heterogeneity. The organic matter supply is also highly variable, either from autochthonous or allochthonous origin. Consequently, a wide diversity of soil layers and types occurs. In this context, the animal and plant communities follow topographic gradients where pedological changes occur from months to several hundred years (Petts and Amoros 1996; Gurnell 2007). The parental mineral material, as its granulometry (slab, blocks, stones,sand, etc.) and its nature (mineralogy, weathering resistance), play a crucial role in the formation and evolution of such humipedons. To improve the current classification, we propose to add new intergrades humipedons being the transition between Para, Terrestrial and Anthropogenic systems, called *Litho* (from Greek "λίθος", rock), which decline in *Psammo* ("ψάμμος", sand), *Peyro* (Πέτρος, stone, pebble) and *Stereo* (στερεός, hard). Discussion and thinking are still under investigation, and the systematic position will soon be clarified and published. In floodplains, most of the humipedons belong to these *Litho* intergrades, especially in the vicinity of the river bed where coarse sediments predominate (Fig. 9). In addition, Para systems such as Crusto and Bryo (Zanella et al. 2018d) and the current Mull, Moder and Amphi systems also coexist in floodplains. As for soils, terraces' topography variability correlates to a gradient of humus systems from the river bed to the mature alluvial forest.

In this context, earthworms and plant roots are the primary ecosystem engineers that govern the formation of aggregates (Le Bayon et al. 2017; Schomburg et al. 2018b; Schomburg et al. 2018a, 2019a; Le Bayon et al. 2021a, b) .The earthworm communities' distribution and composition depend on soil organic matter content, soil texture, altitude and flooding characteristics (Emmerling 1995; Salomé et al.

Fig. 9 Three soil types along an elevation gradient from the direct vicinity of the Sarine River to the mature forest situated onto a terrace at a distance of 150 m (FR, Grandvillard, Switzerland). Left: willow bushes, a Litho intergrade (Psammo Mull) and a Leptosol (Fluvic), with a J horizon (from French "jeune", young, where biogenic aggregates consist of organo-mineral aggregates, however less stable than those observed in an A layer; Baize and Girard 2009); middle: alder forest, an Eumull and a Calcaric Fluvisol; right: ash and beech mature forest, an Oligomull and a Fluvic Cambisol. Organic matter and sand particles are first juxtaposed (right), then the integration begins with the formation of aggregates (middle), and finally, resistant and stable aggregates are formed (right). The formation of aggregates and the integration of the organic matter progressively increase under the influence of earthworms. The dominant species of earthworms in the willow bushes is *L. rubellus*; in the alder forests, *O. tyrtaeum*, *A. rosea*, *A. caliginosa* and *L. terrestris*; and in the beech mature forest, *L. terrestris*, *A. longa*, *A. nocturna*, *O. cyaneum* and *O. tyrtaeum*

2011). Habitat research projects on earthworm communities in floodplains are few. Most of them were conducted at the hill level and focused on flooded meadows in northern Germany (Plum and Filser 2005), grasslands in river valleys in Estonia (Ivask et al. 2007) or short grass and herbaceous vegetation in the Netherlands (Zorn et al. 2008). In 2013, Le Bayon et al. synthesised several studies on earthworm communities in floodplains along an altitudinal gradient from the hill (350 m) to the alpine level (2300 m). Thus, overall in several Swiss near-natural floodplains, Guenat et al. (1999), Bullinger-Weber et al. (2007, 2012) and Salomé et al. (2011) found 27 species and subspecies from subalpine to hill levels (Table 2). This record corresponds to two-thirds of all inventoried earthworm species and subspecies in Switzerland, confirming that floodplains are among the most diverse terrestrial ecosystems. The earthworm distribution is widely heterogeneous within the same floodplain (Bullinger-Weber et al. 2007) and in the same vegetation unit (Bullinger-Weber et al. 2012). The texture and the organic matter are the main drivers of the presence and activity of earthworms. Epigeics are often associated with a coarse sandy texture in contrast to anecics that prefer deep soils and mature forest stages providing the highest carbon content and the finest soil texture (Salomé et al. 2011).

Table 2 Species and subspecies of earthworms recovered in near-natural floodplains in Switzerland at several altitudinal levels. For an update of genus and species names, see the world database DriloBASE on http://taxo.drilobase.org

Epigeic
Bimastos eiseni (Levinsen, 1884)
Dendrobaena octaedra (Savigny, 1826)
Dendrobaena pygmaea cognetti (Michaelsen, 1903)
Dendrobaena pygmaea pygmaea (Savigny, 1826)
Dendrodrilus rubidus rubidus (Savigny, 1826)
Dendrodrilus subrubicundus (Eisen, 1874)
Eiseniella tetraedra tetraedra (Savigny, 1826)
Eisenia andrei (Bouché, 1972)
Lumbricus castaneus (Savigny, 1826)
Lumbricus meliboeus (Rosa, 1884)
Lumbricus rubellus (Hoffmeister, 1843)
Octodrilus argoviensis (Bretscher, 1899)

Anecic
Aporrectodea caliginosa nocturna (Evans, 1946)
Aporrectodea giardi giardi (Ribaucourt, 1901)
Aporrectodea longa longa (Ude, 1885)
Aporrectodea longa ripicola (Bouché, 1972)
Aporrectodea longa ripicola viridis (Bouché, 1972)
Lumbricus terrestris (Linnaeus, 1758)

Endogeic
Allolobophora chlorotica chlorotica (Savigny, 1826)
Aporrectodea caliginosa alternitosa (Bouché 1972)
Aporrectodea caliginosa caliginosa (Savigny, 1826)
Aporrectodea handlirschi handlirschi (Rosa, 1905)
Aporrectodea icterica icterica (Savigny, 1826)
Aporrectodea rosea rosea (Savigny, 1826)
Octolasion cyaneum (Savigny, 1826)
Octolasion tyrtaeum lacteum (Oerley, 1885)
Octolasion tyrtaeum tyrtaeum (Savigny, 1826)

From Le Bayon et al. (2013)

This enormous diversity and vast abundance of earthworms in floodplains provide ideal conditions to study the very first steps of soil pedogenesis, especially aggregate formation. This is particularly the case in calcareous conditions where the presence of calcium ions enhances the binding of mineral particles to organic matter to form the clay-humus complex. At the hill and the mountain level, near the Thur

River and the Sarine River, we studied the implication of vegetation and earthworms on the soil structure in the context of how soils may prevent flood events. As a result, the more the distance and the elevation from the river bed occur, the more the earthworm activity supplants those of roots, but both of them are still involved in the structuring processes (Bullinger-Weber et al. 2007, 2012; Salomé et al. 2011; Schomburg et al. 2018a). In the presence of coarse sand, plant roots dominate the aggregate building. Still, when the texture becomes finer and in width layers of silt, anecics and endogeics activities contribute primarily to creating aggregates and burrows (Schomburg et al. 2018b, 2019a, b).

At the mountain level near Grandvillard (altitude: 750 m a.s.l., FR, Switzerland), earthworms are active for more than 8 months per year from March to December. The pedoclimatic conditions allow maintain their bioturbation activities: the tree canopy, leaves on the soil surface and the constant humidity in the soil create an ideal habitat for earthworms, which may explain the high diversity and abundance encountered (Fig. 10).

These optimal conditions favour the rapid integration of the organic matter and the formation of aggregates. Approximately every 2 years, we go into the field with students at the same place to monitor the evolution of vegetation, humus forms and soils. In one of our study plots, a permanent soil profile is described as a Calcaric Skeletic Pantofluvic Fluvisol (Geoabruptic) according to the IUSS Working Group

Fig. 10 Top of the figure, from left to right, several earthworms recovered in floodplains and belonging to different ecological categories: *L. rubellus* (epigeic), *A. chlorotica* (endogeic), *A. caliginosa* (endogeic), *O. cyaneum* (endogeic) and *A. longa* (anecic). Bottom of the figure: earthworm surface casts that may serve as nutrient pools for plants, here a young beech sprout

WRB (IUSS Working Group WRB 2022), and the humus form is an Eumull (Zanella et al. 2018g) . The A layer mainly comprises macroaggregates (maA) formed by earthworms and plant roots. In 2015, a flood occurred, and a sandy deposition of several centimetres was observed on the soil surface (Fig. 11, top left). After only a few days, earthworms built up surface casts. In 2016, no trace of the flood was visible anymore, and the thickness of the A has increased by at least 1 cm. In 2020, the maA showed a high density of plant roots highlighting the richness of nutrients and the water supply in this layer due to the intensive earthworm bioturbation activities.

The case of the Sarine River highlights how fast the formation of aggregates can be. An organo-mineral layer may be formed in only several years, and the evolution from sandy and silty raw deposits to soil macroaggregates occurs as one moves away from the influence of the river. The very first steps of aggregates' production in such a context are still under research. The X-ray micro-computed tomography helped us to show that the volume ratio of mineral grains within the aggregates is significantly different according to earthworm species (Le Bayon et al. 2021a, b). The organic matter can also be discriminated, and we showed in controlled mesocosms that plant roots could develop in sandy layers. In contrast, earthworms preferentially selected the organic matter and the silt layers (Schomburg et al. 2019a).

Fig. 11 The temporal evolution (2015–2020) of the humipedon is influenced by earthworms and deposits from the Sarine River (FR, Grandvillard, Switzerland). The sediment deposition did not alter the earthworm activity (see the surface cast and the presence of *L. rubellus*). The humus form is an Eumull following the key of Zanella et al. (2018a, b, c, d, e, f, g). Horizons sequence is [nOL-vOL]/maAca/Dca/IIMca-Dca/IIDca according to Baize and Girard (2009). The soil is a FLUVIOSOL TYPIQUE carbonaté, leptique, polylithique, according to the classification of Baize and Girard (2009), and a Calcaric Skeletic Pantofluvic Fluvisol (Geoabruptic), according to the IUSS Working Group WRB (2022)

Fig. 12 From left to right, progressive colonisation by earthworms of the organo-mineral A horizon of a stream bank (Grandvillard, FR, Switzerland). On the left, plant roots dominate in the formation of organo-mineral aggregates close to the river bed (J horizon, see Fig. 9, left); on the right, an A horizon very rich in earthworm aggregates

In floodplains, there is obviously a step between sediment and stable biogenic aggregates. After discussions and debates with soil and humus forms specialists, the J horizon was created in the French classification (Baize and Girard 2009) to describe the initial Fluviosols better (Fluvisols, according to the IUSS Group 2022). In a J horizon (from French "jeune", young), biogenic aggregates are less stable than those observed in an A horizon (Fig. 9). Consequently, we can consider the J horizon as a functional soil layer that usually evolves towards a zoogenic (earthworms) A layer over time (Fig. 12).

This is typically the case we observed at the Sarine floodplain with the succession of soils, from the Leptosol (Fluvic) near the river to the Fluvic Cambisol in the mature and climacic forest. As these processes occur in the humipedon, the ultimate frontier would be to integrate this J layer in the classification key of humus forms, using a specific codification for the aggregates (jA in Psammo forms?). This would be very useful to describe better the transition between Prahumus systems and Terrestrial ones in environments containing vast amounts of sand and silt, as in floodplains. Ellen Desie (2020) encountered the same issue during her fieldwork in Belgium and the Netherlands. She found an accumulation of O layers (OL/OF/OH) directly deposited on sandy mineral layers and highlighted the need for intergrades.

In less active alluvial ecosystems, earthworms are implied in the formation of the humipedon, but they can be restricted to specific areas due to the permanent presence of water. Near the lake of Neuchâtel (St Blaise, altitude 440 m a.s.l., NE, Switzerland), a transect from the forest to the lake showed the influence of the water table on the vegetation and earthworm communities (Fig. 13). In the alluvial forest, all humus forms belong to the Mull system, most of them varying from Eumull to Oligomull. The biodiversity of earthworms is noticeable, with mainly endogeics and *L. terrestris* as the dominant anecic species. Going towards the lake, the

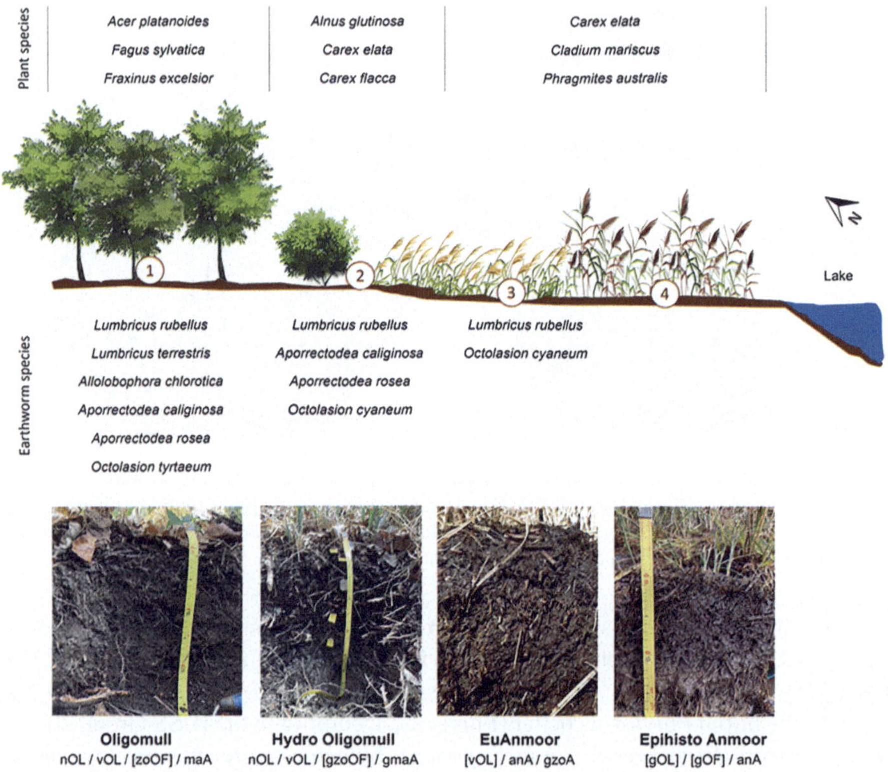

Fig. 13 Transect from the forest to the lake of Neuchâtel showing the humus forms and the horizons sequences, the main earthworm species and the dominant vegetation (St Blaise, NE, Switzerland). The humus forms indicate the progressive influence of the water (total distance: 60 m). The transect was made in the Spring of 2022. Photos credit: Maeder, Bovet, Bulliard and Guerra; scheme: Le Bayon and Semeraro

influence of water begins, the presence of *Carex flacca* being an indication. Earthworms are still numerous and active, and their casts can be easily seen in the A layer of the Hydromull. The increasing effect of water limits the presence and activity of earthworms in the third sampling point: *L. rubellus* (a pioneer species) and *O. cyaneum* (known for its affinity with humid soils) are active when the water level gets down due to the fluctuations of the lake, especially in drier periods of the year. Biogenic aggregates compose the A horizon and are especially observable in May and June. In the vicinity of the lake, no earthworms were observed because of the permanent presence of water. Such transects are pedagogic to students (and researchers!) to highlight how necessary humus forms are for explaining the complex relationships between soils and vegetation.

5 Earthworm Functional Domains as Extended Phenotypes: Eco-evolutionary Aspects

Earthworms, as soil engineers, transform the soil within their functional domain (Lavelle 2000, 2002), also called drilosphere (Brown et al. 2000) or in a few instances vermisphere (Hamilton and Dindal 1983). They create galleries (Amossé et al. 2015), incorporate organic matter to mineral particles (Stout and Goh 1980) and modify the composition of microbial (Dempsey et al. 2013), animal (Migge-Kleian et al. 2006) and plant populations (Hale et al. 2006), favouring or disfavouring species in the space of soil where they live as multispecies assemblages in spite of intense competition (Decaëns et al. 2008). As such they can be considered as agents of the building of ecosystems and communities (Ponge 2021), ensuring multiple ecosystem services far beyond their direct interactions with plants, animals and microbes (Liu et al. 2019). Here we suggest that the concept of extended composite phenotypes (Phillips 2009), covering the environmental properties durably modified by an organism, could be put in synonymy with the functional domain. This has strong consequences for eco-evolutionary issues (Bailey 2012). Considering the functional domain of an earthworm species as an extended composite phenotype means it has evolved and may continue to evolve with it.

This is reminiscent of the concept of niche construction (Odling-Smee et al. 2013), meaning that the niche durably created by an organism (in fact an ecosystem engineer) may, even after its death, be considered as a driving force of evolution. Like coral reefs of the Jurassic Period, the variety of which created various calcareous substrates influencing the composition and spatial distribution of present-day plant communities (Ricci et al. 2018), earthworm activities modify soil properties to an extent that survive their collapse, and even take part to global climate changes. A good example is the case of Mollisols (highly fertile soils with a deep homogeneous mineral-organic horizon), which accompanied the rise of Poaceae in the US Great Plains during the Neogene Period (Retallack 1997) and were thought to be responsible for further global cooling (Retallack 2013). According to what we can deduct from the micromorphology of mollic epipedons (Pawluk and Bal 2015), Mollisols (also called Chernozems) result from the mixing activity of soil engineers, mainly earthworms, although the presence in North America of digging and burying earthworm species (mostly lumbricids) before the first European settlements is still a matter of conjecture (Hendrix et al. 2008), contrary to the European framework (Dreibrodt et al. 2022).

The still postulated (because of the absence of fossils) coevolution of grasses with earthworms is mirrored in the present-day rich and diversified earthworm populations prevailing in permanent grasslands (Ponge et al. 2013). The well-known association of earthworms with the Mull humus system and arthropods with the Moder humus systems and the paucity of fauna in the Mor humus systems led Ponge (2003) to hypothesise that these three main terrestrial humus systems were linked by evolutionary relationships, from Mor (before the rise of arthropods) to Moder (before the rise of terrestrial annelids) and then to Mull. Much more research is needed on the role played by earthworms in the evolution of terrestrial biomes, like this has been fruitfully done in the coevolution of hosts with their parasites (May and Anderson 1990) and plants with their pollinators (Kiester et al. 1984). Here, the postulated processes resort to diffuse coevolution (Janzen 1980), including as a phenotypic set of traits the environment modified by earthworms in addition to signalling strategies (Puga-Freitas and Blouin 2015).

References

Adhikari K, Hartemink AE (2016) Linking soils to ecosystem services—a global review. Geoderma 262:101–111. https://doi.org/10.1016/j.geoderma.2015.08.009

Amossé J, Turberg P, Kohler-Milleret R, Gobat JM, Le Bayon RC (2015) Effects of endogeic earthworms on the soil organic matter dynamics and the soil structure in urban and alluvial soil materials. Geoderma 243–244:50–57. https://doi.org/10.1016/j.geoderma.2014.12.007

Andreetta A, Ciampalini R, Moretti P, Vingiani S, Poggio G, Matteucci G, Tescari F, Carnicelli S (2011) Forest humus forms as potential indicators of soil carbon storage in Mediterranean environments. Biol Fertil Soils 47:31–40. https://doi.org/10.1007/s00374-010-0499-z

Arim M, Abades SR, Neill PE, Lima M, Marquet PA (2006) Spread dynamics of invasive species. Proc Natl Acad Sci U S A 103:374–378. https://doi.org/10.1073/pnas.0504272102

Bailey NW (2012) Evolutionary models of extended phenotypes. Trends Ecol Evol 27(10):561–569 https://doi.org/10.1016/j.tree.2012.05.011

Bais HP, Weir TL, Perry LG, Gilroy S, Vivanco JM (2006) The role of root exudates in rhizosphere interactions with plants and other organisms. Annu Rev Plant Biol 57:233–266. https://doi.org/10.1146/annurev.arplant.57.032905.105159

Baize D, Girard M-C (2009) Référentiel pédologique 2008. Quae, Versailles, France, 405 p. https://www.afes.fr/wpcontent/uploads/2023/10/Referentiel_Pedologique_2008.pdf

Blanchart A, Séré G, Cherel J, Warot G, Marie S, Noël CJ, Louis MJ, Christophe S (2018) Towards an operational methodology to optimize ecosystem services provided by urban soils. Landsc Urban Plan 176:1–9. https://doi.org/10.1016/j.landurbplan.2018.03.019

Bocock KL (1964) Changes in the amounts of dry matter, nitrogen, carbon and energy in decomposing woodland leaf litter in relation to the activities of the soil fauna. J Ecol 52:273–284. https://doi.org/10.2307/2257595

Bohlen PJ, Edwards CA (1995) Earthworm effects on N dynamics and soil respiration in microcosms receiving organic and inorganic nutrients. Soil Biol Biochem 27:341–348. https://doi.org/10.1016/0038-0717(94)00184-3

Bouché MB (2014) Des vers de terre et des hommes. Actes Sud, Arles, 336 p.

Brêthes A, Brun J, Jabiol B, Ponge J, Toutain F (1995) Classification of forest humus forms: a French proposal. Annales des Sciences Forestières 52:535–546. https://doi.org/10.1051/forest:19950602

Brown GG, Barois I, Lavelle P (2000) Regulation of soil organic matter dynamics and microbial activity in the drilosphere and the role of interactions with other edaphic functional domains. Eur J Soil Biol 36:177–198. https://doi.org/10.1016/S1164-5563(00)01062-1

Bullinger-Weber G, Le Bayon RC, Guenat C, Gobat JM (2007) Influence of some physicochemical and biological parameters on soil structure formation in alluvial soils. Eur J Soil Biol 43:57–70. https://doi.org/10.1016/j.ejsobi.2006.05.003

Bullinger-Weber G, Guenat C, Salomé C, Gobat JM, Le Bayon RC (2012) Impact of flood deposits on earthworm communities in alder forests from a subalpine floodplain (Kandersteg, Switzerland). Eur J Soil Biol 49:5–11. https://doi.org/10.1016/j.ejsobi.2011.08.001

Calabrese MS, Mancabelli A, Nicolini G, Sartori G, Zanella A (1996) Humus forestali del Trentino/ Forest humus in Trentino (Italy). Report del Centro di Ecologia Alpina 9:1–53

Cameron EK, Proctor HC, Bayne EM (2013) Effects of an ecosystem engineer on below-ground movement of microarthropods. PLoS One 8:e62796. https://doi.org/10.1371/journal.pone.0062796

Chauvel A, Grimaldi M, Barros E, Blanchart E, Desjardins T, Sarrazin M, Lavelle P (1999) Pasture damage by an Amazonian earthworm. Nature 398 :32–33. https://doi.org/10.1038/17946

Corenblit D, Steiger J, Gurnell AM, Tabacchi E, Roques L (2009) Control of sediment dynamics by vegetation as a key function driving biogeomorphic succession within fluvial corridors. Earth Surf Process Landf 34:1790–1810. https://doi.org/10.1002/esp.1876

Curry JP, Cotton DCF (1983) Earthworms and land reclamation. In: Satchell JE (ed) Earthworm ecology. Springer, Dordrecht, The Netherlands, pp 215–228

Decaëns T, Margerie P, Aubert M, Hedde M, Bureau F (2008) Assembly rules within earthworm communities in North-Western France—a regional analysis. Appl Soil Ecol 39:321–335. https://doi.org/10.1016/j.apsoil.2008.01.007

Dempsey MA, Fisk MC, Yavitt JB, Fahey TJ, Balser TC (2013) Exotic earthworms alter soil microbial community composition and function. Soil Biol Biochem 67:263–270. https://doi.org/10.1016/j.soilbio.2013.09.009

de Nicola C, Zanella A, Testi A, Fanelli G, Pignatti S (2014) Humus forms in a Mediterranean area (Castelporziano Reserve, Rome, Italy): Classification, functioning and organic carbon storage. Geoderma 235–236:90–99. https://doi.org/10.1016/j.geoderma.2014.06.033

Department of Economic and Social Affairs of the United Nations (2022) https://population.un.org/wpp/

Desie E (2020) Litter effects on structure and functions of the belowground forest ecosystem. PhD thesis, KU Leuven, Faculty of Bioscience Engineering, Heverlee, Belgium, 241 pp. Bioscience Engineering, Heverlee, Belgium,

Dominati E, Patterson M, Mackay A (2010) A framework for classifying and quantifying the natural capital and ecosystem services of soils. Ecol Econ 69 :1858–1868. https://doi.org/10.1016/j.ecolecon.2010.05.002

Domínguez A, Jiménez JJ, Ortíz CE, Bedano JC (2018) Soil macrofauna diversity as a key element for building sustainable agriculture in Argentine Pampas. Acta Oecologica 92:102–116. https://doi.org/10.1016/j.actao.2018.08.012

Dreibrodt S, Hofmann R, Dal Corso M, Bork HR, Duttmann R, Martini S, Saggau P, Schwark L, Shatilo L, Videiko M, Nadeau MJ, Grootes PM, Kirleis W, Müller J (2022) Earthworms, Darwin and prehistoric agriculture-Chernozem genesis reconsidered. Geoderma 409:115607. https://doi.org/10.1016/j.geoderma.2021.115607

Edwards CA, Arancon NQ (2022) Biology and ecology of earthworms, 4th edn. Springer, New York, 567 p. https://doi.org/10.1007/978-0-387-74943-3

Eisenhauer N (2010) The action of an animal ecosystem engineer: Identification of the main mechanisms of earthworm impacts on soil microarthropods. Pedobiologia 53:43–352. https://doi.org/10.1016/j.pedobi.2010.04.003

Emmerling C (1995) Long-term effects of inundation dynamics and agricultural land-use on the distribution of soil macrofauna in fluvisols. Biol Fertil Soils 20:130–136. https://doi.org/10.1007/BF00336592

Fonte SJ, Quintero DC, Velásquez E, Lavelle P (2012) Interactive effects of plants and earthworms on the physical stabilization of soil organic matter in aggregates. Plant Soil 359:205–214. https://doi.org/10.1007/s11104-012-1199-2

Galvan P, Ponge J-F, Chersich S, Zanella A (2008) Humus components and soil biogenic structures in norway spruce ecosystems. Soil Sci Soc Am J 72:548–557. https://doi.org/10.2136/sssaj2006.0317

Guenat C, Bureau F, Weber G, Toutain F (1999) Initial stages of soil formation in a riparian zone: Importance of biological agents and lithogenic inheritance in the development of the soil structure. Eur J Soil Biol 35:153–161. https://doi.org/10.1016/S1164-5563(10)70001-7

Gurnell AM (2007) Analogies between mineral sediment and vegetative particle dynamics in fluvial systems. Geomorphology 89:9–22. https://doi.org/10.1016/j.geomorph.2006.07.012

Hale CM, Frelich LE, Reich PB (2006) Changes in hardwood forest understory plant communities in response to European earthworm invasions. Ecology 87:1637–1649. https://doi.org/10.1890/0012-9658(2006)87[1637:CIHFUP]2.0.CO;2

Hamilton W, Dindal D (1983) The vermisphere concept: earthworm activity and sewage sludge. Biocycle 24:54–55. https://archive.org/details/hamilton-dindal-1983

Hamilton WE, Sillman DY (1989) Influence of earthworm middens on the distribution of soil microarthropods. Biol Fertil Soils 8:279–284. https://doi.org/10.1007/BF00266491

Hartmann F (1970) Gli humus forestali—Diagnosi degli humus forestali su basi biomorfologiche. CEDAM, Padova, Italy, pp 190–275

Hendrix PF, Baker GH, Callaham MA, Damoff GA, Fragoso C, González G, James SW, Lachnicht SL, Winsome T, Zou X (2006) Invasion of exotic earthworms into ecosystems inhabited by native earthworms. Biol Invasions 8:1287–1300. https://doi.org/10.1007/s10530-006-9022-8

Hendrix PF, Callaham MA, Drake JM, Huang CY, James SW, Snyder BA, Zhang W (2008) Pandora's box contained bait: The global problem of introduced earthworms. Annu Rev Ecol Evol Syst 39:593–613. https://doi.org/10.1146/annurev.ecolsys.39.110707.173426

IUSS Working Group WRB (2022) World Reference Base for Soil Resources. International soil classification system for naming soils and creating legends for soil maps. In: International Union of Soil Sciences (IUSS), 4th edn, Vienna, Austria. https://wrb.isric.org/files/WRB_fourth_edition_2022-12-18.pdf

Ivask M, Truu J, Kuu A, Truu M, Leito A (2007) Earthworm communities of flooded grasslands in Matsalu, Estonia. Eur J Soil Biol 43:71–76. https://doi.org/10.1016/j.ejsobi.2006.09.009

Janzen DH (1980) When is it Coevolution? Evolution 34:611–612. https://doi.org/10.1111/j.1558-5646.1980.tb04849.x

Jouquet P, Dauber J, Lagerlöf J, Lavelle P, Lepage M (2006) Soil invertebrates as ecosystem engineers: Intended and accidental effects on soil and feedback loops. Appl Soil Ecol 32:153–164. https://doi.org/10.1016/j.apsoil.2005.07.004

Kiester AR, Lande R, Schemske DW (1984) Models of coevolution and speciation in plants and their pollinators. Am Nat 124:220–243. https://doi.org/10.1086/284265

Kubiëna W (1953) The soils of Europe. Illustrated diagnosis and systematics. Thomas Murby, London, 318 p. + 26 plates

Larondelle N, Haase D, Kabisch N (2014) Mapping the diversity of regulating ecosystem services in European cities. Global Environ Change 26:119–129. https://doi.org/10.1016/j.gloenvcha.2014.04.008

Lavelle P (2000) Ecological challenges for soil science. Soil Sci 165:73–86. https://doi.org/10.1097/00010694-200001000-00009

Lavelle P (2002) Functional domains in soils. Ecol Res 17:441–450. https://doi.org/10.1046/j.1440-1703.2002.00509.x

Le Bayon RC, Binet F (1999) Rainfall effect on erosion of earthworm casts and phosphorus transfers by water runoff. Biol Fertil Soils 30:7–13. https://doi.org/10.1007/s003740050580

Le Bayon RC, Moreau S, Gascuel-Odoux C, Binet F (2002) Annual variations in earthworm surface-casting activity and soil transport by water runoff under a temperate maize agroecosystem. Geoderma 106:121–135. https://doi.org/10.1016/S0016-7061(01)00121-5

Le Bayon RC, Bullinger-Weber G, Gobat J-M, Guenat C (2013) Earthworm communities as indicators for evaluating floodplain restoration success. In: Environmental management, restoration and ecological implications. NOVA Science, New York, pp 47–68. https://novapublishers.com/shop/floodplains-environmental-management-restoration-and-ecological-implications/

Le Bayon RC, Bullinger-Weber G, Schomburg A, Turberg P, Schlaepfer R, Guenat C (2017) Earthworms as ecosystem engineers: A review. In: Earthworms: types, roles and research. Nova Science, New York, pp 129–177. https://novapublishers.com/shop/earthworms-types-roles-and-research/

Le Bayon RC, Bullinger G, Schomburg A, Turberg P, Brunner P, Schlaepfer R, Guenat C (2021a) Earthworms, plants, and soils. In: Hunt A, Egli M, Faybishenko B (eds) Hydrogeology, chemical weathering, and soil formation. American Geophysical Union, Washington, DC, pp 81–103. https://doi.org/10.1002/9781119563952.ch4

Le Bayon RC, Guenat C, Schlaepfer R, Fischer F, Luiset A, Schomburg A, Turberg P (2021b) Use of X-ray microcomputed tomography for characterizing earthworm-derived belowground soil aggregates. Eur J Soil Sci 72:1113–1127. https://doi.org/10.1111/ejss.12950

Liu R, Zhu F, Song N, Yang X, Chai Y (2013) Seasonal distribution and diversity of ground arthropods in microhabitats following a shrub plantation age sequence in desertified steppe. PLoS One 8:e77962. https://doi.org/10.1371/journal.pone.0077962

Liu T, Chen X, Gong X, Lubbers IM, Jiang Y, Feng W, Li X, Whalen JK, Bonkowski M, Griffiths BS, Hu F, Liu M (2019) Earthworms coordinate soil biota to improve multiple ecosystem functions. Curr Biol 29:3420–3429.e5. https://doi.org/10.1016/j.cub.2019.08.045

May R, Anderson R (1990) Parasite-host coevolution. Parasitology 100:S89–S101. https://doi.org/10.1017/S0031182000073042

Migge-Kleian S, McLean MA, Maerz JC, Heneghan L (2006) The influence of invasive earthworms on indigenous fauna in ecosystems previously uninhabited by earthworms. Biol Invasions 8:1275–1285. https://doi.org/10.1007/s10530-006-9021-9

Milcu A, Schumacher J, Scheu S (2006) Earthworms (Lumbricus terrestris) affect plant seedling recruitment and microhabitat heterogeneity. Funct Ecol 20:261–268. https://doi.org/10.1111/j.1365-2435.2006.01098.x

Millennium Ecosystem Assessment (2005) Living beyond our means: natural assets and human well-being statement from the board. Washington, D.C. : World Bank Group. http://documents.worldbank.org/curated/en/366241468134713390/Living-beyondour-means-natural-assets-and-human-well-being

Milner AM, Tockner K (2010) River science—what has it contributed to general ecological theory? River Res Appl 26:1–4. https://doi.org/10.1002/rra.1317

Müller PE (1887) Studien über die natürlichen Humusformen und deren Einwirkung auf Vegetation und Boden, mit analytischen Belegen von C.F.A. Tuxen. Springer, Berlin, 324 p. https://doi.org/10.5962/bhl.title.20253

Nuzzo VA, Maerz JC, Blossey B (2009) Earthworm invasion as the driving force behind plant invasion and community change in Northeastern North American forests. Conserv Biol 23:966–974. https://doi.org/10.1111/j.1523-1739.2009.01168.x

Oades JM (1993) The role of biology in the formation, stabilization and degradation of soil structure. Geoderma 56:377–400. https://doi.org/10.1016/0016-7061(93)90123-3

Odling-Smee J, Erwin DH, Palkovacs EP, Feldman MW, Laland KN (2013) Niche construction theory: A practical guide for ecologists. Q Rev Biol 88:4–28. https://doi.org/10.1086/669266

Pawluk S, Bal L (2015) Micromorphology of selected mollic epipedons. In: Douglas LA, Thompson ML (eds) Soil micromorphology and soil classification. Soil Science Society of America, Inc. https://doi.org/10.2136/sssaspecpub15.c4

Pelosi C, Barot S, Capowiez Y, Hedde M, Vandenbulcke F (2014) Pesticides and earthworms. A review. Agron Sustain Dev 34:199–228. https://doi.org/10.1007/s13593-013-0151-z

Petts GE, Amoros C (1996) Fluvial hydrosystems. Chapman & Hall, London

Phillips JD (2009) Soils as extended composite phenotypes. Geoderma 149:143–151. https://doi.org/10.1016/j.geoderma.2008.11.028

Plum NM, Filser J (2005) Floods and drought: Response of earthworms and potworms (Oligochaeta: Lumbricidae, Enchytraeidae) to hydrological extremes in wet grassland. Pedobiologia 49:443–453. https://doi.org/10.1016/j.pedobi.2005.05.004

Ponge JF (2003) Humus forms in terrestrial ecosystems: A framework to biodiversity. Soil Biol Biochem 35:935–945. https://doi.org/10.1016/S0038-0717(03)00149-4

Ponge JF (2021) Communities, ecosystem engineers, and functional domains. Ecol Res 36:766–777. https://doi.org/10.1111/1440-1703.12247

Ponge JF, André J, Zackrisson O, Bernier N, Nilsson MC, Gallet C (1998) The forest regeneration puzzle: Biological mechanisms in humus layer and forest vegetation dynamics. Bioscience 48:523–530. https://doi.org/10.2307/1313314

Ponge JF, Pérès G, Guernion M, Ruiz-Camacho N, Cortet J, Pernin C, Villenave C, Chaussod R, Martin-Laurent F, Bispo A, Cluzeau D (2013) The impact of agricultural practices on soil biota: A regional study. Soil Biol Biochem 67:271–284. https://doi.org/10.1016/j.soilbio.2013.08.026

Puga-Freitas R, Blouin M (2015) A review of the effects of soil organisms on plant hormone signalling pathways. Environ Exp Bot 114:104–116. https://doi.org/10.1016/j.envexpbot.2014.07.006

Retallack GJ (1997) Neogene expansion of the North American prairie. Palaios 12:380–390. https://doi.org/10.2307/3515337

Retallack GJ (2013) Global cooling by grassland soils of the geological past and near future. Annu Rev Earth Planet Sci 41:69–86. https://doi.org/10.1146/annurev-earth-050212-124001

Ricci C, Lathuilière B, Rusciadelli G (2018) Coral communities, zonation and paleoecology of an upper Jurassic reef complex (ellipsactinia limestones, central Apennines, Italy). Rivista Italiana di Paleontologia e Stratigrafia 124:433–508. https://doi.org/10.13130/2039-4942/10611

Ritchie H, Samborska V, Roser M (2024) Urbanization. Published online at OurWorldInData.org. Retrieved from https://ourworldindata.org/urbanization

Salmon S (2001) Earthworm excreta (mucus and urine) affect the distribution of springtails in forest soils. Biol Fertil Soils 34:304–310. https://doi.org/10.1007/s003740100407

Salomé C, Guenat C, Bullinger-Weber G, Gobat JM, Le Bayon RC (2011) Earthworm communities in alluvial forests: Influence of altitude, vegetation stages and soil parameters. Pedobiologia 54:S89–S98. https://doi.org/10.1016/j.pedobi.2011.09.012

Schomburg A, Schilling OS, Guenat C, Schirmer M, Le Bayon RC, Brunner P (2018a) Topsoil structure stability in a restored floodplain: Impacts of fluctuating water levels, soil parameters and ecosystem engineers. Sci Total Environ 639:1610–1622. https://doi.org/10.1016/j.scitotenv.2018.05.120

Schomburg A, Verrecchia EP, Guenat C, Brunner P, Sebag D, Le Bayon RC (2018b) Rock-Eval pyrolysis discriminates soil macro-aggregates formed by plants and earthworms. Soil Biol Biochem 117:117–124. https://doi.org/10.1016/j.soilbio.2017.11.010

Schomburg A, Brunner P, Turberg P, Guenat C, Riaz M, Le Bayon RC, Luster J (2019a) Pioneer plant Phalaris arundinacea and earthworms promote initial soil structure formation despite strong alluvial dynamics in a semi-controlled field experiment. Catena 180:41–54. https://doi.org/10.1016/j.catena.2019.04.001

Schomburg A, Sebag D, Turberg P, Verrecchia EP, Guenat C, Brunner P, Adatte T, Schlaepfer R, Le Bayon RC (2019b) Composition and superposition of alluvial deposits drive macro-biological soil engineering and organic matter dynamics in floodplains. Geoderma 355:113899. https://doi.org/10.1016/j.geoderma.2019.113899

Smith P, Cotrufo MF, Rumpel C, Paustian K, Kuikman PJ, Elliott JA, McDowell R, Griffiths RI, Asakawa S, Bustamante M, House JI, Sobocká J, Harper R, Pan G, West PC, Gerber JS, Clark JM, Adhya T, Scholes RJ, Scholes MC (2015) Biogeochemical cycles and biodiversity as key drivers of ecosystem services provided by soils. Soil 1:665–685. https://doi.org/10.5194/soil-1-665-2015

Sofo A, Zanella A, Ponge JF (2022) Soil quality and fertility in sustainable agriculture, with a contribution to the biological classification of agricultural soils. Soil Use Manag 38:1085–1112. https://doi.org/10.1111/sum.12702

Stout JD, Goh KM (1980) The use of radiocarbon to measure the effects of earthworms on soil development. Radiocarbon 22:892–896. https://doi.org/10.1017/s0033822200010298

Sugden A, Stone R, Ash C (2004) Ecology in the underworld. Science 304:1613–1613. https://doi.org/10.1126/science.304.5677.1613

Zanella A, Berg B, Ponge JF, Kemmers RH (2018a) Humusica 1, article 2: Essential bases—Functional considerations. Appl Soil Ecol 122:22–41

Zanella A, Ponge JF, Briones MJI (2018b) Humusica 1, article 8: Terrestrial humus systems and forms—biological activity and soil aggregates, space-time dynamics. Appl Soil Ecol 122:103–137

Zanella A, Ponge JF, de Waal R, Ferronato C, de Nobili M, Juilleret J (2018c) Humusica 1, article 3: Essential bases—quick look at the classification. Appl Soil Ecol 122:42–55. https://doi.org/10.1016/j.apsoil.2017.05.025

Zanella A, Ponge JF, Fritz I, Pietrasiak N, Matteodo M, Nadporozhskaya M, Juilleret J, Tatti D, Le Bayon RC, Rotschild L, Mancinelli R (2018d) Humusica 2, article 13: Para humus systems and forms. Appl Soil Ecol 122:181–199. https://doi.org/10.1016/j.apsoil.2017.09.043

Zanella A, Ponge JF, Gobat JM, Juilleret J, Blouin M, Aubert M, Chertov O, Rubio JL (2018e) Humusica 1, article 1: Essential bases—vocabulary. Appl Soil Ecol 122:10–21. https://doi.org/10.1016/j.apsoil.2017.07.004

Zanella A, Ponge JF, Jabiol B, Sartori G, Kolb E, Gobat JM, Le Bayon RC, Aubert M, de Waal R, van Delft B, Vacca A, Serra G, Chersich S, Andreetta A, Cools N, Englisch M, Hager H, Katzensteiner K, Brêthes A, de Nicola C, Testi A, Bernier N, Graefe U, Juilleret J, Banas D, Garlato A, Obber S, Galvan P, Zampedri R, Frizzera L, Tomasi M, Menardi R, Fontanella F, Filoso C, Dibona R, Bolzonella C, Pizzeghello D, Carletti P, Langohr R, Cattaneo D, Nardi S, Nicolini G, Viola F (2018f) Humusica 1, article 4: Terrestrial humus systems and forms—Specific terms and diagnostic horizons. Appl Soil Ecol 122:56–74. https://doi.org/10.1016/j.apsoil.2017.07.005

Zanella A, Ponge JF, Jabiol B, Sartori G, Kolb E, Le Bayon RC, Gobat JM, Aubert M, de Waal R, van Delft B, Vacca A, Serra G, Chersich S, Andreetta A, Kõlli R, Brun JJ, Cools N, Englisch M, Hager H, Katzensteiner K, Brêthes A, de Nicola C, Testi A, Bernier N, Graefe U, Wolf U, Juilleret J, Garlato A, Obber S, Galvan P, Zampedri R, Frizzera L, Tomasi M, Banas D, Bureau F, Tatti D, Salmon S, Menardi R, Fontanella F, Carraro V, Pizzeghello D, Concheri G, Squartini A, Cattaneo D, Scattolin L, Nardi S, Nicolini G, Viola F (2018g) Humusica 1, article 5: Terrestrial humus systems and forms—keys of classification of humus systems and forms. Appl Soil Ecol 122:75–86. https://doi.org/10.1016/j.apsoil.2017.06.012

Zhang H, Schrader S (1993) Earthworm effects on selected physical and chemical properties of soil aggregates. Biol Fertil Soils 15:229–234. https://doi.org/10.1007/BF00361617

Zhang T, Li S, Sun X, Zhang Y, Gong X, Fu Y, Jia L (2015) The earthworm eisenia fetida can help desalinate a coastal saline soil in Tianjin, North China. PLoS One 10:e0144709. https://doi.org/10.1371/journal.pone.0144709

Zorn MI, van Gestel CAM, Morrien E, Wagenaar M, Eijsackers H (2008) Flooding responses of three earthworm species, Allolobophora chlorotica, Aporrectodea caliginosa and Lumbricus rubellus, in a laboratory-controlled environment. Soil Biol Biochem 40:87–593. https://doi.org/10.1016/j.soilbio.2007.06.028

Earthworms and Soil Nutrients in Agricultural Ecosystems

Joann K. Whalen and Zhor Abail

Abstract Earthworms are abundant invertebrates in agricultural soils that have a myriad of well-documented effects on soil functions and plant production. Earthworm activities are linked to changes in hydrological flows and accelerated biogeochemical cycling, which readily connects earthworms to the buffering and regulatory processes at an ecosystem level. In this chapter, we explore the relationship between the actions of earthworms in agricultural soils and their relation to biogeochemical cycling, from the perspective of producing an abundance of nutritious food and modulating the nutrient flows in cultivated ecosystems.

1 Introduction

Many people consider earthworms to be an indicator of soil fertility. Earthworm-worked soil is porous and friable, supporting the growth of a deep, extensive plant root system for better anchorage and acquisition of water and nutrients. Earthworms also increase the decomposition of nutrient-rich organic residues through litter comminution. Fragmentation of plant debris by earthworms disrupts cells and releases soluble compounds from vacuoles and other membrane-bound organelles. Some of these soluble compounds are nutrient ions (NO_3^-, $H_2PO_4^-$, SO_4^{2-}, K^+, Ca^{2+}, and Mg^{2+}) along with organic acids, sugars, proteins, and enzymes (Wink 1993). Earthworms bury organic residues in the soil profile, where they are further reduced

J. K. Whalen (✉)
Department of Natural Resource Sciences, Macdonald Campus of McGill University, Sainte-Anne-de-Bellevue, QC, Canada

Chair of Soil Sciences, Mohammed VI Polytechnic University, Ben Guerir, Morocco
e-mail: joann.whalen@mcgill.ca

Z. Abail
Regional Center of Agricultural Research of Kenitra, National Institute of Agricultural Research, Rabat, Morocco
e-mail: zhor.abail@inra.ma

Y. Kooch, Y. Kuzyakov (eds.), *Earthworms and Ecological Processes*,
https://doi.org/10.1007/978-3-031-64510-5_7

in size by soil mesofauna and colonized by soil microorganisms that enzymatically hydrolyze proteins, phospholipids, and other compounds, releasing NH_4^+, $H_2PO_4^-$, and SO_4^{2-} as by-products. These nutrient ions, together with the soluble ions and molecules of boron, chloride, copper, iron, manganese, molybdenum, nickel, and zinc, are absorbed by growing plants. Thus, earthworms increase the quantity of plant-available nutrients for crops.

Earthworms consume plant debris along with sand grains and agglomerated organo-minerals containing silt and clay. In the gizzard, muscular contractions grind organics with mineral grit. The combined crushing and abrasive forces reduce the size of the organic fragments and mineral agglomerates. Organic and mineral particles then move into the digestive tract, where they are mixed with water and intestinal mucus, activating the growth of ingested microorganisms. As the mixture passes through the digestive tract, microorganisms that were ingested or reside in the gut will release enzymes that hydrolyze nutrients from organic polymers and produce acidic compounds that solubilize nutrients bound to mineral surfaces. Some nutrients are assimilated into earthworm tissues for cellular repair, maintenance, and growth. Unassimilated nutrients transit the intestinal tract with partially digested organic matter that binds to minerals. These nutrient-enriched aggregates (casts) are defecated in the soil by the earthworm. Casts of *Amynthas khami* had eight times more NH_4^+, seven times more NO_3^-, 12 times more extractable K, and two times more extractable P than soil aggregates of the same size that were not consumed by earthworms (leached with water; Jouquet et al. 2011). Mineralization and biological weathering in the earthworm gut increase the nutrient concentration of casts, relative to the bulk soil.

Earthworm engineering creates hotspots of soil fertility. Middens are one such hotspot, created when earthworms collect plant and animal debris from across the soil surface and stockpile it in discrete locations. Middens like those constructed by *Lumbricus terrestris* (Fig. 1) contain stem and leaf litter that is mixed with casts in a pile of 10–12 cm in diameter above their burrow opening. Midden-building is typical of the anecic earthworms and reflects their abundance in agroecosystems. About 5–31 middens m^{-2} covered 3–16% of the soil surface in a bean-lupin agroecosystem (Schrader and Seibel 2001) and middens occupied from 15% to 25% of the soil surface in a no-till corn agroecosystem (Subler and Kirsch 1998). In a minimum-tillage winter wheat agroecosystem, middens containing fresh leaf and farmyard manure were a hotspot for mesofauna and endogeic earthworms; these middens had more plant-available nutrients than the reference soil, with at least four times more P, 1.5-fold higher K, six times more S, and about seven times greater Mn concentration (extracted with 0.01 $CaCl_2$ solution; Stroud et al. 2016).

Burrowing is another example of earthworm engineering that affects soil fertility. Earthworms build permanent and semi-permanent burrows that act as macropores (diameter > 6–12 μm; Gorres and Amador 2010), in which gravity exceeds capillarity with respect to water movement. Compared with the bulk soil, earthworm burrows drain quickly and are important for soil aeration. *L. terrestris* burrows are wetter, with higher pH, around twofold more organic carbon, and 1.3–2.2 times greater total nitrogen in the burrow lining than the nearby bulk soil (Tiunov

Fig. 1 Earthworm midden created by *Lumbricus terrestris* in a cultivated maize field is approximately 4 cm (height) by 20 cm (diameter). It contains maize stem and leaf residues, mixed with soil and casts, located above the opening of the earthworm burrow. Photo by the authors

and Scheu 1999). Earthworms have two main strategies for burrow construction: they either ingest and excavate the soil by depositing it on the soil surface, or they push excess soil into nearby cracks, channels, and unused burrows (Fig. 2). Soil ingestion and excavation is the preferred burrowing method of large, epi-anecic *Aporrectodea longa,* whereas smaller endogeic *A. caliginosa* pushes the soil aside until it encounters significant soil mechanical resistance and then it shifts to ingestion to extend its burrow (Arrázola-Vásquez et al. 2022). Burrowing creates new macropores with enhanced water flow and oxygen diffusion, but it simultaneously reduces water movement and air diffusion in the voids that were backfilled with soil by earthworms. The burrow periphery could be a nutrient-rich zone, but this is not always the case. The organic C, total N, and available P concentrations were the same in burrows and topsoil (0–10 cm), whereas fresh casts produced by the anecic *A. giardi* had greater organic C, total N, and available P concentrations than topsoil (Le Bayon et al. 2011).

This chapter explains how earthworms affect soil nutrients in agricultural ecosystems. We explain how earthworm feeding and engineering activities change the amounts of essential macro- and micronutrients that are available for crop growth (i.e., plant-available nutrients). Examples are selected from field and pot-based studies with mineral soil from agricultural ecosystems. We acknowledge that

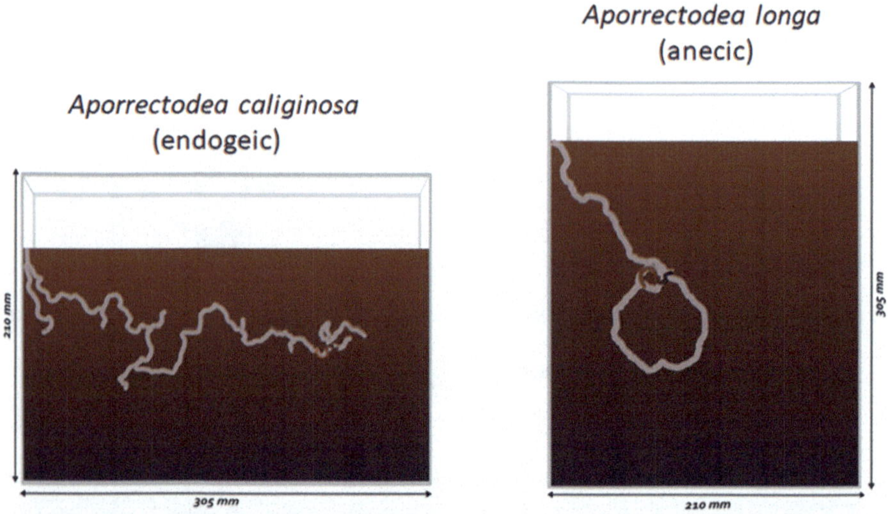

Fig. 2 Earthworm burrows are created by digging and soil displacement. Burrows of the endogeic *Aporrectodea caliginosa* are generally made through cavity expansion (i.e., by pushing soil aside or moving it into soil cracks). In contrast, the epi-anecic *Aporrectodea longa* tends to ingest and excavate the soil, depositing it on the soil surface. Earthworms can change their burrowing mode to the ingestion-excavation method in compacted soils that have limited space for cavity expansion. Adapted from Arrázola-Vásquez et al. (2022)

earthworms often increase the plant-available nutrients in organic substrates such as vermicompost and in artificial soil (e.g., sand-clay-peat mixture; OECD 2004), but the earthworm species used in vermicomposting and artificial soil studies do not represent the earthworm communities of realistic agricultural ecosystems and will not be discussed here.

2 Earthworm Feeding Activity and Nutrient Fluxes

Charles Darwin referred to earthworms as the "intestines of the earth" after observing earthworms ingest vast quantities of plant debris, other organic residues, and soil. In agricultural ecosystems, earthworms consume an estimated 10% of the crop residues and topsoil every year (Edwards and Arancon 2022). Organic residues and mineral particles give earthworms the energy and nutrients they need for physiological functions, growth, and reproduction. Nutrients assimilated into earthworm tissues are released into the soil gradually, in metabolic by-products and upon the death of the earthworm, while the unassimilated nutrients are defecated in casts within 1–2 days. The nutrient flux through earthworms is quantified using an ecological energetics approach (Fig. 3; Whalen and Parmelee 2000). Interestingly, earthworms maintain strict homeostasis in their body elemental composition,

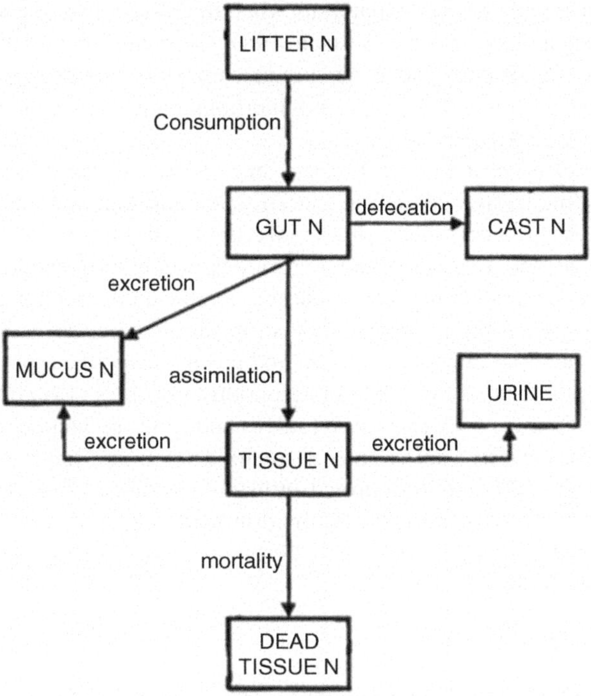

Fig. 3 Ecological energetics of nitrogen flux through earthworms

regardless of the chemical composition of the organic substrates they consume. The C:N ratio of earthworm tissues is around 4–5:1 (Tiunov and Scheu 2004; Chen 2013; Abail and Whalen 2018), with a constant C:N:P ratio of 127:26:1 in the tissue of the endogeic *Pontoscolex corethrurus* (Marichal et al. 2011). In part, this is because earthworms choose what they eat.

Earthworms select portions of organic residues based on the physical, chemical, and biological properties of the residue. Particle size is the most decisive factor, but feeding is also influenced by the shape, hardness, and hairiness of the organic residue (Curry 2004). For instance, *A. caliginosa* ate more barley (*Hordeum vulgare*) straw that was finely ground <0.2 mm than in 1-mm fragments, and the finely ground straw supported nearly two times more weight gain, probably because it was more easily digested to obtain energy and nutrients (Bostrom and Lofs-Holmin 1986). Geophagous species, such as *A. chlorotica*, grow more quickly when given finely-ground crop residues rather than coarse fragments, but *L. terrestris* showed little preference, probably because it prefers to stockpile residues, mix them with casts, and allow them to partially decompose before consuming the residue–cast–microbial mixture (Lowe and Butt 2003). Earthworms generally avoid fresh plant litter, instead choosing to eat partially degraded residues. Bostrom (1987) noted that fresh lucerne litter was toxic to earthworms, but it became palatable after the litter was leached and colonized by microorganisms. Likewise, earthworms ate more straw and other lignin-rich plant materials when the residues were partially decomposed rather than undecomposed (Bostrom 1987; Curry and Byrne 1992). Microbial populations are important because they produce enzymes that degrade organic substrates, making them more digestible, and microbial biomass is a source of nutrition for earthworms (Brown 1995).

In agricultural ecosystems, earthworms have access to many types of organic residues, including fresh plant debris, partially decomposed farmyard manure, and particulate organic matter. Diverse microbial communities in agricultural fields continually colonize and degrade organic residues while biosynthesizing mineral-associated organic matter to regenerate the soil organic matter pool (Gul and Whalen 2022). However, microbially-mediated decomposition depends on agronomic factors such as the amount of residue present and its chemistry, the type and frequency of cultivation, and fertilizer use. These agronomic variables also affect earthworm abundance and their feeding activity. Weather conditions are another consideration. Earthworms become quiescent or enter aestivation when they are exposed to higher temperatures (>25 °C) and sub-optimal moisture conditions (too dry or too wet; Diop et al. 2023). Consequently, the earthworm feeding activity and nutrient fluxes in agricultural fields are likely limited by agronomic and climatic conditions, rather than by a lack of food.

3 Soil Nutrient Enrichment in Earthworm Casts and Engineered Structures

When they are defecated, fresh earthworm casts are moist and enriched in organic matter and nutrients relative to the bulk soil or soil aggregates of a similar size. For example, casts of the anecic *A. khami* contained about 1.8 times more total C and total N, with about 10% more total P and total K than comparable soil macroaggregates from 0.05 to 2 mm diameter (Jouquet et al. 2011). Like other anecic earthworms, *A. khami* selectively ingests fresh organic residues and defecates the undigested portion in globular cast units, which have a larger proportion of SOM and higher nutrient content but the same particle size distribution as the surrounding soil. Feeding on organic residues at the soil surface is common among anecic and epigeic earthworms, but endogeic earthworms are likely to ingest partially fragmented, buried litter within the soil profile. This is evidenced by the fact that casts of *L. rubellus* (epigeic) contained visible strands of partially digested lucerne hay particles, but this was not the case in the casts of the endogeic *A. caliginosa* (Haynes et al. 2003). The selective feeding behavior of earthworms is responsible for the higher plant-available nutrient concentration and microbial activity in fresh casts than in bulk soil (Fig. 4).

Digestive processes that influence the plant-available nutrient concentration in fresh casts include intestinal mucus secretion and the gut transit time. Earthworms flood the gut with intestinal mucus to stimulate the activity of ingested and resident

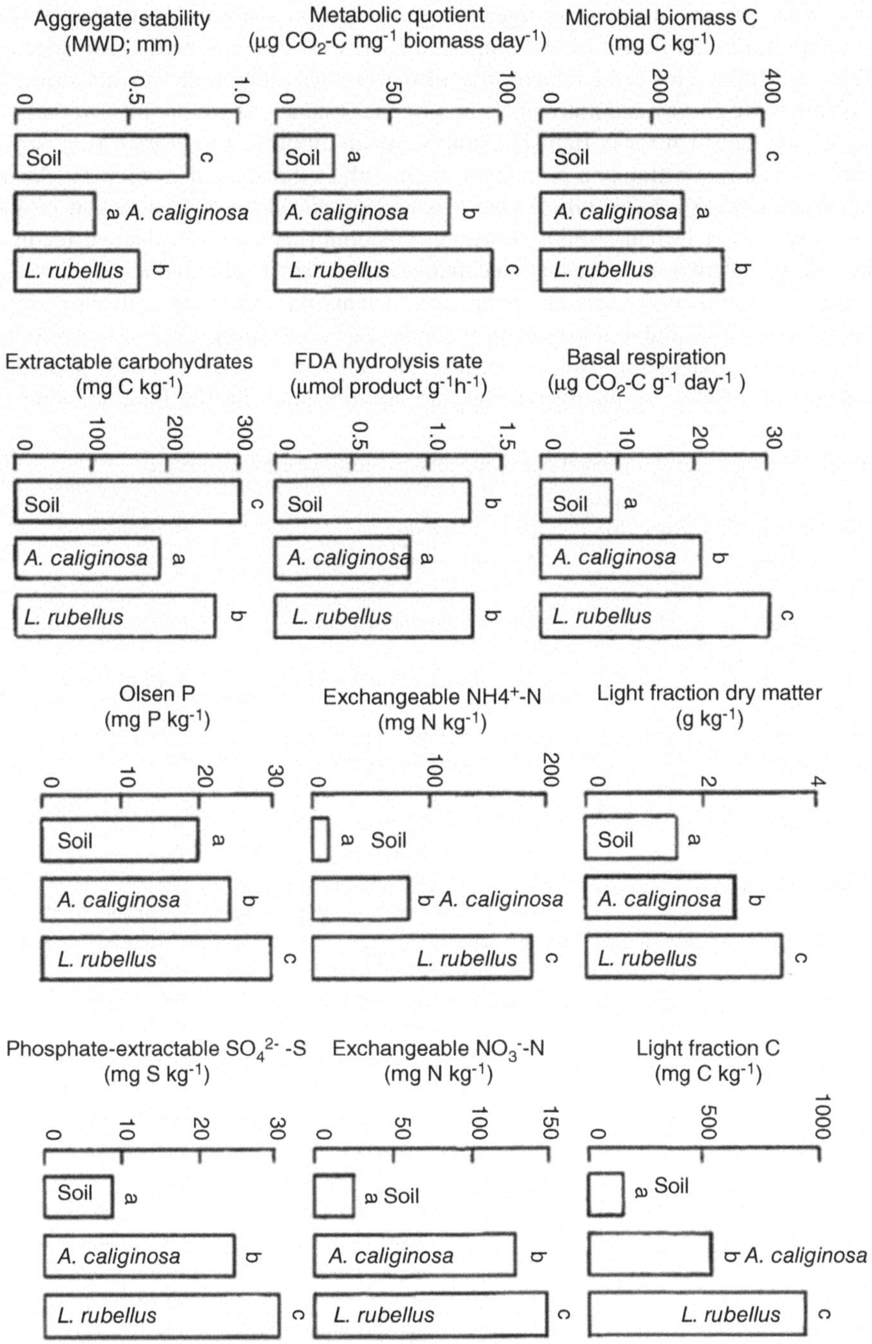

Fig. 4 Fresh earthworm casts are less stable and have smaller microbial biomass, but at least double the microbial activity (i.e., basal respiration, metabolic quotient) of soil. Plant-available N, P, and S concentrations are greater in casts than bulk soil, and casts contain more organic fragments (i.e., light fraction dry matter and light fraction C) than soil. From Haynes et al. (2003)

microbiota, which enhances the digestion of nutrient-poor substrates (Barois 1992). In addition, the gut transit time of ingested food varies among earthworm species (Table 1). This seems to be related to earthworm physiology rather than a strategy to obtain more energy and nutrients from organic residues, based on the observation that it took 21 ± 1 h for N-rich (i.e., soil–soybean mixture) and N-poor (i.e., soil–wheat mixture) substrates to pass from the mouth to the anus of *A. turgida* (Abail and Whalen 2018). As *A. turgida* ate less and produced fewer casts when provisioned with N-rich than N-poor substrate, this confirms that the selective feeding behavior of earthworms is the main mechanism for nutrient enrichment in their casts.

Because earthworm casts are richer in nutrients than the bulk soil, they are a source of plant-available nutrients in the short term. However, the nutrient supply from casts is ephemeral. This is because the nutrient content of casts gradually changes with time, eventually reaching an equilibrium with the bulk soil that is

Table 1 Gut transit time (in hours) of soil ingested by various earthworm species

Earthworm species	Gut transit time (h)	T (°C)	Method	References
Aporrectodea caliginosa	20	15	Non-marked soil	Barley (1959)
Aporrectodea caliginosa	12–24	n/a	Non-marked soil	Piearce (1972)
Aporrectodea caliginosa	0.8–0.9	20	Non-marked soil	Martin (1982)
Aporrectodea caliginosa	0.9	15	Estimated from maximum gut content mass	Scheu (1987)
Aporrectodea caliginosa	2.5	15	Soil marked with dye	Curry and Baker (1998)
Aporrectodea caliginosa	9.6	15	Soil marked with fungal spores	Taylor and Taylor (2014)
Aporrectodea turgida	21	16	Soil marked with glass beads	Abail and Whalen (2018)
Allolobophora rosea	1–2.5	10	Estimated from feces units and egestion rate	Bolton and Phillipson (1976)
Drawida calebi	10	26	Non-marked soil	Dash et al. (1984)
Drawida willsi	2.3	28	Non-marked soil	Dash et al. (1986)
Eisenia fetida	7	15	Soil marked with dye	Hartenstein et al. (1981)
Eisenia fetida	2.5	25	Soil marked with dye	Hartenstein et al. (1981)
Hormogaster elisae	3.6–5.3	18	Soil marked with dye	Cosin et al. (2002)
Lampito maurito	3.5	28	Non-marked soil	Dash et al. (1986)
Lumbricus terrestris	12	15	Soil marked with fungal spores	Taylor and Taylor (2014)
Octalasion lacteum	1.8	15	Estimated from maximum gut content mass	Scheu (1987)
Octochaetona surensis	2.9	28	Non-marked soil	Dash et al. (1986)

detected with soil chemical and near-infrared spectral analysis (Aira et al. 2005; Zangerlé et al. 2014). During the aging of *A. caliginosa* casts, three stages in the nutrient content were observed with near-infrared spectroscopy: (1) rapid changes during the first 48 h that probably correspond to leaching of water-soluble nutrients, (2) a maturation period of 3–30 days that is associated with decomposition and mineralization of the organic C and N in casts, followed by (3) a stabilization period of 45–90 days where cast nutrient levels converged toward those of the control soil (Zangerlé et al. 2014). In the absence of plants, the convergence in the nutrient content of casts and soil can be explained by the immobilization of water-soluble nutrients in microbial biomass, which gradually adds nutrient-rich biomolecules (i.e., microbial necromass) to the soil organic matter pool (Gul and Whalen 2022).

Another reason that casts should be considered an ephemeral source of plant-available nutrients is that casts are spatially associated with the soil that is actively worked by earthworms. Earthworm casts are not uniformly distributed across the soil surface (Fig. 5) nor in layers of the soil profile, and they must be produced by active earthworms. Earthworm casting is controlled primarily by environmental conditions because surface cast production declines to practically zero when the soil temperature is below 5 °C or when it exceeds 25 °C (Whalen et al. 2004), and when the soil is too wet or too dry for earthworm activity (Diop et al. 2023). In temperate regions, surface cast production is most visible in the spring and autumn periods when organic residues are abundant, and the temperature and moisture conditions are favorable to earthworms. These are the times of year when earthworm casts could supply plant-available nutrients to crops and other vegetation.

Burrows produced by anecic earthworms are also nutrient-rich microenvironments, relative to the bulk soil. The distinctive features and relatively large size of these semi-permanent structures have allowed for in-depth study, whereas the smaller, temporary burrows and litter mounds created by endogeic and epigeic earthworms received less attention under field conditions. Burrows of anecic earthworms are macropores that extend vertically to a depth of 1 m or deeper in the soil profile. The burrow is occupied by its builder, but juvenile anecics and up to ten other earthworm species of the epigeic and endogeic groups are observed within and directly beneath middens (Butt and Lowe 2007). Burrow walls, also referred to as the drilosphere, support greater microbial activity and are nutrient-rich relative to the surrounding soil. Plant-available nutrients in the burrow lining could come from organic residues that were buried or mixed into the burrow walls by earthworms, or they could be water-soluble nutrients from decomposing residues, fertilizer, and other sources that infiltrate the burrow (macropore) through preferential flow (Fig. 6). Earthworm bodily secretions, including urine and cutaneous mucus, add urea, NH_4^+, glucopolysaccharides, amino acids, and proteins that can be transformed into plant-available N. Mucus also contains P in the form of organophosphorus compounds and K in electrolytes (Bityutskii et al. 2012). These compounds probably stimulate microbial growth, contributing to the mineralization and solubilization of plant-available nutrients from organic substrates in the drilosphere. Earthworm burrows and middens are discrete units in fields, and the number found depends on the activity of anecic earthworms that build and maintain these structures. Consequently, the presence of burrows and middens, along with casts, is

a)

b)

Fig. 5 Earthworm casts in a temperate riparian zone in early spring, showing the uneven spatial distribution of earthworm activity across the soil surface in (**a**) a 10 cm × 10 cm frame and (**b**) a 3 m × 5 m frame. Photo by the authors

Fig. 6 Leaf residue around the opening and inside a burrow built by *Lumbricus terrestris* in a minimum tillage wheat field. Organic residues were either transported by earthworm activity or were carried passively in water that infiltrated the burrow. From Stroud et al. (2016)

responsible for some of the spatial variability in soil nutrient content across ecosystems and landscapes.

4 Earthworms and Soil Nutrient Dynamics

4.1 Soil Nitrogen Cycling

Nitrogen (N) is the most limiting nutrient element needed to sustain earthworm populations. Every earthworm needs a continual supply of N throughout their lifespan to synthesize proteins, DNA and RNA, vitamins and hormones, as well as protein-rich cutaneous and intestinal mucus that lubricates their movement and aids their digestive functions. Earthworms meet their N requirements by feeding preferentially on N-rich organic materials over N-poor materials (Bostrom 1987; Shipitalo et al. 1988). In nature, organic N is associated with C in amino acids, proteins, purines, and other organic N compounds. The C:N ratio of organic residues is a key indicator of food quality for earthworms. Earthworms grow and reproduce better when they ingest organic residues having a C:N ratio <20 than residues with higher C:N ratios. For instance, earthworms gained on average 68% and 5% of their weight, respectively, when offered alfalfa (C:N ratio of 14) and corn leaves (C:N ratio of 21), while they lost up to 41% of their weight when fed with bromegrass leaves

(C:N ratio of 26; Shipitalo et al. 1988). Bromegrass leaves were also more lignified, as indicated by the aromatic-C content, and appeared to be less palatable for earthworms (Shipitalo et al. 1988). The C:N ratio reflects the energy and N supply that can support the earthworm metabolic process and, consequently, the size of earthworm populations.

The N intake by earthworms is offset by daily N loss in undigested organic residues, which are defecated in the casts, as well as N lost from earthworm tissues. Protein breakdown during digestion and cellular regeneration releases urea and NH_4^+ that are excreted via the nephridium. From 20 to 127 μg N g^{-1}, earthworm (fresh weight) is excreted per day by *A. caliginosa* in a soil-free environment (Needham 1957; Christensen 1987). Earthworms also produce copious amounts of mucus, a mucoprotein with a C:N ratio of 4–5. Intestinal mucus is mostly resorbed within the gut, but some adhere to casts that are defecated, plus mucus representing up to 20% of earthworm fresh body weight is secreted through the cuticle (Abail and Whalen 2018). Urine plus mucus released from epigeic, endogeic, and anecic earthworms is from 278 to 744 μg N g^{-1} earthworm fresh weight per day, depending on the species and their diet (Whalen et al. 2000; Abail and Whalen 2018). The N

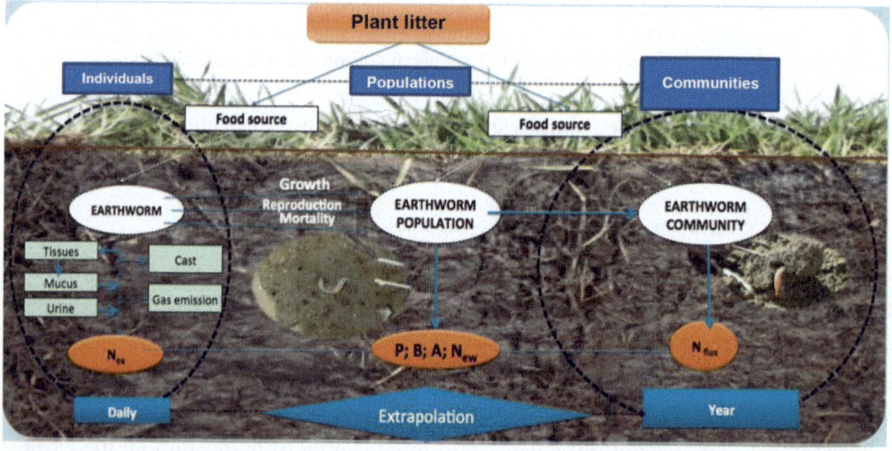

Fig. 7 An ecological energetics view of the direct N flux from plant litter through earthworms in agroecosystems. At the individual level, physiological measurements describe the proportion of N from consumed food that is assimilated in earthworm tissues. The unassimilated N is defecated in casts. The N in earthworm tissues may be used for metabolism, including mucus production, and N in excess of metabolic requirements is excreted in urine or denitrified as N_2O and N_2 gases. N_{ex} represents the amount of N excreted in urine and secreted in mucus. Secondary production of the earthworm population in the field, estimated from biomass accumulation (including reproduction), is also influenced by the quality of the food resource (plant litter and other substrates) and environmental variables. The size, age structure, and dynamics of earthworm populations depends upon access to assimilable energy and nutrients from the available food resources. P is the secondary production, B is the mean biomass of the earthworm population, A is the number of days active during the frost-free period, and N_{ew} is the N concentration of the earthworm body. Physiological and secondary production values are used to estimate the direct N flux (N_{flux}) released by earthworm communities

lost from living earthworms, along with mortality, contributes to the annual N flux from earthworm populations in agricultural fields and natural ecosystems according to ecological energetics. Scaling up earthworm physiological processes through populations and communities (Fig. 7), the annual N flux through earthworms is estimated at 18–105 kg N ha^{-1} $year^{-1}$ (Christensen 1987; Whalen and Parmelee 2000; Abail and Whalen 2021). Considering that N-demanding crops such as maize are fertilized with from 100 to 200 kg N ha^{-1} $year^{-1}$ in temperate agroecosystems, earthworms can make an important direct contribution to the soil N supply.

In addition to releasing N from their tissues, earthworms also stimulate the N cycling of soil microorganisms is earthworm feeding activities and biostructures. This is due to the litter comminution, fragmentation, mixing, and moistening of residues that encourage microbial colonization and growth. For example, the drilosphere of *L. terrestris* contains two to five times greater microbial biomass and bacterial volume than the surrounding soil (Tiunov and Scheu 1999). Still, there are many reports that found similar or smaller microbial biomass in earthworm biostructures as the bulk soil, but this was compensated by greater biological activity in the earthworm-worked soil (Haynes et al. 2003; Vidal et al. 2023). This is probably because earthworms enhance porosity from the microscale (i.e., within their casts; Aira et al. 2005) to the macroscale in their burrows (Arrázola-Vásquez et al. 2022), which favors aerobic processes such as N mineralization, ammonia oxidation, and nitrification. The indirect contribution of earthworms to N mineralization is

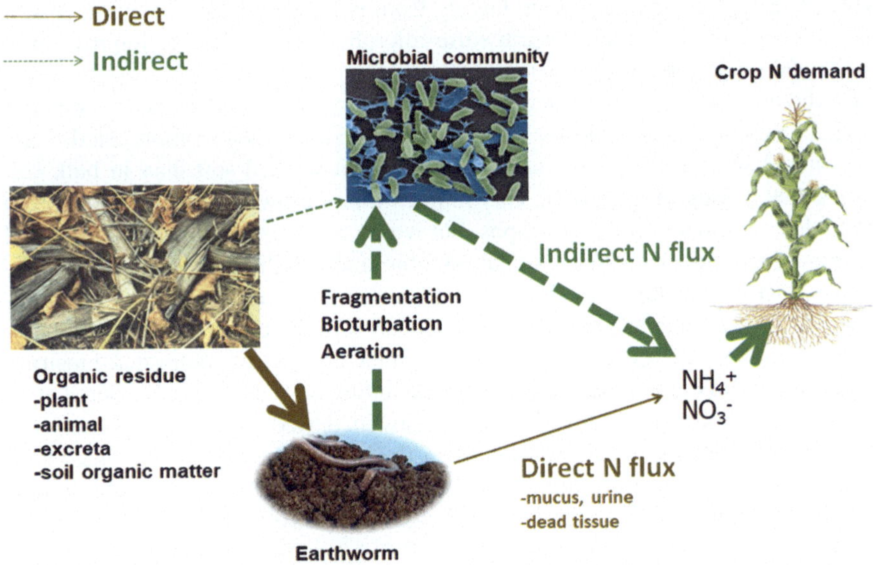

Fig. 8 The direct and indirect N flux through earthworms contributes to the plant-available N supply in the soil. When earthworms are actively feeding and building biostructures, they facilitate the microbial-mediated decomposition and N mineralization processes, effectively accelerating the N transfer from organic residues to crops (illustrated by larger arrows)

estimated at 11–267 kg N ha^{-1} $year^{-1}$ (Marinissen and De Ruiter 1993). This is attributed to earthworm engineering activities that physically and chemically modify the soil habitat, improve plant root growth, and stimulate microbial activity, orchestrating the microbially-mediated transfer of N to crops, as illustrated in Fig. 8.

Several microbially-mediated reactions in the N cycle are influenced by earthworm activities. The ecological energetics method estimates the N flux resulting from decomposition and N mineralization, which degrades organic N compounds and releases NH_4^+ into the soil pore water. Although NH_4^+ is part of the plant-available N pool because it is readily absorbed by plant roots, NH_4^+ is also used in microbial metabolism and tends to be rapidly transformed into NO_3^- in soil. Ammonia oxidation is the first, rate-limiting step of the three-part reaction that transforms NH_4^+ to NO_3^-. This reaction is mediated by ammonia-oxidizing bacteria, ammonia-oxidizing archaea, and complete ammonia oxidizer (comammox, amoA gene of clade-A and clade-B). The earthworm gut is rich in these ammonia oxidizers, which represent about 30–80% of the metabolic functions of the gut microbiome (Thakur et al. 2021). The activation of the ammonia oxidizers in the earthworm gut, coupled with their impact on soil aeration and mixing, can have measurable impacts on the soil N supply. When the endogeic *Pheretima guillelmi* was present in a. rice-fallow rotation, the potential nitrification rate was about 1.6 times greater than without earthworms (Chu et al. 2022). By stimulating microbial processes that produce plant-available N, earthworms can improve crop N nutrition. Besides increasing crop yield by 20–25%, earthworms also increased the N content in plant tissues by 11%, according to a global meta-analysis (van Groenigen et al. 2014; Xiao et al. 2018). Although these findings are largely from pot and small-plot experiments, it does illustrate the potential of earthworm-microbial interactions to improve crop growth and quality in agroecosystems.

Earthworms are also implicated in fixation and loss pathways in the N cycle. For instance, bacteria responsible for the biological nitrogen (N_2) fixation reaction are more abundant in earthworm casts and earthworm-worked soil than in bulk soil (Simek 1993; Jang et al. 2022). This implies that earthworm biostructures are a favorable environment for diazotrophs, but we do not know if this improves the N nutrition of legume plants that associate symbiotically with N_2 fixers, or of the non-legumes that host endophytic diazotrophic bacteria in their leaf, stem and root tissues. Earthworm interactions with denitrifiers, which reduce NO_3^- in partially oxygenated to anoxic environments such as their gut and in compacted biostructures, cause N loss in gaseous by-products, including NO_x, N_2O, and N_2 (Vidal et al. 2023). However, the cumulative N_2O emissions increased by 0.2 mg N_2O kg^{-1} soil after 45 days when the anecic *Metaphire guillelmi* was present in soil planted with lettuce, which was insignificant in the N balance (Na et al. 2022). Of greater importance in this study was N leaching, which was around 10% of the ^{15}N added, with the balance of the ^{15}N remaining in the soil (~70%) and the lettuce crop (~20%). The water flow through earthworm burrows can transport N out of the root zone and into deeper soil layers, which are inaccessible to plant roots. Vertical burrows formed by anecic earthworms are preferential flow pathways for the movement of plant-available N, soluble organic N compounds (i.e., dissolved organic N), and particulate-bound N to deeper soil layers. Leachate collected from the lysimeter had similar inorganic N and dissolved organic N concentrations, but greater water infiltration led to 2.5-fold greater N flux with increased earthworm populations than in those with decreased populations (Domínguez et al. 2004). Since the earthworm interactions with diazotrophs and denitrifiers, as well as the spatial extent of their burrow system are scale-dependent (Chen and Whalen 2016), additional study is needed to extrapolate these relationships and to quantify the soil biota contribution to the N cycle of agroecosystems.

4.2 Earthworms and the Biogeochemical Transformations of Soil Phosphorus and Sulfur

Earthworm activities affect the biological mineralization of organic phosphorus (P) and the biological weathering (bioweathering) of inorganic P. Soil organic P is dominated by phosphate monoesters (occurring mainly as inositol phosphates) plus phosphate diesters (mainly as DNA) and phosphonates, regardless of the parent material, vegetation, and climatic conditions (Huang et al. 2017). Mineralization of organic P compounds by hydrolytic phosphatase enzymes cleaves phosphate ions ($H_2PO_4^-$/HPO_4^{2-}) that are covalently bound to organic molecules into the soil pore water, adding to the plant-available P pool. Soil inorganic P refers to phosphate ions that are electrostatically bound to calcium oxides, iron (hydr)oxides, and aluminum (hydr)oxides, which are associated with clay surfaces and primary minerals. Bioweathering occurs when organic acids and their anions (collectively referred to as organic acids) solubilize the bound inorganic P or desorb phosphates like $H_2PO_4^-$ from anion exchange sites. This is an additional source of P that enters the plant-available P pool. Through their interaction with soil microorganisms, earthworms contribute to the biogeochemical reactions that transform the organically-bound and mineral-bound P into soluble forms.

Earthworms facilitate organic phosphorus mineralization. Organic residues and soil organic matter ingested by earthworms are exposed to a moist gut microenvironment that contains phosphatase enzymes. Acid and alkaline phosphatases are ubiquitous in the peripheral chloragog, epithelium, and typhlosole tissues in the mid-gut of *L. terrestris*, suggesting that earthworms produce phosphatases to meet a physiological P need (Prentø 1987). Phosphatases should include isoenzymes of phosphomonoesterases, phosphodiesterases, and phospholipases, which were detected in *L. terrestris* and *E. fetida* (Park et al. 1992; Hanusova et al. 1999). These enzymes are synthesized in earthworm gut cells and produced by the ingested and resident microorganisms in the earthworm gut. Extracellular phosphatases are often detected in fresh casts, which may have approximately two times more phosphomonoesterase activity than the bulk soil (Kizilkaya and Hepşen 2004; Le Bayon and Binet 2006). Still, an increase in phosphatase activity is not always responsible for the higher plant-available P concentration in earthworm casts. For example, specific phosphatase activity was unchanged or slightly increased above the level in soil when diets of soil mixed with lupin (*Lupinus* sp.), sugar beet (*Beta vulgaris*), maize (*Zea mays*), and barley residue were fed to the endogeic *Octolasion cyaneum* and the anecic *L. terrestris* (Buck et al. 1999). As phosphatase activity could not account for the consistent net gain in plant-available P (from ~5 to 25 μg water-extractable P g^{-1} soil) in the *O. cyaneum* and *L. terrestris* casts, this means that plant-available P was derived from a non-enzymatic process, likely bioweathering.

The earthworm gut is favorable for bioweathering of P minerals due to mechanical grinding, organic acids released from ingested and resident bacteria in the gut, and a pH >7 in the earthworm intestinal tract. Minerals ingested by earthworms are abraded in the gizzard, which should expose occluded P compounds to the gut microenvironment. This idea is supported by the accelerated weathering of anorthite, biotite, smectite, and kaolinite, including the transformation of smectite to illite and the formation of a new mineral phase from kaolinite, after 120 days with the epigeic *E. veneta* (Carpenter et al. 2007). Degradation of organic residues and lysis of microbial cells releases protein and ribonucleic acid into the earthworm gut, which are used by fermentative bacteria to produce formate, lactate, succinate, and propionate (Zeibich et al. 2019). Organic acids are not only a carbon source for earthworms, but they also accelerate mineral weathering by altering the pH, chelation, and complexation reactions, and solubility of ions like Al^{3+} that are involved in the weathering process (Drever and Stillings 1997). Organic acids may be responsible for HPO_4^{2-} displacement from the surface of metal (hydr)oxides through competitive adsorption and desorption reactions in the gut of *L. terrestris*, which had alkaline pH and elevated dissolved organic carbon concentration (Ros et al. 2017). This suggests that bioweathering of the mineral fraction lreleases a considerable amount of plant-available P in earthworm casts.

Earthworm casts contain more total and plant-available P than the bulk soil. In a global meta-data analysis, casts had about 40% more total P relative to the bulk soil and the available P concentration was 84% greater than the bulk soil (van Groenigen et al. 2019). This is due to a combination of preferential feeding and enrichment processes in the digestive process, as illustrated in Fig. 9. While earthworm casts are discrete biostructures associated with earthworm feeding, soil engineering by earthworms could make an even larger contribution to the P biogeochemical cycle. For example, earthworm burrows in the Terra Preta soils of Western Amazonia contained substantial amounts of apatite-rich residues, charcoal, and organic residues that were presumably incorporated by earthworms, as well as considerable pedobiological mixing at the boundary of the A and B soil horizons (Schaefer et al. 2004).

One way to quantify the effect of earthworm feeding and engineering on plant-available P concentration is to incubate soil with earthworms in pot studies. In agricultural soil, plant-available P doubled from ~25 μg P g^{-1} soil to ~50 μg P g^{-1}soil ($NaHCO_3$-extractable P) and the water-extractable P increased from 1.5 μg P g^{-1} soil to 3.3 μg P g^{-1} soil after 14–21 days of ingestion and bioturbation by the endogeic *Pheretima guillelmi* or the epigeic *E. fetida* (Wan and Wong 2004). As the plant-available P extracted from the soil with a chemical solution or water approximates P acquisition by plants, it is more informative to measure plant P uptake directly.

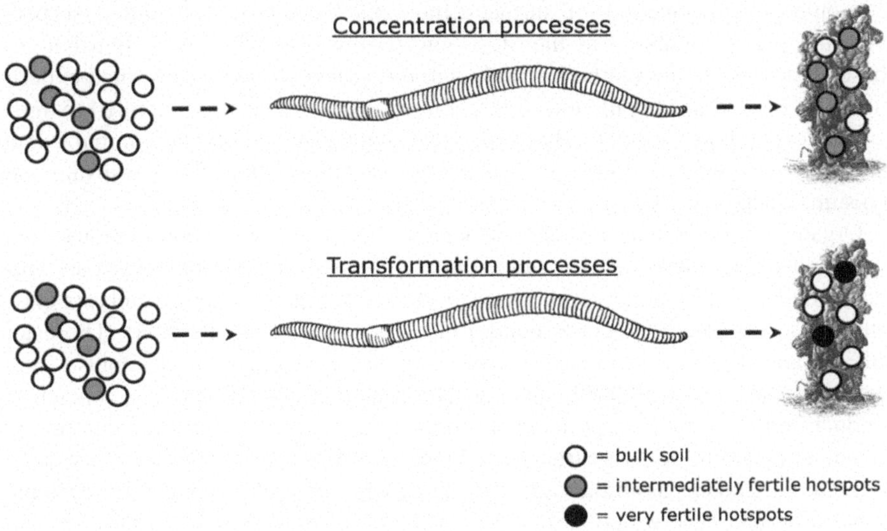

Fig. 9 Earthworm casts can become more fertile than bulk soil through concentration of existing nutrient-rich materials as a result of preferential feeding, or through transformation processes during gut passage leading to new nutrient enrichment. Higher nutrient concentration is represented in the fertile hotspots. From van Groenigen et al. (2019)

Ryegrass (*Lolium perenne*) grew better in pots with earthworms (epigeic *L. rubellus*, anecic *L. terrestris*, and endogeic *A. caliginosa*), but greater P uptake was only observed in the presence of *L. terrestris* after 75 days (Vos et al. 2014). Similarly, the improved P uptake and ryegrass growth in field mesocosms were associated with the presence of the anecics *A. longa* and *L. terrestris* (Vos et al. 2022), suggesting that earthworm engineering activities are possibly more important than feeding activities in delivering plant-available P to plants. In a mesocosm experiment with rice (*Oryza sativa*), the endogeic *P. corethrurus* significantly increased rice shoot biomass (+26%), total biomass (+15%), and rice P nutrition (+65%) by increasing the pool of soluble P from ingested soil and by microbial recovery and transfer to plants of P from dead earthworm tissues (Trap et al. 2021). These reports are consistent with the expectation that earthworms increase soil fertility, and one of the ways they do this is by increasing the plant-available P concentration in soil.

4.3 Earthworms and the Biogeochemical Transformations of Soil Sulfur

Soil sulfur (S) exists in inorganic and organic forms that are transformed by soil biota, including earthworms. In well-aerated agricultural soils, inorganic S exists as moderately water-soluble salts (e.g., $CaSO_4$) and as SO_4^{2-} adsorbed to exchange sites on negatively charged soil surfaces. Inorganic S can be dissolved or desorbed, releasing plant-available S as the SO_4^{2-} ion. Earthworms will likely contribute to these processes via bioweathering, although this still needs to be confirmed experimentally. Earthworm populations tolerate inorganic S amendments applied to agricultural ecosystems, based on the earthworm abundance and biomass in maize and hayfields receiving 2.2–20 Mg ha^{-1} of flue gas desulfurization (FGD) gypsum and gypsum (Chen et al. 2014).

Organic S compounds include sulfonates, sulfolipids, sulfocarbohydrates, and the S-containing amino acids cysteine and methionine originating from plant, animal, and microbial residues (Eriksen 2009). Broadly, organic S compounds are classified as ester sulfates (C–O–S bonds) or carbon–bound sulfur (C–S). The ester sulfates are cleaved enzymatically from organic polymers by arylsulfatases and alkylsulfatases, extracellular hydrolytic enzymes that release SO_4^{2-} as a reaction product. Sulfur-containing proteins are first depolymerized and deaminated before S-containing amino acids are transported into microbial cells for further processing. Soil microorganisms assimilate added cysteine within minutes and degrade cysteine to pyruvate and H_2S, which is then oxidized and released as SO_4^{2-} within 3 h and resorbed by the next microbial generation within 4 days (Ma et al. 2021). Similarly, microbial assimilation of methionine occurs within minutes, followed by hydrolysis to α-ketobutyrate and S-CH_3 before releasing SO_4^{2-} within 24 h, with microbial immobilization occurring within 4 days (Ma et al. 2021). Microorganisms involved in these reactions could be heterotrophic sulfide oxidizers that link S-amino acid metabolism with sulfide production followed by oxidation to sulfite and thiosulfate. Another key group are soil thermophilic bacteria of the Firmicutes Phylum (e.g., genera *Geobacillus*, *Parageobacillus*, *Ureibacillus*, and *Brevibacillus*) that use a dissimilatory organic S mineralization process to produce SO_4^{2-} (Santana et al. 2021). Regardless of what microbial groups are involved, the dynamics of organic S mineralization-immobilization depends upon microbially-controlled stoichiometry.

Earthworms assimilate S into their tissues and defecate the unassimilated S compounds. Casts of *A. caliginosa* and *L. terrestris* contained 1.5–7 times more S derived from ^{34}S-labeled leaf litter from birch (*Betula* sp.) and beech (*Fagus* sp.) than the reference soil-litter mixture without earthworms (Grethe et al. 1996). Casts of *L. terrestris* had higher arylsulfatase activity than agricultural soil mixed with up to 50 g biosolids kg^{-1}, but higher biosolids rates inhibited the arylsulfatase activity in mesocosms with and without earthworms during a 30-day incubation (Kizilkaya and Hepşen 2004). Hence, earthworm-feeding activities will create S-rich casts and support microbial activities for organic S cycling, but it is difficult to predict how earthworms will impact the plant-available S in agricultural ecosystems.

4.4 *Earthworms and Soil Potassium, Calcium, Magnesium, and Micronutrients*

Earthworms accelerate the weathering of silicate minerals, which liberates nutrient ions such as K^+, Ca^{2+}, Mg^{2+}, and micronutrient cations that are sorbed to or occluded within minerals. As discussed in Sect. 4.2, earthworm-induced bioweathering results from the mechanical grinding, pH change, and microbial activity in the earthworm gut. When *Pheretima carnosa* ingested soil mixed with potassium-bearing rock powder (predominantly feldspar) for 10 days, this released water-soluble and nitric acid-extractable K, Ca, Fe, and Al due to bioweathering by a unique, diverse bacteria community in the earthworm gut (Liu et al. 2011). Variation in bioweathering is expected, based on soil texture, mineralogy, earthworm feeding preferences, and the distinct gut microbiota of each earthworm species. For example, the exchangeable K^+ concentration increased by 5–28 mg K^+ kg^{-1} in the casts of *A. caliginosa* and reached 22–42 mg K^+ kg^{-1} in the casts of *L. rubellus* from four agricultural soils after 30 days (Basker et al. 1993). This may be related to the distinctive gut microbiome of *L. rubellus* and *A. caliginosa* that is independent of the soil environment, based on direct counts of bacteria, actinomycetes, and micromycetes (Krišrtůek et al. 1992), specific enzyme activity, and DNA-based analysis (Nebert et al. 2011). Thus, we expect bioweathering to depend upon the earthworm species, their feeding preferences, and gut microbiome.

The exchangeable cations in earthworm casts and earthworm-worked soil are expected to be plant-available for crops in agricultural ecosystems. For example, an alley cropping system with the leguminous leucaena shrub (*Leucaena leucocephala*) and a maize-cassava (*Manihot esculenta*) intercrop supported an active earthworm community that produced casts with more than twice the extractable K^+, Ca^{2+}, and Mg^{2+} concentration of the bulk soil (Asawalam 2006). However, earthworm activity could not sustain the soil fertility in this agricultural ecosystem, as continuous cropping for 6 years significantly reduced the organic carbon, total nitrogen, available P, and exchangeable K, Ca, and Mg concentrations in the topsoil (0–5 cm; Asawalam 2006). Although earthworm casts are rich in plant-available nutrient cations, the nutrients released by earthworm activity cannot overcome the inherently low soil fertility in a humid tropical region, which needs to be managed through judicious use of organic and inorganic fertilizers plus carbon farming practices.

Bioweathering by earthworms and their associated gut microbiota may be a strategy to maintain the necessary stoichiometry of nutrient cations. For instance, intracellular calcium concentration regulates the innate and adaptive immune signaling in earthworm celomocytes, which rely on Ca^{2+} ATPase activity to maintain their calcium homeostasis (Engelmann et al. 2011). The physiological demand for Ca^{2+} leads earthworms to preferentially feed upon Ca-rich materials, which support larger earthworm populations in forests with Ca-rich litter than Ca-poor litter (Reich et al. 2005). Excess Ca^{2+} may combine with anions such as carbonate, sulfate, phosphate, and oxalate in the calciferous glands, leading to the secretion of inorganic salts like calcium carbonate. Granules of calcium carbonate are produced by

Prosellodrilus sp., *L. terrestris*, *A. longa*, *A. trapezoides*, and *Anisochaetae* sp. (García-Montero et al. 2013) and dissolve at a similar rate as other biogenic and inorganic calcites. A population of 20 individuals m^{-2} of *L. terrestris* could produce more than 3000 mol ($CaCO_3$) ha^{-1} $year^{-1}$ (Lambkin et al. 2011), allowing earthworms to buffer soil pH and modulate biogeochemical processes such as decomposition, net N mineralization, and solubilization of essential nutrients and trace metals.

Micronutrient cations are solubilized during earthworm digestion. Casts have more water-soluble Fe, Mn, and Zn than bulk soil, regardless of the soil moisture content, and this effect was more pronounced when *A. caliginosa* consumed soil mixed with rye (*Cecale cereale*) straw from rye (Bityutskii et al. 2012). Cucumber (*Cucumis sativus*) shoot biomass was greater, and there was more Mn and Zn in xylem sap after 30 days when cucumber was grown in earthworm-worked soil than control soil without earthworms. Earthworm casts were also enriched in Fe, Mn, and Zn in no-till cropping systems, which could improve crop nutrition in a rice-wheat rotation (Sharma and Dhaliwal 2021). While this is further evidence that earthworm bioweathering increases plant-available micronutrients, there is a knowledge gap regarding the impact of earthworms on boron, copper, chloride, molybdate, and nickel concentrations in agricultural ecosystems. As earthworms and micronutrients were often studied in vermicompost and artificial soil, targeted studies are needed to measure the earthworm impact on plant-available micronutrients in agricultural fields.

5 Conclusions

Earthworms are involved in soil nutrient transformations through their feeding and engineering activities. These activities release plant-available nutrients that contribute to rop nutrition. In natural ecosystems, it is likely that earthworms co-evolved with plants over millions of years, and this is supported by the frequent presence of earthworms around plant roots and in litter. Still, it remains challenging to integrate this obvious contribution of earthworms to soil fertility into agronomic practice. Soil fertility assessment is based on soil chemical analyses taken once every few years. In contrast, earthworm activities are temporally dynamic and spatially uneven, and our understanding of their involvement in nutrient cycling comes from controlled studies and small-plot measurements that are challenging to extrapolate to an agroecosystem scale. The development of agricultural technologies such as digital sensors is one possible way to detect, monitor, and quantify earthworm activities that are known to affect nutrient cycling. Earthworms must be included in the coming Agriculture 4.0 revolution, to assure that their important contribution to the soil nutrient supply in agroecosystems is fully acknowledged.

References

Abail Z, Whalen JK (2018) Selective ingestion contributes to the stoichiometric homeostasis in tissues of the endogeic earthworm *Aporrectodea turgida*. Soil Biol Biochem 119:121–127

Abail Z, Whalen JK (2021) Earthworm contributions to the soil nitrogen supply in corn-soybean agroecosystems in Quebec, Canada. Pedosphere 31:405–412

Aira M, Monroy F, Domínguez J (2005) Ageing effects on nitrogen dynamics and enzyme activities in casts of *Aporrectodea caliginosa* (Lumbricidae). Pedobiologia 49:467–473

Arrázola-Vásquez E, Larsbo M, Capowiez Y, Taylor A, Sandin M, Iseskog D, Keller T (2022) Earthworm burrowing modes and rates depend on earthworm species and soil mechanical resistance. Appl Soil Ecol 178:104568

Asawalam DO (2006) Influence of cropping intensity on the production and properties of earthworm casts in a leucaena alley cropping system. Biol Fertil Soils 42(6):506–512

Barley K (1959) The influence of earthworms on soil fertility. II. Consumption of soil and organic matter by the earthworm *Allolobophora caliginosa* (Savigny). Crop Pasture Sci 10:179–185

Barois I (1992) Mucus production and microbial activity in the gut of two species of *Amynthas* (Megascolecidae) from cold and warm tropical climates. Soil Biol Biochem 24:1507–1510

Basker A, Macgregor AN, Kirkman JH (1993) Exchangeable potassium and other cations in non-ingested soil and casts of two species of pasture earthworms. Soil Biol Biochem 25:1673–1677

Bityutskii NP, Kaidun PI, Yakkonen KL (2012) The earthworm (*Aporrectodea caliginosa*) primes the release of mobile and available micronutrients in soil. Pedobiologia 55:93–99

Bolton P, Phillipson J (1976) Burrowing, feeding, egestion and energy budgets of *Allolobophora rosea* (Savigny) (Lumbricidae). Oecologia 23:225–245

Bostrom U (1987) Growth of earthworms (*Allolobophora caliginosa*) in soil mixed with either barley, lucerne or meadow fescue at various stages of decomposition. Pedobiologia 30:311–321

Bostrom U, Lofs-Holmin A (1986) Growth of earthworms (*Allolobophora caliginosa*) fed shoots and roots of barley, meadow fescue and lucerne. Studies in relation to particle size, protein, crude fiber content and toxicity. Pedobiologia 29:1–12

Brown GG (1995) How do earthworms affect microfloral and faunal community diversity? Plant Soil 170:209–231

Buck C, Langmaack M, Schrader S (1999) Nutrient content of earthworm casts influenced by different mulch types. Eur J Soil Biol 35:23–30

Butt KR, Lowe CN (2007) Presence of earthworm species within and beneath *Lumbricus terrestris* (L.) middens. Eur J Soil Biol 43:S57–S60

Carpenter D, Hodson ME, Eggleton P, Kirk C (2007) Earthworm induced mineral weathering: Preliminary results. Eur J Soil Biol 43:S176–S183

Chen C (2013) Earthworm interactions with denitrifying bacteria: significance for nitrogen dynamics from the physiological to field scales. PhD thesis, Department of Natural Resource Sciences, McGill University, Montreal, Canada

Chen C, Whalen JK (2016) Earthworm interactions with denitrifying bacteria are scale-dependent: evidence from physiological to riparian ecosystem scales. Basic Appl Ecol 17:1–10

Chen L, Kost D, Tian Y, Guo X, Watts D, Norton D, Wolkowski RP, Dick WA (2014) Effects of gypsum on trace metals in soils and earthworms. J Environ Qual 43:263–272

Christensen O (1987) The effect of earthworms on nitrogen cycling in arable soil. In: Proceedings of the 9th international colloquium on soil zoology, Nauka, Moscow, pp 106–118

Chu X, Bai N, Zheng X, Wang Q, Pan X, Li S, Zhang J, Zhang H, He W, Zhong F, Lv W, Zhang H (2022) Effects of straw returning combined with earthworm addition on nitrification and ammonia oxidizers in paddy soil. Front Microbiol 13:1069554

Cosin DJD, Ruiz MP, Garvín MH, Ramajo M, Trigo D (2002) Gut load and transit time in Hormogasterelisae (Oligochaeta, Hormogastridae) in laboratory cultures. Eur J Soil Biol 38:43–46

Curry JP (2004) Factors affecting the abundance of earthworms in soils. In: Edwards CA (ed) Earthworm ecology. CRC, Boca Raton, pp 3–11

Curry J, Baker G (1998) Cast production and soil turnover by earthworms in soil cores from South Australian pastures. Pedobiologia 42:283–287

Curry J, Byrne D (1992) The role of earthworms in straw decomposition and nitrogen turnover in arable land in Ireland. Soil Biol Biochem 24:1409–1412

Dash M, Satpathy B, Behera N, Dei C (1984) Gut load and turnover of soil, plant and fungal material by Drawidacalebi, a tropical earthworm. Revue d'Ecologie et de Biologie du Sol 21:387–393

Dash H, Beura B, Dash M (1986) Gut load, transit time, gut microflora and turnover of soil, plant and fungal material by some tropical earthworms. Pedobiologia 29:13–20

Diop H, Ejack L, Elmi AA, Whalen JK (2023) Earthworm distribution around tile drainage in agricultural fields of southwest Quebec. Can J Soil Sci 102:377–380

Domínguez J, Bohlen P, Parmelee R (2004) Earthworms increase nitrogen leaching to greater soil depths in row crop agroecosystems. Ecosystems 7:672–685

Drever JL, Stillings LL (1997) The role of organic acids in mineral weathering. Colloids Surf A Physicochem Eng Asp 120:167–181

Edwards CA, Arancon NQ (2022) Biology and ecology of earthworms, 4th edn. Springer, New York. 567 p

Engelmann P, Opper B, Nemeth P (2011) Interactions of intracellular calcium and immune response in earthworms. Invertebr Surviv J 8:78–84

Eriksen J (2009) Soil sulfur cycling in temperate agricultural systems. Adv Agron 102:55–89

García-Montero LG, Valverde-Asenjo I, Grande-Ortíz MA, Menta C, Hernando I (2013) Impact of earthworm casts on soil pH and calcium carbonate in black truffle burns. Agrofor Syst 87:815–826

Gorres JH, Amador JA (2010) Partitioning of habitable pore space in earthworm burrows. J Nematol 42:68–72

Grethe S, Schrader S, Giesemann A, Larink O, Weigel HJ (1996) Influence of earthworms on the sulfur turnover in the soil. Isot Environ Health Stud 32:211–217

Gul S, Whalen JK (2022) Perspectives and strategies to increase the microbial-derived soil organic matter that persists in agroecosystems. Adv Agron 175:347–401

Hanusova R, Bilej M, Brys L, De-Baetselier P, Beschin A (1999) Identification of a coelomic mitogenic factor in *Eisenia foetida* earthworm. Immunol Lett 65:203–211

Hartenstein F, Hartenstein E, Hartenstein R (1981) Gut load and transit-time in the earthworm *Eisenia-foetida*. Pedobiologia 22:5–20

Haynes RJ, Fraser PM, Piercy JE, Tregurtha RJ (2003) Casts of *Aporrectodea caliginosa* (Savigny) and *Lumbricus rubellus* (Hoffmeister) differ in microbial activity, nutrient availability and aggregate stability. Pedobiologia 47:822–887

Huang LM, Jia XX, Zhang GL, Shao MA (2017) Soil organic phosphorus transformation during ecosystem development: A review. Plant Soil 415:17–42

Jang JW, Xiong XY, Liu C, Yoo KX, Ishii S (2022) Invasive earthworms alter forest soil microbiomes and nitrogen cycling. Soil Biol Biochem 171:108724

Jouquet P, Thi PN, Hong HN, Henry-des-Tureaux T, Chevallier T, Duc TT (2011) Laboratory investigation of organic matter mineralization and nutrient leaching from earthworm casts produced by *Amynthas khami*. Appl Soil Ecol 47:24–30

Kizilkaya R, Hepşen S (2004) Effect of biosolid amendment on enzyme activities in earthworm (*Lumbricus terrestris*) casts. J Plant Nutr Soil Sci 167:202–208

Krišrtůek V, Ravasz K, Pižl V (1992) Changes in densities of bacteria and microfungi during gut transit in *Lumbricus rubellus* and *Aporrectodea caliginosa* (Oligochaeta: Lumbricidae). Soil Biol Biochem 24:1499–1500

Lambkin DC, Gwilliam KH, Layton C, Canti MG, Piearce TG, Hodson ME (2011) Production and dissolution rates of earthworm-secreted calcium carbonate. Pedobiologia 54:S119–S129

Le Bayon RC, Binet F (2006) Earthworms change the distribution and availability of phosphorous in organic substrates. Soil Biol Biochem 38:235–246

Le Bayon RC, Matera V, Kohler-Milleret R, Degen C, Gobat J-M (2011) Earthworm activity alters geogenic arsenic and soil nutrient dynamics. Pedobiologia 54:S193–S201

Liu D, Lian B, Wang B, Jiang G (2011) Degradation of potassium rock by earthworms and responses of bacterial communities in its gut and surrounding substrates after being fed with mineral. PLoS One 6:e28803

Lowe CN, Butt KR (2003) Influence of food particle size on inter-and intra-specific interactions of *Allolobophora chlorotica* (Savigny) and *Lumbricus terrestris*. Pedobiologia 47:574–577

Ma Q, Kuzyakov Y, Pan W, Tang S, Chadwick DR, Wen Y, Hill PW, Macdonald A, Ge T, Si LL, Wu LH, Jones DL (2021) Substrate control of sulphur utilisation and microbial stoichiometry in soil: Results of ^{13}C, ^{15}N, ^{14}C, and ^{35}S quad labelling. ISME J 15:3148–3158

Marichal R, Mathieu J, Couteaux M-M, Mora P, Roy J, Lavelle P (2011) Earthworm and microbe response to litter and soils of tropical forest plantations with contrasting C: N: P stoichiometric ratios. Soil Biol Biochem 43:1528–1535

Marinissen J, De Ruiter P (1993) Contribution of earthworms to carbon and nitrogen cycling in agro-ecosystems. Agric Ecosyst Environ 47:59–74

Martin N (1982) Interaction between organic matter in soil and the burrowing activity of three species of earthworms (Oligochaeta: Lumbricidae). Pedobiologia 24:185–190

Na LP, Abail Z, Whalen JK, Liang B, Hu CX, Hu RG, Wu YP (2022) Earthworms increase nitrogen uptake by lettuce and change short-term soil nitrogen dynamics. Appl Soil Ecol 176:104488

Nebert LD, Bloem J, Lubbers IM, van Groenigen JW (2011) Association of earthworm-denitrifier interactions with increased emission of nitrous oxide from soil mesocosms amended with crop residue. Appl Environ Microbiol 77:4097–4104

Needham A (1957) Components of nitrogenous excreta in the earthworms *Lumbricus terrestris*, L. and *Eisenia foetida* (Savigny). J Exp Biol 34:425–446

Organization for Economic Co-operation and Development (OECD) (2004) Test 222, earthworm reproduction test (*Eisenia fetida/Eisenia andrei*). OECD guidelines for the testing of chemicals, section 2. OECD, Paris, France

Park SC, Smith TJ, Bisesi MS (1992) Activities of phosphomonoesterase and phosphodiesterase from *Lumbricus terrestris*. Soil Biol Biochem 24:873–876

Piearce T (1972) The calcium relations of selected Lumbricidae. J Anim Ecol 41:167–188

Prentø P (1987) Distribution of 20 enzymes in the midgut region of the earthworm, *Lumbricus terrestris* L., with particular emphasis on the physiological role of the chloragog tissue. Comp Biochem Physiol A Physiol 87:135–142

Reich PB, Oleksyn J, Modrzynski J, Mrozinski P, Hobbie SE, Eissenstat DM, Chorover J, Chadwick OA, Hale CM, Tjoelker MG (2005) Linking litter calcium, earthworms and soil properties: a common garden test with 14 tree species. Ecol Lett 8:811–818

Ros MBH, Hiemstra T, van Groenigen JW, Chareesri A, Koopmans GF (2017) Exploring the pathways of earthworm-induced phosphorus availability. Geoderma 303:99–109

Santana MM, Dias T, Gonzalez JM, Cruz C (2021) Transformation of organic and inorganic sulfur—adding perspectives to new players in soil and rhizosphere. Soil Biol Biochem 160:108306

Schaefer CEGR, Lima HN, Gilkes RJ, Mello JWV (2004) Micromorphology and electron microprobe analysis of phosphorus and potassium forms of an Indian Black Earth (IBE) Anthrosol from Western Amazonia. Aust J Soil Res 42:401–409

Scheu S (1987) The influence of earthworms (Lumbricidae) on the nitrogen dynamics in the soil litter system of a deciduous forest. Oecologia 72:197–201

Schrader S, Seibel C (2001) Impact of cultivation management in an agroecosystem on hot spot effects of earthworm middens. Eur J Soil Biol 37:309–313

Sharma S, Dhaliwal SS (2021) Conservation agriculture based practices enhanced micronutrients transformation in earthworm cast soil under rice-wheat cropping system. Ecol Eng 163:106195

Shipitalo M, Protz R, Tomlin A (1988) Effect of diet on the feeding and casting activity of *Lumbricus terrestris* and *L. rubellus* in laboratory culture. Soil Biol Biochem 20:233–237

Simek M (1993) Biological nitrogen-fixation (nitrogenase activity) as a measure of soil biological-activity. Rostlinna Vyroba 39:789–796

Stroud JL, Irons DE, Carter JE, Watts CW, Murray PJ, Norris SL, Whitmore AP (2016) *Lumbricus terrestris* middens are biological and chemical hotspots in a minimum tillage arable ecosystem. Appl Soil Ecol 105:31–35

Subler S, Kirsch AS (1998) Spring dynamics of soil carbon, nitrogen, and microbial activity in earthworm middens in a no-till cornfield. Biol Fertil Soils 26:243–249

Taylor AR, Taylor AF (2014) Assessing daily egestion rates in earthworms: using fungal spores as a natural soil marker to estimate gut transit time. Biol Fertil Soils 50:179–183

Thakur SS, Lone AR, Tiwari N, Jain SK, Yadav S (2021) Metagenomic exploration of bacterial community structure of earthworms' gut. J Pure Appl Microbiol 15:1156–1172

Tiunov AV, Scheu S (1999) Microbial respiration, biomass, biovolume and nutrient status in burrow walls of *Lumbricus terrestris* L. (Lumbricidae). Soil Biol Biochem 31:2039–2048

Tiunov AV, Scheu S (2004) Carbon availability controls the growth of detritivores (Lumbricidae) and their effect on nitrogen mineralization. Oecologia 138:83–90

Trap J, Blanchart E, Ratsiatosika O, Razafindrakoto M, Becquer T, Andriamananjara A, Morel C (2021) Effects of the earthworm *Pontoscolex corethrurus* on rice P nutrition and plant-available soil P in a tropical Ferralsol. Appl Soil Ecol 160:103867

van Groenigen JW, Lubbers IM, Vos HM, Brown GG, De Deyn GB, van Groenigen KJ (2014) Earthworms increase plant production: a meta-analysis. Sci Rep 4:6365

van Groenigen JW, van Groenigen KJ, Koopmans GF, Stokkermans L, Vos HMJ, Lubbers IM (2019) How fertile are earthworm casts? A meta-analysis. Geoderma 338:525–535

Vidal A, Blouin M, Lubbers I, Capowiez Y, Sanchez-Hernandez JC, Calogiuri T, van Groenigen JW (2023) The role of earthworms in agronomy: Consensus, novel insights and remaining challenges. Adv Agron 181:1–78

Vos HJJ, Ros MBH, Koopmans GF, van Groenigen JW (2014) Do earthworms affect phosphorus availability to grass? A pot experiment. Soil Biol Biochem 79:34–42

Vos HJJ, Koopmans GF, Ferron LME, Oenema O, van Groenigen JW (2022) Do earthworms increase grass biomass production and phosphorus uptake under field conditions? Appl Soil Ecol 180:104598

Wan JHC, Wong MH (2004) Effects of earthworm activity and P-solubilizing bacteria on P availability in soil. J Plant Nutr Soil Sci 167:209–213

Whalen JK, Parmelee RW (2000) Earthworm secondary production and N flux in agroecosystems: a comparison of two approaches. Oecologia 124:561–573

Whalen JK, Parmelee RW, Subler S (2000) Quantification of nitrogen excretion rates for three lumbricid earthworms using ^{15}N. Biol Fertil Soils 32:347–352

Whalen JK, Sampedro L, Waheed T (2004) Quantifying surface and subsurface cast production by earthworms under controlled laboratory conditions. Biol Fertil Soils 39:287–291

Wink M (1993) The plant vacuole: a multifunctional compartment. J Exp Bot 44:231–246

Xiao Z, Wang X, Koricheva J, Kergunteuil A, Le Bayon RC, Liu M, Hu F, Rasmann S (2018) Earthworms affect plant growth and resistance against herbivores: A metaanalysis. Funct Ecol 32:150–160

Zangerlé A, Hissler C, Blouin M, Lavelle P (2014) Near infrared spectroscopy (NIRS) to estimate earthworm cast age. Soil Biol Biochem 70:47–53

Zeibich L, Staege M, Schmidt O, Drake HL (2019) Amino acids and ribose: Drivers of protein and RNA fermentation by ingested bacteria of a primitive gut ecosystem. Appl Environ Microbiol 85:e01297–e01219

City Dwellers: Earthworms in Urban Ecosystems

Olaf Schmidt

Abstract Urban ecosystems are often overlooked environments where earthworms thrive as hidden underground dwellers. This chapter explores the occurrence, diversity and ecological roles of earthworms in various urban settings, including parks, lawns, home gardens, buildings, and green infrastructures. It examines how earthworm species adapt to urban habitats and investigates the influence of human activities – such as soil compaction, chemical pollution, and habitat fragmentation – on earthworm populations. By comparing earthworm populations across rural–urban gradients, the review highlights how urbanization influences species distribution, abundance, and ecological functions. Integrating dispersed studies, reports and observation from diverse geographic regions, this chapter provides a comprehensive overview of how urbanization affects earthworm communities. It underscores the importance of maintaining earthworm populations in urban soils to promote sustainable urban green spaces, enhance soil quality, and support other urban biodiversity, including earthworm predators. Finally, the chapter emphasizes the educational value of earthworm biodiversity in cities and its usefulness for increased public awareness and engagement in urban ecological initiatives, conservation and sustainable urban planning. This chapter contributes to the broader understanding of urban ecology and biodiversity, as well as of urban soils and their functions.

1 Introduction

Earthworms are usually studied in rural land uses for primary production, such as agriculture and forestry, or they are of interest from a biodiversity and nature conservation perspective, in that case almost always in exurban, (semi)natural ecosystems. However, earthworms dwell in a remarkable number of urban habitats and

O. Schmidt (✉)
UCD School of Agriculture and Food Science, University College Dublin, Dublin 4, Ireland

UCD Earth Institute, University College Dublin, Dublin 4, Ireland
e-mail: olaf.schmidt@ucd.ie

Y. Kooch, Y. Kuzyakov (eds.), *Earthworms and Ecological Processes*,
https://doi.org/10.1007/978-3-031-64510-5_8

micro-locations, often close to housing, in gardens, greenhouses, and even inside buildings. Of the 6000 or so currently described earthworm species, a small number (87) are known as cosmopolitan or peregrine species (Misirlioglu et al. 2023). These anthropochore (or synanthropic) species have spread widely over the centuries through human trade, transport, and movement, and they usually benefit from habitat modification and soil disturbance by farming, gardening, planting, and related human activities (Terhivuo 1991). It is therefore not surprising that earthworms have adapted to environments in towns, cities, and urban infrastructures that offer basic conditions for survival (Pilipiuk 1981; Toth et al. 2020). Due to its island geography and well-documented history of human settlement and trade over the last 1000 years or so, Iceland is a particularly fine case study for understanding these anthropochore species and their connections to humans (Rundgren 2007). Five species of the family Lumbricidae that are common and widespread in Iceland (including *Dendrodrilus rubidus, Dendrobaena octaedra*, and *Lumbricus rubellus*) are presumed to have arrived and spread naturally. On the contrary, species such as *Aporrectodea rosea, Lumbricus terrestris,* and *Octolasion cyaneum* only occur in human-modified soils, richer in organic matter and nutrients, and are usually closely associated with urban areas where people live, garden, landscape and create improved patches of soil (Rundgren 2007).

Most people who live in cities are unaware of their hidden underground companions. Occasionally, people might notice earthworms on pavements that have emerged after heavy rains (Schwert 1980), and rarely, they might even witness mass migration and "worm circling" events (Zaborski and Gittenger 2001; Weisberger 2021). Earthworms are often called "the gardener's best friend," and people with access to gardens or recycling spaces might see compost worms in their composting bins (Appelhof 2018; Burton and Cameron 2021). In contrast to such benign real-life encounters, the list of fictional "worms" that urban consumers know from literature, movies, and video games is long and mostly encompasses monstrous, destructive, parasitic, and otherwise unpleasant creatures (Wikipedia Contributors 2023). It is therefore important to document the presence, ecology, and importance of earthworms in urban habitats and to provide a factual account of the published body of research, for other researchers, interested lay people, citizen scientists, and educators. The occurrence in and adaptation to urban environments have been studied and reviewed for many animal taxa, most notably birds and mammals (McDonnell and Hahs 2015). Recently, Szlavecz et al. (2020) have produced an overview of urban soil fauna (chiefly macrofauna), with a focus on ecological theories related to invasive species, disturbance, habitat fragmentation, and biotic homogenization. Taking a narrower perspective and descriptive approach, this chapter deals exclusively with earthworms because such a focused literature review is lacking.

This chapter will collate, describe, and review the scattered information on earthworms in cities, the environments where they occur, their adaptations to urban stressors including pollution, and their importance for soil functions and also as food for many vertebrate animals. In most cases, earthworms are beneficial, but there are instances of disservices (e.g., on runway greens in airports). The chapter will conclude with opportunities for the use of earthworms in citizen science, public awareness building, and education in general. The use of earthworms in composting and organic waste management will not be considered here because this topic has been covered extensively, in popular (Appelhof 2018), educational (Melear and Lunsford 2007), and scientific publications (Bhat et al. 2018).

2 Parks, Lawns, and Other Green Space

It is widely accepted that our cities and their urban/suburban habitats are home to diverse and locally abundant animal species (McKinney 2002, 2006) and also soil arthropods (Szlavecz et al. 2020). This is no different for earthworms. Wherever soil sealing leaves some natural soils in place, earthworms will colonize. This may include larger areas such as parks, woodlands, open green spaces, graveyards, sports turf pitches, and allotments, as well as smaller green areas such as grassy roadside strips, small lawns, and even tiny front gardens. One such inner-city front garden in Dublin (Ireland) is the type locality for the earthworm *Lumbricus friendi*, described in 1893 by the naturalist Reverend Hilderic Friend. The site was revisited by local researchers exactly 120 years later to the day of the original discovery and the presence of that species was confirmed (Boots et al. 2012). This is just one example of the fact that urban habitats and fragments of habitats can have a special earthworm fauna (see Fig. 1).

Reflecting the growing interest in two converging topics—urban biodiversity and urban soils—many published studies have reported the abundance and species richness of earthworms of urban habitats. In one of the earliest studies, Pilipiuk (1981) analyzed earthworm records from Warsaw and the surrounding Mazovia region (Poland) and concluded that, while species numbers were similar, the actual species composition differed. Urban habitats lacked stenotopic species (e.g., wetlands species), had additional, presumably recently introduced species (e.g., *Dendrobaena veneta*), and were dominated by eurytopic, cosmopolitan species that can tolerate drier soils, raking/removal of surface litter, and higher soil pH values, all properties caused by human management activities. Recently, Toth et al. (2020) published a comparative international multi-city study involving Baltimore (USA), Budapest (Hungary), Helsinki and Lahti (Finland), and Potchefstroom (South Africa). The authors uncovered a relatively small number of species (19) and a high degree of similarity (dominated by peregrine Lumbricidae species) and interpreted these findings as a sign of biotic homogenization. Across sites, earthworm communities were structured by habitat type, while the few measured soil properties (including soil organic matter and pH) did not explain structures or abundance.

The fact that urban habitats harbor earthworms has now been corroborated by many studies from different parts of the world. To give some examples from Europe, Schlaghamersky and Pizl (2009) described lumbricid communities (eight species in total) from differently sized parks in Brno (Czechia); Glasstetter (2012) included understudied habitats (e.g., river edges and flood plain forests) in her survey of 60 locations in Basel (Switzerland) and, as a result, recorded 22 Lumbricidae species; Francini et al. (2018) undertook a large soil survey of 41 urban parks in southern

Fig. 1 The author at the type locality of *Lumbricus friendi*, a small front garden in inner-city Dublin (Ireland). Individual worms were expelled from their burrows using a diluted mustard oil solution. The species originally discovered in 1892 was confirmed at the site 120 years later (see Boots et al. 2012). Photographs: Aidan Keith

Finland and discovered that earthworm communities were similar to those in control forests; also in Finland (Helsinki), old parks and reference forests showed the same outcome of having significantly larger earthworm biomass under coniferous trees than deciduous trees, inspite of much larger N inputs in the urban parks (Lu et al. 2022); in Brussels (Belgium), Tiho and Josens (2000) characterized earthworm communities in lawn "islets" formed by a heavily trafficked road and found surprisingly large populations (89–253 individuals m^{-2}) with between four and seven species; finally, Butt et al. (2014) recorded nine lumbricid species in a graveyard in England (UK), a green space type studied very rarely.

In South America, Ferreira et al. (2018) surveyed five parks in Curitiba (Brazil) and recorded mostly exotic species (eight) but also native species (four) able to survive in such urban habitats; Dudas et al. (2023) reported very similar findings from a study of a university campus and a public green space in the same city. A number of recent studies from China have provided new data and insights, including Xie et al. (2018) who showed that earthworm communities in radial residential areas of Beijing are associated with soil properties (e.g., bulk density, moisture content, soil organic carbon), but also the age of the residential area (negative correlation), and landscape indices (negative correlation with percentage impervious surface; positive correlation with percentage green space area); recently, Xie et al. (2022) conducted a similar analysis of 16 parks in Beijing.

Land does not have to be "green" to have earthworms, the so-called brownfield (post-industrial) sites can also harbor earthworms. In Kunming City (China), Xie et al. (2022) found that abandoned land (34% vegetation cover) still had four earthworm species (43 individuals m^{-2}); however, this was comparatively much lower than population sizes in forest and nursery land. A documented brownfield example is a former steelworks site near Glasgow, UK, which now has 16 lumbricid species at the rehabilitated, afforested site (Butt and Quigg 2020). Constructed soils will be considered in Sect. 6.

3 Rural–Urban Comparisons and Gradients

To compare earthworm communities in rural and urban contexts, researchers either select the same habitat types in the two contexts and conduct a straight comparison, or they sample spatially along a transect or gradient, with an increasing degree of urbanization. An example of the former approach is the previously mentioned analysis by Pilipiuk (1981) who compared species lists from Warsaw (urban) and the Mazovia region (rural). The advantage of the gradient approach is that sometimes thresholds can be identified beyond which drastic community changes or species additions/disappearances occur. An example is the study by Pizl and Josens (1995a) who sampled earthworms in six grassy park areas in Brussels, Belgium, and found that earthworm biomass increased significantly with distance from the city center. In that study, all sites were located close to busy roads, and earthworm density or biomass was negatively correlated with heavy metal concentrations in soil, including Cd, Zn, Mg, and Cu. In Paris, France, Pelosi et al. (2021) studied 15 grassland patches on a gradient of urbanized land cover around them (the proportion of urbanized land varied between 0.05 and 92%), and they did not observe a significant change in earthworm communities. In Finland, Lu et al. (2022) compared "old" (>70 years) urban parks in Helsinki with similarly aged, exurban reference forests; total earthworm biomass was lower in parks but not significantly so. Related to the expansion of urbanization, Smetak et al. (2007) compared differently aged residential turf grass sites in Moscow, Idaho (USA), and found substantially larger earthworm populations in 75-year-old (121 individuals m^{-2}) compared with 10-year-old properties (26 individuals m^{-2}), suggesting that soil disturbance, higher initial soil bulk density, and lower initial soil C and N contents are detrimental to earthworm populations. In a large survey of 85 gardens along a gradient of urban density (based on position and sealed surface area) in Zürich, Switzerland, Tresch et al. (2018) discovered that earthworm biomass and species richness were driven by garden management factors (such as vegetable growing, compost addition, and bare soil) and not by factors related to urbanization such as proximity to industry.

The picture is confounded in places where exotic species still invade and spread, with cities often acting as the origin (Szlavecz et al. 2020). These studies observed larger earthworm populations in forest soils in urban contexts; in one study from oak forests in New York, Steinberg et al. (1997) found ten times larger earthworm populations in urban than in rural settings, and the three suburban sites were intermediate. The authors ascribed this finding to the close association of introduced exotic species with humans, an interpretation echoed by Baxter et al. (1999). Using microsatellite markers, Klein et al. (2017) detected decreased genetic variance in the exotic *L. terrestris* away from the center of Calgary, Canada, probably reflecting single introduction events (from fish bait) at the most distant locations.

It is usually assumed that urbanization creates gradients of disturbance that impact biological diversity (McKinney 2002, 2006). Taken together, the rural–urban gradient studies mentioned here suggest that earthworm communities are not always impacted systematically by such physical and chemical disturbances. Rather, local soil properties and land management practices are the ultimate drivers.

4 Home Gardens

Home gardens and allotments are obvious places where earthworms are known and appreciated, and no popular publication on gardening misses to mention them (Barter 2016; RHS 2023). Charles Darwin himself studied earthworms and their activities on the grounds of his family home, Down House, London Borough of Bromley, where he lived from 1842 to 1882. Remarkably, Butt et al. (2008) discovered in a survey of that estate that Darwin's Kitchen Garden had the largest earthworm abundance and biomass (715 individuals m^{-2}, 261 g m^{-2}) of all sampled habitats.

Gardens have been included in a number of other scientific studies on urban soils and their biodiversity. For example, in the aforementioned study of 85 gardens in Zürich, Switzerland, Tresch et al. (2018) showed convincingly that soil quality in urban gardens is explained by management practices, with vegetable growing, compost addition, and percentage of bare soil all associated with larger earthworm species richness and abundance, driven mostly by endogeic worms. In a survey of 50 home gardens spread throughout the humid tropical lowlands of Tabasco (Mexico), Huerta and van der Wal (2012) found a strong positive correlation ($r^2 = 0.5$) between soil organic matter and earthworm abundance, and high soil clay and loam contents as well as tree cover in the gardens were also associated with higher earthworm abundances. While this study was not exclusively on urban soils, it identified some drivers of earthworm populations in an understudied part of the world. Domestic horticultural land management even leaves a legacy. Keplin and Broll (1997) compared abandoned lands (for >10 years) in the Ruhr industrial zone (Germany) and found that the sites previously used as gardens had larger earthworm populations (170–224 individuals m^{-2}) than former agricultural sites (57–86 individuals m^{-2}).

5 Botanic Gardens and Greenhouses

The fact that botanic gardens and especially their heated greenhouse are an entry point and a living repository for exotic earthworms was well-known even to the earliest earthworm researchers. For instance, Boettger (1930) reported cases from Italian cities, and Gates (1963) summarized 70 years' of earthworm records from greenhouses in Maine, USA. More recently, Reynolds (2011) pointed out that arboreta, such as the one associated with the University of Guelph, Ontario (Canada), are also prime sources for imported and exotic earthworms, introduced with root balls and soil. For domestic gardens and landscaping, commercial garden centers and plant traders used to be the prime route. To give just one number to support this statement, Gates (1982) reported that, when working for the US Bureau of Quarantine from 1950 to 1982, he intercepted 50 different earthworm species at US ports imported with plant material shipments.

The earthworm checklist for Britain contains about 30 species known from outdoor habitats and about 35 additional, exotic species only recorded from botanic gardens, greenhouses, and similar protected habitats (Sherlock and Carpenter 2009; Sims and Gerard 1999). Since Britain has acquired a full set of earthworm species since the last glaciation, there is less concern about potential new introductions via the greenhouse route. However, the New Zealand flatworm (*Arthurdendyus triangulatus*), a predator of earthworms, has taken exactly that route and is now widespread, especially in northern Britain and Northern Ireland (Boag et al. 1994; Boag and Yeates 2001) (see Fig. 2). In the city of Basel (Switzerland), Glasstetter (2012) surmised that the Zoological Garden boosted such a large earthworm species richness (18) because it had many green spaces "built up from imported soil." Concern about new species introductions is much heightened in parts of North America where glaciation left large land masses devoid of native earthworms (Chang et al. 2021).

Fig. 2 The New Zealand flatworm (*Arthurdendyus triangulatus*), an earthworm predator, has been introduced to Britain and Ireland and spread widely through commercial garden plant trade and domestic exchange. Photograph: Jane Shackleton

6 Technosols and Green Infrastructures

As was mentioned earlier (Sect. 2), land does not have to be "green" and under mature vegetation cover to harbor earthworms, and in urban areas, many soils are in the early phases of development after initial construction. There is a body of research on these so-called "technosols," i.e., soils created through and dominated by human activity, typically in urban and industrial settings (Rossiter 2007). The presence of earthworm communities in such soils is of interest either because they are seen as useful indicators of general soil development and biological status, or because they are known to make important contributions to soil functions and soil development itself. In a comprehensive study of differently aged (1–13 years) technosols in Neuchâtel (Switzerland), Amosse et al. (2016) observed that soil depth and bulk density correlated with earthworm community data (redundancy analysis), and soil depth, in turn, was positively correlated with age. In a companion paper, Amosse et al. (2015) showed that endogeic earthworms had higher burrowing activity and were associated with larger C and N mineralization in an urban technosol than in a natural alluvial soil, highlighting the roles earthworms fulfill in such urban soils with low organic matter contents. Scharenbroch and Johnston (2011) showed that the large-bodied, anecic *L. terrestris* had various stimulating effects on microbial biomass and activity as well as nutrient availability in a designed soil (the Cornell University Structural Soil); however, the effects were short-lived because earthworm survival was very low after 140 days in this material.

That latter study highlights the importance of assessing technosols and the materials used for their suitability as a habitat for earthworms and other soil biota, so that for future projects biological activity can be considered at the design stage. For that purpose, bioassays are commonly used, mainly with respect to chemical suitability, e.g., Hankard et al. (2005) (see Sect. 9). In a field-based survey of 20 technosols aged between 2 and 64 years located in parks in Paris (France), Vergnes et al. (2017) observed that earthworm abundances increased with soil age only when the initial design included natural topsoil additions over mineral backfill, but in general all communities were dominated by a few common species. Also in Paris, Marechal et al. (2021) did not find effects on earthworm communities of different road engineering approaches 20–30 years after construction; however, they found some evidence for significantly reduced colonization by earthworms of highly isolated roadside soils (surrounded by large asphalted roads). These authors concluded that the preservation of reservoir areas and maintenance of habitat patch connections is important to ensure earthworm colonization. Targeted inoculation is an alternative approach (e.g., Butt et al. 2022).

Once technosols are in situ, management and especially amendments can also influence the development of earthworm communities. This has been studied extensively in former mining and urban brownfield sites, for example in a study around Liverpool (UK) where earthworm counts were used as one of several soil biological indicators (Hartley et al. 2008). In that study, soil inversion (to 1 m depth), application of dredged contaminated canal sediment, and use of a former metalliferous site all resulted in very small earthworm populations, while planting of trees and establishment of amenity grassland helped boost earthworm populations at a number of sites. Using experimental grass plots to compare soil cover options popular in residential landscapes in the USA, Byrne et al. (2008) discovered that bark mulch sustained the largest earthworm populations, whereas gravel cover had the lowest, with lawn and unmown treatments being intermediate. Active management in order to suppress earthworm populations, and especially the surface casting species, is a topic in the context of sports turfs and similarly constructed soils. Lowering the soil pH, adding grit or irritants, and changing grass-cutting regimes are examples of management practices that have been trialed (Boyle et al. 2019). A study of a park in Dnipro (Ukraine) found that park "reconstruction" works (including path restoration, shrub removal, removal of old trees, and replanting of new trees) resulted in increased soil compaction and disturbed soil moisture regimes, which in turn resulted in reduced earthworm populations in the short term (Zhukov et al. 2023). In a survey of soccer pitches near Moscow (Russia), Kutovaya et al. (2014) discovered that all-year-round artificial heating, from a heating system located 25–30 cm below the soil surface, as well as on-demand artificial irrigation facilitates the development of "abundant" earthworm populations (no data given) comprising three species including the cosmopolitan endogeic *A. caliginosa* and the epigeic *L. rubellus*.

Urban planners and managers make more and more use of "nature-based solutions" and "green infrastructures" to tackle problems we face in cities, including (storm)water management, heat island effects, and biodiversity loss. Green infrastructures such as green roofs and stormwater planters combine physical engineering with nature and often contain soil or soil-like substrates, but we know little about the fauna in such habitats. For example, Jusselme et al. (2019) reviewed the published literature on rain gardens and related stormwater infrastructures and concluded that, while soil macrofauna including earthworms potentially have important functions in these systems, almost no studies have been published yet. Ayers and Kangas (2018) surveyed ten bioretention cells in Maryland and Washington DC (USA) and found substantial numbers of earthworms in all of them (equivalent to between 72 and 805 individuals m^{-2}). The PhD thesis (Ayers 2009) contains more details and suggests that, first, the age of the retention cells (between 1 and 10 years) was not correlated with earthworm abundance (i.e., colonization occurred rapidly), and second, the identified communities comprised both endemic and exotic species. More recently, Fernandes et al. (2022) investigated a large (2 ha) infiltration basin in Chassieu (France) and recorded between 0 and 246 earthworms m^{-2}, with common endogeic and epigeic species accounting for half of the identifiable adults. As all sample locations in the basin with a moisture content <80% had earthworms, the authors speculate that higher water saturation of the sediment prevents earthworm survival. These findings clearly suggest that such deep soils/sediments in green infrastructures can harbor substantial earthworm populations, which should be investigated further.

7 Buildings

The presence of living invertebrate animals, or "creepy-crawlies," in people's houses is universally unwanted. Many arthropods show up in domestic dwellings and often even live there. For example, in a monumental survey of 50 houses in Raleigh (USA), Bertone et al. (2016) were astonished to find between 32 and 211 morphospecies and between 24 and 128 families of arthropods per house, with true flies (Diptera), spiders (Araneae), beetles (Coleoptera), and ants/bee/wasps (Hymenoptera) accounting for 23%, 19%, 16%, and 15%, respectively, of all recorded species. By contrast, earthworms are only exceptionally rarely recorded inside dwellings, and they can only survive in micro-sites that meet their basic needs including those for moisture and food. In fact, elongated arthropods (centipedes and millipedes) that are much more likely to invade houses are probably often mistakenly identified as earthworms (Back 1939). Even in countries with subtropical monsoon climates such as Bangladesh, house incursions by earthworms are very rare relative to other animal taxa and are usually linked with blocked drainage, low-lying areas, and flooding events (Kabir et al. 2020).

One notable species in this regard is *Dichogaster bolaui* (family Octochaetidae), originally from eastern Africa and now widespread throughout the tropics. This species has been termed "domicole" by Csuzdi et al. (2008), on account of it having been recorded as a resident in sewage pipes, toilet bowls, sinks, and similar indoor places in dispersed parts of the world including India, Finland, Sweden, and Israel. There is one case report from Ireland where the managers of a heated swimming pool instigated an investigation and eradication of a worm infestation in the shower areas and between tiles that had offended clients (Rota and Schmidt 2006). Archeological reports of thriving earthworm populations in latrines dating from medieval (Judas and Büchner 1989) and even Roman (Dobson and Satchell 1956) times suggested that there may be an older synanthropic connection with some specialized species (such as *Dendrobaena hortensis*) that exploit human wastes.

Terhivuo (1991) had reported *D. bolaui* from domestic houses in southern Finland from sanitary units connected to the sewer system and also from potted house plants, and that author pointed out that the latter likely are the source for the former. We know from Gates' record keeping (Gates 1963) that plant containers with growth media conceal earthworms and are an ideal mode of transportation. Given that more stringent plant health and biosecurity regulations have been introduced in many parts of the world since then, this route is perhaps less important, but it still exists internationally (Sugiura 2008).

8 Airports

A well-documented "ecosystem disservice" that large earthworm populations potentially provide relates to hazards for aviation from birds that feed on earthworms in vegetated areas around runways. For instance, in an early study at Christchurch Airport, New Zealand Moeed (1976) found that black-backed gulls (*Larus dominicanus*) fed exclusively on earthworms, while starlings and magpies fed opportunistically on earthworms. To reduce airport bird-strike hazards, efforts were initially focused on synthetic pesticides for chemical control, for example, using the fungicide benomyl, a carbamate that was known to reduce earthworm populations in orchards (Tomlin et al. 1981). Later, attention was given to natural products for chemical control including saponin-rich tea oil byproducts (Potter et al. 2010), fertilizers that reduce soil pH and thus make soil less hospitable for earthworms (Seamans et al. 2015), and also fungal endophytes (*Epichloë* sp.) of grasses that, likely, affect litter palatability via alkaloidal secondary metabolites (Pennell et al. 2018). Ecological management and long-term monitoring (20 years) of earthworm populations at Manchester airport were reported recently by Butt et al. (2022).

9 Earthworms and Urban Pollution

Analogous to the use of earthworms to assess the biological status of technosols (Sect. 6), earthworms are chosen as study organisms in relation to chemical soil pollution because they are seen as useful indicators of general soil biological status (Bargagli et al. 2019), or because they are known to make important contributions to soil functions and to be a food source for other animals (Sect. 10). A very voluminous literature exists on the use of earthworms in assessing the impacts of toxic chemicals, with approaches ranging from short-term, acute toxicity assays; medium-term field-based toxicity tests; to long-term risk prediction and modeling (Reinecke and Reinecke 2004). Broadly speaking, many more studies are conducted on agri-chemicals, which are primarily used on rural soils, while fewer studies focus on pollutants (especially heavy metals) that dominate in urban systems. That reflects the wider context of rural and urban soils having very different pollution profiles (Shifaw 2018). Cities are characterized by the concentration of polluting human activities (including road transport, industries, construction, and households), and consequently, many studies have investigated the effects on and response of earthworms in urban soils polluted with heavy metals.

Standardized laboratory toxicity assays are used routinely to screen urban soils for chemical toxicity (e.g., Hankard et al. 2005; Santorufo et al. 2012). Such bioassays are also useful for materials used in technosol production (Sect. 6). Some field studies found correlations between the concentrations of at least some heavy metals in soils and the earthworms inhabiting them (e.g., Koeckritz et al. 1999; Pizl and Josens 1995b), but other studies concluded that the soil–worm metal relationships depend on other factors, for example, the exact layering of metals in the soil and the species-specific feeding behaviors (Suthar et al. 2008). Much effort has also been expended on the question of whether earthworms can act as "bioaccumulators" of certain metals and thus potentially endanger predator species higher up the food chain. For instance, Scheifler et al. (2006) estimated that 63% of lead intake in blackbirds (*Turdus merula*) comes from the consumption of earthworms and they concluded that lead persistence in urban soils and its transfer through food chains continues to pose a problem. A similar conclusion was reached by Pouyat et al. (2015) who studied metal burdens in the Baltimore Washington area (USA). In general, the heavy metal geochemistry of cities is complicated because it is driven not only by known anthropogenic inputs but also by the prevailing parent material, age of settlement, and local peculiarities such as the usage of lead-rich paints on local houses (Martin et al. 2023).

Other contaminants studied include polycyclic aromatic hydrocarbons (PAH) (Pizl et al. 2009), which had no measurable effects, and, more recently, microplastics and particulate matter (Holzinger et al. 2022). The latter authors concluded from bioassays with *Eisenia fetida* that vehicle break dust was the most harmful. Using a local earthworm species (*Alma nilotica*, Microchaetidae), Nfor et al. (2022) applied bioassays in Douala (Cameroon) to test soils from e-waste management sites. These authors found significant growth and reproduction inhibition in waste site soils that had much higher heavy metal concentrations than the two control soils.

10 Earthworm Predators in Cities

A diverse range of vertebrates that live in cities prey on earthworms, and for some of them, including moles and foxes, earthworms are the principal food all year round or seasonally (Macdonald 1983). For example, earthworms were the second largest identifiable food item (after scavenged food) in the fecal analysis of urban foxes in Oxford (UK) (Doncaster et al. 1990). Stomach content analysis showed that the diet of both urban and rural badgers in Norway was dominated by earthworms in spring. In forest fragments in Auckland (New Zealand), earthworms were found in the stomachs of 43% of 44 hedgehogs analyzed (Nottingham et al. 2019). Staying with the same species, earthworm availability was one of the predictors of hedgehog abundance in a study in north-eastern France, where hedgehog density estimates were about nine times higher in urban than in rural environments (Hubert et al. 2011). In urban parks in central Tokyo (Japan), earthworms were detected in 49% of fecal samples produced by raccoon dogs in the winter season (Enomoto et al. 2018). A recent study even showed that earthworms constitute a major dietary component for most of the year for domestic cats in Parisian suburbs (Castañeda et al. 2020).

Of the various bird species that feed on earthworms, gulls have been studied often in cities because of their increasing presence and, sometimes, divisive feeding behavior giving rise to human–wildlife conflict. In Quebec (Canada), Brousseau et al. (1996) detected a shift in the diet of ring-billed gulls from 1978 to 1993 in that they relied more on refuse and earthworms and less on fish, presumably reflecting changes in waste management and urbanization. Coulson and Coulson (2008) concluded from the analysis of regurgitated pellets from lesser black-backed gulls in the town of Dumfries (UK) that this species feeds extensively on earthworms and never on fish or fish offal, a finding that differed from most previous studies.

Cities are good locations for studying predator behavior. For example, Macdonald (1983) provided a fascinating description of a fox hunting for earthworms and surmised that a fox detects an earthworm (usually *L. terrestris*) by "the scraping sound made by its chaetae" (p. 404). In the Dutch city of Groningen, Nienhuis (2022) accumulated observations on black-headed gulls that suggest this bird species, which does not normally feed at night, uses artificial light from streetlamps for nocturnal foraging for earthworms. Large numbers of gulls engaged in this behavior (up to 300 individuals per site at the same time) and foraging activity peaked between 20 and 22 o'clock in the winter.

Fewer reports exist on invertebrate earthworm predators. A subterranean, earthworm-eating slug (*Selenochlamys ysbryda*) from an urban garden in Cardiff (UK) was described as new to science by Rowson and Symondson (2008). Although it belongs to a molluscan family (Trigonochlamydidae) endemic to the Caucasus region, the authors noted its similarity in earthworm-predation behavior and deep soil-dwelling habits to the west-palearctic Testacellidae family. The close association of the aforementioned New Zealand flatworm (*A. triangulatus*) (see Fig. 2), an invasive species and exclusive predator of earthworms, with (urban) people and especially urban gardens has been documented in great detail from the UK (Boag et al. 1994; Boag and Yeates 2001). For example, in their early survey of Scotland (conducted 1991–1993), Boag et al. (1994) recorded this species 348 times from domestic gardens, 56 from botanic gardens and garden centers/nurseries, but just 13 times from farms.

11 Education and Citizen Science in Cities

Earthworms are readily recognized by most people, including children, and their large size, ubiquitous occurrence, and easy handling make them ideal study objects for educational events about urban biodiversity and conservation, but also soil functions, soil sealing, flooding, and so on. Earthworms are perfect as an easily accessible, locally available, but somewhat hidden, resource for conveying and illustrating biodiversity concepts. Their use especially in primary education has been relayed in several publications (Baker et al. 1997; Martay and Pearce-Higgins 2018; Varelas et al. 2014). Earthworms can be used in simple observations, measurements, experiments, and projects suitable for education at all levels and public awareness. Collecting and sampling earthworms is very easy and at its most basic (hand-sorting of soil) only requires a ruler, spade, and plastic sheet. However, more sophisticated methods are available, for instance, electrical extraction that has the big advantage of being non-destructive and thus suitable for sensitive and public areas where soil cannot be disturbed (see Fig. 3).

Based on many years of experience, Butt and Grigoropoulou (2010) recounted and described a range of simple and inexpensive methods for studying the life cycle, behavior, and functions of earthworms. Ecological roles of earthworms on the one hand, and impacts of human activities on earthworms on the other hand, can also be demonstrated and investigated with learners. One illustrative case on ecological roles that urban citizens can relate to is the disappearance and decomposition of fallen leaf litter, raked together and heaped up around trees (Cothrel et al. 1997). Some other functions, such as improvement of soil structure by earthworm burrowing and stimulation of nutrient cycling, have already been mentioned in Sect. 6; however, they are usually more demanding as a topic of investigation. An example of a visible human impact is trampling, caused by pedestrians walking along unofficial paths in urban parks; this greatly reduced earthworm populations due to soil compaction and a reduction in soil moisture (Pizl and Schlaghamersky 2007). The species richness, ecology, and functions of earthworms can be explained, demonstrated, and investigated indoors and outdoors. As shown in this chapter, urban habitats often have many species and different ecological groups, and different habitats can be close together in cities. For example, on an urban educational farm in Dublin (Ireland) of 13.5 ha, 11 habitats were surveyed and they each boasted between 5 and 12 earthworm species (Melody and Schmidt 2012).

Earthworms can often be observed in cities at night, especially anecic individuals that live in cracks, under walls, and between paving stones. For example, Griffith et al. (2013) observed (and filmed) *L. terrestris* surfacing from gravel driveways and stony garden areas to forage, in Sequim WA (USA). The surfacing behavior of these large species at night offers great practical opportunities for direct observation, counting, and photography/videoing (see Fig. 4).

Compost worms are, of course, a wonderful subject matter when teaching and demonstrating composting, vermicomposting, and related organic waste management topics to an urban audience. This was not covered here because plenty of

Fig. 3 Sampling earthworms and estimating population sizes can be done by digging and hand-sorting of soil (top), but non-destructive methods such as the electrical "octet" method are also available (bottom). Photographs: Ganna Shutenko (top), Olaf Schmidt (bottom)

Fig. 4 Semi-emerged *Lumbricus terrestris* foraging (top) and mating (bottom) on the soil surface. Photographed at night in a brightly lit traffic isle in inner-city Halle/Saale (Germany). The worms showed no escape response to torchlight. Photographs: Olaf Schmidt

publications already exist in this space (Appelhof 2018; Bhat et al. 2018), also for school classroom-based learning (Melear and Lunsford 2007), and for young curious audiences (Burton and Cameron 2021).

Citizen science approaches to earthworm surveys have been attempted, including in very large projects in the UK as part of the OPAL (Open-Air laboratories) program (Davies et al. 2011), in Australia (Baker et al. 1997), with school pupils in

Norway (Anon 2010), and in the USA, with a focus on earthworms as invasive species (Ziter et al. 2021).

For the scientific community, urban ecology is also a thriving scientific research field. Expert knowledge was compiled recently to formulate no fewer than 62 research hypotheses considered important for urban ecology (Lokatis et al. 2023). It is essential that below-ground systems and organisms, such as earthworms, are also included in this endeavor (Szlavecz et al. 2020).

12 Conclusions

Earthworms rarely make it into mainstream media, apart from the odd report on "worm tornados" (Weisberger 2021). It is clear from the studies reviewed here that diverse and abundant earthworm communities occur in parks, lawns, gardens, and other green, and sometimes brown, areas in the world's cities. They often comprise cosmopolitan and/or exotic species but can also include native species. Their species composition and abundance are mainly driven by habitat type and, sometimes, soil properties, management, and urbanization factors. Earthworms are an important food source for many other animals in urban habitats, and they are exposed to many stressors including chemical pollutants. Earthworms are an easily accessible animal group for education and citizen science in various topics including urban ecology, biodiversity, and the importance of soils.

Acknowledgments The author acknowledges research funding under the EPA Research Program 2021–2030, a Government of Ireland initiative funded by the Department of the Environment, Climate and Communications (Research Grant 2023-NE-1199), and funding from the European Union (Grant Agreement 101112831). Views and opinions expressed are those of the author only and do not necessarily reflect those of the EPA or the European Union.

References

Amosse J, Turberg P, Kohler-Milleret R, Gobat J-M, Le Bayon R-C (2015) Effects of endogeic earthworms on the soil organic matter dynamics and the soil structure in urban and alluvial soil materials. Geoderma 243:50–57

Amosse J, Dozsa-Farkas K, Boros G, Rochat G, Sandoz G, Fournier B, Mitchell EAD, Le Bayon R-C (2016) Patterns of earthworm, enchytraeid and nematode diversity and community structure in urban soils of different ages. Eur J Soil Biol 73:46–58

Anon (2010) An earthworm is never the same. Report from the Research Campaign 2010 [in Norwegian]. Norges forskningsråd, St. Hanshaugen, Norway. https://www.miljolare.no/kampanjer/forskningskampanjen/2010/FK2010_meitemark_sluttrapport.pdf

Appelhof M (2018) Worms eat my garbage: how to set up and maintain a worm composting system (35th anniversary edition). Storey, North Adams, MA

Ayers EM (2009) Pedogenesis in rain gardens: The role of earthworms and other organisms in long-term soil development. PhD thesis, University of Maryland, USA

Ayers EM, Kangas P (2018) Soil layer development and biota in bioretention. Water 10:article 1587

Back EA (1939) Centipedes and millipedes in the house. U.S. Department of Agriculture (Leaflet No. 192), pp 1–6

Baker GH, Thumlert TA, Meisel LS, Carter PJ, Kilpin GP (1997) "Earthworms Downunder": A survey of the earthworm fauna of urban and agricultural soils in Australia. Soil Biol Biochem 29:589–597

Bargagli R, Ancora S, Bianchi N, Rota E (2019) Deposition, abatement and environmental fate of pollutants in urban green ecosystems: Suggestions from long-term studies in Siena (Central Italy). Urban For Urban Green 46:article 126483

Barter G (2016) RHS how do worms work? A gardener's collection of curious questions and astonishing answers. Mitchell Beazley, London

Baxter J, Pickett STA, Carreiro MM, Dighton J (1999) Ectomycorrhizal diversity and community structure in oak forest stands exposed to contrasting anthropogenic impacts. Can J Bot - Rev Can Bot 77:771–782

Bertone MA, Leong M, Bayless KM, Malow TLF, Dunn RR, Trautwein MD (2016) Arthropods of the great indoors: characterizing diversity inside urban and suburban homes. PeerJ 4:article e1582

Bhat SA, Singh J, Vig AP (2018) Earthworms as organic waste managers and biofertilizer producers. Waste Biomass Valorization 9:1073–1086

Boag B, Yeates GW (2001) The potential impact of the New Zealand flatworm, a predator of earthworms, in western Europe. Ecol Appl 11:1276–1286

Boag B, Palmer LF, Neilson R, Chambers SJ (1994) Distribution and prevalence of the predatory planarian *Artioposthia triangulata* (Dendy) (Tricladida, Terricola) in Scotland. Ann Appl Biol 124:165–171

Boettger CR (1930) Untersuchungen über die Gewächshausfauna Unter- und Mittelitaliens. Z Wiss Biol — Abt A Morph Okol Tiere 19:534–590

Boots B, Keith A, Schmidt O (2012) Looking out for an old friend. Biodiversity Ireland – Bull Natl Biodivers Data Centre 10(Winter):8–9

Boyle PE, Richardson MD, Savin MC, Karcher DE, Potter DA (2019) Ecology and management of earthworm casting on sports turf. Pest Manag Sci 75:2071–2078

Brousseau P, Lefebvre J, Giroux JF (1996) Diet of ring-billed gull chicks in urban and non-urban colonies in Quebec. Colonial Waterbirds 19:22–30

Burton V, Cameron E (2021) Learning more about earthworms with Citizen Science. Front Young Minds 9:article 548525

Butt KR, Grigoropoulou N (2010) Basic research tools for earthworm ecology. Appl Environ Soil Sci 2010, article 562816. https://doi.org/10.1155/2010/562816

Butt KR, Quigg SM (2020) Soils and earthworms as a final chapter in the narrative of a steelworks. Glasgow Nat 27(2):1–10

Butt KR, Lowe CN, Beasley T, Hanson I, Keynes R (2008) Darwin's earthworms revisited. Eur J Soil Biol 44:255–259

Butt KR, Lowe CN, Duncanson P (2014) Earthworms of an urban cemetery in Preston: General survey and burrowing of Lumbricus terrestris. Zeszyty Naukowe 17:23–30

Butt KR, Gilbert JA, Kostecka J, Lowe CN, Quigg SM, Euteneuer P (2022) Two decades of monitoring earthworms in translocated grasslands at Manchester Airport. Eur J Soil Biol 113:article 103443

Byrne LB, Bruns MA, Kim KC (2008) Ecosystem properties of urban land covers at the aboveground-belowground interface. Ecosystems 11:1065–1077

Castañeda I, Zarzoso-Lacoste D, Bonnaud E (2020) Feeding behaviour of red fox and domestic cat populations in suburban areas in the south of Paris. Urban Ecosyst 23:731–743

Chang C-H et al (2021) The second wave of earthworm invasions in North America: biology, environmental impacts, management and control of invasive jumping worms. Biol Invasions 23:3291–3322

Cothrel SR, Vimmerstedt JP, Kost DA (1997) In situ recycling of urban deciduous litter. Soil Biol Biochem 29:295–298

Coulson JC, Coulson BA (2008) Lesser Black-backed Gulls *Larus fuscus* nesting in an inland urban colony: the importance of earthworms (Lumbricidae) in their diet. Bird Study 55:297–303

Csuzdi C, Pavlicek T, Nevo E (2008) Is *Dichogaster bolaui* (Michaelsen, 1891) the first domicole earthworm species? Eur J Soil Biol 44:198–201

Davies L et al (2011) Open air laboratories (OPAL): A community-driven research programme. Environ Pollut 159:2203–2210

Dobson R, Satchell J (1956) *Eophila oculata* at Verulamium: a Roman earthworm population? Nature 177:796–797

Doncaster CP, Dickman CR, MacDonald DW (1990) Feeding ecology of red foxes (*Vulpes vulpes*) in the city of Oxford, England. J Mammal 71:188–194

Dudas RT, Tavares AAT, Ercole C, Lara BLD, Carlos EDS, Torres JLM, Smokanit M, Guaranha RM, Brown GG, Bartz MLc. (2023) Urban green areas as earthworm species maintainers in Curitiba, Parana, Brazil. Zootaxa 5255:336–346

Enomoto T, Saito MU, Yoshikawa M, Kaneko Y (2018) Winter diet of the raccoon dog (*Nyctereutes procyonoides*) in urban parks, central Tokyo. Mammal Study 43:275–280

Fernandes G, Roques O, Lassabatere L, Sarles L, Venisseau A, Marchand P, Bedell JP (2022) Influence of edaphic conditions and persistent organic pollutants on earthworms in an infiltration basin. Environ Pollut 304:article 119192

Ferreira T et al (2018) Earthworm species in public parks in Curitiba, Parana, Brazil. Zootaxa 4496:535–547

Francini G, Hui N, Jumpponen A, Kotze DJ, Romantschuk M, Allen JA, Setala H (2018) Soil biota in boreal urban greenspace: Responses to plant type and age. Soil Biol Biochem 118:145–155

Gates GE (1963) Miscellanea Megadrilogica. VII. Greenhouse earthworms. Proc Biol Soc Washington 76:9–18

Gates GE (1982) Farewell to North American megadriles. Megadrilogica 4:12–77

Glasstetter M (2012) Earthworm diversity in urban habitats of Basel (Northwestern Switzerland) (Oligochaeta: Lumbricidae). Zool Middle East 58 (Issue Sup 4):95–102

Griffith B, Tuerke M, Weisser WW, Eisenhauer N (2013) Herbivore behavior in the anecic earthworm species *Lumbricus terrestris* L.? Eur J Soil Biol 55:62–65

Hankard PK, Bundy JG, Spurgeon DJ, Weeks JM, Wright J, Weinberg C, Svendsen C (2005) Establishing principal soil quality parameters influencing earthworms in urban soils using bioassays. Environ Pollut 133:199–211

Hartley W, Uffindell L, Plumb A, Rawlinson HA, Putwain P, Dickinson NM (2008) Assessing biological indicators for remediated anthropogenic urban soils. Sci Total Environ 405:358–369

Holzinger A, Mair MM, Luecker D, Seidenath D, Opel T, Langhof N, Otti O, Feldhaar H (2022) Comparison of fitness effects in the earthworm *Eisenia fetida* after exposure to single or multiple anthropogenic pollutants. Sci Total Environ 838:article 156387

Hubert P, Julliard R, Biagianti S, Poulle M-L (2011) Ecological factors driving the higher hedgehog (*Erinaceus europeaus*) density in an urban area compared to the adjacent rural area. Landsc Urban Plan 103:34–43

Huerta E, van der Wal H (2012) Soil macroinvertebrates' abundance and diversity in home gardens in Tabasco, Mexico, vary with soil texture, organic matter and vegetation cover. Eur J Soil Biol 50:68–75

Judas M, Büchner S (1989) *Dendrobaena hortensis* (Michaelsen) (Annelida, Oligochaeta, Lumbricidae) inhabiting medieval sewers in Göttingen. Göttinger Naturkundl Schr 1:71–76

Jusselme MD, Pruvost C, Motard E, Giusti-Miller S, Frechault S, Alphonse V, Balland-Bolou-Bi C, Dajoz I, Mora P (2019) Increasing the ability of a green roof to provide ecosystem services by adding organic matter and earthworms. Appl Soil Ecol 143:61–69

Kabir A, Ali S, Hawkeswood TJ (2020) Invader animals in Bangladesh houses and their control. Calodema 780:1–14

Keplin B, Broll G (1997) Earthworms and dehydrogenase activity of urban biotopes. Soil Biol Biochem 29:533–536

Klein A, Cameron EK, Heimburger B, Eisenhauer N, Scheu S, Schaefer I (2017) Changes in the genetic structure of an invasive earthworm species (*Lumbricus terrestris*, Lumbricidae) along an urban - rural gradient in North America. Appl Soil Ecol 120:265–272

Koeckritz T, Irmler U, Weppen P (1999) Heavy metal content in *Aporrectodea caliginosa* (Oligochaeta, Lumbricidae) in the city of Kiel (Germany, Schleswig-Holstein). J Plant Nutr Soil Sci 162:477–482

Kutovaya OV, Zamotaev IV, Belobrov VP (2014) Communities of microorganisms and invertebrates in soil-like bodies of soccer fields in Moscow oblast. Eurasian Soil Sci 47:1107–1115

Lokatis S et al (2023) Hypotheses in urban ecology: building a common knowledge base. Biol Rev 98:1530–1547

Lu C, Kotze DJ, Setala HM (2022) Plant functional type affects nitrogen dynamics in urban park soils similarly to boreal forest soils. Plant Soil 479:573–587

Macdonald DW (1983) Predation on earthworms by terrestrial vertebrates. In: Satchell JE (ed) Earthworm ecology: from Darwin to vermiculture. Chapman & Hall, London, pp 393–414

Marechal J, Hoeffner K, Marie X, Cluzeau D (2021) Response of earthworm communities to soil engineering and soil isolation in urban landscapes. Ecol Eng 169:article 106307

Martay B, Pearce-Higgins JW (2018) Using data from schools to model variation in soil invertebrates across the UK: The importance of weather, climate, season and habitat. Pedobiologia 67:1–9

Martin AP, Lim C, Kah M, Rattenbury MS, Rogers KM, Sharp EL, Turnbull RE (2023) Soil pollution driven by duration of urbanisation and dwelling quality in urban areas: An example from Auckland, New Zealand. Appl Geochem 148:article 105518

McDonnell MJ, Hahs AK (2015) Adaptation and adaptedness of organisms to urban environments. Annu Rev Ecol Evol Syst 46:261–280

McKinney ML (2002) Urbanization, biodiversity, and conservation. Bioscience 52:883–890

McKinney ML (2006) Urbanization as a major cause of biotic homogenization. Biol Conserv 127:247–260

Melear CT, Lunsford E (2007) Worms cultivate our curriculum: A long-term, theme-based unit. Sci Act 44:48–54

Melody C, Schmidt O (2012) Northward range extension of an endemic soil decomposer with a distinct trophic position. Biol Lett 8:956–959

Misirlioglu M et al (2023) Earthworms (Clitellata, Megadrili) of the world: an updated checklist of valid species and families, with notes on their distribution. Zootaxa 5255:417–438

Moeed A (1976) Birds and their food resources at Christ Church International Airport, New Zealand. NZ J Zool 3:373–390

Nfor B, Fai PBA, Fobil JN, Basu N (2022) Effects of electronic and electrical waste-contaminated soils on growth and reproduction of earthworm (*Alma nilotica*). Environ Toxicol Chem 41:287–297

Nienhuis J (2022) Night-time foraging for earthworms by Black-headed *Gulls Larus ridibundus* [In Dutch]. Limosa 95:41–45

Nottingham CM, Glen AS, Stanleyl MC (2019) Snacks in the city: the diet of hedgehogs in Auckland urban forest fragments. N Z J Ecol 43:article 3374

Pelosi C, Baudry E, Schmidt O (2021) Comparison of the mustard oil and electrical methods for sampling earthworm communities in rural and urban soils. Urban Ecosyst 24:355–364

Pennell CGL, Rolston MP, van Koten C, Mace WJ, Hume DE, Card SD (2018) The effect of grass endophytes on earthworms and slugs under a turf mowing regime. N Z Plant Prot 71:240–247

Pilipiuk I (1981) Earthworms (Oligochaeta, Lumbricidae) of Warsaw and Mazovia. Memorabilia Zoologica 34:69–77

Pizl V, Josens G (1995a) Earthworm communities along a gradient of urbanization. Environ Pollut 90:7–14

Pizl V, Josens G (1995b) The influence of traffic pollution on earthworms and their heavy-metal contents in an urban ecosystem. Pedobiologia 39:442–453

Pizl V, Schlaghamersky J (2007) The impact of pedestrian activity on soil annelids in urban greens. Eur J Soil Biol 43:S68–S71

Pizl V, Schlaghamersky J, Triska J (2009) The effects of polycyclic aromatic hydrocarbons and heavy metals on terrestrial annelids in urban soils. Pesquisa Agropecuaria Brasileira 44:1050–1055

Potter DA, Redmond CT, Meepagala KM, Williams DW (2010) Managing earthworm casts (Oligochaeta: Lumbricidae) in turfgrass using a natural byproduct of tea oil (*Camellia* sp.) manufacture. Pest Manag Sci 66:439–446

Pouyat RV, Szlavecz K, Yesilonis ID, Wong CP, Murawski L, Marra P, Casey RE, Lev S (2015) Multi-scale assessment of metal contamination in residential soil and soil fauna: A case study in the Baltimore-Washington metropolitan region, USA. Landsc Urban Plan 142:7–17

Reinecke AJ, Reinecke SA (2004) Earthworms as test organisms in ecotoxicological assessment of toxicant impacts on ecosystems. In: Edwards CA (ed) Earthworm ecology, 2nd edn. CRC, Boca Raton, FL, pp 299–320

Reynolds JW (2011) The earthworms (Oligochaeta: Lumbricidae) of the University of Guelph Arboretum, Wellington County, Ontario, Canada. Megadrilogica 15:15–20

RHS (2023) Wild about lawns—bring your lawn to life. Royal Horticultural Society, London

Rossiter DG (2007) Classification of urban and industrial soils in the world reference base for soil resources. J Soils Sediments 7:96–100

Rota E, Schmidt O (2006) *Dichogaster bolaui* (Oligochaeta : Octochaetidae), an unusual invader in a swimming pool in Ireland. J Nat Hist 40:161–167

Rowson B, Symondson WOC (2008) *Selenochlamys ysbryda* sp nov from Wales, UK: A Testacella-like slug new to Western Europe (Stylommatophora : Trigonochlamydidae). J Conchol 39:537–552

Rundgren S (2007) Lumbricidae in Iceland. Insect Syst Evol Suppl 64:121–159

Santorufo L, Van Gestel CAM, Maisto G (2012) Ecotoxicological assessment of metal-polluted urban soils using bioassays with three soil invertebrates. Chemosphere 88:418–425

Scharenbroch BC, Johnston DP (2011) A microcosm study of the common night crawler earthworm (*Lumbricus terrestris*) and physical, chemical and biological properties of a designed urban soil. Urban Ecosyst 14:119–134

Scheifler R et al (2006) Lead concentrations in feathers and blood of common blackbirds (*Turdus merula*) and in earthworms inhabiting unpolluted and moderately polluted urban areas. Sci Total Environ 371:197–205

Schlaghamersky J, Pizl V (2009) Enchytraeids and earthworms (Annelida: Clitellata: Enchytraeidae, Lumbricidae) of parks in the city of Brno, Czech Republic. Soil Organ 81:145–173

Schwert DP (1980) Active and passive dispersal of lumbricid earthworms. In: Dindal DL (ed) Soil biology as related to land use practices. EPA, Washington, DC, pp 182–189

Seamans TW, Blackwell BF, Bernhardt GE, Potter DA (2015) Assessing chemical control of earthworms at airports. Wildl Soc Bull 39:434–442

Sherlock E, Carpenter D (2009) An updated earthworm list for the British Isles and two new 'exotic' species to Britain from Kew Gardens. Eur J Soil Biol 45:431–435

Shifaw E (2018) Review of heavy metals pollution in china in agricultural and urban soils. J Health Pollut 8:article 180607

Sims RW, Gerard BM (1999) Earthworms. Notes for the identification of British species. Field Studies Council, Shrewsbury, UK

Smetak KM, Johnsn-Maynard JL, Lloyd JE (2007) Earthworm population density and diversity in different-aged urban systems. Appl Soil Ecol 37:161–168

Steinberg DA, Pouyat RV, Parmelee RW, Groffman PM (1997) Earthworm abundance and nitrogen mineralization rates along an urban-rural land use gradient. Soil Biol Biochem 29:427–430

Sugiura S (2008) Hot water tolerance of soil animals: utility of hot water immersion in preventing invasions of alien soil animals. Appl Entomol Zool 43:207–212

Suthar S, Singh S, Dhawan S (2008) Earthworms as bioindicator of metals (Zn, Fe, Mn, Cu, Pb and Cd) in soils: Is metal bioaccumulation affected by their ecological category? Ecol Eng 32:99–107

Szlavecz K, Csuzdi C, Hornung E, Korsós Z (2020) Urban soil fauna. In: Douglas I, Anderson PML, Goode D, Houck MC, Maddox D, Nagendra H, Tan PY (eds) The Routledge handbook of urban ecology, 2nd edn. Routledge, London, pp 425–438

Terhivuo J (1991) *Dichogaster bolaui* Michaelsen Octochaetidae an allochthonous oligochaete invading urban sewer system. Memoranda Soc Fauna Flora Fenn 67:61–65

Tiho S, Josens G (2000) Earthworm populations of Roosevelt Avenue (Brussels, Belgium): composition, density and biomass. Belgian J Zool 130:131–138

Tomlin AD, Tolman JH, Thorn GD (1981) Suppression of earthworm (*Lumbricus terrestris*) populations around an airport by soil application of the fungicide, benomyl. Prot Ecol 2:319–323

Toth Z et al (2020) Earthworm assemblages in urban habitats across biogeographical regions. Appl Soil Ecol 151:article 103530

Tresch S, Moretti M, Le Bayon R-C, Mader P, Zanetta A, Frey D, Fliessbach A (2018) A gardener's influence on urban soil quality. Front Environ Sci 6:article 25

Varelas M, Pieper L, Arsenault A, Pappas CC, Keblawe-Shamah N (2014) How science texts and hands-on explorations facilitate meaning making: learning from Latina/o third graders. J Res Sci Teach 51:1246–1274

Vergnes A, Blouin M, Muratet A, Lerch TZ, Mendez-Millan M, Rouelle-Castrec M, Dubs F (2017) Initial conditions during Technosol implementation shape earthworms and ants diversity. Landsc Urban Plan 159:32–41

Weisberger M (2021) Bizarre 'worm tornado' in New Jersey has scientists baffled. Live Science, 30 Mar 2021. Online document. https://www.livescience.com/squirming-worm-tornado-new-jersey.html. Accessed 29 Aug 2023

Wikipedia Contributors (2023) List of fictional worms. Online document. Wikipedia, The Free Encyclopedia. https://en.wikipedia.org/wiki/List_of_fictional_worms. Accessed 29 Aug 2023

Xie T, Wang M, Chen W, Uwizeyimana H (2018) Impacts of urbanization and landscape patterns on the earthworm communities in residential areas in Beijing. Sci Total Environ 626:1261–1269

Xie T, Li X, Wang M, Chen W, Faber JH (2022) Factors influencing earthworm fauna in parks in megacity Beijing, China: An application of a Synthetic and Simple Index (ESI). Sustainability 14:article 6054

Zaborski ER, Gittenger LAS (2001) *Amynthas hupeiensis* (Michaelsen, 1895) (Oligochaeta: Megascolecidae) in Illinois, USA, with observations on worm circling. Megadrilogica 8:13–16

Zhukov O, Kunakh O, Yorkina N, Tutova A (2023) Response of soil macrofauna to urban park reconstruction. Soil Ecol Lett 5:article 220156

Ziter CD, Herrick BM, Johnston MR, Turner MG (2021) Ready, set, go: community science field campaign reveals habitat preferences of nonnative Asian earthworms in an urban landscape. Bioscience 71:280–291

The Role of Earthworms in Grasslands

Eduardo Vázquez, Jacobo Arango, Mirjam Pulleman, and Nikola Teutscherová

Abstract Grasslands are among the most important ecosystems for human livelihoods. Besides their irreplaceable role in human food production, grasslands provide a wide range of ecosystem services at different scales. Earthworms, being soil ecosystem engineers, can play a key role in affecting the capacity of grassland soils to provide ecosystem services. Earthworms are known to improve soil physical properties, nutrient availability, plant biomass production, and soil water balance. Specifically, a high earthworm abundance and diversity in grasslands is often related to high soil porosity and water retention, low soil compaction, formation of soil biogenic aggregates with great stability, high availability of nutrients, and accelerated soil organic matter cycling, leading to high plant biomass production. This, in turn, encourages the maintenance of large and diverse earthworm populations in grasslands.

Most of the world's managed grasslands sustain livestock production (as opposed to natural grasslands without human intervention). Earthworms are sensitive to agricultural management. For instance, fertilization, grazing and manure management, soil compaction, and modification of the plant community composition (i.e., introduction of improved forages, trees, or legumes) can affect earthworm communities

E. Vázquez (✉)
Escuela Técnica Superior de Ingeniería Agronómica, Alimentaria y de Biosistemas (ETSIAAB), Universidad Politécnica de Madrid (UPM), Madrid, Spain
e-mail: eduardo.vazquez@upm.es

J. Arango
Alliance Bioversity-CIAT, Palmira, Colombia
e-mail: j.arango@cgiar.org

M. Pulleman
Alliance Bioversity-CIAT, Palmira, Colombia

Soil Biology Group, Wageningen University, Wageningen, The Netherlands
e-mail: m.pulleman@cgiar.org

N. Teutscherová
Czech University of Life Sciences, Prague, Czech Republic
e-mail: teutscherova@ftz.czu.cz

Y. Kooch, Y. Kuzyakov (eds.), *Earthworms and Ecological Processes*,
https://doi.org/10.1007/978-3-031-64510-5_9

in direct and indirect (and often interrelated) ways. Here, we summarize the potential of different grassland management practices and their possible effects on earthworm populations and communities, as well as the subsequent soil properties and functions that underpin sustainable intensification.

Based on their importance for soil functions and high sensitivity to grassland management, we encourage the inclusion of the earthworm communities among the studied factors for the assessment of grassland management practices' effects on soil health. Furthermore, we suggest that using earthworms as bioindicators of soil health in grassland is practically accessible and easily interpretable by farmers and agriculture services. The development of a standardized method of earthworm assessment, adapted to different edaphoclimatic conditions and types of grassland management, could help to transfer scientific knowledge generated during the last decades to final users and managers of grasslands.

1 Grasslands

1.1 Importance of Grasslands

Grasslands cover about 40% of the ice-free area on Earth and are one of the most important ecosystems (Suttie et al. 2005; Lemaire et al. 2011; Bengtsson et al. 2019). The grassland biome is characterized by the dominance of graminoids and the presence of broadleaves forbs, in addition to less than 10% coverage by tree canopy in temperate regions or up to 40% in tropical regions (Suttie et al. 2005; Jones 2019). Grasslands were formed by processes related to climate, fire, and wildlife grazing, and they occur naturally in temperate regions (steppes and prairies) and along desert borders. In more humid regions, grasslands have been generated anthropogenically by forest clearance for grazing domesticated animals (Jones 2019; Bengtsson et al. 2019). Globally, grassland areas have decreased during the last century because of increasing food demand and the related land use change to cropland (Manuel-Navarrete et al. 2009; Peeters 2009). Nevertheless, in some regions of Europe and North America, the abandonment of marginal croplands with low productivity contributes to grassland and prairie restoration (Cramer et al. 2008; Török et al. 2011).

The importance of grasslands for human livelihood is primarily related to fodder and forage production for the livestock sector (Herrero et al. 2013; Rao et al. 2015; Bengtsson et al. 2019; Notenbaert et al. 2021). Grasslands produce plant biomass that is the basis for meat and dairy production, particularly in the Global South where the livestock sector is based on mixed crop-livestock systems and family farms (Herrero et al. 2013; Rao et al. 2015; Notenbaert et al. 2021). The most important role of grasslands, nevertheless, lies in their ecological functions. Grasslands provide a wide array of ecosystem services affecting ecological processes and human welfare at the landscape, regional, and global scale (Lamarque et al. 2011;

Bengtsson et al. 2019). At the landscape scale, grasslands are considered reservoirs for the maintenance of above and belowground biodiversity, which is also linked to other ecosystem services such as providing floral support for pollinators and biological pest control (Lamarque et al. 2011; Bengtsson et al. 2019). At the regional scale, grasslands provide water services by storing, purifying, and regulating the water flows and reservoirs. They also limit erosion and deliver cultural services, such as tourism; recreation and sport areas; hunting; landscape heritage; and social cohesion (Lamarque et al. 2011; Bengtsson et al. 2019; Notenbaert et al. 2021). Finally, at the global scale, grassland ecosystems are crucial for climate regulation because their soils store about 20% of global carbon (C) stocks and can become a sink or source of CO_2 depending on the grassland status (Lal 2004; Wang and Fang 2009; Bengtsson et al. 2019). However, all of these ecosystem services are endangered by land use change, climate change, and unsustainable grassland management practices (Egoh et al. 2016; Bengtsson et al. 2019; Notenbaert et al. 2021). Therefore, the understanding, conservation, and restoration of grasslands in an agroecology framework are crucial to ensure biodiversity conservation, food security, and sustainable livelihoods at the global scale (Bengtsson et al. 2019; Notenbaert et al. 2021).

1.2 Natural Versus Managed Grasslands

Grasslands generally require a certain degree of management to support the increasing demand for livestock production worldwide (Herrero et al. 2013). Natural and managed grasslands differ in the intensity of human intervention. Natural or semi-natural grasslands have a low level of human intervention and are only grazed by wild herbivores or sparsely by domestic herbivores (Bengtsson et al. 2019; Chang et al. 2021). In contrast, managed grasslands or pasturelands have been anthropogenically modified to support livestock production (Bengtsson et al. 2019; Chang et al. 2021). Grassland management includes practices such as tree and shrubs clearance; construction of farm infrastructures including fences, shelters, or watering places; introduction of improved grasses or legumes to replace the natural floral composition; fertilization with chemical fertilizers or manure; and/or periodical burnings. Within the managed grasslands, we could distinguish between pastures (permanent grasslands under grazing by domestic herbivores) and meadows (permanent grasslands used for mechanical harvesting of grass by mowing), but in this chapter, we will refer indistinctly to pastures and meadows as managed grasslands, except when differences between types of managed grasslands are expected. The trend of globally increasing demand for livestock products has been accelerating the management intensification of these ecosystems, resulting in the replacement of wild herbivores and natural grasslands with managed grasslands grazed by domestic herbivores (Smith et al. 2016; Chang et al. 2021).

2 Earthworms in Grasslands

2.1 Occurrence of Earthworms in Grasslands

Earthworms are probably the most studied soil invertebrates in grasslands because of their pivotal role in providing or maintaining many soil ecosystem services and their relative ease of evaluation (Phillips et al. 2021). Earthworms are more abundant in grasslands than in any other ecosystem (Decaëns et al. 1994; Lavelle et al. 1997; Paoletti 1999). Earthworm biomass can be as high as 200 g m^{-2} in temperate grasslands (Curry et al. 2008; Hoeffner et al. 2021b) or up to 150 g m^{-2} in tropical pastures (Decaëns et al. 1994; Feijoo et al. 2011), exceeding the biomass of any other invertebrate. At the global scale, the average biomass of earthworms in pastures and grasslands is approximately 100 g m^{-2} (Paoletti 1999). Similarly, earthworm abundance in grasslands and pastures is approximately 300 individuals m^{-2}, although it can range between 0 and 500 individuals m^{-2}, while the earthworm species diversity commonly ranges between 6 and 15 species (Lavelle et al. 1997; Paoletti 1999; Feijoo et al. 2011; Hoeffner et al. 2021b; van de Logt et al. 2023). Earthworms in grasslands compared with those in croplands or forests benefit particularly from higher and continuous availability of food (organic matter inputs to soil, including animal manure) (compared with croplands), absence of soil tillage (compared with croplands), and high-quality litter (compared with forests), although exceptions can be found in both cases.

2.2 Impacts of Earthworms on Grassland Soils

Earthworm activity is crucial for grassland soil functioning: they regulate litter decomposition and nutrient availability; maintain soil structure and water retention; and affect the greenhouse gas (GHG) balance (Lavelle 1988; Lavelle et al. 1997; Eisenhauer and Scheu 2008; Pulleman et al. 2012; Blouin et al. 2013). In addition, due to their impact on soil properties and nutrient availability, earthworms affect the ecosystem's primary production or even the competition between different plant functional groups (Lavelle 1988; Brown et al. 2004; Eisenhauer and Scheu 2008; Blouin et al. 2013). Therefore, the study of earthworm communities in grasslands can help to understand the soil health status and ecosystem functioning (Pulleman et al. 2012; Blouin et al. 2013). In this section, we will summarize the main impacts of earthworms on grassland soils as well as on plant community composition and biomass production. The main relationships among soil properties, earthworms, plant biomass production, grazing animals, and anthropogenic inputs are summarized in Fig. 1.

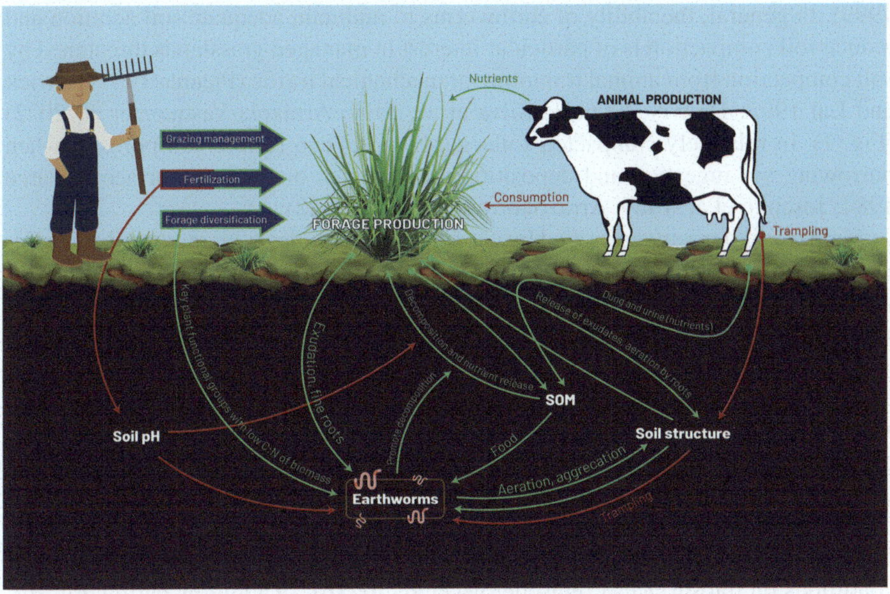

Fig. 1 The reciprocal impacts of grazing management, grassland fertilization, and forage diversification (blue arrows) on forage production and earthworms mediated by effects on soil pH, soil organic matter (SOM), and soil structure. Positive and negative effects are represented by green and red color, respectively (figure made by José Luis Urrea and Anny Isabella Yedra (Alliance Bioversity-CIAT))

2.2.1 Soil Physical Properties

The life strategy of earthworms affects the physical properties of soil in many ways. They dig burrows and tunnels, deposit casts with a particular physical structure, and contribute to the mixing of soil horizons, which subsequently incorporate plant litter into deeper soil layers (Lavelle 1988). In this way, earthworms improve soil porosity and aeration, water infiltration and storage, and soil aggregation and structural stability (Lavelle 1988; Spurgeon et al. 2013) (Fig. 1).

Many studies have found a higher soil porosity in grasslands with a higher earthworm abundance (Drewry 2006; Lamandé et al. 2011; Velásquez et al. 2012; Schon et al. 2017; Vázquez et al. 2020; Teutscherová et al. 2021b). Earthworms can increase macropores by burrowing, while meso- and microporosity is mainly affected by casting. Variation occurs between different earthworm species in the extent to which they impact soil porosity (Brown et al. 2004). Surface-casting species, like *Lumbricus terrestris*, are recognized for their prolific production of surface casts in compacted soils, which can alleviate soil compaction (Piearce 1984; Joschko et al. 1989). Although exceptional, some invasive species, such as *Pontoscolex corenthrurus*, have been observed to colonize grasslands established after Amazonian Forest clearance leading to soil compaction as a result of their uncontrolled casting activity that reduces the soil macroporosity (Chauvel et al.

1999). In general, the ability of earthworms to maintain adequate soil aeration and reduce soil compaction is of particular interest in managed grasslands threatened by soil compaction from animal trampling or mechanical traffic (Piearce 1984; Herrick and Lal 1995; Drewry 2006; da Silva et al. 2019; Arrázola-Vásquez et al. 2022) (Fig. 1). In relatively compacted soils, earthworms can restore soil porosity when burrowing via ingestion and deposition the soil cast on the soil surface (Piearce 1984; Joschko et al. 1989; Arrázola-Vásquez et al. 2022).

Higher soil porosity generated by earthworms coincides with increased soil aeration and soil hydraulic conductivity, thereby affecting some important ecosystem services provided by grasslands, such as plant biomass production, water fluxes regulation, and soil erosion control (Blouin et al. 2013; Spurgeon et al. 2013) (Fig. 1). The relationship between higher earthworm abundance and water infiltration and storage has been observed in experimental grasslands (Clements et al. 1991; Bouché and Al-Addan 1997; Velásquez et al. 2012; Fischer et al. 2014) and real farm grasslands (Teutscherová et al. 2021b). Higher water infiltration is mainly caused by the continuous macropores created by earthworm burrowing activity, primarily by anecic species. This can contribute to reduced soil erosion through superficial water runoff (Weiler and Naef 2003; Blouin et al. 2013) or waterlogging conditions on flat surfaces (Teutscherová et al. 2021b). In addition, earthworm surface casting can reduce soil erosion by increasing surface roughness (Blouin et al. 2013); however, this seems to be more important in agricultural soils with abundant bare soil areas than in permanent grasslands.

Earthworms are well known for increasing soil aggregation in grasslands (Guggenberger et al. 1996; Fonte et al. 2012; Velásquez et al. 2012; Vázquez et al. 2020; Lavelle et al. 2020; Garcia-Franco et al. 2021; Rodríguez et al. 2021b). The positive effect of earthworms on soil aggregation can be caused directly by the production of highly stable casts or indirectly through the promotion of the root colonization of nutrient-rich casts (Pulleman et al. 2005a, b; Fonte et al. 2012). Generally, earthworm casts are very rich in C and other nutrients, and these structures remain stable upon drying (Guggenberger et al. 1996; Pulleman et al. 2005a, b; Jouquet et al. 2008; Fonte et al. 2012; Lavelle et al. 2020).

2.2.2 SOM and Nutrient Cycling

Among the most studied and accepted impacts of earthworm activity in grasslands is, without a doubt, their capacity to enhance soil C sequestration by improving soil aggregation and organic matter mineralization (Lavelle et al. 2020) (Fig. 1). Earthworm casts are formed by resistant aggregates enriched in C stabilized with soil particles (Lavelle and Martin 1992; Guggenberger et al. 1996; Pulleman et al. 2005b). Plant debris is decomposed in the earthworm gut and mixed with mucus and soil particles that promote the physical stabilization of organic matter (Lavelle and Martin 1992; Guggenberger et al. 1996). After release, casts are enriched in nutrients and show a high microbial activity; however, after a few days/weeks, the microbial activity decreases, resulting in organic matter that is stabilized and protected from further mineralization within the aggregates (Lavelle and Martin 1992; Guggenberger et al. 1996). This occlusion of organic matter within stable aggregates created by earthworms suggests that casts may influence positively long-term C sequestration in soils (Lavelle and Martin 1992; Guggenberger et al. 1996; Pulleman et al. 2005a, b; Lavelle et al. 2020). Yet, some studies have suggested that earthworms only increase the soil C sequestration in the long term when organic matter inputs are also increased compared with the previous management (Don et al. 2008; Fonte and Six 2010).

2.2.3 Plant Biomass Productivity and Plant Community

The positive effect of earthworms on plant biomass productivity (Fig. 1) was already described by Darwin in 1881. Since then, many studies, reviews, and meta-analyses have confirmed such observation (Curry 1987; Brown et al. 1999; Scheu 2003; Blouin et al. 2013; van Groenigen et al. 2014) and at least five mechanisms have been identified to justify the observed effects of earthworms on plant biomass productivity (Brown et al. 1999; Scheu 2003; Blouin et al. 2013): (1) higher nutrient availability due to accelerated organic matter mineralization, (2) improved water retention and soil structure, (3) pest control, (4) increased microbial activity, and (5) promotion of symbiotic relationships. In addition, earthworms can affect the plant diversity of grassland by modifying the competition between different plant functional groups, such as grasses and legumes (Eisenhauer and Scheu 2008; Eisenhauer et al. 2009a) or by affecting the seed bank dispersion (Decaëns et al. 2003; Zaller and Saxler 2007). However, the understanding of these mechanisms and their relevance under real grassland conditions remains a challenge because earthworms both affect and are affected by plant biomass and diversity in a positive feedback loop (Brussard 1999). In Sect. 3, the different management practices that simultaneously promote plant biomass or diversity and earthworm abundance in managed grasslands are discussed.

3 Earthworms in Managed Grasslands

As the earthworm abundance and distribution in soil are primarily driven by the presence of food supply, earthworm communities are affected by, for instance, sward composition and diversity; the quantity and quality of produced plant biomass; the type and intensity of grazing or cutting events; the dung deposition; and the application of both organic and inorganic fertilizers (Lavelle et al. 2001; Curry 2004; Curry et al. 2008; Singh et al. 2021). Therefore, all grassland management practices affecting these grassland properties can affect the earthworm community (Lavelle et al. 2001; Curry 2004). Not only does grassland management affect organic matter supply in soil, but it also affects earthworm communities by changing the soil's physical structure, microclimatic conditions, and/or chemical composition (which includes toxic substances) (Ma et al. 1990; Drewry 2006; Curry et al. 2008; Birkhofer et al. 2008; Schon et al. 2017; Vázquez et al. 2020).

To better understand the earthworm ecology and to predict potential impacts of grassland management on earthworm communities, several studies have provided interesting information on the different drivers of earthworm abundance and activity using data collected from experimental plots with different levels of plant richness (Milcu et al. 2008; Eisenhauer et al. 2009b; Velásquez et al. 2012) or N fertilization type/dose (Ma et al. 1990; King et al. 2007; van Eekeren et al. 2009; Hoeffner et al. 2021a). Nevertheless, under real farm conditions, several management practices are generally applied simultaneously to optimize the productivity and/or sustainability of a managed grassland (intensification or sustainable intensification of a grassland) (Dumont et al. 2018). For instance, grassland fertilization is commonly accompanied by increased livestock stocking rates or the frequency of cuttings, both possibly influencing soil earthworms. Thus, the response of earthworm communities to grassland management is more complex in comparison to effects caused by a single treatment applied in experimental plots (Curry et al. 2008; Singh et al. 2021). In addition, suitable grassland management practices are strongly dependent on the region, the climatic conditions, the type of farming, and/or the aim of the farmer. The principal aim of this section is to provide an overview of the potential effects of different management practices on earthworm communities and their interactions under real farm conditions. In the next four sections, different effects are evaluated separately, and at the end, the combination of different common management practices in temperate and tropical regions is discussed, aiming to provide an overview of the potential impact of sustainable intensification of grasslands on earthworm communities.

3.1 *Impact of Grassland Establishment*

Land use change has a profound impact on soil ecosystem functioning including the earthworm community and its functions in soil (Decaëns et al. 2004; Postma-Blaauw et al. 2012; Spurgeon et al. 2013). In the last century, the growing human population and food demand have forced the intensification of agricultural production, which, on one hand, has led to the transformation of grasslands into agricultural fields, and on the other hand, the establishment of newly managed grasslands in formerly forested areas (Fragoso et al. 1999; Ellis and Ramankutty 2008; Peres et al. 2010; Graesser et al. 2015; Dumont et al. 2018; Notenbaert et al. 2021). The latter land use change is particularly true in Latin America, where the growing demand for meat has dramatically moved the deforestation barrier into the tropical forests and savannas to establish new grasslands or man-made pasturelands (Rodrigues et al. 2009; Rao et al. 2015). In contrast, in temperate regions of Europe and North America, a trend of land management de-intensification to restore grasslands and prairies in abandoned agricultural areas (Cramer et al. 2008; Török et al. 2011; Postma-Blaauw et al. 2012; Spurgeon et al. 2013; Wodika et al. 2014) or introduction of temporal grass leys in crop rotations (Karwat et al. 2017; Rodríguez et al. 2020; Hoeffner et al. 2021a) has been observed.

The effect of grassland establishment on earthworm communities depends on the previous land use, particularly on the degree to which there is a similarity between the previous land use and vegetation and the newly established grassland. For instance, the impact of grassland establishment on earthworm communities will drastically differ in grasslands formed in previously forested areas versus grasslands that were previously savannas (Decaëns et al. 1994, 2004; Fragoso et al. 1999). In tropical savannas (one of the predominant ecosystems in Latin America), the man-made pastures used for intensive livestock production are established by replacing the native grasses with African grasses and forage legumes, hence simplifying the plant community but increasing the biomass production (Fragoso et al. 1999; Notenbaert et al. 2021). Nevertheless, if the vegetation structure and functionality remain similar, the taxonomic richness and earthworm community composition can be preserved (Decaëns et al. 1994, 2004; Fragoso et al. 1999). In fact, earthworm abundance and biomass can even augment after such intervention as a result of higher organic matter inputs from plant biomass, higher quality plant litter (especially in case of legume inclusion), absence of periodic burning, and higher animal dung deposition (Decaëns et al. 1994, 2004; Jiménez et al. 1998; Fragoso et al. 1999).

In contrast, earthworm communities are usually drastically altered when tropical rainforests (such as in the case of the Amazon Basin) are substituted with pasturelands, as native earthworm species are commonly depleted and replaced by a single exotic peregrine species (Fragoso et al. 1999; Decaëns et al. 2004; Marichal et al. 2010). The exotic replacement species seem to benefit from the following changes to the soil environment, the removal of the soil litter layer (affecting in particular anecic and epigeic species) and high soil organic C availability in managed pasturelands (Fragoso et al. 1999; Decaëns et al. 2004; Mathieu et al. 2005; Peres et al. 2010; Marichal et al. 2010). Such a strong reduction of earthworm diversity inevitably leads to the loss of healthy soil structure, as the role of earthworms as soil engineers is lacking. This is reflected in a rapid degradation of soils as commonly observed across Latin America pasturelands (Chauvel et al. 1999; Marichal et al. 2014). This highlights the importance of considering earthworm communities during the assessment of the impact of land use change on ecosystem services and sustainability.

Unlike in the tropics, temperate grassland areas have tended to diminish during the last century because of the increasing demand for agricultural lands (Manuel-Navarrete et al. 2009; Peeters 2009). This agriculturalization process of grasslands, which is still in progress in some regions (e.g., the Argentinian pampas), has led to the reduction of earthworm abundance and the invasion of exotic or opportunistic earthworms and, therefore, soil degradation (Manuel-Navarrete et al. 2009; Domínguez et al. 2010; Falco et al. 2015). In contrast, in many regions of Europe and North America, land with low agricultural productivity and marginal croplands have been abandoned, which contributes to grassland and prairie restoration (Cramer et al. 2008; Török et al. 2011). Even though the success of the grassland plant community recovery in such abandoned agricultural fields may require some human support to reach the native vegetation diversity status (the success seems to be dependent on the intensity of previous agricultural management) (Cramer et al. 2008; Török et al. 2011), a shift toward grasslands at the expense of an agricultural field generally leads to an increase in earthworm abundance (van Eekeren et al. 2008; Postma-Blaauw et al. 2012; Spurgeon et al. 2013; Wodika et al. 2014). This effect derives from the higher organic matter input in grasslands compared with croplands, lack of tillage or pesticide application, and the continuous soil cover that buffers the soil microclimatic conditions (van Eekeren et al. 2008; Domínguez et al. 2010; Postma-Blaauw et al. 2012; Spurgeon et al. 2013; Wodika et al. 2014). It has been observed that earthworm abundance can react to land use change as early as 0–3 years after a transition (van Eekeren et al. 2008; Spurgeon et al. 2013). However, the earthworm species composition in restored grasslands is not always the same as in a native grassland. For instance, in restored prairies of Illinois (USA), the earthworm community was dominated by exotic European earthworm species, which may affect the soil and grassland functioning (Wodika et al. 2014). Similarly, van Eekeren et al. (2008) described a recovery of earthworm abundance and biomass in Belgium after 3 years of grassland establishment, but the dominance of anecic species, as in older grasslands, was not achieved. In general, earthworm communities in croplands are dominated by endogeic (feeding on deeper soil layers) species, which are likely the first earthworm group that reacts positively to grassland conversion as they can recruit new progeny from the standing arable crop community. Anecic and epigeic species, on the contrary, likely need a longer period of time after grassland restoration until their microhabitat is recovered (especially surface litter) and they can recolonize the area (Spurgeon et al. 2013).

In recent years, the introduction of temporal grass ley phase in crop rotations has gained attention as an alternative to continuous cropping systems (particularly in organic farming) that can reduce soil degradation, loss of biodiversity, and N losses and increase the soil C sequestration (Tsiafouli et al. 2015; Lemaire et al. 2015; Karwat et al. 2017; Hoeffner et al. 2021a). The grassland period, which commonly ranges between a season to several years, temporally modifies the soil ecosystem and subsequently can affect the earthworm community (van Eekeren et al. 2008; Postma-Blaauw et al. 2012; Rodríguez et al. 2020; Hoeffner et al. 2021a). Hoeffner et al. (2021a) observed that the increase in earthworm abundance and shift in community composition depended on the time since grassland establishment. In particular, they observed that anecic species benefited from grassland establishment only after 6 years, indicating that the effects on soil earthworm communities and their ecosystem services may differ depending on the extent of the grassland period in the rotation. Therefore, the introduction of temporal grasslands in crop rotations can have a positive effect on soil earthworm communities compared with continuous cropping; however, the positive effects would not reach the level of permanent grasslands due to the short term of the grassland period and the consequent returned to cropland including tillage and other agricultural practices. Further studies should evaluate possible legacy effects on the earthworm communities established during the grassland period in the next cropping periods of the rotation.

In summary, a shift toward managed grasslands depends on the previous land use type and can have a positive or negative impact on earthworm communities. When a grassland is established on a recently deforested area, expansion of exogenic species and soil degradation are often observed. Nevertheless, the substitution of native grasslands, savannas, or agricultural areas with managed grasslands may have a positive impact on earthworms due to an increase in organic inputs to the soil via higher plant biomass production or livestock dung deposition and/or the reduction of periodical burnings, tillage, and pesticide application. However, the composition and diversity of the earthworm community can differ compared with the native community, particularly when the ecosystem functioning of the former land use is different, which can have strong implications for soil properties and soil health. Therefore, we encourage consideration of the potential responses of earthworm communities to grassland establishment in order to minimize negative impacts on newly established soil earthworm communities.

3.2 Grassland Management and Intensification

3.2.1 Introduction of Improved Grasses

During the last decades, native vegetation of natural tropical savannas has been replaced in large areas by improved tropical grasses with higher potential biomass production in order to increase livestock production systems (Rao et al. 2015; Notenbaert et al. 2021). We use the term "improved grasses" to refer to grass genotypes that have been selected from wild relatives or have gone through a breeding and selection program to obtain desirable traits (Rao et al. 2015; Notenbaert et al. 2021). In Latin America and the Caribbean, and to a lesser extent in Africa, improved cultivars of *Urochloa* or *Megathyrsus* grasses are commonly sown due to their high productivity that can sustain higher livestock stocking rates (Notenbaert et al. 2021). Besides the exceptionally high aboveground biomass production, these cultivars are resistant to pests and provide an array of environmental benefits (Rao et al. 2015; Notenbaert et al. 2021). Several of the improved grasses are characterized by a dense and deep rooting system that contributes to the accumulation of SOM (Fisher et al. 1994; Trujillo et al. 2006) and soil structure build-up. This in turn is often reflected in a higher abundance of soil earthworms feeding on the SOM (Decaëns et al. 1994, 2004; Jiménez et al. 1998; Webster et al. 2019). In addition, the introduction of improved grasses is also an alternative to the common practice of burning pastures (to temporarily increase their fertility), which is also beneficial to earthworm communities because the negative effects of burning are avoided (Decaëns et al. 1994, 2004; Fragoso et al. 1999). However, the potential effects of improved grasses with different traits (such as contrasting above-belowground biomass allocation) on earthworm communities have not yet been evaluated, to our knowledge. On one hand, the denser rooting system of some improved grasses may enhance the earthworm abundance due to the higher root exudation (Decaëns et al. 1994), but, on the other hand, a very dense rooting system may also hamper the development of earthworm communities (Eisenhauer et al. 2009b). We suggest that the effect of the selected grass on the earthworm community should be evaluated and considered when designing the sustainable intensification strategy of a grassland.

3.2.2 Impact of Grassland Diversification

Earthworms benefit from a high plant biomass production (i.e., higher organic matter inputs) and from high-quality organic matter (i.e., a low C/N ratio); hence, the management practices leading to a higher biomass production and organic matter quality can positively affect the earthworm community (Lavelle et al. 2001; Curry 2004; Eisenhauer et al. 2009b; Hoeffner et al. 2024).

Grasslands with higher plant species diversity generally support a higher abundance of earthworms (Spehn et al. 2000; Milcu et al. 2008; Hoeffner et al. 2024), which is often explained by (1) a higher biomass production, (2) a more diverse offering of litter residues with different chemical qualities, and (3) variable timing in the organic matter supply to soil (Spehn et al. 2000). However, more recent studies have suggested that earthworms benefit more from the presence of specific plant functional groups, such as legumes, than strictly from high plant diversity per se (Eisenhauer et al. 2009b; Piotrowska et al. 2013; Singh et al. 2021). Hence, effects on the earthworm population assumed to be a result of higher plant diversity might actually be caused by the increasing likelihood of legumes being present in highly diverse grasslands (Eisenhauer et al. 2009b). This consideration is critical in a scenario of global loss of legumes and their N_2 fixation capacity due to anthropogenic impacts (Tognetti et al. 2021; Vázquez et al. 2022) and may, therefore, also indirectly affect earthworm communities.

Legumes play a key role in grassland ecosystems by increasing plant biodiversity, providing floral support to pollinators, and fixing atmospheric N via symbiotic N_2 fixation (Lamarque et al. 2011; Lüscher et al. 2014). Fixed N is not only used for growth by legumes but also contributes to the nutrition of other grasses and forbs in the sward, which increases the overall biomass productivity and the N content of plant biomass (Herridge et al. 2008). Therefore, legumes are promoted in managed grasslands aiming to boost their N_2 fixation capacity and reduce the use of synthetic fertilizers (Lüscher et al. 2014; Rao et al. 2015). Because earthworms are benefited by organic matter inputs rich in N, the promotion of legumes in grasslands can have a positive effect on earthworm abundance, an observation that has been largely presented in the literature (Milcu et al. 2008; Eisenhauer et al. 2009b; Velásquez et al. 2012; Piotrowska et al. 2013; Vázquez et al. 2020; Hoeffner et al. 2024). In particular, anecic species benefit more from a high legume abundance for several reasons, their proximity to the high-quality residues entering the soil, the abundance of N-rich seeds, and legumes have a less dense root system compared with grasses (Eisenhauer et al. 2009b; Hoeffner et al. 2024). Similarly, the introduction of forage legume grasslands into crop rotations increased the earthworm abundance more than the introduction of grasses into Argentinian pampas as a consequence of the lower C/N ratio of legume residues (Rodríguez et al. 2020). Therefore, a positive effect on earthworm communities can be considered another benefit of legume promotion in grasslands.

A particular case of grassland diversification is the establishment of silvopastoral systems, which consists of the introduction/maintenance of woody perennials in a pastoral system. The *dehesa* (or *montado*) ecosystem covers 3.1 million ha in the Iberian Peninsula (Moreno and Pulido 2009) and is the most emblematic example of a managed silvopastoral system. Dehesa is a silvopastoral system based on livestock rearing and oak exploitation after the partial clearing of oak woodlands (Moreno and Pulido 2009). Dehesa and other silvopastoral systems are characterized by high biodiversity, including earthworms, as a result of the existence of many different microhabitats (Moreno and Pulido 2009; Moreno et al. 2016; Marsden et al. 2020). Similar positive effects on earthworm communities have also been observed in the tropics (Fragoso et al. 1999; Barros et al. 2003; Lavelle et al. 2016; Vázquez et al. 2020; Rodríguez et al. 2021a), where the greatest benefits can be expected in cases where legume woody perennials provide high-quality litter (Murgueitio et al. 2011; Webster et al. 2019).

Altogether, earthworm communities benefit from highly productive grasslands characterized by a high plant diversity including legumes and in the case of silvopastoral systems from high spatial heterogeneity. Thus, grassland management should be directed toward supporting the highest biodiversity and legume abundance, without penalizing biomass production.

3.2.3 Grassland Fertilization

Fertilized grasslands have higher plant biomass production and, therefore, by providing more food supply, can generally sustain a higher biomass of earthworms (Curry 1976; Curry et al. 2008; van Eekeren et al. 2009; Murchie et al. 2015; Schon et al. 2017; Hoeffner et al. 2021a). The positive impact of N fertilization on earthworm abundance is particularly strong when N fertilization is combined with irrigation, which also increases plant biomass growth (Schon et al. 2017). The addition of organic fertilizers can be even more beneficial than synthesized N fertilizers due to the simultaneous supply of organic matter (Birkhofer et al. 2008; van Eekeren et al. 2009; Murchie et al. 2015; Deru et al. 2023). However, in other cases, low or no effect of fertilization on earthworm abundance has been detected due to the already high soil organic content of many grasslands (Edwards and Lofty 1982; Timmerman et al. 2006). Besides the effects on earthworm abundance, fertilization can also alter the earthworm community composition (Edwards and Lofty 1982; Murchie et al. 2015). For example, epigeic earthworms generally benefit more from manure or cattle slurry addition because they feed directly upon surface organic matter (Edwards and Lofty 1982; Murchie et al. 2015; Deru et al. 2023).

However, the positive effects on earthworm abundance through increased plant biomass production can be hindered by changes in soil properties, especially in the case of continuous additions of synthesized N fertilizers, which can strongly reduce soil pH (Ma et al. 1990; van Eekeren et al. 2009; Schon et al. 2017). Similarly, the addition of high doses of manure slurry has been found to be detrimental to earthworm populations due to the high presence of heavy metals or other toxic substances in earthworms (Curry 1976; Edwards and Lofty 1982; Murchie et al. 2015; Segat et al. 2020). Grassland N fertilization can also affect earthworms indirectly by changing plant community composition: grasses and non-N-fixing forbs frequently displace legumes due to their lower competitiveness under high N availability (Tognetti et al. 2021; Vázquez et al. 2022). Therefore, the effect of fertilization depends on the balance between the positive effects of higher biomass production and SOM and the detrimental effects of soil acidification, reduced legume abundance, and potential toxic effects.

3.2.4 Increasing Livestock Stocking Rate and Mowing Frequency

Efforts to increase plant biomass production are almost exclusively directed to higher animal production, e.g., to the increasing stocking rates in grazed pastures or mowing frequency in meadows. Higher stocking rates can negatively affect earthworm communities by increasing the risk of soil compaction and poaching caused by excessive animal trampling (Cluzeau et al. 1992; Drewry 2006; Curry et al. 2008; Schon et al. 2017; Vázquez et al. 2020; Hoeffner et al. 2021b, 2024). In particular, earthworm species living near the soil surface are more affected by the negative effects of excessive trampling (Piearce 1984; Cluzeau et al. 1992). Both direct mortality due to crushing and poor conditions of compacted soils can contribute to earthworm decline and downward shifts in vertical distribution (Piearce 1984; Cluzeau et al. 1992).

The detrimental impacts of trampling on soil properties and earthworm communities depend on the treading pressure; thus, more negative impacts can be observed in grasslands grazed by cattle than by sheep (Schon et al. 2017). It can also be localized to specific areas of the grassland (i.e., under the trees or close to the water sources) due to preferential animal behavior (Vázquez et al. 2020). As earthworms often play a key role in reducing soil compaction, the disappearance of earthworms further stimulates soil physical degradation in a negative feedback loop (Piearce 1984; Herrick and Lal 1995; Drewry 2006).

Nevertheless, it should also be noted that increasing stocking rates can, in some cases, stimulate earthworms by increasing the dung deposition and availability of fresh organic matter rich in nutrients and microorganisms (Decaëns et al. 1994, 2004; Curry et al. 2008; Bacher et al. 2018; Singh et al. 2021; Teutscherová et al. 2021b). Therefore, the impact of a higher stocking rate will depend on the balance between the negative effect of higher trampling and the positive effect of increased dung deposition and likely plant biomass production (Curry et al. 2008; Schon et al. 2017; Teutscherová et al. 2021b; Hoeffner et al. 2024). For example, Curry et al. (2008) described a higher earthworm abundance with an increasing stocking rate and N fertilization as a consequence of higher plant biomass and dung deposition counterbalancing the negative effects of higher soil trampling. In this sense, a correct grazing management that ensures an adequate period of rest for the grassland by rotational grazing seems to be a promising strategy to avoid the negative impacts on earthworm communities from high stocking rates (Teutscherová et al. 2021b). Otherwise, unsustainable grazing management can lead to plant community degradation, an increase in the areas of bare soil prone to erosion, soil structure degradation, and, finally, a lower earthworm abundance (Teague et al. 2011; Teutscherová et al. 2021b).

Intensified meadows, which are not directly grazed, can be mowed more frequently in comparison to extensively managed grasslands. A recent study showed a negative impact of mowing on earthworm abundance compared with no-mowing (Hyvönen et al. 2021) due to the forage biomass removal. In contrast, Hoeffner et al. (2024) found an increase in earthworm abundance and diversity in mowed compared with abandoned grasslands. The impact of mowing on earthworm communities is a priori lower than the impact of grazing due to the absence of physical disturbance by animal trampling (Schlaghamerský et al. 2007; Tälle et al. 2016). Yet, other studies have shown no differences between mowed and grazed grasslands, likely because less organic matter was incorporated into the soil (no dung deposition and forage removal from the plot) in the mowed than in the grazed system (Schon et al. 2010, 2017; Hoeffner et al. 2024). However, more frequent mowing may have negative impacts on soil earthworm communities by affecting the plant diversity or through soil compaction if the mowing is performed by machinery under suboptimal soil moisture conditions, although more studies should evaluate in detail the impact of mowing timing and frequency on soil earthworm communities.

3.2.5 Trends of Grassland Sustainable Intensification Strategies

Intensification of forage-based agriculture (in terms of increased plant biomass and animal productivity) can have detrimental effects on ecosystems unless appropriate measures to maintain or increase pasture (and soil) health are applied. The most commonly observed negative impacts include soil degradation, higher GHG emissions, plant community degradation (Dumont et al. 2018), and finally, a decrease in earthworm abundance and diversity (Birkhofer et al. 2008; Piotrowska et al. 2013; Manning et al. 2015; Siebert et al. 2019; Singh et al. 2021). It is, therefore, imperative to follow the agroecological framework adopting sustainable intensification practices to cover the increasing demand for livestock products while improving livelihoods and environmental co-benefits: land sparing; GHG mitigation; soil health and plant; and faunal biodiversity (Rao et al. 2015; zu Ermgassen et al. 2018; Dumont et al. 2018; Notenbaert et al. 2021). Only in that case, an abundant and diverse earthworm community that supports the sustainability of grasslands can be maintained (Decaëns et al. 1994; Birkhofer et al. 2008; Manning et al. 2015; Siebert et al. 2019; Vázquez et al. 2020; Singh et al. 2021; Teutscherová et al. 2021b).

In the tropical regions, a very common sustainable intensification strategy is the establishment of silvopastoral systems including improved grasses and legumes (both forages and shrubs/trees) combined with an intensive short-durational grazing period (Alfaro-Arguello et al. 2010; Rao et al. 2015; Lerner et al. 2017; zu Ermgassen et al. 2018; Notenbaert et al. 2021). These systems are characterized by a high plant biomass production, soil rich in N due to the presence of legumes, low or no input of synthesized fertilizers, high deposition of dung, and no soil disturbance in the form of tillage or periodical burning (Rao et al. 2015; Lerner et al. 2017; Notenbaert et al. 2021). The combination of these management practices can increase the stocking rates while reducing the negative environmental impacts of livestock production (Murgueitio et al. 2011; Ferguson et al. 2013; Lerner et al. 2017; Vázquez et al. 2020; de Freitas et al. 2020; Rodríguez et al. 2021a; Notenbaert et al. 2021). These diversified and well-managed systems enhance the earthworm abundance compared with grassland monocultures due to several reasons, including higher plant diversity and biomass production, higher spatial heterogeneity, the presence of legumes, higher dung deposition (as a consequence of the higher stocking rate), and the rotational management (Webster et al. 2019; Vázquez et al. 2020; Teutscherová et al. 2021b; Rodríguez et al. 2021a). This higher abundance of earthworms is linked to improved soil C storage, soil aggregation, and water infiltration, which confirms the sustainability of the systems and the important role of earthworms in soil health. However, the spatial heterogeneity of the paddock, as a result of tree presence, can also cause heterogeneity in the spatial distribution of earthworms. Vázquez et al. (2020) observed in a silvopastoral system that the continuous trampling of cattle under the rows of legume trees (better microclimatic conditions and more palatable food under the trees) leads to a decrease in earthworm abundance. This finding encourages a further evaluation of different silvopastoral arrangements with the aim of finding the most appropriate paddock design and management practices to optimize the sustainability of intensive short-durational grazing silvopastoral systems.

In the case of temperate regions, the sustainable intensification points toward the substitution of intensive meadows managed using high doses of fertilizers, high cutting frequency, and reduced plant diversity with extensively grazed grasslands with a high plant diversity (particularly legumes) and manure instead of synthesized fertilizers addition (Oberson et al. 2013; Manning et al. 2015; Dumont et al. 2018; Siebert et al. 2019). The adoption of those practices can enhance earthworm abundance due to the higher plant diversity and biomass production, the presence of legumes, the dung deposition, and the reduction or absence of synthesized fertilizers (Birkhofer et al. 2008; Singh et al. 2021). Similarly, the establishment or protection of silvopastoral systems, such as the Dehesa ecosystem, can enhance the abundance and diversity in earthworm communities (Moreno et al. 2016; Marsden et al. 2020).

Therefore, we consider that the sustainable intensification of grassland management could meet the increasing livestock production demands while enhancing earthworm communities compared with traditionally managed grasslands. Similarly, due to the sensitivity of earthworm communities to different management practices, the monitoring of earthworm communities may help to understand the sustainability of management practices, as we will discuss in the next section.

4 Earthworms as an Early Indicator of Grassland Management

The potential of earthworms to serve as bioindicators of the land use sustainability has been proposed by many researchers in the last decades (Paoletti et al. 1991; Muys and Granval 1997; Paoletti 1999; Tondoh et al. 2007; Pérès et al. 2011). In grasslands, multiple scientific studies have confirmed strong correlations between earthworm abundance and/or diversity and many important ecosystem services (see Sect. 2). Furthermore, there is strong evidence that earthworms are sensitive to different grassland management practices (see Sect. 3). This evidence, together with the low cost and relatively easy assessment, makes earthworms perfect candidates for being used as grassland health bioindicators. Some authors have already proposed earthworms to indicate the health status of pastures in Australia and New Zealand (King et al. 2007; Schon et al. 2022). Schon et al. (2022) proposed a specific assessment method and flowchart to provide a tool for farmers to assess the health status of their pastures based on the abundance and diversity of earthworms. This progressing trend of earthworm inclusion into ready-to-use methodology for farmers to assess the health status of their grasslands can be also observed in the visual soil evaluation methodologies (Ball et al. 2017) which are well studied in temperate grasslands. Recently, both visual assessment of soils and earthworm counting showed a great correlation also in the tropics (Colombian pastures), and in addition, both proved to be suitable for distinguishing among contrasting grazing management practices (Teutscherová et al. 2021a).

To be successfully used by farmers and agricultural services, the evaluation must be easily performed and include earthworm counting and only basic classification. In addition, flowcharts with different thresholds could help to assess the health status of grasslands. Furthermore, a recent database compiled hundreds of studies in earthworm communities from the last decades (Phillips et al. 2021), which could help to adapt this methodology to different climates, soils, or types of grasslands (pastures or meadows). In this sense, the development of a digital tool, such as an app for mobile phones, could boost the real application of earthworms as bioindicators of grassland health status. This app could help farmers assess the soil health status of their grasslands based on the observed abundance and type of earthworms in combination with edaphoclimatic and management parameters.

5 Outlook and Future Perspectives

With the increasing population and global demand for meat products, the importance of grasslands has become more apparent than ever. These grassland ecosystems are crucial for human livelihood because they (1) play a crucial role in livestock-based agriculture worldwide and (2) provide many ecosystem services related to soil, water, and air quality. The latter is largely facilitated by earthworms, which greatly affect grassland soil physical properties, nutrient availability, plant biomass production, and water and climate regulation. However, earthworm communities and their functionality are very sensitive to different grassland management practices as confirmed in many studies and summarized in this chapter. It is precisely their importance and sensitivity that make earthworms suitable candidates to assess the health status of grasslands and help to understand the potential impacts of grassland management on soil functioning. We advise considering the impacts on earthworm communities in grassland establishment, management, and other potential land-use changes (e.g., grassland establishment after forest clearance) to better predict the long-lasting impacts of such activities.

We suggest that future research should comprise both mechanistic studies in experimental plots and mesocosms and on-farm observations to increase our understanding of different management interactions, as the simultaneous application of techniques can have individually contrasting effects on earthworms. However, we recognize that the comparison of real farms (as opposed to strictly monitored and controlled experimental grasslands) brings some challenges in data quality and interpretation due to potential baseline confounding bias. In general, the evaluation of real farms is based on the comparison of adjacent farms with contrasting managements. These across-the-fence comparisons make it difficult to distinguish between the effects caused by the current management, the legacy of the previous management, and/or other potential micro-edaphoclimatic differences or confounding factors. This problem could be solved by sampling both farms in a time series and then comparing the development of earthworm communities over time under different management strategies. Another problem arises with the lack of independent replications because it is not realistic to find different farms that are replicating the exact same practices under the same conditions. Farmers generally adapt management practices to the local edaphoclimatic conditions and resource possibilities. Then with time series data, comparisons can be made between farms following management practices with the same inspiration, even though the management is not the same (i.e., rotational management practices vs continuous grazing, organic vs conventional, or silvopastures vs no-tree pastures). Despite these limitations, we consider that more real farm studies can help to better understand earthworm communities and their functionality in grasslands in a scenario of continuous changes in grassland management practices.

Finally, we encourage the development of accessible and easy methods to assess the earthworm community health status in grasslands similar to what the visual assessment has done for soil health (Ball et al. 2017). Such methods could be a tool for farmers and agricultural services to assess the impact of their management on the earthworm community and help them adjust their practices. These earthworm assessment methods, adapted to different regions and grasslands, can be developed based on the work performed in the last decades by thousands of scientists, farmers, and agronomists who have evaluated and studied earthworms in grasslands worldwide. Thus, it is time to further develop the potential of earthworms for healthier and more sustainable grasslands.

Acknowledgments We thank José Luis Urrea and Anny Isabella Yedra for graphic design assistance, and Erica Packard for language editing.

References

Alfaro-Arguello R, Diemont SAW, Ferguson BG et al (2010) Steps toward sustainable ranching: An emergy evaluation of conventional and holistic management in Chiapas, Mexico. Agric Syst. https://doi.org/10.1016/j.agsy.2010.08.002

Arrázola-Vásquez E, Larsbo M, Capowiez Y et al (2022) Earthworm burrowing modes and rates depend on earthworm species and soil mechanical resistance. Appl Soil Ecol 178:104568. https://doi.org/10.1016/J.APSOIL.2022.104568

Bacher MG, Fenton O, Bondi G et al (2018) The impact of cattle dung pats on earthworm distribution in grazed pastures. BMC Ecol 18:1–12. https://doi.org/10.1186/S12898-018-0216-6/TABLES/2

Ball BC, Guimarães RML, Cloy JM et al (2017) Visual soil evaluation: A summary of some applications and potential developments for agriculture. Soil Tillage Res 173:114–124. https://doi.org/10.1016/J.STILL.2016.07.006

Barros E, Neves A, Blanchart E et al (2003) Development of the soil macrofauna community under silvopastoral and agrosilvicultural systems in Amazonia. Pedobiologia (Jena) 47:273–280. https://doi.org/10.1078/0031-4056-00190

Bengtsson J, Bullock JM, Egoh B et al (2019) Grasslands—more important for ecosystem services than you might think. Ecosphere 10. https://doi.org/10.1002/ECS2.2582

Birkhofer K, Bezemer TM, Bloem J et al (2008) Long-term organic farming fosters below and aboveground biota: Implications for soil quality, biological control and productivity. Soil Biol Biochem 40:2297–2308. https://doi.org/10.1016/J.SOILBIO.2008.05.007

Blouin M, Hodson ME, Delgado EA et al (2013) A review of earthworm impact on soil function and ecosystem services. Eur J Soil Sci 64:161–182. https://doi.org/10.1111/EJSS.12025

Bouché MB, Al-Addan F (1997) Earthworms, water infiltration and soil stability: Some new assessments. Soil Biol Biochem 29:441–452. https://doi.org/10.1016/S0038-0717(96)00272-6

Brown GG, Pashanasi B, Villenave C et al (1999) Effects of earthworms on plant production in the tropics. In: Lavelle P, Brussard L, Hendrix PF (eds) Earthworm management in tropical agroecosystems. CABI, Wallingford, pp 87–148

Brown GG, Edwards CA, Brussard L (2004) How earthworms affect plant growth: burrowing into the mechanisms. In: Edwards CA (ed) Earthworm ecology, 2nd edn. CRC, Boca Raton, pp 13–49

Brussard L (1999) On the mechanisms of interactions between earthworms and plants. Pedobiologia (Jena) 43:880–885. https://doi.org/10.1016/S0031-4056(24)00626-7

Chang J, Ciais P, Gasser T et al (2021) Climate warming from managed grasslands cancels the cooling effect of carbon sinks in sparsely grazed and natural grasslands. Nat Commun 12:1–10. https://doi.org/10.1038/s41467-020-20406-7

Chauvel A, Grimaldi M, Barros E et al (1999) Pasture damage by an Amazonian earthworm. Nature 398:32–33. https://doi.org/10.1038/17946

Clements RO, Murray PJ, Sturdy RG (1991) The impact of 20 years' absence of earthworms and three levels of N fertilizer on a grassland soil environment. Agric Ecosyst Environ 36:75–85. https://doi.org/10.1016/0167-8809(91)90037-X

Cluzeau D, Binet F, Vertes F et al (1992) Effects of intensive cattle trampling on soil-plant-earthworms system in two grassland types. Soil Biol Biochem 24:1661–1665. https://doi.org/10.1016/0038-0717(92)90166-U

Cramer VA, Hobbs RJ, Standish RJ (2008) What's new about old fields? Land abandonment and ecosystem assembly. Trends Ecol Evol 23:104–112. https://doi.org/10.1016/J.TREE.2007.10.005
Curry JP (1976) Some effects of animal manures on earthworms in grassland. Pedobiologia (Jena) 16:425–438
Curry JP (1987) The invertebrate fauna of grassland and its influence on productivity. III. Effects on soil fertility and plant growth. Grass Forage Sci 42:325–341. https://doi.org/10.1111/J.1365-2494.1987.TB02121.X
Curry JP (2004) Factors Affecting the Abundance of Earthworms in Soils. In: Edwards CA (ed) Earthworm ecology, 2nd edn. CRC, Boca Raton, pp 91–113
Curry JP, Doherty P, Purvis G, Schmidt O (2008) Relationships between earthworm populations and management intensity in cattle-grazed pastures in Ireland. Appl Soil Ecol 39:58–64. https://doi.org/10.1016/J.APSOIL.2007.11.005
da Silva HMS, Dubeux JCB, Silveira ML et al (2019) Soil and root attributes in pastures managed under different stocking rates and nitrogen fertilization levels. Agrosyst Geosci Environ 2:1–9. https://doi.org/10.2134/AGE2018.08.0031
Decaëns T, Lavelle P, Jimenez Jaen JJ et al (1994) Impact of land management on soil macrofauna in the oriental Llanos of Colombia. Eur J Soil Biol 30:157–168
Decaëns T, Mariani L, Betancourt N, Jiménez JJ (2003) Seed dispersion by surface casting activities of earthworms in Colombian grasslands. Acta Oecologica 24:175–185. https://doi.org/10.1016/S1146-609X(03)00083-3
Decaëns T, Jiménez JJ, Barros E et al (2004) Soil macrofaunal communities in permanent pastures derived from tropical forest or savanna. Agric Ecosyst Environ 103:301–312. https://doi.org/10.1016/J.AGEE.2003.12.005
de Freitas IC, Ribeiro JM, Araújo NCA et al (2020) Agrosilvopastoral systems and well-managed pastures increase soil carbon stocks in the Brazilian Cerrado. Rangel Ecol Manag 73:776–785. https://doi.org/10.1016/j.rama.2020.08.001
Deru JGC, Bloem J, de Goede R et al (2023) Effects of organic and inorganic fertilizers on soil properties related to the regeneration of ecosystem services in peat grasslands. Appl Soil Ecol 187:104838. https://doi.org/10.1016/J.APSOIL.2023.104838
Domínguez A, Bedano JC, Becker AR (2010) Negative effects of no-till on soil macrofauna and litter decomposition in Argentina as compared with natural grasslands. Soil Tillage Res 110:51–59. https://doi.org/10.1016/J.STILL.2010.06.008
Don A, Steinberg B, Schöning I et al (2008) Organic carbon sequestration in earthworm burrows. Soil Biol Biochem 40:1803–1812. https://doi.org/10.1016/J.SOILBIO.2008.03.003
Drewry JJ (2006) Natural recovery of soil physical properties from treading damage of pastoral soils in New Zealand and Australia: A review. Agric Ecosyst Environ 114:159–169. https://doi.org/10.1016/J.AGEE.2005.11.028
Dumont B, Groot JCJ, Tichit M (2018) Review: Make ruminants green again—how can sustainable intensification and agroecology converge for a better future? Animal 12:s210–s219. https://doi.org/10.1017/S1751731118001350
Edwards CA, Lofty JR (1982) Nitrogenous fertilizers and earthworm populations in agricultural soils. Soil Biol Biochem 14:515–521. https://doi.org/10.1016/0038-0717(82)90112-2
Egoh BN, Bengtsson J, Lindborg R et al (2016) The importance of grasslands in providing ecosystem services: opportunities for poverty alleviation. In: Potschin M, Haines-Young R, Robert F, Kerry Turner R (eds) Routledge handbook of ecosystem services. Routledge, pp 421–441. https://doi.org/10.4324/9781315775302-37
Eisenhauer N, Scheu S (2008) Earthworms as drivers of the competition between grasses and legumes. Soil Biol Biochem 40:2650–2659. https://doi.org/10.1016/J.SOILBIO.2008.07.010
Eisenhauer N, Milcu A, Nitschke N et al (2009a) Earthworm and belowground competition effects on plant productivity in a plant diversity gradient. Oecologia 161:291–301. https://doi.org/10.1007/S00442-009-1374-1/TABLES/4

Eisenhauer N, Milcu A, Sabais ACW et al (2009b) Plant community impacts on the structure of earthworm communities depend on season and change with time. Soil Biol Biochem 41:2430–2443. https://doi.org/10.1016/J.SOILBIO.2009.09.001
Ellis EC, Ramankutty N (2008) Putting people in the map: anthropogenic biomes of the world. Front Ecol Environ 6:439–447. https://doi.org/10.1890/070062
Falco LB, Sandler R, Momo F et al (2015) Earthworm assemblages in different intensity of agricultural uses and their relation to edaphic variables. PeerJ 2015:e979. https://doi.org/10.7717/PEERJ.979/SUPP-1
Feijoo A, Carvajal AF, Zúñiga MC et al (2011) Diversity and abundance of earthworms in land use systems in central-western Colombia. Pedobiologia (Jena) 54:S69–S75. https://doi.org/10.1016/J.PEDOBI.2011.09.016
Ferguson BG, Diemont SAW, Alfaro-Arguello R et al (2013) Sustainability of holistic and conventional cattle ranching in the seasonally dry tropics of Chiapas, Mexico. Agric Syst. https://doi.org/10.1016/j.agsy.2013.05.005
Fischer C, Roscher C, Jensen B et al (2014) How do earthworms, soil texture and plant composition affect infiltration along an experimental plant diversity gradient in grassland? PLoS One 9:e98987. https://doi.org/10.1371/JOURNAL.PONE.0098987
Fisher MJ, Rao IM, Ayarza MA et al (1994) Carbon storage by introduced deep-rooted grasses in the South American savannas. Nature 371:236–238. https://doi.org/10.1038/371236a0
Fonte SJ, Six J (2010) Earthworms and litter management contributions to ecosystem services in a tropical agroforestry system. Ecol Appl 20:1061–1073. https://doi.org/10.1890/09-0795.1
Fonte SJ, Quintero DC, Velásquez E, Lavelle P (2012) Interactive effects of plants and earthworms on the physical stabilization of soil organic matter in aggregates. Plant Soil 359:205–214. https://doi.org/10.1007/s11104-012-1199-2
Fragoso C, Lavelle P, Blanchart E et al (1999) Earthworm communities of tropical agroecosystems: origin, structure and influence of management practices. In: Lavelle P, Brussaard L, Hendrix P (eds) Earthworm management in tropical agroecosystems. CAB, Wallingford, pp 27–55
Garcia-Franco N, Walter R, Wiesmeier M et al (2021) Biotic and abiotic controls on carbon storage in aggregates in calcareous alpine and prealpine grassland soils. Biol Fertil Soils 57:203–218. https://doi.org/10.1007/S00374-020-01518-0/FIGURES/2
Graesser J, Aide TM, Grau HR, Ramankutty N (2015) Cropland/pastureland dynamics and the slowdown of deforestation in Latin America. Environ Res Lett 10:034017. https://doi.org/10.1088/1748-9326/10/3/034017
Guggenberger G, Thomas RJ, Zech W (1996) Soil organic matter within earthworm casts of an anecic-endogeic tropical pasture community, Colombia. Appl Soil Ecol 3:263–274. https://doi.org/10.1016/0929-1393(95)00081-X
Herrero M, Havlík P, Valin H et al (2013) Biomass use, production, feed efficiencies, and greenhouse gas emissions from global livestock systems. Proc Natl Acad Sci U S A 110:20888–20893. https://doi.org/10.1073/PNAS.1308149110/SUPPL_FILE/SAPP.PDF
Herrick JE, Lal R (1995) Soil physical property changes during dung decomposition in a tropical pasture. Soil Sci Soc Am J 59:908–912. https://doi.org/10.2136/SSSAJ1995.03615995005900030040X
Herridge DF, Peoples MB, Boddey RM (2008) Global inputs of biological nitrogen fixation in agricultural systems. Plant Soil 311:1–18. https://doi.org/10.1007/s11104-008-9668-3
Hoeffner K, Hotte H, Cluzeau D et al (2021a) Effects of temporary grassland introduction into annual crop rotations and nitrogen fertilisation on earthworm communities and forage production. Appl Soil Ecol 162:103893. https://doi.org/10.1016/J.APSOIL.2021.103893
Hoeffner K, Santonja M, Monard C et al (2021b) Soil properties, grassland management, and landscape diversity drive the assembly of earthworm communities in temperate grasslands. Pedosphere 31:375–383. https://doi.org/10.1016/S1002-0160(20)60020-0
Hoeffner K, Louault F, Lerner L, Pérès G (2024) Management of grassland: A necessary tool to maintain plant and earthworm diversity. Eur J Soil Biol 120:103589. https://doi.org/10.1016/J.EJSOBI.2023.103589

Hyvönen T, Huusela E, Kuussaari M et al (2021) Aboveground and belowground biodiversity responses to seed mixtures and mowing in a long-term set-aside experiment. Agric Ecosyst Environ 322:107656. https://doi.org/10.1016/J.AGEE.2021.107656

Jiménez JJ, Moreno AG, Decaëns T et al (1998) Earthworm communities in native Savannas and man-made pastures of the Eastern Plains of Colombia. Biol Fertil Soils 28:101–110. https://doi.org/10.1007/S003740050469

Jones MB (2019) Projected climate change and the global distribution of grasslands. Grasslands Clim Change 67–81. https://doi.org/10.1017/9781108163941.006

Joschko M, Diestel H, Larink O (1989) Assessment of earthworm burrowing efficiency in compacted soil with a combination of morphological and soil physical measurements. Biol Fertil Soils 8:191–196. https://doi.org/10.1007/BF00266478/METRICS

Jouquet P, Bottinelli N, Podwojewski P et al (2008) Chemical and physical properties of earthworm casts as compared to bulk soil under a range of different land-use systems in Vietnam. Geoderma. https://doi.org/10.1016/j.geoderma.2008.05.030

Karwat H, Moreta D, Arango J et al (2017) Residual effect of BNI by Brachiaria humidicola pasture on nitrogen recovery and grain yield of subsequent maize. Plant Soil 420:389–406. https://doi.org/10.1007/S11104-017-3381-Z/TABLES/3

King KL, Hutchinson KJ, King KL, Hutchinson KJ (2007) Pasture and grazing land: assessment of sustainability using invertebrate bioindicators. Aust J Exp Agric 47:392–403. https://doi.org/10.1071/EA05270

Lal R (2004) Soil carbon sequestration impacts on global climate change and food security. Science (1979) 304:1623–1627. https://doi.org/10.1126/science.1097396

Lamandé M, Labouriau R, Holmstrup M et al (2011) Density of macropores as related to soil and earthworm community parameters in cultivated grasslands. Geoderma 162:319–326. https://doi.org/10.1016/J.GEODERMA.2011.03.004

Lamarque P, Tappeiner U, Turner C et al (2011) Stakeholder perceptions of grassland ecosystem services in relation to knowledge on soil fertility and biodiversity. Reg Environ Change 11:791–804. https://doi.org/10.1007/s10113-011-0214-0

Lavelle P (1988) Earthworm activities and the soil system. Biol Fertil Soils 6:237–251. https://doi.org/10.1007/BF00260820/METRICS

Lavelle P, Martin A (1992) Small-scale and large-scale effects of endogeic earthworms on soil organic matter dynamics in soils of the humid tropics. Soil Biol Biochem 24:1491–1498. https://doi.org/10.1016/0038-0717(92)90138-N

Lavelle P, Bignell D, Lepage M et al (1997) Soil function in a changing world: The role of invertebrate ecosystem engineers. Eur J Soil Biol 33:159–193

Lavelle P, Barros E, Blanchart E et al (2001) SOM management in the tropics: Why feeding the soil macrofauna? In: Martius C, Tiessen H, Vlek PLG (eds) Managing organic matter in tropical soils: scope and limitations. Springer, Dordrecht, pp 53–61. https://doi.org/10.1007/978-94-017-2172-1_6

Lavelle P, Dolédec S, de Sartre XA et al (2016) Unsustainable landscapes of deforested Amazonia: An analysis of the relationships among landscapes and the social, economic and environmental profiles of farms at different ages following deforestation. Global Environ Change 40:137–155. https://doi.org/10.1016/j.gloenvcha.2016.04.009

Lavelle P, Spain A, Fonte S et al (2020) Soil aggregation, ecosystem engineers and the C cycle. Acta Oecologica 105. https://doi.org/10.1016/j.actao.2020.103561

Lemaire G, Hodgson J, Chabbi A (2011) Grassland productivity and ecosystem services. CABI, Wallingford

Lemaire G, Gastal F, Franzluebbers A, Chabbi A (2015) Grassland–cropping rotations: an avenue for agricultural diversification to reconcile high production with environmental quality. Environ Manag 56:1065–1077. https://doi.org/10.1007/S00267-015-0561-6/FIGURES/3

Lerner AM, Zuluaga AF, Chará J et al (2017) Sustainable cattle ranching in practice: moving from theory to planning in Colombia's livestock sector. Environ Manag 60:176–184. https://doi.org/10.1007/s00267-017-0902-8

Lüscher A, Mueller-Harvey I, Soussana JF et al (2014) Potential of legume-based grassland-livestock systems in Europe: A review. Grass Forage Sci 69:206–228. https://doi.org/10.1111/gfs.12124

Ma WC, Brussaard L, de Ridder JA (1990) Long-term effects of nitrogenous fertilizers on grassland earthworms (Oligochaeta: Lumbricidae): Their relation to soil acidification. Agric Ecosyst Environ 30:71–80. https://doi.org/10.1016/0167-8809(90)90184-F

Manning P, Gossner MM, Bossdorf O et al (2015) Grassland management intensification weakens the associations among the diversities of multiple plant and animal taxa. Ecology 96:1492–1501. https://doi.org/10.1890/14-1307.1

Manuel-Navarrete D, Gallopín GC, Blanco M et al (2009) Multi-causal and integrated assessment of sustainability: The case of agriculturization in the Argentine Pampas. Environ Dev Sustain 11:621–638. https://doi.org/10.1007/S10668-007-9133-0/FIGURES/7

Marichal R, Martinez AF, Praxedes C et al (2010) Invasion of Pontoscolex corethrurus (Glossoscolecidae, Oligochaeta) in landscapes of the Amazonian deforestation arc. Appl Soil Ecol 46:443–449. https://doi.org/10.1016/J.APSOIL.2010.09.001

Marichal R, Grimaldi M, Feijoo MA et al (2014) Soil macroinvertebrate communities and ecosystem services in deforested landscapes of Amazonia. Appl Soil Ecol 83:177–185. https://doi.org/10.1016/J.APSOIL.2014.05.006

Marsden C, Martin-Chave A, Cortet J et al (2020) How agroforestry systems influence soil fauna and their functions—a review. Plant Soil 453:29–44. https://doi.org/10.1007/S11104-019-04322-4/TABLES/3

Mathieu J, Rossi JP, Mora P et al (2005) Recovery of soil macrofauna communities after forest clearance in Eastern Amazonia, Brazil. Conserv Biol 19:1598–1605. https://doi.org/10.1111/J.1523-1739.2005.00200.X

Milcu A, Partsch S, Scherber C et al (2008) Earthworms and legumes control litter decomposition in a plant diversity gradient. Ecology 89:1872–1882. https://doi.org/10.1890/07-1377.1

Moreno G, Pulido FJ (2009) The functioning, management and persistence of Dehesas. Agroforest Europe 127–160. https://doi.org/10.1007/978-1-4020-8272-6_7

Moreno G, Gonzalez-Bornay G, Pulido F et al (2016) Exploring the causes of high biodiversity of Iberian dehesas: the importance of wood pastures and marginal habitats. Agroforest Syst 90:87–105. https://doi.org/10.1007/S10457-015-9817-7/TABLES/5

Murchie AK, Blackshaw RP, Gordon AW, Christie P (2015) Responses of earthworm species to long-term applications of slurry. Appl Soil Ecol 96:60–67. https://doi.org/10.1016/J.APSOIL.2015.07.005

Murgueitio E, Calle Z, Uribe F et al (2011) Native trees and shrubs for the productive rehabilitation of tropical cattle ranching lands. For Ecol Manag 261:1654–1663. https://doi.org/10.1016/j.foreco.2010.09.027

Muys B, Granval P (1997) Earthworms as bio-indicators of forest site quality. Soil Biol Biochem 29:323–328. https://doi.org/10.1016/S0038-0717(96)00047-8

Notenbaert AMO, Douxchamps S, Villegas DM et al (2021) Tapping into the environmental co-benefits of improved tropical forages for an agroecological transformation of livestock production systems. Front Sustain Food Syst 5:434. https://doi.org/10.3389/FSUFS.2021.742842/BIBTEX

Oberson A, Frossard E, Bühlmann C et al (2013) Nitrogen fixation and transfer in grass-clover leys under organic and conventional cropping systems. Plant Soil 371:237–255. https://doi.org/10.1007/s11104-013-1666-4

Paoletti MG (1999) The role of earthworms for assessment of sustainability and as bioindicators. Agric Ecosyst Environ 74:137–155. https://doi.org/10.1016/S0167-8809(99)00034-1

Paoletti MG, Favretto MR, Stinner BR et al (1991) Invertebrates as bioindicators of soil use. Agric Ecosyst Environ 34:341–362. https://doi.org/10.1016/0167-8809(91)90120-M

Peeters A (2009) Importance, evolution, environmental impact and future challenges of grasslands and grassland-based systems in Europe. Grassl Sci 55:113–125. https://doi.org/10.1111/J.1744-697X.2009.00154.X

Peres CA, Gardner TA, Barlow J et al (2010) Biodiversity conservation in human-modified Amazonian forest landscapes. Biol Conserv 143:2314–2327. https://doi.org/10.1016/J.BIOCON.2010.01.021

Pérès G, Vandenbulcke F, Guernion M et al (2011) Earthworm indicators as tools for soil monitoring, characterization and risk assessment. An example from the national Bioindicator programme (France). Pedobiologia (Jena) 54:S77–S87. https://doi.org/10.1016/J.PEDOBI.2011.09.015

Phillips HRP, Bach EM, Bartz MLC et al (2021) Global data on earthworm abundance, biomass, diversity and corresponding environmental properties. Sci Data 8:1–12. https://doi.org/10.1038/s41597-021-00912-z

Piearce TG (1984) Earthworm populations in soils disturbed by trampling. Biol Conserv 29:241–252. https://doi.org/10.1016/0006-3207(84)90101-0

Piotrowska K, Connolly J, Finn J et al (2013) Evenness and plant species identity affect earthworm diversity and community structure in grassland soils. Soil Biol Biochem 57:713–719. https://doi.org/10.1016/J.SOILBIO.2012.06.016

Postma-Blaauw MB, de Goede RGM, Bloem J et al (2012) Agricultural intensification and de-intensification differentially affect taxonomic diversity of predatory mites, earthworms, enchytraeids, nematodes and bacteria. Appl Soil Ecol 57:39–49. https://doi.org/10.1016/J.APSOIL.2012.02.011

Pulleman MM, Six J, Uyl A et al (2005a) Earthworms and management affect organic matter incorporation and microaggregate formation in agricultural soils. Appl Soil Ecol 29:1–15. https://doi.org/10.1016/J.APSOIL.2004.10.003

Pulleman MM, Six J, Van Breemen N, Jongmans AG (2005b) Soil organic matter distribution and microaggregate characteristics as affected by agricultural management and earthworm activity. Eur J Soil Sci 56:453–467. https://doi.org/10.1111/J.1365-2389.2004.00696.X

Pulleman M, Creamer R, Hamer U et al (2012) Soil biodiversity, biological indicators and soil ecosystem services-an overview of European approaches. Curr Opin Environ Sustain 4:529–538. https://doi.org/10.1016/J.COSUST.2012.10.009

Rao I, Peters M, Castro A et al (2015) LivestockPlus—The sustainable intensification of forage-based agricultural systems to improve livelihoods and ecosystem services in the tropics. Trop Grasslands Forrajes Tropicales. https://doi.org/10.17138/TGFT(3)59-82

Rodrigues ASL, Ewers RM, Parry L et al (2009) Boom-and-bust development patterns across the amazon deforestation frontier. Science (1979) 324:1435–1437. https://doi.org/10.1126/SCIENCE.1174002/SUPPL_FILE/RODRIGUES.SOM.PDF

Rodríguez MP, Domínguez A, Ferroni MM et al (2020) The diversification and intensification of crop rotations under no-till promote earthworm abundance and biomass. Agronomy 10:919. https://doi.org/10.3390/AGRONOMY10070919

Rodríguez L, Suárez JC, Pulleman M et al (2021a) Agroforestry systems in the Colombian Amazon improve the provision of soil ecosystem services. Appl Soil Ecol 164. https://doi.org/10.1016/j.apsoil.2021.103933

Rodríguez L, Suárez JC, Rodriguez W et al (2021b) Agroforestry systems impact soil macroaggregation and enhance carbon storage in Colombian deforested Amazonia. Geoderma 384:114810. https://doi.org/10.1016/J.GEODERMA.2020.114810

Scheu S (2003) Effects of earthworms on plant growth: patterns and perspectives: The 7th international symposium on earthworm ecology · Cardiff · Wales · 2002. Pedobiologia (Jena) 47:846–856. https://doi.org/10.1078/0031-4056-00270

Schlaghamerský J, Šídová A, Pižl V (2007) From mowing to grazing: Does the change in grassland management affect soil annelid assemblages? Eur J Soil Biol 43:S72–S78. https://doi.org/10.1016/J.EJSOBI.2007.08.054

Schon NL, Mackay AD, Yeates GW, Minor MA (2010) Separating the effects of defoliation and dairy cow treading pressure on the abundance and diversity of soil invertebrates in pastures. Appl Soil Ecol 46:209–221. https://doi.org/10.1016/J.APSOIL.2010.08.011

Schon NL, Dennis S, Fraser PM et al (2017) Intensification of pastoral systems influences earthworm populations. N Z J Agric Res 60:423–436. https://doi.org/10.1080/00288233.2017.1363788

Schon NL, Fraser PM, Mackay AD (2022) Earthworms for inclusion as an indicator of soil biological health in New Zealand pastures. N Z J Agric Res. https://doi.org/10.1080/00288233.2022.2041676

Segat JC, Sousa JP, Baretta D, Klauberg-Filho O (2020) Increasing level of liquid pig manure reduces Eisenia andrei and Enchytraeus crypticus reproduction in subtropical soils. Sci Rep 10:1–6. https://doi.org/10.1038/s41598-020-67360-4

Siebert J, Thakur MP, Reitz T et al (2019) Extensive grassland-use sustains high levels of soil biological activity, but does not alleviate detrimental climate change effects. In: Resilience in complex socio-ecological systems. Advances in ecological research. Academic, London, pp 25–58

Singh J, Cameron E, Reitz T et al (2021) Grassland management effects on earthworm communities under ambient and future climatic conditions. Eur J Soil Sci 72:343–355. https://doi.org/10.1111/EJSS.12942

Smith FA, Hammond JI, Balk MA et al (2016) Exploring the influence of ancient and historic megaherbivore extirpations on the global methane budget. Proc Natl Acad Sci U S A 113:874–879. https://doi.org/10.1073/PNAS.1502547112/SUPPL_FILE/PNAS.1502547112.SAPP.PDF

Spehn EM, Joshi J, Schmid B et al (2000) Plant diversity effects on soil heterotrophic activity in experimental grassland ecosystems. Plant Soil 224:217–230. https://doi.org/10.1023/A:1004891807664/METRICS

Spurgeon DJ, Keith AM, Schmidt O et al (2013) Land-use and land-management change: Relationships with earthworm and fungi communities and soil structural properties. BMC Ecol 13:1–13. https://doi.org/10.1186/1472-6785-13-46/FIGURES/7

Suttie JM, Reynolds SG, Batello C (2005) Grasslands of the world. Food and Agriculture Organization of the United Nations, Rome

Tälle M, Deák B, Poschlod P et al (2016) Grazing vs. mowing: A meta-analysis of biodiversity benefits for grassland management. Agric Ecosyst Environ 222:200–212. https://doi.org/10.1016/J.AGEE.2016.02.008

Teague WR, Dowhower SL, Baker SA et al (2011) Grazing management impacts on vegetation, soil biota and soil chemical, physical and hydrological properties in tall grass prairie. Agric Ecosyst Environ. https://doi.org/10.1016/j.agee.2011.03.009

Teutscherová N, Vazquez E, Baquero D et al (2021a) Visual assessment of soil structure as an early indicator of soil quality in response to intensive rotational grazing. In: International grassland congress proceedings

Teutscherová N, Vázquez E, Sotelo M et al (2021b) Intensive short-duration rotational grazing is associated with improved soil quality within one year after establishment in Colombia. Appl Soil Ecol 159:103835. https://doi.org/10.1016/J.APSOIL.2020.103835

Timmerman A, Bos D, Ouwehand J, de Goede RGM (2006) Long-term effects of fertilisation regime on earthworm abundance in a semi-natural grassland area. Pedobiologia (Jena) 50:427–432. https://doi.org/10.1016/J.PEDOBI.2006.08.005

Tognetti PM, Prober SM, Báez S et al (2021) Negative effects of nitrogen override positive effects of phosphorus on grassland legumes worldwide. Proc Natl Acad Sci U S A 118. https://doi.org/10.1073/pnas.2023718118

Tondoh JE, Monin LM, Tiho S, Csuzdi C (2007) Can earthworms be used as bio-indicators of land-use perturbations in semi-deciduous forest? Biol Fertil Soils 43:585–592. https://doi.org/10.1007/S00374-006-0144-Z/FIGURES/3

Török P, Vida E, Deák B et al (2011) Grassland restoration on former croplands in Europe: An assessment of applicability of techniques and costs. Biodivers Conserv 20:2311–2332. https://doi.org/10.1007/S10531-011-9992-4/TABLES/3

Trujillo W, Fisher MJ, Lal R (2006) Root dynamics of native savanna and introduced pastures in the Eastern Plains of Colombia. Soil Tillage Res 87:28–38. https://doi.org/10.1016/J.STILL.2005.02.038

Tsiafouli MA, Thébault E, Sgardelis SP et al (2015) Intensive agriculture reduces soil biodiversity across Europe. Glob Chang Biol 21:973–985. https://doi.org/10.1111/GCB.12752

van de Logt R, van der Sluijs T, van Eekeren N (2023) Lumbricus terrestris abundance in grasslands on sandy soils in relation to soil texture, hydrology and earthworm community. Eur J Soil Biol 119:103545. https://doi.org/10.1016/J.EJSOBI.2023.103545

van Eekeren N, Bommelé L, Bloem J et al (2008) Soil biological quality after 36 years of ley-arable cropping, permanent grassland and permanent arable cropping. Appl Soil Ecol 40:432–446. https://doi.org/10.1016/J.APSOIL.2008.06.010

van Eekeren N, de Boer H, Bloem J et al (2009) Soil biological quality of grassland fertilized with adjusted cattle manure slurries in comparison with organic and inorganic fertilizers. Biol Fertil Soils 45:595–608. https://doi.org/10.1007/S00374-009-0370-2/FIGURES/3

van Groenigen JW, Lubbers IM, Vos HMJ et al (2014) Earthworms increase plant production: a meta-analysis. Sci Rep 4:1–7. https://doi.org/10.1038/srep06365

Vázquez E, Teutscherova N, Lojka B et al (2020) Pasture diversification affects soil macrofauna and soil biophysical properties in tropical (silvo)pastoral systems. Agric Ecosyst Environ 302:107083. https://doi.org/10.1016/j.agee.2020.107083

Vázquez E, Schleuss PM, Borer ET et al (2022) Nitrogen but not phosphorus addition affects symbiotic N2 fixation by legumes in natural and semi-natural grasslands located on four continents. Plant Soil 478:689–707. https://doi.org/10.1007/s11104-022-05498-y

Velásquez E, Fonte SJ, Barot S et al (2012) Soil macrofauna-mediated impacts of plant species composition on soil functioning in Amazonian pastures. Appl Soil Ecol 56:43–50. https://doi.org/10.1016/j.apsoil.2012.01.008

Wang W, Fang J (2009) Soil respiration and human effects on global grasslands. Glob Planet Change 67:20–28. https://doi.org/10.1016/J.GLOPLACHA.2008.12.011

Webster E, Gaudin ACM, Pulleman M et al (2019) Improved pastures support early indicators of soil restoration in low-input agroecosystems of Nicaragua. Environ Manag. https://doi.org/10.1007/s00267-019-01181-8

Weiler M, Naef F (2003) An experimental tracer study of the role of macropores in infiltration in grassland soils. Hydrol Process 17:477–493. https://doi.org/10.1002/HYP.1136

Wodika BR, Klopf RP, Baer SG (2014) Colonization and recovery of invertebrate ecosystem engineers during prairie restoration. Restor Ecol 22:456–464. https://doi.org/10.1111/REC.12084

Zaller JG, Saxler N (2007) Selective vertical seed transport by earthworms: Implications for the diversity of grassland ecosystems. Eur J Soil Biol 43:S86–S91. https://doi.org/10.1016/J.EJSOBI.2007.08.010

zu Ermgassen EKHJ, de Alcântara MP, Balmford A et al (2018) Results from on-the-ground efforts to promote sustainable cattle ranching in the Brazilian Amazon. Sustainability 10:1301. https://doi.org/10.3390/SU10041301

Earthworm Contributions to Agricultural Sustainability

Céline Pelosi, Lucas Petit-Dit-Grezeriat, Onja Ratsiatosika, and Eric Blanchart

1 Introduction

Soils are a major component of ecosystems and they provide many benefits to humanity. Located in the Earth's critical zone, at the interface between the atmosphere, the hydrosphere, the lithosphere, and the biosphere, soil biogeochemical cycles are essential for the sustenance of life in terrestrial ecosystems at a planetary scale. Soil is therefore one of the richest habitats for biological species. However, soils have for decades been subjected to many pressures that cause degradation and jeopardize their ability to provide services to humanity. Agriculture, in its capacity to ensure sufficient food for a growing human population, is mainly concerned with the soil degradation that lowers crop yields. Sustainable management practices are designed to produce higher crop yields and other services, including the regeneration of soil ecological functions, i.e., the intensification of all functions performed by soil organisms. This improvement in soil functions is recognized as a basis for

C. Pelosi
National Research Institute for Agriculture, Food and the Environment, Paris, France
e-mail: celine.pelosi@inrae.fr

L. Petit-Dit-Grezeriat
Mediterranean Institute of Marine and Terrestrial Biodiversity and Ecology, National Research Institute for Agriculture, Food and the Environment, Avignon University, Avignon, France
e-mail: lucas.petit-dit-grezeriat@inrae.fr

O. Ratsiatosika
French National Research Institute for Sustainable Development, Montpellier, France

University of Antananarivo, Antananarivo, Madagascar

E. Blanchart (✉)
French National Research Institute for Sustainable Development, Montpellier, France
e-mail: eric.blanchart@ird.fr

Y. Kooch, Y. Kuzyakov (eds.), *Earthworms and Ecological Processes*,
https://doi.org/10.1007/978-3-031-64510-5_10

sustainable agriculture (FAO, ITPS, GSBI, SCBD and EC 2020; Larbodière et al. 2020; FAO 2021, 2022). Soil biodiversity, seen through its taxonomic diversity, but also and above all through its functional diversity, is therefore at the heart of sustainable soil management. Among this soil biodiversity, earthworms, a major contributor to the animal biomass in soil, perform many ecological functions of interest. Often perceived as beneficial for the functioning of soils and indicators of their quality, we are now poised with a greater understanding of the contribution of earthworms to the sustainability of agriculture (e.g., Lavelle et al. 2006; Jouquet et al. 2014; Bertrand et al. 2015). In this chapter, we will briefly review the processes and functions provided by earthworms in soils, particularly those necessary to achieve agricultural sustainability. Next, we will analyze the effect of agricultural practices on earthworm communities, considering those that have either a negative or a positive impact on their communities. Finally, through case studies and knowledge from temperate and tropical environments, and from laboratory and field experiments, we will discuss the possibility of using earthworms to improve the productivity and sustainability of agricultural systems.

2 Effects of Earthworms on Soil Processes

Earthworms are well-known for their contribution to agricultural production and sustainability. Earthworms influence soil functions primarily through their feeding and burrowing activities. Together, these activities lead (1) to chemical, physical, and biological modifications in the detritus and minerals ingested, which is seen when comparing the uningested materials and the fecal material (casts) and (2) to the redistribution of detritus, minerals, casts, residues, and living organisms in the soil profile. By modifying soil's physical, chemical, and biological properties, earthworms affect the whole functioning of the soil and regulate the activity of other soil organisms through resource availability: this is why they are called "soil engineers" (Jones et al. 1994). Earthworm activities in the soil contribute to four major ecological functions (*sensu* Kibblewhite et al. 2008) that are the basis of the ecosystem services provided by soil and control the sustainability of agricultural systems, namely: (1) carbon transformations, (2) nutrient cycling, (3) maintenance of soil structure, and (4) regulation of pests and diseases (Fig. 1). Improving or intensifying these four ecological functions is required when adopting new practices for sustainable agriculture (Trap and Blanchart 2023).

2.1 Carbon Transformation

Earthworms assimilate part of the organic matter contained in their food: it can be the organic matter contained in the soil for geophagous species, plant litter for detritivorous species, or living organisms (mainly microorganisms) contained in the soil

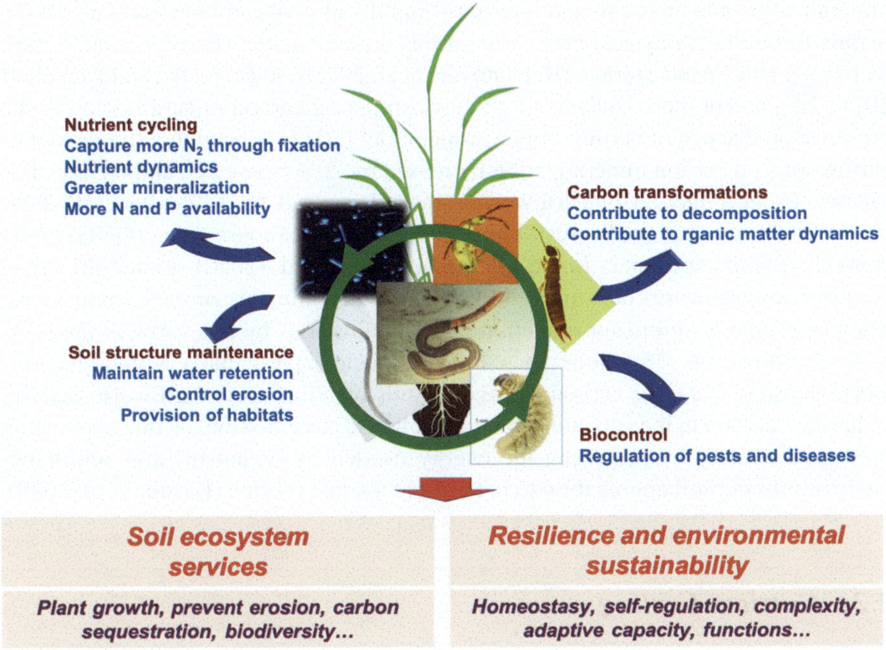

Fig. 1 Earthworm interactions with other soil organisms (including roots, bacteria, fungi, and fauna) are essential for the main ecological functions that support the ecosystem services, resilience, and environmental sustainability of agricultural soil. Photo credits: bacteria (Laetitia Bernard, IRD), nematodes (Jean Trap, IRD), white grubs (Richard Randriamanantsoa, FOFIFA, and Henri-Pierre Aberlenc, CIRAD), earthworm, Dermaptera, and collembola (Eric Blanchart, IRD)

or plant litter. Part of the organic matter is assimilated through a mutualistic interaction with the gut microbiome, as earthworms have insufficient enzymatic capacity to fully digest organic matter in their intestinal tract. In doing so, earthworms move large quantities of fresh and partially decomposed organic matter within the soil; this action is mainly carried out by epi-anecic and strict anecic earthworms. Epi-anecic earthworms (such as *Lumbricus rubellus* or *Lumbricus terrestris*) especially consume fresh litter at the soil surface, which is then buried in the soil, while strict anecic earthworms (such as *Aporrectodea giardi*) ingest a mixture of litter and soil organic matter (Hoeffner et al. 2019). The same authors measured surface litter mass loss by earthworm activity in mesocosms from ca. 5 to 23 mg g^{-1} day^{-1}. Transformations of organic matter start as soon as the earthworm ingests the material and continue during intestinal transit, in connection with the gut microbiome mentioned above, but also in fresh excreta that is defecated by earthworms. Fresh casts thus constitute, for a certain time, hotspots of microbial activity in the soil (Bernard et al. 2012).

How earthworms globally influence the fate of carbon in terrestrial ecosystems is a relevant question for scientific research with significant impacts on carbon sequestration and greenhouse gas emissions. Earthworms promote short-term mineralization and medium- and long-term stabilization of carbon (Coq et al. 2007), mainly through the physical protection of this organic matter (Bossuyt et al. 2005), increasing soil carbon storage (Ratsiatosika et al. 2021b; Ratsiatosika and Blanchart 2021). In a recent meta-analysis of studies comparing carbon mineralization in the presence or absence of earthworms, Garnier et al. (2022) noticed that the impact of earthworms on carbon mineralization increased by 20% between 0 and 60 days following the introduction of earthworms, then decreased by 14% after 350 days. Earthworm activity may also result in a decrease in greenhouse gases (GHG) emissions. In a field experiment in the Philippines, John et al. (2020) studied the effect of different earthworm densities and rice straw amendments on GHG emissions. They measured a significant reduction of CH4 emissions by rice straw in the presence of earthworms (*Metaphire* sp., endogeic species), particularly with high earthworm density. Thus, the carbon storage in soils resulting from earthworm activity sequesters carbon in the soil and avoids greenhouse gas emissions, while supporting biological activity by providing the energy needed by organisms and simultaneously improving soil aggregation to better fight against erosion (Lavelle et al. 2020).

2.2 Nutrient Cycling

Earthworm casts contain higher concentrations of available nutrients than uningested soil. Earthworm casts therefore represent a microenvironment that is rich in available nutrients (van Groenigen et al. 2019). This outcome is the result of microbial activity stimulated in the digestive tract and continuing in the casts. Microbially mediated decomposition and mineralization release some of the nutrients contained in the organic matter, making them available to plants and other soil organisms. This was documented for nitrogen and phosphorus, which is positively related to greater plant growth and nutrient uptake in the presence of earthworms (Chapuis-Lardy et al. 2011; Ros et al. 2017), and earthworm casts also contain more available calcium, magnesium, and silicon than bulk soil (Bityutskii et al. 2016). Mobilization of available phosphorus by earthworms is of particular importance, specifically in the P-fixing Ferralsol from Madagascar. A recent greenhouse experiment using labeled-P showed that earthworms (*Pontoscolex corethrurus*) significantly increased rice shoot biomass (+26%) and P nutrition (+65%) (Trap et al. 2021). This increase in plant-available P was mainly explained by the solubilization of native soil P during gut transit. Other mechanisms to explain earthworm-enhanced phosphate solubility have been recently proposed (Vos et al. 2022).

2.3 *Maintenance of Soil Structure*

Soil structure describes the three-dimensional arrangement of solids and pores that allows for gas and solution transfer in the profile, root growth, and water retention. Soil structure is thus a fundamental condition for plant growth. Earthworms modify this structure, mainly by digging into the soil, ingesting soil, and redistributing it in the form of aggregates. In addition to quantifying the biostructures (galleries and aggregates) produced by earthworms, researchers note the link between the production of biostructures and the functioning of the soil: does aggregation or the digging of galleries improve soil ecological functions, including plant growth? Earthworm biostructures, in particular water-stable globular casts, are important in regulating the hydrological properties of the soil (Blanchart et al. 2004; Jouquet et al. 2012). On the other hand, surface casts can increase soil erosion in some situations, as it was shown by Jouquet et al. (2013) for the granular casts of *Metaphire posthuma*, in Vietnam. Earthworms can be grouped as decompacting species and compacting species (Blanchart et al. 1997), who act sequentially in time and space to always maintain a well-structured soil, suitable for plant development in the long term (Barot et al. 2007). If one of the two groups is missing, through loss of biodiversity, loss of soil functions can occur, as shown in Brazilian Amazonia after the conversion of forest to pasture (Chauvel et al. 1999).

2.4 *Regulation of Pests and Diseases*

Earthworm activity regulates the diversity and activity of other soil organisms. All soil organisms are affected by the upheavals created in the soil by earthworms, including free-living (Bernard et al. 2012) and symbiotic (Meng et al. 2022) microorganisms, nematodes (Villenave et al. 2010), mesofauna (Loranger et al. 1998), plant seeds (Clause et al. 2015), and plant roots (Wang et al. 2020). If the diversity of these organisms is modified, so is their activity. These functional interactions between earthworms and other soil biota hold promise for improving plant production and other ecosystem services, but still need more investigation to understand their role, the functioning of ecosystems and agrosystems (Liu et al. 2019). The challenge remains to explain the complexity of the interactions that link the different taxa, the importance of the abiotic environment, and also the self-organization capacity of the soil, which remains poorly understood. Within the framework of these biological interactions, it is above all the regulation of pests and diseases that has been the subject of numerous studies. Earthworms are able to regulate soilborne pathogens and airborne pathogens (Blouin et al. 2005; Wurst 2010; Loranger-Merciris et al. 2012; Blanchart et al. 2020). An interesting mechanism by which earthworms can reinforce plant resistance to disease is via their ability to increase the availability of silicon (Bityutskii et al. 2016), which is known to increase plant defense against pathogens (Coskun et al. 2019).

3 Effects of Agricultural Practices on Earthworm Communities

Agriculture is a process by which humans manage their ecosystems and control the biological cycle of domesticated species, to produce food and other useful resources for their societies. It designates all the know-how and activities relating to the cultivation of soils, and, more generally, all work on the natural environment (not only terrestrial) allowing the cultivation and collection of living beings (plants, animals, fungi, and bacteria) that are useful to humans. Agriculture thus involves changes in land use and management with consequences for the exploited ecosystems and in particular on soil biodiversity.

In theory, all the elements of the cropping system (i.e., the crop rotation and cropping techniques such as the nature of the species grown) can affect the size of earthworm populations and their specific diversity and activity. Many studies have been carried out to assess the effects of different agricultural practices on earthworm populations: e.g., soil tillage (review in Chan 2001; Briones and Schmidt 2017), pesticide use (Kula and Kokta 1992; Pelosi et al. 2014), and fertilizer use (Whalen et al. 1998; Betancur-Corredor et al. 2022). Tillage, the use of pesticides, and compaction negatively affect earthworm communities, whereas fertilization, liming, and irrigation are generally favorable to them. However, earthworm diversity, abundance, and biomass are often lower in cultivated areas than in undisturbed lands (Smeaton et al. 2003). This finding can be explained by three major effects: a lower carbon resource, direct mortality linked to agricultural machinery or the use of chemicals, and environmental modifications linked to soil tillage and compaction.

3.1 Reduction of the Carbon Resource

The carbon resource is less abundant and uniform in agricultural fields compared to natural grasslands and forests due to plant diversity. Crop residues are the major food source for earthworms in agroecosystems. Some agricultural lands receive animal-based organic matter (e.g., manure, biosolids) but this does not occur in most arable cropping systems. In general, the crop residues are only available at certain times of the year (after harvest) and are not suitable (size, palatability) for all ecological categories of earthworms. In most cases, crop residues are buried within the soil profile. This is favorable for soil-dwelling endogeic earthworms but does not provide for the epigeic and anecic earthworms that feed at the soil surface. In some cases, and mainly in direct seeding systems, crop residues are left on the surface, which is a disadvantage to the endogeic species. Only species adapted to these conditions and to the quality and quantity of organic matter available will therefore be able to survive and develop in cultivated environments. Consequently, agricultural environments harbor less earthworm diversity, density, and biomass than non-agricultural systems (Blakemore 2018). Furthermore, the presence of other species and/or individuals of the same species can induce inter- and/or intra-specific competition for the food resource. Earthworm competition changes the length, depth, and size of the galleries (Jegou et al. 2001) as well as the growth of individuals (Lowe and Butt 2002; Eriksen-Hamel and Whalen 2007). Lowe and Butt (1999) showed that *Lumbricus rubellus'* reproductive rate was negatively affected by interspecific competition with *L. terrestris* for surface organic matter.

3.2 *Tillage*

Conventional tillage (involving plowing and harrowing) is responsible for mechanically mixing a deep layer of soil, incorporating and fragmenting plant debris, exposing soil pests to sunshine for biocontrol, disrupting large clods, and leveling the ground (De Medeiros Barbosa 2015). Conventional tillage negatively influences earthworm populations due to mechanical damage, exposure to predators, and desiccation due to soil inversion (Edwards and Bohlen 1996; Chan 2001; Briones and Schmidt 2017). In addition, plowing destroys earthworm galleries (habitat for certain species) and buries organic matter while changing physical conditions such as temperature and soil humidity.

Plowing affects the abundance, biomass, composition, and diversity of earthworm communities (Chan 2001; Kladivko 2001). However, the different categories of earthworms are not affected in the same way and to the same magnitude. Anecic earthworms are the most affected by plowing because they are more susceptible to mechanical damage due to their large size (Briones and Schmidt 2017). In addition, the vertical galleries in which they live are destroyed by the passage of the plow. Finally, after plowing, these earthworms no longer have access to their preferred food resource because it is buried within the profile. Epigeic species that live at the soil surface are rarely found in conventional agricultural systems, perhaps because they are disadvantaged by the removal of their food resource and habitat by plowing. Endogeic earthworms are less affected by plowing and may even be favored by the burial of fragmented crop residues in the soil profile, which eases their access to the resource (Wyss and Glasstetter 1992; Nuutinen 1992; Chan 2001; Pelosi et al. 2009). Despite this, most authors emphasize the negative impact of plowing on earthworm communities. Evans and Guild (1948) found no decrease in earthworm populations in the first 6 months after plowing but described a 70% reduction in earthworm numbers and 80% decline in earthworm biomass after 5 years.

Shallow or minimum tillage has less impact on soil structure than plowing because it does not invert the soil mass and is less deep (i.e., generally <10 cm while plowing is 20–30 cm deep). However, it can still reduce earthworm populations, particularly through the destruction of habitats. Direct seeding, on the other hand, promotes the development of earthworms, and in particular the return and increase in anecic species. After 5 years of changing tillage practices, the earthworm biomass was 180 kg ha^{-1} for plowing, 500 kg ha^{-1} for shallow tillage, and 1500 kg ha^{-1} for direct seeding (Tebrügge and Düring 1999).

3.3 Pesticides

Worldwide, the diversity and quantity per hectare of synthetic pesticides used are increasing, along with an increase in the area of treated surfaces (Bernhardt et al. 2017; DiBartolomeis et al. 2019). Global pesticide use (in tons of active ingredients) increased by 80% worldwide between 1990 (2,285,881 tons) and 2017 (4,113,591 tons) (FAOSTAT 2019; Zhang et al. 2011), and the total sales of pesticides remained constant in Europe between 2011 and 2018, revealing that there was no reduction in reliance on pesticides during this period (Eurostat 2020; FAOSTAT 2019). In 2021, 355,175 tons of pesticides were sold in the EU, which represents a moderate increase of 2.7% compared with 2020 (346,000 tons), and a further rebound back toward the medium-term annual average between 2011 and 2021. More than 450 active ingredients are currently used in several thousand commercial pesticides authorized for use in the European Union. Despite the precautions taken by farmers to limit pesticide contact with nontarget organisms, their application causes contamination of the waterways and soil. The presence of active ingredients and byproducts is detected in the soils of cultivated fields directly treated with pesticides, sometimes at high concentrations (Chiaia-Hernandez et al. 2017; Karasali et al. 2016; Silva et al. 2019; Zhao et al. 2018). A study in central France revealed the presence of at least one pesticide ingredient in 100% of the 180 sampled soils and in 92% of the 155 earthworms (*Allolobophora chlorotica*) collected from both pesticide-treated crops and untreated habitats (hedgerows, permanent grasslands, and cereals under organic farming) (Pelosi et al. 2021). Mixtures of at least one insecticide, one herbicide, and one fungicide, at concentrations above the limit of quantification, contaminated 90% of the soils and 54% of the earthworms at levels that were sometimes dangerous. About 46% of samples were assessed at a high risk of chronic toxicity to earthworms, both in treated winter cereals and nontreated habitats considered as refugees.

The pesticides generally used in field crops are herbicides, fungicides, insecticides, nematicides, and molluscicides. Their impact on earthworm populations depends on the applied dose, the type of application (spreading of granules, spraying, etc.), the period of application, the active ingredient they contain, the frequency, and the application intensity but also on climatic conditions and earthworm behavior and habitat (Edwards and Bohlen 1996; Maggi and Tang 2021). However, Pelosi et al. (2014) highlighted in a review that earthworms were impacted by pesticides at all levels of biological organization, inducing modifications in enzymatic activities, individual mortality, decrease in fecundity and growth, or change in individual behavior.

3.4 Soil Compaction

Soil compaction in arable farming systems is a consequence of the passage of agricultural machinery. It corresponds to a decrease in air volume and porosity in the soil, which reduces the water- and air-filled pore space and impedes plant root growth. This has negative consequences on yields (Hargreaves et al. 2019) and increases the release of greenhouse gases (Pulido-Moncada et al. 2022) while also impacting soil organisms. Mechanization of agriculture, particularly due to the increase in size and weight of agricultural machinery, increases the risk of soil compaction. These risks depend on the bearing capacity of a soil which is its ability to support wheel loads. A soil is more or less load-bearing depending on its texture, structure, and moisture content.

Soil compaction can kill earthworms by crushing and destroying their habitat, thus reducing the size and diversity of earthworm populations (Langmaack et al. 1999). The horizontal burrows of endogeic earthworms that are close to the soil surface are more sensitive to the effects of compaction than the vertical or oblique and deeper burrows of anecic species (Jegou et al. 2002). However, Beylich et al. (2010) concluded from a literature review that no general effect of soil compaction was found, due to the high variability of experimental situations and conditions. Compaction may impact more the biological processes (e.g., burrowing) than the species abundance (Buck et al. 2000; Beylich et al. 2010; Ampoorter et al. 2011).

To rehabilitate compacted soils, farmers are tempted to increase soil tillage frequency and/or depth. This short-term solution can move the compaction to a deeper soil layer (i.e., a plow pan) and is thus not sustainable. In conservation agriculture systems that rely on shallow tillage or no-tillage, soil decompaction occurs gradually through natural processes such as alternating freeze/thaw, wetting/drying, and biological activity (roots and earthworms). It is important to consider other agronomic levers that can be used to promote earthworm communities and improve soil biological functions.

4 Agronomic Levers to Promote Earthworm Communities and Improve Biological Soil Functioning

Postwar modern agriculture was built on a productivist model in which the increase in yields was the main objective. An important development of mechanization took place, with increasingly heavy machines and the use of chemical inputs (pesticides and mineral fertilizers) often derived from the petroleum industry. During this period, soils were seen only as inert supports for production, neglecting most of the biological processes that took place in this complex environmental compartment. Since the end of the last century, these productivist conventional systems have been seriously questioned due to their negative externalities such as depletion in organic matter, compaction, contamination, and loss of biodiversity. A more sustainable agriculture is needed and new agricultural practices and cropping systems have been proposed in order to decrease the dependency on external inputs and to increase the system sustainability.

4.1 Conservation Tillage

Conservation tillage is any tillage practice (e.g., minimum tillage, no-till, and direct-seeding) that builds up crop residues on the soil surface to minimize the impact of water and wind erosion (Sullivan 2003). Conservation tillage generally increases earthworm abundance and biomass and modifies species diversity and composition (Holland 2004). Actually, no-tillage and minimum tillage that increase the quantity of organic matter on the soil surface are favorable to earthworms, which can live in an undisturbed biotope with abundant surface residues (Edwards and Bohlen 1996; Chan 2001; Henneron et al. 2015). Cropping systems without plowing are known to rapidly promote earthworm communities, especially the larger individuals (Chan 2001; De Oliveira et al. 2012). In the review of Chan (2001), different studies are reported showing that between 3 and 5 years are needed for the recovery of earthworm populations after mouldboard ploughing to the level observed under direct drilling.

4.2 Reduce Pesticide Use

Pesticide effects on earthworms are mostly studied under laboratory conditions (Pelosi et al. 2014). Field ecotoxicological effects of pesticides have been mostly assessed by comparing earthworm communities in organic and conventional systems. These studies often show variable results (Hole et al. 2005; Nuutinen and Haukka 1990), mainly due to (1) confounding factors such as differences in tillage, organic inputs, or crop types (Hole et al. 2005) and to (2) variable differentiation time (i.e., the time between stopping pesticides and collecting earthworms) between cropped fields. To overcome the first issue (confounding factors), Pelosi et al. (2013) used a 5-year experimental database including conventional and organic cropping systems with practices other than pesticide use as similar as possible (conventional plowing, same amount of organic matter inputs, same crops, i.e., winter wheat). The authors established simple relationships between the Treatment Frequency Index—a phytosanitary indicator of pesticide pressure—and the abundance of three important earthworm species variably in contact with the soil surface and belonging to three different ecological categories. Insecticides had greater potential toxicity to earthworm species than herbicides and fungicides. These chemicals should thus be managed as a priority to prevent harm to earthworm populations. Moreover, species living in close contact with the soil surface were the most affected by pesticides (the epigeic *Lumbricus castaneus* > the epi-anecic *L. terrestris* > the surface endogeic

A. chlorotica) (Fig. 2). The authors highlighted that the abundance of these earthworm species could drastically increase if pesticides were stopped. They could be multiplied by a factor 1.5–4 if the Treatment Frequency Index was halved, as it is required by some European regulations.

To overcome the second mentioned issue (i.e., differentiation time in cropped fields), ecotoxicological effects of pesticides on earthworms can be assessed in medium or long-term trials or farmers' fields. In the literature, the effects on earthworm communities are often assessed from a single sampling date, in trials whose duration is not always specified or which were set up only a few years before the sampling (e.g., Nuutinen and Haukka 1990). Few studies have been done on medium (8–10 years) and long-term (more than 15 years) trials, with sampling intervals of several years. In a trial initiated in 1997 near Paris (France), species composition, abundance, and biomass of earthworms were measured at two periods in fall (six times in total): once a year between 2005 and 2007 (period 1) and once a year between 2011 and 2013 (period 2) (Pelosi et al. 2015). A conventional and an organic cropping system were compared. In period 1 (after 8–10 years of differentiation), earthworm abundance did not differ between the two cropping systems, being even higher in the conventional system at some sampling times. Mean total abundance of earthworms was 122 and 149 individuals m^{-2} in period 1 and 408 and 216 individuals m^{-2} in period 2 in the organic and conventional systems, respectively. While earthworm abundance and biomass increased slightly in the conventional system between the two periods, they were more than tripled in the organic system. This was mainly due to the species *Aporrectodea caliginosa, L. castaneus, L. terrestris,* and *Aporrectodea longa* that greatly increased between the two periods in the organic system. After around 15 years of system differentiation, organic farming system contained almost two times more earthworms than the conventional

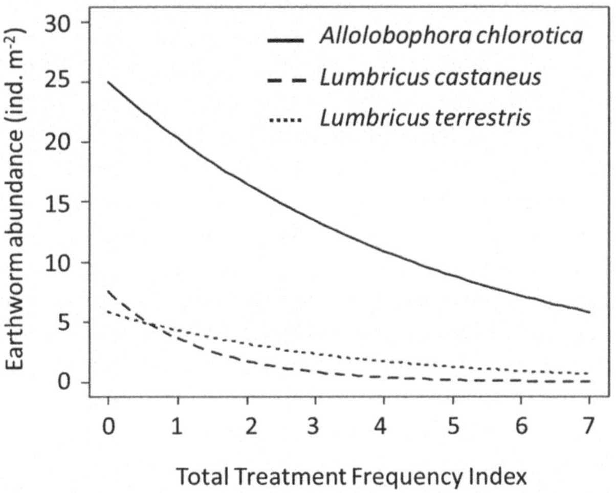

Fig. 2 Poisson regressions of densities of three earthworm species, i.e., *L. castaneus*, *L. terrestris*, and *A. chlorotica* as a function of total Treatment Frequency Index (adapted from Pelosi et al. 2013)

system. The time needed for the increase in earthworm communities (abundance, biomass, and diversity) in the organic system may be due to the remanence (including metabolites) of harmful pesticides applied before the set-up of the trial, and to a slow progressive recolonization (external recovery, EFSA Scientific Committee 2016) of the organic fields by earthworms.

Finally, in field trials, another difficulty arises for the assessment of the effects based on the application date of pesticides because they are subject to spatially heterogeneous degradation through light, chemical and biological activity. Consequently, an accurate assessment of the exposure would require direct measurement of the actual dose of active ingredients that the earthworms are exposed to in the soil-litter-plant system.

4.3 Mineral and Organic Fertilization

Mineral fertilization increases the quantity of biomass produced and leaves more unharvested residue in the agroecosystem, which supports larger earthworm populations (Lofs-Holmin 1983). However, a decrease in earthworm numbers and biomass is observed when ammonium-based nitrogen fertilizers cause soil acidification (e.g., Hansen and Engelstad 1999). Liming (i.e., treating soil with lime to reduce acidity and improve fertility or oxygen levels) is generally beneficial to earthworms, which prefer neutral soil pH conditions (McCallum et al. 2016). Moreover, the spreading of decomposed organic matter on cultivated soils can avoid the negative impacts of soil acidification that are associated with the use of certain mineral fertilizers in the absence of lime applications. According to Cluzeau et al. (1994), adding urban compost over 12 years to an experimental vineyard of the 'Comité Interprofessionnel des Vins de Champagne' resulted in a 30% increase in earthworm biomass, particularly of the anecic species, mainly due to the organic matter input, i.e., more food for earthworms. Moreover, Anderson et al. (1983) showed that adding slurry or manure rapidly increased earthworm populations. However, some ammonia-rich slurries have harmful effects on earthworms (Curry 1976). In general, applications of organic materials were found to have beneficial effects on earthworm communities and their activities, therefore promoting related properties and functions such as soil porosity and water retention (Riley et al. 2008). For instance, digestates obtained after anaerobic digestion of organic wastes to produce biogas are considered a valuable organic fertilizer and soil amendment (Bachmann et al. 2014; Coelho et al. 2020). Earthworm abundance was 150% higher in the fields treated for 2 years with digestates or cattle effluents than the field treated with chemical fertilizers (Moinard et al. 2021). However, microcosm experiments showed that digestate can have a negative impact on earthworm behavior in the short term, either through avoidance (a strategy that earthworms use to limit their exposure to potentially harmful substances) or reduced surface foraging (that triggers mortality).

4.4 Crop Rotations

Cover crops are an essential element of sustainable agriculture and can benefit earthworm populations by adding more organic residues and by keeping the soil covered, thus moderating the temperature and moisture conditions at the soil surface (Clapperton et al. 2007; Korucu et al. 2018; Scavo et al. 2022). Although it is not fully understood how C:N ratio and cover crops affect earthworm populations (Euteneuer et al. 2020), monocropping and rotations with cereals are associated with a long-term decline in earthworm populations (Schmidt et al. 2003; Bogužas et al. 2022). Torppa and Taylor (2022) found that earthworm abundance and activities like bioturbation can be increased by a diversified crop rotation. Legume cover crops have been described as a good way to restore and/or to conserve soil fertility by controlling weeds and erosion and by enhancing carbon and nitrogen stocks in soil (Blanchart et al. 2006; Rodríguez et al. 2020). Because legume residues are an easily palatable resource with high nitrogen content, crop rotations that include legumes (e.g., pea, soybean, white clover) were found to increase the abundance and biomass of earthworm communities (Hubbard et al. 1999; Rodríguez et al. 2020; Schmidt et al. 2003).

4.5 Irrigation

Earthworms contain 65–90% water by mass, depending on their state of hydration (Lee 1985). They rely on cutaneous respiration and must stay moist to breathe. Earthworm strategies to survive dry conditions include aestivation, diapause, or migration to deeper soil layers. Irrigation creates a favorable environment for their activity, increasing earthworm populations during drought (Curry 1998) and allowing certain species to remain active in otherwise unfavorable soil moisture conditions (Edwards 1998). However, the effects of irrigation on earthworm communities depend on agricultural practices and can also be unfavorable, linked for instance to soil compaction due to the passage of agricultural machinery on wet soils in autumn (Radics et al. 2022). In the current context of climate change, which is projected to have significant impacts on agricultural production, irrigation requirements are increasing, which can influence earthworm presence and activity.

4.6 Alternative Cropping Systems

Alternative cropping systems such as conservation, organic, biodynamic, or regenerative agriculture include some of the principles and practices mentioned above and have been proposed to go toward a more sustainable agriculture (Médiène et al. 2011; Pelosi et al. 2015). They are all considered as "agroecological" systems, and agroecology is considered as the ultimate and most comprehensive solution to the many challenges of the agricultural and food system (Ewert et al. 2023). When switching from one system to another, earthworm communities will change over time. However, once earthworm populations are affected, their recovery in the field can be long (Kattwinkel et al. 2015) as it depends on the species reproductive and recolonization traits (e.g., life cycle, number of offsprings, dispersal capacity, and distribution) (Liess and von der Ohe 2005) for internal recovery, and on the field surroundings for external recovery (EFSA Scientific Committee 2016). To accelerate this process and to ensure that earthworm populations will help the system to recover, earthworms can be inoculated in soils. These key soil organisms, highly recognized and used for decades as bioindicators of soil quality (Paoletti 1999), can also be used as actors in soils to improve agrosystem sustainability and to support agricultural transition.

5 Use of Earthworms to Improve Agrosystem Sustainability

Earthworms are widely recognized as an essential organism in agricultural systems in the temperate (Scheu 2003) and tropical regions (Lavelle et al. 1998; Senapati et al. 1999). Their beneficial effects on plant growth and productivity are well documented in the literature. Besides the plant production, earthworms also have positive effects on other agroecosystem services through their impact on soil ecological processes (Blanchart et al. 2020; Trap et al. 2021; Ratsiatosika et al. 2021b). Earthworm manipulation is suggested as a key measure to successfully restore terrestrial ecosystems (Jouquet et al. 2014; Wubs et al. 2016; Ratsiatosika and Blanchart 2021). Earthworm inoculation was first experienced in the early 1950s by a New Zealander farmer who found the abundance of earthworms was related to the fertility of his soil (Nielsen 1951). Since then, earthworm introduction was studied for its impact on soil sustainability (Fig. 3), i.e., to maintain soil functional integrity (Ludwig et al. 2018). Several techniques are proposed to inoculate earthworms and restore degraded soils (Butt et al. 1995; Piearce et al. 2003). Inoculation of earthworms, called "in-soil" technology, appears to have better effects on ecosystem functioning than "off-soil" technology that uses earthworms to provide compost (vermicomposting) because a large amount of carbon is lost that might have been used to sustain mechanical activities of earthworms and other invertebrates in the soil (Senapati et al. 1999). Inoculation can be done by adding cocoons, juveniles, adults, or applying microenvironments containing earthworms (sod transplantation method) (Butt et al. 1997). Before inoculation, earthworms can be bought in a commercial unit, collected from neighboring areas, or produced in large quantities in culture beds (Senapati et al. 1999). In the following paragraphs, we will analyze how earthworms can affect the provision of ecosystem services, not only plant growth but also carbon sequestration, and resistance to erosion under temperate and tropical climates.

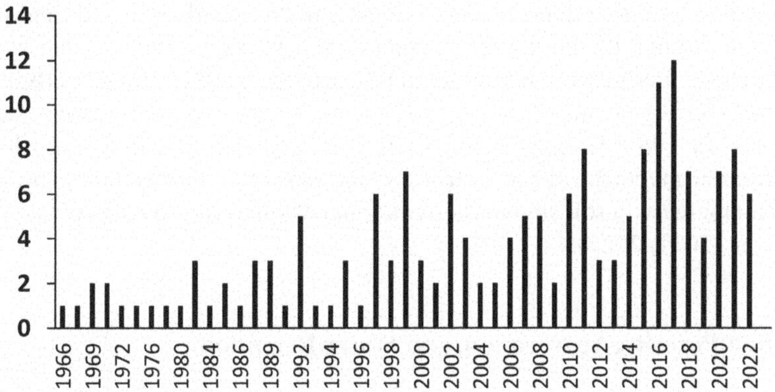

Fig. 3 Numbers of references dealing with earthworm inoculation, based on data collected from the ISI Web of Knowledge from January 1996 to December 2022. Bibliometric results were obtained using "All Databases," and querying the "Title" with these keywords: "(earthworm* or lumbricid* or aporrectodea or lumbricus or allolobophora) and (inoculat* or introduc* or restor* or bioremediat* or biostimulat* or biofertilizat* or revitalizat*)"

5.1 Earthworm Inoculation in Temperate Climates

Earthworm inoculations under temperate climate generally revealed positive impacts on different soil characteristics. With an objective of soil rehabilitation/restoration, most of these studies were performed in degraded and/or polluted soils such as in mines (Marashi and Scullion 2004; Scullion and Malik 2000), landfill sites (Butt et al. 1999; Forey et al. 2018), urban roofs (Jusselme et al. 2019) or constructed Technosols (Pey 2010), and a few were performed in agricultural systems. Among them, some "only" assessed earthworm populations or communities after inoculation initiatives (e.g., Frazao et al. 2019; Nuutinen et al. 2006; van der Werff et al. 1998). We here give examples of the effects of earthworm inoculation on different soil functions in agricultural systems.

5.1.1 Soil Fertility (Nutrients) and Plant Growth

In agroecosystems, the presence of earthworms led to a 25% increase in crop yield and 23% increase in aboveground biomass (van Groenigen et al. 2014). The magnitude of these effects depends on crop residues, soil texture, earthworm density, and type and rate of fertilization. Since then, no studies have been published that evaluate the effects of earthworm inoculation on soil nutrients and plant growth in agricultural systems under temperate climate. Some can be found in tropical areas and will be presented in the next Sect. 5.2 (e.g., Ratsiatosika et al. 2021a).

The positive effect of earthworm inoculation on plant growth occurs predominantly through increasing available nitrogen and other nutrients for plants (Bityutskii et al. 2016; Ros et al. 2017; van Groenigen et al. 2014). Indeed, earthworms are highly involved in the fragmentation of organic matter residues in soils and in their degradation through the promotion of microbial activity. By moving into and onto the soil and by mixing mineral and organic particles, earthworms directly interact with other soil organisms belonging to macro-, meso-, and microfauna, but also with roots (Lee 1985; Jacquiod et al. 2020). Dual inoculation with both earthworms and arbuscular mycorrhizal fungi strongly magnifies the response on plant growth through an increase in soil enzyme activities and changes in soil nutrient availability (Meng et al. 2021, 2022).

5.1.2 Soil Porosity, Aggregation, and Water Retention

The effects of earthworm inoculation on soil porosity were assessed by Ligthart and Peek (1997) in an earthworm-free pasture in the Netherlands. More than 7 years after the inoculation of a mix of four species (the epigeic *L. rubellus*, the endogeic *A. caliginosa*, and the anecics *A. longa* and *L. terrestris*), burrow systems were mapped and recorded at 46 m of the inoculation front, at a maximum depth of 0.9 m. They found *A. longa* to be the fastest colonizer, while *L. terrestris* dispersed slowly. The abundance of earthworms, along with the burrow systems, was the highest 4 years after the inoculation, and decreased then. The study of earthworm burrow system not only allows assessing the impact of inoculation on porosity, but also it is a nondestructive technique to follow the dispersion of inoculated individuals. Brais (1989) showed on a clay loam soil of southwest Quebec that inoculation of *L. terrestris* individuals increased the macroporosity in untilled plots but had no effect in tilled soils (in which the porosity is mainly driven by soil tillage). Blackwell and Blackwell (1989) showed no difference in the macroporosity between earthworm inoculated and control plots in southeast Australia, despite an increase in air permeability in the inoculated areas. These results should be interpreted in regards to the success of earthworm inoculation (e.g., by assessing the abundance or biomass of inoculated individuals) since inoculation initiatives can fail under unfavorable soil conditions for earthworm establishment.

Regarding aggregation, Van Rhee (1969) introduced *A. caliginosa* and *L. terrestris* in orchards in new polder soils in Holland. They found that the casts produced by earthworms tended to increase the aggregate stability of the top 15 cm of soil and the water available for plant growth. Similarly, after an inoculation of 70 individuals m^{-2} (species mix), Scullion and Malik (2000) showed a clear improvement in aggregate stability of upper layer aggregates in the >2 and >63 μm size range due to earthworms.

Earthworm inoculation initiatives have also been used in forest management, especially for soil structure restoration. Ampoorter et al. (2011) studied the effect of *L. terrestris* inoculation on soil compaction caused by wheel tracks at four study sites in Belgium. A small decrease in bulk density was observed due to earthworms on the upper layer beside the tracks but no effect inside tracks. The authors suggested that the level of compaction was too high for this species, reducing the activity of the inoculated individuals (similar abundance of *L. terrestris* inside and beside tracks). Besides, Hamilton and Dindal (1989) highlighted a positive correlation between the density of inoculated *L. terrestris* and the percentage of 4 mm water-stable soil aggregates at four sites in a mixed hardwood forest and in an herbaceous old field. However, the 1 mm water-stable aggregates tended to decrease in the presence of *L. terrestris*, highlighting a redistribution of the pore and aggregate systems in soils under earthworm influence (van Rhee 1969).

5.1.3 Greenhouse Gas Storage and Emission

In a forest soil, in Germany, Borken et al. (2000) inoculated *L. terrestris* in PVC columns (two individuals per column, equivalent to 113 individuals m^{-2}). Despite an increase in CO_2 emission rates in the first 3–4 weeks after earthworm inoculation, there was no effect on a 120-day period. However, *L. terrestris* reduced the cumulative CH_4 oxidation rate by 53% (thus potentially increasing CH_4 emissions) and increased the cumulative N_2O emissions by 57% compared to the columns without earthworms. In 2013, Lubbers et al. concluded that earthworms increased net soil greenhouse gas emissions. Since then, studies have been published confirming that earthworm activity can increase the intensity of N_2O emission (Maslov et al. 2022), but other publications highlighted that the time is of great importance for concluding about positive or negative effects of earthworms on greenhouse gas emissions. For instance, despite a short-term (3 months) increase in mean N_2O fluxes in the presence of artificially elevated earthworm populations, Ejack et al. (2021) observed that this effect disappeared at longer terms (>1 year).

5.2 *Earthworm Inoculation in Tropical Climates*

The management of earthworms in tropical agroecosystems was reviewed by Lavelle et al. (1999). The effects of earthworms on plant production, and sometimes on other ecosystem services have been studied at different levels, spatially and temporally: from short-term experiments (few days to several months) in micro- and mesocosms, to medium-term experiments in field plots with or without enclosures, and to long-term experiments after earthworm rearing and inoculation (Brown et al. 1999). This last approach, to our knowledge, has only been implemented on two occasions: in a tea plantation in India (Giri 1995) and in a corn crop in Peru (Pashanasi et al. 1996). In a review gathering 246 data points taken from 28 experiments in both greenhouse and field scales performed in eight tropical countries, Brown et al. (1999) revealed the beneficial effect of earthworms on plant production in the tropics. Indeed, a positive effect of earthworms on the plant production was recorded in 75% of cases, with a mean increase of 63% (Fig. 4). Particularly, the shoot biomass was enhanced, on average, by 57% in the presence of earthworms; this positive effect was observed in 75% of the experiments. For the grain yields, an increase of 36% was noted in 72% of cases. With regard to the root biomass, a mean

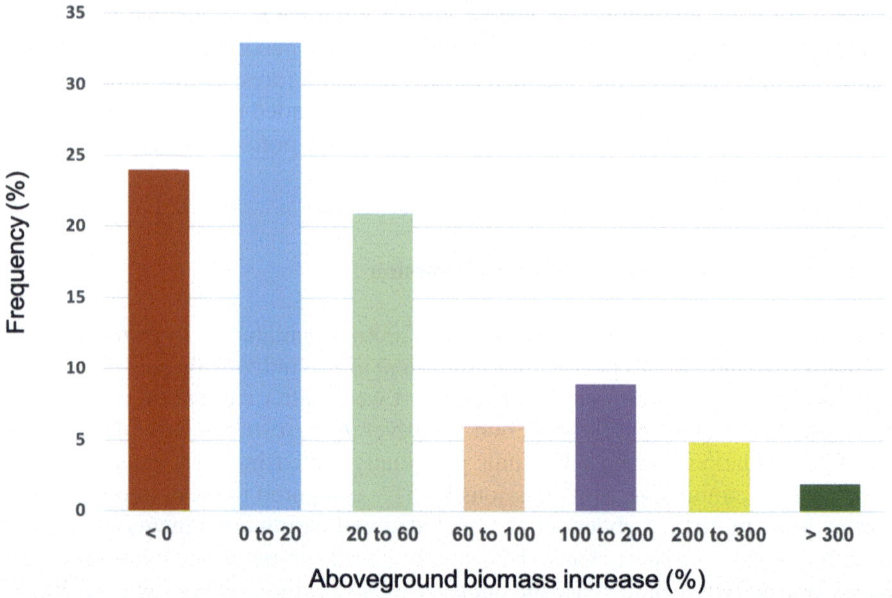

Fig. 4 Frequency that earthworm presence had a negative effect (<0%) or increased (from 0 to > 300%) the above-ground (shoot) plant biomass (n = 246 data points). Adapted from Brown et al. (1999, CABI Publishing)

increase of 66% was induced by earthworm presence in 59% of the experiments. Specifically for the tea plantation in India, the restoration of soil fertility by trenching of tea prunings, organic materials, and earthworm inoculation between tea rows strongly increased the tea yields (+239%) and profitability (US$ 5500 ha^{-1}) (Senapati et al. 1999) compared to conventional treatment. More largely, the global effect of earthworms on plant production was more recently reviewed by van Groenigen et al. (2014).

Here, we present some case studies on the effect of earthworm management and inoculation on agrosystem services in rice cropping systems in the tropics. Rice is among the most studied crops in terms of the assessment of the effect of earthworm inoculation in agrosystems (van Groenigen et al. 2014). Different study scales were considered to better understand the processes through which earthworms affect agrosystem services: (1) microcosm and mesocosm scales and (2) field scale. It is important to keep in mind that earthworms have significant effects on plant growth from minimal density or biomass (Spain et al. 1992).

5.2.1 Microcosm and Mesocosm Experiments

Many laboratory experiments have compared the presence and absence of earthworms on plant growth and nutrition, in order to understand their ability to provide healthy conditions for rice growth (Noguera et al. 2011; Ratsiatosika et al. 2019; Blanchart et al. 2020; Trap et al. 2021). Ferralsols that are predominant in the Highlands of Madagascar are characterized by a multiple nutrient limitation (P, Ca, N, and Mg) and a low organic matter content (Raminoarison et al. 2020). Given the important role of earthworms in increasing the soil nutrient availability, especially phosphorus in Ferralsols (Trap et al. 2021), the inoculation of earthworms seems a relevant strategy to enhance agrosystem services in order to meet the context of sustainable agriculture. Two mesocosm experiments gathering 16 data points (comparisons between treatments with and without earthworms) were conducted in the Laboratoire des Radio-Isotopes (University of Antananarivo, Madagascar) to test the effect of inoculation of earthworms (*P. corethrurus*, Rhinodrilidae) on rice growth and nutrition. In the first experiment, the response of six rice cultivars mostly used by Malagasy farmers to the presence of earthworms was studied. The objective of the second experiment was to assess the effect of earthworms on rice growth and nutrition under ten fertilization treatments. The duration of each experiment was 8 weeks. At the end of the experiments, results showed that rice positively responded to the presence of earthworms compared to situations without earthworms, irrespective of the experiment. Among the growth studied traits (i.e., shoot biomass, root biomass, plant biomass, and shoot:root ratio), the positive effect of earthworms was mainly observed on the shoot biomass, with a mean increase of 78% ± 30%. However, it is important to emphasize that the magnitude of the positive effect of earthworms varied among rice cultivars as earthworms increased the shoot biomass from 40 to 130% according to cultivars (Ratsiatosika et al. 2021a). Likewise, the effect of earthworms on the rice nutrition was highly variable, especially for nitrogen (from 116 to 355%) and phosphorus (from 48 to 147%). Previous mesocosm experiments conducted by Noguera et al. (2011) in Columbia already showed the variability of the effect of earthworms according to rice cultivars. Another point to emphasize is that earthworms enhance the effect of organic amendments in improving plant growth and nutrition. Nevertheless, our study revealed that the magnitude of increase induced by earthworm presence was highly variable according to the types of organic amendments applied. For the shoot biomass, the increase ranged from 37 to 117%, and for the shoot phosphorus amount, it varied from 33 to 122%.

5.2.2 Field Experiment

Despite the widely recognized potential of earthworms under controlled conditions, only a few studies report the effect of earthworm inoculation in field experiments. Here, we present some field case studies showing the effect of earthworm inoculation on agrosystem services in tropical areas.

- Plant growth and production
- In the highlands of Madagascar, the effect of earthworm inoculation was tested on agrosystem services in upland rice cropping systems during four cropping seasons (Ratsiatosika et al. 2021b). The endogeic species *P. corethrurus* were inoculated each year at a density of 75 individuals m^{-2}. After 3 years of earthworm inoculation, results showed that earthworms increased the rice height at the end of the tillering and at the flowering stages. Also, an increase of 26% in the rice shoot biomass was induced by earthworm inoculation. In terms of production, the grain yield was enhanced by 44.6% in treatments where earthworms were inoculated. Similarly, the earthworm inoculation enhanced rice nutrition; grain N and P amounts were increased by 43% and 28%, respectively.
- Mitigation of greenhouse gas emission
- In the Philippines, farmers shift from traditionally flooded double rice cropping systems to non-flooded crops such as upland rice due to increasing water scarcity and rapid socio-economic development. However, this conversion may have a negative impact not only on the soil organic matter contents because of soil nutrient mining but also on the greenhouse gas balance (Haefele et al. 2013). In this situation, farmers incorporate rice straw in rice agroecosystems to offset nutrient mining. Nevertheless, residue restitution is known to increase CO_2 emissions and CH_4 release. As earthworms have a pivotal role in organic matter decomposition, the presence of earthworms may increase the benefits of rice straw restitution. Recently, John et al. (2020) carried out an in situ mesocosm experiment to assess both the potential and the risks associated with earthworm effects on GHG emissions and crop productivity. The authors showed that the inoculation of earthworms contributes to the mitigation of GHG emissions, especially CH_4. This study underlines that the manipulation of earthworms appears as a promising way to substitute the crop residues burning in the field and to reduce the straw-induced CH_4 emissions.
- C sequestration and soil aggregation
- Besides plant growth and nutrition, earthworm inoculation of earthworms had also a beneficial effect on C sequestration and soil aggregation. In the highlands of Madagascar, earthworm inoculation resulted in a slight increase in C stock equal to 0.61 t C ha^{-1} $year^{-1}$ in the upper 10 cm of soil after 3 (Ratsiatosika et al. 2021b; Ratsiatosika and Blanchart 2021). Likewise, the proportion of soil macroaggregates to the total soil was 11% higher in treatments where earthworms were inoculated than without earthworm inoculation. In Lamto savannas in Côte d'Ivoire, earthworm inoculation induced an additional C storage of 0.3 t C ha^{-1} $year^{-1}$ after 3 years in the upper 10 cm of soil (Gilot 1997). In Peru, Duboisset (1995) and Hallaire et al. (2000) showed through the analysis of soil thin sections, that the introduction of *P. corethrurus* in the soil induced a strong modification of soil structure in the 0–2 cm depth: compact aggregates became dominant, at the expense of soil porosity and especially of macroporosity. This superficial layer became very compact and impermeable due to the coalescence of casts.

6 Conclusions and Recommendations

We saw in this chapter that earthworms can be used as bioindicators and also as actors in soils for improving soil's status and fertility. The activities they carry out in the soil have a strong impact on the provision by soil of ecosystem services through the main ecological functions, which are the carbon transformations, the recycling of nutrients, the maintenance of the soil structure, and the control of pathogens. It therefore appears fundamental to implement cultural practices that promote their presence, diversity, and activities. This can be done thanks to numerous agronomic levers that have shown their capacity to promote earthworm communities in cultivated soils, i.e., conservation tillage, reduction of pesticide use, organic fertilization, rotations, etc. Another alternative consists of inoculating earthworms as has been done in different agrosystems around the world, with often positive consequences for soil, plant, and agrosystem health.

An important and last point to consider is to involve stakeholders in this knowledge and in the development of innovative, more agroecological, agricultural practices allowing the better use of this potential offered by earthworms and other soil organisms. A co-design practice involving both agricultural and scientific knowledge is of great importance to ensure agricultural sustainability. Another point to emphasize is the sensitization of decision-makers to make them recognize the beneficial role of earthworms in agriculture and to encourage them to facilitate the dissemination of earthworm inoculation to farmers.

References

Ampoorter E, De Schrijver A, De Frenne P, Hermy M, Verheyen K (2011) Experimental assessment of ecological restoration options for compacted forest soils. Ecol Eng 37:1734–1746

Anderson JM, Ineson P, Huish SA (1983) Nitrogen and cation mobilization by soil fauna feeding on leaf litter and soil organic matter from deciduous woodlands. Soil Biol Biochem 15:463–467

Bachmann S, Gropp M, Eichler-Lobermann B (2014) Phosphorus availability and soil microbial activity in a 3-year field experiment amended with digested dairy slurry. Biomass Bioenergy 70:429–439. https://doi.org/10.1016/j.biombioe.2014.08.004

Barot S, Rossi J-P, Lavelle P (2007) Self-organization in a simple consumer–resource system, the example of earthworms. Soil Biol Biochem 39:2230–2240. https://doi.org/10.1016/j.soilbio.2007.03.021

Bernard L, Chapuis-Lardy L, Razafimbelo T, Razafindrakoto M, Pablo AL, Legname E, Poulain J, Brüls T, O'Donohue M, Brauman A, Chotte JL, Blanchart E (2012) Endogeic earthworms shape bacterial functional communities and affect organic matter mineralization in a tropical soil. ISME J 6:213–222

Bernhardt ES, Rosi EJ, Gessner MO (2017) Synthetic chemicals as agents of global change. Front Ecol Environ 15:84–90. https://doi.org/10.1002/fee.1450

Bertrand M, Barot S, Blouin M, Whalen J, de Oliveira T, Roger-Estrade J (2015) Earthworm services for cropping systems. A review. Agron Sustain Dev. https://doi.org/10.1007/s13593-014-0269-7

Betancur-Corredor B, Lang B, Russell DJ (2022) Organic nitrogen fertilization benefits selected soil fauna in global agroecosystems. Biol Fertil Soils 59:1–16

Beylich A, Oberholzer H-R, Schrader S, Höper H, Wilke B-M (2010) Evaluation of soil compaction effects on soil biota and soil biological processes in soils. Soil Tillage Res 109:133–143

Bityutskii N, Kaidun P, Yakkonen K (2016) Earthworms can increase mobility and bioavailability of silicon in soil. Soil Biol Biochem 99(2016):47–53. https://doi.org/10.1016/j.soilbio.2016.04.022

Blackwell PS, Blackwell J (1989) The introduction of earthworms to an ameliorated, irrigated duplex soil in south-eastern Australia and the influence on macropores. Aust J Soil Res 27:807–814

Blakemore RJ (2018) Critical decline of earthworms from organic origins under intensive, humic SOM-depleting agriculture. Soil Syst 2(2):33

Blanchart E, Lavelle P, Braudeau E, Le Bissonnais Y, Valentin C (1997) Regulation of soil structure by geophagous earthworm activities in humid savannas of Côte d'Ivoire. Soil Biol Biochem 29:431–439

Blanchart E, Albrecht A, Brown GG, Decaëns T, Duboisset A, Lavelle P, Mariani L, Roose E (2004) Effects of tropical endogeic earthworms on soil erosion: a review. Agric Ecosyst Environ 104:303–315

Blanchart E, Villenave C, Viallatoux A, Barthes B, Girardin C, Azontonde A, Feller C (2006) Long-term effect of a legume cover crop (Mucuna pruriens var. utilis) on the communities of soil macrofauna and nematofauna under maize cultivation, in southern Benin. Eur J Soil Biol 42:S136–S144

Blanchart E, Ratsiatosika O, Raveloson H, Razafimbelo T, Razafindrakoto M, Sester M, Becquer T, Bernard L, Trap J (2020) Nitrogen supply reduces the earthworm-silicon control on rice blast disease in a Ferralsol. Appl Soil Ecol 145

Blouin M, Zuily-Fodil Y, Pham-Thi A-T, Laffray D, Reversat G, Pando A, Tondoh J, Lavelle P (2005) Belowground organism activities affect plant aboveground phenotype, inducing plant tolerance to parasites. Ecol Lett 8:202–208. https://doi.org/10.1111/j.1461-0248.2004.00711.x

Bogužas V, Skinulienė L, Butkevičienė LM, Steponavičienė V, Petrauskas E, Maršalkienė N (2022) The effect of monoculture, crop rotation combinations, and continuous bare fallow on soil CO2 emissions, earthworms, and productivity of winter rye after a 50-year period. Plants 11:431. https://doi.org/10.3390/plants11030431

Borken W, Gründel S, Beese F (2000) Potential contribution of *Lumbricus terrestris* L. to carbon dioxide, methane and nitrous oxide fluxes from a forest soil. Biol Fertil Soils 32:142–148

Bossuyt H, Six J, Hendrix PF (2005) Protection of soil carbon by microaggregates within earthworm casts. Soil Biol Biochem 37:251–258. https://doi.org/10.1016/j.soilbio.2004.07.035

Brais S (1989) Effets de différentes pratiques sylvicoles et de l'introduction de vers de terre sur la structure du sol en plantations de feuillus. Can J Soil Sci 69:705–709

Briones MJI, Schmidt O (2017) Conventional tillage decreases the abundance and biomass of earthworms and alters their community structure in a global meta-analysis. Glob Chang Biol 23:4396–4419. https://doi.org/10.1111/gcb.13744

Brown GG, Pashanasi B, Villenave C, Patron JC, Senapati BK, Giri S, Barois I, Lavelle P, Blanchart E, Blakemore RJ, Spain AV, Boyer J (1999) Effects of earthworms on plant production in the tropics. In: Lavelle P, Brussaard L, Hendrix P (eds) Earthworm management in tropical agroecosystems. CABI, Wallingford, pp 87–148

Buck C, Langmaack M, Schrader S (2000) Influence of mulch and soil compaction on earthworm cast properties. Appl Soil Ecol 14:223–229

Butt KR, Frederickson J, Morris RM (1995) An earthworm cultivation and soil inoculation technique for land restoration. Ecol Eng 4:1–9

Butt KR, Frederickson J, Morris RM (1997) The earthworm inoculation unit (EIU) technique: an integrated system for cultivation and soil inoculation of earthworms. Soil Biol Biochem 29:251–257

Butt KR, Frederickson J, Lowe CN (1999) Colonisation, survival and spread of earthworms on a partially restored landfill site. Pedobiologia 43:684–690

Chan KY (2001) An overview of some tillage impacts on earthworm population abundance and diversity—implications for functioning in soils. Soil Tillage Res 57:179–191

Chapuis-Lardy L, Le Bayon RC, Brossard M, Lopez-Hernandez D, Blanchart E (2011) Role of soil macrofauna in P cycling. In: Bünemann EK, Oberson A, Frossard E (eds) Phosphorus in action—biological processes in soil phosphorus cycling. Springer soil biology series 26. Springer, New York, pp 199–213

Chauvel A, Grimaldi M, Barros E, Blanchart E, Desjardins T, Sarrazin M, Lavelle P (1999) Pasture damage by an Amazonian earthworm. Nature 398:32–33

Chiaia-Hernandez AC, Keller A, Wächter D, Steinlin C, Camenzuli L, Hollender J, Krauss M (2017) Long-term persistence of pesticides and TPs in archived agricultural soil samples and comparison with pesticide application. Environ Sci Technol 51:10642–10651. https://doi.org/10.1021/acs.est.7b02529

Clapperton MJ, Chan KY, Larney FJ (2007) Managing the soil habitat for enhanced biological fertility. In: Abbott LK, Murphy DV (eds) Soil biological fertility. Springer, Dordrecht. https://doi.org/10.1007/978-1-4020-6619-1_10

Clause J, Forey E, Lortie CJ, Lambert AM, Barot S (2015) Non-native earthworms promote plant invasion by ingesting seeds and modifying soil properties. Acta Oecol 64:10–20

Cluzeau D, Guo ZT, Chaussod D, Fedoroff N, Normand M, Perraud A (1994) Interaction between soil, biological activities and organic matter enrichments in Champagne soils. In: Transactions of the XV world congress of soil science. INEG and CNA, Mexico, 4b, pp 149–150

Coelho JJ, Hennessy A, Casey I, Woodcock T, Kennedy N (2020) Biofertilisation with anaerobic digestates: effects on the productive traits of ryegrass and soil nutrients. J Soil Sci Plant Nutr 20:1665–1678. https://doi.org/10.1007/s42729-020-00237-7

Coq S, Barthès BG, Oliver R, Rabary B, Blanchart E (2007) Earthworm activity affects soil aggregation and soil organic matter dynamics according to the quality and localization of crop residues—an experimental study (Madagascar). Soil Biol Biochem 39:2119–2128

Coskun D, Deshmukh R, Sonah H, Menzies JG, Reynolds O, Ma J, Kronzucker HJ, Bélanger RR (2019) The controversies of silicon's role in plant biology. New Phytol 221:67–85

Curry JP (1976) Some effects of animal manures on earthworms in grassland. Pedobiology 16:425–438

Curry JP (1998) Factors affecting earthworm abundance in soils. In: Edwards CA (ed) Earthworm ecology. St. Lucie Press, Boca Raton, p 389

De Medeiros Barbosa VFA (2015) Chapter 3—Planting. In: Santos F, Borém A, Caldas C (eds) Sugarcane. Academic, San Diego, pp 35–51

De Oliveira T, Bertrand M, Roger-Estrade J (2012) Short-term effects of ploughing on the abundance and dynamics of two endogeic earthworm species in organic cropping systems in northern France. Soil Tillage Res 119:76–84

DiBartolomeis M, Kegley S, Mineau P, Radford R, Klein K (2019) An assessment of acute insecticide toxicity loading (AITL) of chemical pesticides used on agricultural land in the United States. PLoS One 14:e0220029. https://doi.org/10.1371/journal.pone.0220029

Duboisset A (1995) Caractérisation et quantification par analyse d'image des modifications structurales engendrées par *Pontoscolex corethrurus* (Glossoscolecidae Oligochaeta) sur un Sol Ferrallitique Cultivé (Pérou). DEA thesis, The University of Nancy 1

Edwards CA (1998) The use of earthworms in the breakdown and management of organic wastes. In: Edwards CA (Ed) Earthworm Ecology, CRC Press, Boca Raton, 1998, pp 327–354

Edwards CA, Bohlen PJ (1996) Biology and ecology of earthworms, 3rd edn. Chapman and Hall, London. 426 pp

EFSA Scientific Committee (2016) Recovery in environmental risk assessments at EFSA. EFSA J 14:4313. https://doi.org/10.2903/j.efsa.2016.4313

Ejack L, Kernecker ML, Prieto R, Chen C, Gul S, Bradley RL, Whalen JK (2021) Earthworms did not increase long-term nitrous oxide fluxes in perennial forage and riparian buffer ecosystems. Pedobiologia 85–86:150727

Eriksen-Hamel NS, Whalen JK (2007) Competitive interactions affect the growth of *Aporrectodea caliginosa* and *Lumbricus terrestris* (Oligochaeta: Lumbricidae) in single- and mixed-species laboratory cultures. Eur J Soil Biol 43:142–150

Eurostat (2020). https://appsso.eurostat.ec.europa.eu/nui/show.do?dataset=aei_fm_salpest09&lang=en

Euteneuer P, Wagentristl H, Steinkellner S, Fuchs M, Zaller JG, Piepho H-P, Butt KR (2020) Contrasting effects of cover crops on earthworms: Results from field monitoring and laboratory experiments on growth, reproduction and food choice. Eur J Soil Biol 100:103225

Evans AC, Guild WJM (1948) Studies on the relationships between earthworms and soil fertility; some effects of earthworms on soil structure. Ann Appl Biol 35:471–484

Ewert F, Baatz R, Finger R (2023) Agroecology for a sustainable agriculture and food system: from local solutions to large-scale adoption. Ann Rev Resour Econ 15:351–381

FAO (2021) Keep soil alive, protect soil biodiversity. In: Global symposium on soil biodiversity 19–22 April 2021. Proceedings, Rome. https://doi.org/10.4060/cb7374en

FAO (2022) Soils for nutrition: state of the art. Rome. https://doi.org/10.4060/cc0900en

FAO, ITPS, GSBI, SCBD and EC (2020) State of knowledge of soil biodiversity—Status, challenges and potentialities. Report 2020. FAO, Rome. https://doi.org/10.4060/cb1928en

FAOSTAT (2019). http://www.fao.org/faostat/en/#data/

Forey E, Chauvat M, Coulibaly SFM, Langlois E, Barot S, Clause J (2018) Inoculation of an ecosystem engineer (Earthworm: *Lumbricus terrestris*) during experimental grassland restoration: Consequences for above and belowground soil compartments. Appl Soil Ecol 125:148–155

Frazao J, de Goede RGM, Salanki TE, Brussaard L, Faber JH, Hedde M, Pulleman MM (2019) Responses of earthworm communities to crop residue management after inoculation of the earthworm *Lumbricus terrestris* (Linnaeus, 1758). Appl Soil Ecol 142:177–188

Garnier P, Makowski D, Hedde M, Bertrand M (2022) Changes in soil carbon mineralization related to earthworm activity depend on the time since inoculation and their density in soil. Sci Rep 12:13616. https://doi.org/10.1038/s41598-022-17855-z

Gilot C (1997) Effects of a tropical geophageous earthworm, M. anomala (Megascolecidae), on soil characteristics and production of a yam crop in Ivory Coast. Soil Biol Biochem 29(3–4), 353–359

Giri S (1995) Short term input operational experiment in tea garden with application of organic matter and earthworm. MPhil thesis, Sambalpur University, India. 254 pp

Haefele SM, Banayo NPM, Amarante ST, Siopongco JDLC, Mabesa RL (2013) Characteristics and management options for rice-maize systems in the Philippines. Field Crop Res 144:52–61

Hallaire V, Curmi P, Duboisset A, Lavelle P, Pashanasi B (2000) Soil structure changes induced by the tropical earthworm Pontoscolex corethrurus and organic inputs in a Peruvian ultisol. Eur J Soil Biol 36:35–44

Hamilton WE, Dindal DL (1989) Impact of landspread sewage sludge and earthworm introduction on established earthworms and soil structure. Biol Fertil Soils 8:160–165

Hansen S, Engelstad F (1999) Earthworm populations in a cool and wet district as affected by tractor traffic and fertilisation. Appl Soil Ecol 13:237–250

Hargreaves PR, Baker KL, Graceson A, Bonnett S, Ball BC, Cloy JM (2019) Soil compaction effects on grassland silage yields and soil structure under different levels of compaction over three years. Eur J Agron 109:125916

Henneron L, Bernard L, Hedde M, Pelosi C, Villenave C, Chenu C, Bertrand M, Girardin C, Blanchart E (2015) Fourteen years of evidence for positive effects of conservation agriculture and organic farming on soil life. Agron Sustain Dev 35:169–118

Hoeffner K, Santonja M, Cluzeau D, Monard C (2019) Epi-anecic rather than strict-anecic earthworms enhance soil enzymatic activities. Soil Biol Biochem 132:93–100

Hole DG, Perkins AJ, Wilson JD, Alexander IH, Grice PV, Evans AD (2005) Does organic farming benefit biodiversity? Biol Conserv 122:113–130

Holland JM (2004) The environmental consequences of adopting conservation tillage in Europe: reviewing the evidence. Agric Ecosyst Environ 103:1–25

Hubbard VC, Jordan D, Stecker JA (1999) Earthworm response to rotation and tillage in a Missouri claypan soil. Biol Fertil Soils 29:343–347. https://doi.org/10.1007/s00374005056

Jacquiod S, Puga-Freitas R, Spor A, Mounier A, Monard C, Mougel C, Philippot L, Blouin M (2020) A core microbiota of the plant-earthworm interaction conserved across soils. Soil Biol Biochem 144:107754

Jegou D, Capowiez Y, Cluzeau D (2001) Interactions between earthworm species in artificial soil cores through the 3D reconstruction of the burrow systems. Geoderma 102:123–137

Jegou D, Brunotte J, Rogasik H, Capowiez Y, Diestel H, Schrader S, Cluzeau D (2002) Impact of soil compaction on earthworm burrow systems using X-ray computed tomography: preliminary study. Eur J Soil Biol 38:329–336

John K, Janz B, Kiese R, Wassmann R, Zaitsev AS, Wolters V (2020) Earthworms offset straw-induced increase of greenhouse gas emission in upland rice production. Sci Total Environ 710(2020):136352. https://doi.org/10.1016/j.scitotenv.2019.136352

Jones CG, Lawton JH, Shachak M (1994) Organisms as ecosystem engineers. Oikos 69:373–386

Jouquet P, Janeau JL, Pisano A, Tran H, Orange D, Luu Thi Nguyet M, Valentin C (2012) Influence of earthworms and termites on runoff and erosion in a tropical steep slope fallow in Vietnam: A rainfall simulation experiment. Appl Soil Ecol 61(2012):161–168. https://doi.org/10.1016/j.apsoil.2012.04.004

Jouquet P, Bottinelli N, Kerneis G, Henry-des-Tureaux T, Doan TT, Planchon O, Tran TD (2013) Surface casting of the tropical Metaphire posthuma increases soil erosion and nitrate leaching in a laboratory experiment. Geoderma 204–205:10–14. https://doi.org/10.1016/j.geoderma.2013.04.003

Jouquet P, Blanchart E, Capowiez Y (2014) Utilization of earthworms and termites for the restoration of ecosystem functioning. Appl Soil Ecol 73:34–40

Jusselme MD, Pruvost C, Motard E, Giusti-Miller S, Frechault S, Alphonse V, Balland-Bolou-Bi C, Dajoz I, Mora P (2019) Increasing the ability of a green roof to provide ecosystem services by adding organic matter and earthworms. Appl Soil Ecol 143:61–69

Karasali H, Marousopoulou A, Machera K (2016) Pesticide residue concentration in soil following conventional and low-input crop management in a Mediterranean agro-ecosystem, in Central Greece. Sci Total Environ 541:130–142. https://doi.org/10.1016/j.scitotenv.2015.09.016

Kattwinkel M, Liess M, Arena M, Bopp S, Streissl F, Roembke J (2015) Recovery of aquatic and terrestrial populations in the context of European pesticide risk assessment. Environ Rev 23:382–394

Kibblewhite MG, Ritz K, Swift MJ (2008) Soil health in agricultural systems. Philos Trans R Soc B Biol Sci 363(1492):685–701

Kladivko EJ (2001) Tillage systems and soil ecology. Soil Tillage Res 61:61–76

Korucu T, Shipitalo MJ, Kaspar TC (2018) Rye cover crop increases earthworm populations and reduces losses of broadcast, fall-applied, fertilizers in surface runoff. Soil Tillage Res 180:99–106. https://doi.org/10.1016/j.still.2018.03.004

Kula H, Kokta C (1992) Side effects of selected pesticides on earthworms under laboratory and field conditions. Soil Biol Biochem 24:1711–1714

Langmaack M, Schrader S, Rapp-Bernhardt U, Kotzke K (1999) Quantitative analysis of earthworm burrow systems with respect to biological soil-structure regeneration after soil compaction. Biol Fertil Soils 28:219–229

Larbodière L, Davies J, Schmidt R, Magero C, Vidal A, Arroyo Schnell A, Bucher P, Maginnis S, Cox N, Hasinger O, Abhilash PC, Conner N, Westerburg V, Costa L (2020) Common ground: restoring land health for sustainable agriculture. IUCN, Gland, Switzerland

Lavelle P, Barios I, Blanchart E, Brown G, Brussaard L, Fragoso C, Jimenez JJ, Kajondo KK, Martinez MA, Moreno AG, Poshanasi B, Senapati BK, Villenave C (1998) Earthworms as a resource in tropical agroecosystems. Nat Resour 34:26–41

Lavelle P, Brussaard L, Hendrix P (1999) Earthworm management in tropical agroecosystems. CAB, Wallingford, UK

Lavelle P, Decaëns T, Aubert M, Barot S, Blouin M, Bureau F, Margerie F, Mora P, Rossi J-P (2006) Soil invertebrates and ecosystem services. Eur J Soil Biol 42(2006):S3–S15. https://doi.org/10.1016/j.ejsobi.2006.10.002

Lavelle P, Spain A, Fonte S, Camilo Bedano J, Blanchart E, Galindo V, Grimaldi M, Jimenez JJ, Velasquez E, Zangerlé A (2020) Soil aggregation, ecosystem engineers and the C cycle. Acta Oecol 105(2020):103561. https://doi.org/10.1016/j.actao.2020.103561

Lee KE (1985) Earthworms: their ecology and relationship with soils and land use. Academic, New York. 411 pp

Liess M, von der Ohe PC (2005) Analyzing effects of pesticides on invertebrate communities in streams. Environ Toxicol Chem 24:954–965

Ligthart TN, Peek G (1997) Evolution of earthworm burrow systems after inoculation of lumbricid earthworms in a pasture in the Netherlands. Soil Biol Biochem 29:453–462

Liu T, Chen X, Gong X, Lubbers IM, Jiang Y, Feng W, Li X, Whalen JK, Bonkowski M, Griffiths BS, Hu F, Liu M (2019) Earthworms coordinate soil biota to improve multiple ecosystem functions. Curr Biol 29:1–10. https://doi.org/10.1016/j.cub.2019.08.045

Lofs-Holmin A (1983) Influence of agricultural practices on earthworms (Lumbricidae). Acta Agric Scand 33:225–234

Loranger G, Ponge JF, Blanchart E, Lavelle P (1998) Impact of earthworms on the diversity of microarthropods in a Vertisol (Martinique). Biol Fertil Soils 27:21–26

Loranger-Merciris G, Cabidoche Y-M, Deloné B, Quénéhervé P, Ozier-Lafontaine H (2012) How earthworm activities affect banana plant response to nematodes parasitism. Appl Soil Ecol 52(2012):1–8. https://doi.org/10.1016/j.apsoil.2011.10.003

Lowe CN, Butt KR (1999) Interspecific interactions between earthworms: a laboratory-based investigation. Pedobiology 43:808–817

Lowe CN, Butt KR (2002) Growth of hatchling earthworms in the presence of adults: interactions in laboratory culture. Biol Fertil Soils 35:204–209

Ludwig M, Wilmes P, Schrader S (2018) Measuring soil sustainability via soil resilience. Sci Total Environ 626:1484–1493

Maggi F, Tang FHM (2021) Estimated decline in global earthworm population size caused by pesticide residue in soil. Soil Security 5:100014

Marashi A, Scullion J (2004) Porosity and hydrological changes in surface mine soils. In: ISCO 2004—13th international soil conservation organisation conference—Brisbane, July 2004

Maslov M, Astaykina A, Pozdnyakov L (2022) Earthworm *Lumbricus terrestris* contributes nitrous oxide emission from temperate agricultural soil regardless of applied mineral nitrogen fertilizer doses. Agronomy (Basel) 12

McCallum HM, Wilson JD, Beaumont D, Sheldon R, O'Brien MG, Park KJ (2016) A role for liming as a conservation intervention? Earthworm abundance is associated with higher soil pH and foraging activity of a threatened shorebird in upland grasslands. Agric Ecosyst Environ 223:182–189. https://doi.org/10.1016/j.agee.2016.03.005

Médiène S, Valantin-Morison M, Sarthou J-P, de Tourdonnet S, Gosme M, Bertrand M, Roger-Estrade J, Aubertot J-N, Rusch A, Motisi N, Pelosi C, Doré T (2011) Agroecosystem management and biotic interactions: a review. Agron Sustain Dev 31:491–514

Meng L, Srivastava A, Kuca K, Giri B, Rahman M, Wu Q-S (2021) Interaction between earthworms and arbuscular mycorrhizal fungi in plants: a review. Phyton 90:687–699

Meng L-L, Srivastava AK, Kuča K, Wu Q-S (2022) Earthworm (*Pheretima guillelmi*)-mycorrhizal fungi (*Funneliformis mosseae*) association mediates rhizosphere responses in white clover. Appl Soil Ecol 172:104371

Moinard V, Redondi C, Etiévant V, Savoie A, Duchene D, Pelosi C, Houot S, Capowiez Y (2021) Short- and long-term impacts of anaerobic digestate spreading on earthworms in cropped soils. Appl Soil Ecol 168:104149

Nielsen RL (1951) Earthworms and soil fertility. N Z J Agric 83:433–435

Noguera D, Laossi KR, Lavelle P, de Carvalho MHC, Asakawa N, Botero C, Barot S (2011) Amplifying the benefits of agroecology by using the right cultivars. Ecol Appl 21:2349–2356

Nuutinen V (1992) Earthworm community response to tillage and residue management on different soil types in southern Finland. Soil Tillage Res 23:221–239

Nuutinen V, Haukka J (1990) Conventional and organic cropping systems at Suitia. VII: Earthworms. J Agr Sci Finland 62:357–367

Nuutinen V, Nieminen M, Butt KR (2006) Introducing deep burrowing earthworms (*Lumbricus terrestris* L.) into arable heavy clay under boreal conditions. Eur J Soil Biol 42:S269–S274

Paoletti MG (1999) The role of earthworms for assessment of sustainability and as bioindicators. Agric Ecosyst Environ 74:137–155

Pashanasi B, Lavelle P, Alegre J, Charpentier F (1996) Effect of the endogeic earthworm Pontoscolex corethrurus on soil chemical characteristics and plant growth in a low-input tropical agrosystem. Soil Biol Biochem 28:801–810

Pelosi C, Bertrand M, Roger-Estrade J (2009) Earthworm community in conventional, organic and direct seeding with living mulch cropping systems. Agron Sustain Dev 29:287–295

Pelosi C, Toutous L, Chiron F, Dubs F, Hedde M, Muratet A, Ponge J-F, Salmon S, Makowski D (2013) Reduction of pesticide use can increase earthworm populations in wheat crops in a European temperate region. Agric Ecosyst Environ 181:223–230

Pelosi C, Barot S, Capowiez Y, Hedde M, Vandenbulcke F (2014) Pesticides and earthworms. A review. Agron Sustain Dev 34:199–228

Pelosi C, Bertrand M, Thénard J, Mougin C (2015) Earthworms in a 15 years agricultural trial. Appl Soil Ecol 88:1–8

Pelosi C, Bertrand C, Daniele G, Coeurdassier M, Benoit P, Nélieu S, Lafay F, Bretagnolle V, Gaba S, Vulliet E, Fritsch C (2021) Residues of currently used pesticides in soils and earthworms: A silent threat? Agriculture. Ecosyst Environ 305:107167

Pey B (2010) Contribution de la faune du sol au fonctionnement et à l'évolution des Technosols. Sciences agricoles. Institut National Polytechnique de Lorraine

Piearce TG, Budd T, Hayhoe JM, Sleep D, Clasper PJ (2003) Earthworms of a land restoration site treated with paper mill sludge. Pedobiologia 47:792–795

Pulido-Moncada M, Petersen SO, Munkholm LJ (2022) Soil compaction raises nitrous oxide emissions in managed agroecosystems. A review. Agron Sustain Dev 42:38

Radics Z, Dekemati I, Gyuricza C, Simon B, Ibrahim HTM, Vinogradov S, Birkás M (2022) Effects of irrigation and organic mulching on the abundance and biomass of earthworms. Pol J Environ Stud 31:2811–2821

Raminoarison M, Razafimbelo T, Rakotoson T, Becquer T, Blanchart E, Trap J (2020) Multiple-nutrient limitation of upland rainfed rice in ferralsols: a greenhouse nutrient-omission trial. J Plant Nutr 43:270–284

Ratsiatosika O, Blanchart E (2021) Earthworm inoculation. In: FAO and ITPS (eds) Recarbonizing global soils: A technical manual of best management practices, Volume 4. Cropland, grassland, integrated systems and farming approaches—case-studies. Food and Agriculture Organization of the United Nations, Rome, Italy

Ratsiatosika O, Bernard L, Rabary B, Rainihanjarimanana I, Randriamanantsoa R, Razafimbelo T, Razafindrakoto M, Trap J, Blanchart E (2019) Earthworm functional groups, and residue quality and management impact rice biomass allocation, growth and grain yield in a rainfed rice cropping system in Madagascar Highlands. J Exp Agric Int 30(5):1–14

Ratsiatosika O, Blanchart E, Razafimbelo T, Razafindrakoto M, Vom Brocke K, Cao-Hamadou T-V, Rakotomalala Andriamarosata J-M, Ramanantsoanirina A, Trap J (2021a) Does rice breeding affect the ability of plant to interact with earthworms in nutrient-depleted Ferralsols? Appl Soil Ecol 163

Ratsiatosika O, Razafindrakoto M, Razafimbelo T, Rabenarivo M, Becquer T, Bernard L, Trap J, Blanchart E (2021b) Earthworm inoculation improves upland rice crop yield and other agrosystem services in Madagascar. Agriculture 11:60. https://doi.org/10.3390/agriculture11010060

Riley H, Pommeresche R, Eltun R, Hansen S, Korsaeth A (2008) Soil structure, organic matter and earthworm activity in a comparison of cropping systems with contrasting tillage, rotations, fertilizer levels and manure use. Agric Ecosyst Environ 124:275–284

Rodríguez MP, Domínguez A, Moreira Ferroni M, Wall LG, Bedano JC (2020) The diversification and intensification of crop rotations under no-till promote earthworm abundance and biomass. Agronomy 10:919. https://doi.org/10.3390/agronomy10070919

Ros MBH, Hiemstra T, van Groenigen JW, Chareesri A, Koopmans GF (2017) Exploring the pathways of earthworm-induced phosphorus availability. Geoderma 303(2017):99–109. https://doi.org/10.1016/j.geoderma.2017.05.012

Scavo A, Fontanazza S, Restuccia A, Pesce GR, Abbate C, Mauromicale G (2022) The role of cover crops in improving soil fertility and plant nutritional status in temperate climates. A review. Agron Sustain Dev 42:93

Scheu S (2003) Effects of earthworms on plant growth: patterns and perspectives. Pedobiologia 47:846–856

Schmidt O, Clements RO, Donaldson G (2003) Why do cereal-legume intercrops support large earthworm populations? Appl Soil Ecol 22:181–190. https://doi.org/10.1016/S0929-1393(02)00131-2

Scullion J, Malik A (2000) Earthworm activity affecting organic matter, aggregation and microbial activity in soils restored after opencast mining for coal. Soil Biol Biochem 32:119–126

Senapati BK, Lavelle P, Giri S, Pashanasi B, Alegre J, Decaens T, Jiménez JJ, Albrecht A, Blanchart E, Mahieux M, Rousseaux L, Thomas R, Panigrahi PK, Venkatachalam M (1999) In-soil earthworm technologies for tropical agroecosystems. In: Lavelle P, Brussard L, Hendrix PF (eds) Earthworm management in tropical agroecosystems. CAB, Wallingford, pp 199–238

Silva V, Mol HGJ, Zomer P, Tienstra M, Ritsema CJ, Geissen V (2019) Pesticide residues in European agricultural soils—a hidden reality unfolded. Sci Total Environ 653:1532–1545. https://doi.org/10.1016/j.scitotenv.2018.10.441

Smeaton TC, Daly AN, Cranwell JM (2003) Earthworm population responses to cultivation and irrigation in a South Australian soil. Pedobiol. 47:379–385

Spain AV, Lavelle P, Mariotti A (1992) Stimulation of plant growth by tropical earthworms. Soil Biol Biochem 24:1629–1634

Sullivan P (2003) Conservation tillage. National Center for Appropriate Technology ATTRA publication No. CT 105

Tebrügge F, Düring RA (1999) Reducing tillage intensity—a review of results from a long-term study in Germany. Soil Tillage Res 53:15–28

Torppa KA, Taylor AR (2022) Alternative combinations of tillage practices and crop rotations can foster earthworm density and bioturbation. Appl Soil Ecol 175:104460

Trap J, Blanchart E (2023) Intensifying the soil ecological functions for sustainable agriculture: Acting with stakeholders. Curr Res Environ Sustain 5:100225

Trap J, Blanchart E, Ratsiatosika O, Razafindrakoto M, Becquer T, Andriamananjara A, Morel C (2021) Effects of the earthworm Pontoscolex corethrurus on rice P nutrition and plant-available soil P in a tropical Ferralsol. Appl Soil Ecol 160:103867

van der Werff PA, Noordhuis R, Dekkers TBM (1998) Introduction of earthworms into an organic arable farming system. Appl Soil Ecol 9:311–317

van Groenigen JW, Lubbers IM, Vos HMJ, Brown GG, de Deyn GB, van Groenigen KJ (2014) Earthworms increase plant production: a meta-analysis. Sci Rep 4:6365. https://doi.org/10.1038/srep06365

van Groenigen JW, van Groenigen KJ, Koopmans GF, Stokkermans L, Vos HMJ, Lubbers IM (2019) How fertile are earthworm casts? A meta-analysis. Geoderma 338:525–535

van Rhee JA (1969) Inoculation of earthworms in a newly drained polder. Pedobiologia 9:128–132

Villenave C, Rabary B, Kichenin E, Djigal D, Blanchart E (2010) Earthworms and plant residues modify nematodes in tropical cropping soils (Madagascar): a mesocosm experiment. Appl Environ Soil Sci 2010:323640

Vos HMJ, Hiemstra T, Prado Lopez M, van Groenigen JW, Voegelin A, Mangold S, Koopmans GF (2022) Earthworms affect reactive surface area and thereby phosphate solubility in iron-(hydr) oxide dominated soils. Geoderma 428(2022):116212

Wang P, Hasnain M, Cahill J, Wu D (2020) The multi-response of root foraging strategy to a neighbor, soil heterogeneity and earthworm. Appl Soil Ecol 155:103684. https://doi.org/10.1016/j.apsoil.2020.103684

Whalen JK, Parmelee RW, Edwards CA (1998) Population dynamics of earthworm communities in corn agroecosystems receiving organic or inorganic fertilizer amendments. Biol Fertil Soils 27:400–407

Wubs ERJ, van der Putten WH, Bosch M, Bezemer TM (2016) Soil inoculation steers restoration of terrestrial ecosystems. Nat Plants 2

Wurst S (2010) Effects of earthworms on above- and belowground herbivores. Appl Soil Ecol 45(2010):123–130. https://doi.org/10.1016/j.apsoil.2010.04.005

Wyss E, Glasstetter M (1992) Tillage treatments and earthworm distribution in a Swiss experimental corn field. Soil Biol Biochem 24:1635–1639

Zhang W, Jiang F, Ou J (2011) Global pesticide consumption and pollution: with China as a focus. Proc Int Acad Ecol Environ Sci 1:125–144

Zhao P, Wang Z, Li K, Guo X, Zhao L (2018) Multi-residue enantiomeric analysis of 18 chiral pesticides in water, soil and river sediment using magnetic solid-phase extraction based on amino modified multiwalled carbon nanotubes and chiral liquid chromatography coupled with tandem mass spectrometry. J Chromatogr A 1568:8–21. https://doi.org/10.1016/j.chroma.2018.07.022

Part III
Earthworms, Ecosystem Services and Future Perspectives

The Role of Earthworms in Soil Formation

Jan Frouz

Abstract Despite being an important ecosystem engineer, earthworms cannot form soil alone. The role of earthworms in soil formation occurs through interactions with other soil-forming factors namely parent material, climate, topography, vegetation, and time. I begin by exploring questions about how earthworms interact with these soil-forming factors, namely, which of these affect earthworm activity and how earthworm activities can modify how these factors contribute to soil formation. Then, I examine how earthworms are involved in soil formation at various spatiotemporal scales. Earthworms form biostructures with dimensions of several millimeters (feces) to 1 to several meters (burrows); however, these biostructures affect soil processes that occur at smaller spatiotemporal scales as well as hydrologic and vegetative processes at much larger scales. Since earthworm biostructures persist in soil, they can have an impact on soil processes beyond the lifespan of an earthworm. I conclude with an analysis of the extent of earthworms and soil formation across spatiotemporal scales.

1 Introduction

Earthworms have a tremendous role in soil properties across spatiotemporal scales. In this chapter, I focus on the effect of earthworms on soil formation. I will approach this question from two different views. In the first part of the chapter, I will explore how earthworms interact with key soil-forming factors. How soil-forming factors, namely, geological substrate, climate, topography, biota, and time of soil development may affect earthworms and how earthworms can modify the effect of these soil-forming factors. In the second part of the chapter, we will explore how

J. Frouz (✉)
Biology Centre CAS, Institute of Soil Biology and Biogeochemistry, České Budějovice, Czech Republic

Charles University, Prague, Czech Republic
e-mail: frouz@natur.cuni.cz

Y. Kooch, Y. Kuzyakov (eds.), *Earthworms and Ecological Processes*,
https://doi.org/10.1007/978-3-031-64510-5_11

earthworms affect processes of soil formation from soil aggregates to soil profile and beyond.

2 Earthworm Interacting with Soil-Forming Factors

Soil formation is a process during which mineral material interacts with organic matter under the influence of soil-forming factors to produce new soil materials. Soil-forming factors are parent material, which provides the mineral constituents of the soil, which together with climate, topography, and biota determine the speed and magnitude of soil formation during a period of time (Weil and Brady 2016). In this chapter, I will explore how these soil-forming factors may interact with the soil-forming activity of earthworms, which have engineering capacity. It is also interesting to understand how earthworm activity can be affected by soil formation, which may make the soil profile a more or less habitable environment for earthworms.

2.1 Parent Material

Parent material is the unconsolidated weathered rock that may be transported and deposited, e.g., due to water or wind erosion or glaciers and/or other forces (Weil and Brady 2016). Both character of rock and transformation processes determine the properties of the parent material (due to the character of weathering, leaching, mixing, etc.). The proportions of clay and other particles (sand, silt) in a given mass of soil determine its texture, also the chemical properties of the parent material determine the availability of nutrients and pH. Earthworm presence has also been shown to be closely related to the geological history of the area especially glaciation (Mathieu and Jonathan Davies 2014; Rutgers et al. 2016), but links to particular bedrock are less established. Moreover, also due to the fact that soil can form either directly from bedrock or from saprolite (which is in situ weathered bedrock), but also from rhyolite (which is clastic, partly weathered material, various sources, local, as well as those brought on the place by various transport processes). The fact that weathered material from which soil in originating may not be produced locally, but transported here from various sources, made link between bedrock, soil properties and earthworms more complex. Furthermore, in some soils such as organic peat and fen soil, the effect of mineral is less pronounced, and thus texture is not predictive of earthworms and their activity.

Still, earthworm communities are strongly dependent on soil texture and pH (Hendrix et al. 1992; De Wandeler et al. 2016), which are directly or indirectly dependent on the parent material. Earthworms tend to preffer loamy soil and close to neutral pH. Also, the effect of soil fauna and earthworms in particular on soil properties strongly depends on substrate texture and pH (Józefowska et al. 2017).

Finally, various bedrock types can produce soils with similar pH and texture, which makes the relationship between bedrock and earthworms even less tight.

The weathering of parent material is affected by the presence of earthworms. For example, by creating preferential flow pathways, earthworms accelerate water infiltration from the soil surface to the subsoil and in the soil bedrock interface, which can indirectly affect weathering. Direct weathering of clastic minerals by earthworms is well-known, based on reports dating back Darwin (1881), who describes that the red color of red-oxide sand disappeared after passing through earthworm intestines. There are some studies that explore the effect of earthworms on clay minerals using X-ray diffraction (Hušák et al. 2005; Carpenter et al. 2007). These rather preliminary studies report earthworms increase the weathering of anorthite, biotite, smectite, and kaolinite. These results also suggest the transformation of smectite to illite and the formation of a new mineral phase from kaolinite. Similarly Pop (1998) found that *Octodrilus* earthworms affect the clay mineralogy and formation of illite in the soil. However, the mechanisms behind these changes are not quite clear. Earthworms accelerate basalt weathering and increase the availability of potassium and phosphorus from minerals bearing these elements (Mackay et al. 1983; Liu et al. 2011), suggesting earthworms may speed up the chemical weathering of these minerals. On the contrary, earthworms may reduce metal mobility in the soil in certain circumstances (Sizmur and Hodson 2009). The effect of earthworms on mineral weathering and hence nutrient availability in soil may be ecologically significant; unfortunately, the effect of earthworms on minerals and rock weathering is highly underexplored.

2.2 Climate

Climate is a key soil-forming factor. Climate affects earthworm distribution (Phillips et al. 2019; Fourcade and Vercauteren 2022) and consequently their function in the ecosystem. This is the consequence of the geographic distribution of earthworms, along with the temporal variation in earthworm activity and population dynamics in response to ongoing climate change (Singh et al. 2019; Phillips et al. 2019; Fourcade and Vercauteren 2022). Earthworms are unlikely to modify local soil thermal conditions but may modify the moisture factor of climate that influences soil formation. For example, one of the most important effects of climate on soil formation is the degree to which water percolates through the soil profile, as this alters clay movement in the soil profile and triggers the soil-forming processes associated with eluviation and illimerization. Earthworms may affect these processes by the creation of preferential flows, which affect water movement in soil profile or by upward movement of clays due to bioturbation (Sauzet et al. 2023). For example, Jačka et al. (2021) indicate that infiltration (K) was about threefold higher in alder plantation, densely colonized by earthworms, compared to neighboring spruce where earthworms were almost absent.

2.3 Topography

Topography is another important soil-forming factor. Topography causes variation in environmental conditions and hence also conditions for soil formation but perhaps more importantly topography triggers many transport processes, such as landslides, erosion, and solifluction that cause soil-forming materials to be transported from one habitat to another. Already Darwin (1881) pointed out that earthworms may affect these processes by depositing casts on the soil surface that subsequently become susceptible to erosion. Also, other studies show that casts or droppings produced by soil fauna may be subject to more erosion than the surrounding soil matrix (Yair 1995; Le Bayon and Binet 2001). However, the interaction of earthworms and erosion may be more complex, as discussed below. For example, earthworms increase infiltration (Ehlers 1975; Francis and Fraser 1998; Jačka et al. 2021), which may potentially reduce surface runoff and water erosion. Eroded material in many cases gets deposited in other places in the landscape. Since Darwin's times (Darwin 1881), it is known that earthworms can bury the objects in the soil. Earthworms may also cause the burial of the sedimented layers and its mixing into the soil underneath.

2.4 Biota

Vegetation is the most important soil-forming factor that influences the earthworm community (Mathieu and Jonathan Davies 2014; Rutgers et al. 2016). This is because vegetation provides substrates and habitat for earthworms. The effect of vegetation on soil formation is largely mediated by the response of soil biota, particularly earthworms, to the presence of particular plant communities (Frouz et al. 2013a, 2013b; Ponge 2003, 2013; Józefowska et al. 2017; Frouz 2018). The evidence supporting this claim comes from several sources: (1) correlation between certain vegetation traits, earthworm abundance (and/or activity), and soil properties (Frouz et al. 2013a), (2) manipulation experiments (Frouz et al. 2006, 2007a), and (3) from the significant change in soil properties after earthworm invasion into a new area previously devoid of earthworms (Hendrix and Bohlen 2002; Bohlen et al. 2004) or in a location where earthworms were eradicated, e.g., by mining activity (Frouz et al. 2008, 2013b). By changing soil properties, earthworms affect other soil biota and change fungal:bacterial ratio, and composition of the soil food web (Frouz et al. 2013b), which, in turn, may affect other soil processes such as aggregate formation or the quantity of carbon stored in soil. This relationship between vegetation, litter traits, and soil faunal activity can be conveniently described by the concept of humus forms (Ponge 2003, 2013; Frouz 2018).

The formation of humus form is affected by litter quality and also by the properties of the soil-forming substrate determining nutrients, pH, and texture (Fig. 1), which determine the composition of soil biota among them earthworms play a

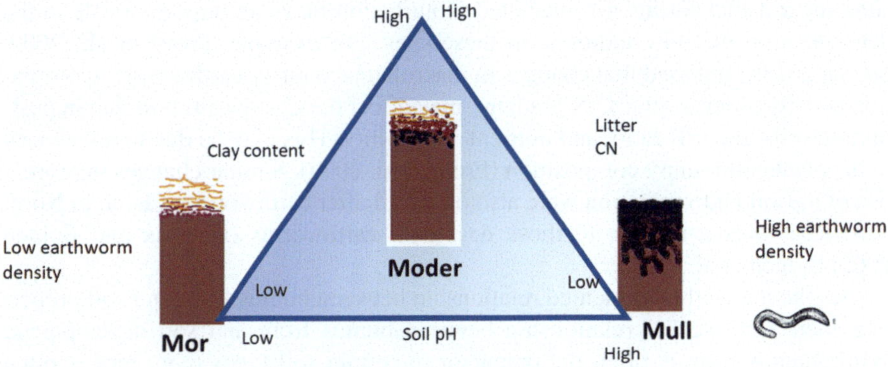

Fig. 1 Humus forms in undisturbed ecosystems are affected by the soil properties (e.g., clay content, pH) and characteristics of the vegetation, here described as the C:N ratio of the litter layer from senescent leaves and other plant debris

principal role. Soil fauna, namely earthworms, then determine the intensity of bioturbation and the way how bioturbation is performed, which consequently yields in the formation of specific humus form affecting other soil properties that generate a feedback by supporting vegetation with specific leaf traits (Frouz 2018). In soils supplied with litter that has a high C:N ratio and decomposes slowly, namely on substrates that are sandy and acidic, which limit earthworm activity, a mor type of humus develops. The bioturbation effect of fauna in this case is small. Litter is either not eaten by fauna or is consumed by soil mesofauna and transformed into holoorganic mesofauna feces. As a consequence, most litter is decomposed on the soil surface by fungi, leading to a high fungal-to-bacterial ratio (F:B). The activity of soil fauna, namely earthworms, is more pronounced in soil supplied with easily decomposable litter with a low C:N ratio namely in substrates that are close to neutral pH and loamy, resulting in the incorporation of a substantial part of the litter into the soil and in the formation of a mull type of humus (Frouz et al. 2009; Ponge 2003). In a mull humus, macrofauna biomass is high, and there are abundant anecic earthworms that forage for litter on the surface but live deeper in the soil. Bioturbation by macrofauna, and by earthworms in particular, plays a principal role in the formation of the soil profile. Because bacteria dominate in a mull, the F:B ratio is low. Frouz et al. (2013a, 2013b) found that the F:B ratio is more closely related to the intensity of earthworm bioturbation than to the litter C:N ratio. In soils supplied with litter of an intermediate quality, a moder type of humus develops. This type of humus is characterized by an intermediate F:B ratio and an intermediate level of bioturbation that is maintained by litter-feeding macroarthropods (such as millipedes, isopods, and dipteran larvae) and epigeic and endogeic earthworms (Frouz 2018). This basic outline of humus form can be expanded to a more complex classification of humus form (Zanella et al. 2011).

The presence of specific humus forms corresponds with the vegetation and nutrient status of the soil. Also, observations made after the invasion of earthworm-free sites suggest that earthworm-mediated bioturbation plays an important role in the determination of soil conditions in these sites. For example, Frouz et al. (2009, 2013a, 2013b) reported that changes in macrofauna, namely earthworms' bioturbation activity along a litter C:N gradient, may affect the C sequestration rate in post-mining soils, the F:B ratio, and nutrient availability. This may be due to a feedback in the plant community composition (Frouz et al. 2009). Similar changes in properties of topsoil and vegetation were also observed after earthworm invasion in North American forests, relative to those devoid of earthworms (Hendrix and Bohlen 2002; Bohlen et al. 2004).

Despite the well-documented relationship between humus form and soil properties, there is a weaker relationship between humus from and soil development. While humus form is often the precursor to certain soil types (e.g., mor is often associated with podsols or other spodosol, while mull is related to mollisols, and so on), variable humus form could be associated with the same soil type (Frouz et al. 2013a). There are two reasons for that: first, the formation of a specific soil type is affected by a broader set of soil-forming factors and second, the humus form affects the soil profile close to the surface, specifically the topsoil and overlying organic horizons. These changes can occur quickly and affect processes in deeper soil layers, but it typically takes longer before changes in the deeper layers of the soil profile become visible. More detailed studies are needed to explain how soil fauna activity contributes to the formation of the humus form and evaluate the connection between humus form and particular soil types.

Finally, all soil-forming factors can be important when a sufficiently long timescale is considered. As the time passes and soil develops, its suitability for earthworms will change. Earthworms colonize sites after soil conditions become suitable for their survival, following initial soil development (Frouz et al. 2006, 2007a, 2007b). At the same time, earthworm activity affects the continued soil development in a site-specific manner. For example, Toyota et al. (2013) compared the effect of earthworms on N use efficiency in soils in a tillage–no tillage experiment; soils used in this experiment came from the same field divided into tillage and no-tillage treatment decades ago. In the meantime, earthworms' density increased in no-tillage treatment, which contributed to soil changes. Then earthworms were experimentally removed from both soils and reintroduced in some treatments. Earthworms' presence in tillage soil (with a low-soil fauna legacy) increased N loss by leaching and decreased N use efficiency by plants. In contrast, fauna presence in no-till soil (higher soil fauna legacy) decreased N leaching and increased N use efficiency by plants. Similarly, the effect of earthworms on soil carbon storage changes as the soil accumulates more carbon (Frouz 2017). In soil without any organic matter, the rate of accumulation is slow and increases as organic matter increases, but as the soil approaches saturation, it slows down and eventually stops. More similar examples can be found in Frouz (2024).

3 Earthworms and Soil Formation across Spatiotemporal Scales

Earthworms have a direct effect on soil processes that are proportional to their size and lifespan; however, the soil environment maintains traces of earthworm activity, and the cumulative effect of individual earthworms persists across space and time. Consequently, earthworms affect soil processes across spatiotemporal scales. Earthworms produce soil aggregates through their digestive and soil agglomeration activities. The corridors formed by earthworms in the drilosphere allow for gas exchange and the flow of water, dissolved solutes, and particulate transport. The small-scale reactions inside these biostructures can be important at short time scales. Still, the cumulated effects of earthworm-created microstructures contribute to hydrologic and biogeochemical cycling in the whole soil profile, potentially shaping soil formation at the landscape scale.

3.1 Drilosphere and Microscale Effect

Earthworms feed on organic material at the soil surface and inside the soil profile and egest casts (aggregated fecal material) in which organic and mineral particles are mixed (Fig. 2). These feces are deposited either on the soil surface or in cavities inside the soil. Moreover, earthworms move by pushing soil on sides, this creates a layer of compacted soil around earthworm corridors, which can be enriched by earthworm mucus and excrement. Drilosphere differs from the surrounding soil (Frouz et al. 2011). It typically has higher organic matter content than surrounding soil, greater aggregate stability, and elevated microbial activity. Microbial community change as litter and mineral matrix gets mixed into earthworm cast, namely fewer fungi and bacteria prevail (Frouz 2018). Microbial activity changes as well, these changes vary in time, microbial activity may be enhanced in young earthworm casts and suppressed in older ones. Mixing of organic matter in the aggregates can contribute to soil carbon storage as reviewed by Angst et al. (2019, 2022). Earthworms mix small fragments of particulate organic matter (POM) in the mineral matrix inside aggregates. Decomposition of this occluded POM gradually declines without substantial chemical changes (Angst et al. 2017). The reason is that occluding organic matter in clay, which is typically water-saturated, reduces the access of oxygen, which has to move inside by diffusion. Diffusion is a slow process, and part of the oxygen can be consumed by microbes in the surface layer of aggregates, this can yield in anaerobic conditions inside aggregates, which slows down decomposition. Also, the removal of the product of metabolism from occluded organic matter can be reduced by the same mechanism, which further restrains the decomposition of occluded organic matter. Moreover, microbes grow massively in occluded organic matter, namely, in their vicinity, using dissolved organic matter that leaches out of occluded POM. This hotspot of microbial biomass can eventually

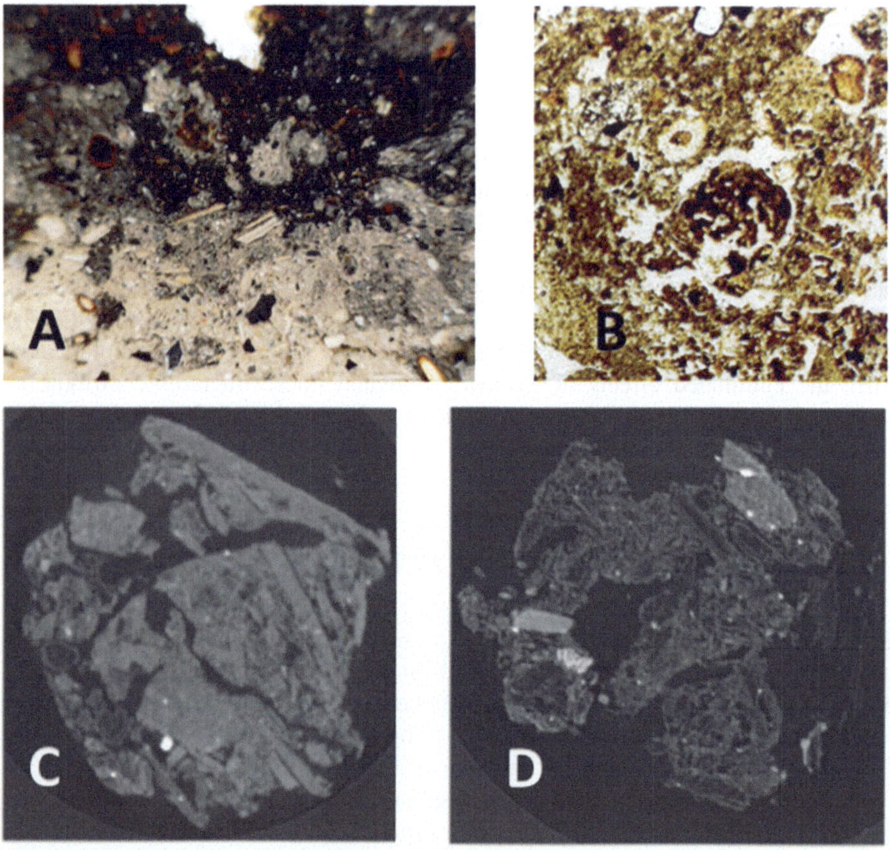

Fig. 2 Comparison of X-ray of aggregate that was formed without being affected by earthworms (**a**), with earthworm cast (**b**) and view of earthworm cast in the immature soil profile (**c**), and a detailed view of the earthworm cast from the same location (**d**)

turn into microbial necromass following the lysis of microbial cells. The microbial necromass remains in the soil and forms a substantial part of mineral-associated soil organic matter. The association of mineral particles and microbial necromass stimulates the formation of microaggregates inside soil aggregates (Six et al. 2002, 2004). The organic matter bound in soil aggregates, and specially in microaggregates inside aggregates, is more stable and resistant against decomposition (Marín-Spiotta et al. 2008). Despite the remarkable difference in the drilosphere and the surrounding soil, Hedenec et al. (2020) show that the colonization of soil by earthworms leads to significant changes of soil microbial processes not only in drilosphere but also in bulk soil, which was a few cm away from earthworm created structures. This suggests that the effect of earthworms can be much broader than one may have thought.

3.2 Bioturbation and Topsoil Formation

The effect of individual earthworm cast is low, but earthworm cast may accumulate in soil that changes soil condition. Not only by changes that happen inside drilosphere but also the way how individual structures produced by earthworms are moved and organized in soil, in other words by earthworm bioturbation. Bioturbation is a movement of soil by biological processes. In moist soils that do not freeze, earthworms that inhabit that soil can move 10–50 t/ha/y in some tropical locations, rates exceed 100 t/ha/y (e.g., Watanabe and Ruaysoongnern 1984; Krishnamoorthy 1985). Earthworm bioturbation can move soil from deeper soil layers upward but also horizontally in soil. Due to this movement, objects on the soil surface can be buried in the soil, because they are undermined by the earthworm corridor and at the same time covered by earthworm cast, which causes them to sink into the soil. Formation of corridor may affect water movement in soil and root growth (Cheng et al. 2017), which can have implications for soil development. Moreover, bioturbation does not mean only mixing of soil but it also causes the incorporation of organic matter in the soil (Jegou et al. 1998, 2000) and modification of soil properties. Another consequence of simultaneous activities of many earthstorms over time is that certain layers of soil are enriched by earthworm casts that contribute to the formation of soil horizons. Frouz et al. (2007a, 2007b) used thin soil sections to describe how accumulation of earthworm cast corresponds to the formation of the A horizon in the initial stage of soil development in post-mining sites. These results show that, in suitable conditions, almost 10 cm of A horizon can develop in 40 years (Frouz et al. 2009). Most of that layer consists of earthworm casts (Frouz et al. 2009). In addition, soil development in a common garden experiment, where various tree species were planted close to each other on the same post-mining overburden, shows that the increase in A horizon depth is positively correlated with earthworm abundance; furthermore, the A horizon depth and earthworm abundance are negatively correlated with the litter C:N. So, in initial soil development, earthworms strongly affect the formation of the A horizon but are controlled by the litter quality. When C:N ratio is low, such as under alder or lime, almost 10 cm of A horizon can be formed in 40 years, but almost no A layer formation was observed during the same time under spruce and pine with high C:N ratio of litter, despite the fact that these sites were just a few dozen meters apart (Frouz et al. 2013a). Spruce has also low earthworm density, and over time, a thick Oe layer formed by partly decomposed litter accumulated on the soil surface. This clearly shows that earthworms are responsible for litter mixing in soil and fermion of A layer. If earthworms are absent, the litter decomposes on the soil surface without being mixed into the soil. This can be nicely illustrated in unclaimed plots where due to migration constrains and a bit slower development of vegetation lead to slower colonization of sites, by earthworm (Fig. 3). Frouz et al. (2008) show that before earthworms colonize these sites (during the first 15–20 years of soil development), there is no visible formation of A horizon and instead a thick Oe layer of pertly decomposed litter develops on soil surface. When earthworms colonize sites, A layers start to develop.

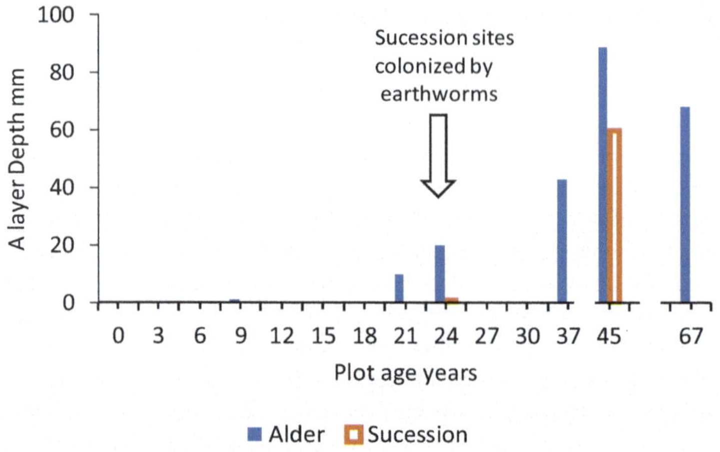

Fig. 3 Development of the A horizon in post-mining soils near Sokolov (Czech Republic). The alder plantations "Alder" were colonized by earthworms from Year 0 of the study. In contrast, earthworms do not colonize sites with unassisted natural succession "Succession" until 20 years after mining activities cease (based on Frouz et al. 2001, 2008, 2009)

A horizon thickness may reach about 5 cm within 20 years after earthworms colonize the sites. Moreover, the growth of A horizon in these sites is closely correlated to the proportion of earthworm casts in soil volume. All these data strongly support the idea that A horizon formation in the initial stages of soil development may be strongly affected by earthworm bioturbation. However, the formation of A horizon is not linear with time. The growth of A horizon is fast initially and later it slows down (Fig. 3) and eventually stops. This is consistent with the degree of bioturbation decreasing exponentially with time as was mentioned earlier. Studies from post-mining sites (Frouz et al. 2001, 2007a, 2009) show that about 50% of soil volume, or in other words, the majority of solid state of the soil depicted in Fig. 3, is formed by earthworm casts. One may expect that the depth of A horizon corresponds somehow with the depth of bioturbation, not only by earthworms but also by other organisms (Ruiz et al. 2023), but there is little data directly comparing those two.

Formation of A horizon can strongly affect soil development. Soil horizons of young developing soils, where earthworm bioturbation dominates, tend to accumulate carbon (Fig. 4) and have higher microbial biomass and F:B ratio shifted toward bacteria (Frouz et al. 2009, 2013a, 2013b). Shift of fungal-to-bacterial ratio leads to a shift in the whole food web more to bacterial driven (Frouz et al. 2013b). Soil layers that have been largely affected by earthworm activity have greater water retention (Frouz and Kuráž 2014) and also larger porosity and water infiltration (Jačka et al. 2021). These changes in soil condition may yield into changes in plant community composition (Frouz et al. 2008; Mudrák and Frouz 2018), which has been observed during decade after the sites have been colonized by earthworms. Mudrák and Frouz (2018) showed that the changes in plant community arise from the

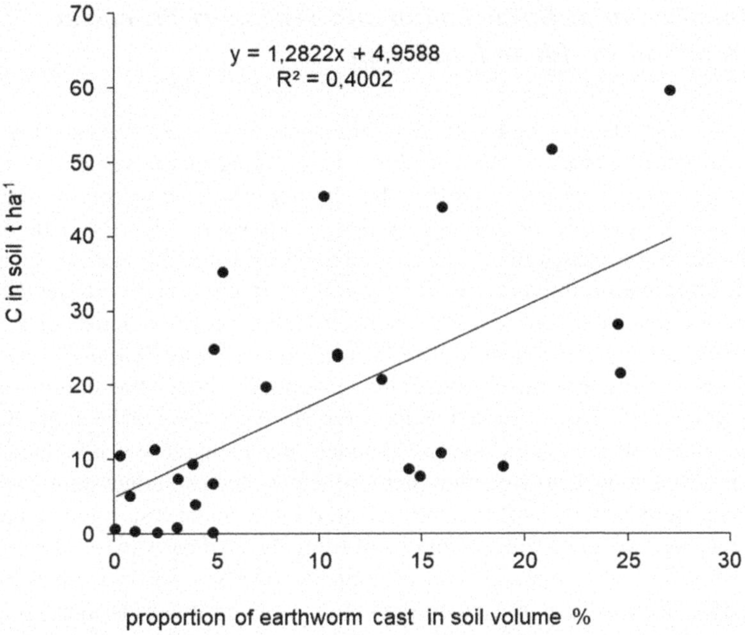

Fig. 4 Relationships between the earthworm casts in topsoil (0 to X cm) and the soil C concentration in the A horizon developing in post-mining soils near Sokolov (Czech Republic). Based on Frouz et al. (2009), modified

combination of immediate effects fauna had on plants arising, namely, from nutrient availability, but also from long-term changes in soil structure, which develop over a decade or two. In the early stages of development, the effect on soil structure is more pronounced; once soil structure has been established, the immediate effect of earthworms such as effect on nutrient availability becomes more important (Mudrák and Frouz 2018). Earthworms may also change the mycorrhizal association with plants roots. Data from post-mining soils in Czech Republic suggest that earthworms may support arbuscular mycorrhizal fungi preferentially, to the detriment of ectomycorrhizal fungi, which may support changes in vegetation composition mentioned earlier (Frouz et al. 2019). However, it should be highlighted that all these examples come from the initial stages of soil development. With time, the effect of earthworms becomes less pronounced, until it cannot be discerned detected in a developed soil. This is indirectly supported by the observation that in soils that are in the initial stages of soil development (after only a few decades of soil development, such as post-mining heap) growing plants producing litter with low C:N that support earthworm activity support soil carbon storage. The opposite was observed for mature soils where soil development already occurred continuously for several millennia (Hublova and Frouz 2021). More examples on interactions between immediate effects and long-term legacy can be found in Frouz (2024).

3.3 *Interaction of Bioturbation and Transport Processes from Soil Profile to Landscape*

Bioturbation can also interact with other transport processes, such as clay movement in soil profile, erosion and deposition. Concerning the movement of material in soil profile bioturbation may partly offer eluviation as earthworms are transporting soil from deeper soil including clay particles toward soil surface. In absolute terms, bioturbation brings more clay particles upward than eluviation downward; however, bioturbation intensity strongly non-linearly decreases with depth, which forms an area in deeper layers of soil where eluviation prevails (Sauzet et al. 2023). Bioturbation affects erosion as well. On one hand, as was demonstrated above, earthworms increase infiltration, which may potentially reduce the amount of water moving on the soil surface—and thus the strength of erosion (Jačka et al. 2021; Le Mer et al. 2021). At the same time, cast production may increase soil surface roughness, which may reduce erosion (Jouquet et al. 2008). Other studies show that earthworm casts brought to the surface are more erodible than the rest of the soil and can support water erosion and movement of soil from elevated parts to the slopes downhill where sedimented material can accumulate (Darwin 1881; Román-Sánchez et al. 2019). Still, there is not much data about how earthworms may potentially contribute to mixing this newly sedimented material to existing soil and doing that contribute to increasing A horizon thickness in these areas. However, in areas where large, active earthworm populations exist, they are probably contributing to soil hydrological processes and corresponding intensity of erosion (Bottinelli et al. 2021; Schneider et al. 2018; van Schaik et al. 2014). Along slope catena, bioturbation can thus increase erosion in the upper parts of the catchment. This brings soil-forming substrate closer to the soil surface, which may potentially increase the rate of soil formation. At the same time, eroded material accumulates in the bottom part of the catchment, which can promote the formation of deeper soil there. These processes, happening over large areas, may affect land denudation as already pointed out by Darwin (1881) and more recently confirmed by modern techniques (Johnson et al. 2014; Román-Sánchez et al. 2019). Wilkinson et al. (2009) even postulate that bioturbation-induced soil formation and erosion process can be as rapid as sustained maximum rates of tectonic uplift (Fig. 5).

4 Conclusion

Earthworms interact with the key soil-forming factors, at various spatiotemporal scales, which contribute to soil formation. In highly disturbed soils after mining, earthworms are involved with soil formation in the A horizon. Importantly, these earthworm–soil interactions have a nearly instantaneous impact on soil processes such as nutrient release and microbial community structure, which are beneficial for the establishment of vegetation and other biota. While earthworm-mediated

Fig. 5 Two soil profiles in post-mining soils near Sokolov (Czech Republic). On the left panel is an alder plantation with over 250 earthworms m^{-2}, while the right panel is under a pine forest with less than 50 earthworms m^{-2}

formation of soil aggregates and preferential flow paths created by their burrowing may have a limited immediate effect, the persistence of these biostructures can become enhanced across space and time. This magnification of an initially small but persistent effect creates a substantial legacy of earthworm engineering that persists even after organisms that started the whole process are already gone.

References

Angst S, Mueller CW, Cajthaml T, Angst G et al (2017) Stabilization of soil organic matter by earthworms is connected with physical protection rather than with chemical changes of organic matter. Geoderma 289:29–35. https://doi.org/10.1016/j.geoderma.2016.11.017

Angst G, Mueller CW, Prater I, Angst S et al (2019) Earthworms act as biochemical reactors to convert labile plant compounds into stabilized soil microbial necromass. Commun Biol 2:441. https://doi.org/10.1038/s42003-019-0684-z

Angst G, Frouz J, van Groenigen JW, Scheu S et al (2022) Earthworms as catalysts in the formation and stabilization of soil microbial necromass. Glob Chang Biol 28(16):4775–4782. https://doi.org/10.1111/gcb.16208

Bohlen PJ, Scheu S, Hale CM, McLean MA et al (2004) Non-native invasive earthworms as agents of change in northern temperate forests. Front Ecol Environ 2(8):427–435. https://doi.org/10.2307/3868431

Bottinelli N, Maeght JL, Pham RD, Valentin C et al (2021) Anecic earthworms generate more topsoil than they contribute to erosion—evidence at catchment scale in northern Vietnam. Catena 201:105186. https://doi.org/10.1016/j.catena.2021.105186

Carpenter D, Hodson ME, Eggleton P, Kirk C (2007) Earthworm induced mineral weathering: preliminary results. Eur J Soil Biol 43(Supplement 1):S176–S183. https://doi.org/10.1016/j.ejsobi.2007.08.053

Cheng Y, Ogden FL, Zhu J (2017) Earthworms and tree roots: a model study of the effect of preferential flow paths on runoff generation and groundwater recharge in steep, saprolitic, tropical lowland catchments. Water Resour Res 53:5400–5419. https://doi.org/10.1002/2016WR020258

Darwin C (1881) The formation of vegetable mould, through the action of worms, with observations on their habits. John Murray, London

De Wandeler H, Sousa-Silva R, Ampoorter E, Bruelheide H et al (2016) Drivers of earthworm incidence and abundance across European forests. Soil Biol Biochem 99:167–178. https://doi.org/10.1016/j.soilbio.2016.05.003

Ehlers W (1975) Observations on earthworm channels and infiltration on tilled and untilled loess soil. Soil Sci 119(3):242–249. https://doi.org/10.1097/00010694-197503000-00010

Fourcade Y, Vercauteren M (2022) Predicted changes in the functional structure of earthworm assemblages in France driven by climate change. Divers Distrib 28(5):1050–1066. https://doi.org/10.1111/ddi.13505

Francis GS, Fraser PM (1998) The effects of three earthworm species on soil macroporosity and hydraulic conductivity. Appl Soil Ecol 10(1–2):11–19. https://doi.org/10.1016/S0929-1393(98)00045-6

Frouz J (2017) Effects of soil development time and litter quality on soil carbon sequestration: assessing soil carbon saturation with a field transplant experiment along a post-mining chronosequence. Land Degrad Dev 28:664–672. https://doi.org/10.1002/ldr.2580

Frouz J (2018) Effects of soil macro- and mesofauna on litter decomposition and soil organic matter stabilization. Geoderma 332:161–172. https://doi.org/10.1016/j.geoderma.2017.08.039

Frouz J (2024) Plant-soil feedback across spatiotemporal scales from immediate effects to legacy. Soil Biol Biochem 189:109289. https://doi.org/10.1016/j.soilbio.2023.109289

Frouz J, Kuráž V (2014) Soil fauna and soil physical properties. In: Frouz J (ed) Soil biota and ecosystem development in post mining sites. CRC press, Boca Raton, pp 265–278

Frouz J, Keplin B, Pižl V, Tajovský K et al (2001) Soil biota and upper soil layers development in two contrasting post-mining chronosequences. Ecol Eng 17:275–284. https://doi.org/10.1016/S0925-8574(00)00144-0

Frouz J, Elhottová D, Kuráž V, Šourková M (2006) Effects of soil macrofauna on other soil biota and soil formation in reclaimed and unreclaimed post mining sites: results of a field microcosm experiment. Appl Soil Ecol 33(3):308–320. https://doi.org/10.1016/j.apsoil.2005.11.001

Frouz J, Elhottová D, Pižl V, Tajovský K et al (2007a) The effect of litter quality and soil faunal composition on organic matter dynamics in post-mining soil: a laboratory study. Appl Soil Ecol 37(1–2):72–80. https://doi.org/10.1016/j.apsoil.2007.04.001

Frouz J, Pižl V, Tajovský K (2007b) The effect of earthworms and other saprophagous macrofauna on soil microstructure in reclaimed and un-reclaimed post-mining sites in Central Europe. Eur J Soil Biol 43(Supplement 1):S184–S189. https://doi.org/10.1016/j.ejsobi.2007.08.033

Frouz J, Prach K, Pižl V, Háněl L et al (2008) Interactions between soil development, vegetation and soil fauna during spontaneous succession in post mining sites. Eur J Soil Biol 44:109–121. https://doi.org/10.1016/j.ejsobi.2007.09.002

Frouz J, Pižl V, Cienciala E, Kalčík J (2009) Carbon storage in post-mining forest soil, the role of tree biomass and soil bioturbation. Biogeochemistry 94:111–121. https://doi.org/10.1007/s10533-009-9313-0

Frouz J, Krištůfek V, Livečková M, van Loo D et al (2011) Microbial properties of soil aggregates created by earthworms and other factors: spherical and prismatic soil aggregates from unreclaimed post-mining sites. Folia Microbiol 56:36–43. https://doi.org/10.1007/s12223-011-0011-7

Frouz J, Livečková M, Albrechtová J, Chroňáková A (2013a) Is the effect of trees on soil properties mediated by soil fauna? A case study from post-mining sites. For Ecol Manag 309:87–95. https://doi.org/10.1016/j.foreco.2013.02.013

Frouz J, Thébault E, Pižl V, Adl S (2013b) Soil food web changes during spontaneous succession at post mining sites: a possible ecosystem engineering effect on food web organization? PLoS One 8(11):e79694. https://doi.org/10.1371/journal.pone.0079694

Frouz J, Moradi J, Pueschel D, Rydlova J (2019) Earthworms affect growth and competition between ectomycorrhizal and arbuscular mycorrhizal plants. Ecosphere 10:e02736. https://doi.org/10.1002/ecs2.2736

Hedenec P, Cajthaml T, Pizl V, Marialigeti K et al (2020) Long-term effects of earthworms (*Lumbricus rubellus* Hoffmeister, 1843) on activity and composition of soil microbial community under laboratory conditions. Appl Soil Ecol 150:103463. https://doi.org/10.1016/j.apsoil.2019.103463

Hendrix PF, Bohlen PJ (2002) Exotic earthworm invasions in North America: ecological and policy implications: expanding global commerce may be increasing the likelihood of exotic earthworm invasions, which could have negative implications for soil processes, other animal and plant species, and importation of certain pathogens. Bioscience 52(9):801–811. https://doi.org/10.1641/0006-3568(2002)052[0801:EEIINA]2.0.CO;2

Hendrix PF, Mueller BR, Bruce RR, Langdale GW, Parmelee RW (1992) Abundance and distribution of earthworms in relation to landscape factors on the Georgia Piedmont, U.S.A. Soil Biol Biochem 24(12):1357–1361. https://doi.org/10.1016/0038-0717(92)90118-H

Hublova L, Frouz J (2021) Contrasting effect of coniferous and broadleaf trees on soil carbon storage during reforestation of forest soils and afforestation of agricultural and post-mining soils. J Environ Manag 290:112567. https://doi.org/10.1016/j.jenvman.2021.112567

Hušák M, Frouz J, Had J (2005) Can earthworm gut passage modify clay mineral structure: preliminary results. In: Tajovský K, Schlaghamerský J, Pižl V (eds) Contribution to soil zoology in Central Europe. Ústav půdní biologie AV ČR, České Budějovice, pp 37–39

Jačka L, Walmsley A, Kovář M, Frouz J (2021) Effects of different tree species on infiltration and preferential flow in soils developing at a clayey spoil heap. Geoderma 403:115372. https://doi.org/10.1016/j.geoderma.2021.115372

Jégou D, Cluzeau D, Balesdent J, Tréhen P (1998) Effects of four ecological categories of earthworms on carbon transfer in soil. Appl Soil Ecol 9(1–3):249–255. https://doi.org/10.1016/S0929-1393(97)00057-7

Jégou D, Cluzeau D, Hallaire V, Balesdent J, Tréhen P (2000) Burrowing activity of the earthworms Lumbricus terrestris and Aporrectodea giardi and consequences on C transfers in soil. Eur J Soil Biol 36(1):27–34. https://doi.org/10.1016/S1164-5563(00)01046-3

Johnson MO, Mudd SM, Pillans B, Spooner NA et al (2014) Quantifying the rate and depth dependence of bioturbation based on optically-stimulated luminescence (OSL) dates and meteoric ^{10}Be. Earth Surf Process Landf 39:1188–1196. https://doi.org/10.1002/esp.3520

Jouquet P, Podwojewski P, Bottinelli N, Mathieu J, Ricoy M, Orange D, Duc Tran T, Valentin C (2008) Above-ground earthworm casts affect water runoff and soil erosion in northern Vietnam. Catena 74(1):13–21

Józefowska A, Pietrzykowski M, Wos B, Cajthaml T, Frouz J (2017) The effects of tree species and substrate on carbon sequestration and chemical and biological properties in reforested post-mining soils. Geoderma 292:9–16. https://doi.org/10.1016/j.geoderma.2017.01.008

Krishnamoorthy RV (1985) A comparative study of wormcast production by earthworm populations from grassland and woodland sites near Bangalore, India. Rev Ecol Biol Sol 22:209–219

Le Bayon R-C, Binet F (2001) Earthworm surface casts affect soil erosion by runoff water and phosphorus transfer in a temperate maize crop. Pedobiologia 45(5):430–442. https://doi.org/10.1078/0031-4056-00097

Le Mer G, Jouquet P, Capowiez Y, Maeght J et al (2021) Age matters: dynamics of earthworm casts and burrows produced by the anecic Amynthas khami and their effects on soil water infiltration. Geoderma 382:114709. https://doi.org/10.1016/j.geoderma.2020.114709

Liu D, Lian B, Wang B, Jiang G (2011) Degradation of potassium rock by earthworms and responses of bacterial communities in its gut and surrounding substrates after being fed with mineral. PLoS One 6(12):e28803. https://doi.org/10.1371/journal.pone.0028803

Mackay AD, Springett JA, Syers JK, Gregg PEH (1983) Origin of the effect of earthworms on the availability of phosphorus in a phosphate rock. Soil Biol Biochem 15(1):63–73. https://doi.org/10.1016/0038-0717(83)90121-9

Marín-Spiotta E, Swanston CW, Torn MS, Silver WL, Burton SD (2008) Chemical and mineral control of soil carbon turnover in abandoned tropical pastures. Geoderma 143(1–2):49–62. https://doi.org/10.1016/j.geoderma.2007.10.001

Mathieu J, Jonathan Davies T (2014) Glaciation as an historical filter of below-ground biodiversity. J Biogeogr 41:1204–1214. https://doi.org/10.1111/jbi.12284

Mudrák O, Frouz J (2018) Earthworms increase plant biomass more in soil with no earthworm legacy than in earthworm-mediated soil, and favour late successional species in competition. Funct Ecol 32:626–635. https://doi.org/10.1111/1365-2435.12999

Phillips HRP, Guerra CA, Bartz MLC, Briones MJI et al (2019) Global distribution of earthworm diversity. Science 366(6464):480–485. https://doi.org/10.1126/science.aax4851. Erratum in: Science 2020 Jul 31;369(6503): PMID: 31649197; PMCID: PMC7335308

Ponge J-F (2003) Humus form in ter restrial ecosystem: a framework to biodiversity. Soil Biol Biochem 35:935–945. https://doi.org/10.1016/S0038-0717(03)00149-4

Ponge J-F (2013) Plant–soil feedbacks mediated by humus forms: a review. Soil Biol Biochem 57:1048–1060. https://doi.org/10.1016/j.soilbio.2012.07.019

Pop VV (1998) Earthworm biology and ecology–a case study: the genus Octodrilus omodeo, 1956 (Oligochaeta, Lumbricidae), from the carpathians in FAO Earthworm ecology. St. Lucie Press, Boca Raton, pp 65–100

Román-Sánchez A, Reimann T, Wallinga J, Vanwalleghem T (2019) Bioturbation and erosion rates along the soil-hillslope conveyor belt, part 1: insights from single-grain feldspar luminescence. Earth Surf Process Landf 44:2051–2065. https://doi.org/10.1002/esp.4628

Ruiz S, Hallett P, Or D (2023) Bioturbation—physical processes. Reference module in earth systems and environmental sciences. Elsevier, p B9780128229743002000. https://doi.org/10.1016/B978-0-12-822974-3.00180-4

Rutgers M, Orgiazzi A, Gardi C, Römbke J et al (2016) Mapping earthworm communities in Europe. Appl Soil Ecol 97:98–111. https://doi.org/10.1016/j.apsoil.2015.08.015

Sauzet O, Cammas C, Gilliot J-M, Montagne D (2023) Long-term quantification of the intensity of clay-sized particles transfers due to earthworm bioturbation and eluviation/illuviation in a cultivated Luvisol. Geoderma 429:116251. https://doi.org/10.1016/j.geoderma.2022.116251

Schneider A-K, Hohenbrink TL, Reck A, Zangerlé A et al (2018) Variability of earthworm-induced biopores and their hydrological effectiveness in space and time. Pedobiologia 71:8–19. https://doi.org/10.1016/j.pedobi.2018.09.001

Singh J, Schädler M, Demetrio W, Brown GG, Eisenhauer N (2019) Climate change effects on earthworms – a review. Soil Org 91(3):114–138. https://doi.org/10.25674/so91iss3pp114. PMID: 31908681; PMCID: PMC6944501

Six J, Conanat RT, Paul EP, Paustian K (2002) Stabilization mechanism of soil organic matter: implication for C-saturation of soil. Plant Soil 241:155–176. https://doi.org/10.1023/A:1016125726789

Six J, Bossuyt H, Degryze S, Denef K (2004) A history of research on the link between (micro) aggregates, soil biota, and soil organic matter dynamics. Soil Tillage Res 79(1):7–31. https://doi.org/10.1016/j.still.2004.03.008

Sizmur T, Hodson ME (2009) Do earthworms impact metal mobility and availability in soil? A review. Environ Pollut 157(7):1981–1989. https://doi.org/10.1016/j.envpol.2009.02.029

Toyota A, Hynšt J, Cajthaml T, Frouz J (2013) Soil fauna increase nitrogen loss in tilled soil with legume but reduce nitrogen loss in non-tilled soil without legume. Soil Biol Biochem 60:105–112. https://doi.org/10.1016/j.soilbio.2013.01.017

van Schaik L, Palm J, Klaus J, Zehe E, Schröder B (2014) Linking spatial earthworm distribution to macropore numbers and hydrological effectiveness. Ecohydrology 7(2):401–408. https://doi.org/10.1002/eco.1358

Watanabe H, Ruaysoongnern S (1984) Cast production by the megascolecid earthworm *Pheretima* sp. in northeastern Thailand. Pedobiologia 26:37–44

Weil RR, Brady NC (2016) Nature and properties of soils, 15th edn. Pearson Education

Wilkinson MT, Richards PJ, Humphreys GS (2009) Breaking ground: pedological, geological, and ecological implications of soil bioturbation. Earth-Sci Rev 97(1–4):257–272. https://doi.org/10.1016/j.earscirev.2009.09.005

Yair A (1995) Short and long term effects of bioturbation on soil erosion, water resources and soil development in an arid environment. Geomorphology 13(1–4):87–99. https://doi.org/10.1016/0169-555X(95)00025-Z

Zanella A, Jabiol B, Ponge J-F, Sartori G et al. (2011) European humus forms reference base. https://hal.science/hal-00541496v2

Earthworms and Soil Biogeochemical Processes

Katharina H. E. Meurer and Klaus Schützenmeister

Abstract Earthworms inhabit almost all ecosystems and they play an important role in the soil's functioning. Through their burrowing activities, i.e., the ingestion and egestion of mineral soil and organic material, they actively change their surrounding environment, consequently altering the hydrological and biogeochemical properties of the soil. In addition to the physical burrowing, the physical and chemical properties of earthworm casts differ from the surrounding soil material, thus affecting the water retention properties, as well as the soil's density. However, differences exist between earthworm species in to which regard and degree their bioturbation influences the soil structure. More specifically, different species can have opposite effects, highlighting the impact of the loss of species, e.g., through soil and land management. Overall, when assessing the presence of earthworms, it is recommended to consider the soil as a system in order to include the interactions between individual parts. Mathematical models are powerful tools in this regard, as earthworm processes, such as bioturbation, are highly dynamic and influence on different parts of the soil system.

1 Introduction

Earthworms are the most abundant biomass in terrestrial ecosystems (Lavelle and Spain 2001) and are therefore considered the most important components of the soil biota (Edwards 2004). By changing fundamental soil properties like soil structure

K. H. E. Meurer (✉)
Department of Soil & Environment, Swedish University of Agricultural Sciences, Uppsala, Sweden
e-mail: katharina.meurer@slu.se

K. Schützenmeister
Institute for Environmental Sciences, University of Kaiserslautern-Landau (RPTU), Landau, Germany
e-mail: klaus.schuetzenmeister@rptu.de

Y. Kooch, Y. Kuzyakov (eds.), *Earthworms and Ecological Processes*,
https://doi.org/10.1007/978-3-031-64510-5_12

and spatiotemporal distribution of nutrients, earthworms influence basic soil processes, such as water infiltration and storage and nutrient cycling (Frelich et al. 2006). Their influence can be of both a purely physical (alteration of pathways for air and water movement in the soil profile and enhanced preferential flow) or a more biophysical nature (increased soil respiration due to own respiration and alteration of soil litter, as well as increased microbial respiration). The earthworm movement through the soil involves displacement of soil particles, as well as compression of the surrounding pore space (e.g., Binet and Curmi 1992; Schrader et al. 2007; Koestel and Schlüter 2019). More precisely, the soil ingested by the earthworm remains in the gut for some time and gets usually egested at some distance from the site of ingestion, which can be both in along the soil profile, but also at the soil surface (Meurer et al. 2020). This directly alters the bulk density in the soil profile (e.g., Jarvis et al. 2010) and has in some cases been shown to add to the loosening of compacted soil (e.g., Zund et al. 1997; Francis et al. 2001). In addition to that, the casts egested by the earthworms have properties different to the ingested soil both in terms of density and pore size distribution (e.g., Blanchart et al. 1993; Lipiec et al. 2015). In general, the passage of ingested soil through earthworm guts accelerates the decomposition of organic material by mineralization and increases the microbial activity, which usually would be expected to an increased loss of soil carbon (C). However, the study of Angst et al. (2017) found that the C loss was lower in treatments where earthworms were present compared to those where soil and litter were mixed mechanically. The authors found no differences in the chemical composition of the organic matter between the treatments, which made them suggest that the lower C loss and, consequently, greater sequestration of organic matter in the earthworm treatment is caused by the physical protection of soil organic matter inside the cast aggregates. Besides the obvious creation of macropores through burrowing (e.g., Capowiez et al. 1998), earthworm casting affects the water retention properties of the soil matrix due to their physical properties that differ from the surrounding soil (Meurer et al. 2020). This will further be explained in Sect. 2.1.

Differences exist between ecosystems with regard to the benefits and the way earthworms change their surrounding environment: while the invasion of forest soils by earthworms can lead to the loss of the organic horizon, earthworms generally have beneficial effects on soil processes and, consequently, ecosystem services, in cultivated and grassland systems (e.g., Keith and Robinson 2012). These processes include decomposition, nutrient cycling, soil pore creation, as well as soil aggregate formation. The importance of bioturbation has been highlighted in several studies, but Constanza et al. (1998) even assigned a value to it. According to their study, soil-forming processes account for more than one-third of the average value of the total ecosystem services.

Earthworms can be found in almost all ecosystems. Besides the fact that agricultural fields are unstable biotopes for earthworms (Edwards and Bohlen 1996), positive impact on, among others plant growth have been reported. Brown et al. (1999) distilled five possible pathways through which earthworms can positively affect plant growth: (i) biocontrol of pests and diseases, (ii) stimulation of microbial plant symbionts, (iii) production of plant growth regulating substances, (iv) soil structure changes, and (v) increased nutrient availability. In a meta-analysis, Van Groenigen et al. (2014) concluded that the main benefit of earthworm bioturbation was an enhanced nutrient mineralization and not only soil structure improvement. The role of earthworms in nutrient cycling is, among others, due to the large amounts of organic matter they ingest. Whalen et al. (1999) summarized a number of studies and reported that 2–15 Mg of organic matter ha^{-1} is being consumed by earthworms per year. In addition to that, they estimated that earthworms are responsible for 5–30% of total heterotrophic respiration. For agroecosystems, the authors also found that the nitrogen (N) flux through earthworms has been estimated to range between 10 and 83 kg ha^{-1} $yr.^{-1}$.

In the following chapters, we will look closer into the influence of earthworm bioturbation on soil structure development (Sect. 2.1), as well as nutrient cycling (Sect. 2.2; here with emphasis on the emission of greenhouse gases (GHGs), i.e., carbon dioxide (CO_2), nitrous oxide (N_2O), and methane (CH_4)). Given that results from field experiments are still scarce, extrapolation of data in space and time is difficult. In Chap. 3, we will therefore briefly discuss the development of mathematical models as a tool to increase process understanding and the consideration of the soil as an interlinked system rather than individual disciplines.

2 Earthworms Change the Soil's Structure and Functioning

2.1 Soil Structure

Earthworms are considered a keystone species due to their contribution to soil-forming processes (Sharma et al. 2017). This contribution consists primarily of the way they reorganize the soil and the soil's structure while making their way through the soil, i.e., by ingesting and egesting soil and organic material. While ingestion creates additional (macro-)pores, egestion not only fills parts of these pores but also modifies the soil bulk density through the incorporation of organic material into the soil (Sharma et al. 2017). The presence of earthworms and their bioturbation in the soil are mostly associated with an improvement of the soil structure. The positive effect of bioturbation on soil porosity and pore size distribution, however, depends a lot on the variability of the earthworm species present, as some make the soil more compact, while others have the expected effect of making it looser. Casts are usually more compact than the surrounding soil (Blanchart et al. 1993, 1999; Jouquet et al. 2008), consequently increasing the soil's bulk density, and, depending on the species, covered by a cortex made of fine organic and mineral particles, reducing aeration and inhibiting microbial activity at the scale of weeks to months (Blanchart et al. 1993). The cast stability can be attributed to the incorporation of organic matter and, consequently, the earthworm's diet. For example, Guggenberger et al.

(1996) related the enrichment of the cast of *Martiodrilus* sp. in carbohydrates and lignin to the carbohydrate-rich plant debris these earthworms ingest.

The compacting or decompacting effect of endogeic earthworm bioturbation, i.e., small- to medium-sized earthworms moving horizontally in the upper mineral soil, has been observed in several studies and both in the laboratory and under field conditions (Table 1). For example, in a 3-year study in the Peruvian Amazon, Alegre et al. (1996) found that bulk density was significantly higher in treatments where *Pontoscolex corethrurus* was present (an increase from 1.12 to 1.23 g cm^{-3}) compared to treatments without earthworm inoculation. Moreover, soil porosity decreased significantly in earthworm treatments over the course of the study (from 58 to 53%). The authors further found changes in the proportion of aggregates of different sizes with the proportion of macroaggregates (> 10 mm) increasing from 25.1 to 32.7%, while microaggregates (< 2 mm) decreased from 32.2 to 26.1%. The changes in the proportion of aggregates can be explained by the fact that *Pontoscolex corethrurus* ingests aggregates smaller than the diameter of its mouth (Blanchart et al. 1989), while it egests larger aggregates. This means that smaller aggregates are

Table 1 Overview of studies reporting compacting or decompacting influence of endogeic earthworm species studied individually or together with another species

Earthworm species	Family	Functional attribute re soil structure	Field or lab study	Reference
Reginaldia omodeoi/Millsonia omodeoi	Acanthodrilidae	Compacting	Field, Ivory Coast	Blanchart et al. (1999); Lavelle et al. (2004) Guei et al. (2012)
Pontoscolex corethrurus	Glossoscolecidae	Compacting	Field, Peruvian Amazon	Alegre et al. (1996)
	Eudrilidae	Decompacting	Field, Ivory Coast	Blanchart et al. (1999)
Millsonia anomala	Megascolecidae	Compacting	Field, savannah	Blanchart et al. (1997)
Chuniodrilus zielae	Eudrilidae	Decompacting	Sieved soil, Ivory Coast Field, Ivory Coast	Blanchart et al. (1989) Rossi (2003)
Stuhlmannia porifera	Eudrilidae	Decompacting	Sieved soil, Ivory Coast Field, Ivory Coast	Blanchart et al. (1989) Rossi (2003)
Hyperiodrilus africanus	Eudrilidae	Decompacting	Field, Ivory Coast	Guei et al. (2012)
Dichogaster terraenigrae	Acanthodrilidae	Compacting	Field, Ivory Coast	Guei et al. (2012)
Metaphire posthuma	Megascolecidae	Compacting (after rainfall)	Laboratory, Hanoi, Vietnam	Botinelli et al. (2010)

continuously transformed into larger aggregates (Alegre et al. 1996). Increased aggregation when earthworms were present has already been reported by Blanchart et al. (1989), where two earthworms, *Chuniodrilus zielae* and *Stuhlmannia porifera (Eudrilidae)*, were included in the study. The authors found that the rather small filiform earthworms contributed to the formation of medium-sized aggregates. Overall, their presence has been reported to rather decompact the soil (e.g., Blanchart et al. 1999). The same group of authors (Blanchart et al. 1989) showed significantly higher percentages of aggregates >2 mm at the expense of smaller aggregates in the presence of the medium-sized species *Millsonia anomala*. In general, the fact that the size of the aggregates that are being formed depends on the size of the earthworm is comprehensible. Blanchart et al. (1997) compared *Millsonia anomala* with smaller eudrilid earthworms in a field experiment over a period of 28 months. Similar to the outcomes presented above, the authors found that treatments with only *Millsonia anomala* showed an increase in macroaggregates, whereas smaller aggregates were observed when only eudrilid earthworms were present. With this study, Blanchart et al. (1997) actually developed the concept of compacting and decompacting species (Guei et al. 2012). It is important to note that this has primarily been studied on endogeic species, i.e., small- to medium-sized earthworms living in the mineral soil and burrowing the soil in a horizontal fashion.

In addition to the aggregate distribution, Haynes and Fraser (1998) found increased aggregate stability in the cast of *Aporrectodea caliginosa* compared to the surrounding soil whenever organic amendments were applied. The authors attributed the cast stability to the mixing of organic material from the amendments and the soil with the organic material becoming embedded within the soil material, as well as fungal hyphae emanating from within the casts.

Rossi (2003) went a step further and studied the spatiotemporal dynamics of different functional groups in a field experiment, given their importance for soil structure development. The author included two decompacting (*Chuniodrilus zielae* and *Stuhlmannia porifera*) and one compacting (*Millsonia anomala*) species. Even though the study found an overall highly heterogenous, patchy, and structured assemblage in the earthworm community, the study confirmed the previous results and found an association between low bulk density and high eudrilid density, as well as between high bulk density and high *Millsonia anomala* density. The studies mentioned above were limited to investigating earthworm species individually—however, in nature, there is often a variability of earthworm species abundant, even though different ecological groups inhabit different habitats and feed on different food resources. Nevertheless, depending on the composition of this variability, interactions of different species can balance out the effects of individual species. For example, *Millsonia omodeoi* and *Dichogaster terraenigrae* have a compacting effect when observed individually, while *Hyperiodrilus africanus* presence decreases compaction (Guei et al. 2012). This impact of *Hyperiodrilus africanus* was observed even when studied together with either *Millsonia omodeoi* or *Dichogaster terraenigrae* or with both of them (Guei et al. 2012). Gui et al. (2012) further showed that together, *Millsonia omodeoi* or *Dichogaster terraenigrae* have been found to have a significant compacting effect. As for the aggregate sizes present in the soil, treatment including *Millsonia omodeoi* had a higher proportion of macroaggregates (> 2 mm), while this proportion decreased to the benefits of small- (0.2–0.5 mm) and medium-sized (0.5–1 mm) aggregates when *Hyperiodrilus africanus* was present. Less large aggregates were found in the presence of *Dichogaster terraenigrae* compared to treatments where compacting and decompacting species interacted. In general, larger compacting species, such as *Millsonia anomala,* feed on smaller aggregates (< 2 mm) while excreting larger, stable, and compact cast (Blanchart et al. 1999). Since earthworms do not reingest their own casts (this has at least been shown for *M. anomala* (Lavelle 1978)), these macroaggregates are left in the soil, decreasing the porosity. Decompacting species, such as *Hyperiodrilus africanus* or eudrilid earthworms, feed on the cast of the larger earthworms, eventually egesting smaller casts, which, in turn, are being ingested by the larger species (Blanchart et al. 1997). Consequently, the interaction between species with different functional traits is of high importance for soil structure development and functioning. On top of that, contrasting diversity and composition of invertebrate assemblage can lead to differences in soil structure and aboveground plant communities Cifuentes-Croqueville et al. (2020).

2.2 Soil Nutrient Cycling

In addition to the mechanical impact on the soil structure, earthworms drive nutrient dynamics in the soil ecosystem (Crooks 2002) as they feed on litter and mix nutrients into the soil (Blouin et al. 2013). Doing this, they can stimulate nutrient mobilization and plant nutrient uptake (Wardle et al. 2004). Similar to the soil compaction state that has been discussed previously (Sect. 2.1), changes in the bioturbation activities and the community composition, as influenced by, e.g., soil management practices, can be used as indicators of soil fertility and quality (Lemtiri et al. 2014).

Casts and burrow walls have specific physical, chemical, and biological properties that are different from the initial material (e.g. Amador and Görres 2007; Jouquet et al. 2008) and are considered hot spots for microbial activity (e.g., Martin and Marinissen 1993; Marhan et al. 2007). For example, Don et al. (2008) showed that even though the sequestration of C in burrows can be fast (~22 g C m^{-2} yr^{-1} in burrow linings), the accumulated C can be mineralized rather rapidly with turnover times of 3–5 years. In that study, the authors concluded that earthworm bioturbation does not necessarily lead to more stabilized organic C, but that the burrows can serve as fast ways for fresh organic C transport into deeper soil layers.

Their role in changing their surrounding environment becomes apparent in areas where earthworms are invasive species. In particular, in forests, earthworms have been shown to mix the litter layer with the underlying mineral soil (Lyttle et al. 2011). A striking and illustrative example for the effect of earthworms on soils was shown in North America (Burtelow et al. 1998; Hendrix and Bohlen 2002; Hale et al. 2005): thick layers of organic matter overlaying soils considerably diminished within decades. Besides the risk of increased nitrogen leaching (Bohlen et al.

2004a), higher denitrification enzyme activity (Burtelow et al. 1998) and genes related to denitrification and nitrogen fixation have been found in soils that were more heavily invaded by earthworms (Jang et al. 2022).

Apart from situations where earthworms invade new sites, they are suspected to increase greenhouse gas (GHG) emissions from soils (Lubbers et al. 2013). Being heterotrophic organisms, earthworms do not add to the overall carbon (C) and nitrogen (N) supply but alter the turnover rates of organic matter (Burtelow et al. 1998). Nevertheless, altered fluxes of GHG from soils in the presence of earthworms were reported by several studies and they generally agree on increased CO_2 and N_2O emissions (e.g., Karsten and Drake 1997; Matthies et al. 1999; Borken et al. 2000; Speratti and Whalen 2008; Charpuis-Lardy et al. 2010; Nieminen et al. 2015). Exceptions exist and Fisk et al. (2004) reported no influence of earthworms on total CO_2 efflux from soils, while Borken and Brumme (1997) reported lower N_2O emissions from limed forest soils. In the latter study, the authors related the resulting pH increase to the activation of surviving earthworms and the creation of more aerobic conditions and O_2 diffusion into the mineral soil. The impact on CH_4 fluxes has received less attention so far, but Borken et al. (2000) and Borken and Brumme (1997) found a reduced sink function of temperate forest soils in the presence of earthworms. Lubbers et al. (2013) concluded from a meta-analysis that experimental duration matters and earthworm-induced CO_2 emissions increased during a relatively short experimental period, but decreased significantly when studies lasted longer than 200 days. Most likely earthworms accelerate initial C decomposition, but do not increase the total amount of decomposed material in the longer term (Lubbers et al. 2013). The opposite was found for N_2O and fluxes induced by earthworms increased over time. This may be related to the anaerobic conditions in the earthworm gut, which favors denitrification, as well as the labile C and N found there (Drake and Horn 2006). However, most of these studies were realized in the absence of plants that otherwise would take nitrogen from the soil.

The inclusion of earthworms and plants, which also influence the soil environment through, among others, root growth, has received less attention in field and laboratory experiments. Even though the C sink capacity of forest and grassland soils is higher than that of cropland soils (Schulze et al. 2010), most studies are focused on arable species rather than natural ecosystems. With regard to the soil's biophysical processes, both trees and earthworms significantly contribute to nutrient cycling. In the presence of both, the large burrows created by the earthworms are preferably used by plant roots to grow in and the organic compounds and mineral nutrients left by the earthworms after digestion are taken up and reused by plants (Blouin et al. 2013). For example, increases in both N microbial biomass and N concentration in total grain-N due to the presence of earthworms have been reported for soybean plants, while soil mineral N correlated positively with earthworm density (Ericksen-Hamel and Whalen 2007). Thus, the emergent behavior controlled by the impact of plants and earthworms on the net GHG fluxes between the atmosphere and soil ecosystem cannot be excluded.

In a mesocosm study, emissions of CO_2, N_2O, and CH_4 were measured under the combination of two earthworm species (*Lumbricus terrestris* and *Aporrectodea caliginosa*) and ash (*Fraxinus excelsior* L) saplings and compared to emissions observed when the two were considered individually. The estimated effect of the manipulation was calculated by subtracting the control from the other treatments, i.e., Earthworm (E) – Control (C), Ash (A) – C, and Ash + Earthworm (AE) – C. The combined effect of E and A was calculated as $AE_{add} = (E - C) + (A—C)$.

As it turned out, there was a clear effect of the presence of both earthworms and ash, but the reaction differed between the gases. Ash (A) and ash + earthworms (AE) showed significantly higher total CO_2–C emissions (4.7 and 5.3 g m^{-2}, respectively) compared to the treatments without plants C (2.8 g m^{-2}) and E (3.3 g m^{-2}) (Table 1, Fig. 1a). Moreover, the combined effect of earthworms and ash sapling (AE) on cumulative CO_2–C fluxes was very close to the prediction given by the individual treatments E and A with the difference being only 0.1 g m^{-2} (Table 2). For N_2O–N, fluxes were slightly stimulated by earthworms alone (E) and reduced by ash (A) (Fig. 1b, Table 2). The latter has also been reported by Schuetzenmeister et al. (2020) and might be related to photosynthesis. Increased N_2O emissions have been shown by different studies (e.g., Karsten and Drake 1997; Borken et al. 2000; Tianxiang et al. 2008) and can be explained by altered denitrification rates in the burrows, casts with high organic content, and the gut homogenates, which are rich in anaerobic microorganisms (e.g., Han and Wang 2003). Karsten and Drake (1997) showed nicely the role of earthworms in soil-borne N_2O emissions, making clear that earthworms can and should be considered "moveable hotspots" of N_2O

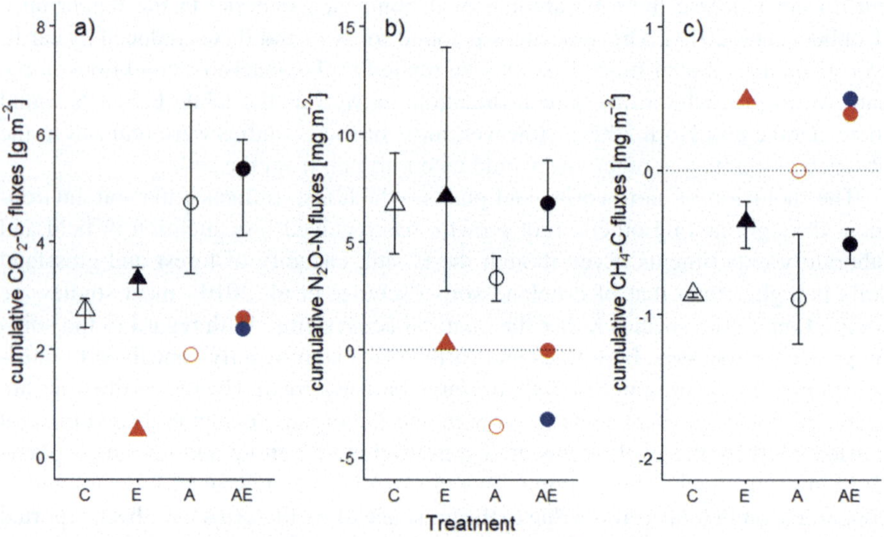

Fig. 1 Treatment-specific cumulative average fluxes of CO_2–C [g m^{-2}], N_2O–N [mg m^{-2}], and CH_4–C [mg m^{-2}]. Red triangles and circles represent the estimated fluxes after the deduction of C (third column in Table 1). The blue circle shows the estimated flux after the addition of the individual fluxes (after subtraction of C) (AE_{add}, fourth column in Table 1). C = control (soil and ash litter), E = ash litter and earthworms, A = planted ash sapling, and AE = planted ash sapling and earthworms. Error bars show minimum and maximum values. $n = 4$

Table 2 Treatment-specific cumulative average emissions for CO_2–C [g m^{-2}], N_2O-N [mg m^{-2}], and CH_4–C [mg m^{-2}] fluxes. Minimum and maximum values are presented in parenthesis. Different letters behind the values represent significant differences between the treatments (rows). The estimated effect of the manipulation (calculated as E – C, A – C, and AE – C, respectively) is presented in the third column (Manipulation effect). The fourth column (AE_{add}) contains the estimated fluxes after summarizing E and A (more precisely: (E – C) + (A – C)). C = Control (soil and ash litter), E = ash litter and earthworms, A = planted ash sapling, and AE = planted ash sapling and earthworms

Treatment	Mean (min; max)	Manipulation effect	AE_{add}
CO_2–C [g m^{-2}]			
C	2.8 (2.4; 3.0) *a*		
E	3.3 (3.0; 3.5) *a*	0.5	
A	4.7 (3.4; 6.5) *b*	1.9	
AE	5.3 (4.1; 5.9) *b*	2.5	2.4
N_2O–N [mg m^{-2}]			
C	6.8 (4.4; 9.1) *a*		
E	7.1 (2.8; 14.0) *a*	0.3	
A	3.3 (2.3; 4.4) *a*	−3.5	
AE	6.8 (5.6; 8.8) *a*	0.0	−3.2
CH_4–C [mg m^{-2}]			
C	−0.9 (−0.9; −0.8) *a*		
E	−0.4 (−0.5; −0.2) *b*	0.5	
A	−0.9 (−1.2; −0. 5) *a*	0.0	
AE	−0.5 (−0.6; −0.4) *b*	0.4	0.4

emissions. In the experiment presented here, the combined effect of earthworms and ash could not be predicted by the individual treatments: If added up (Fig. 1b), predicted fluxes were – 3.2 mg m^{-2}, while there were no fluxes found solely due to the manipulation treatment AE (0.0 mg m^{-2}) (Table 2). CH_4–C fluxes were negative from all treatments (Fig. 1c), implying that CH_4 was oxidized in the soil. Earthworms showed a decelerating effect on the CH_4–C uptake and both treatments E and AE showed significantly lower uptake rates than their counterpart without earthworms (Table 1). This corresponds to the findings of Borken et al. (2000) who reported that *Lumbricus terrestris* had a reducing effect on CH_4 oxidation in forest soil. The authors related this to anaerobic conditions in the earthworm gut caused by the feeding substrate and soil moisture leading to CH_4 production. The combined treatment AE and the prediction by adding the individual treatments E and A resulted in similar CH_4–C fluxes (0.4 and 0.5 mg m^{-2}, respectively). In contrast to the observed fluxes, estimated CH_4–C fluxes were positive after the deduction of the control treatment C.

The results suggest that it is important to consider systems (in this case the soil–plant–earthworm system) toward a sounder process understanding rather than disconnecting individual processes. However, besides the significant role of earthworms on nutrient cycling and, more specifically, greenhouse gas emissions, the results presented here should be taken with care, given the general setup, i.e., laboratory conditions. As highlighted by Lavelle et al. (2004), it is important to consider both spatial and temporal scales when addressing the roles of earthworms in nutrient cycling processes. Extrapolating from rather short-term laboratory microcosm experiments is a delicate issue when aiming at assessing long-term effects at the field or landscape level (e.g., Bohlen et al. 2004b). Lubbers et al. (2013) concluded that short-term experiments are not sufficient for identifying meaningful effects. The authors also recommended an experimental duration of at least 200 days. A promising step into that direction is the application and improvement of process-oriented models (see Chap. 3), which, in turn, rely on data from both laboratory and field studies in order to improve the underlying processes and improve their mathematical description.

3 Examples for Mathematical Modelling of Earthworm Bioturbation

The significant role of earthworm bioturbation on soil structure and nutrient cycling has been mentioned in previous chapters. The complexity of the processes occurring in the soil, which moreover are governed by environmental conditions, makes it difficult to understand and predict how parts of the system react to changes and disturbances and which impacts this will have on other parts. Mathematical models can help to understand and analyze these processes. One of the first attempts to simulate the processes by which earthworms consume, accumulate, and turn over C and N in soil has been made by Whalen et al. (1999). In their energetics-based model, the authors described the transfer between different pools using first-order kinetics. Besides the simplicity of the model, it successfully reproduced the growth and biomass gain of *Aporrectodea tuberculata* in relation to the litter quality and under different soil temperatures and moisture conditions as observed in the laboratory. This model was, however, purely focused on the ingestion and egestion of organic material by earthworms, and the potential impact of the burrowing itself on C and N turnover was not considered.

This has been taken up in more recent studies, e.g., Pelosi et al. (2008), Blanchart et al. (2009), and Meurer et al. (2020). This list is by far not extensive but rather aims at providing an overview on different modeling approaches of earthworm bioturbation targeting soil structure dynamics and, eventually, organic matter turnover. However, the three approaches taken by the models are quite different, as they target different properties of earthworm bioturbation. The population dynamics model WORMDYN presented by Pelosi et al. (2008) simulates the population density of a single earthworm species (*Lumbricus terrestris*) in the topsoil ($\leq$ 30 cm) as a function of changes in soil temperature and water content. Even though migration between soil layers in different development stages of the earthworm is included in the model, changes to the soil structure due to bioturbation are not considered.

In contrast to that, the SWORM model (Blanchart et al. 2009), a multi-agent (informatics) model, has been shown to successfully reproduce the effects of compacting and decompacting endogeic earthworms. In this model, bioturbation is considered through ingestion and egestion of soil and organic material, and this process creates biopores. Surprisingly, the model assumes a constant bulk density, i.e., no changes in bulk density occur due to the earthworm bioturbation, which is not what is usually being observed in the field. More specifically, it has been shown that changes in the physical and mechanical properties of soil as induced by ingestion and egestion can have consequences for soil aggregation and structural stability (e.g., Barré et al. 2009; Larink et al. 2001; Jouquet et al. 2012). This implies that the temporal evolution of the soil bulk density, as well as the porosity, is related to earthworm activity and bioturbation, respectively. Moreover, the fact that the model is limited to endogeic earthworms ignores the fact of casts left at the soil surface rather than in the burrows.

Both issues, i.e., the dynamically changing bulk density and the consideration of surface casting, have been addressed in a modeling approach presented by Meurer et al. (2020). In a case study, the authors applied a modeling framework to simulate soil structure recovery from severe traffic compaction in a field experiment at Zurich in Switzerland (see the experiment described by Keller et al. 2017). Given that the chosen treatment was a bare soil, the influence of plant roots was ignored, and the main agent for soil recovery was considered to be faunal bioturbation. For the modeling, the authors focused on earthworm bioturbation as a means to recover degraded soil structure. The experimental data showed that the topsoil porosity of the compacted bare soil had increased over a 4-year period, which could in part be attributed to soil deposition at the soil surface, i.e., earthworm casting. In order to account for this, Meurer et al. (2020) assumed that only a fraction of the ingested soil is egested into macropores, while the remaining fraction (f_{surf}) is cast at the soil surface. This, in turn, implies an increase in the total soil volume. In order to satisfactorily reproduce the observed values for soil bulk density, macro-, meso-, and micropores over 4 years of soil recovery, the authors had to assume larger surface casting rates in more compact soil—a fact which has been found in earlier studies (e.g., Joschko et al. 1989; Kretzschmar 1991; Zund et al. 1997; Buck et al. 2000; Jouquet et al. 2012). After calibration, the model simulations suggested that topsoil macro- and mesoporosity had largely recovered from compaction within 3 years as a result of earthworm bioturbation (Fig. 2). This result assumes, however, a rate of soil ingestion of 2.79 g soil g^{-1} biomass day^{-1}. A value, that is at the high end of the range reported by Curry and Schmidt (2007) from field experiments for temperate geophagous, i.e., soil-eating, species.

The modeling approach presented by Meurer et al. (2020) clearly demonstrated the impact of earthworm bioturbation on different parts of the soil structure (bulk density, but also pore size distribution). Given the fact that both the burrowing and the casting cause changes in the physical, chemical, and biological properties of the soil (e.g., as hot spot for microbial activity), the dynamic approach by Meurer et al. (2020) would be a valuable addition to soil organic matter models. In that case, the fact that the cast also provides physical protection of the organic matter from

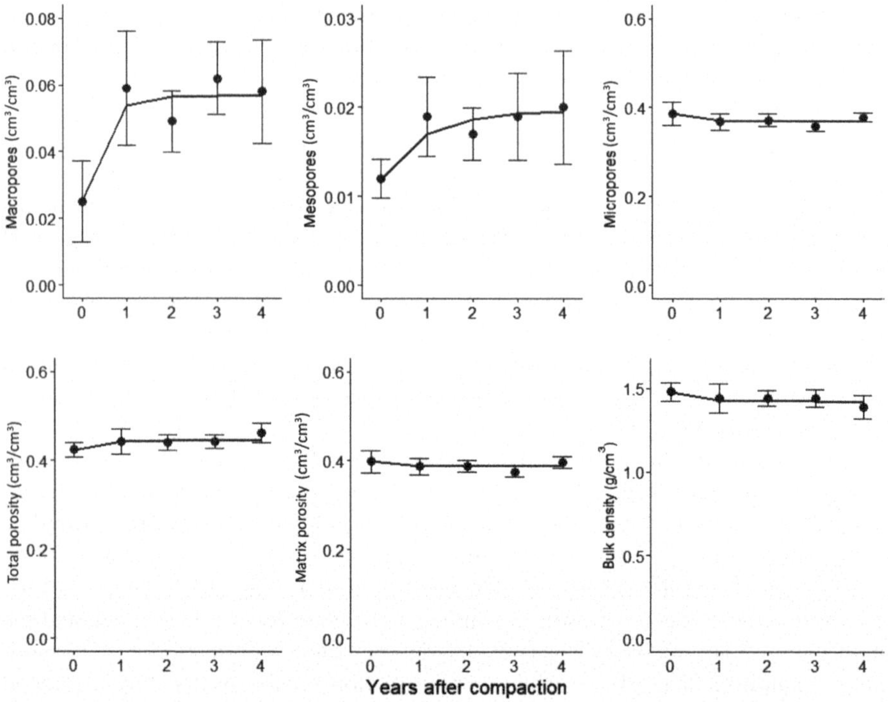

Fig. 2 Comparison of model simulations (lines) with observed soil physical properties on bare soil plots at the compaction recovery experiment (Keller et al. 2017; bars are standard deviations; Meurer et al. 2020)

mineralization, as has been suggested by Angst et al. (2017) is a process that has not received attention so far, but should be incorporated in the model.

4 Conclusions

Earthworms and, more specifically, bioturbation are vital for the functioning of the different aspects of the soil system, such as soil structure development and nutrient cycling. Through ingestion and egestion of mineral soil and organic material, earthworms alter the soil structure and pores size distribution and create new hotspots for microbial activities, eventually resulting in changes in the carbon and nitrogen turnover. Studies have shown that this effect differs between different earthworm species and groups. Endogeic earthworms have been studied in particular, as their presence has been related to both compaction and decompaction of soil. Here, differences have been pointed out between earthworm species and, more importantly, the size and material they feed on, as well as the structure of their cast. Besides structural changes, earthworm cast is considered a hotspot for microbial activity and has been shown to alter the denitrification rates in the burrows. The high content in organic material and anaerobic microorganisms lead to higher emissions of nitrous oxide (N_2O) when earthworms are present, making them "moveable hotspots." This fact is hardly considered in mathematical biogeochemical models. However, the mathematical description of earthworm bioturbation in soil has been reported, opening up the way for a better inclusion of these important soil and ecosystem engineers in future assessments of soil degradation and quality as a result of land management.

References

Alegre JC, Pashanasi B, Lavelle P (1996) Dynamics of soil physical properties in Amazonian agroecosystems inoculated with earthworms. Soil Sci Soc Am J 60:1522–1529

Amador JA, Görres JH (2007) Microbiological characterization of the structures built by earthworms and ants in an agricultural field. Soil Biol Biochem 39:2070–2077

Angst Š, Mueller CW, Cajthaml T, Angst G, Lhotáková Z, Bartuška M, Spaldonova A, Frouz J (2017) Stabilization of soil organic matter by earthworms is connected with physical protection rather than with chemical changes of organic matter. Geoderma 289:29–35

Barré P, McKenzie B, Hallett P (2009) Earthworms bring compacted and loose soil to a similar mechanical state. Soil Biol Biochem 41:656–658

Binet F, Curmi P (1992) Structural effects of *Lumbricus terrestris* (Oligochaeta: Lumbricidae) on the soil-organic matter system: micromorphological observations and autoradiographs. Soil Biol Biochem 24:1519–1523

Blanchart E, Lavelle P, Spain AV (1989) Effects of two species of tropical earthworms (Oligochaeta: Eudrilidae) on the size distribution of aggregates in an African soil. Revue d'Ecologie et de Biologie du Sol 26(4):417–425

Blanchart E, Bruand A, Lavelle P (1993) The physical structure of casts of *Millsonia anomala* (Oligochaete: Megascolecidae) in shrub savannah soils (Côte d'Ivoire). Geoderma 56:119–132

Blanchart E, Lavelle P, Braudeau E, Le Bissonnais Y, Valentin C (1997) Regulation of soil structure by geophagous earthworm activities in humid savannas of cote D'Ivoire. Soil Biol Biochem 29(4/4):431–439

Blanchart E, Albrecht A, Alegre J, Duboisset A, Gilot C, Pashanasi B, Lavelle P, Brussaard L (1999) Effects of earthworms on soil structure and physical properties. In: Lavelle P, Brussaard L, Hendrix PF (eds) Earthworm management in tropical agroecosystems. CABI Publishing, pp 87–147

Blanchart E, Marilleasu N, Chotte J-L, Drogoul A, Perrier E, Cambier C (2009) SWORM: an agent-based model to simulate the effect of earthworms on soil structure. Eur J Soil Sci 60:13–21

Blouin M, Hodson ME, Delgado EA, Baker G, Brussaard L, Butt KR, Dai J, Dendooven L, Peres G, Tondoh JE, Cluzeau D, Brun J-J (2013) A review of earthworm impact on soil function and ecosystem services. Eur J Soil Sci 64:161–182

Bohlen PJ, Pelletier DM, Groffman PM, Fahey TJ, Fisk MC (2004a) Influence of earthworm invasion on redistribution and retention of soil carbon and nitrogen in northern temperate forests. Ecosystems 7:13–27

Bohlen PJ, Parmelee RW, Blair JM (2004b) Integrating the effects of earthworms on nutrient cycling across spatial and temporal scales. In: Edwards CA (ed) Earthworm ecology, pp 161–180

Borken W, Brumme R (1997) Liming practices in temperate forest ecosystems and the effects on CO_2, N_2O and CH_4 fluxes. Soil Use Manag 13:251–257

Borken W, Grundel S, Beese F (2000) Potential contribution of *Lumbricus terrestris* L. to carob dioxide, methane and nitrous oxide fluxes from a forest soil. Biol Fertil Soils 32:142–148

Botinelli N, Henry-des-Tureaux T, Hallaire V, Mathieu J, Benard Y, Duc Tran T, Jouquet P (2010) Earthworms accelerate soil porosity turnover under watering conditions. Geoderma 156:43–47

Brown GG, Pashanasi B, Villenave C, Patrón JC, Senapati BK, Giri S, Barois I, Lavelle P, Blanchart E, Blakemore RJ, Spain AV, Boyer J (1999) Effects of earthworms on plant production in the tropics. In: Lavelle P, Brussaard L, Hendrix PF (eds) Earthworm management in tropical agroecosystems. CABI Publishing, pp 87–147

Buck C, Langmaack M, Schrader S (2000) Influence of mulch and soil compaction on earthworm cast properties. Appl Soil Ecol 14:223–229

Burtelow AE, Bohlen PJ, Groffman PM (1998) Influence of exotic earthworm invasion on soil organic matter, microbial biomass and denitrification potential in forest soils in the Northeast United States. Appl Soil Ecol 9(1–3):197–202

Capowiez Y, Pierret A, Daniel O, Monestiez P, Kretzschmar A (1998) 3D skeleton reconstructions of natural earthworm burrow systems using CAT scan images of soil cores. Biol Fertil Soils 27:51–59

Charpuis-Lardy L, Brauman A, Bernard L, Pablo AL, Toucet J, Mano MJ, Weber J, Brunet D, Razafimbelo T, Chotte JL, Blanchart E (2010) Effect of the endogeic earthworm Pontoscolex corethrurus on the microbial structure and activity related to CO_2 and N_2O fluxes from a tropical soil (Madagascar). Appl Soil Ecol 45(3):201–208

Cifuentes-Croqueville C, Stanton DE, Armesto J (2020) Soil invertebrate diversity loss and functional changes in temperate forest soils replaced by excotic pine plantations. Sci Rep 10:7762

Constanza R, dÁrge R, de Groot R, Farber S, Grasso M, Hannon B, Limburg K, Naeem S, O'Neill RV, Paruelo J, Raskin RG, Sutton P, van den Belt M (1998) The value of the world's ecosystem services and natural capital. Ecol Econ 25:3–15

Crooks JA (2002) Characterizing ecosystem-level consequences of biological invasions: the role of ecosystem engineers. Oikos 97:153–166

Curry J, Schmidt O (2007) The feeding ecology of earthworms—A review. Pedobiologia 50:463–477

Don A, Steinberg B, Schöning I, Pritsch K, Joschko M, Gleixner G, Schulze ED (2008) Organic carbon sequestration in earthworm burrows. Soil Biol Biochem 40:1803–1812

Drake HL, Horn MA (2006) Earthworms as a transient heaven for terrestrial denitrifying microbes: a review. Eng Life Sci 6(3):261–265

Edwards CA, Bohlen PJ (1996) Biology and ecology of earthworms, 3rd edn. Chapman and Hall, London, p 426

Edwards CA (2004) Earthworm ecology, 2nd edn. CRC Press, 456P

Ericksen-Hamel NS, Whalen JK (2007) Impacts of earthworms on soil nutrients and plant growth in soybean and maize agroecosystems. Agric Ecosyst Environ 120:442–448

Fisk MC, Fahey TJ, Groffman PM, Bohlen PJ (2004) Earthworm invasion, fine-root distributions, and soil respiration in north temperate forests. Ecosystems 7(1):55–62

Francis GS, Tabley FJ, Butler RC, Fraser PM (2001) The burrowing characteristics of three common earthworm species. Aus J Soil Res 39(6):1453–1465

Frelich LE, Hale CM, Scheu S, Holdsworth AR, Heneghan L, Bohlen PJ, Reich PB (2006) Earthworm invasion into previously earthworm-free temperate and boreal forests. Biol Invasions 8:1235–1245

Guéi AM, Badai Y, Tondoh JE, Huising J (2012) Functional attributes: compacting vs decompacting earthworms and influence on soil structure. Curr Zool 58(4):556–565

Guggenberger G, Thomas RT, Zech W (1996) Soil organic matter within earthworm casts of an anecic-endogeic tropical pasture community, Colombia. Appl Soil Ecol 3:263–274

Hale CM, Frelich LE, Reich PB, Pastor J (2005) Effects of European earthworm invasion on soil characteristics in northern hardwood forests of Minnesota, USA. Ecosystems 8:911–927

Han XG, Wang ZP (2003) Soil biodiversity and trace gases (CO_2, CH_4, N_2O) metabolism: a review. Biodivers Sci 11(4):322–332

Haynes RJ, Fraser PM (1998) A comparison of aggregate stability and biological activity in earthworm casts and uningested soil as affected by amendment with wheat or Lucerne straw. Eur J Soil Sci 49:629–636

Hendrix PF, Bohlen PJ (2002) Exotic earthworm invasions in North America: ecological and policy implications. Bioscience 52(9):801–811

Jang J, Xiong X, Liu C, Yoo K, Ishii S (2022) Invasive earthworms alter forest soil microbiomes and nitrogen cycling. Soil Biol Biochem 171:108724

Jarvis NJ, Taylor A, Larsbo M, Etana A, Rosén K (2010) Modelling the effects of bioturbation on the re-distribution of 137Cs in an undisturbed grassland soil. Eur J Soil Sci 61(1):24–34

Joschko M, Diestel H, Larink O (1989) Assessment of earthworm burrowing efficiency in compacted soil with a combination of morphological and soil physical measurements. Biol Fertil Soils 8:191–196

Jouquet P, Bottinelli N, Podwojewski P, Hallaire V, Duc TT (2008) Chemical and physical properties of earthworm casts as compared to bulk soil under a range of different land-use systems in Vietnam. Geoderma 146:231–238

Jouquet P, Huchet G, Bottinelli N, Thu T-D, Duc T-T (2012) Does the influence of earthworms on water infiltration, nitrogen leaching and soil respiration depend on the initial soil bulk density? A mesocosm experiment with the endogeic species *Metaphire posthuma*. Biol Fertil Soils 48:561–567

Karsten GR, Drake HL (1997) Denitrifying bacteria in the earthworm gastrointestinal tract and in vivo emission of nitrous oxide (N_2O) by earthworms. Appl Environ Microbiol 63(5):1878–1882

Keith AM, Robinson D (2012) Earthworms as natural capital: ecosystem service providers in agricultural soils. Economol J 2(2):91–99

Keller T, Colombi T, Ruiz S, Manalili MP, Rek J, Stadelmann V, Wunderli H, Breitenstein D, Reiser R, Oberholzer H, Schymanski S, Romero-Ruiz A, Linde N, Weisskopf P, Walter W, Or D (2017) Long-term soil structure observatory for monitoring post-compaction evolution of soil structure. Vadose Zone J 16(4):1–16

Koestel J, Schlüter S (2019) Quantification of the structure evolution of a garden soil over the course of two years. Geoderma 338:597–609

Kretzschmar A (1991) Burrowing ability of the earthworm Aporrectodea longa limited by soil compaction and water potential. Biol Fertil Soils 11:48–51

Larink O, Werner D, Langmaack M, Schrader S (2001) Regeneration of compacted soil aggregates by earthworm activity. Biol Fertil Soils 33:395–401

Lavelle P (1978) Les vers de terre de la savane de Lamto (Cote d'Ivoire): Peulements, populations et fonctions dans l'ecosystbme. Publications du Laboratoire de Zoologic de l'I~cole Normale Superieure, Paris, p 12

Lavelle P, Spain A (2001) Soil ecology. Springer Science & Business Media, 654P

Lavelle P, Pashanasi B, Charpentier F, Gilot C, Rossi J-P, Derouard L, Andre J, Ponge JF, Bernier N (2004) Effects of earthworms on soil organic matter and nutrient dynamics at a landscape scale over decades. In: Edwards CA (ed) Earthworm ecology, pp 145–160

Lemtiri A, Colinet G, Alabi T, Cluzeau D, Zirbes L, Haubruge E, Francis E (2014) Impacts of earthworm on soil components and dynamics. A review. Biotechnol Agron Soc Environ 18(1):1–13

Lipiec J, Turski M, Hajnos M, Świeboda R (2015) Pore structure, stability and water repellency of earthworm casts and natural aggregates in loess soil. Geoderma 243(244):124–129

Lubbers IM, Van Groenigen KJ, Fonte SJ, Six J, Brussard L, Groengen WV (2013) Greenhouse-gas emissions from soils increased by earthworms. Nat Clim Chang 3:187–194

Lyttle A, Yoo K, Hale C, Aufdenkampe A, Sebestyen S (2011) Carbon-mineral interactions along an earthworm invasion gradient at a sugar maple forest in Northern Minnesota. Appl Geochem 26:S85–S88

Marhan S, Kandeler E, Scheu S (2007) Phospholipid fatty acid profiles and xylanase activity in particle size fractions of forest soil and casts of *Lumbricus terrestris* L. (Oligochaeta, Lumbricidae). Appl Soil Ecol 35:412–422

Martin A, Marinissen JCY (1993) Biological and physico-chemical processes in excrements of soil animals. Geoderma 56:331–347
Matthies C, Griesshammer A, Schmittroth M, Drake HL (1999) Evidence for involvement of gut-associated denitrifying bacteria in emission of nitrous oxide (N_2O) by earthworms obtained from garden and forest soils. Appl Environ Microbiol 65(8):3599–3604
Meurer K, Barron J, Chenu C, Coucheney E, Fielding M, Hallett P, Herrmann AM, Keller T, Koestel J, Larsbo M, Lewan E, Or D, Parsons D, Parvin N, Taylor A, Vereecken H, Jarvis N (2020) A framework for modelling soil structure dynamics induced by biological activity. Glob Chang Biol 26:5382–5403
Nieminen M, Hurme T, Mikola J, Regina K, Nuutinen V (2015) Impact of earthworm *Lumbricus terrestris* living sites on the greenhouse gas balance of no-till arable soil. Biogeosciences 12:5481–5493
Pelosi C, Bertrand M, Makowski D, Roger-Estrade J (2008) Wormdyn: a model of *Lumbricus terrestris* population dynamics in agricultural fields. Ecol Model 218(3–4):219–234
Rossi J-P (2003) The spatiotemporal pattern of a tropical earthworm species assemblage and its relationship with soil structure. Pedobiologia 47:497–503
Schrader S, Rogasik H, Onasch I, Jégou D (2007) Assessment of soil structural differentiation around earthworm burrows by means of X-ray computed tomography and scanning electron microscopy. Geoderma 137:378–387
Schuetzenmeister K, Meurer KHE, Gronwald M, Hartmann ABD, Gansert D, Jungkunst HF (2020) N_2O emissions from plants are reduced under photosynthetic activity. Plant-Environ Interact 1(1):48–56
Schulze ED, Ciais P, Luyssaert S, Schrumpf M, Janssens IA, Thiruchittampalam B, Theloke J, Saurat M, Bringezu S, Lelieveld J, Lohila A, Rebmann C, Jung M, Bastviken D, Abril G, Grassi G, Leip A, Freibauer A, Kutsch W, Don A, Nieschulze J, Börner A, Gash JH, Dolman AJ (2010) The European carbon balance. Part 4: integration of carbon and other trace-gas fluxes. Glob Chang Biol 16:1451–1469
Sharma DK, Tomar S, Chakraborty D (2017) Role of earthworm in improving soil structure and functioning. Curr Sci 113(6):1064–1071
Speratti AB, Whalen JK (2008) Carbon dioxide and nitrous oxide fluxes from soil as influenced by anecic and endogeic earthworms. Appl Soil Ecol 38:27–33
Tianxiang L, Huixin L, Tong W, Feng H (2008) Influence of nematodes and earthworms on the emissions of soil trace gases (CO_2, N_2O). Acta Ecol Sin 28(3):993–999
van Groenigen JW, Lubbers IM, Vos HMJ, Brown GG, De Deyn GB, van Groenigen KJ (2014) Earthworms increase plant production: a meta-analysis. Sci Rep 4:6365
Wardle DA, Bardgett RD, Klironomos JN, Setälä H, van der Putten WH, Wall DH (2004) Ecological linkages between aboveground and belowground biota. Science 304:1629–1633
Whalen JK, Paustian KH, Parmelee RW (1999) Simulation of growth and flux of carbon and nitrogen through earthworms. Pedobiologia 43:537–546
Zund P, Pillai-McGarry U, McGarry D, Bray S (1997) Repair of a compacted Oxisol by the earthworm *Pontoscolex corethrurus* (Glossoscolecidae, Oligochaeta). Biol Fertil Soils 25:202–208

Earthworm Effects on Microbial and Enzyme Activities in Soil

Yahya Kooch and Yakov Kuzyakov

Abstract Earthworms have well-documented beneficial influences on plant growth and soil processes. The earthworm's capacity to decomposition of organic matter has been associated with the microbial communities residing in their digestive tracts or constructed structures, which serve as hotspots for microbial and enzymatic activities. Nevertheless, it is still unknown how the structure of soil microbial communities is altered by earthworms and how these alternations impact microbial functions in soil. Do earthworms feed on microorganisms and thereby lower microbial abundance and activity, or do they choose and promote certain microbial associations? The purpose of this chapter is to furnish a comprehensive overview of the effects of earthworms on the structure and functions of soil microbial communities. Additionally, this chapter aims to elucidate the impact of earthworm-induced shifts in the microbiota on soil functions, especially those that occur in the rhizosphere and are associated with plant health and growth. Finally, a framework for earthworm, microbial, and enzyme activity connections will be provided, along with recommendations for further research.

Y. Kooch (✉)
Faculty of Natural Resources & Marine Sciences, Tarbiat Modares University, Noor, Mazandaran, Iran
e-mail: yahya.kooch@modares.ac.ir

Y. Kuzyakov
Department of Soil Science of Temperate Ecosystems, Department of Agricultural Soil Science, University of Göttingen, Göttingen, Germany

Institute of Environmental Sciences, Kazan Federal University, Kazan, Russia

Peoples Friendship University of Russia (RUDN University), Moscow, Russia
e-mail: kuzyakov@gwdg.de

Y. Kooch, Y. Kuzyakov (eds.), *Earthworms and Ecological Processes*,
https://doi.org/10.1007/978-3-031-64510-5_13

1 Introduction

Earthworms, counting class N800, are an imperative group of soil fauna common in soils. Since their activities and interactions control soil biogeochemistry, earthworms are frequently alluded to as soil environment engineers (Blouin et al. 2013). Earthworms are one of the foremost vital categories of soil organisms that are essential in the decomposition of the plant residues in the topsoil. Earthworms keep up numerous interrelationships with microorganisms. They rely on microorganisms as their primary nourishment source, and by breaking down organic matter (OM) and inoculating it with microorganisms, they back microbial actions within the rot of natural matter and scatter microorganisms within the soil and litter (Edwards and Arancon 2022). Earthworms transfer enormous massive quantities of dead plant materials with a more carbon (C):nitrogen (N) ratio through their intestine system and digest more microbial protoplasm (Xiao et al. 2022). The microbial flora is enhanced by the soil earthworms that comminute these materials and, since the soil fauna uses such large amounts of organic matter, it regularly reuses what was once plant tissue (Whalen and Benslim 2021). The burrowing exercises and defecation of earthworms support the functioning of soil microbiota (Edwards and Bohlen 1996), and nutrient-rich earthworm casts are a great medium for advancing microbial development (Lee 1985). The digestive tracts of earthworms host various microorganisms. Earthworm molds contain more microbes than bulk soils. Earthworm castings include many more microorganisms, in expansion to showing more increased enzyme activities than the surrounding soil. The casts of distinctive species of earthworms have diverse characteristics based on the species and the inalienable features interior of their intestine.

Recent research has revealed that earthworm casts harbor a greater abundance of bacteria compared to bulk soils (Vidal et al. 2023). Furthermore, earthworm castings exhibit significantly higher levels of bacterial, actinomycete, and fungal populations, as well as greater enzymatic activity than the surrounding soil. It has been noted that the attributes of earthworm casts exhibit variability contingent upon the earthworm species and the inherent properties of their gut. In addition to their role in the dissemination of microorganisms through the ingestion and subsequent release of food in different locations, earthworms also facilitate the transport of microorganisms by carrying them on their bodies. This phenomenon has been documented by Edwards and Arancon (2022), highlighting the crucial role of earthworms in shaping microbial communities in soil ecosystems. Earthworms accelerate the turnover of soil organic matter (SOM) by excreting nutrients and moving within and between soil horizons. Their gastrointestinal tract has a huge number of organisms that frame a complicated interdependency with the host and recreate critical regulatory roles within the host's immune system, nutrient metabolism, and other host physiological processes (Liu et al. 2018). Earthworms are environment engineers, who change the physical, chemical, and biological soil properties through the movement of soil organisms, counting alters in microbial biomass, breath, or enzymatic action (Aira and Dominguez 2009). Earthworms can increase or decrease microbial abundance. Earthworm burrows give a habitat for other spineless creatures such as nematodes (Andriuzzi et al. 2016) and arthropods (Boivin et al. 2006), these together give a drilosphere layer containing organic carbon (C). The input of labile C and non-sustainable energy affect microbial community abundance and enzyme activities within the drilosphere and casts formed by earthworms in soil (Valchovski 2011; Lipiec et al. 2015) (Fig. 1). The present chapter endeavors to investigate the impact of earthworms on the activities of soil microbes and enzymes.

2 Earthworms and Hotspots for Soil Microbial and Enzyme Activities

Microbial activities play a crucial role in facilitating biogeochemical processes. Despite the abundance of microbial organisms, ranging from 10^7 to 10^{12} per gram of soil (Watt et al. 2006), their habitat is confined to small microhabitats, which constitute less than 1% of the total soil volume (Young et al. 2008). Furthermore, their presence is only observed on a minuscule fraction, approximately 10–6%, of the soil surface (Young and Crawford 2004). Numerous soil microbiota form biofilms and colonies covering entire microbial hotspots (Ekschmitt et al. 2005). Subsequently, environmentally pertinent biogeochemical forms frequently happen in small-volume soil hotspots. Hotspots are characterized as small soil volumes that exhibit

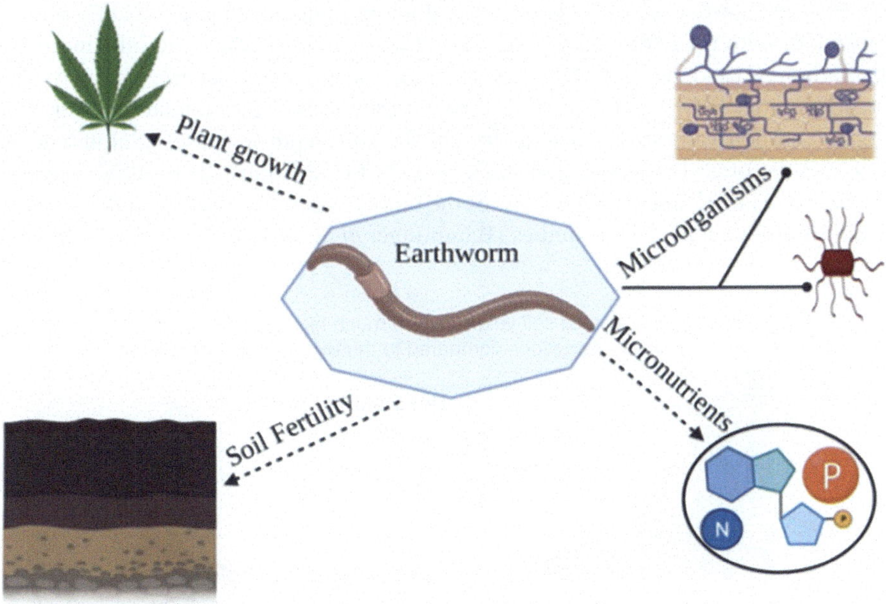

Fig. 1 The role of earthworms in soil fertility, as well as their interaction with microorganisms

significantly accelerated process rates and heightened interactions in comparison to the average soil conditions (Fig. 2) (Kuzyakov and Blagodatskaya 2015). Hence, hotspots and hotspot moments are defined by rates and energetic properties, specifically the use of the process. According to a survey conducted by Kuzyakov and Blagodatskaya (2015), the majority of research efforts have concentrated on microbial hotspots found in the rhizosphere and detritusphere. Limited findings are available regarding other hotspots, such as the earthworm drilosphere and castings. However, it is important to note that the impact of these hotspots on soil and ecosystem performance serves as an indicator for the decomposition of soil organic matter.

Biopores are critical hotspots with high rates of microbial processes (Hoang et al. 2016). The biopores are described as tubular soil macropores (Nakamoto 2000) or voids (Kautz 2015) created by earthworms (drilosphere), roots (the pores staying after root decomposition) (detritusphere), or persistent colonization of old root troughs by earthworms (detritus- and drilosphere) (Nakamoto 2000; Sanaullah et al. 2011). Root litter accumulation and earthworm burrows intensely enhance soil microbiota biomass (Sanaullah et al. 2011). The significance of subsoil biopores in facilitating nutrient supply to plants is noteworthy, as prolonged development and expulsion of litter from the field may lead to the depletion of nutrients and organic C in the topsoil (Kautz et al. 2013). Biopores impact environmental conditions by expanding the circulation of air and water in the subsoil layers of the soil (Kautz 2015). In addition, they also alter the biochemical features of the soil by modifying the microbial biomass and enzyme activities implicated in the cycles of phosphorus (P), nitrogen (N), carbon (C), and other nutrients (Jégou et al. 2001). However, numerous biopore research has concentrated on surface layers (Menichetti et al. 2014). The significance of biopores in facilitating plant development is often overlooked, despite their prominent role. Microbial activities, which are influenced by the carbon (C) and nitrogen (N) contents, are regulated by the decomposition of soil organic matter (SOM) (Hodge et al. 2000). In the bulk soil, the total nitrogen and organic carbon values exhibit a decline as the soil depth increases (Salomé et al. 2011). This forces biopores an essential source of C and nutrients in subsoil layers. Increased N and C stocks in the walls of the biopores compared to the bulk soil have been documented in several studies (Bamminger et al. 2014). The solubilization of

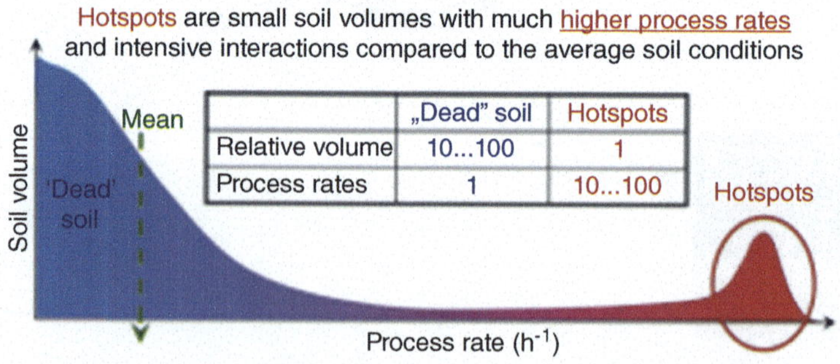

Fig. 2 Concept of microbial hotspots in soil (Kuzyakov and Blagodatskaya 2015)

P in soil is enhanced by earthworms through various mechanisms, including the reduction of pH, desorption of P from soil particles, and acceleration of microbial activities (Lopez-Hernandez et al. 1993). The composition and dynamics of the microbial community, in turn, may be influenced by factors such as the labile nature of OM, the availability of C and N, and the persistence of biopores originating from diverse sources (Kuzyakov and Blagodatskaya 2015).

3 Earthworm Impacts on Microbial Functions and Soil Structure

The impact of earthworms on microbial functions has been a subject of enduring discussion within the field of soil ecology. The potential consequences of the interplay between earthworms and microbes are dependent on the nature of their association and can lead to either positive or negative impacts. Given their relatively large size, earthworms are capable of preying upon smaller soil organisms, particularly microbes. As such, they are regarded as elevated-level consumers within detritivorous networks (De Ruiter et al. 1994; Hunt et al. 1987), and it is reasonable to anticipate that they would result in a reduction of microbial biomass. It should be noted that a reduction in biomass does not necessarily contradict an increase in microbial production, as the rate of recycling may be heightened (Barot et al. 2007). Consequently, the interactions between earthworms and microbes have been characterized as mutualistic. Earthworms are recognized as significant components of the soil ecosystem due to their ability to produce the largest biomass and their interaction with soil bacteria, which contributes to crucial processes of decomposition and nutrient cycling in the soil. As both dung and soil feeders, earthworms aim to enhance rates of microbial activity by changing the composition and structure of the soil microbial community during digestion (Aira et al. 2019). Earthworms play an essential function in the decomposition of plant materials by increasing the rate of decay of OM and modifying microbial processes (Butenschoen et al. 2007; Ernst and Emmerling 2009). Earthworms indirectly support microbial populations by pulverizing organic matter, thus creating an extensive surface area for further decomposition and microbial colonization (Seeber et al. 2008). The soil transmission by earthworms is indicative of the shift from aerobic to anaerobic environments (Drake and Horn 2007). Consequently, certain microorganisms traverse the earthworm's digestive tract and are assimilated into the earthworm's diet, whereas others, predominantly fermenters, proliferate. (Wüst et al. 2011). Microorganisms are commonly identified through the biochemical breakdown of soil organic matter. Nonetheless, soil earthworms have the ability to directly consume microorganisms and surrounding areas subsequent to the decomposition of SOM, thereby influencing microbial activity and consequently, the functioning of microbial decomposers. In certain instances, microorganisms may serve as a food source for soil earthworms. The number of microorganisms consumed and the capacity of earthworms to assimilate and digest biota are contingent upon the type of earthworm, nutrient substrate, and ecosystem in which they reside, ultimately impacting the rate of absorption and assimilation by earthworms (Gómez-Brandón et al. 2011). Also, soil aeration, the fragmentation in addition to the organic matter turnover via the activities of earthworms (burrowing and feeding), along with microbial activities ongoing within their guts, speeds up the humification process. Also, the higher cellulose and peroxidase activities level in the intestine of earthworms demonstrates their vital role in the condensation and decomposition levels of the humification process.

4 Microbial Activity in Biopores of Earthworms

Controlling the rate of humification and interactions with soil microflora through feeding, burrowing, and casting activities is probably one of the most essential functions of earthworms. This mainly happens to control carbon input into the soil from buried litter and to increase the rate of decomposition, to control microbial activity in drilosphere (the soil directly covering earthworm burrows) and castings, and to conserve carbon in stable aggregates such as carbon (Brown et al. 2000). Earthworm biopores are one of the main hotspots of microbial activity in soil (Fig. 3) (Kuzyakov and Blagodatskaya 2015) because of the input of (1) allocation and processing of unstable and recalcitrant OM in the hindgut of earthworms (Brown et al. 2000; Schrader et al. 2007), (2) earthworm casts and animal feces containing labile and recalcitrant organics (Tiunov and Scheu 2004), (3) higher content of N, C, and P, and (4) good aeration and water infiltration.

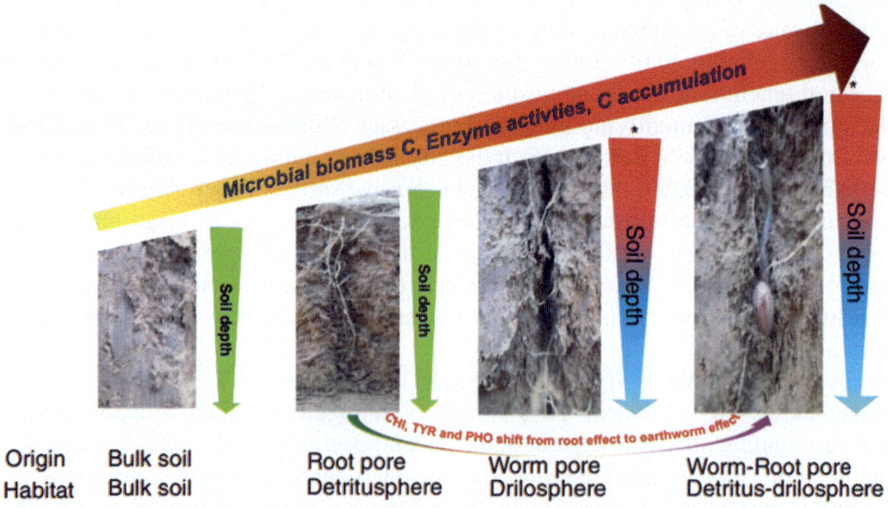

Fig. 3 The increment of MBC, enzyme activities, and C accumulation from bulk soil to biopores with a maximum in earthworm–root pores (Hoang et al. 2016)

Earthworms' activities (mechanical and biological) regulate microbial biomass and mobility, catalyzing C and N mineralization, SOM degradation, and nutrient cycles (Tiunov and Scheu 1999). Diverse earthworm species are apprehended to extend soil microbial respiration (Scheu 1987) or enzyme activities connected to N, C, and P turnover, particularly within the drilosphere (burrow and casts dividers, Aira et al. 2003). The body fluid secreted by earthworms is additionally a nutrients source for microbial action (Martin 1991). Regardless, consuming microbial biomass can reduce the overall microbial biomass while raising the specific activity of its residues (e.g., extracellular enzyme activity, Aira and Dominguez 2009). Higher microbial activity in biopores includes (1) the respiration of microorganisms because of mineralization of organic compounds, (2) a higher activity of a broad range of enzymes (Fig. 4, Hoang et al. 2016) leading to hydrolysis of plant-derived polymers, and (3) a larger area of hotspot percentage. All these lead to accelerated SOM turnover, e.g., priming effects (Fig. 5). Despite the accelerated SOM decomposition, the amount of C and N in biopores is more than in the bulk soil, because of much larger C input (1) from the topsoil allocated by anecic earthworms, (2) by growing roots and mycorrhiza hyphae, and (3) leached from litter horizon. Higher

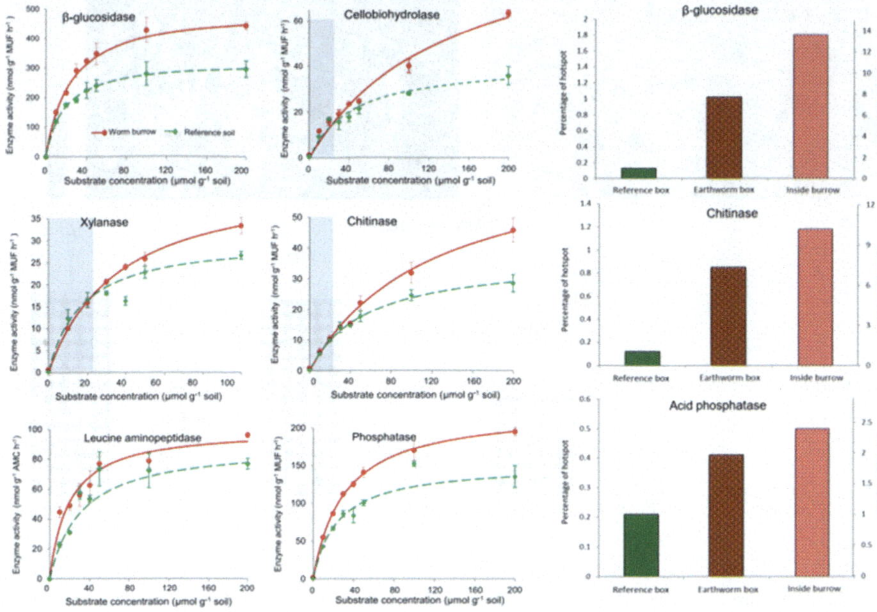

Fig. 4 On the left side, the kinetics of six enzymes (β-glucosidase, cellobiohydrolase, xylanase, chitinase, leucine aminopeptidase, and phosphatase) in earthworm biopores are presented through red points and curves, in comparison to bulk soil represented by green points and curves. The color-shaded sections highlight the concentration range where earthworm burrow effects were absent. On the right side, a comparison is made between the percentage of hotspot area in reference soil, earthworm boxes, and inside burrows (left axis) and the increment of hotspots (right axis) (modified from Hoang et al. 2016)

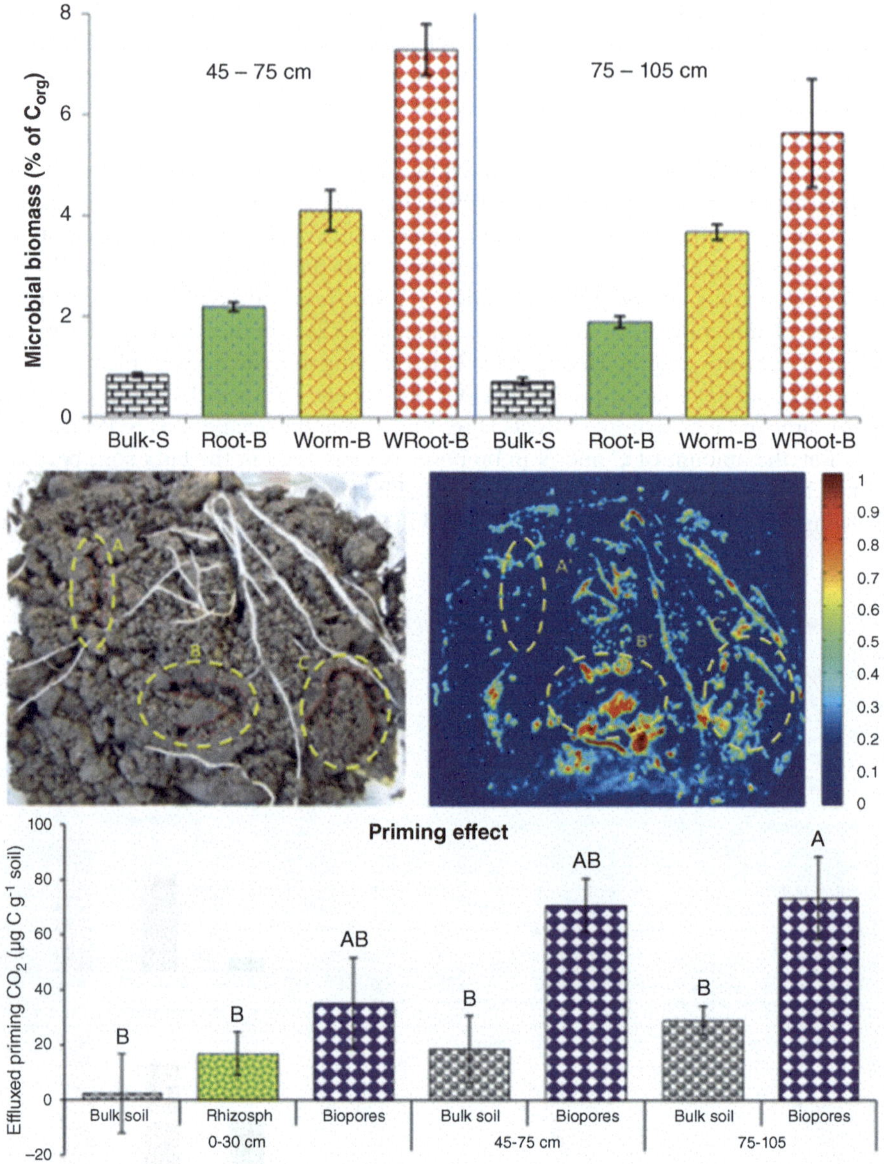

Fig. 5 Microbial biomass (top), enzyme activity hotspots (middle), and priming effects in bulk soil, rhizosphere soil, and biopores of various origins (bottom) (modified from Hoang et al. 2016)

microbial and enzyme activities as well as higher O_2 content, organic toxicants, e.g., pesticides, are decomposed faster in biopores compared to the bulk soil (Badawi et al. 2013).

5 Impact on Soil Nutrient Cycling Through Earthworms by Altering Soil Microbial Communities

The casts produced by earthworms are considered to be areas of high microbial activity due to their crucial involvement in the nutrient dynamics of soil ecosystems (Aira et al. 2019). Furthermore, they exert a notable influence on microbial communities, regulating nutrient cycles and promoting plant growth through their interactions (Medina-Sauza et al. 2019.(Earthworms are essential regulators of soil functions and processes such as nutrient dynamics, OM decomposition, microbial biomass and decomposition, soil structure, and plant growth. Earthworms, which participate in nutrient cycling processes involving P (Chapuis-Lardy et al. 1998), N (Lavelle et al. 1992), are important decomposers of soil. The influence of soil earthworms on soil structure has been extensively reported, and these organisms are widely regarded as the primary drivers of biological functions. Earthworms exert a significant influence on diverse soil features, including pH, organic matter value, N levels, and particle size, through their involvement in soil formation and destruction, as well as the conversion of OM. The interactions between earthworms, microbes, other organisms, and soil enzymes result in the production of nutrients. As earthworms consume soil, the ingested matter undergoes significant chemical and microbial transformations as it traverses their digestive tracts. This process results in the breakdown of organic materials, which subsequently leads to an increase in both pH levels and microbial activity within the stomach contents (Edwards and Bohlen 1996; Lukkari et al. 2006).

The acceleration of N mineralization in OM by earthworms is contingent upon the species and its interplay with soil properties, OM source, and soil microbiota, as illustrated in Fig. 6 (Butenschoen et al. 2009). Given the proclivity of organisms from diverse biological groups toward distinct diets, they can exert an impact on the mineral content of food. Anecic earthworms facilitate the conversion of litter nutrient items into soil minerals, thereby rendering them available for soil nutrients (Bossuyt et al. 2006). In contrast to other earthworm species, endogeic earthworms predominantly feed on soil and humified organic matter located in the upper layer of the mineral soil. The process of organic particle transformation in the soil is predominantly facilitated by soil microbiota, which handles the responsibility for this phenomenon. As such, their activity is a crucial factor in the cycling of soil organic matter (SOM) (Coq et al. 2007). The expansion of SOM mineralization in earthworm casts, when compared to the bulk soil, is associated with the enhancement of diverse compounds and subsequent increase in microbial activity. This phenomenon can be attributed to the digestive processes of earthworms themselves, as well as alterations in the intestinal microbiome (Abail et al. 2017).

Devliegher and Verstraete (1997) sought to separate between the nutrient enrichment function linked to the incorporation of SOM and the intestinal process associated with the migration of soil and SOM from the gut of *Lumbricus terrestris*. Their findings indicated that nutrient-enrichment forms, rather than intestinal processes, were responsible for the observed increase in microbial biomass and typical plant

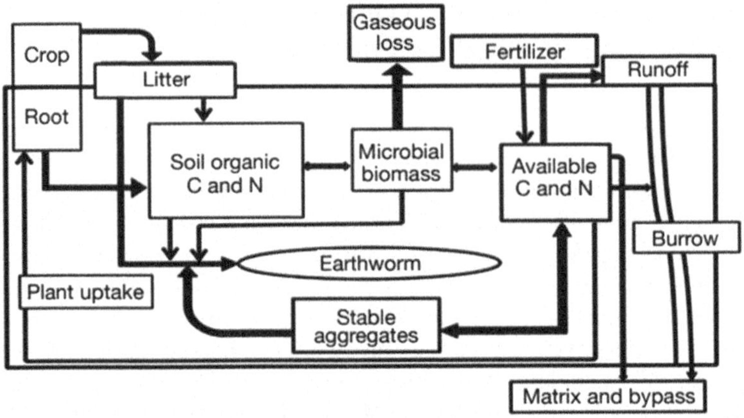

Fig. 6 The ecosystem budget model was utilized by Bohlen et al. (2004) to examine the various pools and fluxes of carbon and nitrogen in the presence of earthworms

movement. Earthworms change SOM dynamics since they are capable of affecting microbial activities and communities of other soil-dwelling invertebrates. Microorganisms heavily depend on larger organisms such as earthworms and roots to access fresh substrates due to their restricted motility. It is noteworthy that microorganisms are responsible for the majority of carbon turnover in soil, accounting for approximately 85–90% of the process (Gómez-Brandón et al. 2011). This was called the "sleeping beauty paradox" because roots and soil invertebrates stimulated the dormant microbial communities. As a result of bioturbation by earthworms, soils are biologically reworked and organic components are introduced into them, enhancing microbial activity. According to Vidal et al, a parallel buildup of microbial-dominated organic matter occurred through the decay of organic matter within litter, both in terms of living microorganisms' biomass and dead microorganisms' residues. Earthworms also boost the microbial biomass by adding large amounts of labile C through mucus during the organic matter ingestion. A major contributing factor to the C sequestration in soil is the increased presence of earthworms in the organic matter mineral-associated and tiny occluded particulate in the necromass stored in the soil. This is at the cost of losing plant combinations in less stable fractions.

Therefore, based on Ludwig et al., it is recognized that strong microbial activity supports C sequestration by microbial necromass production, creating a significant source of organic matter. Due to bacteria's large protein content, they act as stabilizing agents by increasing aggregate stability. The major populations of bacteria, fungi, actinomycetes, and also the related enzyme activities are found in the casts of earthworms thanks to the greater nutrient availabilities and higher structural stabilities of the casts. As organic C conservation is intimately linked with soil particles, it is further bound by microbial fungi hyphae and mucilage, resulting in cast stability and organic carbon conservation. According to Aira et al., while the biomass of the microbial community of the casts remains the same with age, the activity of the bacteria decreases resulting in a decreased organic matter mineralization. The microbial population reduction may decrease the CO_2 emission through microbial respiration. The bioturbation activities of earthworms, along with the occlusion of carbon and nitrogen within macroaggregates, have been found to play a crucial role in mitigating CO_2 emissions in non-flooded paddy fields that receive nitrogen fertilizer supplementation. The involvement of microorganisms in these processes is fundamental, and therefore, the influence of earthworms on microbial communities is a crucial factor in forecasting and comprehending the effects of earthworms on ecosystem functionality (Hong et al. 2011; Ahmed and Al-Mutairi 2022). Therefore, the influence of earthworms on the growth of plants and the cycling of nutrients is primarily indirect rather than direct, as it is primarily facilitated through alterations to the microbial community (Fujii et al. 2012; Shan et al. 2013).

6 Relationships Between Soil Earthworms and Enzymes

Soil enzymes, primarily produced and released from plant roots and soil microbiota and animals, recreate a critical function in mediating biochemical process exchanges, including nutrient cycling and organic residue decomposition in soil (McLatchey and Reddy 1998). Therefore, soil enzymes have the potential to function as a highly responsive indicator for the assessment of alterations in microbial activity and soil fertility (Parthasarathi and Ranganathan 2000). This is due to their strong association with the decomposition of essential nutrient elements, including nitrogen (N), phosphorus (P), and carbon (C). The role of earthworms in the process of litter decomposition is of paramount importance, particularly in their interaction with soil microorganisms. Additionally, earthworms act as soil engineers, altering the composition and enzymatic activities of microbial communities (Hoeffner et al. 2019). Earthworms that rely on microbes, grit present, and litter in the soil for their feeding and survival are expected to possess a diverse range of enzymes. The excrement produced by earthworms, commonly known as castings, is widely acknowledged as a valuable reservoir of plant growth-enhancing substances, encompassing enzymes, growth hormones, and vitamins (Kiyasudeen et al. 2016). Earthworm castings are known to harbor a diverse array of advantageous microorganisms, including those capable of nitrogen fixation, phosphorous solubilization, and cellulose decomposition, all of which contribute to the enhancement of soil productivity. Additionally, earthworms possess a self-contained reservoir of enzymes, such as acid phosphatase, nitrate reductase, and alkaline phosphatase, which play a crucial role in the metabolic processes associated with nitrogen and phosphate materials present within the soil (Kiyasudeen et al. 2016). The occurrence of mucus, which is formed by gut fluids with elevated levels of C and N and a reduced C/N ratio, has been observed in both excreta and earthworm burrows (Kizilkaya and Hepsen 2004). According to Prabha et al. (2005), earthworms accelerate the process of composting and convert waste materials into castings that are rich in nutrients through the assistance of enzymes. Consequently, the microbial and enzyme activities observed in earthworm excreta and biopores are generally higher in comparison to those present in the surrounding soil (Tiunov and Scheu 1999). The consumption patterns of earthworms are primarily demarcated by their ecological classification, which includes epigeic, endogeic, and anecic species. These organisms predominantly consume soil organic matter, which encompasses decaying plant remains and soil. Epigeic species primarily feed on soil organic matter, while endogeic and anecic species exhibit a preference for the composition of soil organic matter. Additionally, earthworms may consume soil microorganisms, including fungi and bacteria (Doube et al. 1997; Schonholzer et al. 1999). Therefore, the composition of materials and enzyme activities found in the intestines or excreta of earthworms undergo changes depending on their feeding habits. In the vicinity of earthworms, fluctuations in the quantity of organic matter give rise to variations in the enzyme activities of their excreta, which are further influenced by the nature and provenance of the enzymes involved.

7 Conclusions

The significance of earthworms in facilitating participation in biological cycles, promoting root growth, and augmenting the soil volume for C sequestration cannot be overstated. Microorganisms play a predominant role in the biochemical degradation of organic matter (OM). Earthworms are known to play a crucial role in the process of decomposition by acting as significant agents that can exert an influence on the activity of microbial decomposers through the direct consumption of microorganisms. Earthworms support soil microbial activity, nutrient transport, and improve physico-chemical properties, and finally provide beneficial conditions for important soil functions. Earthworms allocate organic matter and nutrients far below the tillage horizon and below the frequently dense Bt horizon of natural soils and extend the soil depth for all biological processes. Earthworms affect the structure of the soil microbial community directly through ingestion or indirectly via the priming effect caused by increased sensitivity. Microbial activities and biomass are much higher in earthworm-originated biopores as compared to bulk soil and even to the rhizosphere. Therefore, earthworms' abundance and activities promote soil enzyme activity, which allows more and faster biochemical processes to govern C, N, and P mineralization and cycling in soil. The aforementioned activities hold significant importance in the conversion of OM and plant nutrients into forms that are readily available and accessible to both plants and soil microorganisms, ultimately leading to an enhancement in soil quality and fertility. It is imperative to comprehend the impact of earthworms on soil fertility via their activities, particularly their influence on soil microbes and enzymes. These findings are important for knowing how nutrients move through the environment and the health of the soil.

References

Abail Z, Sampedro L, Whalen JK (2017) Short-term carbon mineralization from endogeic earthworm casts as influenced by properties of the ingested soil material. Appl Soil Ecol 116:79–86

Ahmed N, Al-Mutairi KA (2022) Earthworms effect on microbial population and soil fertility as well as their interaction with agriculture practices. Sustain For 14(13):7803

Aira M, Domínguez J (2009) Microbial and nutrient stabilization of two animal manures after the transit through the gut of the earthworm *Eisenia fetida* (Savigny, 1826). J Hazard Mater 161(2–3):1234–1238

Aira M, Monroy F, Domínguez J (2003) Effects of two species of earthworms (Allolobophora spp.) on soil systems: a microfaunal and biochemical analysis: the 7th international symposium on earthworm ecology· Cardiff·Wales·2002. Pedobiologia 47(5–6):877–881

Aira M, Pérez-Losada M, Domínguez J (2019) Microbiome dynamics during cast ageing in the earthworm Aporrectodea caliginosa. Appl Soil Ecol 139:56–63

Andriuzzi WS, Ngo PT, Geisen S, Keith AM, Dumack K, Bolger T, Bonkowski M, Brussaard L, Faber JH, Chabbi A, Rumpel C (2016) Organic matter composition and the protist and nematode communities around anecic earthworm burrows. Biol Fertil Soils 52:91–100

Badawi N, Johnsen AR, Brandt KK, Sørensen J, Aamand J (2013) Hydraulically active biopores stimulate pesticide mineralization in agricultural subsoil. Soil Biol Biochem 57:533–541

Bamminger C, Zaiser N, Zinsser P, Lamers M, Kammann C, Marhan S (2014) Effects of biochar, earthworms, and litter addition on soil microbial activity and abundance in a temperate agricultural soil. Biol Fertil Soils 50:1189–1200

Barot S, Ugolini A, Brikci FB (2007) Nutrient cycling efficiency explains the long-term effect of ecosystem engineers on primary production. Funct Ecol 21:1–1

Blouin M, Hodson ME, Delgado EA, Baker G, Brussaard L, Butt KR, Dai J, Dendooven L, Pérès G, Tondoh JE, Cluzeau D (2013) A review of earthworm impact on soil function and ecosystem services. Eur J Soil Sci 64(2):161–182

Bohlen PJ, Scheu S, Hale CM, McLean MA, Migge S, Groffman PM, Parkinson D (2004) Non-native invasive earthworms as agents of change in northern temperate forests. Front Ecol Environ 2(8):427–435

Boivin G, Kölliker-Ott UM, Bale J, Bigler F (2006) Assessing the establishment potential of inundative biological control agents. In: Environmental impact of invertebrates for biological control of arthropods: methods and risk assessment. CABI Publishing, Wallingford UK, pp 98–113

Bossuyt H, Six J, Hendrix PF (2006) Interactive effects of functionally different earthworm species on aggregation and incorporation and decomposition of newly added residue carbon. Geoderma 130(1–2):14–25

Brown GG, Barois I, Lavelle P (2000) Regulation of soil organic matter dynamics and microbial activity in the drilosphere and the role of interactions with other edaphic functional domains. Eur J Soil Biol 36:177–198

Butenschoen O, Poll C, Langel R, Kandeler E, Marhan S, Scheu S (2007) Endogeic earthworms alter carbon translocation by fungi at the soil–litter interface. Soil Biol Biochem 39(11):2854–2864

Butenschoen O, Marhan S, Langel R, Scheu S (2009) Carbon and nitrogen mobilisation by earthworms of different functional groups as affected by soil sand content. Pedobiologia 52(4):263–272

Chapuis-Lardy L, Brossard M, Lavelle P, Schouller E (1998) Phosphorus transformations in a ferralsol through ingestion by *Pantoscolex corethrurus*, a geophagous earthworm. Eur J Soil Biol 34(2):61–67

Coq S, Barthès BG, Oliver R, Rabary B, Blanchart E (2007) Earthworm activity affects soil aggregation and organic matter dynamics according to the quality and localization of crop residues—an experimental study (Madagascar). Soil Biol Biochem 39(8):2119–2128

de Ruiter PC, Neutel AM, Moore JC (1994) Modelling food webs and nutrient cycling in agro-ecosystems. Trends Ecol Evol 9(10):378–383

Devliegher W, Verstraete W (1997) The effect of Lumbricus terrestris on soil in relation to plant growth: effects of nutrient-enrichment processes (NEP) and gut-associated processes (GAP). Soil Biol Biochem 29(3–4):341–346

Doube BM, Schmidt O, Killham K, Correll R (1997) Influence of mineral soil on the palatability of organic matter for lumbricid earthworms: a simple food preference study. Soil Biol Biochem 29(3–4):569–575

Drake HL, Horn MA (2007) As the worm turns: the earthworm gut as a transient habitat for soil microbial biomes. Ann Rev Microbiol 61:169–189

Edwards CA, Arancon NQ (2022) The role of earthworms in organic matter and nutrient cycles. In: Biology and ecology of earthworms. Springer US, New York, NY, pp 233–274

Edwards CA, Bohlen PJ (1996) Biology and ecology of earthworms, vol 3. Springer Science & Business Media

Ekschmitt K, Liu M, Vetter S, Fox O, Wolters V (2005) Strategies used by soil biota to overcome soil organic matter stability—why is dead organic matter left over in the soil? Geoderma 128(1–2):167–176

Ernst G, Emmerling C (2009) Impact of five different tillage systems on soil organic carbon content and the density, biomass, and community composition of earthworms after a ten-year period. Eur J Soil Biol 45(3):247–251

Fujii K, Ikeda K, Yoshida S (2012) Isolation and characterization of aerobic microorganisms with cellulolytic activity in the gut of endogeic earthworms. Int Microbiol 15(3):121–130

Gómez-Brandón M, Lazcano C, Lores M, Domínguez J (2011) Short-term stabilization of grape marc through earthworms. J Hazard Mater 187(1–3):291–295

Hoang DT, Pausch J, Razavi BS, Kuzyakova I, Banfield CC, Kuzyakov Y (2016) Hotspots of microbial activity induced by earthworm burrows, old root channels, and their combination in subsoil. Biol Fertil Soils 52:1105–1119

Hodge A, Stewart J, Robinson D, Griffiths BS, Fitter AH (2000) Plant N capture and microfaunal dynamics from decomposing grass and earthworm residues in soil. Soil Biol Biochem 32(11–12):1763–1772

Hoeffner K, Santonja M, Cluzeau D, Monard C (2019) Epi-anecic rather than strict-anecic earthworms enhance soil enzymatic activities. Soil Biol Biochem 132:93–100

Hong SW, Lee JS, Chung KS (2011) Effect of enzyme-producing microorganisms on the biomass of epigeic earthworms (*Eisenia fetida*) in vermicompost. Bioresour Technol 102(10):6344–6347

Hunt HW, Coleman DC, Ingham ER, Ingham RE, Elliott ET, Moore JC, Rose SL, Reid CPP, Morley CR (1987) The detrital food web in a shortgrass prairie. Biol Fertil Soils 3:57–68

Jégou D, Capowiez Y, Cluzeau D (2001) Interactions between earthworm species in artificial soil cores assessed through the 3D reconstruction of the burrow systems. Geoderma 102(1–2):123–137

Kautz T (2015) Research on subsoil biopores and their functions in organically managed soils: a review. Renew Agr Food Syst 30(4):318–327

Kautz T, Amelung W, Ewert F, Gaiser T, Horn R, Jahn R, Javaux M, Kemna A, Kuzyakov Y, Munch JC, Pätzold S (2013) Nutrient acquisition from arable subsoils in temperate climates: a review. Soil Biol Biochem 57:1003–1022

Kiyasudeen SK, Ibrahim MH, Quaik S, Ahmed Ismail S, Ibrahim MH, Quaik S, Ismail SA (2016) Important digestive enzymes of earthworm. In: Prospects of organic waste management and the significance of earthworms, pp 105–122

Kızılkaya R, Hepsen S (2004) Effect of biosolid amendment on enzyme activities in earthworm (Lumbricus terrestris) casts. J Plant Nutr Soil Sci 167(2):202–208

Kuzyakov Y, Blagodatskaya E (2015) Microbial hotspots and hot moments in soil: concept & review. Soil Biol Biochem 83:184–199

Lavelle P, Melendez G, Pashanasi B, Schaefer R (1992) Nitrogen mineralization and reorganization in casts of the geophagous tropical earthworm Pontoscolex corethrurus (Glossoscolecidae). Biol Fertil Soils 14:49–53

Lee KE (1985) Earthworms: their ecology and relationships with soils and land use. Academic Press Inc.

Lipiec J, Turski M, Hajnos M, Świeboda R (2015) Pore structure, stability and water repellency of earthworm casts and natural aggregates in loess soil. Geoderma 243:124–129

Liu D, Lian B, Wu C, Guo P (2018) A comparative study of gut microbiota profiles of earthworms fed in three different substrates. Symbiosis 74:21–29

Lopez-Hernandez D, Lavelle P, Fardeau JC, Niño M (1993) Phosphorus transformations in two P-sorption contrasting tropical soils during transit through Pontoscolex corethrurus (Glossoscolecidae: Oligochaeta). Soil Biol Biochem 25(6):789–792

Lukkari T, Teno S, Väisänen A, Haimi J (2006) Effects of earthworms on decomposition and metal availability in contaminated soil: microcosm studies of populations with different exposure histories. Soil Biol Biochem 38(2):359–370

Martin A (1991) Short- and long-term effects of the endogeic earthworm *Millsonia anomala* (Omodeo) (Megascolecidæ, Oligochæta) of tropical savannas, on soil organic matter. Biol Fert Soils 11:234–238

McLatchey GP, Reddy KR (1998) Regulation of organic matter decomposition and nutrient release in a wetland soil (Vol. 27, No. 5, pp. 1268–1274). American Society of Agronomy, Crop Science Society of America, and Soil Science Society of America

Medina-Sauza RM, Álvarez-Jiménez M, Delhal A, Reverchon F, Blouin M, Guerrero-Analco JA, Cerdán CR, Guevara R, Villain L, Barois I (2019) Earthworms building up soil microbiota, a review. Front Environ Sci 7:81

Menichetti L, Ekblad A, Kätterer T (2014) Contribution of roots and amendments to soil carbon accumulation within the soil profile in a long-term field experiment in Sweden. Agric Ecosyst Environ 200:79–87

Nakamoto T (2000) The distribution of wheat and maize roots as influenced by biopores in a subsoil of the Kanto loam type. Plant Prod Sci 3(2):140–144

Parthasarathi K, Ranganathan LS (2000) Aging effect on enzyme activities in pressmud vermicasts of Lampito mauritii (Kinberg) and Eudrilus eugeniae (Kinberg). Biol Fertil Soils 30:347–350

Prabha LM, Jayraaj IA, Jeyaraaj R (2005) Macro and micronutrient changes in vermicomposting of vegetable wastes using Eudrilus eugeniae. South Asian J Soc Polit Stud 2(129–130):156

Salomé C, Guenat C, Bullinger-Weber G, Gobat JM, Le Bayon RC (2011) Earthworm communities in alluvial forests: influence of altitude, vegetation stages and soil parameters. Pedobiologia 54:S89–S98

Sanaullah M, Chabbi A, Leifeld J, Bardoux G, Billou D, Rumpel C (2011) Decomposition and stabilization of root litter in top-and subsoil horizons: what is the difference? Plant Soil 338:127–141

Scheu S (1987) Microbial activity and nutrient dynamics in earthworm casts (Lumbricidae). Biol Fertil Soils 5:230–234

Schönholzer F, Hahn D, Zeyer J (1999) Origins and fate of fungi and bacteria in the gut of Lumbricus terrestris L. studied by image analysis. FEMS Microbiol Ecol 28(3):235–248

Schrader S, Rogasik H, Onasch I, Jégou D (2007) Assessment of soil structural differentiation around earthworm burrows by means of X-ray computed tomography and scanning electron microscopy. Geoderma 137(3–4):378–387

Seeber J, Seeber GUH, Langel R, Scheu S, Meyer E (2008) The effect of macro-invertebrates and plant litter of different quality on the release of N from litter to plant on alpine pastureland. Biol Fertil Soils 44:783–790

Shan J, Liu J, Wang Y, Yan X, Guo H, Li X, Ji R (2013) Digestion and residue stabilization of bacterial and fungal cells, protein, peptidoglycan, and chitin by the geophagous earthworm *Metaphire guillelmi*. Soil Biol Biochem 64:9–17

Tiunov AV, Scheu S (1999) Microbial respiration, biomass, biovolume and nutrient status in burrow walls of Lumbricus terrestris L.(Lumbricidae). Soil Biol Biochem 31(14):2039–2048

Tiunov AV, Scheu S (2004) Carbon availability controls the growth of detritivores (Lumbricidae) and their effect on nitrogen mineralization. Oecologia 138:83–90

Valchovski H (2011) Microbiological and agrochemical characteristics of earthworm midden casts of Lumbricus terrestris. J Balkan Ecol 14(4):383–389

Vidal A, Blouin M, Lubbers I, Capowiez Y, Sanchez-Hernandez JC, Calogiuri T, van Groenigen JW (2023) The role of earthworms in agronomy: consensus, novel insights and remaining challenges. Adv Agron 181:1–78

Watt M, Hugenholtz P, White R, Vinall K (2006) Numbers and locations of native bacteria on field-grown wheat roots quantified by fluorescence in situ hybridization (FISH). Environ Microbiol 8(5):871–884

Whalen JK, Benslim H (2021) Earthworm populations are stable in temperate agricultural soils receiving wood-based biochar. Pedosphere 31(3):398–404

Wüst PK, Horn MA, Drake HL (2011) Clostridiaceae and Enterobacteriaceae as active fermenters in earthworm gut content. ISME J 5(1):92–106

Xiao R, Ali A, Xu Y, Abdelrahman H, Li R, Lin Y, Bolan N, Shaheen SM, Rinklebe J, Zhang Z (2022) Earthworms as candidates for remediation of potentially toxic elements contaminated soils and mitigating the environmental and human health risks: a review. Environ Int 158:106924

Young IM, Crawford JW (2004) Interactions and self-organization in the soil-microbe complex. Science 304(5677):1634–1637

Young IM, Crawford JW, Nunan N, Otten W, Spiers A (2008) Microbial distribution in soils: physics and scaling. Adv Agron 100:81–121

Earthworms for Soil Organic Matter Mineralization and Carbon Sequestration

Yahya Kooch and Yakov Kuzyakov

Abstract Earthworms are known to modify soil structure and alter the physico-chemical and biological regimes of soils. Both laboratory and field experiments confirmed the effects of earthworms on soil characteristics, which may be physical, chemical, biological, or a combination of all three. Earthworms through their vital activities such as feeding, burrowing, and casting play important roles in the transformations of primary and secondary minerals and nutrients to available and accessible forms for plants and soil microorganisms and thus increase soil fertility and quality. Many challenges remain in determining the operational role of earthworms in Soil organic matter (SOM) dynamics, carbon sequestration, and aggregate formation. After moving through the stomachs of earthworms, soil particles are entirely rearranged and form aggregates, and consequently, contribute to mineral weathering and SOM formation. Earthworms partially decompose SOM, whereas a portion of it is physically protected within microaggregates. Complementary studies are needed to assess the impact of earthworms on SOM dynamics, the conditions under which physical protection develops, and the impact of earthworms on microbial community structure.

Y. Kooch (✉)
Faculty of Natural Resources & Marine Sciences, Tarbiat Modares University, Noor, Mazandaran, Iran
e-mail: yahya.kooch@modares.ac.ir

Y. Kuzyakov
Department of Soil Science of Temperate Ecosystems, Department of Agricultural Soil Science, University of Göttingen, Göttingen, Germany

Institute of Environmental Sciences, Kazan Federal University, Kazan, Russia

Peoples Friendship University of Russia (RUDN University), Moscow, Russia
e-mail: kuzyakov@gwdg.de

Y. Kooch, Y. Kuzyakov (eds.), *Earthworms and Ecological Processes*,
https://doi.org/10.1007/978-3-031-64510-5_14

1 Introduction

Soil formation is a long-term process that is partially dependent on climate and parent material (Chesworth 1992). These include the weathering of primary minerals and the stabilization of organic matter (OM). Darwin (1881) was one of the first to include organisms, particularly earthworms, in his list of factors that make up the soil through the process of accumulation and mixing of soil organisms excreta. The importance and vital role of earthworms in chemical weathering were first reported by Darwin (1881). The role of earthworms in generating SOM has been studied in more detail than in mineral weathering (Edwards et al. 2011). Earthworms are heterotrophs and are often involved in the breakdown of SOM and molecules produced not only by plants but also by other heterotrophs. One of the most important functions of earthworms is to dig, feed, water, and maintain SOM by interacting with organisms. Earthworms not only contribute to mineral weathering and SOM formation but also excavate organic matter from the surface and allocate them deep into the soil, as well as carry soil particles from deep layers to the surface.

Erosion is crucial for soil formation, and earthworms perform a vital role in erosion, especially through the formation of casts on the soil surface. Earthworms accelerate SOM decomposition by increasing the level of organic matter through grinding (Seeber et al. 2008). Organic compounds in the form of small monomers are released into the environment after being digested by earthworms or converted into nutrients. These nutrients, especially nitrogen (N), are used by plants and promote plant growth (Lee 1985; Bityutskii et al. 2002). Earthworms increase soil organic matter uptake and macroaggregate formation through digging, feeding, and excretion activities (Blanchart et al. 1997). This suggests that earthworms also play a serious role in soil carbon sequestration, as storing C in compact and stable aggregates is a key stabilization process and prevents its rapid release as a greenhouse gas (Lavelle et al. 2006). Therefore, this review aims to discuss the impact of earthworms on the network of decomposers, SOM dynamics, decomposition processes, plant growth, and soil C sequestration.

2 Earthworms and Decomposition Processes

Earthworms are the dominant community of large decomposers in temperate deciduous grasslands and forests (Regulska and Kołaczkowska 2015; Kooch et al. 2021). Since the ground-breaking work of Darwin (1881), the earthworm's impacts on the soil physicochemical properties of materials have been well documented. There has been extensive research on the earthworm casts enrichment with plant nutrients such as N, P, K, Ca, and Mg, as well as their role in incorporating plant residues in pastures and cropping systems, as well as in forest ecosystems (Schon et al. 2017). Earthworms are generally referred to as ecosystem engineers because of their ability to modify soil and plant communities (Lavelle et al. 1997; Hale et al. 2005).

Through the incorporation of surface litter, casting, burrowing, and other activities alter soil nutrient content and availability (Edwards and Bohlen 1996), soil physical properties (Edwards et al. 1989), microbial communities (Doube and Brown 1998), and above-ground plant communities (Piearce et al. 1994; Wurst et al. 2005). However, there is still a demand for studies on the mechanisms through which earthworms improve soil fertility. Through feeding on organic substances, earthworms decrease the C:N ratio in plant residues. By digesting organic matter and assimilating a small portion into the tissues of earthworms, some C is removed as CO_2 through the respiration process, therefore decreasing the C:N ratio, while the larger fractions of ingested materials are excreted (Regulska and Kołaczkowska 2015).

There is a need for much more quantifiable work in this context, i.e., that earthworms can have a more vital role in decomposition processes than is commonly known (Sapkota et al. 2020). About 80–90% of net primary production above ground enters the soil food web as dead plants in most terrestrial ecosystems (Bardgett et al. 2005). Bacteria and fungi, which account for up to 90% of soil microbial biomass, are the major litter decomposers (Berg and Laskowski 2005), but the food webs of decomposers are very complex and include protozoa, nematodes, micronodes, foot animals, and earthworms (Wardle 2006). Earthworms play an important role in degrading organic matter, greatly accelerating decomposition rates and nutrient turnover (Lavelle and Spain 2001a, 2001b, 2001c). Earthworms influence the decomposition of organic matter through gut-related processes (direct effect). They are released into earthworm feces via ingestion, digestion, and assimilation effects of organic and microbial matter (Monroy et al. 2008; Aira et al. 2009). Such indirect effects derive from direct effects and include processes such as aging (weeks to months) of material treated by earthworms and mixing this material with a substrate not yet treated by earthworms (Aira et al. 2010).

Earthworms constitute the majority of soil fauna biomass in most temperate terrestrial ecosystems and have an important impact on SOM turnover, nutrient cycling, and ecosystem function and structure (Tamartash and Ehsani 2021). Lavelle et al. (2004) described the "drilosphere" to comprise populations of the earthworm and all the soil volume, populations of microbes and invertebrates expressed by their activity, including (i) the inner microenvironment of the earthworm intestine system (internal rumen), (ii) the surface of earthworm in contact with soil and the external made earthworm structures, (iii) belowground and surface casts, (iv) middens, and (v) burrows, diapauses chambers or galleries (closed and open). The impact of "drilosphere" processes and structures (external and internal) on SOM dynamics and microbial activities at various scales of time and space, from the earthworm intestine to the soil profile and from hours to decades (see Fig. 1) is described by Brown et al. (2000). Earthworms play a critical role in the dynamics of C and N and agricultural management influences the direction and magnitude of their impact (Kooijman et al. 2008). Even though this position seems to be unique, which is, particular mechanisms for every type of earthworm community for each type of soil in peculiar states (Lavelle et al. 2004).

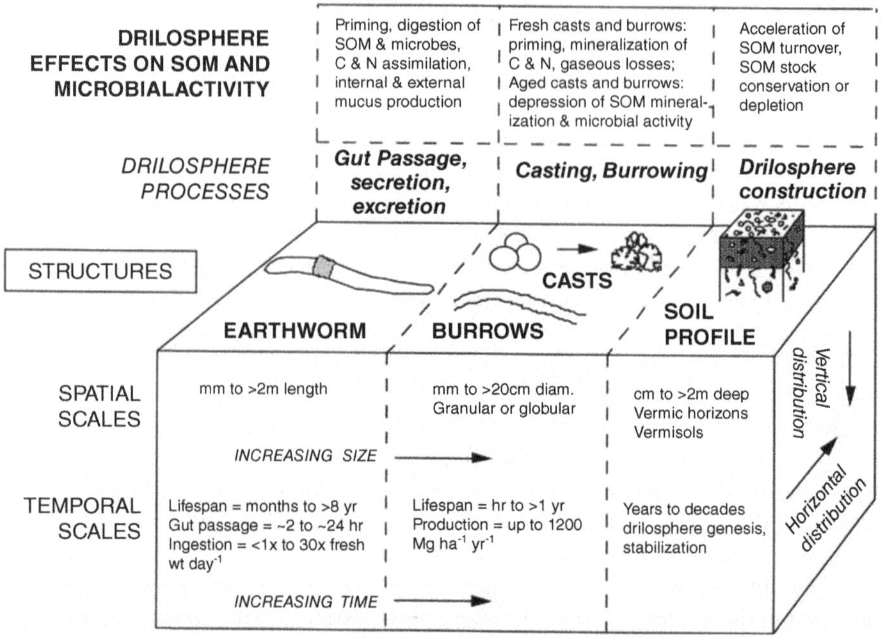

Fig. 1 The influence of drilosphere structures and processes (internal and external) on the dynamics of SOM and soil microbial activity at different spatial and temporal scales, from the earthworm gut to the soil profile and from several hours to several decades (Brown et al. 2000)

3 Groups and Localization of Earthworms and Their Living Story

Earthworm groups include multiple functional associations and groups that have different ecologies and effects on the environment. In the early 1970s, soil biologist Marcel Boucher classified French earthworm species into three ecological categories. This classification system has been very successful and is widely used outside of Europe to describe functional groups of earthworms. Bouché used morphoanatomical features to distinguish three main categories (Bottinelli et al. 2020). Bouché (1972) described the various life strategies of earthworms known as ecological groups as (i) *anecic* (feeding on the surface litter, yet otherwise residing at depth soil), (ii) *epigeous* (dwelling and feeding under bark or on surface litter), and (iii) *endogeic* (feeding and living in horizons of the soil underneath the litter). This classification has supplied effective mechanisms to classify earthworms' activities and their impacts on soil (see Table 1). Anecic earthworms usually move on the soil surface, while endogeic earthworms typically move horizontally in the top 20 cm of soil (Manlay et al. 2007). About the dynamics of soil water, for instance, the anecics vertical burrows increase the rapid preferential flow, while the activity of epigeic earthworms increases topsoil moisture and the activity of endogeic earthworms increases the subsoil water discharge and infiltration rates. Through the process of

Table 1 Ecological groups of soil earthworms (Bouché 1972)

Ecological group	Endogeic	Anecic	Epigeic
Physical appearance Size Color	Medium 1–20 cm Pink or light grey	Large 10–110 cm Dark grey, or red or brown	Small 1–5 cm Dark red
Picture	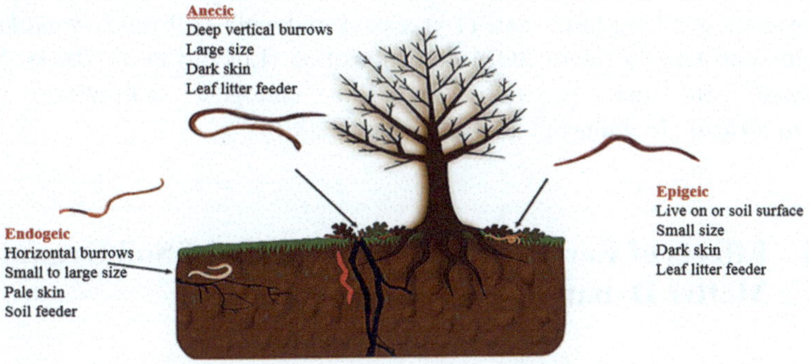		

composting, from the viewpoint of ecological engineering, organic matter is transformed via epigeic earthworms.

4 Effects Earthworms on Soil Organic Matter

Organic matter and litter are utilized in a process known as vermicomposting to produce organic materials, which have better nutritional, physical, and biochemical features than those in traditional composts (Kumar et al. 2021). Aside from ingesting large quantities of soil, endogeic earthworms also assimilate a portion of SOM, which is probably why they are regularly called "geophagous." Microorganisms are moved to new substrates during the transfer in the endogeic intestine (Lassauce et al. 2011) and the presence of mucus (readily assimilated organic matter) and fresh organic residue in ingested soil stimulate their activity. The endogeic earthworm's function in agroecosystems intensification is probably critical. In agronomical systems, endogeic earthworms are more universal in particular, while anecic and epigeic earthworms depend on agricultural practices with the litter layer management to survive. As ecosystem engineers, they interact with microflora, selectively digest soil particles, produce casts, burrows, galleries, and chambers, and can influence the dynamics of organic matter, nutrients, and pedological processes through mutualistic interactions with microflora, selective ingestion of soil particles, and high ingestion rates (Law and Lai 2021). Endogeic earthworms have been given comparably small attention given their presumptive significance in soil functioning.

Ecological groups of earthworms impact nutrient turnover and dynamics of SOM. As far as the carbon cycle is concerned, studies have shown that there is a higher carbon (C) content in soil casts, and the mineralization rate of this carbon is slower when compared to the bulk soil around them at time scales ranging from days to months (Londo 2001). Since Fresh organic matter (FOM) from the surface of litter ingested by anecic earthworms and blended to bulk soil can accumulate C content compared to bulk soil, the higher C content can be rational. The low mineralization in casts made by intestinal earthworms mucopolysaccharides could occasionally be explained as a result of a physically protected C pool (Lubbers et al. 2013). Some other research conducted with endogeic species concluded that in savannas at a long-term scale (1 yr. casts), endogeic *Millsonia anomala* populations can considerably retard SOM decomposition (Lukkari et al. 2004). Nevertheless, based on the research on the endogeic earthworm *Pontoscolex corethrurus* (Mabuhay et al. 2003; Londo 2001).

5 Effects of Earthworms in the Models of Soil Organic Matter Dynamics

In temperate agricultural systems, earthworm species are the greatest effective macroorganisms. In pedobiological models that describe soil dynamics, earthworm activities are not represented, despite their importance as regulators of soil physical structure and dynamics of SOM. Few studies have attempted to include the effects of micro- and macroorganisms of the soil biota in the models of SOM mineralization conducted for modeling of soil C: Chertov and Komarov (1997) modeled SOM considering soil organisms as biological factors and not as distinct pools. A network of soil food as a "super-organism" that processes the SOM was described by Fu et al. (2000). A long-term period of earthworm's effects was modeled by Lavelle et al. (2004). According to their simulation, the activity of earthworms in the dynamics of SOM could decrease by 10% in 30 years based on the various pools of the "CENTURY" model. A fractal approach was utilized for modeling a complex environment and combining it into a multi-agent system (MAS) which simulates earthworms in a virtual world as located factors (Marilleau et al. 2008; Blanchart et al. 2009). They also recommended a SWORM (for 'Simulated Worms') model helping to describe the effects of earthworms on processes of physics and biology in soil functioning; however, a confirmed mathematical model of SOM turnover was lacking. A simulation model to assess the possible effects of invasions of earthworms on the dynamics of carbon in forest ecosystems was developed by Huang et al. (2010). In temperate ecosystems, earthworms form the major part of the biomass of macrofauna in soil. The earthworm's significance in influencing organic matter processing, nutrient cycling, and soil structure has been accepted for a long time (McCay and Scull 2019). With their casting and burrowing activities, earthworms can turn over enormous amounts of soil annually (up to 10% of volume in topsoil of some natural Chernozems and phaeozems) and can cause significant changes in soil structure, which is significant for the soil development (Medina-Sauza et al. 2019). Soil aggregates are destroyed during the earthworm intestine passage, creating organic materials that intimately mix the soil matrix. The organic matter interaction with the soil matrix during the earthworm's intestine system can promote carbon stabilization by enhancing the linkage with soil clay minerals (Mosavi and Raisi 2009). Gut-associated processes (GAPs) include processes that force the organic matter to be mobilized during the gut passage of earthworms (Nachtergale et al. 2002). By contrast, the presence of "cast-associated processes" (CAPs) in earthworm casts can stabilize organic matter (Neirynck et al. 2000).

6 Seasonal Dynamics of Earthworm Populations

Seasonal dynamics of earthworm populations strongly depend on climate and as consequence on plant development and litter input. Despite its critical role, the temporal heterogeneity of communities of earthworms is still understudied. Considering that earthworms are sensitive to negligible changes in soil temperature and moisture, the dynamics of earthworm communities are controlled by the weather and have seasonality. There is a variation in the length of the growing season in temperate ecosystems, and the earthworm abundance differs during the year in croplands (McCay and Scull 2019). Nonetheless, the population curves form and the temporal change reasons are yet not defined completely. Whilst the smallest number of earthworms are visible in the driest cropping seasons (late summer to early fall), the most numerous earthworms are recorded during and after peak precipitation (late spring to summer) (Medina-Sauza et al. 2019). In a temperate climate, populations of earthworms follow a different seasonal pattern than in other climates. The genesis of cocoons tends to be higher in the spring and summer seasons in a temperate climate. In spring, however, a high amount of cocoons hatches and produces a huge juvenile community because of cocoon accumulation during the chillier months (fall and spring) (Fu et al. 2000). Cocoons are produced by juveniles who survive until the late summer season and mature to become adults. Then a lot of those individuals die because of a lack of food or frost during the winter season. Cocoons, however, are kept protected from these disturbances, tend to be safe during winter, and then start the new cycle by hatching in the spring. The mentioned life cycle is usually observed with a large population dominated by juveniles in spring and a smaller and more uniform population in the late summer (Mosavi and Raisi 2009). Larger populations of earthworms were discovered in the fall season than in summer and spring in some long-season prairies in Kansas and Europe. Populations of earthworms were four to ten times higher in fall than in late spring and early summer in either conventional or organic farmlands of Pennsylvania (McCay and Scull 2019). In the case of humid conditions in summer, it is possible for cocoons laid in spring to hatch and produce the juvenile's next cohort in fall (Mosavi and Raisi 2009). Also, harsh climatic conditions, such as extended freezing of soil surface or drought, can impact their populations. Populations of earthworms, which were initially huge in spring and fall, dropped after drought in the late summer and fall and did not get back to normality in the following year (Power 2010).

7 Effects of Earthworms on Plant Growth

The importance of earthworms in plant growth has been known for more than 100 years since the publication of Charles Darwin's book in 1881. Over the last century, the effects of earthworms on crop production were described in the field mainly in the temperate zone and small (pot) levels. These earthworms exert primarily beneficial effects on plant growth, although in a few cases, negative or no effects could be induced. Despite the earthworms are useful for the growth of plants, there is no convincing scientific evidence. The earthworm's impact on nutrients and the growth of plants is not stable and is extremely reliant on soil type, plant species, and earthworm species. The earthworm's relationships, plant nutrient uptake, and yields, as well as soil N pools, are also influenced by soil type, species of earthworm, and fertility treatments. Brown et al. (2004) proposed five ways in which earthworms can affect plant growth: (i) biological control of pets and disease, (ii) stimulation of microbial plant symbiotes, (iii) production of plant growth regulators, (iv) soil structure alteration, and (v) Increase the availability of nutrients (Fig. 2).

7.1 Effects of Earthworms on Biocontrol of Pests and Diseases

Soil fauna influences plant health and growth through direct and indirect interactions in the rhizosphere (Schrader et al. 2013). In particular, earthworms are known to interact with soil biota (Gómez-Brandón et al. 2012). Soil microorganisms are the most important food source for many types of earthworms (Byzov et al. 2007). Therefore, the consumption of plant pathogens in soil by earthworms has many positive effects on plant health (Bi et al. 2018). For saprophytic pathogens, earthworms can actively consume all pathogen-infected crop residues, thereby reducing the pressure of the disease on the plant. Also, incorporating such plant matter into the soil reduces disease pressure (Wolfarth et al. 2011). Therefore, the effects of pathogens in plants can be achieved both indirectly by consuming dormant spores or tissues and by increasing plant resistance. Thus, the main mechanisms by which earthworms control plant diseases may be the result of both direct and indirect effects.

7.2 Effects of Earthworms on Stimulation of Microbial Plant Symbionts

Earthworms are important environmental mediators and can influence soil functions and microbial activity by inducing priming effects. Earthworms can change the dynamics of SOM since they are capable of affecting soil microbial activities and communities of other soil-dwelling invertebrates. The microbes, because of their

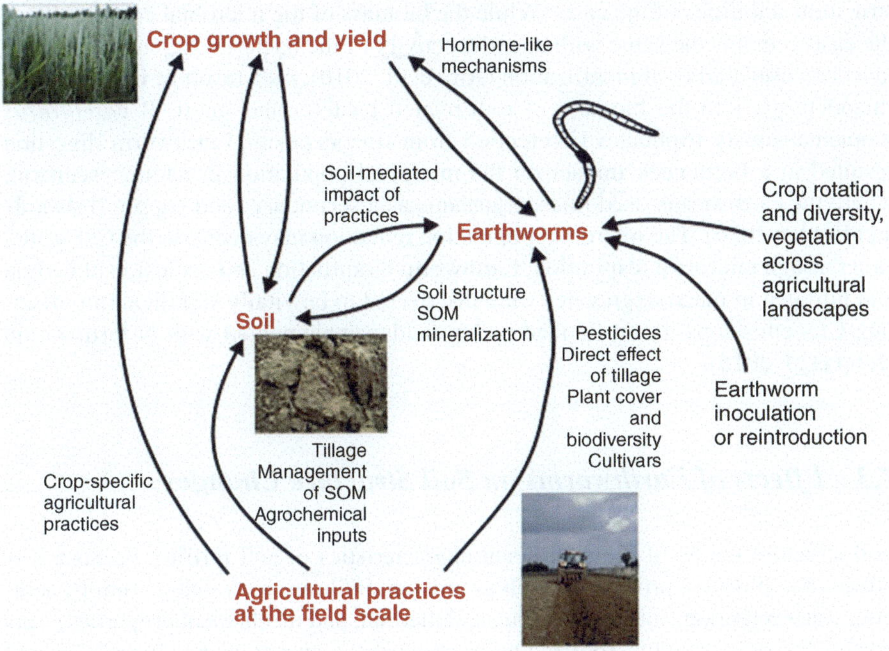

Fig. 2 Effects of earthworms on plant growth and health (Bertrand et al. 2015); magnitude of changes caused by earthworms in soil structure, water and nutrient availability, hormonal changes in crop health as a result of land use and activities, and other factors affecting the ability of earthworms can improve soil fertility and crop demand

limited motility, extremely rely on large organisms such as earthworms and roots to reach the fresh substrates, although most of the carbon turnover (approximately 85–90%) in the soil is mediated by microorganisms (Gómez-Brandón et al. 2012). This was called the "sleeping beauty paradox" because roots and soil invertebrates stimulated the dormant microbial communities. As a result of bioturbation by earthworms, soils are biologically reworked, and organic components are introduced into them, enhancing soil microbial activity (Sheehan et al. 2008). A parallel buildup of microbially dominated organic matter occurred through the decay of organic matter within litters, both in terms of living microorganisms biomass and dead microorganisms residues (Vidal et al. 2019). Earthworms also boost the microbial biomass in mineral soils by adding large amounts of labile C through mucus during the organic matter ingestion. The major contributing factor to C sequestration is the increased presence of earthworms in the organic matter mineral-associated and tiny occluded particulates in the necromass stored in the soil. This is at the expense of plant compounds in the less stable fractions (Angst et al. 2019a, 2019b). The strong microbial activity supports C sequestration with microbial necromass, creating a significant source of organic matter in soils (Ludwig et al. 2015). Soil's major populations of bacteria, fungi, actinomycetes, and also the related enzyme activities are found in the casts of earthworms thanks to the greater nutrient availabilities and higher structural stabilities of the casts. While the biomass of the microbial community of the casts remains the same with age, the activity of the bacteria decreases resulting in a decreased carbon mineralization (Aira et al. 2010). As a result of immobilizing rhizodeposits into the biomass of root-related bacteria and fungi, *P. corethrurus* momentaneously triplicates C retention from rhizodeposits. Earthworm digestion resulted in a bottleneck impact on the microbial population in another scenario, where the earthworms used microorganisms as a secondary food supply (Edwards and Bohlen 2020). The microbial population reduction may decrease the CO_2 emission through microbial respiration. Earthworm bioturbation and occlusion of carbon and nitrogen in macroaggregates were discovered to be vitally significant in offsetting CO_2 emissions in not flooded paddy fields supplemented with N fertilization (John et al. 2015).

7.3 Effects of Earthworms on Soil Structure Changes

Soil structure is one of the significant characteristics of soil fertility because it is effective in physical processes (such as erosion, infiltration, drainage, runoff, aeration, water retention, soil evaporation, soil thermal, and mechanical properties), carbon cycle (organic inputs, root and microorganisms respiration, decomposition, and localization of organic matter), nutrient cycling (mineralization, ion exchange, and immobilization), and biological properties (soil fauna and activity of microorganisms) (Dindal 1985; Elliott and Coleman 1988). Soil structure can be directly affected by earthworms through the creation of burrows and casts (Fig. 3). Casting increases soil aggregation, while burrowing improves the structure of the soil. Earthworm activity reduces soil bulk density by creating large pores. These large pores allow water (Chan and Heenan 1993) and gases to enter the soil. Earthworms translocate surface OM to deeper soils (Shuster et al. 2001) and produce organo-mineral casts (Wilcox et al. 2002). When an earthworm moves through the soil, it encounters two opposing mechanical effects. First, the soil is broken down, and second, the ingested soil is compacted as it passes through the intestine before being excreted. Earthworms begin to transform loose, compacted soil into moderate conditions that are better for structural stability and root development.

7.4 Effect of Earthworms on Nutrient Availability

Earthworms are one of the most serious detritivores in the soil food webs (Blouin et al. 2013a, 2013b). Through activities such as burrowing and casting, earthworms lead to increased SOM mineralization and/or humification, modification of soil porosity and aggregation, and stimulation of soil microbiota, thereby improving soil nutrient availability (Bertrand et al. 2015; Cunha et al. 2016). Through their feeding and burrowing habits, earthworms promote litter decomposition, nitrogen (N)

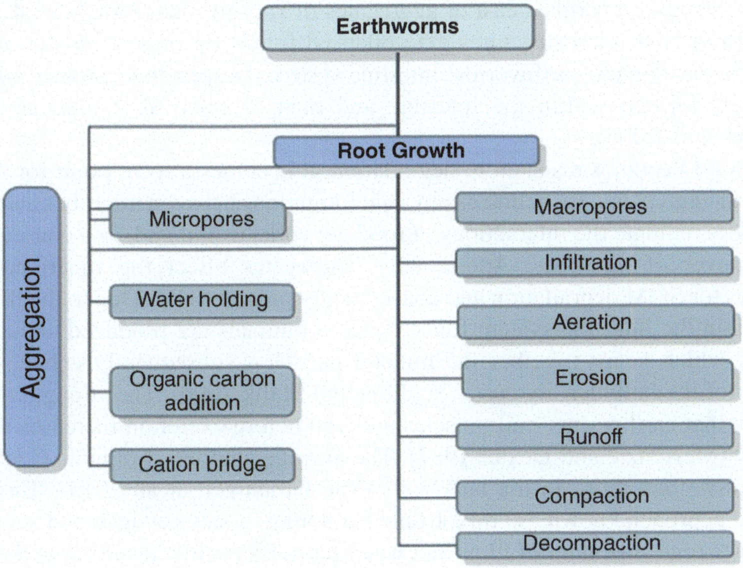

Fig. 3 Effects of earthworms on soil structure

mineralization, and water infiltration (Baker 2007) and have a significant impact on soil properties (Hättenschwiler and Gasser 2005). Earthworms transport nutrients (N, P, K, and Ca) through their physical properties and nutrient cycling. By forming aggregates and pores (i.e., biostructures), they play an important role in the supply of nutrients for plant growth (Mora et al. 2005). The major impact of earthworms is on the availability of nutrients and material cycles, and their interaction with soil microbes in the decomposition of organic matter.

8 Earthworms and Soil Carbon Sequestration

Earthworms play an essential role in soil carbon sequestration (Gilot 1997; Bossuyt et al. 2005). A review of 237 findings from 57 recently published research indicated that earthworm communities brought about 33% boost in soil CO_2 emission (Lubbers et al. 2013). Earthworms can generate a "carbon trap" and facilitate C sequestration in soils (Zhang et al. 2013). The three fundamental mechanisms for SOM stabilization are: (1) physical protection, (2) chemical stabilization, and (3) biochemical stabilization (Christensen 1996). Many studies link SOM stabilization to physical protection within cast aggregates (e.g., Lavelle and Martin 1992; Six et al. 2004; Bossuyt et al. 2005). Earthworms positively influence physical soil properties such as aeration, water retention, and infiltration (Bossuyt et al. 2005). Therefore, increased earthworm density supports soil quality, resulting in higher amounts of sequestered C. As soil and organic matter (OM) pass through an earthworm's gut, it forms a cast of aggregates of varying sizes (Angst et al. 2017). The existing soil microstructures get collapsed following organic matter and soil transition through the earthworms' intestine system. Despite this, another microaggregate is formed within the intestine and then excreted as a macroaggregate (Bossuyt et al. 2004).

As SOM becomes resistant to degradation, it becomes a focal point for the formation of aggregates. By acting as physical limitations between microbes and SOM, aggregates regulate the interactions of food networks in soil and, as a consequence, the turnover of microbes. Additionally, aggregates affect the functionality of enzymes for SOM degradation and consequently protect SOM. Due to the flocculation within the intestine system folds, calcium humates are produced and act as a cement, which keeps together the mineral particles, subsequently enhancing the stability of the bulk density casts (Scullion and Malik 2000). The strength of compression that earthworms' tail muscles undergo before excretion increases the cast stability (McKenzie and Dexter 1987). The average amount of organic C is higher by 40–48% than those in the bulk soils (Van Groenigen et al. 2019). Through a physical approach known as "thixotropic hardening," casts get aged and stabilized, whereby a rearrangement of films and water particles occurs, resulting in the edge-to-edge connection of the clay particles which is re-established without losing water. In the long-term, as aggregates age and dry out, they shrink and close up, i.e., they form strong connections among flora and polysaccharides (found inside the organic matter) and clay particles (Bossuyt et al. 2005). A cortex area (25 μm thick layer of fine portions) shown by a thin sectional analysis of the casts hampers the movement of water and air, creating anaerobic conditions unfavorable for microbes and, consequently, reducing OM decomposition. The long-term performance of organic matter conservation in aggregates relies mostly on two factors: (1) the highest quantity of aggregates feasible at a certain location and (2) the aggregate's life span (Lavelle et al. 2004). The casts depend on many factors according to a wide set of research, including the earthworms' digestion patterns, the behavior of species feedings, and the related microbes inside their intestine, as well as the ingested food nature and soil types (Abail et al. 2017). In the short term, the earthworm's biological features had no impact on the C mineralization extent; in the long term, these features were the keys to lowering C availability and therefore soil C stabilization regulation (Zhang et al. 2013). A conceptual model for carbon sequestration regulation utilizing earthworms is described by Zhang et al. (2013) in Fig. 4.

8.1 Epigeic Earthworms and Soil Carbon Sequestration

Epigeic earthworms' role in forming aggregates mostly relies on detritus contact with minerals. Surprisingly, epigeic earthworms dwelling on litter can transfer carbon (C) into the soil through their feeding activities, where it is stored in a more stabilized form within soil aggregates (Wu et al. 2017). The epigeic earthworms combine the C obtained from the litter in addition to that of soil, into aggregates

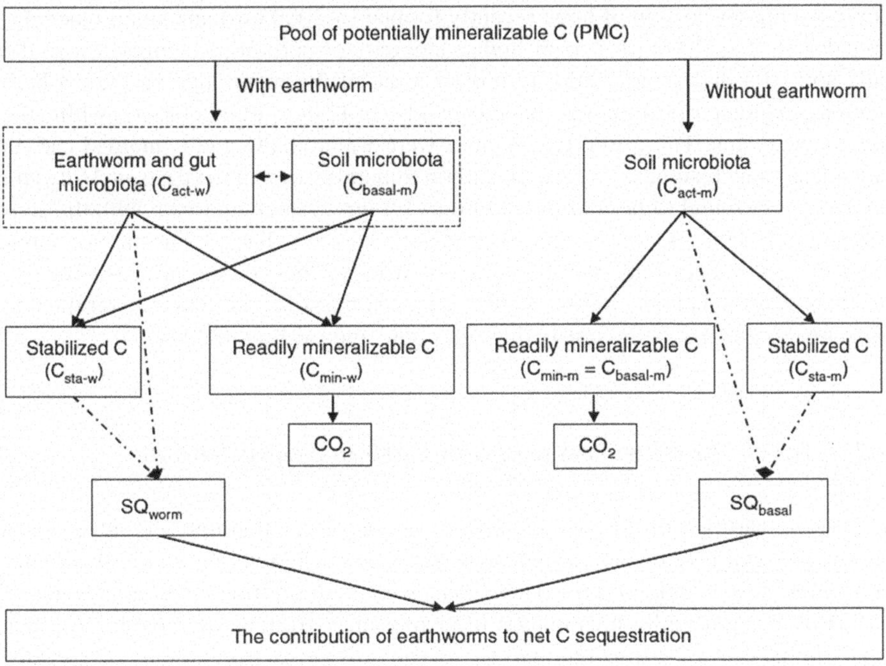

Fig. 4 Carbon sequestration regulation by earthworms (conceptual model). Abbreviations [Cact-w: earthworm-activated; Cact-m: soil microbiota activated mineralizable C; Cmin-w: net earthworms-induced pool of readily mineralizable C; Csta-w: pool of stabilized C in the presence of soil earthworms; Cmin-m = Cbasal-m: pool of readily mineralizable C in the absence of soil earthworms; Csta-m: the pool of stabilized C in the absence of soil earthworms; SQworm: soil C sequestration in the presence of soil earthworms; SQbasal: soil C sequestration in the absence of soil earthworms]. Single-ended dot dash and the double-ended dotted lines, respectively, refer to the major components of the SQ values and feasible interactions (Zhang et al. 2013)

(Wu et al. 2017). C stored in soil is increased by earthworms. Labile C was transformed into non-labile forms by earthworms (Angst et al. 2019a, 2019b). In particular, epigeic earthworms may influence microbial decomposer activity by directly grazing on microorganisms and by increasing the surface area available for microbial attack after the organic matter is broken down.

8.2 *Endogeic Earthworms and Soil Carbon Sequestration*

Endogeic earthworms cause the stabilization of OC forming large macroaggregates and transforming labile C into recalcitrant forms (Sánchez de León et al. 2014). Endogeic earthworms then were defined as "soil main aggregation agents and organic matter stabilizers" (Lavelle and Spain 2001a, 2001b, 2001c). The *Aporrectodea caliginosa* endogenic earthworms (worked soils) were discovered to possess a higher portion of large recently formed (> 2000 μm) and small macroaggregates (> 250–2000 μm), with higher amounts of carbon in comparison to the bulk soil (Bossuyt et al. 2005). In the presence of the mentioned species, a high microaggregate proportion was shaped quickly (in 12 days of incubation) within the macroaggregates. These microaggregates were water-stable and contained carbon taken from new residues. The rate of carbon mineralization in the casts of *Millsonia anomala* was found to be four times lower (3 %/yr.$^{-1}$) in comparison with the control soil (11 %/yr.$^{-1}$). Anecic earthworms have burrows that are constant or semi-constant and are extended inside the soil deeply; they commonly use organic residues as food that decompose on the surface of the soil, and then transfer them to the burrows they have previously dug (Shipitalo and Le-Bayon 2004).

8.3 Anecic Earthworms and Soil Carbon Sequestration

The anecic earthworms known as *Amynthas khami* formed simply organized casts with higher stabilized water, which are capable to reach a height up to 20 cm after deposition for a couple of days. These casts were built up from small aggregates of less than 500 μm with an estimated life expectancy from a few months to a few years. In the casts, the time required for a drop of water to penetrate the aggregates is considerably higher than that of the control soil. These casts contained a higher portion of OC and N within the class 50–250 μm. The plausible reason for the physical protection of organic matter from the mineralization that occurs in the casts could be the lack of large pores and voids in them. The CO_2 emissions from cast aggregates smaller than 250 μm in size were significantly lower than those in aggregates larger than 500 μm. The casts of *Lumbricus terrestris* (anecic species) contained greater amounts of organic carbon in comparison with those of *A. caliginosa* (endogeic species). In incubation research (21 days) using earthworms, plants, and compost compounds, their relative significance in forming aggregate and stabilizing soil structure was evaluated (Deeb et al. 2017). The aggregate distribution was also affected by the system management activities. Forming the small macroaggregates can be fostered by using a mix of *A. caliginosa, L. terrestris,* and *Lumbricus rubellus* species (Frazão et al. 2019). The positive impact of earthworms on soil aggregation occurred precisely thanks to the presence of plants or the contributions of inputs from litter (Fonte and Six 2010) or of carbon in the roots (Fonte et al. 2012). Also, the earthworms' impact on the stabilization of aggregates differs according to the variation in the added residues (Coq et al. 2007). Although earthworms macerate lignocelluloses to make them accessible to microorganisms' intestine, combining the stabilizing agents with the materials ingested in the anterior intestine system, followed by the latter biochemical processes in the hindgut, along with the water reabsorption, causes a stable organ mineral complex (Vidal et al. 2019).

9 Earthworms and Pedogenic Carbonates

Organic matter bound to clay is stable and slow to form, but earthworms speed up the creation of these crawling pools by feeding on them. As a consequence of the relatively higher pH, cation exchange capacity, and divalent cations (Mg^{2+} and Ca^{2+}) in the casts compared to the surrounding soil, the electrostatic bonds between particles in the casts are increased, causing better stability. Polysaccharides as well as mineral properties, produced by living microorganisms during decomposition, play a vital role in controlling the interactions between OM and minerals (Jouquet et al. 2008). In casts, this increases inter particles bonds, which increases the mineral's adsorption and consequently strengthens the organo-mineral bonds. While casts are wet, the bonds among complexes of organo-minerals had been determined to be restrained to particularly cation bridges and water. The casts' aggregate stability was improved by way of the bonds between clay–polyvalent cations–organic matters (C–PC–OM). In the majority of cases, Ca^{2+} together with Mg^{2+} (to a lesser extent) are considered as parts of the C- PC- OM binding. A sizable part of that in dried casts was coordination complexes composites, which were discovered to be completely strong. Compared to the surrounding undigested soil, casts contained greater amounts of amorphous Fe, sesquioxides related to organic complexes, and polysaccharides associated with heavy fractions. Microaggregate stability is primarily associated with lighter fractions of organic matter and amorphous Al and Fe in casts, whereas the stability of macroaggregate was found to be primarily associated with heavier fractions of that (Oyedele et al. 2006). SOC content increases in the presence of reactive minerals such as Al and Fe (Baldock and Skjemstad 2000). As evidenced by the analysis of 60 earthworm casts, organo-mineral complexes have been confirmed to be formed from mineral and organic particles. The mineral OM absorption is the first way of SOC conservation. Earthworm calciferous glands, or esophageal glands, produce a carbonate trait that is excreted in earthworm casts (Zamanian et al. 2016). The majority of earthworm species, especially those in the *Lumbricidae* family, are true biomineralizers because they produce $CaCO_3$ through their calciferous glands. These granules are mostly made of calcite, aragonite, vaterite, and a small amount of amorphous $CaCO_3$ (Lehmann and Kleber 2015). Amorphous $CaCO_3$, which is then transformed into granule calcite via a mechanism of dissolution precipitation, occurs in calciferous glands by dissolving CO_2 into carbonate and bicarbonate (Briones et al. 2008; Zamanian et al. 2016). Earthworms process calcite granules in their calciferous glands and then move them into their esophagus and intestine before finally excreting them in the soil (Hodson et al. 2015). These granules can act as long-term carbon pools in the $CaCO_3$ formation since their life expectancy is much more than 300,000 years (Versteegh et al. 2014). Thus, the earthworms' calciferous glands supply a regulation mechanism for the level of carbon dioxide in tissues in addition to restoring the metabolic and environmental carbon dioxide. Acting as a cementing factor for the organic matter that moves through the intestine, $CaCO_3$ is capable to affect positively the SOC stabilization and structure (Rowley et al. 2018) (Fig. 5).

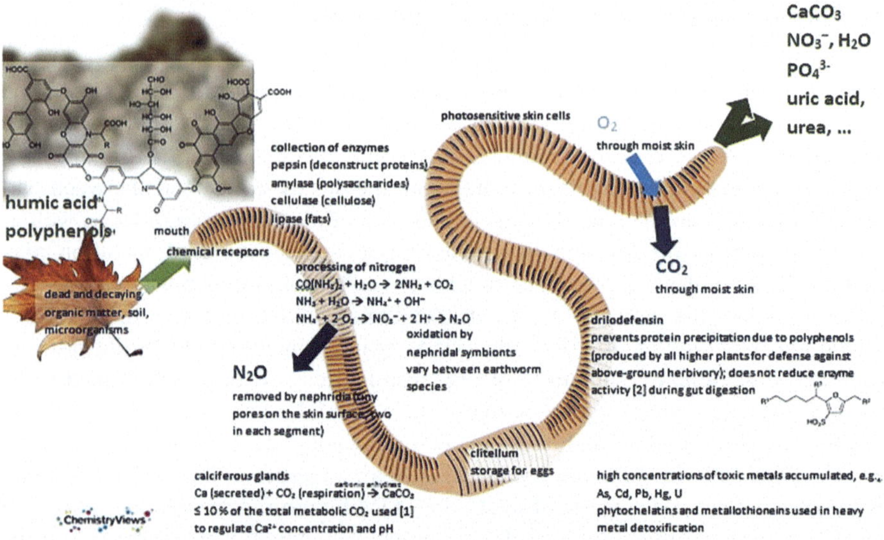

Fig. 5 Earthworms play a role in the pedogenic carbonates (Liebeke et al. 2015)

10 Phytolith-C by Earthworms

Biogeochemical carbon sequestration by phytoliths offers a long-term opportunity to sequester atmospheric CO_2 (Song et al. 2016). Soil properties (such as bulk density), biological turbulence (such as earthworm peristalsis), and climatic conditions (linked by temperature and precipitation) can affect the movement and distribution of soil phytoliths (Yang et al. 2018). Casts of earthworms include a higher water extractable mono silicic acid ($Si(OH)_4$) and Si compared to the bulk soil because of the silicate solubilizing bacteria in their intestine. Soil bioturbation by earthworms enhances the xylem Si transportation within maize and cucumber cultivation (Hu et al. 2018). While roots of plants absorb the mono silicic acid, Si is deposited in a no-crystalline mineral form within the walls of cells, inter-cellular and lumen places of plants, forming phytoliths of Si. These phytoliths further sequester a portion of organic carbon (1–6%) during plant growth, which is assimilated during the photosynthesis process from atmospheric carbon dioxide. After plant organic matter decomposition, the phytolith-occluded carbon (PhytOC), which is very resistant to degradation, deposits on the surface of the soil, therefore playing a role as a potential sink for carbon. Phytolith occluded carbon can be responsible for approximately 82% of the total carbon in well-drained soils, even after thousands of years of organic matter decomposition because of its high resistance to degradation (Li et al. 2015). Additionally, earthworms' bioturbation is highly capable of translocating the protected phytolith vertically and horizontally (Zangerlé et al. 2016). Most of the organic carbon in litter in terrestrial ecosystems moves rapidly into the turnover of organic matter pools. This includes only short-term carbon sequestration.

11 Conclusions

Earthworms affect SOM mineralization and sequestration. Earthworms adapt to cultivation techniques and can be used as bioindicators of soil health. Accordingly, earthworms contribute to soil biology and the reduction of CO_2 emissions, which affect the carbon cycle. Along with facilitating C input into the soil, earthworms also support the (micro) biological activities. In the short term, mineralization increases, and mineral nutrients become more accessible to plants. Earthworms emit some substances similar to those produced by plant growth. Therefore, earthworms help maintain and improve soil structure as well as aerate the soil and incorporate dead plants into the soil's organic matter. The fact that earthworms can specifically affect soil organic matter mineralization is of great importance for increasing sustainable land use in degraded soils and agroecosystems. Nevertheless, earthworm research will be addressed in future work on soil properties, vegetation types, climate details, species-level identification of earthworms, and the presence of other soil microfauna.

Acknowledgments The valuable and constructive comments/suggestions provided by Dr. Rosa Francaviglia (Council for Agricultural Research and Economics, Research Centre for Agriculture and Environment, Rome, Italy) are highly appreciated. We thank also the Strategic Academic Leadership Program "Priority 2030" of the Kazan Federal University and the RUDN University Strategic Academic Leadership Program.

References

Abail Z, Sampedro L, Whalen JK (2017) Short-term carbon mineralization from endogeic earthworm casts as influenced by properties of the ingested soil material. Appl Soil Ecol 116:79–86

Aira M, Monroy F, Domínguez J (2009) Changes in bacterial numbers and microbial activity of pig manure during gut transit of epigeic and anecic earthworms. J Hazard Mat 162:1404–1407

Aira M, Lazcano C, Gómez-Brandón M, Domínguez J (2010) Ageing effects of casts of Aporrectodea caliginosa on soil microbial community structure and activity. Appl Soil Ecol 46:143–146

Angst Š, Mueller CW, Cajthaml T, Angst G, Lhotáková Z, Bartuška M, Špaldoňová A, Frouz J (2017) Stabilization of soil organic matter by earthworms is connected with physical protection rather than with chemical changes of organic matter. Geoderma 289:29–35

Angst G, Mueller KE, Eissenstat DM, Trumbore S, Freeman KH, Hobbie SE, Chorover J, Oleksyn J, Reich PB, Mueller CW (2019a) Soil organic carbon stability in forests: distinct effects of tree species identity and traits. Glob Chang Biol 25(4):1529–1546

Angst G, Mueller CW, Prater I, Angst Š, Frouz J, Jílková V, Peterse F, Nierop KGJ (2019b) Earthworms act as biochemical reactors to convert labile plant compounds into stabilized soil microbial necromass. Commun Biol 2:1–7

Baker G (2007) Differences in nitrogen release from surface and incorporated plant residues by two endogeic species of earthworms (Lumbricidae) in a red–brown earth soil in southern Australia. Eur J Soil Biol 43:S165–S170

Baldock JA, Skjemstad JO (2000) Role of the soil matrix and minerals in protecting natural organic materials against biological attack. Org Geochem 31:697–710

Bardgett RD, Usher MB, Hopkins DW (2005) Biological diversity and function in soils. Cambridge University Press

Berg B, Laskowski R (2005) Decomposers: soil microorganisms and animals. Adv Ecol Res 38:73–100

Bertrand M, Barot S, Blouin M, Whalen J, de Oliveira T, Roger-Estrade J (2015) Earthworm services for cropping systems. A review. Agron Sustain Dev 35(2):553–567

Bi Y-M, Tian G-L, Wang C, Zhang Y, Wang D-N, Zhang F-F, Zhang L-S, Sun Z-J (2018) Differential effects of two earthworm species on fusarium wilt of strawberry. Appl Soil Ecol 126:174–181

Bityutskii NP, Lapshina IN, Lukina EI, Solov'eva AN, Patsevich VG, Vygovskaya AA (2002) Role of earthworms in mineralization of organic nitrogen compounds in soil. Eurasian Soil Sci 35:1100–1107

Blanchart E, Lavelle P, Braudeau E, Le Bissonnais Y, Valentin C (1997) Regulation of soil structure by geophagous earthworm activities in humid savannas of Côte D'ivoire. Soil Biol Biochem 29:431–439

Blanchart E, Marilleau N, Chotte JL, Drogoul A, Perrier E, Cambier C (2009) SWORM: an agent-based model to simulate the effect of earthworms on soil structure. Eur J Soil Sci 60:13–21

Blouin M, Hodson ME, Delgado EA, Baker G, Brussaard L, Butt KR, Dai J, Dendooven L, Pérès G, Tondoh JE, Cluzeau D (2013a) A review of earthworm impact on soil function and ecosystem services. Eur J Soil Sci 64(2):161–182

Blouin M, Sery N, Cluzeau D, Brun JJ, Bédécarrats A (2013b) Balkanized research in ecological engineering revealed by a bibliometric analysis of earthworms and ecosystem services. Environ Manag 52(2):309–320

Bossuyt H, Six J, Hendrix PF (2004) Rapid incorporation of carbon from fresh residues into newly formed stable microaggregates within earthworm casts. Eur J Soil Sci 55:393–399

Bossuyt H, Six J, Hendrix PF (2005) Protection of soil carbon by microaggregates within earthworm casts. Soil Biol Biochem 37:251–258

Bottinelli N, Hedde M, Jouquet P, Capowiez Y (2020) An explicit definition of earthworm ecological categories–Marcel Bouché's triangle revisited. Geoderma 372:114361

Bouché MB (1972) Lombriciens de France. Ecologie et Systématique. INRA, Paris

Briones MJI, López E, Méndez J, Rodríguez JB, Gago-Duport L (2008) Biological control over the formation and storage of amorphous calcium carbonate by earthworms. Mineral Mag 72:227–231

Brown GG, Barois I, Lavelle P (2000) Regulation of soil organic matter dynamics and microbial activity in the drilosphere and the role of interactions with other edaphic functional domains. Eur J Soil Biol 36:177–198

Brown GG, Edwards CA, Brussaard L (2004) How earthworms affect plant growth: burrowing into the mechanisms. Earthworm Ecol 2:13–49

Byzov BA, Khomyakov NV, Kharin SA, Kurakov AV (2007) Fate of soil bacteria and fungi in the gut of earthworms. Eur J Soil Biol 43:S149–S156

Chan KY, Heenan DP (1993) Surface hydraulic properties of a red earth under continuous cropping with different management practices. Soil Res 31(1):13–24

Chertov OG, Komarov AS (1997) SOMM: a model of soil organic matter dynamics. Ecol Model 94:177–189

Chesworth W (1992) Weathering systems. In: Martini IP, Chesworth W (eds) Weathering, soils and Paleosols. Elsevier, Amsterdam, pp 19–40

Christensen BT (1996) Carbon in primary and secondary organo-mineral complexes. In: Carter MR, Stewart BA (eds) Structure and organic matter storage in agricultural soils. CRC Press Inc., Boca Raton, Florida, pp 97–165

Coq S, Barthès BG, Oliver R, Rabary B, Blanchart E (2007) Earthworm activity affects soil aggregation and organic matter dynamics according to the quality and localization of crop residues—an experimental study (Madagascar). Soil Biol Biochem 39:2119–2128

Cunha L, Brown GG, Stanton DW, Da Silva E, Hansel FA, Jorge G, McKey D, Vidal-Torrado P, Macedo RS, Velasquez E, James SW (2016) Soil animals and pedogenesis: the role of earthworms in anthropogenic soils. Soil Sci 181(3/4):110–125

Darwin C (1881) The formation of vegetable mould through the action of worms, with observations on their habits. John Murray, London

Deeb M, Desjardins T, Podwojewski P, Pando A, Blouin M, Lerch TZ (2017) Interactive effects of compost, plants and earthworms on the aggregations of constructed Technosols. Geoderma 305:305–313

Dindal DL (1985) Soil animals and soil fabric production: facts and perceptions. Quaestiones Entomologicae 21:587–594

Doube BM, Brown GG (1998) Life in a complex community: functional interactions between earthworms, organic matter, microorganisms, and plants

Edwards WM, Shipitalo MJ, Owens LB, & Norton LD (1989) Water and nitrate movement in earthworm burrows within long-term no-till cornfields. Journal of Soil and Water Conservation 44(3):240–243

Edwards CA, Bohlen PJ (1996) Biology and ecology of earthworms, vol 3. Springer Science & Business Media

Edwards CA, Bohlen PJ (2020) Biology and ecology of earthworms, 4th edn. Springer, US, p 426

Edwards CA, Arancon NQ, Sherman R (eds) (2011) Vermiculture technology: earthworms, organic wastes, and environmental management. CRC Press, Boca Raton

Elliott ET, Coleman DC (1988) Let the soil work for us. Ecol Bull 39:23–32

Fonte SJ, Six J (2010) Earthworms and litter management contributions to ecosystem services in a tropical agroforestry system. Ecol Appl 20:1061–1073

Fonte SJ, Quintero DC, Velásquez E, Lavelle P (2012) Interactive effects of plants and earthworms on the physical stabilization of soil organic matter in aggregates. Plant Soil 359:205–214

Frazão J, de Goede RGM, Capowiez Y, Pulleman MM (2019) Soil structure formation and organic matter distribution as affected by earthworm species interactions and crop residue placement. Geoderma 338:453–463

Fu S, Cabrera ML, Coleman DC, Kisselle KW, Garrett CJ, Hendrix PF, Crossley DA Jr (2000) Soil carbon dynamics of conventional tillage and no-till agroecosystems at Georgia Piedmont—HSB-C models. Ecol Model 131:229–248

Gilot C (1997) Effects of a tropical geophageous earthworm, M. Anomala (Megascolecidae), on soil characteristics and production of a yam crop in Ivory Coast. Soil Biol Biochem 29:353–359

Gómez-Brandón M, Lores M, Domínguez J (2012) Species-specific effects of Epigeic earthworms on microbial community structure during first stages of decomposition of organic matter. PLoS One 7(2):e31895

Hale CM, Frelich LE, Reich PB, Pastor J (2005) Effects of European earthworm invasion on soil characteristics in northern hardwood forests of Minnesota, USA. Ecosystems 8(8):911–927

Hättenschwiler S, Gasser P (2005) Soil animals alter plant litter diversity effects on decomposition. Proc Natl Acad Sci 102(5):1519–1524

Hodson ME, Benning LG, Demarchi B, Penkman KEH, Rodriguez-Blanco JD, Schofield PF, Versteegh EAA (2015) Biomineralisation by earthworms – an investigation into the stability and distribution of amorphous calcium carbonate. Geochem Trans 16:1–16

Hu L, Xia M, Lin X, Xu C, Li W, Wang J, Zeng R, Song Y (2018) Earthworm gut bacteria increase silicon bioavailability and acquisition by maize. Soil Biol Biochem 125:215–221

Huang CY, Hendrix PF, Fahey TJ, Bohlen PJ, Groffman PM (2010) A simulation model to evaluate the impacts of invasive earthworms on soil carbon dynamics. Ecol Model 221:2447–2457

John K, Jauker F, Marxsen J, Zaitsev AS, Wolters V (2015) Earthworm bioturbation stabilizes carbon in non-flooded paddy soil at the risk of increasing methane emissions under wet soil conditions. Soil Biol Biochem 91:127–132

Jouquet P, Bottinelli N, Podwojewski P, Hallaire V, Duc TT (2008) Chemical and physical properties of earthworm casts as compared to bulk soil under a range of different land-use systems in Vietnam. Geoderma 146:231–238

Kooch Y, Parsapour MK, Egli M, Moghimian N (2021) Forest floor and soil properties in different development stages of oriental beech forests. Appl Soil Ecol 161:103823

Kooijman AM, Kooijman-Schouten MM, Martinez-Hernandez GB (2008) Alternative strategies to sustain N-fertility in acid and calcaric beech forests: low microbial N-demand versus high biological activity. Basic Appl Ecol 9:410–421

Kumar S, Tripathi G, Mishra GV (2021) A comparative study on earthworm biodiversity and species habitat-relationship of hilly and plain areas of Sirohi District of Rajasthan, India. Appl Ecol Environ Sci 9:419–439

Lassauce A, Paillet Y, Jactel H, Bouget C (2011) Deadwood as a surrogate for forest biodiversity: meta-analysis of correlations between deadwood volume and species richness of saproxylic organisms. Ecol Indic 11:1027–1039

Lavelle P, Martin A (1992) Small-scale and large-scale effects of endogeic earthworms on soil organic matter dynamics in soils of the humid tropics. Soil Biol Biochem 24(12):1491–1498

Lavelle P, Spain AV (2001a) Soil ecology. Springer, Netherlands, p 654

Lavelle P, Spain AV (2001b) Soil organisms. Soil Ecol:201–356

Lavelle P, Spain AV (2001c) Soil ecology. Kluwer Academic Publishers, London

Lavelle P, Bignell D, Lepage M et al (1997) Soil function in a changing world: the role of invertebrate ecosystem engineers. Eur J Soil Biol 33:159–193

Lavelle P, Charpentier F, Villenave C, Rossi JP, Derouard L, Pashanasi B, André J, Ponge JF, Bernier N (2004) Effects of earthworms on soil organic matter and nutrient dynamics at a landscape scale over decades. In: Edwards CA (ed) Earthworm ecology. CRC Press, Boca Raton, Florida, pp 145–160

Lavelle P, Decaens T, Aubert M, Barot S, Blouin M, Bureau F et al (2006) Soil invertebrates and ecosystem services. Eur J Soil Biol 42:3–15

Law MMS, Lai DYF (2021) Impacts of wetting-drying cycles on short-term carbon and nitrogen dynamics in Amynthas earthworm casts. Pedosphere 31:423–432

Lee KE (1985) Earthworms their ecology and relationships with soils and land use. Academic Press, Sydney

Lehmann J, Kleber M (2015) The contentious nature of soil organic matter. Nature 528:60–68

Li B, Song Z, Li Z, Wang H, Gum R, Song R (2015) Phylogenetic variation of phytolith carbon sequestration in bamboos. Sci Rep 4:1–6

Liebeke M, Strittmatter N, Fearn S, Morgan AJ, Kille P, Fuchser J, Wallis D, Palchykov V, Robertson J, Lahive E, Spurgeon DJ (2015) Unique metabolites protect earthworms against plant polyphenols. Nat Commun 6(1):1–7

Londo AJ (2001) Bucket mounding as a mechanical site preparation technique in wetlands. NIAF 18:7–13

Lubbers IM, Van Groenigen KJ, Fonte SJ, Six J, Brussaard L, Van Groenigen JW (2013) Greenhouse-gas emissions from soils increased by earthworms. Nat Clim Chang 3:187–194

Ludwig M, Achtenhagen J, Miltner A, Eckhardt KU, Leinweber P, Emmerling C, Thiele-Bruhn S (2015) Microbial contribution to SOM quantity and quality in density fractions of temperate arable soils. Soil Biol Biochem 81:311–322

Lukkari T, Taavitsainen M, Vaisanen A, Haimi J (2004) Effects of heavy metals on earthworms along contamination gradients in organic rich soils. Ecotoxicol Environ Saf 59:340–348

Mabuhay JA, Nakagoshi N, Horikoshi T (2003) Microbial biomass and abundance after forest fire in pine forests in Japan. Ecol Res, Springer Science & Business Media BV 18:431–441

Manlay RJ, Feller C, Swift MJ (2007) Historical evolution of soil organic matter concepts and their relationships with the fertility and sustainability of cropping systems. Agric Ecosyst Environ 119:217–233

Marilleau N, Cambier C, Drogoul A, Chotte JL, Perrier E, Blanchart E (2008) Multiscale MAS modelling to simulate the soil environment: application to soil ecology. Simul Model Pract Theory 16:736–745

McCay TS, Scull T (2019) Invasive lumbricid earthworms in northeastern north American forests and consequences for leaf-litter fauna. Biol Invasions 21:2081–2093

McKenzie BM, & Dexter AR (1987) Physical properties of casts of the earthworm Aporrectodea rosea. Biology and Fertility of Soils 5:152–157

Medina-Sauza RM, Álvarez-Jiménez M, Delhal A, Reverchon F, Blouin M, Guerrero-Analco JA, Cerdán CR, Guevara R, Villain L, Barois I (2019) Earthworms building up soil micro biota, a review. Front Environ Sci 7:81. https://doi.org/10.3389/fenvs.2019.00081

Monroy F, Aira M, Domínguez J (2008) Changes in density of nematodes, protozoa and total coliforms after transit through the gut of four epigeic earthworms (Oligochaeta). Appl Soil Ecol 39:127–132

Mora P, Miambi E, Jiménez JJ, Decaëns T, Rouland CJSB (2005) Functional complement of biogenic structures produced by earthworms, termites and ants in the neotropical savannas. Soil Biol Biochem 37(6):1043–1048

Mosavi F, Raisi F (2009) The effect of earthworm and organic matter on carbon mineralization and microbial biomass, Articles collection of 11th soil sciences congress in Iran, pp 112–113 (In Persian)

Nachtergale L, Ghekiere K, De Shriver A, Muys B, Luyssaert S, Lust N (2002) Earthworm biomass and species diversity in wind throw sites of a temperate lowland forest. Pedobiologia 46:440–451

Neirynck J, Mirtcheva S, Sioen G, Lust N (2000) Impact of *Tilia platyphyllos* Scop., *Fraxinus excelsior* L., *Acer pseudoplatanus* L., *Quercus robur* L., and *Fagus sylvatica* L. on earthworm biomass and physico -chemical properties of loamy topsoil. For Ecol Manag 133:275–286

Oyedele DJ, Schjønning P, Amusan AA (2006) Physicochemical properties of earthworm casts and uningested parent soil from selected sites in southwestern Nigeria. Ecol Eng 28:106–113

Piearce TG, Roggero N, Tipping R (1994) Earthworms and seeds. J Biol Educ 28(3):195–202

Power AG (2010) Ecosystem services and agriculture: tradeoffs and synergies. Philos Trans R Soc B: Biol Sci 365(1554):2959–2971

Regulska E, Kołaczkowska E (2015) Landscapes patch pattern effect on relationships between soil properties and earthworm assemblages: a comparison of two farmlands of different spatial structure. Pol J Ecol 63:549–558

Rowley MC, Grand S, Verrecchia EP (2018) Calcium-mediated stabilization of soil organic carbon. Biogeochemistry 137:27–49

Sánchez-de León Y, Lugo-Pérez J, Wise DH, Jastrow JD, González-Meler MA (2014) Aggregate formation and carbon sequestration by earthworms in soil from a temperate forest exposed to elevated atmospheric CO_2: a microcosm experiment. Soil Biol Biochem 68:223–230

Sapkota R, Santos S, Farias P, Krogh PH, Winding A (2020) Insights into the earthworm gut multi-kingdom microbial communities. Sci Total Environ 727:13801

Schon NL, MacKay AD, Gray RA, Van Koten C, Dodd MB (2017) Influence of earthworm abundance and diversity on soil structure and the implications for soil services throughout the season. Pedobiologia 62:41–47

Schrader S, Wolfarth F, Oldenburg E (2013) Biological control of soil-borne phytopatogenic fungi and their mycotoxins by soil fauna. Bull UASMV Agric 70:291–298

Scullion J, Malik A (2000) Earthworm activity affecting organic matter, aggregation and microbial activity in soils restored after opencast mining for coal. Soil Biol Biochem 32:119–126

Seeber J, Seeber GUH, Langel R, Scheu S, Meyer E (2008) The effect of macro-invertebrates and plant litter of different quality on the release of N from litter to plant on alpine pastureland. Biol Fertil Soils 44:783–790

Sheehan C, Kirwan L, Connolly J, Bolger T (2008) The effects of earthworm functional diversity on microbial biomass and the microbial community level physiological profile of soils. Eur J Soil Biol 44:65–70

Shipitalo MJ, Le Bayon RC (2004) Quantifying the effects of earthworms on soil aggregation and porosity. In: Edwards CA (ed) Earthworm ecology. CRC Press, Boca Raton, Florida, pp 183–200

Shuster WD, Subler S, McCoy EL (2001) Deep-burrowing earthworm additions changed the distribution of soil organic carbon in a chisel-tilled soil. Soil Biol Biochem 33(7–8):983–996

Six J, Bossuyt H, Degryze S, Denef K (2004) A history of research on the link between (micro) aggregates, soil biota, and soil organic matter dynamics. Soil Tillage Res 79(1):7–31

Song Z, McGrouther K, Wang H (2016) Occurrence, turnover and carbon sequestration potential of phytoliths in terrestrial ecosystems. Earth Sci Rev 158:19–30

Tamartash R, Ehsani SM (2021) The effect of earthworms on plant diversity and soil properties under different land uses. Acta Ecol Sin 41:171–176

Van Groenigen JW, Van Groenigen KJ, Koopmans GF, Stokkermans L, Vos HMJ, Lubbers IM (2019) How fertile are earthworm casts? A meta-analysis. Geoderma 338:525–535

Versteegh EAA, Black S, Hodson ME (2014) Environmental controls on the production of calcium carbonate by earthworms. Soil Biol Biochem 70:159–161

Vidal A, Watteau F, Remusat L, Mueller CW, Nguyen Tu TT, Buegger F, Derenne S, Quenea K (2019) Earthworm cast formation and development: a shift from plant litter to mineral associated organic matter. Front Environ Sci 7:1–15

Wardle DA (2006) The influence of biotic interactions on soil biodiversity. Ecol Lett 9:870–886

Wilcox C, Dominguez J, Parmelee R, McCartney D (2002) Soil carbon and nitrogen dynamics in *Lumbricus terrestris*. L. middens in four arable, a pasture, and a forest ecosystems. Biol Fertil Soils 36:26–34

Wolfarth F, Schrader S, Oldenburg E, Weinert J, Brunotte J (2011) Earthworms promote the reduction of fusarium biomass and deoxynivalenol content in wheat straw under field conditions. Soil Biol Biochem 43:1858–1865

Wu J, Li H, Zhang W, Li F, Huang J, Mo Q, Xia H (2017) Contrasting impacts of two subtropical earthworm species on leaf litter carbon sequestration into soil aggregates. J Soils Sediments 17:1672–1681

Wurst S, Langel R, Scheu S (2005) Do endogeic earthworms change plant competition? A microcosm study. Plant Soil 271(1):123–130

Yang X, Song Z, Liu H, Van Zwieten L, Song A, Li Z, Hao Q, Zhang X, Wang H (2018) Phytolith accumulation in broadleaf and conifer forests of northern China: implications for phytolith carbon sequestration. Geoderma 312:36–44

Zamanian K, Pustovoytov K, Kuzyakov Y (2016) Pedogenic carbonates: forms and formation processes. Earth Sci Rev 157:1–17

Zhang W, Hendrix PF, Dame LE, Burke RA, Wu J, Neher DA, Li J, Shao Y, Fu S (2013) Earthworms facilitate carbon sequestration through unequal amplification of carbon stabilization compared with mineralization. Nat Commun 4:2576. https://doi.org/10.1038/ncomms3576

Zangerlé A, Renard D, Iriarte J, Suarez Jimenez LE, Adame Montoya KL, Juilleret J, & McKey D (2016) The surales, self-organized earth-mound landscapes made by earthworms in a seasonal tropical wetland. PLoS One 11(5):e0154269

Role of Earthworms on C and N Biogeochemical Cycles and Potential Links to Greenhouse Gas Emissions

Maria J. I. Briones

Abstract Earthworms play a critical role in many essential soil processes (e.g., soil organic matter (SOM) decomposition and nutrient cycling, improved soil structure, better drainage and aeration, etc.), which have important implications for the sequestration/emissions of carbon (C) and nitrogen (N) in soils. This is the result of a wide array of activities, including feeding, casting, burrowing, and interacting with other above- and below-ground organisms. The soil mixing, the incorporation of fresh organic residues into the soil, the formation of micro and macroaggregates, and the changes in soil structure promote both SOM mineralization and stabilization. Therefore, several studies have tried to elucidate whether the positive effects of earthworms on mineralization and nitrification/denitrification processes would make them responsible for increased greenhouse emissions (mainly CO_2 and N_2O) or to be the mediators for trapping nutrients and thus helping to reduce C and N emissions instead. Part of the problem is the difficulties in measuring the "direct" earthworm contribution to the dynamic transformations (mobilization vs. stabilization) of SOM and their separation from their "indirect" effects (via plant and microbial interactions). Proving that earthworms are strengthening the greenhouse effect requires a more complete understanding of the functional dissimilarity among species and the modulating effects of the soil abiotic conditions. Here, I will synthesize the current knowledge on the role of earthworms on biogeochemical cycles and how earthworm biogenic structures (casts, burrows, and pores) could act as potential hotspots for C and N mineralization/retention. The final implications for C and N stocks will also need to take into account experimental biases and research gaps to fully integrate all controlling factors so that the role of earthworms on C and N cycling and dynamics can be unambiguously assessed.

M. J. I. Briones (✉)
Departamento. Ecología y Biología Animal, Universidad de Vigo, Vigo, Spain
e-mail: mbriones@uvigo.gal

Y. Kooch, Y. Kuzyakov (eds.), *Earthworms and Ecological Processes*,
https://doi.org/10.1007/978-3-031-64510-5_15

1 Earthworm Contribution to C and N Cycling

Earthworm feeding and casting, burrowing, and soil mixing together with their metabolic activities (respiration) and their interactions with other soil organisms (namely, plants and microbial populations) are responsible for changing the physical and chemical conditions of soils and hence, the dynamic of soil processes (Fig. 1). By consuming litter and soil, earthworms accelerate the decomposition of soil organic matter (SOM). In tropical savannahs, earthworm communities are dominated by geophagous endogeic species, with average population densities and biomass of 202 individuals and 39.7 g m^{-2} (respectively), and can ingest up to 1000–1250 t dry soil ha^{-1} $year^{-1}$ (Lavelle et al. 1998). In temperate systems, the maximum amount of ingested soil is estimated to reach 500 t dry soil ha^{-1} $year^{-1}$ (Lavelle 1988). However, a higher number of earthworm species with different feeding strategies, in particular in milder areas, can have greater effects on litter consumption (Bohlen et al. 1997; Heděnec et al. 2022) and its distribution (Nuutinen and Butt 2019). For example, in agroecosystems, it has been shown that they can consume 840 kg ha^{-1} $yr.^{-1}$ of surface litter (Bohlen et al. 1997) and increase litter dispersal by 25–37% (Nuutinen and Butt 2019).

Since earthworms feed preferentially on N-rich litter (i.e., low carbon (C)/nitrogen (N) ratio), N efflux as ammonium released in feces and mucus and as dead worms is an important pathway to return nutrients to the soil (Fig. 1) and can range from 21 to 50 kg ha^{-1} $year^{-1}$ (Lavelle et al. 1998). Indeed, it has been shown that this release of N retained in plant residues and SOM is the main mechanism (and not their positive effect on soil structure) by which earthworms stimulate plant growth (van Groenigen et al. 2014). However, there are other concomitant processes that also play a key role in stimulating plant growth, such as their interactions with mycorrhizas (Wurst et al. 2004; Yu et al. 2005; Milleret et al. 2009; Zaller et al.

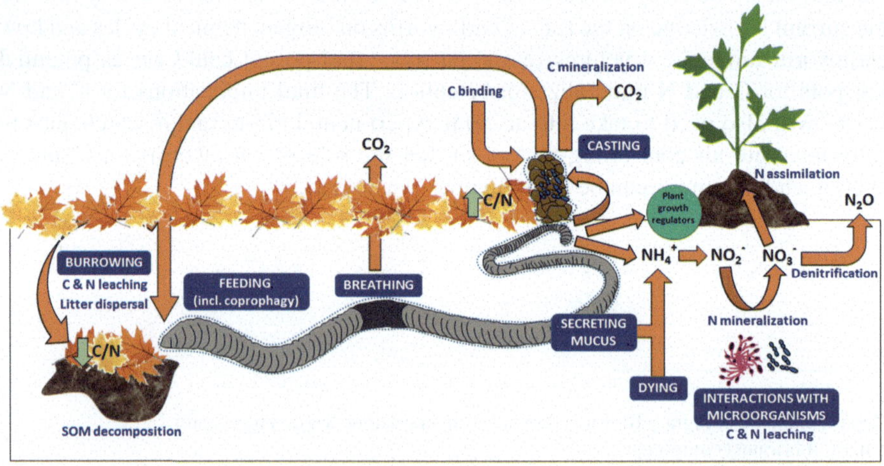

Fig. 1 Earthworm activities influencing soil processes, including direct and indirect effects

2011; Li et al. 2013a, b; Wang et al. 2020) and the release of plant growth regulators (Fig. 1) (Tomati et al. 1988; Canellas et al. 2002; Puga-Freitas et al. 2012).

Interestingly, the plant litter accumulated by some anecic worms in middens, which has been specifically collected by these earthworms, has a lower C/N ratio and greater microbial activity than litter not associated with these structures and thus, increasing the overall C/N ratio of the remaining surface litter (Bohlen et al. 1997) (Fig. 1). As these biogenic structures act as an "external rumen" (Lavelle et al. 2007; see also Sect. 2), preliminary microbial attack could also facilitate the degradation of less palatable litter, which could have important implications for nutrient cycling in nutrient-poor systems. Reingestion of their feces or coprophagy is another mechanism to increase nutritional assimilation from suboptimal foods (Ponge 1991; Mariani et al. 2001).

The effects of earthworms on N cycling have been further expanded with a focus on microbial N functional genes. Accordingly, recent studies have shown a decrease in the abundance of nitrification genes (amoA and nxrB), but an increase of those associated with denitrification (napA, nirK, nirS, norB, and nosZ) and nitrogen fixation processes (nifH) in the presence of earthworms (Jang et al. 2022). However, a wider comparison among published articles evidenced that the direct and indirect effects of earthworms on N cycling are modulated by their feeding behavior, with nitrification rates decreasing more in the presence of anecic worms and denitrification ones increasing when endogeic worms were present (Xue et al. 2022). Furthermore, N release from soils can also be enhanced through the leaching of soluble forms (namely, nitrate, but also dissolved organic nitrogen) due to a combination of ammonium excretion, casting, and burrowing by earthworms (Subler et al. 1997; Domínguez et al. 2004; Costello and Lamberti 2008), which also points toward the importance of ecological categories in understanding the earthworm regulatory roles on N pools.

On the other hand, earthworms can either enhance C mineralization, by stimulating microbial activities (Angst et al. 2020), or C sequestration by physically protecting organic C within micro- and macroaggregates (Bossuyt et al. 2005; Pulleman et al. 2005; Barthod et al. 2020) and through binding it to mineral surfaces (Shan et al. 2013). Because earthworms consume microbial biomass present in soils but also promote microbial growth, the overall effects of earthworms on microbial biomass have led to contradictory results in the available literature (reviewed by Medina-Sauza et al. 2019). Nevertheless, enhanced microbial attack is likely to be more pronounced in the oxic litter layer under the action of epigeic earthworms, which have shown to alter the soil microbial composition more than geophagous species (Pawlett et al. 2009). Accordingly, not only Actinobacteria and Firmicutes (Koubová et al. 2015) but also Gammaproteobacteria (Fjøsne et al. 2018) increase their abundances after epigeic gut passage. In contrast, anecic species influence microbial populations in the deeper layers through burrowing and induce a shift in bacterial communities from anaerobic Firmicutes to aerobic or facultative-aerobic saprotrophic Bacteroidetes and Proteobacteria (Menezes et al. 2018). A similar impact on bacterial community composition following the introduction of endogeic earthworms was observed by Bernard et al. (2012), who found that these soil feeders strongly promoted the growth of Flavobacteriaceae (Bacteroidetes phylum) and Paenibacillaceae (Firmicutes phylum). The fact that these bacterial families can degrade more recalcitrant compounds, including phenol derivatives (Xu et al. 2010; Gopal et al. 2017) and chitin (Wieczorek et al. 2019), could indicate that, by stimulating specific microbial degraders, earthworms could have access to nutrients that are locked away in the SOM complex matrix. These bacterial taxa are also predominant in the earthworm guts (Sapkota et al. 2020), and specifically harbored from the surrounding soil by different species/ecological groups, which explains the contrasting results observed in the literature depending on the habitat/soil type (Thakuria et al. 2010; Menezes et al. 2018) or species/ecological group present (Scheu et al. 2002; Thakuria et al. 2010).

Similar to N, C losses as dissolved organic carbon (DOC) can also be enhanced by earthworms due to a combination of casting and burrowing (Fig. 1) and by stimulating microbial processes (Song et al. 2020). However, the earthworm contribution to C release through leaching has been difficult to prove due to the fact that microorganisms use DOC as a nutrient source (Perelo and Munch 2005), and from this, it was concluded that the earthworm effects on DOC production diminish with time (Amador et al. 2003; Guo et al. 2019). Furthermore, the earthworm control on the release of soluble C is modulated by plant community composition and moisture conditions, with earthworms enhancing C losses when soils are drier and shrubs are absent, indicating that their effect depended on their interaction with soil moisture and microbial activity (Peach et al. 2021). In addition, initial soil C content (Bohlen et al. 2004) and its lability (Lago et al. 2020) also determine DOC concentrations in the leachates, and therefore, land use history and management can also have a stronger influence on DOC exports from soils.

Increased binding of soil organic C inside soil aggregates by earthworms has been demonstrated in several studies (Bossuyt et al. 2005; Pulleman et al. 2005; Fonte and Six 2010; Zhang et al. 2013; Song et al. 2020; Zhu et al. 2021) (see also Sect. 2), with their different feeding strategies having an effect on the origin of the C being retained in these biogenic structures. Accordingly, more plant- and root-derived C was incorporated by epigeic worms, whereas more soil-derived C was found in those macroaggregates produced by endogeic worms (Sánchez-de León et al. 2014). However, although epigeic and anecic species induced a larger incorporation of litter-derived C into microaggregates within large macroaggregates, they have less effect on soil aggregation than endogeic species, and therefore, the combination of this group with the other two groupings is crucial for a larger earthworm effect on C retention (Bossuyt et al. 2006).

In view of this, it is essential to include the interplay of the biotic interactions (between earthworm species and ecological groups but also with plants and microbes) and abiotic factors in the system investigated. As being climate-sensitive, their temporal dynamics also play an important role in understanding their effects on C and N mineralization/storage, with extreme conditions reducing or even impeding surface feeding activities as well as the vertical redistribution of SOM from the organic layers down the soil profile (Guidi et al. 2022).

2 Biogenic Structures as Potential Hotspots for C and N Mineralization and Retention: Implications for C and N Stocks

Earthworms produce two types of organo-mineral structures, burrows and casts, which result from feeding while navigating through the soil and from the chemical transformations after gut transit. Because they are impregnated with mucus, a source of labile nutrients, they are microsites that attract high microbial activities leading to increased C and N mineralization (Fig. 2). In addition, earthworms also create biopores that can be used by plant roots and other organisms and that promote aerobic decomposition processes, but also SOM retention (Fig. 2). Finally, those species with active calciferous glands that secrete amorphous calcium carbonate can mediate C (and perhaps N) stabilization (Briones et al. 2008a).

The estimated production of earthworm casts in temperate regions ranges between 5.8 and 100 tons per ha, but it can reach as much as 2600 tons per ha in

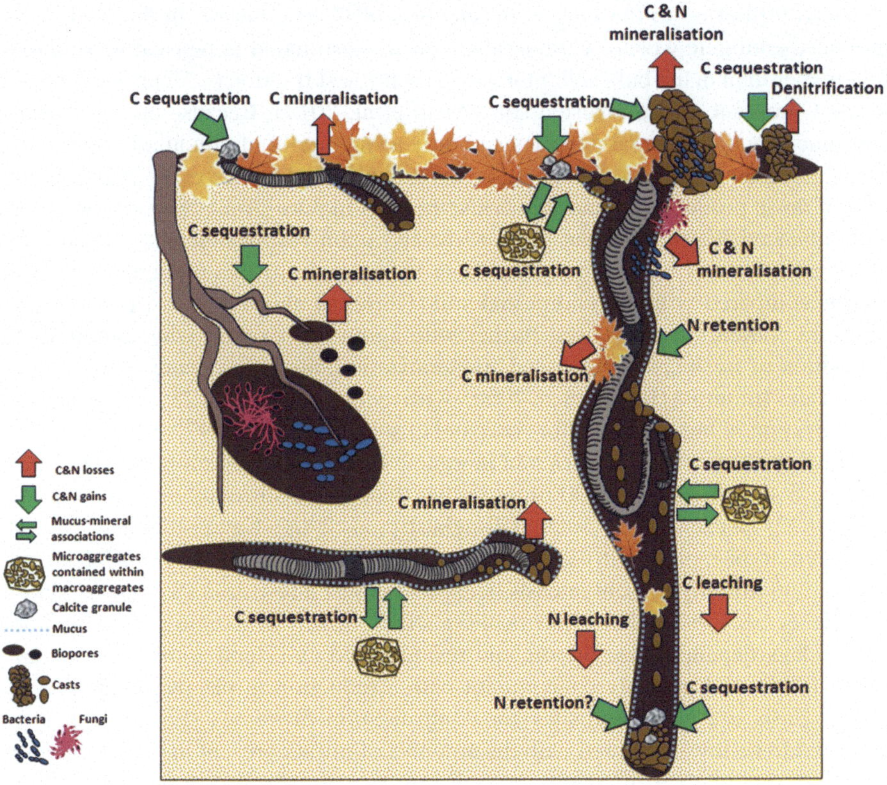

Fig. 2 Earthworm biogenic structures (burrows, casts, and pores) and secretions (mucus and calcium carbonate concretions) play a crucial role in determining the dynamic of soil processes

tropical regions (Edwards and Arancon 2022). However, this might be an underestimation as earthworms also deposit casts underground and burrowing linings (Fig. 2) and casting activity exhibits temporal fluctuations (Wang et al. 2021). Casts are richer in amino-sugars, phenols and proteins (Mora et al. 2003), calcium, potassium, and magnesium (Kim et al. 2022), as well as total amounts of organic C, P, and N (van Groenigen et al. 2019; Wang et al. 2021) when compared to the surrounding soil. They are also stable structures and the proposed stabilization mechanisms include bacterial gums, fungal hyphae, and organo-mineral bonds (Tomlin et al. 1995), although strongly influenced by the clay and carbonate contents of the parent soil (Schrader and Zhang 1997) and the presence of organic residues (Shipitalo and Protz 1988). However, the potential of earthworms to store C in casts relies on the balance between C mineralization and stabilization (Fig. 2). Physical protection of C occurs via its incorporation into microaggregates (53–250 μm) contained within the macroaggregates (>250 μm) (Six et al. 2004; Bossuyt et al. 2005), which are resistant to degradation. Thixotropic hardening, which involves the rearrangement of particles and water films leading to stronger chemical bonds, increases the strength of this physico-chemical protection even further (Thomas et al. 2020), more so over time, as result of drying and aging (Marinissen and Dexter 1990) and hence, contributing to the long-term retention of C. In contrast, in the fresh casts produced endogeic worms, C mineralization was estimated to be twofold to threefold higher than in the bulk soil, more so in their finest fraction (< 53 μm) and linked to the C content of the ingested soil (Abail et al. 2017). Because epigeic worms consume less fine mineral particles and only a small pool of potentially mineralizable C is egested, C mineralization is less magnified (Thomas et al. 2020). In the case of those produced by anecic worms, the shift in pore structure to smaller pores and enhanced water retention seem to be the main responsible mechanisms for observed increases in C mineralization in the casts from this particular group of worms compared with the undigested soil (Görres et al. 2001). Cast porosity and moisture content decrease even further with aging and explain why, although the deposition of fresh casts increases C mineralization, the rates decrease over longer time scales (months) (Brown et al. 2000), with reductions of up to 28% after 30 years of cultivation being predicted for Amazonian soils (Lierman 1997).

Fresh earthworm casts are rich in inorganic N, in particular ammonium (Fig. 1), ranging between 10.9 and 700 times higher than in the surrounding soils (Lavelle et al. 1992; Decaëns et al. 1999; Tillinghast et al. 2001; Kawaguchi et al. 2011). However, it is rapidly nitrified by microorganisms resulting in 42–87% of the ammonium initially present in casts being lost in just a few days (Syers et al. 1979; Lavelle et al. 1992). Then nitrification slows down with aging, usually after 28–30 days depending on the organic residue used as a food source (Parkin and Berry 1994), and possibly concomitant with a reduction in pH (Kawaguchi et al. 2011). Casts are also microsites for denitrification processes (Fig. 2), but rates appeared to be low compared with the amounts of nitrate present in these structures and hence, not likely to affect nitrate pools in the short term (Parkin and Berry 1994).

Earthworm burrowing activities also have profound effects on soil structure and chemistry, with important implications for water drainage and C and N dynamics

(Devliegher and Verstraete 1997; Heemsbergen et al. 2004). By reusing the burrows, dragging litter down, and selecting the finest soil particles anecic worms ensure a continuous input of C belowground and provide adsorption sites at the burrow walls that favor the formation of mucus–mineral associations (Guhra et al. 2020) (Fig. 2). These modifications of the soil environment not only contribute to nutrient retention but also to mineralization, as mucus acts as a priming microbial agent. Burrow linings contain higher C (5–500%) and inorganic N levels (21–78%) than non-worm-worked soil (Stehouwer et al. 1993; Görres et al. 1997; Jégou et al. 2000; Hagedorn and Bundt 2002). Measurements of C and N mineralization rates in burrows indicate that they are also high, ranging from 2.3 to 7.5 times and 0.42 to 1.8 times higher, respectively, than in the bulk soil (Görres et al. 1997, 2001). Furthermore, since earthworm burrows can function as preferential flow pathways, they are also important structures for increasing water drainage and, in turn, the leaching of dissolved elements. Soluble forms of C and N are susceptible to be exported from the soils due to burrowing activity (Fig. 2) and previous studies have shown that the total amounts of N leached in agricultural soils can be 2.5-fold greater in plots with increased earthworm populations than in those with decreased populations (Domínguez et al. 2004). The attribution of these infiltration effects to earthworm burrows came from earlier work by Edwards et al. (1992) who found that distilled water poured directly into the surface openings of *Lumbricus terrestris* burrows became enriched with up to 40 mg of NO_3-N L^{-1}, a value greater than that measured in water samples collected from natural rainstorms. Nevertheless, these C and N losses can be counteracted by stabilization mechanisms, and thus, C sequestration rates of about 22 g C m^{-2} $yr.^{-1}$ have been measured in burrow linings of anecic worms (Don et al. 2008). No similar estimates have been produced for N retention rates in earthworm burrows, although the available evidence indicates that anecic worms can accumulate N in their burrows, which can extend to the surrounding soil (Amador et al. 2006; Andriuzzi et al. 2013) and is facilitated by increased soil moisture (Amador et al. 2005). Further research is needed to quantify if these elevated C and N accumulation rates in the burrowing linings can lead to an overall increase in C and N pools in the long term. Retention processes could be enhanced if earthworm continuous input of labile C sources that promote microbial priming became reduced, which is the main argument in favor of the idea that the positive effects of earthworm burrowing activity on C and N retention are mostly indirect through supporting root growth (Banfield et al. 2017).

Finally, it is known that some earthworm species (namely, epigeic and anecic worms) can secrete amorphous calcium carbonate before finally stabilizing into calcite (Fig. 2), with the excess being expelled into the soil as calcite granules (Briones et al. 2008b; Gago-Duport et al. 2008). In those species with active calciferous glands, such as *L. terrestris,* granule production rate ranges between 0.8 to 2.9 mg per earthworm per day (Canti 2007; Lambkin et al. 2011; Versteegh et al. 2013). This is a possible binding agent (Tisdall and Oades 1982) and could help to enhance the chemical stabilization of CO_2 in certain ecosystems (Briones et al. 2008a; Rowley et al. 2018, 2021). Calcium can also act as a cementing agent for the SOM that passes through the gut leading to the formation of "Ca humates" (Satchell

1967), and indeed, clay-polyvalent cation-soil organic matter (C-P-SOM) linkages have been suggested to be responsible for bonding plant and microbial polysaccharides and other organic compounds to clay (Shipitalo and Protz 1988). Organo-mineral associations mediated by clay are also a well-known mechanism to stabilize organic N (Vogel et al. 2014); however, direct links between earthworm calcium carbonate secretion and N sequestration have not yet been established.

3 Earthworm Contribution to C and N Greenhouse Gas Emissions

The previous two sections provided evidence for the crucial roles of earthworms in mediating SOM decomposition and nutrient immobilization processes, which are both also modulated by abiotic (namely, soil temperature and moisture as well as the chemical composition of soils) and biotic interactions (plants and microorganisms). Oxidation of SOM that results in the production of CO_2 emissions is an aerobic process, whereas CH_4 is the end product of the anaerobic decomposition, and N_2O emissions can be the result of either aerobic (nitrification) or anaerobic processes (denitrification). From this, it would be expected that the anoxic environment in the earthworm gut would enhance denitrification, whereas earthworm biopores/burrows and casts would represent hotspots for nitrification and C mineralization, but also for C and N sequestration (as seen above). However, based on the latest review work (Singh and Singh 2019), earthworms increase CO_2, CH_4, and N_2O emissions, and this greenhouse gas (GHG) generation can be amplified even further depending on the soil physicochemical properties, the type and amount of residues added to the soils, earthworm abundance, and the feeding strategy of different earthworm species. Because the direct earthworm contribution to either GHG emissions or C and N sequestration rates is difficult to measure, the published literature shows very different results and the subject remains controversial. Resolving this issue would require a more detailed investigation of all components, factors, and mechanisms modulating earthworm-induced GHG production.

That earthworms accelerate SOM decomposition and stimulate microbial activities has long been acknowledged and consequently, it is well accepted that the outcome of these earthworm–microorganisms interactions is the increase in CO_2 production (Šimek and Pižl 2010; Lubbers et al. 2013; Zheng et al. 2017). In agreement with this, the results from a meta-analysis using 237 observations of 57 articles published between 1990 and 2011 concluded that earthworms increased CO_2 emissions by 33% (Lubbers et al. 2013). However, the fact that nearly all studies included in this review were performed under laboratory conditions and without plants provides only a partial picture of whether earthworms are good or bad in terms of converting soils into either a net source or a sink of GHG. For example, the duration of the incubation experiment proved to be crucial in determining the overall earthworm effect on CO_2 emissions, and with increasing experimental duration

(> 200 days) less CO_2 was emitted and hence, it could be considered to be a transient response (Lubbers et al. 2013; Chang et al. 2016). Other controlling factors of earthworm contribution to CO_2 fluxes highlighted in Lubbers' study included: (i) the different earthworm ecological groups, with anecic worms inducing the highest increase in CO_2 emissions (50%), (ii) population sizes, evidenced by a 41% increase when earthworm densities were greater than 150 individuals m^{-2}, (iii) type of nutrient inputs, with earthworm induced effects increasing if mineral fertilizers were added (61%), (iv) soil organic carbon content, with low values (<2%) corresponding with greater emissions of CO_2, (v) C/N ratio, with positive effects if the values were below 12.5, and (vi) ecosystem simulated in the experimental design, which showed significantly higher increases in agroecosystems (45%) compared to natural systems (18%). According to this analysis, those agricultural practices that promote earthworm populations (e.g., no-till, ground-cover), but with reduced residue input and regular applications of mineral N (rendering lower C/N ratios), are more prone to stimulate C emissions from soils due to a higher anecic earthworm activity. However, some field studies have shown that crop residue additions shift the balance between C mineralization and stabilization processes, from the dominance of the former to the dominance of the latter at low earthworm densities (Lubbers et al. 2017; John et al. 2020), but they compensate each other at higher densities (John et al. 2020) and when a greater functional diversity is present (Xia et al. 2011). Ultimately, they could even offset the observed stimulation of CO_2 emissions via enhancing plant growth (Van Groenigen et al., 2014) and cast production (Dewi and Senge 2015).

The potential effects of earthworms on CH_4 fluxes have received considerably less attention, with the majority of the observations coming from laboratory incubations of epigeic earthworms. The outcomes from this experimental work show very contrasting results, and while some studies suggest that earthworm activities consistently lowered net CH_4 production (Kumaresan et al. 2011; Bradley et al. 2012; Kernecker et al. 2015), others concluded that they promote CH_4 generation (Koubová et al. 2012; John et al. 2020; Schulz et al. 2015). The earthworm gut and casts are the main microsites for methanogenic bacteria (Kammann et al. 2009); however, casts exhibit more CH_4 oxidation potential associated with a higher number of methanotrophs and presence of oxygen (Park et al. 2008; Kim et al. 2011) (Fig. 3) and hence, earthworm casts have been proposed as a promising tool for methane removal in landfills (Moon et al. 2010). Similarly, earthworm burrows by increasing soil porosity and oxygen concentrations in the soil profile also enhance methane oxidation, as it has been shown that an indirect constraint on methane oxidation is a reduced earthworm burrowing activity due to soil acidification of a former earthworm-rich soil (Brumme and Borken 1999). In contrast, the earthworm gut provides the anaerobic conditions and the methanogenic communities for methane generation (Fig. 3). For example, some earthworm species such as *Eudrilus eugeniae* (Eudrilidae) display higher methanogenic capacities than other species belonging to the Megascolecidae and Glossoscolecidae families (Depkat-Jakob et al. 2012), thanks to the ingestion of denitrifying and fermenting microbes (Egert et al. 2004; Horn et al. 2005; Wüst et al. 2009, 2011). Other studies, however, have indicated

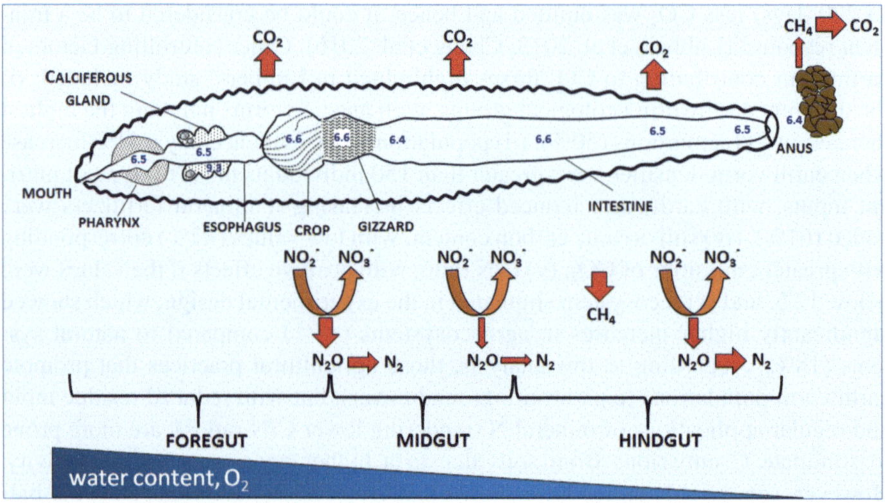

Fig. 3 Potential GHG generation processes "directly" attributable to earthworms based on the literature. The relative importance of each gas is indicated by the size of the arrows while differentiating between external (above) and internal diffusion (below). Gradients of water and oxygen contents and pH values (blue font) are also shown

that methanogens are not metabolically significant in the earthworm guts (Karsten and Drake 1995). Despite the fact that some earthworms can emit important amounts of H_2 (e.g., *L. terrestris* can emit approximately 6 nmol H_2 g (fresh weight)$^{-1}$ h^{-1}; Wüst et al. 2009), which forms important trophic links between acetogenic, methanogenic, and sulfate-reducing processes (Zinder 1993), the consumption of H_2 by the earthworm gut microflora under anaerobic conditions appeared to be minimal (Karsten and Drake 1995) (Fig. 3).

The earthworm gut also provides the optimum conditions for N_2O-producing microorganisms (Fig. 3) by the continuous supply of labile substrates (in the form of a mixture of low-molecular-weight mucus with ingested SOM), the anaerobic environment, neutral pH (Robertson 1936), and high moisture content (Ihssen et al. 2003; Lemtiri et al. 2014). Specifically, certain Protobacteria belonging to the Rhodocyclales (e.g., *Dechloromonas denitrificans*) appear to be the primary agents producing N_2O in the earthworm gut (Horn et al. 2005), as well as other nitrate and nitrite-reducing bacteria, such as members of the Rhizobiales, that are not only responsible for the symbiotic N_2-fixation in nature but also for N emissions in earthworms (Depkat-Jakob et al. 2012). Thanks to this diversity of denitrifiers, lumbricid earthworms (unlike Megascolecidae) can emit up to 11 and on average 1.5 nmol N_2O h^{-1} g^{-1} (fresh weight) as a result of denitrification processes, but also via non-denitrifying, dissimilatory nitrate reducers, which convert the nitrite in the earthworm gut to N_2O. In contrast, fresh casts only produce N_2O only under anoxic conditions, which indicates that anaerobic conditions are needed for the gut-associated denitrifying bacteria to generate N_2O (Matthies et al. 1999). Several studies have shown that the presence of earthworms increases soil N_2O emissions up to 42% (reviewed by Lubbers et al. 2013), especially over longer periods (> 30 days), when epigeic species are present and extra N is applied to the system (Rizhiya et al. 2007; Giannopoulos et al. 2010, 2011; Nebert et al. 2011; Lubbers et al. 2011, 2013). In contrast, endogeic species do not seem to contribute significantly to N_2O emissions unless they have access to N-derived plant litter, especially when it is incorporated into the soil by those earthworm ecological groupings capable of doing so (Rizhiya et al. 2007; Speratti and Whalen 2008; Giannopoulos et al. 2010, 2011). According to these studies, the generation of N_2O was associated with a corresponding increase in the turnover rate of macroaggregates, which indicates that microbial priming effects might be behind this response. It should be noted, however, that although the earthworm crop/gizzard and hindgut are the regions where the largest amounts of N_2O can be emitted, complete denitrification (i.e., the reduction of nitrate or nitrite to N_2) also takes place in these regions at high rates (Horn et al. 2006; Wüst et al. 2009) (Fig. 3). This could make the overall earthworm contribution to N_2O emissions not to exceed those of N_2 (Wüst et al. 2009).

Therefore, it seems clear that earthworms can simultaneously increase and decrease CO_2 and N_2O emissions depending on the influence of plants and/or if organic residues are added to the soils (and where they are deposited), soil parameters, and earthworm feeding strategy. This strongly suggests that earthworm-induced GHG emissions are an inevitable side effect of increased soil C and N mineralization processes that are fueled by feeding, burrowing, casting, soil mixing, mucus formation, and interactions with microorganisms (as seen before). Temporality is another important modulating factor in the outcome of these activities, and experimental observations have shown that over longer periods of time (> 30 days), earthworms induced an increase in N_2O emissions, but diminished those of CO_2 (Lubbers et al. 2013). If the earthworm effect is actually a microbial priming effect (via increasing substrate availability to microbes, as suggested above), then long-term studies with and without organic additions will allow to determine the actual “direct” earthworm contribution to GHG generation. In support of this, it has been suggested that earthworm-induced C emissions will diminish over time as a result of inducing an unequal amplification of C stabilization compared with C mineralization, and hence, mediating a “C trap” (Zhang et al. 2013). From this, it is possible to suggest that a similar mechanism could also operate in the case of the earthworm-induced N emissions and that they could also mediate an “N trap” under suitable conditions and over long periods of time.

4 Conclusions, Way Forward and Research Needs

In view of all the evidence described above, it is possible to suggest that earthworm casts can be considered to be C and N sinks, as they facilitate the long-term protection of these two elements in the soil, whereas earthworm guts are sources of GHG (thanks to anaerobic microbial activity), and burrows could act as either sinks or sources depending on the species that produced them and its abundance. However, the fact that a number of modulating factors can determine the balance between mineralization and sequestration means that under contrasting climatic conditions, plant and microbial interactions, land uses and management, organic additions and their chemistry, and the soil physico-chemical properties, the same or different earthworm species/ecological groupings could have opposite effects on GHG generation/mitigation. To quantify the magnitude of unequal C processes, Zhang et al. (2013) coined the sequestration quotient (SQ, Csta/Cact), which is calculated separately for systems with earthworms (SQworm) and without (SQbasal). A similar quotient should also be developed to introduce the concept of earthworm-mediated "N trap." Although most published literature highlights the effects of earthworms on increasing N availability for plants (e.g., van Groenigen et al. 2014), a few studies have indicated that earthworms are also capable of mediating the increase of N in the soil, at least in the deeper layers (10–30 cm; Wironen and Moore 2006). The fact that earthworm casts are more enriched in mineral N than the surrounding soil (80–100 mg nitrate-N kg^{-1}) has been long known (Parkin and Berry 1994), but also confirmed in more recent work (e.g., Johnson et al. 2012).

The role of vegetation in earthworm-mediated effects on C and N mineralization/retention also deserves more attention. In addition to the chemical quality of the food sources (litter and SOM) available to earthworms, which strongly influences their community structure and activities, above-ground plant composition has also an important effect in controlling the direction and magnitude of plant–earthworm interactions. Again, although most research efforts have focused on the effects of earthworms on plant growth and not vice versa, some studies have indicated that habitat cover greatly influences earthworm abundance and biomass (Phillips et al. 2019) and that species diversity is inversely correlated to productivity (Gudeta et al. 2022), which suggest that vegetation is an important environmental filter driving their community composition. Land use systems with vegetation that produces high amounts of good quality litter could attract certain species (e.g., belonging to the epigeic and anecic groupings) and determine the extent of their bioturbation effects (Briones 2018). In support of this, Kuznetsova et al. (2021) found that the richer the forest litter is, the higher the biomass of earthworms, and the faster the organic horizon decomposes, resulting in decreases of C stocks in the organic horizon. In contrast, C stocks increased in the mineral layers and were positively related to the biomass of those worms whose life activity is related to the mineral layer (endogeic, epi-endogeic, and anecic groups). Similar work on the effects of different land uses on different earthworm ecological groupings and the consequences for N retention is also needed. It is noteworthy that the little evidence available seems to point in the same direction, at least when considering invasive earthworms, and it shows that they can decrease ammonium stocks without affecting nitrate stocks in the organic soil, but increase nitrate stocks but not ammonium stocks in the mineral soil (Ferlian et al. 2020).

Furthermore, in the particular case of forest and woodland ecosystems, certain trees also exert control over the amount of light/shade and volume of precipitation that penetrates under their canopy (Lukina et al. 2022), which could have a significant effect on earthworm activities and hence on soil C and N stocks. Therefore, in agreement with previous findings, climate and habitat are crucial in shaping earthworm communities (Phillips et al. 2019), but local abiotic and biotic factors also exert a strong influence in determining their ecological groupings' assemblages (Hoeffner et al. 2021) and the way they affect soil processes and hence, advocates for more field research under a wide variety of plant communities (e.g., grass vs. woody plant cover) to fully decipher their role in C and N cycling.

Further insights into the fate of SOM due to the presence of earthworms come from investigating the molecular composition of SOM, which has helped to increase our understanding of how different earthworm species/ecological groups affect the stability of the different C and N pools. By using cutin- and suberin-derived lipids as specific markers for leaf- and root-derived SOM in combination with their ^{13}C composition and soil physical fractionations, Angst et al. (2020) concluded that earthworms favor the decomposition of leaf- over root-derived SOM. A finer refinement of this approach can be obtained from pyrolysis gas chromatography/mass spectrometry (Py-GC/MS), which enables the link between SOM molecules to specific plant and microbial origins and to the degree of SOM decomposition. This methodology has recently proven to be useful in identifying changes in the decomposition processes in the top and subsoil after forest conversions (Nikolaus et al. 2023). More advances in this line of research are foreseen, as it has the potential to trace the original plant material entering the soil into the different SOM fractions and uncover the earthworm effects on their decomposition in the different organic and mineral compartments.

References

Abail Z, Sampedro L, Whalen JK (2017) Short-term carbon mineralization from endogeic earthworm casts as influenced by properties of the ingested soil material. Appl Soil Ecol 116:79–86

Amador JA, Gorres JH, Savin MC (2003) Carbon and nitrogen dynamics in *Lumbricus terrestris* (L.) burrow soil: relationship to plant residues and macropores. Soil Sci Soc Am J 67:1755–1762

Amador JA, Görres JH, Savin MC (2005) Role of soil water content in the carbon and nitrogen dynamics of *Lumbricus terrestris* L. burrow soil. Appl Soil Ecol 28:15–22

Amador JA, Gorres JH, Savin MC (2006) Effects of *Lumbricus terrestris* L. on nitrogen dynamics beyond the burrow. Appl Soil Ecol 33:61–66

Andriuzzi WS, Bolger T, Schmidt O (2013) The drilosphere concept: fine-scale incorporation of surface residue-derived N and C around natural *Lumbricus terrestris* burrows. Soil Biol Biochem 64:136–138. https://doi.org/10.1016/j.soilbio.2013.04.016

Angst G, Angst Š, Frouz J, Peterse F, Nierop KCJ (2020) Preferential degradation of leaf- vs. root-derived organic carbon in earthworm-affected soil. Geoderma 372:114391. https://doi.org/10.1016/j.geoderma.2020.114391

Banfield CC, Dippold MA, Pausch J, Hoang DTT, Kuzyakov Y (2017) Biopore history determines the microbial community composition in subsoil hotspots. Biol Fertil Soils 53:573–588. https://doi.org/10.1007/s00374-017-1201-5

Barthod J, Dignac MF, Le Mer G, Bottinelli N, Watteau F, Kögel-Knabner J, Rumpel C (2020) How do earthworms affect organic matter decomposition in the presence of clay-sized minerals? Soil Biol Biochem 143:107730. https://doi.org/10.1016/j.soilbio.2020.107730

Bernard L, Chapuis-Lardy L, Razafimbelo T, Razafindrakoto T, Pablo A-L, Legname E, Poulain J, Brüls T, O'Donohue M, Brauman A, Chotte J.-L-, Blanchart E. (2012) Endogeic earthworms shape bacterial functional communities and affect organic matter mineralization in a tropical soil. ISME J 6:213–222. https://doi.org/10.1038/ismej.2011.87

Bohlen PJ, Parmelee RW, McCartney DA, Edwards CA (1997) Effects on carbon and nitrogen dynamics of surface litter in corn agroecosystems. Ecol Appl 7:1341–1349

Bohlen PJ, Pelletier DM, Groffman PM, Fahey TJ, Fisk MC (2004) Influence of earthworm invasion on redistribution and retention of soil carbon and nitrogen in northern temperate forests. Ecosystems 7:13–27

Bossuyt H, Six J, Hendrix PF (2005) Protection of soil carbon by microaggregates within earthworm casts. Soil Biol Biochem 37:251–258

Bossuyt H, Six J, Hendrix PF (2006) Interactive effects of functionally different earthworm species on aggregation and incorporation and decomposition of newly added residue carbon. Geoderma 130:14–25

Bradley RL, Chronakova A, Elhottova D, Šimek M (2012) Interactions between land-use history and earthworms control gross rates of soil methane production in an overwintering pasture. Soil Biol Biochem 53:64–71

Briones MJI (2018) The serendipitous value of soil fauna in ecosystem functioning: the unexplained explained. Front Environ Sci 6:149. https://doi.org/10.3389/fenvs.2018.00149

Briones MJI, Ostle NJ, Piearce TG (2008a) Stable isotopes reveal that the calciferous gland of earthworms is a CO2-fixing organ. Soil Biol Biochem 40:554–557

Briones MJI, López E, Méndez J, Rodríguez JB, Gago-Duport L (2008b) Biological control over the formation and storage of amorphous calcium carbonate by earthworms. Mineral Mag 72:227–231

Brown GG, Barois I, Lavelle P (2000) Regulation of soil organic matter dynamics and microbial activity in the drilosphere and the role of interactions with other edaphic functional domains. Eur J Soil Biol 36:177–198

Brumme R, Borken W (1999) Site variation in methane oxidation as affected by atmospheric deposition and type of temperate forest ecosystem. Glob Biogeochem Cycles 13:493–501

Canellas LP, Olivares FL, Okorokova-Facanha AL, Facanha AR (2002) Humic acids isolated from earthworm compost enhance root elongation, lateral root emergence, and plasma membrane H^{+}–ATPase activity in maize roots. Plant Physiol 130:1951–1957. https://doi.org/10.1104/pp.007088

Canti MG (2007) Deposition and taphonomy of earthworm granules in relation to their interpretative potential in Quaternary stratigraphy. J Quat Sci 22:111–118

Chang CH, Szlavecz K, Buyer JS (2016) Species-specific effects of earthworms on microbial communities and the fate of litter-derived carbon. Soil Biol Biochem 100:129–139. https://doi.org/10.1016/j.soilbio.2016.06.004

Costello DM, Lamberti GA (2008) Non-native earthworms in riparian soils increase nitrogen flux into adjacent aquatic ecosystems. Oecologia 158:499–510. https://doi.org/10.1007/s00442-008-1149-0

Decaëns T, Rangel AF, Asakawa N, Thomas RJ (1999) Carbon and nitrogen dynamics in ageing earthworm casts in grasslands of the eastern plains of Colombia. Biol Fertil Soils 30:20–28

Depkat-Jakob PS, Hunger S, Schulz K, Brown GG, Tsai SM, Drake HL (2012) Emission of methane by *Eudrilus eugeniae* and other earthworms from Brazil. Appl Environ Microbiol 78:3014–3019

Devliegher W, Verstraete W (1997) Microorganisms and soil physico-chemical conditions in the drilosphere of *Lumbricus terrestris*. Soil Biol Biochem 29:1721–1729

Dewi WS, Senge M (2015) Earthworm diversity and ecosystem services under threat. Rev Agric Sci 3:25–35. https://doi.org/10.7831/ras.3.25

Domínguez J, Bohlen P, Parmelee R (2004) Earthworms increase nitrogen leaching to greater soil depths in row crop agroecosystems. Ecosystems 7:672–685. https://doi.org/10.1007/s10021-004-0150-7

Don A, Steinberg B, Schöning I, Pritsch K, Joschko M, Gleixner G, Schulze E-D (2008) Organic carbon sequestration in earthworm burrows. Soil Biol Biochem 40:1803–1812
Edwards CA, Arancon NK (2022) Biology and ecology of earthworms. Springer, New York
Edwards WM, Shipitalo MJ, Traina SJ, Edwards CA, Owens LB (1992) Role of *Lumbricus terrestris* (L.) burrows on quality of infiltrating water. Soil Biol Biochem 24:1555–1561
Egert M, Marhan S, Wagner B, Scheu S, Friedrich MW (2004) Molecular profiling of 16S rRNA genes reveals diet-related differences of microbial communities in soil, gut, and casts of *Lumbricus terrestris* L. (Oligochaeta: Lumbricidae). FEMS Microbiol Ecol 48:187–197
Ferlian O, Thakur M, Castañeda GA, San Emeterio LM, Marr S, da Silva Rocha B, Eisenhauer N (2020) Soil chemistry turned upside down: a meta-analysis of invasive earthworm effects on soil chemical properties. Ecology 101:e02936. https://doi.org/10.1002/ecy.2936
Fjøsne T, Myromslien FD, Wilson RC, Rudi K (2018) Earthworms are associated with subpopulations of Gammaproteobacteria irrespective of the total soil microbiota composition and stability. FEMS Microbiol Lett 365:fny071. https://doi.org/10.1093/femsle/fny071
Fonte SJ, Six J (2010) Earthworms and litter management contributions to ecosystem services in a tropical agroforestry system. Ecol Appl 20:1061–1073
Gago-Duport L, Briones MJI, Rodríguez JB, Covelo B (2008) Amorphous calcium carbonate biomineralization in the earthworm's calciferous gland: pathways to the formation of crystalline phases. J Struct Biol 162:422–435
Giannopoulos G, Pulleman MM, Van Groenigen JW (2010) Interactions between residue placement and earthworm ecological strategy affect aggregate turnover and N_2O dynamics in agricultural soil. Soil Biol Biochem 42:618–625
Giannopoulos G, Van Groenigen JW, Pulleman MM (2011) Earthworm-induced N_2O emissions in a sandy soil with surface-applied crop residues. Pedobiologia 54:S103–S111
Gopal M, Bhute SS, Gupta A, Prabhu SR, Thomas GV, Whitman WB, Jangid K (2017) Changes in structure and function of bacterial communities during coconut leaf vermicomposting. Antonie Leeuwenhoek 110:1339–1355. https://doi.org/10.1007/s10482-017-0894-7
Görres JH, Savin MC, Amador JA (1997) Dynamics of carbon and nitrogen mineralization, microbial biomass, and nematode abundance within and outside the burrow walls of anecic earthworms (*Lumbricus terrestris*). Soil Sci 162:666–671
Görres JH, Savin MC, Amador JA (2001) Soil micropore structure and carbon mineralization in burrows and casts of an anecic earthworm (*Lumbricus terrestris*). Soil Biol Biochem 33:1881–1887. https://doi.org/10.1016/S0038-0717(01)00068-2
Gudeta K, Bhagat A, Julka JM, Bhat SA, Sharma GK, Bantihun G, Amarowicz R, Belina M (2022) Impact of aboveground vegetation on abundance, diversity, and biomass of earthworms in selected land use systems as a model of synchrony between aboveground and belowground habitats in mid-Himalaya, India. Soil Syst 6:76. https://doi.org/10.3390/soilsystems6040076
Guhra T, Stolze K, Schweizer S, Totsche KU (2020) Earthworm mucus contributes to the formation of organo-mineral associations in soil. Soil Biol Biochem 145:107785. https://doi.org/10.1016/j.soilbio.2020.107785
Guidi C, Frey B, Brunner I, Meusburger K, Vogel ME, Chen X, Stucky T, Gwiazdowicz DJ, Skubała P, Bose AK, Schaub M, Rigling A, Hagedorn F (2022) Soil fauna drives vertical redistribution of soil organic carbon in a long-term irrigated dry pine forest. Glob Chang Biol 28:3145–3160. https://doi.org/10.1111/gcb.16122
Guo Y, Zhang X, Zhang Y, Wu D, McLaughlin N, Zhang S, Chen X, Jia S, Liang A (2019) Temporal variation of earthworm impacts on soil organic carbon under different tillage systems. Int J Environ Res Public Health 16:1908. https://doi.org/10.3390/ijerph16111908
Hagedorn F, Bundt M (2002) The age of preferential flow paths. Geoderma 108:119–132
Heděnec P, Jiménez JJ, Moradi J, Domene X, Heackenberger D, Barot S, Oktaba L, Filser J, Kindlmann P, Frouz J (2022) Global distribution of soil fauna functional groups and their estimated litter consumption across biomes. Sci Rep 12:17362. https://doi.org/10.1038/s41598-022-21563-z

Heemsbergen DA, Berg MP, Loreau M, van Hal JR, Faber JH, Verhoef HA (2004) Biodiversity effects on soil processes explained by interspecific functional dissimilarity. Science 306:1019–1020. https://doi.org/10.1126/science.1101865

Hoeffner K, Santonja M, Monard C, Barbe L, Le Moing M, Cluzeau D (2021) Soil properties, grassland management, and landscape diversity drive the assembly of earthworm communities in temperate grasslands. Pedosphere 31:375–383

Horn MA, Ihssen J, Matthies C, Schramm A, Acker G, Drake HL (2005) *Dechloromonas denitrificans* sp. nov., *Flavobacterium denitrificans* sp. nov., *Paenibacillus anaericanus* sp. nov., and *Paenibacillus terrae* strain MH72, N_2O-producing bacteria isolated from the gut of the earthworm *Aporrectodea caliginosa*. Int J Syst Evol Microbiol 55:1255–1265

Horn MA, Mertel R, Kästner M, Gehre M, Drake HL (2006) In vivo emission of dinitrogen by earthworms via denitrifying bacteria in the gut. Appl Environ Microbiol 72:1013–1018

Ihssen J, Horn MA, Matthies C, Gössner A, Schramm A, Drake HL (2003) N_2O-producing microorganisms in the gut of the earthworm *Aporrectodea caliginosa* are indicative of ingested soil bacteria. Appl Environ Microbiol 69:1655–1661

Jang J, Xiong X, Liu C, Yoo K, Ishii S (2022) Invasive earthworms alter forest soil microbiomes and nitrogen cycling. Soil Biol Biochem 171:108724. https://doi.org/10.1016/j.soilbio.2022.108724

Jégou D, Cluzeau D, Hallaire V, Balesdent J, Trehen P (2000) Burrowing activity of the earthworms *Lumbricus terrestris* and *Aporrectodea giardi* and consequences on C transfers in soil. Eur J Soil Biol 36:27–34

John K, Janz B, Kiese R, Wassmann R, Zaitsev AS, Wolters W (2020) Earthworms offset straw-induced increase of greenhouse gas emission in upland rice production. Sci Total Environ 710:136352

Johnson S, Bose A, Snaddon J, Moss B (2012) The role of earthworms in nitrogen and solute retention in a tropical forest in Sabah, Malaysia: A pilot study. J Trop Ecol 28:611–614. https://doi.org/10.1017/S0266467412000624

Kammann C, Hepp S, Lenhart K, Müller C (2009) Stimulation of methane consumption by endogenous CH_4 production in aerobic grassland soil. Soil Biol Biochem 41:622–629

Karsten GR, Drake HL (1995) Comparative assessment of the aerobic and anaerobic microfloras of earthworm guts and forest soils. Appl Environ Microbiol 61:1039–1044

Kawaguchi T, Kyoshima T, Kaneko N (2011) Mineral nitrogen dynamics in the casts of epigeic earthworms (*Metaphire hilgendorfi*: Megascolecidae). Soil Sci Plant Nutr 57(3):387–395. https://doi.org/10.1080/00380768.2011.579879

Kernecker M, Whalen JK, Bradley RL (2015) Endogeic earthworms lower net methane production in saturated riparian soils. Biol Fertil Soils 51:271–275. https://doi.org/10.1007/s00374-014-0965-0

Kim TG, Moon KE, Lee EH, Choi SA, Cho KS (2011) Assessing effects of earthworm cast on methanotrophic community in a soil biocover by concurrent use of microarray and quantitative real-time PCR. Appl Soil Ecol 50:52–55

Kim G, Jo H, Kim HS, Kwon M, Son Y (2022) Earthworm effects on soil biogeochemistry in temperate forests focusing on stable isotope tracing: a review. Appl Biol Chem 65:88. https://doi.org/10.1186/s13765-022-00758-y

Koubová A, Goberna M, Šimek M, Chronakova A, Pizl V, Insam H, Elhottova D (2012) Effects of the earthworm *Eisenia andrei* on methanogens in a cattle-impacted soil: a microcosm study. Eur J Soil Biol 48:32–40

Koubová A, Chronáková A, Pižl V, Sánchez-Monedero MA, Elhottová D (2015) The effects of earthworms *Eisenia* spp. on microbial community are habitat dependent. Eur J Soil Biol 68:42–55. https://doi.org/10.1016/j.ejsobi.2015.03.004

Kumaresan D, Héry M, Bodrossy L, Singer AC, Stralis-Pavese N, Thompson IP, Murrell JC (2011) Earthworm activity in a simulated landfill cover soil shifts the community composition of active methanotrophs. Res Microbiol 162:1027–1032. https://doi.org/10.1016/j.resmic.2011.08.002

Kuznetsova AI, Geraskina AP, Lukina NV, Smirnov VE, Tikhonova EV, Shevchenko NE, Gornov AV, Ruchinskaya EV, Tebenkova DN (2021) Linking vegetation, soil carbon stocks, and earthworms in upland coniferous–broadleaf forests. Forests 12:1179. https://doi.org/10.3390/f12091179

Lago MCF, Barreal ME, Gallego PP, Briones MJI (2020) Legacy effects of agricultural practices override earthworm control on C dynamics in kiwifruit orchards. Front Environ Sci 8:545609. https://doi.org/10.3389/fenvs.2020.545609

Lambkin DC, Gwilliam KH, Layton C, Canti MG, Piearce TG, Hodson ME (2011) Production and dissolution rates of earthworm-secreted calcium carbonate. Pedobiologia 54:S119–S129

Lavelle P (1988) Earthworm activities and the soil system. Biol Fertil Soils 6:237–251. https://doi.org/10.1007/BF00260820

Lavelle P, Melendez G, Pashanasi B, Schaefer R (1992) Nitrogen mineralization and reorganization in casts of the geophagous tropical earthworm *Pontoscolex corethrurus* (Glossoscolecidae). Biol Fertil Soils 14:49–53. https://doi.org/10.1007/BF00336302

Lavelle P, Pashanasi B, Charpentier FR, Gilot C, Rossi J, Derouard L, André J, Ponge J-P, Bernier N (1998) Large-scale effects of earthworms on soil organic matter and nutrient dynamics. In: Edwards CA (ed) Earthworm ecology. St. Lucie Press, Boca Raton, pp 103–122

Lavelle P, Barot S, Decaëns T, Blouin M (2007) Earthworms as key actors in self-organised soil systems. Theor Ecol Ser 4:77–106. https://doi.org/10.1016/S1875-306X(07)80007-4

Lemtiri A, Colinet G, Alabi T, Cluzeau D, Zirbes L, Haubruge É, Francis F (2014) Impacts of earthworms on soil components and dynamics. A review. Biotechnol Agron Soc Environ 18:121–133

Li H, Wang C, Li X, Christie P, Dou Z, Zhang J, Xiang D (2013a) Impact of the earthworm *Aporrectodea trapezoides* and the arbuscular mycorrhizal fungus *Glomus intraradices* on N-15 uptake by maize from wheat straw. Biol Fertil Soils 49:263–271. https://doi.org/10.1007/s00374-012-0716-z

Li H, Wang C, Li X, Xiang D (2013b) Inoculating maize fields with earthworms (*Aporrectodea trapezoides*) and an arbuscular mycorrhizal fungus (*Rhizophagus intraradices*) improves mycorrhizal community structure and increases plant nutrient uptake. Biol Fertil Soils 49:1167–1178. https://doi.org/10.1007/s00374-013-0815-5

Lierman C (1997) Simulation de l'effet de vers de terre tropicaux sur la dynamique de la matière organique et la production végétale, DEA d'Écologie. Université de Paris XI, VI, Institut National Agronomique Paris-Grignon, Paris

Lubbers IM, Brussaard L, Otten W, Van Groenigen JW (2011) Earthworm-induced N mineralization in fertilized grassland increases both N2O emission and crop-N uptake. Eur J Soil Sci 62:152–161. https://doi.org/10.1111/j.1365-2389.2010.01313.x

Lubbers IM, Van Groenigen KJ, Fonte SJ, Six J, Brussaard L, Van Groenigen JW (2013) Greenhouse-gas emissions from soils increased by earthworms. Nat Clim Chang 3:187–194

Lubbers IM, Pulleman MM, van Groenigen JW (2017) Can earthworms simultaneously enhance decomposition and stabilization of plant residue carbon? Soil Biol Biochem 105:15–24. https://doi.org/10.1016/j.soilbio.2016.11.008

Lukina NV, Kuznetsova AI, Geraskina AP, Smirnov VE, Ivanova VN, Teben'kova DN, Gornov AV, Shevchenko NE, Tikhonova EV (2022) Unaccounted factors determining carbon stocks in forest soils. Russ Meteorol Hydrol 47:791–803. https://doi.org/10.3103/S1068373922100077

Mariani L, Bernier N, Jiménez Jaén JJ, Decaëns T (2001) Diet of an anecic earthworms from the Colombian savannas: A question about ecological categories. In: Jiménez Jaén JJ, Thomas RJ (eds) Nature's plow: soil macroinvertebrate communities in the neotropical savannas of Colombia. Centro Internacional de Agricultura Tropical (CIAT), Cali, CO, pp 137–148. (CIAT publication no. 324)

Marinissen JCY, Dexter AR (1990) Mechanisms of stabilization of earthworm casts and artificial casts. Biol Fertil Soils 9:163–167. https://doi.org/10.1007/BF00335801

Matthies C, Griesshammer A, Schmittroth M, Drake HL (1999) Evidence for involvement of gut-associated denitrifying bacteria in emission of nitrous oxide (N_2O) by earthworms obtained from garden and forest soils. Appl Environ Microbiol 65:3599–3604

Medina-Sauza RM, Álvarez-Jiménez M, Delhal A, Reverchon F, Blouin M, Guerrero-Analco JA, Cerdán CR, Guevara R, Villain L, Barois I (2019) Earthworms building up soil microbiota, a review. Front Environ Sci 7:81. https://doi.org/10.3389/fenvs.2019.00081

Menezes AB, Prendergast-Miller MT, Macdonald LM, Toscas P, Baker G, Farrell M, Wark T, Richardson AE, Thrall PH (2018) Earthworm-induced shifts in microbial diversity in soils with rare versus established invasive earthworm populations. FEMS Microbiol Ecol 94:fiy051. https://doi.org/10.1093/femsec/fiy051

Milleret R, Le Bayon RC, Gobat JM (2009) Root, mycorrhiza and earthworm interactions: their effects on soil structuring processes, plant and soil nutrient concentration and plant biomass. Plant Soil 316:1–12. https://doi.org/10.1007/s11104-008-9753-7

Moon KE, Lee SY, Lee SH, Ryu HW, Cho KS (2010) Earthworm cast as a promising filter bed material and its methanotrophic contribution to methane removal. J Hazard Mater 176:131–138

Mora P, Seugé C, Chotte JL, Rouland C (2003) Physico-chemical typology of the biogenic structures of termites and earthworms: a comparative analysis. Biol Fertil Soils 37:245–249. https://doi.org/10.1007/s00374-003-0592-7

Nebert LD, Bloem J, Lubbers IM, van Groenigen JW (2011) Association of earthworm-denitrifier interactions with increased emissions of nitrous oxide from soil mesocosms amended with crop residue. Appl Environ Microbiol 77:4097–4104. https://doi.org/10.1128/AEM.00033-1

Nikolaus K, Schellekens J, Mols S, Jansen B, Briones MJI, Desie E, Cornelis J-T, Absalah S, Muys B, Vancampenhout K (2023) Seeing the forest for the fractions—comparing soil organic matter fractionation methods using molecular features after forest stand conversion. Geoderma 430:116280. https://doi.org/10.1016/j.geoderma.2022.116280

Nuutinen V, Butt KR (2019) Earthworm dispersal of plant litter across the surface of agricultural soils. Ecology 100:1–3

Park S, Lee I, Cho C, Sung K (2008) Effects of earthworm cast and powdered activated carbon on methane removal capacity of landfill cover soils. Chemosphere 70:1117–1123

Parkin TB, Berry EC (1994) Nitrogen transformations associated with earthworm casts. Soil Biol Biochem 26:1233–1238

Pawlett M, Hopkins DW, Moffett BF, Harris JA (2009) The effect of earthworms and liming on soil microbial communities. Biol Fertil Soils 45:361–369. https://doi.org/10.1007/s00374-008-0339-6

Peach ME, Hicks PC, Friedland AJ (2021) Plants and earthworms control soil carbon and water quality trade-offs in turfgrass mesocosms. Sci Total Environ 753:141884

Perelo LW, Munch JC (2005) Microbial immobilization and turnover of ^{13}C labeled substrates in two arable soils under field and laboratory conditions. Soil Biol Biochem 37:2263–2272. https://doi.org/10.1016/j.soilbio.2005.02.039

Phillips H.R.P., Guerra C.A., Bartz M.L.C., Briones M.J.I., Brown G.G., Ferlian O., Gongalsky K.B., Krebs J., Orgiazzi A., Schwarz B., Bach E.M., …, Eisenhauer N. (2019) Global distribution of earthworm diversity. Science 366: 480–485. https://doi.org/10.1126/science.aax4851

Ponge JF (1991) Food resources and diets of soil animals in a small area of Scots pine litter. Geoderma 49:33–62. https://doi.org/10.1016/0016-7061(91)90090-G

Puga-Freitas R, Barot S, Taconnat L, Renou J-P, Blouin M (2012) Signal molecules mediate the impact of the earthworm *Aporrectodea caliginosa* on growth, development and defence of the plant *Arabidopsis thaliana*. PLoS One 7:e49504. https://doi.org/10.1371/journal.pone.0049504

Pulleman MM, Six J, Van Breemen N, Jongmans AG (2005) Soil organic matter distribution and microaggregate characteristics as affected by agricultural management and earthworm activity. Eur J Soil Sci 56:453–467

Rizhiya E, Bertora C, van Vliet PCJ, Kuikman PJ, Faber JH, van Groenigen JW (2007) Earthworm activity as a determinant for N2O emission from crop residue. Soil Biol Biochem 39:2058–2069. https://doi.org/10.1016/j.soilbio.2007.03.008

Robertson JD (1936) The function of the calciferous glands of earthworms. J Exp Biol 13:279–297. https://doi.org/10.1242/jeb.13.3.279
Rowley MC, Grand S, Verrecchia EP (2018) Calcium-mediated stabilisation of soil organic carbon. Biogeochemistry 137:127–149
Rowley MC, Grand S, Spangenberg JE, Verrecchia EP (2021) Evidence linking calcium to increased organo-mineral association in soils. Biogeochemistry 153:223–241. https://doi.org/10.1007/s10533-021-00779-7
Sánchez-de León Y, Lugo-Pérez J, Wise DH, Jastrow JD, González-Meler MA (2014) Aggregate formation and carbon sequestration by earthworms in soil from a temperate forest exposed to elevated atmospheric CO_2: A microcosm experiment. Soil Biol Biochem 68:223–230
Sapkota R, Santos S, Farias P, Krogh PH, Winding A (2020) Insights into the earthworm gut multi-kingdom microbial communities. Sci Total Environ 727:138301. https://doi.org/10.1016/j.scitotenv.2020.138301
Satchell JE (1967) Lumbricidae. In: Burgess A, Raw F (eds) Soil biology. Academic Press, London, pp 259–322
Scheu S, Schlitt N, Tiunov A, Newington JE, Jones TH (2002) Effects of the presence and community composition of earthworms on microbial community functioning. Oecologia 133:254–260. https://doi.org/10.1007/s00442-002-1023-4
Schrader S, Zhang H (1997) Earthworm casting: stabilization or destabilization of soil structure? Soil Biol Biochem 29:469–475. https://doi.org/10.1016/S0038-0717(96)00103-4
Schulz K, Hunger S, Brown GG, Tsai SM, Cerri CC, Conrad R, Drake HL (2015) Methanogenic food web in the gut contents of methane-emitting earthworm *Eudrilus eugeniae* from Brazil. ISME J 9:1778–1792. https://doi.org/10.1038/ismej.2014.262
Shan J, Liu J, Wang Y, Yan X, Guo H, Li X, Ji R (2013) Digestion and residue stabilization of bacterial and fungal cells, protein, peptidoglycan, and chitin by the geophagous earthworm *Metaphire guillelmi*. Soil Biol Biochem 64:9–17
Shipitalo MJ, Protz R (1988) Factors influencing the dispersibility of clay in worm casts. Soil Sci Soc Am J 52:764–769. https://doi.org/10.2136/sssaj1988.03615995005200030030x
Šimek M, Pižl V (2010) Soil CO_2 flux affected by *Aporrectodea caliginosa* earthworms. Cent Eur J Biol 5:364–370. https://doi.org/10.2478/s11535-010-0017-1
Singh A, Singh GS (2019) Is earthworm a protagonist or an antagonist in greenhouse gas (GHG) emissions from the soil? Int J Environ Sci Technol 16:1145–1158. https://doi.org/10.1007/s13762-018-1922-5
Six J, Bossuyt H, Degryze S, Denef K (2004) A history of research on the link between (micro) aggregates, soil biota, and soil organic matter dynamics. Soil Tillage Res 79:7–31. https://doi.org/10.1016/j.still.2004.03.008
Song K, Sun L, Lv W, Zheng X, Sun Y, Terzaghi W, Qin Q, Xue Y (2020) Earthworms accelerate rice straw decomposition and maintenance of soil organic carbon dynamics in rice agroecosystems. PeerJ 8:e9870. https://doi.org/10.7717/peerj.9870
Speratti AB, Whalen JK (2008) Carbon dioxide and nitrous oxide fluxes from soil as influenced by anecic and endogeic earthworms. Appl Soil Ecol 38:27–33
Stehouwer RC, Dick WA, Traina SJ (1993) Characteristics of earthworm burrow lining affecting atrazine sorption. J Environ Qual 22:181–185
Subler S, Baranski CH, Edwards CA (1997) Earthworm additions increased short-term nitrogen availability and leaching in two grain-crop agroecosystems. Soil Biol Biochem 29:413–421. https://doi.org/10.1016/S0038-0717(96)00173-3
Syers JK, Sharpley AN, Keeney DR (1979) Cycling of nitrogen by surface-casting earthworms in a pasture ecosystem. Soil Biol Biochem 11:181–185
Thakuria D, Schmidt O, Finan D, Egan D, Doohan FM (2010) Gut wall bacteria of earthworms: a natural selection process. ISME J 4:357–366. https://doi.org/10.1038/ismej.2009.124
Thomas E, Prabha VS, Kurien VT, Thomas AP (2020) The potential of earthworms in soil carbon storage: a review. Environ Exp Biol 18:61–75. https://doi.org/10.22364/eeb.18.06

Tillinghast EK, O'Donnell R, Eves D, Calvert E, Taylor J (2001) Water-soluble luminal contents of the gut of the earthworm *Lumbricus terrestris* L. and their physiological significance. Comp Biochem Physiol A Mol Integr Physiol 129:345–353. https://doi.org/10.1016/S1095-6433(00)00352-4

Tisdall JM, Oades JM (1982) Organic matter and water-stable aggregates in soils. Eur J Soil Sci 33:141–163. https://doi.org/10.1111/j.1365-2389.1982.tb01755.x

Tomati U, Grappelli A, Galli E (1988) The hormone-like effect of earthworm casts on plant growth. Biol Fertil Soils 5:288–294. https://doi.org/10.1007/BF00262133

Tomlin AD, Shipitalo MJ, Edwards WM, Protz R (1995) Earthworms and their influence on soil structure and infiltration. In: Hendrix PF (ed) Earthworm ecology and biogeography in North America. Lewis Publishers Inc., Boca Raton, pp 159–183

van Groenigen J, Lubbers I, Vos H, Brown GG, De Deyn GB, van Groeningen KJ (2014) Earthworms increase plant production: a meta-analysis. Sci Rep 4:6365. https://doi.org/10.1038/srep06365

Van Groenigen JW, Van Groenigen KJ, Koopmans GF, Stokkermans L, Vos HMJ, Lubbers IM (2019) How fertile are earthworm casts? A meta-analysis. Geoderma 338:525–535. https://doi.org/10.1016/j.geoderma.2018.11.001

Versteegh EAA, Black S, Canti MG, Hodson ME (2013) Earthworm-produced calcite granules: a new terrestrial palaeothermometer? Geochim Cosmochim Acta 123:351–357

Vogel C, Mueller CW, Hoeschen C, Buegger F, Heister K, Schulz S, Schloter M, Kögel-Knabner I (2014) Submicron structures provide preferential spots for carbon and nitrogen sequestration in soils. Nat Commun 5:2947. https://doi.org/10.1038/ncomms3947

Wang G, Wang L, Ma F, Yang D, You Y (2020) Earthworm and arbuscular mycorrhiza interactions: strategies to motivate antioxidant responses and improve soil functionality. Environ Pollut 272:115980. https://doi.org/10.1016/j.envpol.2020.115980

Wang X, Fu S, Wang X, Li Z, Li J, Zhang W (2021) One-year monitoring of daily earthworm cast production: surface cast contribution to soil fertility in a subtropical forest. Forests 12:865. https://doi.org/10.3390/f12070865

Wieczorek AS, Schmidt O, Chatzinotas A, von Bergen M, Gorissen A, Kolb S (2019) Ecological functions of agricultural soil bacteria and microeukaryotes in chitin degradation: a case study. Front Microbiol 10:1293. https://doi.org/10.3389/fmicb.2019.01293

Wironen M, Moore TR (2006) Exotic earthworm invasion increases soil carbon and nitrogen in an old-growth forest in southern Quebec. Can J For Res 36:845–854. https://doi.org/10.1139/x06-016

Wurst S, Dugassa-Gobena D, Langel R, Bonkowski M, Scheu S (2004) Combined effects of earthworms and vesicular–arbuscular mycorrhizas on plant and aphid performance. New Phytol 163:169–176. https://doi.org/10.1111/j.1469-8137.2004.01106.x

Wüst PK, Horn MA, Drake HL (2009) In situ hydrogen and nitrous oxide as indicators of concomitant fermentation and denitrification in the alimentary canal of the earthworm *Lumbricus terrestris*. Appl Environ Microbiol 75:1852–1859. https://doi.org/10.1128/AEM.02745-08

Wüst PK, Horn MA, Drake HL (2011) Clostridiaceae and Enterobacteriaceae as active fermenters in earthworm gut content. ISME J 5:92–106

Xia L, Szlavecz K, Swan CM, Burgess JL (2011) Inter- and intra-specific interactions of *Lumbricus rubellus* (Hoffmeister, 1843) and *Octolasion lacteum* (Örley, 1881) (Lumbricidae) and the implication for C cycling. Soil Biol Biochem 43:1584–1590. https://doi.org/10.1016/j.soilbio.2011.04.009

Xu L, Teng Y, Li ZG, Norton JM, Luo YM (2010) Enhanced removal of polychlorinated biphenyls from alfalfa rhizosphere soil in a field study: the impact of a rhizobial inoculum. Sci Total Environ 408:1007–1013

Xue R, Wang C, Liu X, Liu M (2022) Earthworm regulation of nitrogen pools and dynamics and marker genes of nitrogen cycling: A meta-analysis. Pedosphere 32:131–139. https://doi.org/10.1016/S1002-0160(21)60063-2

Yu X, Cheng J, Wong MH (2005) Earthworm-mycorrhiza interaction on Cd uptake and growth of ryegrass. Soil Biol Biochem 37:195–201

Zaller JG, Saccani F, Frank T (2011) Effects of earthworms and mycorrhizal fungi on the growth of the medicinal herb *Calendula officinalis* (Asteraceae). Plant Soil Environ 57:499–504. https://doi.org/10.17221/189/2011-PSE

Zhang W, Hendrix PF, Dame LE, Burke RA, Wu J, Neher DA, Li J, Shao Y, Fu S (2013) Earthworms facilitate carbon sequestration through unequal amplification of carbon stabilization compared with mineralization. Nat Commun 4:2576. https://doi.org/10.1038/ncomms3576

Zheng Y, Wang S, Bonkowski M, Chen X, Griffiths B, Hu F, Liu M (2017) Litter chemistry influences earthworm effects on soil carbon loss and microbial carbon acquisition. Soil Biol Biochem 123:105–114. https://doi.org/10.1016/j.soilbio.2018.05.012

Zhu X, Hu Y, Li Z, Wu D (2021) Earthworms promote the accumulation of rhizodeposit carbon to soil macroaggregate in a Mollisol of Northeast China, primarily in long-term no-till soil. Soil Ecol Lett 3:84–93. https://doi.org/10.1007/s42832-020-0062-2

Zinder SH (1993) Physiological ecology of methanogens. In: Ferry JG (ed) Methanogenesis. Chapman and Hall, Inc., New York, pp 128–206

Earthworm Biopores for Transport and Nutrient Cycling

Yakov Kuzyakov and Yahya Kooch

Abstract Earthworms build a complex network of stable biopores with diameters of up to 0.5 cm existing over years and decades in soil. Besides the direct habitat and pathway functions for the earthworms, the biopores strongly accelerate the diffusion and migration of gases, solutions, nutrients, and organic substances, as well as mineral particles within the soil profile and link the top with subsoil. After a short introduction to pore classification and formation mechanisms, we focus on the functions of biopores produced by earthworms as relate not only to the transport processes. We conclude that biopores are real hotspots of transport, and cycling of elements, where the turnover of C, N, and P is intensively ongoing. Concludingly, biopores formed by earthworms are important for the fast transport of gases, solutions, and nutrients, provide highways for root growth into the subsoil for nutrient and water acquisition, contribute to C sequestration, and therefore are one of the ways for the amelioration of soil by natural processes. The higher C input into biopores enriches C and N in the subsoil. In general, earthworms increase SOM turnover and cycling of elements by stimulating microbial activity, especially in the subsoil. Consequently, it is very crucial to assess the spatial and temporal distribution of the density of biological voids and their effectiveness for the reliable representation of the coarse flow and the transfer of elements in the soil.

Y. Kuzyakov (✉)
Department of Soil Science of Temperate Ecosystems, Department of Agricultural Soil Science, University of Göttingen, Göttingen, Germany

Institute of Environmental Sciences, Kazan Federal University, Kazan, Russia

Peoples Friendship University of Russia (RUDN University), Moscow, Russia
e-mail: kuzyakov@gwdg.de

Y. Kooch
Faculty of Natural Resources & Marine Sciences, Tarbiat Modares University, Noor Mazandaran, Iran
e-mail: yahya.kooch@modares.ac.ir

Y. Kooch, Y. Kuzyakov (eds.), *Earthworms and Ecological Processes*,
https://doi.org/10.1007/978-3-031-64510-5_16

1 Introduction

Plant roots extend into the soil to access resources such as nutrients and water. However, many soils hinder deep root growth (Schneider et al. 2021). Therefore, an important resource, mainly subsoil (soil depth below 0.3 m), containing more than half of soil nutrients and moisture, remains untapped (Gocke et al. 2021). Plants can use soil biopores for root growth and resource utilization, which is the best way for roots to grow in pores due to less resistance (Valentine et al. 2012). Earthworms and plant roots mainly lead to the construction of biological pores (biopores) in the soil (Fig. 1). Biopores occupy about 1–10% of the total soil volume and are found primarily in the subsoil of the croplands because of tillage in the topsoil (Kuzyakov and Blagodatskaya 2015). The activity of roots and animals enriches the inner walls of the biopores with C, N, and P and therefore more C and N content compared to the bulk soil (Hoang et al. 2016). Biopores play an essential but poorly known function in soils. They are hotspots for microbial function and nutrient mineralization and characterize the preferred flow path for root growth, water infiltration, and gas exchange. Although biopores are produced by many types of soil organisms (e.g., bees, ants, and spiders nesting in soil), most reports focus on earthworm biopores.

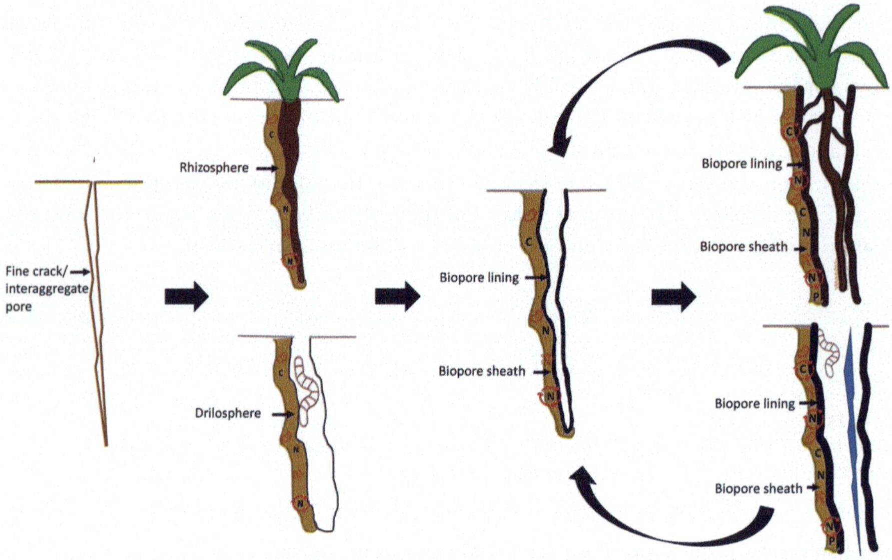

Fig. 1 The following is a schematic timeline depicting the interactions between roots and biopores. Beginning from the left, when biopores are accessible, roots and earthworms utilize the fine cracks in the soil due to the reduced penetration resistance. This results in the development of a rhizosphere or a drilosphere, respectively, which exhibit enhanced nutrient cycling and microbial biomass and activity in comparison to the bulk soil. Even after the roots have died or the earthworms have exited or perished, nutrients continue to be stored along the lining and sheaths of the biopores

Therefore, this review aims to discuss the impact of earthworm biopores on the microbial activities, transport, and cycling of elements.

1.1 Pores and Biopores in Soil: Origin, Formation, and Stability

Soil structure generally refers to the arrangement of voids and soil solids in various scales (Yudina and Kuzyakov 2023). It can be explained in terms of both solid phase and pore volume (Rabot et al. 2018). Pores are soil voids that play a significate role in the storage, retention, and movement of soil. Therefore, the relationship between porosity and soil hydraulics is very close. The soil of the root zone maintains the air and water necessary for plant growth and is directly connected to the distribution of pore sizes (Reynolds et al. 2002). Porosity and particle size distribution in soil affect many hydraulic properties of soil such as electrical conductivity, infiltration, water retention, and water quality. Water and air fluxes, organic matter (OM) decomposition, and plant availability are straightly related to the arrangement and geometry of the pore in the soil (Pires et al. 2017).

Pores (voids) are crucial for all functions of mass, energy, and information exchange in the soil as well as all biochemical reactions and localization of (micro) organisms (Yudina and Kuzyakov 2023). Pores occupy usually 35–55% of the volume of mineral soils and up to 90% of the volume of organic soils. Pore morphology is of great importance because it has direct effects on root growth and air and water transport through the soil. According to the diameter, all pores are divided into nano- (<0.1 μm), micro- (0.1–60/75 μm), and macropores (>60/75 μm) (Fig. 2, note however that various approaches and classifications use various borders). Four other pore size classes have been identified, namely, macropores (greater than 1000 μm), large mesopores (1000–60 μm), fine mesopores (60–10 μm), and micropores (10 μm). Macropores usually make up a small fraction of the total soil volume but are the main controller of soil water (Dhaliwal and Kumar 2020). In particular, soil fauna burrowing and root activity may cause the formation of prolate pores in soil (biopores) (Carducci et al. 2014). The existence of these long pores, which improve root penetration and water and air transfer in the soil, has a favorable impact on plant growth (Pagliai et al. 2004).

These pores are those larger than ~10 μm in diameter and permit (1) quick diffusion of air and water vapor, (2) rapid water infiltration and drainage, and (3) access to nutrients (Bauke et al. 2017). However, exposed surfaces of these pores dry quickly, and this would restrict the activity of organisms living within water films. Macropores may be created as spaces between sand grains and coarse silt grains in seepages with a coarse texture and as pore spaces accumulated in well-integrated loam and clay texture seepages. Roots, earthworms, insects, and other soil organisms can also create soil macropores. Soil pores less than 10 μm in diameter are known as micropores and are critical for water retention for plants and for creating

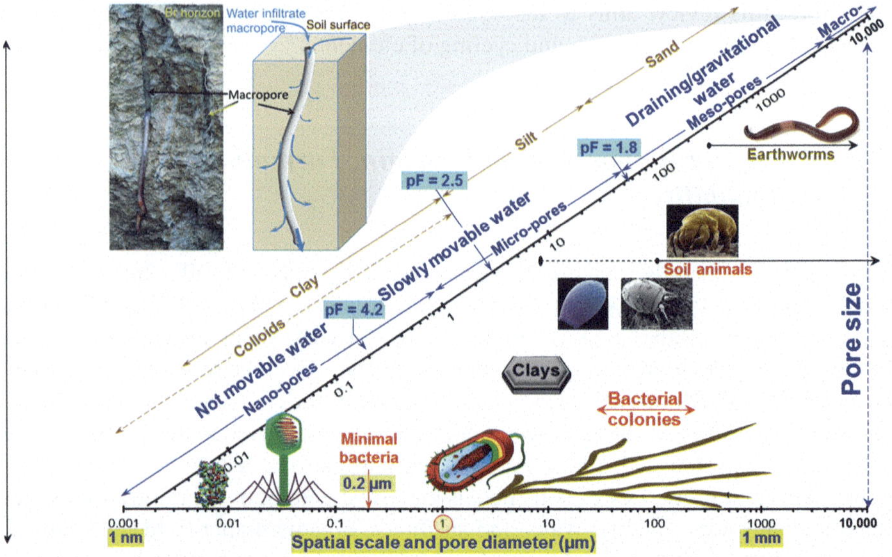

Fig. 2 Schematic representation of the physical properties (upslope scale) and biological presence (*x* scale) of soil pores. Note that the logarithmic scale shows a 10^8-fold change in pore size across the scale. Pore size and water resistance are shown in blue (where ψ is soil water potential). Biota is shown in red. The size of the texture particles (clay, silt, sand) and colloids appears in brown (modified from Kuzyakov and Mason-Jones 2018)

habitats with sufficient water layer thickness for the development of microbial biofilms. Water flow and diffusion of gases and nutrients in micropores are slow, typically <0.025 cm h^{-1}, and because of this, the immediate environment for soil microorganisms remains relatively constant. Larger micropores and smaller macropores can accommodate root hairs, micropores, and microbiota, while pores smaller than ~5 µm cannot accommodate most microorganisms, but only viruses and some bacteria. They may reduce or even inhibit the uptake of potential substrates by microorganisms by limiting the release and diffusion of exoenzymes and nutrients. Topsoil typically contains about 50% of the pore space, while only one-fourth to half of this pore space is available for microorganisms. Larger pores drain and dry very quickly, while smaller pores are a major limitation.

Restricting pore size in small aggregates can slow the penetration of organic matter to restrict interactions and microbial decay of soil organisms, thus preventing predation by fauna. Larger aggregates benefit water infiltration and soil aeration and remain intact as long as the soil is not disturbed by intensive tillage or water erosion. The pore space within small aggregates that make up the larger aggregates is called the inter-aggregate pore space and is important in ensuring soil retention. The pore space around the aggregates (Yudina and Kuzyakov 2023), collectively known as the pore aggregate area, is where plant roots and large plant fragments of plant debris are observed, whose wide channels facilitate water entry and aeration. Aggregation is vital for controlling SOM circulation and microbial activity in surface soils because the interpore space allows the exchange of gases (CO_2, CH_4, and N_2O) produced by microorganisms and plant roots with oxygen and other gases in the atmosphere. These pores are also the sites receiving annual deposits of plant residues. However, the larger pores in aggregates dry quickly so they would only be a habitat for organisms able to live outside of thin water films. In this way, they also control the availability of organic matter to be decomposed by microorganisms.

By demarcating the nature and pore size distribution in the soil, aggregates can classify the pore area characteristics of fine-textured, clay, and loam-textured soils and sandy soil to facilitate aeration, water infiltration, and drainage. Although total pore space is essential, pore size and connectivity are key to determining soil habitability. This feature of the soil, or structure, describes the soil microarchitecture and is largely a function of texture, aggregation, and location within the soil profile. Physical (weathering, drying-wetting, freezing-thawing, etc.) and chemical processes (weathering, solution, etc.) may build pores of all size classes. Soil organisms and roots, however, cannot produce nano-pores, may produce only a very limited amount of micropores (e.g., by fungal hyphen), and are mainly responsible for meso- and macropore formation. Despite other physical and chemical processes also responsible for the formation of these large pores, here we focus only on the part of biopores (Fig. 3).

2 Definition and Origin of Biopores

Within the second half of the nineteenth century, Hugo Thiel was one of the primary researchers to consider root development within the existent channels. He demonstrated that these channels arose from the roots of previously growing plant or soil animals such as nematodes, earthworms, or beetles (Thiel 1865). Subsequently, the term "biopore" was used to describe pores resulting from biological functions. *Biopores* are pores produced by various organisms and roots in the soil. Because this book is aimed at earthworms, this section describes the origin, properties, and functions solely of biopores produced by and/or with the contribution of earthworms – the earthworm biopores. Please note that the term biopores is used below for pores produced by earthworms, despite roots also producing biopores. Earthworm biopores originate mainly from anecic and endogeic earthworms but partly also from epigeic earthworms. These biopores differ in their size, form, orientation, and connectivity (Table 1) and consequently in functions for matter transport and element cycling. The density and volume of biopores originating from anecic and endogeic earthworms increase from topsoil, reach the maximum between 70 and 120 cm, and then decrease again (Jegou et al. 1999).

Many studies were focused on the lifestyle of the anecic, endogeic, and epigeic earthworms, their food preferences, and their direct contribution to the soil bioturbation and so, the mixing of horizons and their contained elements. The earthworm biopores, however, remain in the soil also after the earthworm death (or outmigration) and function not only as a habitat for the successor earthworms and other

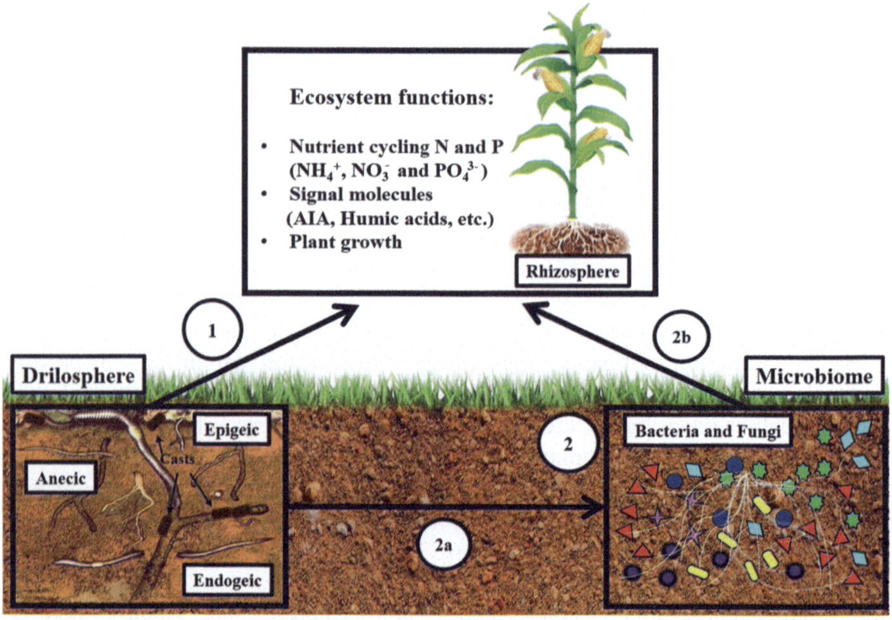

Fig. 3 Consequences of earthworms on nutrient cycle and plant growth. The effects of earthworms are either direct (1) or indirect, through microbial functions (2). Soil bacteria alter the microbial community (2a), and earthworms affect how bacteria affect ecosystem function (2b) (Medina-Sauza et al. 2019)

organisms as well as roots but also mediate a full range of transport and turnover processes. This is especially the case of the large stable permanent biopores produced and maintained by anecic earthworms. Therefore, this section is devoted to the overview of the functions of biopores remaining after the activities of the earthworms.

3 Properties of Biopores

Biopores with high microbial process rates are known as important hotspots (Hoang et al. 2016). Tubular macropores or cavities (Kautz et al. 2013) as (1) roots (pores left after root decomposition) (detritosphere), (2) earthworms (drilosphere), or (3) persistent colonization of old roots (detritus- or drilosphere) by earthworms are interpreted. The microbial functions of the soil increase significantly with the addition of root litter and earthworms (Sanaullah et al. 2011). Subsoil biopores are important for plant access to nutrients, as arable topsoil can be depleted in nutrients and organic matter after prolonged cultivation and waste dumping from fields. Bioporosity affects soil structure through keeping air and water below the surface (Kautz et al. 2013).

Table 1 Main differences of biopores formed by anecic, endogeic, and epigeic earthworms[a]

Properties	Anecic earthworms	Endogeic earthworms	Epigeic earthworms
Diameter	Up to 1 cm	<0.3 cm	<0.5 cm
Depth	Deeper than 2 m	Mainly in the top 30–40 cm	Top 10 cm
Orientation	Vertical	Mainly horizontal	Mainly horizontal
Connectivity	One main vertical pore; very few galleries	Many connected pores of the same size	Some connected pores
Walls	A lot of organic materials; dark	Few organic materials	A lot of organic materials
Burrows	Permanent, reusable	Temporary	Not permanent, partly reusable
Age	Very old, decades; inherited by the successor	Temporary, weeks	Not permanent, weeks–months
Stability	Very stable	Unstable	Stable
Casts	Many	Many	Few
Typical examples	*Lumbricus terrestris* *Aporrectodea giardia* *Dendrobaena octaedra* *Bimastos parvus*	*Aporrectodea caliginosa* *Aporrectodea trapezoids* *Aporrectodea tuberculata* *Lumbricus terrestris*	*Aporrectodea longa* *Aporrectodea icteria* *Perelia kaznakovi* *Octolasion cyaneum*

[a]For the main representatives of groups of anecic, endogeic, and epigeic earthworms as well as their lifestyle

Biopores alter soil biochemical features by changing microbial biomass and the enzyme's activity implicated in the cycles of phosphorus (P), nitrogen (N), carbon (C), and other nutrients (Jégou et al. 2001). Most studies of biopores have concentrated on processes and their effects on the topsoil (Menichetti et al. 2014). The role of subsoil biopores on plant growth, although important, has yet to be predicted. Organic carbon and total nitrogen contents reduce with soil depth (Salome et al. 2010). This problem leads to the biological pores being an essential source of carbon and nutrients for the subsoil. The investigation of Bamminger et al. (2014) recorded the increase of N and C storage in burrow walls compared to bulk soil (Fig. 4).

4 Functions of Biopores

4.1 Transport

Biopores may function for long-distance transport of gases, solutions, and particles as well as root growth to availability of nutrients and water in the subsoil (Table 2). From all transport mechanisms (Table 2), the downward gravitational movement with water after intensive rainfall is the fastest process, and the diffusion is the slowest process.

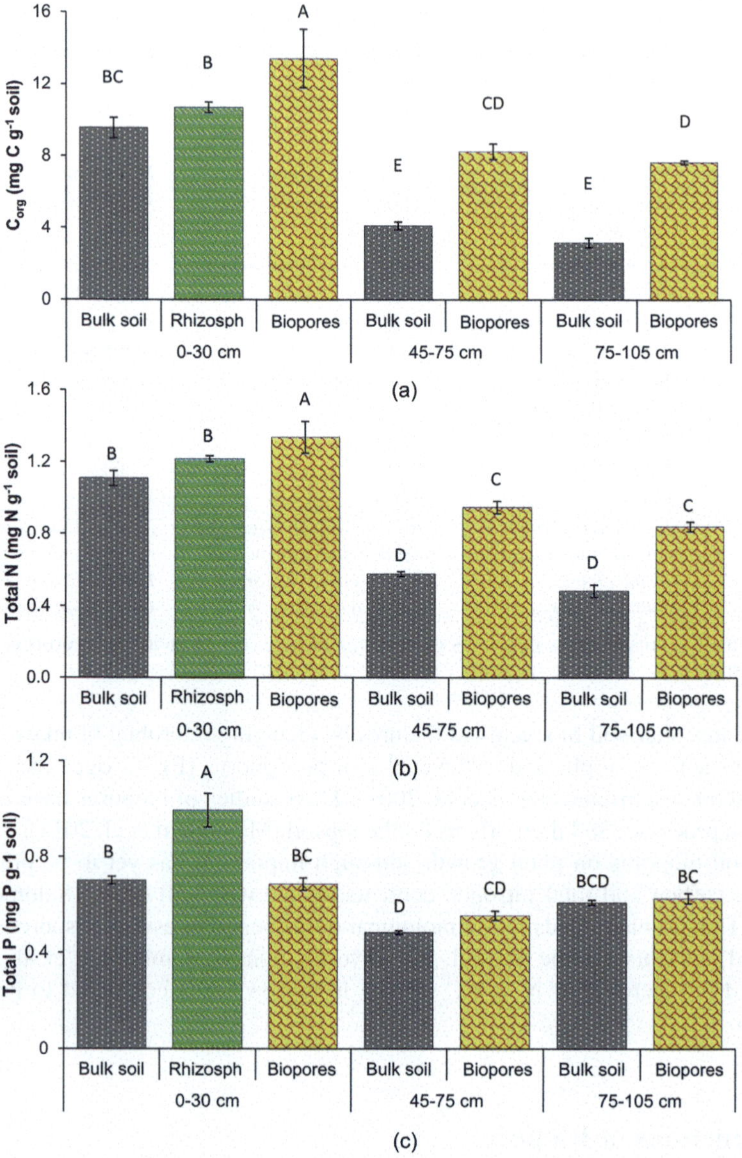

Fig. 4 A comparative analysis was conducted on the levels of organic carbon (Corg), total nitrogen, and total phosphorus in earthworm burrows, rhizosphere, and bulk soil. The results indicate that earthworms significantly enhance the content of Corg and total nitrogen in biological cavities ($p < 0.05$) by 1.2–2.4 times when compared to bulk soil. Furthermore, the disparity between the total nitrogen of Corg and biopores and subsoil is as high as 1.7 times that of the topsoil, underscoring the crucial role of soil in the associated organic matter, particularly in the subsoil (modified from Hoang et al. 2017)

Table 2 Main mechanisms of downward and upward transport of gases, liquids, and solids in biopores

	Downward	Upward
Gases		
– O_2	Diffusion	
– CO_2		Diffusion
– Water vapor	(Diffusion)	Diffusion
Liquids		
– Water	Gravitation •	Root transpiration •
– Nutrients	Gravitation, soluted in water •	Root uptake •
– Colloids	Gravitation, with water •	
Particles		
– Microaggregates	Gravitation with/without water •	In earthworms •
– Litter	+ on or in earthworms •	(on or in earthworms)
Organisms		
– Earthworms	Active migration	Active migration
– Other animals	Active migration and passive	Active migration
– Fungi	Drop	Hyphae growth
– Bacteria	Hyphae growth	Passive on worms and animals
– Viruses	Passive on worms and animals Passive on worms and animals	Passive on worms and animals

•: relative transport rates: very slow, medium, fast

Biopores function not only for the transport of water, air, and solutions by physical processes, but also earthworm-induced bioturbation crucially contributes to the mixing of the soil and allocation of solid compounds within the soil profile. Organic substances and mineral particles will be mechanically allocated into deeper horizons, and the release of organic matter enriched earthworm casts that strongly raise microbial and enzyme activities in biopores (Fig. 5).

4.2 *Nutrient Dynamics in Biopores*

Biopores are usually tubular loops that run vertically through the soil, but there are also horizontal biopores (Kautz 2015). Although the pore volume in the soil can be up to 50% of the total soil volume, biopores are only 1–10% of the total soil volume (Banfeld et al. 2017) and are a part of the macropore system. Evidence for the role of earthworm biopores in influencing plant productivity through the availability of new nutrients in deep soil was displayed by Van Groenigen et al. (2014). Utilizing a meta-analysis, they determined that the existence of earthworms improved aboveground plant biomass by an average of 23%, mainly through nitrogen release from OM. Digestion of OM can affect soil conditions and the distribution of nutrients, as well as activities that can alter aeration, oxygen, and soil moisture, and ultimately increase the activity of microorganisms in the plant growth environment. Biopore

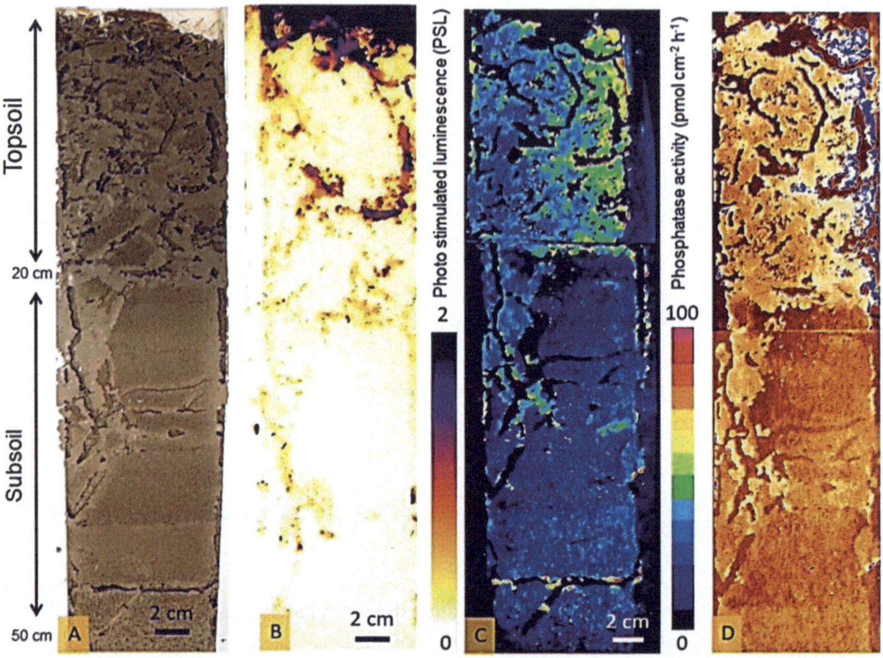

Fig. 5 Bioturbation by earthworm burrowing activities visualized by ^{14}C autoradiography and zymography. (**a**) Burrow channels with plant litter on the open side panel. (**b**) ^{14}C activity, with dark purple and purple color expressing hotspots of ^{14}C allocation, with white meaning no ^{14}C activity signal. (**c**) Expressed acid phosphatase distribution with higher color intensity (red color) concentrated along earthworm burrows. The color bar corresponds to enzyme activity (pmol cm^{-2} h^{-1}). (**d**) Overlapping ^{14}C and enzyme distribution (overlap correlation >68%) (Hoang et al. 2020)

walls are rich in nutrients (Athmann et al. 2013) and have great microbial activity due to rotting roots or soil fauna in the pores (Hoang et al. 2017; Banfeld et al. 2018). Due to this feature, soil pores are accepted as important places for plant roots to absorb nutrients from the soil (Hoang et al. 2016). Regardless, this claim ignores the reality that the contact of the roots with the soil will decrease, particularly in pores with a larger diameter than the root diameter, thus potentially reducing plant nutrient uptake and reducing and narrowing the pores (Landl et al. 2019).

4.3 Root and Hyphae Ingrowth into Biopores

Biological pores created and utilized by earthworms and plant roots supply availability to the subsoil. Biopores can also be regarded as hotspots to obtain nutrients and have high nutritional value than the surrounding soil (Pankhurst et al. 2002). Because the main physical obstacle for roots and fungal hyphae growth is the soil density, all meso- and macropores pores are the predestined pathways for root and hyphae development (Fig. 6) (Athmann et al. 2013). Roots preferably grow into the existing biopores because of (i) low impedance, (ii) higher oxygen availability (Dexter 1991), (iii) better water supply and availability, (iii) high nutrient content, and (iv) a higher abundance of microorganisms including mycorrhiza (Athmann et al. 2013) leading to higher colonization of old compared to new biopores (Pagenkemper et al. 2015). Compared to bacteria, fungi have the best ability to reach long distances with extremely unstable C: hyphae can penetrate the soil film in the pores because the release of hydrophobicity from the hyphae strongly reduces tension. Therefore, hyphae often use biological pores for deep growth (Athmann et al. 2013).

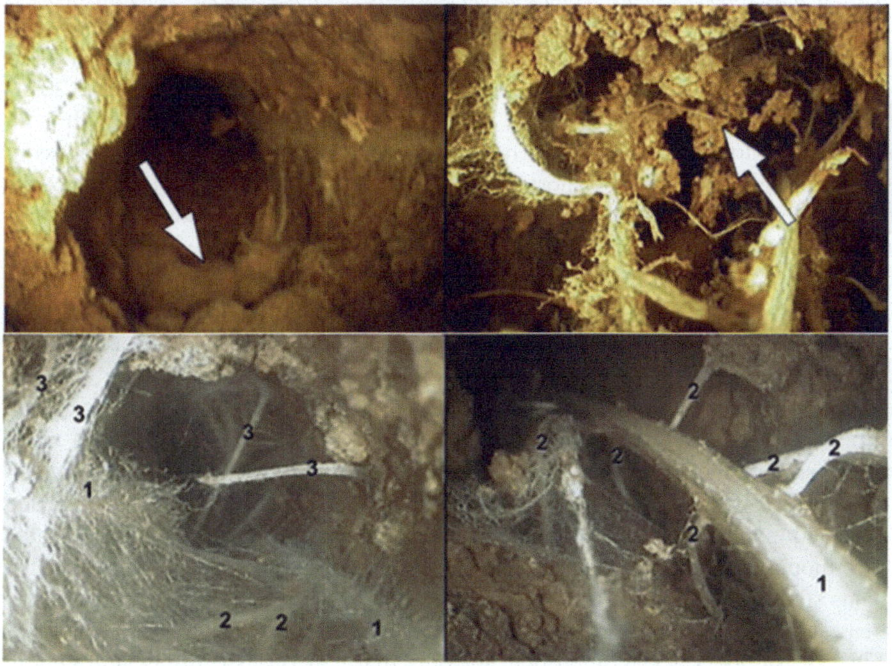

Fig. 6 Fresh earthworm cast lining (top left) and macropore root soil (top right) in biopores. Left picture: 75 cm soil depth, with barley root. Right picture: 65 cm soil depth, with barley roots. Root growth of barley (bottom left) and oilseed rape (bottom right) in biopores: bottom left: 95 cm soil depth; two vertical roots in close contact to the pore wall, also via root hairs (1) with lateral roots (2) and four ingrowing roots (3). Bottom right: 105 cm soil depth; vertical root without direct contact to the pore wall (1) but with several lateral roots (2) (from Athmann et al. 2013, modified)

4.4 Biopores for Plant Growth and Crop Yield

Because the soil pore structure controls the amount of water infiltration and storage in the soil and affects root growth, they are considered a critical indicator of soil quality (Pires et al. 2005). Many soil functions and processes are determined through soil structure, including the arrangement of solids and pores (Rabot et al. 2018; Yudina and Kuzyakov 2023). The main regulators of soil pore structure are plant roots, which affect the pore structure through various mechanisms such as directly creating/changing soil porosity, increasing SOM, mucilage secretion, and water absorption (Gregory 2022). Root growth interacts with the pore structure in various forms: (1) roots can extend from the root system underground; (2) can pass through existent pores, including biological pores; and (3) they can organize an existing pore at some angle before infiltrating the soil (Lucas 2022). During the growth, the root system overcomes the resistance of the soil by compacting existing pores and creating new biological pores. After the root dies and decomposes, the biological pores it creates can then be reused by the new growing roots. White and Kirkegaard (2010) found that 32–47% of rice roots are in biopores at a depth of 0.3–0.6 m, while all roots are in biopores below 1 m. Comprehension of the importance and process of soil pore organization in the short term and its subsequent effect on pore structure, as the crop continues to grow, is important for making decisions about crop management and use. In this regard, Lu et al. (2020) investigated the impact of root modifications on hydraulic soils and found that thick roots increase coarse porosity at the expenditure of small pores. However, root effects depend on total root volume (Lu et al. 2020). The movement of soil macrofauna (e.g., ants, termites, and earthworms) and mesofauna (e.g., enchytraeids) causes pores to form in the soil based on their size. In particular, earthworms form extensive pores and change the pore wall through secretion and compression of the medium (Kautz 2015). Haruna et al. (2020) studied the impacts of biopore density on cover crops.

Plant yield increases with biopore density due to better root development and increased plant biomass. Data from Lucas (2022) showed that roots and soil fauna form biopores of various sizes, reducing the volume of pores <0.2 mm. Few of the plants studied had a root thickness of >1 mm, while large biopores (>1 mm) had the greatest bioporosity and time, and plant species had no effect on biological pores (Fig. 7). Biopores act as microbial hotspots that can furnish further nutrients to crops – but only if their roots extend into the biopores (Banfield et al. 2017). After death, roots grow faster in subsoil through biopores (Kautz et al. 2013) and reach further nutrient resources. One of the most significant benefits is presumably for succeeding crops, the biochemical conditions achieved by root decomposition, and rhizodeposition in the environment (Banfield et al. 2017). In the field, even 2 years after the death and decomposition of the roots, the carbon content in the root biocavities (detritosphere) is still 2.5 times higher than in the bulk soil, resulting in a 5.5 times increase in microbial biomass and enzyme activity (Hoang et al. 2016; Banfield et al. 2017).

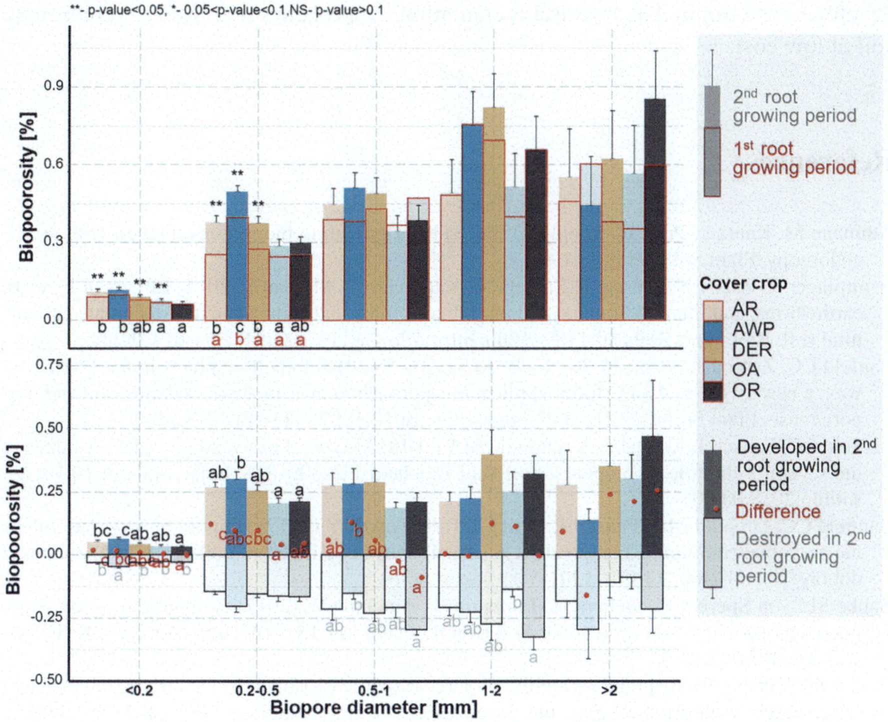

Fig. 7 (**a**) Biopore size distribution in soil remaining after five cover crops; letters indicate significant differences among cover plant species in each size class ($p < 0.05$). (**b**) Biopores (light colors) disappeared and formed (dark colors) during the second growth period. The red dots show the total changes resulting from bioporosity (Lucas et al. 2022)

5 Conclusions

Earthworm biopores are crucial for involving deep soil in biological cycles, deepening root growth, and providing larger soil volume for C sequestration. Earthworms allocate organic matter and nutrients far below the tillage horizon and below the frequently dense Bt horizon of natural soils and extend the soil depth for all biological processes. Despite the accelerated decomposition of OM in biopores, the total C and N contents there are higher than in the bulk soil. This is especially in the subsoil, where all processes take longer periods compared to the topsoil. Earthworm biopores provide important pathways for root penetration of crops and lead to yield increases, because of larger resources in the subsoil. Supporting access to water and nutrients, biopore management follows the principles of organic farming. The density of biopores (number per area) can define the water and nutrient supply, especially under drought conditions, thus contributing to the stability of crops that are other specific targets of organic agriculture. These "natural" methods can be developed and applied as "natural reclamation" approaches and used to rehabilitate soil at low cost.

References

Athmann M, Kautz T, Pude R, Koepke U (2013) Root growth in biopores evaluation with *in situ* endoscopy. Plant Soil 371:179–190

Bamminger C, Zaiser N, Zinsser P, Lamers M, Kammann C, Marhan S (2014) Effects of biochar, earthworms, and litter addition on soil microbial activity and abundance in a temperate agricultural soil. Biol Fertil Soils 50:1189–1200. https://doi.org/10.1007/s00374-014-0968-x

Banfeld CC, Zarebanadkouki M, Kopka B, Kuzyakov Y (2017) Labelling plants in the Chernobyl way: a new 137Cs and 14 C foliar application approach to investigate rhizodeposition and biopore reuse. Plant Soil 417:301–315. https://doi.org/10.1007/s11104-017-3260-7

Banfeld CC, Pausch J, Kuzyakov Y, Dippold MA (2018) Microbial processing of plant residues in the subsoil – the role of biopores. Soil Biol Biochem 125:309–318. https://doi.org/10.1016/j.soilbio.2018.08.004

Banfield CC, Dippold MA, Pausch J, Hoang DT, Kuzyakov Y (2017) Biopore history determines the microbial community composition in subsoil hotspots. Biol Fertil Soils 53:573–588. https://doi.org/10.1007/s00374-017-1201-5

Bauke SL, von Sperber C, Siebers N, Tamburini F, Amelung W (2017) Biopore effects on phosphorus biogeochemistry in subsoils. Soil Biol Biochem 111:157–165. https://doi.org/10.1016/j.soilbio.2017.04.012

Carducci CE et al (2014) Spatial variability of pores in oxidic latosol under a conservation management system with different gypsium doses. Ciência e Agrotecnologia 38(5):445–460. https://doi.org/10.1590/S1413-70542014000500004

Dexter AR (1991) Amelioration of soil by natural processes. Soil Tillage Res 20:87–100. https://doi.org/10.1016/0167-1987(91)90127-J

Dhaliwal JK, Kumar S (2020) Hydro-physical soil properties as influenced by short and long-term integrated crop-livestock agroecosystems. Soil Sci Soc Am J. https://doi.org/10.1002/saj2.20214

Gocke MI, Don A, Heidkamp A et al (2021) The phosphorus status of German cropland – an inventory of top- and subsoils. J Plant Nutr Soil Sci 184:51–64. https://doi.org/10.1002/jpln.202000127

Gregory PJ (2022) RUSSELL REVIEW Are plant roots only "in" soil or are they "of" it? Roots, soil formation and function. Eur J Soil Sci 73(1):e13219. https://doi.org/10.1111/ejss.13219

Haruna SI, Anderson SH, Udawatta RP, Gantzer CJ, Phillips NC, Cui S, Gao Y (2020) Improving soil physical properties through the use of cover crops: a review. Agrosyst Geosci Environ 3(1):e20105. https://doi.org/10.1002/agg2.20105

Hoang DTT, Pausch J, Razavi BS, Kuzyakova I, Banfield CC, Kuzyakov Y (2016) Hotspots of microbial activity induced by earthworm burrows, old root channels and their combination in subsoil. Biol Fertil Soils 52:1105–1119. https://doi.org/10.1007/s00374-016-1148-y

Hoang DTT, Bauke SL, Kuzyakov Y, Pausch J (2017) Rolling in the deep: priming effects in earthworm biopores in topsoil and subsoil. Soil Biol Biochem 114:59–71. https://doi.org/10.1016/j.soilbio.2017.06.021

Hoang DTT, Maranguit D, Kuzyakov Y, Razavi BS (2020) Accelerated microbial activity, turnover and efficiency in the drilosphere is depth dependent. Soil Biol Biochem 147:107852. https://doi.org/10.1016/j.soilbio.2020.107852

Jegou D, Hallaire V, Cluzeau D, Trehen P (1999) Characterization of the burrow system of the earthworms *Lumbricus terrestris* and *Aporrectodea giardi* using X-ray computed tomography and image analysis. Biol Fertil Soils 29:314–318. https://doi.org/10.1007/s003740050558

Jégou D, Schrader S, Diestel H, Cluzeau D (2001) Morphological, physical and biochemical characteristics of burrow walls formed by earthworms. Appl Soil Ecol 17:165–174
Kautz T (2015) Research on subsoil biopores and their functions in organically managed soils: a review. Renew Agr Food Syst 30(4):318–327. https://doi.org/10.1017/S1742170513000549
Kautz T, Amelung W, Ewert E, Gaiser T, Horn R, Jahn R, Javaux M, Kuzyakov Y, Munch JC, Pätzold S, Peth S, Scherer HW, Schloter M, Schneider H, Vanderborght J, Vetterlein D, Wiesenberg G, Köpke U (2013) Nutrient acquisition from the arable subsoils in temperate climates: a review. Soil Biol Biochem 57:1003–1022. https://doi.org/10.1016/j.soilbio.2012.09.014
Kuzyakov Y, Blagodatskaya E (2015) Microbial hotspots and hot moments in soil: Concept & review. Soil Biology & Biochemistry 83:184–199. https://doi.org/10.1016/j.soilbio.2015.01.025
Kuzyakov Y, Mason-Jones K (2018) Viruses in soil: nano-scale undead drivers of microbial life, biogeochemical turnover and ecosystem functions. Soil Biol Biochem 127:305–317. https://doi.org/10.1016/j.soilbio.2018.09.032
Landl M, Schnepf A, Uteau D, Peth S, Athmann M, Kautz T, Perkons U, Vereecken H, Vanderborght J (2019) Modeling the impact of biopores on root growth and root water uptake. Vadose Zone J 18(1):1–20. https://doi.org/10.2136/vzj2018.11.0196
Lu J, Zhang Q, Werner AD, Li Y, Jiang S, Tan Z (2020) Root-induced changes of soil hydraulic properties–a review. J Hydrol 589:125203. https://doi.org/10.1016/j.jhydrol.2020.125203
Lucas M (2022) Perspectives from the Fritz-Scheffer Awardee 2020—the mutual interactions between roots and soil structure and how these affect rhizosphere processes#. J Plant Nutr Soil Sci 185(1):8–18. https://doi.org/10.1002/jpln.202100385
Lucas M, Nguyen LT, Guber A, Kravchenko AN (2022) Cover crop influence on pore size distribution and biopore dynamics: enumerating root and soil faunal effects. Front Plant Sci 13:928569. https://doi.org/10.3389/fpls.2022.928569
Medina-Sauza RM, Álvarez-Jiménez M, Delhal A, Reverchon F, Blouin M, Guerrero-Analco JA, Cerdán CR, Guevara R, Villain L, Barois I (2019) Earthworms building up soil microbiota, a review. Front Environ Sci 7:81
Menichetti L, Landi L, Nannipieri P, Katterer T, Kirchmann H, Renella G (2014) Chemical properties and biochemical activity of colonized and abandoned litter-feeding termite (Macrotermes spp.) mounds in chromic Cambisol area on the Borana plateau, Ethiopia. Pedosphere 24(3):399–407. https://doi.org/10.1016/S1002-0160(14)60026-6
Pagenkemper SA, Athmann M, Uteau D, Kautz T, Peth S, Horn R (2015) The effect of earthworm activity on soil bioporosity – investigated with X-ray computed tomography and endoscopy. Soil Tillage Res 146:79–88. https://doi.org/10.1016/j.still.2014.05.007
Pagliai M, Vignozzi N, Pellegrini S (2004) Soil structure and the effect of management practices. Soil Tillage Res 79(2):131–143. https://doi.org/10.1016/j.still.2004.07.002
Pankhurst CE, Pierret A, Hawke BG, Kirby JM (2002) Microbiological and chemical properties of soil associated with macropores at different depths in a red-duplex soil in NSW Australia. Plant Soil 238:11–20. https://doi.org/10.1023/A:1014289632453
Pires LF, Bacchi OO, Reichardt K (2005) Gamma ray computed tomography to evaluate wetting/drying soil structure changes. Nucl Instrum Methods Phys Res, Sect B 229(3-4):443–456. https://doi.org/10.1016/j.nimb.2004.12.118
Pires LF, Borges JA, Rosa JA, Cooper M, Heck RJ, Passoni S, Roque WL (2017) Soil structure changes induced by tillage systems. Soil Tillage Res 165:66–79. https://doi.org/10.1016/j.still.2016.07.010
Rabot E, Wiesmeier M, Schlüter S, Vogel HJ (2018) Soil structure as an indicator of soil functions: a review. Geoderma 314:122–137. https://doi.org/10.1016/j.geoderma.2017.11.009
Reynolds WD, Bowman BT, Drury CF, Tan CS, Lu X (2002) Indicators of good soil physical quality: density and storage parameters. Geoderma 110(1-2):131–146. https://doi.org/10.1016/S0016-7061(02)00228-8
Salome C, Nunan N, Pouteau V, Lerch TZ, Chenu C (2010) Carbon dynamics in topsoil and in subsoil may be controlled by different regulatory mechanisms. Global Change Biology 16:416–426

Sanaullah M, Blagodatskaya E, Chabbi A, Rumpel C, Kuzyakov Y (2011) Drought effects on microbial biomass and enzyme activities in the rhizosphere of grasses depend on plant community composition. Appl Soil Ecol 48(1):38–44. https://doi.org/10.1016/j.apsoil.2011.02.004

Schneider F, Amelung W, Don A (2021) Origin of carbon in agricultural soil profles deduced from depth gradients of C:N ratios, carbon fractions, δ13C and δ15N values. Plant Soil 460(1–2):123–148. https://doi.org/10.1007/s11104-020-04769-w

Thiel H (1865) De radicum plantarum quarundam ab agricolis praecipue cultarum directione et extensione. Dissertation, University of Bonn

Valentine TA, Hallett PD, Binnie K et al (2012) Soil strength and macropore volume limit root elongation rates in many UK agricultural soils. Ann Bot 110:259–270. https://doi.org/10.1093/aob/mcs118

Van Groenigen JW, Lubbers IM, Vos HM, Brown GG, De Deyn GB, Van Groenigen KJ (2014) Earthworms increase plant production: a meta-analysis. Sci Rep 4(1):6365. https://doi.org/10.1038/srep06365

White RG, Kirkegaard JA (2010) The distribution and abundance of wheat roots in a dense, structured subsoil–implications for water uptake. Plant Cell Environ 33(2):133–148. https://doi.org/10.1111/j.1365-3040.2009.02059.x

Yudina A, Kuzyakov Y (2023) Dual nature of soil structure: The unity of aggregates and pores. Geoderma 434:116478. https://doi.org/10.1016/j.geoderma.2023.116478

Earthworm Role in Soil Food Webs

Josef H. Görres and Timothy McCay

Abstract Within the food web, earthworms are typically regarded as organisms in a mid-level trophic category. However, earthworms' trophic interaction with other organisms is much more complex. As geophagous animals they consume a whole range of organisms in soil aggregates and thus they are sometimes called ecosystemvores. Not all earthworms are the same and their nutrition depends much on their ecological group. Earthworms that live deep in the soil (endogeics) consume older carbon than those that live at the surface (epigeic) earthworms and the organisms that they ingest are also different. In addition, their indirect effects on the food web are also substantial. It is well known that the drilosphere, soil modified by earthworms, consists of microsites where microbivores graze on bacteria and fungi. The microsites are high in organic matter that was either translocated there by the earthworms or is based in earthworm frass. Most earthworms are infected by gregarine parasites that feed on sperm morulae and thus create an "internal" part of the food web. There are many earthworm predators that range from invertebrates, such as flatworms, to snakes and birds. The earthworm's place in the food web is complex and both its indirect effects as an ecosystem engineer and its direct trophic interactions need to be considered when appreciating the ecology of earthworms.

1 Introduction

Like all heterotrophs, earthworms consume organic material and are consumed by other heterotrophs, holding a place within soil food webs. Earthworms are ecosystem engineers which change soil properties considerably. For this reason, they are affecting the food web directly by ingesting other smaller biota, enriching the soil with gut microorganisms, and indirectly by altering the habitat of other soil-dwelling

J. H. Görres (✉)
University of Vermont, Burlington, VT, USA
e-mail: jgorres@uvm.edu

T. McCay
Colgate University, Hamilton, NY, USA
e-mail: tmccay@colgate.edu

Y. Kooch, Y. Kuzyakov (eds.), *Earthworms and Ecological Processes*,
https://doi.org/10.1007/978-3-031-64510-5_17

organisms. They consume massive amounts of soil relative to their body weight and provide an internal milieu for interactions among soil organisms during their passage through the gut (Drake and Horn 2007; Medina-Sauza et al. 2019). Their fecal matter (casts) provides an environment distinct from the surrounding soil for microbial action on undigested organic material (Edwards and Arancon 2022). And earthworms create voids in the earth and accumulations of organic materials in middens, changing the physical and microclimatic conditions for other species (Parkin and Berry 1999). In these and other ways, earthworms may serve as engineers of their ecosystems and likely have many non-consumptive effects on the soil food web (Frelich et al. 2019).

Earthworms are diverse and occupy a variety of niches in the soil (Barois et al. 1999). Bouché's (1977) influential work defined three main ecological groups of earthworms, which since have been refined (Kooch and Jalilvand 2008; Singh et al. 2020; Hsu et al. 2023; Table 1). Ecological categories were established largely on the basis of burrowing behavior and vertical habitat use and habitat use affects diet (Lavelle 1988). Epigeic earthworms live at the soil surface and consume surface litter. Endogeic earthworms live within the mineral soil and excavate burrows as part of their geophagous feeding behavior. Anecic earthworms make vertical, permanent burrows that they line with litter gathered at the surface and whose function is sometimes likened to that of an "external rumen."

Table 1 Ecological groups of earthworms, their habitat, and food preferences (Singh et al. 2020) determine what they ingest. Carbon age of ingested organic matter estimated from $\delta^{13}C$ (Pollierer et al. 2009; Briones et al. 2005). Ecological sub-groups based on Reynolds (1977)

Ecological group	Ecological sub-groups	Habitat	Food preference	Burrows, castings, carbon age
Epigeic	Epigeic	Litter	Leaf litter and soil microbes	No burrows, surface casting Carbon age 0–3 years
	Epi-anecic	Surface soil	Leaf litter and soil	Burrows, casting below- and aboveground
	Epi-endogeic	Interface between the organic matter layer and the layer just below burrows	Leaf litter and soil	Horizontal burrows
	Limnocolus			Wetland soils
	Corticole		Wood	Under the bark of dead trees, litter
Anecic	Anecic	Vertical burrows connecting the surface with subsoil	Leaf litter and soil	Near vertical burrows, lined with litter, carbon age 4–6 years
Endogeic	Polyhumic	Upper 20 cm of soil	Soil with high organic matter	Near horizontal burrows
	Mesohumic	Upper 20 cm of soil	Soil from 0 to 10 cm	
	Endo-anecic	0–50 cm of soil	Soil from 20 to 40 cm	Carbon age >4–8 years
	Oligohumic	15–20 cm	Soil from 0 to 10 cm	

Earthworms are classically described as geophagous and detritivorous, ingesting organic and inorganic soil matter. Yet, they also ingest organisms, including small arthropods and microorganisms (Bonkowski and Schaefer 1997; Drake and Horn 2007; Pokarzhevskii et al. 1997) and even seeds and seedlings (Eisenhauer et al. 2010). Ingested microbiota may be digested by earthworms or remain undigested, potentially participating in an internal gut food web such as the one predicted for the human gut (Wang et al. 2019). The abundance of some ingested soil microfauna increases during gut passage, supporting the idea that they participate in trophic interactions within the gut (Monroy et al. 2008).

Parasites are observed in many families of earthworms (Kundu and Bandyopadhyay 2019), but the level of pathogenicity and the mechanisms by which they affect earthworm fitness are often not clear (Field and Michiels 2005; Weller 2013). Earthworms participate as an intermediate host in the life cycles of some parasites with more well-studied terminal hosts (e.g., Antipov et al. 2018). Earthworms serve as food for a variety of predators. Many consumers use earthworms opportunistically, and a few almost exclusively (Lee 1981; MacDonald 1983; Edwards and Arancon 2022).

Plant growth is an obvious, but not the only, indirect effect of earthworm activities on the food web. Earthworms concentrate resources by moving and transforming soil. These changes may influence the abundance and diversity of nematodes, protozoa, and other microorganisms (Li et al. 2021), which consequently may restructure the food web. In addition, casting and burrowing by earthworms affect the spatial variability of food resources at the microsite scale. It is well known that the drilosphere, i.e., the soils directly affected by earthworm activity, is enriched in microorganisms (Parkin and Berry 1999; Andriuzzi et al. 2013). However, competition between earthworms and other species (Snyder et al. 2013) can also affect the structure of food webs and the turnover of matter in ecosystems.

The position of earthworms in the food web appears unremarkable, neither a top predator nor the base of the food web (Fig. 1, right). Yet, their impact on the food web is profound because they influence each level of the food web in what Frelich

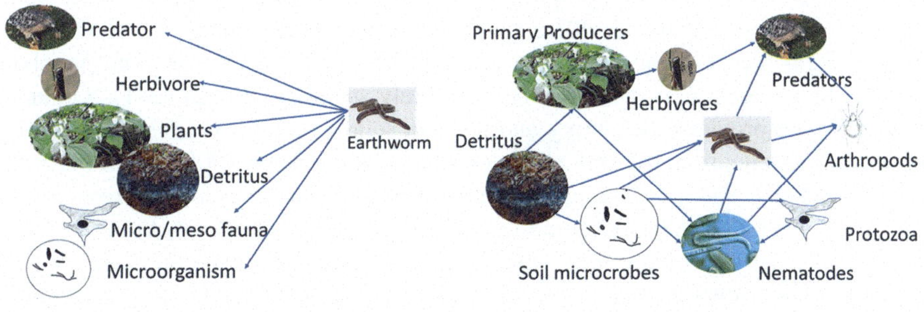

Fig. 1 Lateral influence of earthworms on several trophic levels, affecting the soil food web and causing ecosystem scale cascades (left, redrawn after Frelich et al., 2019). Traditional view that earthworms are a midtrophic level participant in the transfer of energy and matter in an ecosystem but omitting indirect effects (right, redrawn from Das et al., 2021)

et al. (2019) attribute to their "sideways" position (Fig. 1, left). Although most ecologists think of food webs in terms of direct feeding linkages, earthworms and other ecosystem engineers also affect food webs indirectly by modulating and regulating other trophic interactions. For example, the ability of earthworms to accelerate the decomposition and mineralization of organic matter avails nutrients to plants, improving their productivity in both natural and agricultural ecosystems (Van Groenigen et al. 2014).

The resilience of ecosystems to disturbance and changes in abiotic factors such as climate or soils depends on the trophic linkages among taxa (Pinnegar et al. 2005). The belowground soil food web is highly complex and efficient in recycling of nutrients, with more than 90% of nitrogen being recycled in temperate ecosystems (Cleveland et al. 2013). Interactions within the soil food web are difficult to observe because of the opacity of the soil and the large number of organisms and taxa involved. Thus, many uncertainties remain in soil ecology. The effect of high biodiversity and the associated level of functional redundancy (Ettema 1998) and the effect of considering species on ecosystem functioning rather than the more convenient taxonomic abstraction to feeding guilds (Ferris et al. 1997) are two concepts from nematology that need more exploring. These concepts are also relevant to earthworm ecology. While nematode species are organized into trophic groups, i.e., bacterial, fungal, plant feeders, etc., earthworm species are classed into ecological groups. Although it is easy to distinguish species by morphology or molecular methods, it is not straightforward to collect physiological data for each species that would then assist in determining the transfer of energy and matter to and from the earthworm fauna.

2 Direct Interactions

2.1 Earthworms as Consumers

Earthworms ingest soils and detritus. From this standpoint, earthworms could be classified as detritivores and geophages, feeding on the base of the soil food web. However, earthworm species differ ecologically, which determines the food available to them. Epigeic and anecic earthworms are both detritivorous and geophagous, whereas endogeic earthworms are solely geophagous (Table 1). Organic matter, which provides the energy and nutrient source of soil food webs, generally decreases with depth in the soil profile (Jobbágy and Jackson 2000). Similarly, soil microorganisms and microfauna that are ingested by earthworms decrease with depth (Yeates 1977, 2007; Ekelund et al. 2001). The soil strata inhabited by each ecological group thus determine what is ingested and the nutritional value of the ingestate. The deeper the location of the habitat, the less energy and the less ingestible soil biota available and the more substrate that has to be ingested to obtain sufficient nutrition. Endogeic earthworms ingest more matter than the other ecological groups to make up for the poorer nutrition found deeper in the soil profile (Singh et al. 2020).

Whereas epigeic and anecic earthworms feed on litter resources, i.e., "new carbon," endogeic earthworms feed on more recalcitrant "old carbon" locked up in either aggregates (physically protected) or as organo-mineral complexes (chemically protected) (Ferlian et al. 2014), potentially stabilized by earthworms in other ecological groups. Judas (1992) inspected the gut content of six earthworm species for organic and mineral material. They distinguished three ingestate fractions: mineral, particulate organic matter (less decomposed), and non-particulate organic matter (more decomposed). They found that endogeic species have lower particulate organic and higher mineral gut content than *L. terrestris* (anecic) and *L. castaneus* (epigeic). Endogeic species did not ingest soil haphazardly but preferred soil components with high non-particulate detritus content. Whether this purported preference for this substrate was due to lower amounts of particulate organic matter at the habitat depth of endogeic earthworms or a real preference for the older carbon was not determined. Briones et al. (2005) also found that the age of ingested carbon was older in endogeic than epigeic earthworms (Table 1).

Earthworms likely also ingest microorganisms and microfauna with the detritus that they consume. Because this consumption is apparently inadvertent, we will refer to earthworms as grazers rather than predators of these organisms. Because of the wide variety of organisms that they ingest, earthworms might be called omnivorous, and it becomes important to ask what nutrition is derived from dead detrital carbon compounds versus from organisms that have colonized it?

Microbiota are important food for the common composting earthworm, *Eisenia fetida* (an epigeic species). Miles (1963) investigated whether sterilized soils inoculated with bacteria and fungi, or with protozoa, affected *Eisenia fetida* growth. In the sterilized soil enriched with protozoa, *E. fetida* juveniles grew to reproductive maturity, whereas in soils enriched with only bacteria and fungi, they did not. Flack and Hartenstein (1984) investigated various microbiota as food sources for *E. fetida*. They measured the growth of the earthworm when presented with a blend of cellulose, ashed (i.e., only the mineral portion of) loamy soil, and each of 42 species of bacteria, fungi, and protozoa. *E. fetida* grew most consistently on a diet of the bacterial and fungal species, but only on 70% of the fungal species presented to them. Mixtures of cellulose and the mineral soil did not cause *E. fetida* to gain weight even when macronutrients and various vitamins were added. A similar experiment found that fungi and protozoa promote growth and survival in *E. fetida* (Edwards and Fletcher 1988), mirroring the results of Bonkowski and Schaefer (1997), whereas bacteria and nematodes did not sustain *E. fetida*.

Piearce and Phillips (1980) investigated the effect of earthworm intestinal fluids from different sections of the intestinal tract (Fig. 2) on the ciliate *Colpidium campylum*. Whereas fluids taken from the esophagus, pharynx, and the hind gut of *L. terrestris* had little effect on the behavior or the integrity of the ciliates, mid-gut fluids even in low concentrations led to immobilization and disintegration of the ciliate cells. These early studies were confirmed by more current studies (Bonkowski and Schaefer 1997). Not all protozoa are food sources. Bamforth (1988) enumerated both encysted and active protozoa after gut passage. While active protozoa were digested by earthworms, cysts were able to pass the gut unharmed. Piearce and

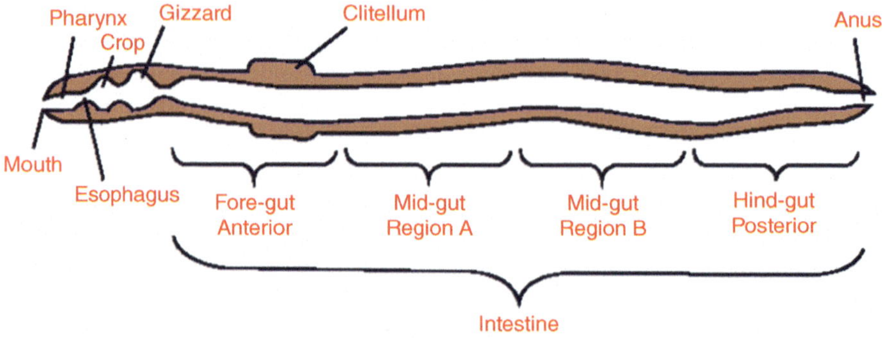

Fig. 2 Representation of the digestive tract of earthworms defining sections of the gut frequently used in ecological studies of the gut microbiome (from Adnan and Joshi 2013)

Philips (1980) observed that some astomate protozoa survived exposure to intestinal fluids. However, they were distributed along the alimentary tract of earthworms in accordance with the availability of nutrients and chemical conditions. Many astomate ciliates are obligate endosymbionts of earthworms (Obert et al. 2021), in which they derive their nutrition from nutrient ions (Nana et al. 2012). Bonkowski and Schaefer (1997) confirmed early work by Miles (1963) and by Piearce and Phillips (1980) that protozoans are an important part of the nutrition of earthworms by analyzing their weight gain when fed on soils with and without protozoans. They also found protozoans were depleted in feces of *Octolasion lacteum (endogeic)*, suggesting that they digested protozoans.

Monroy et al. (2008) fed pig slurry to *Eisenia andrei*, *E. fetida*, *Lumbricus rubellus*, and *Eudrilus eugeniae*, four litter-feeding, epigeic species, to determine how gut passage affected the number of ingested organisms. None of the nematodes offered for ingestion were found after gut passage, and the number of coliforms in the hind gut was less than in pig slurry in all of the four earthworm species. These and epigeic species likely derived nutritional benefits from the nematofauna. In contrast, their results also showed that naked amoeba increased during gut passage, probably feeding on some of the gut microbiome or other ingested cells.

More evidence of the nutritional value of microbiota came from the comparison of amino acid concentrations in litter and soil, with those in earthworms. Earthworms cannot synthesize all essential amino acids (EAA), and thus EAAs may serve as clues to the food sources of earthworms. Pokarzhevskii et al. (1997) found that there were insufficient amino acids in soils and litter to match the concentrations found in earthworms. They concluded that earthworms satisfy amino acid requirements using ingested microbiota along with amino acids synthesized by the gut ecosystem (microbiome).

Like Pokarzhevskii et al. (1997), Larsen et al. (2016) investigated the question whether the food of earthworms came from plant sources (litter), bacterial sources, or fungal sources with essential amino acid and ^{13}C isotope analysis (Fig. 3). For two anecic species, *Lumbricus terrestris* and *Aporrectodea longa*, greater than 70%

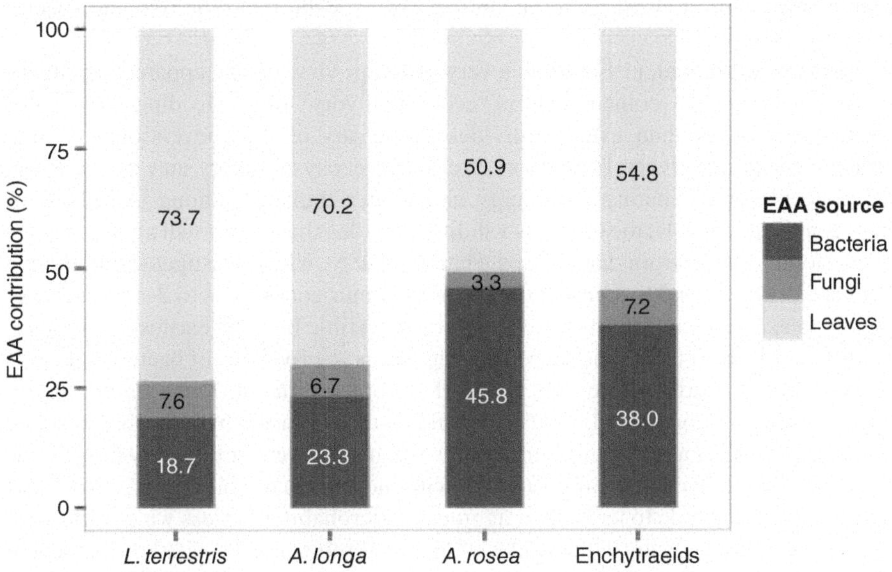

Fig. 3 Food origin of essential amino acids (EAA) in earthworm tissue estimated using ^{13}C analysis (from Larsen et al. 2016)

of their essential amino acids were derived from plant sources, while the only endogeic species, *A. rosea*, derived only 50% of its amino acids from litter. Bacteria were sources of between 16 and 23% of amino acids for the anecic species, but for the endogeic species, bacterial sources accounted for as much as 45% of amino acids. Fungal contributions were less than 10%. This agrees with the idea that endogeic earthworms derive nutrition from older carbon sources than earthworms that feed on plant litter. However, all three sources—i.e., plant litter, bacteria, and fungi—contributed to the essential amino acid nutrition of the earthworms. Further studies of essential amino acid stable isotope analyses by Potapov et al. (2019) expanded on Larsen et al.'s (2016) analysis by investigating two epigeic species (*Dendrobaena octaedra* and *Lumbricus rubellus*), two endogeic species (*Aporrectodea caliginosa* and *Allolobophora chlorotica*), and one wood litter-feeding species (*Dendrobaena rubidus*) with both amino acid stable isotope analyses. The origin of earthworm essential amino acids can be predicted from ^{13}C and ^{15}N isotope signatures of the essential amino acids extracted from earthworm biomass and from potential food organisms (Potapov et al. 2019). Contributions of the three sources of essential amino acids in earthworm food (plants, fungi, bacteria) were found to be much more evenly distributed across these sources than in the study by Larsen et al. (2016) with fungal contributions greater than 20% for all earthworm species. However, epigeic earthworms generally derived more amino acid nutrition from plants and the endogeic earthworms mainly from bacteria. Although *D. rubidus* is classified as an epigeic earthworm, its feeding on decaying wood skewed the analysis of essential amino acids more toward bacterial sources than for the two other epigeic species (Potapov et al. 2019).

Pokarzhevskii et al. (1997) took a very different view of this apparent omnivory by earthworms. They coined the term "ecosystemivory" for the feeding activities of earthworms rather than using terms describing any of the conventional feeding strategies, e.g., fungivory, bacterivory, etc. While ecosystemivory may not be a useful concept for estimating the energy and matter transfers among organisms or between trophic levels, the term has validity. The idea that even small aggregates are ecosystems comes from Vargas and Hattori (1986) who investigated interaction between protozoa and bacteria in the inner and outer zones of 1- to 2-mm-diameter aggregates. Aggregates of that size would be ingestible by most earthworm species. Hattori and Hattori (1993) described aggregates as microcosms of bacteria and protozoan habitat. Furthermore they found that communities differed among aggregates. Similarly, Görres et al. (1999) formulated the enclosure hypothesis that stated that aggregates act as ecological enclosures when soils were sufficiently dry and the aggregates were hydrologically isolated, with motile aquatic biota, e.g., nematodes and protozoa, not able to leave their aggregate microhabitats unless water films connected the aggregates during episodes of high soil moisture. This has consequences for elemental cycling and food web interactions (Görres et al. 1999). Presumably, being confined in aggregates reduces the volume of space in which microfaunal microbivores, such as protozoans and nematodes, quest for food, which in turn intensifies the interaction between microbivorous grazers and their forage. This likely results in an acceleration of elemental cycling (Görres et al. 2001) within aggregates introducing soil moisture as a regulative parameter on detritus-detrivore interaction. As such, geophagy may well be described as ecosystemivory as suggested by Pokarzhevskii et al. (1997).

A wide variety of organisms are found in the earthworm gut. Apart from bacteria, the digestive tract of earthworms contains fungi, nematodes, and protozoa (Dash et al. 1980). However, identifying all ingested species may be difficult as they degrade quickly in the earthworm gut. New molecular methods may improve the precision with which the ingested organisms are identified over the conventional visual examination of gut contents. A prerequisite for this is that one can eliminate the possibility that the PCR primers used in assays misidentify prey species due to interference from earthworm DNA amplifying in the polymer chain reactions. Knowledge of the earthworm DNA is essential to avoid this problem (Admassu et al. 2006) and to advance studies that explore the gut content of earthworms in order to determine what they ingest.

2.2 *Microbial Changes During Gut Passage*

The gut of earthworms presents a different environment to microbes associated with ingested material, including soil and litter (Drake and Horn 2007). Earthworms' digestive tracts tend to be higher in pH, less in aerobic, and more carbon rich than the surrounding soil environment (Brown et al. 2000; Drake and Horn 2007). Studies have consistently demonstrated a difference in microbial communities of earthworm guts as compared to the surrounding soil (Pass et al. 2015). This is likely due to promoted growth of certain microbes and inhibition of others already present in the ingested material, although taxa unique to earthworm digesta have been identified (Monroy et al. 2008). Promotion and inhibition are mostly due to the physical and chemical conditions of the digestive tract of the earthworms. Passage of pathogenic microbes may have implications for dispersal of insect and plant hosts (Campos-Herrera et al. 2006; Boyer et al. 2013). In contrast to the loss of nematodes during gut passage, naked amoebae increase in numbers during gut passage which suggest that there are interactions among the organisms of the earthworm gut (Monroy et al. 2008). The hypothesis that there is an internal gut food web is difficult to test, and to our knowledge, there has not been a study of those interactions for earthworms.

2.3 *Parasites of Earthworms*

Because of the feeding habits of earthworms, they likely ingest parasites and pathogens. There are a large number of gregarine parasites that infect earthworms. Kundu and Bandyopadhyay (2019) list 137 species in 7 genera of these parasitic protozoans. They are found in various earthworm tissues ranging from the coelom to the seminal vesicles. Different life stages may be found in different organs. Some parasites are specific to one species of earthworm; others infect several species.

Frequently, various life stages of gregarines, especially those in the genus *Monocystis*, are found in high density in the seminal vesicles where they feed on sperm morulae (Keller and Schall 2020; Weller 2013), castrating the worms to varying degrees. In the hermaphroditic species *L. terrestris*, the presence of gregarines does not appear to have much of an effect on fitness (Weller 2013). Weller (2013) hypothesized that this may be so because the virulence of the gregarines may not be severe enough to fully castrate the *L. terrestris*. This insures that host populations persist which in turn allows greater parasite prevalence in the wild. There may still be a cost. Field and Michiels (2005) observed a reduction in earthworm growth with parasite infections. Weller (2013) argued that the response of the immune system of *L. terrestris* to *Monocystis* species exacts a cost and thus thinks of *Monocystis* as true parasites.

The parthenogenetic earthworm *Amynthas agrestis* is infected by *Monocystis perplexa*, whose life cycle also mainly occurs in the seminal vesicle (Keller 2018). There, its trophozoite life stage feeds on sperm morulae, a.k.a. sperm forming cells. Even if *M. perplexa* fully castrated the parthenogenetic *A. agrestis*, its reproduction would not be affected by it, allowing high prevalence of the parasite in wild populations (Keller and Schall 2020).

A. agrestis is within a group of very similar earthworms called pheretimoids. Many pheretimoids have been shown to become infected by gregarine parasites, and often the infection presents in their seminal vesicles (Kundu and Bandyopadhyay 2019). Three sympatrically co-occurring parthenogenetic earthworms in the pheretimoid clade, *A. tokioensis*, *A. agrestis*, and *Metaphire hilgendorfi* (Chang et al. 2018), all become infected by gregarines, albeit different ones (Schall 2021). These three earthworms vary with regard to the presence or absence of organs used in the passing and storing of sperm ("male organs"). At one end of the "male organ possession spectrum," hermaphroditic forms, presumably the ancestral forms, are rare. At the other end, fully parthenogenetic pheretimoids (absence of all male organs) are found more frequently. There are also morphologies with different combinations of male organ possession. It is interesting that these parthenogenetic pheretimoids are still producing sperm cells to the benefit of their *Monocystis* parasites. Could there be a mutualistic relationship between these earthworms and the parasite?

2.4 Predators of Earthworms

Because of their large biomass, earthworms are efficient trophic conduits for transferring matter and energy from belowground to aboveground components of terrestrial ecosystems. They feed at or below the soil surface but fall prey to many aboveground predators. Earthworms have high energy and protein content, which makes them desirable food items for many species (Sabine 1983). Earthworms form a portion of the diets of many predators and are exclusively eaten by a much smaller number of predators. One prominent example of an obligate earthworm predator in forested ecosystems of the northeastern USA is the American woodcock that feeds on earthworms introduced from Europe (Owen Jr. and Galbraith 1989).

Earthworm bodies are approximately 80% water when fully hydrated, though this varies among species, and earthworms may tolerate significant dehydration (Roots 1956). Dry masses of earthworms that have been studied are more than 50% protein (e.g., *Lumbricus terrestris*, Fosgate and Babb 1972, Sabine 1983) with a broad profile of amino acids (Taboga 1980; Sabine 1983; Pokarzhevskii et al. 1997; Istiqomah et al. 2009). The amount of fat in earthworm tissue varies among species but has been less than 20% in all species studied to date (e.g., 2.8% in *Lumbricus terrestris*, Fosgate and Babb 1972, and 18.5% in *L. rubellus*, Istiqomah et al. 2009). Thus, earthworms may serve as a nutritional, and possibly complete, food for many animals and have been eaten by humans (Marconi et al. 2002; Sun and Jiang 2017).

Terrestrial vertebrates that rely on earthworms for a large portion of their diet include moles, certain shrews, hedgehogs, badgers, certain birds, many squamates (lizards and snakes), salamanders, anurans, and caecilians (Yalden 1976; MacDonald 1983; Edwards and Arancon 2022). In most cases, these predators are generalists that hunt at the upper layers of the soil and consume earthworms opportunistically or consume night-feeding earthworms when they emerge (MacDonald 1980; Maerz et al. 2005). In a few cases, predators specialize in hunting earthworms and depend on them principally, at least during particular times in particular habitats.

The common eastern mole of North America (*Scalopus aquaticus*) often consumes earthworms, but the portion of the diet that consists of earthworms is variable and is usually inferior to that of insects (West 1910; Hartman et al. 2000). The star-nosed mole (*Condylura cristata*), which is typically found in wet soils, more reliably consumes earthworms (Petersen and Yates 1980). The diet of the masked shrew (*Sorex cinereus*) included about 10% earthworm tissue in a forest in Pennsylvania, USA (McCay and Storm 1997). Badgers are often regarded as earthworm specialists, but Li et al. (2013) found that the diet of Eurasian badger (*Meles meles*) included a variety of non-earthworm prey as well.

The consumption of earthworms by birds is iconic, and earthworms may make up a part of the diets of hundreds of species (MacDonald 1983). Some birds known to often eat earthworms include the American woodcock (*Scolopax minor*, Owen Jr. and Galbraith 1989), European starling (*Sturnus vulgaris*, Moeed 1980), little owl (*Athene noctua*, Hounsome et al. 2004), American robin (*Turdus migratorius*) and other thrushes (*Turdus* spp., Heppner 1965), and various gulls (*Larus* spp., Coulson and Coulson 2008, Pennycott et al. 2020). Martay and Pearce-Higgins (2020) found that the abundance of European robins (*Erithacus rubecula*) and various thrushes (*Turdus* spp.) could be used as an index to the abundance of earthworms, explaining about 20% of the variability in earthworm abundance. Earthworms may be particularly important food items for nestlings due to the ease with which they can be carried intact (Lin et al. 2007). There are few studies on how the availability and abundance of earthworms affect populations of their predator, though Cameron and Bayne (2012) found that the distribution of invasive earthworms was correlated with the distribution of the American robin (*Turdus migratorius*).

It is likely that many generalist lepidosaurs and amphibians consume earthworms at least opportunistically. Snakes in the genus *Thamnophis* (garter snakes) frequently consume earthworms (Hamilton Jr. 1951; Gregory 1978; Crone et al. 2022), and the Mesa Central blotched garter snake (*T. scaliger*) is an earthworm specialist (Reguera et al. 2011). Brown et al. (2011) used molecular diagnostics to detect several species of earthworms of many ecological types in the diets of slow worms (*Anguis fragilis*, a legless lizard). Several species of lizards are known to eat giant earthworms in the family Rhinodrilidae (Strüssmann et al. 2013; Picelli et al. 2019). A variety of amphibians consume earthworms, including red-backed (*Plethodon cinereus*, Maerz et al. 2005), fire (*Salamandra salamandra*, Wang et al. 2021), and seal (*Desmognathus monticola*, Gorsuch and Owen 2014) salamanders and American toads (*Anaxyrus americanus*, Crone et al. 2022).

Interesting negative feedback on earthworm populations may occur when flatworms in the family Geoplanidae are present. Flatworms are predators of earthworms, snails, and slugs. Their impact on earthworm populations has been documented in the UK and other areas in Europe where flatworms were introduced with plant materials via botanical gardens and horticultural trade (Boag and Yeates 2001). The most impactful flatworm invader in the UK, *Arthurdendyus triangulatus*, is from New Zealand. Flatworms more negatively affect anecic than other ecotypes of earthworms in the UK (Murchie and Gordon 2013). A wide variety of earthworm species were similarly attacked and killed by the geoplanid flatworm *Bipalium adventitium* in North America (Zaborski 2002). Preference of flatworms for a particular earthworm group may be due to their burrowing and feeding behavior. For example, the occasional presence of anecic worms at the surface and their permanent burrows allow flatworms better access to this prey. Flatworms may have negative impacts on populations of other earthworm predators such as moles, badgers, shrews, and birds, though this has not been extensively studied (Alford et al. 1995; Boag 2000).

Earthworm species differ in their defensive adaptations. Gorsuch and Owen (2014) demonstrated that a planarian (*Bipalium adventitium*), a salamander (*Desmognathus monticola*), and the ribbon leach (*Nephelopsis obscura*) were more likely to consume two lumbricid earthworms (*Lumbricus rubellus* and *Aporrectodea caliginosa*) than a megascolecid species (*Amynthas agrestis*). The defenses of earthworms against predators have received little attention. Laakso and Setälä (1997) collected evidence that *Bimastos* (*Dendrodrilus*) *rubidus* has a chemical defense against predation by the ants with which it was commonly associated in Finland. Saim (1985) found that benzaldehyde, phenylacetaldehyde, and phenethyl alcohol were emitted by the "green stink worm" (*Amynthas hupeiensis*). These substances may repel predators, though to our knowledge this earthworm has never been presented to predators in a controlled setting.

Earthworms represent a route by which wildlife may be exposed to various soil contaminants (Boykins 1967; Sample et al. 2014). Earthworms may play a major role in mobilizing heavy metals by ingesting contaminated soils, remobilizing, and bioaccumulating contaminants (Morgan and Morgan 1999). Contaminants may thus enter the terrestrial food web (van den Brink et al. 2010; Richardson et al. 2015). In certain regions of the globe that have experienced heavy loads of toxic trace metals such as mercury, zinc, cadmium, selenium, and lead, concentrations in earthworm tissue can exceed levels deemed toxic to small mammals and birds (Richardson et al. 2015; van den Brink et al. 2010). While there is an abundance of studies investigating the ability of different earthworm species to accumulate soil pollutants, very little is known about how much this affects the predators of earthworms. The effect of earthworms on bioavailability of trace metals needs to be considered when addressing the risk to wildlife populations. This may not be limited to toxic trace metals. For example, Vermeulen et al. (2010) investigated the ability of earthworms to mobilize organic compounds from the soil into the food web via hedgehogs. However, they point out that inconclusive results on the accumulation of persistent organic compounds in hedgehogs as a result of consuming earthworms might be due to varying foraging behavior.

2.5 Indirect Effects on Organisms

Direct consumption by earthworms will affect the food web through consumption or promotion of microbiota. However, earthworms also affect the food web indirectly through their engineering activities. The permanent macropores (burrows) created by anecic earthworms may benefit prey organisms that can use the additional porosity for refuge. Salmon et al. (2005) conducted a mesocosm experiment with a collembolan (*Heteromurus nitidus*), which was drawn to earthworms and predatory centipedes. Their hypothesis was that collembola would benefit from the presence of anecic earthworms (*L. terrestris* and *A. giarda*). They found that collembola mortality was less when earthworms were present than when they were not and hypothesized that the collembola were able to avoid predators by using the earthworm galleries as escapes.

Earthworms also may have indirect effects on the food web by competing with other species for resources. Ransom (2012) found that *L. terrestris* affected red-backed salamanders (*Plethodon cinereus*) by consuming leaf litter that was inhabited by decomposer arthropods that salamanders would feed on. Interestingly, the permanent burrows also provided refuges for these salamanders. Here it is important to again realize that generalizing the effect of earthworms on a food web is difficult. Ziemba et al. (2015) found that *Amynthas* spp. (epi-endogeic) affected red-backed salamanders by consuming the litter cover in forest. This had a considerable effect on the salamanders which began to utilize alternative cover, and minimized their movements, and thus consumed fewer prey (Ziemba et al. 2016). Zhang et al. (2010) also found that the invasive earthworm *Amynthas* spp. competed with millipedes indirectly by consuming the litter layer and thus depriving the millipedes of their prey, reducing their numbers. Gao et al. (2017) studied the interaction between *A. agrestis* (epigeic), collembola (*Sinella curviseta*), and a centipede predator in the Cryptopidae family. *A. agrestis* reduced collembola abundance. The centipede predator of collembola, however, did not. Moreover, centipedes lost weight when in mesocosm that contained collembola only. Instead, centipede biomass increased potentially because they preyed on the earthworms.

When looking at the part of the soil food web that includes microbiota living in water films between soil particles, there are two major indirect effects. First, there is the alteration of habitat structure, providing habitable pore spaces for microorganisms. Figure 4 shows the pore size distribution of bulk soil and drilosphere soils of *L. terrestris* and pheretimoid earthworms (genera *Amynthas* and *Metaphire*). These pore size distributions were obtained with mercury intrusion porosimetry which gives the

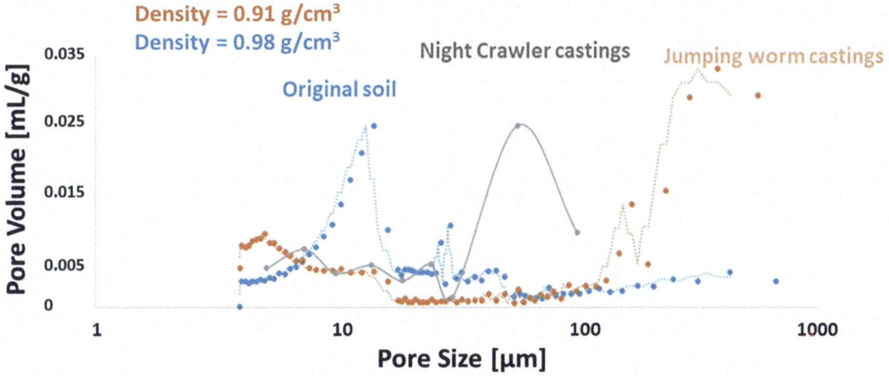

Fig. 4 Pore size distribution of aggregates from an earthworm free soil, created by *L. terrestris* and by *A. agrestis* (unpublished data)

pore volume that is accessible through pore necks of a given diameter per unit dry mass of soil (Görres and Amador 2010). The majority of pore volume in pheretimoid casts can be accessed through pore necks greater than 100 μm. For *L. terrestris*, the majority of the pore volume can be accessed through pores greater than 50 but less than 100 μm. These pore sizes are similar to or greater than body diameters of nematodes. For soil unaltered by *L. terrestris* or the pheretimoid worms, the majority of pore volume can be accessed through pores greater than 10 μm but less than 25 μm. These pore sizes are less than most adult nematode body diameters. The differences in pore structure represent differences in habitable pore space which would promote soil organisms of different body diameters. Second, the drilosphere is enriched in resources which likely promote the growth of decomposers (Burtis et al. 2014). Together, enrichment and pore structure modification have the potential to affect the food web in the drilosphere and possibly the quantity and quality of food resources of earthworms.

To dig deeper into how burrows and burrow wall enrichment of *L. terrestris* might influence the detritus-detritivore link, Savin et al. (2004) conducted a mesocosm experiment. Specifically, they probed the effect of resource enrichment, greater gas diffusion, and presence of earthworms on the drilosphere community. There were four treatments:

1. First, artificial burrows that would allow greater oxygen diffusion without the addition of litter or earthworms (ARTF).
2. Second, artificial burrows with litter pushed against the burrow walls (LEAF).
3. Third, the presence of the anecic earthworm *L. terrestris* and litter resources at the surface (WORM) relying on the earthworm to translocate the litter into the burrow.
4. Fourth, these were compared to a control without burrows, earthworms, or litter (CRTL).

The microbiota in the burrow walls in each treatment and the CRTL soil were examined for bacterial and fungal biomass and nematode and protozoan functional groups. The abundances of all nematode trophic groups were greatest in the artificial burrow walls (ARTF) that also received leaf litter (Fig. 5). In contrast

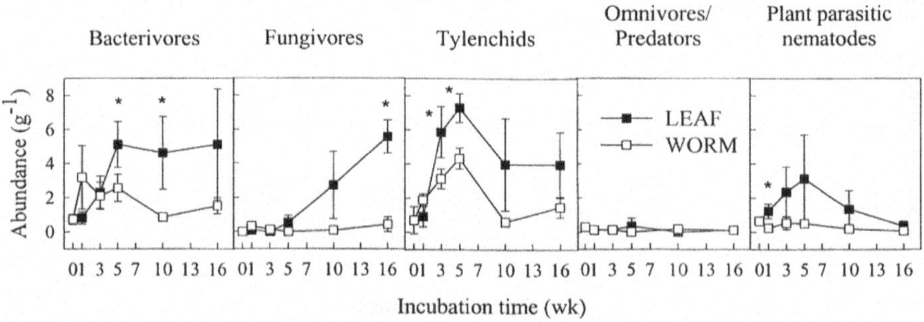

Fig. 5 Distribution of microfauna in artificial burrow soils lined with leaf litter (LEAF) and in *L. terrestris* burrows (WORM) (from Savin et al. 2004)

protozoans were greater in the walls of the occupied burrows (WORM). This would suggest, at first inspection, that earthworms were reducing the numbers of nematodes, which is supported by the findings of Monroy et al. (2008). Promotion of protozoan abundance in the WORM treatment also agrees with the observation by Monroy et al. (2008) that some protozoans increase in numbers during gut passage. However, this would be a simplification because the presence of detrital resources is necessary to promote the growth of microorganisms which in turn is essential for the growth of both microbivorous nematodes and protozoa. In some cases earthworms have been shown to grow better in the presence of protozoans (Bonkowski and Schaefer 1997) suggesting digestion and use of the protozoans. The effect on the bacterial and fungal biomass varied over the course of Savin et al.'s (2004) study with the presence of the earthworm promoting active fungi but not active bacteria, except early in the incubation when bacterial biomass exceeded that in the other treatments.

The dynamics in the burrow walls is evidence that earthworms influence the structure and composition of the food web. But, these effects likely differ among ecological groups and also among species within the same ecological group. For example, Yakkou et al. (2022) compared the abundance of microfauna in the castings of two endogeic *Aporrectodea* species with the bulk soil. Flagellate protozoa and nematodes in the casts of *A. caliginosa* were lower than in the bulk soil. This was opposite to what was found for the cast of *A. molleri*.

Whether modification of soil food web structure within the castings affects the quantity and quality of food resources available to the earthworm may depend on whether an earthworm species is coprophagic. Anecic earthworms line their burrows with detritus that they weld to the burrow walls with their castings, providing an initial inoculation of the detritus. Swift et al. (1979) formulated the external rumen hypothesis and suggested that the drilosphere of anecic earthworms acts like a rumen where detritus may be predigested, improving the nutritional value and palatability of detritus by enriching the drilosphere with microorganisms and their grazers. Given that earthworms cannot synthesize their own essential amino acids, promoting the growth of microorganisms on detritus stored in the drilosphere would enrich the food resources. This also implies that the acceleration of nutrient cycles in the drilosphere is not only an effect of the presence of the earthworm but also of growth of ingestible organisms that participate in the food web.

3 Conclusion

Earthworms are important parts of the food web, through their midlevel position in the food web, and their large biomass influences soil organisms in both directions of the trophic interactions. However, it can be argued that earthworms are geophagous and their effect as ecosystem engineers needs to be considered to appreciate their indirect effect on trophic interactions. Their engineering activities in the soil influence the habitat structure for other smaller organisms such as nematodes and protozoa which in turn would affect their interactions with microorganisms. They consume and derive nutrition from the very base of the detrital food web and microorganisms that inhabit detritus. The large biomass represented by them makes them an important link between belowground consumers and aboveground predators. Further studies are needed to fully appreciate their population effects on predator populations. The indirect effects of earthworms on food web interactions and the concomitant changes in biogeochemical cycling need integration into food web models.

References

Admassu B, Juen A, Traugott M (2006) Earthworm primers for DNA-based gut content analysis and their cross-reactivity in a multi-species system. Soil Biol Biochem 38(6):1308–1315

Adnan M, Joshi N (2013) The uniqueness of microbial diversity from the gut of earthworm and its importance. J Microbiol Biotech Res 3:111–115

Alford DV, Hancocks PJ, Parker WE (1995) The potential impact of New Zealand flatworm (Artioposthia triangulata) on agriculture and the environment in England and Wales. MAFF

Andriuzzi WS, Bolger T, Schmidt O (2013) The drilosphere concept: fine-scale incorporation of surface residue-derived N and C around natural Lumbricus terrestris burrows. Soil Biol Biochem 64:136–138

Antipov, A., Bakhur, T., Feshchenko, D., Romanishina, T., Avramenko, N., Goncharenko, V., Zghozinska, O., Solovyova, L., Koziy, N., Pidborska, R. and Shahanenko, V., 2018. Earthworms (Lumbricidae) as intermidiate hosts of lung nematodes (Metastrongylidae) of swine in Kyiv and Zhytomyr regions of Ukraine

Bamforth SS (1988) Interactions between protozoa and other organisms. Agric Ecosyst Environ 24(1–3):229–234

Barois I, Lavelle P, Brossard M, Tondoh J, Martinez M, Rossi JP, Senapati BK, Angeles A, Fragoso C, Jimenez JJ, Decaëns T (1999) Ecology of earthworm species with large environmental tolerance and/or extended distributions. Earthworm management in tropical agroecosystems, vol 3, pp 57–84

Boag B (2000) The impact of the New Zealand flatworm on earthworms and moles in agricultural land in western Scotland. Asp Appl Biol 62:79–84

Boag B, Yeates GW (2001) The potential impact of the New Zealand flatworm, a predator of earthworms, in western Europe. Ecol Appl 11(5):1276–1286

Bonkowski M, Schaefer M (1997) Interactions between earthworms and soil protozoa: a trophic component in the soil food web. Soil Biol Biochem 29(3–4):499–502

Bouché MB (1977) Strategies lombriciennes. Soil organisms as components of ecosystems. Ecol Bull (Stockholm) 25:122–132

Boyer J, Reversat G, Lavelle P, Chabanne A (2013) Interactions between earthworms and plant-parasitic nematodes. Eur J Soil Biol 59:43–47

Boykins EA (1967) The effects of DDT-contaminated earthworms in the diet of birds. Bioscience 17:37–39

Briones MJI, Garnett MH, Piearce TG (2005) Earthworm ecological groupings based on 14C analysis. Soil Biol Biochem 37(11):2145–2149

Brown GG, Barois I, Lavelle P (2000) Regulation of soil organic matter dynamics and microbial activity in the drilosphere and the role of interactions with other edaphic functional domains. Eur J Soil Biol 36:177–198

Brown DS, Jarman SN, Symondson WOC (2011) Pyrosequencing of prey DNA in reptile faeces: analysis of earthworm consumption by slow worms. Mol Ecol Resour 12:259–266

Burtis JC, Fahey TJ, Yavitt JB (2014) Impact of invasive earthworms on Ixodes scapularis and other litter-dwelling arthropods in hardwood forests, central New York state, USA. Appl Soil Ecol 84:148–157
Cameron EK, Bayne EM (2012) Invasion by a non-native ecosystem engineer alters distribution of a native predator. Divers Distrib 18(12):1190–1198
Campos-Herrera R, Trigo D, Gutiérrez C (2006) Phoresy of the entomopathogenic nematode *Steinernema feltiae* by the earthworm *Eisenia fetida*. J Invertebr Pathol 92:50–54
Chang CH, Johnston MR, Görres JH, Dávalos A, McHugh D, and Szlavecz K (2018) Co-invasion of three Asian earthworms, Metaphire hilgendorfi, Amynthas agrestis and Amynthas tokioensis in the USA. Biological invasions, 20, pp. 843–848.
Cleveland CC, Houlton BZ, Smith WK, Marklein AR, Reed SC, Parton W, Del Grosso SJ, Running SW (2013) Patterns of new versus recycled primary production in the terrestrial biosphere. Proc Natl Acad Sci U S A 110(31):12733–12737
Coulson JC, Coulson BC (2008) Lesser Black-backed Gulls *Larus fuscus* nesting in an inland urban colony: the importance of earthworms (Lumbricidae) in their diet. Bird Study 55:297–303
Crone ER, Sauer EL, Herrick BM, Drake D, Preston DL (2022) Effects of invasive jumping worms (*Amynthas* spp.) on microhabitat and trophic interactions of native herpetofauna. Biol Invasions 24:2499–2512
Das A, Rani K, Trivedi A, Yadav DK (2021) Earthworm: the ecosystem engineer. Harit Dhara 4:16–20
Dash MC, Senapati BK, Mishra CC (1980) Nematode feeding by tropical earthworms. Oikos:322–325
Drake HL, Horn MA (2007) As the worm turns: the earthworm gut as a transient habitat for soil microbial biomes. Ann Rev Microbiol 61:169–189
Edwards CA, Arancon NQ (2022) The role of earthworms in organic matter and nutrient cycles. In: Biology and ecology of earthworms. Springer US, New York, NY, pp 233–274
Edwards CA, Fletcher KE (1988) Interactions between earthworms and microorganisms in organic-matter breakdown. Agric Ecosyst Environ 24(1–3):235–247
Eisenhauer N, Butenschoen O, Radsick S, Scheu S (2010) Earthworms as seedling predators: importance of seeds and seedlings for earthworm nutrition. Soil Biol Biochem 42(8):1245–1252
Ekelund F, Rønn R, Christensen S (2001) Distribution with depth of protozoa, bacteria and fungi in soil profiles from three Danish forest sites. Soil Biol Biochem 33:475–481
Ettema CH (1998) Soil nematode diversity: species coexistence and ecosystem function. J Nematol 30(2):159
Ferlian O, Cesarz S, Marhan S, Scheu S (2014) Carbon food resources of earthworms of different ecological groups as indicated by 13C compound-specific stable isotope analysis. Soil Biol Biochem 77:22–30
Ferris H, Venette RC, Lau SS (1997) Population energetics of bacterial-feeding nematodes: carbon and nitrogen budgets. Soil Biol Biochem 29(8):1183–1194
Field SG, Michiels NK (2005) Parasitism and growth in the earthworm Lumbricus terrestris: fitness costs of the gregarine parasite Monocystis sp. Parasitology 130(4):397–403
Flack FM, Hartenstein R (1984) Growth of the earthworm Eisenia foetida on microorganisms and cellulose. Soil Biol Biochem 16(5):491–495
Fosgate OT, Babb MR (1972) Biodegradation of animal waste by *Lumbricus terrestris*. J Dairy Sci 55:870–872
Frelich LE, Blossey B, Cameron EK, Dávalos A, Eisenhauer N, Fahey T, Ferlian O, Groffman PM, Larson E, Loss SR, Maerz JC (2019) Side-swiped: ecological cascades emanating from earthworm invasions. Front Ecol Environ 17(9):502–510
Gao M, Taylor MK, Callaham MA Jr (2017) Trophic dynamics in a simple experimental ecosystem: Interactions among centipedes, Collembola and introduced earthworms. Soil Biol Biochem 115:66–72
Görres JH, Savin MC, Neher DA, Weicht TR, Amador JA (1999) Grazing in a porous environment: 1. The effect of soil pore structure on C and N mineralization. Plant Soil 212:75–83

Görres JH, Savin MC, Amador JA (2001) Soil micropore structure and carbon mineralization in burrows and casts of an anecic earthworm (Lumbricus terrestris). Soil Biol Biochem 33(14):1881–1887

Gorres JH, Amador JA (2010) Partitioning of habitable pore space in earthworm burrows. J Nematol 42(1), pp. 68–72.

Gorsuch JP, Owen PC (2014) Potential edaphic and aquatic predators of a nonindigenous Asian earthworm (*Amynthas agrestis*) in the Eastern United States. Northeast Nat 21(4):652–661

Gregory PT (1978) Feeding habits and diet overlap of three species of garter snakes (*Thamnophis*) on Vancouver Island. Can J Zool 56:1967–1974

Hamilton WJ Jr (1951) The food and feeding behavior of the garter snake in New York State. Am Midl Nat 46:385–390

Hartman GD, Whitaker JO Jr, Munsee JR (2000) Diet of the mole *Scalopus aquaticus* from the coastal plain region of South Carolina. Am Midl Nat 144:342–351

Hattori R, Hattori T (1993) Soil aggregates as microcosms of bacteria–protozoa biota. In: Soil structure/soil biota interrelationships. Elsevier, pp 493–501

Heppner F (1965) Sensory mechanisms and environmental clues used by the American Robin in locating earthworms. Condor 67(3):247–256

Hounsome T, O'Mahony D, Delahay R (2004) The diet of Little Owls *Athene noctua* in Gloucestershire, England. Bird Study 51:282–284

Hsu G-C, Szlavecz K, Csuzdi C, Bernard M, Chang C-H (2023) Ecological groups and isotopic niches of earthworms. Appl Soil Ecol 181:104655

Istiqomah L, Sofyan A, Damayanti E, Julendra H (2009) Amino acid profile of earthworm and earthworm meal (*Lumbricus rubellus*) for animal feed stuff. J Indones Trop Anim Agric 34:253–257

Jobbágy EG, Jackson RB (2000) The vertical distribution of soil organic carbon and its relation to climate and vegetation. Ecol Appl 10(2):423–436

Judas M (1992) Gut content analysis of earthworms (Lumbricidae) in a beechwood. Soil Biol Biochem 24(12):1413–1417

Keller EL (2018) Uncovering the variable life history traits and strategies of the gregarine parasite, *Monocystis perplexa*, in its invasive earthworm host, *Amynthas agrestis*. M.S. Thesis. The University of Vermont and State Agricultural College, Burlington, VT

Keller EL, Schall JJ (2020) A new species of *Monocystis* (Apicomplexa: Gregarina: Monocystidae) from the Asian invasive earthworm *Amynthas agrestis* (Megascolecidae), with an improved standard for *Monocystis* species descriptions. J Parasitol 106:735–741

Kooch Y, Jalilvand H (2008) Earthworms as ecosystem engineers and the most important detritivors in forest soils. Pak J Biol Sci 11(6):819–825

Kundu B, Bandyopadhyay PK (2019) An annotated list of aseptate gregarine parasites (Protozoa; Apicomplexa; Sporozoa) infecting oligochaete hosts. J Parasit Dis 43:487–497

Laakso J, Setälä H (1997) Nest mounds of red wood ants (*Formica aquilonia*): hot spots for litter-dwelling earthworms. Oecologia 111:565–569

Larsen T, Pollierer MM, Holmstrup M, D'Annibale A, Maraldo K, Andersen N, Eriksen J (2016) Substantial nutritional contribution of bacterial amino acids to earthworms and enchytraeids: a case study from organic grasslands. Soil Biol Biochem 99:21–27

Lavelle P (1988) Earthworm activities and the soil system. Biol Fertil Soils 6(3):237–251

Lee KE (1981) Earthworms (Annelida: Oligochaeta) of Vanua Tu (New Hebrides Islands). Aust J Zool 29(4):535–572

Li F, Luo Z, Li C, Li C, Jiang Z (2013) Biogeographical patterns of the diet of Palearctic badger: Is badger an earthworm specialist predator? Chin Sci Bull 58:2255–2261

Li S, Song M, Jing S (2021) Effects of different carbon inputs on soil nematode abundance and community composition. Appl Soil Ecol 163:103915

Lin R-S, Yao C-T, Lee P-F (2007) The diet of Fairy Pitta *Pitta nympha* nestlings in Taiwan as revealed by videotaping. Zool Stud 46:355–361

Macdonald DW (1980) The red fox, Vulpes vulpes, as a predator upon earthworms, Lumbricus terrestris. Z Tierpsychol 52:171–200

Macdonald DW (1983) Predation on earthworms by terrestrial vertebrates. In: Satchell JE (ed) Earthworm ecology: from Darwin to Vermiculture. Chapman and Hall, London, pp 393–414

Maerz JC, Karuzas JM, Madison DM, Blossey B (2005) Introduced invertebrates are important prey for a generalist predator. Divers Distrib 11:83–90

Marconi S, Manzi P, Pizzoferrato L, Buscardo E, Cerda H, Hernandez DL, Paoletti MG (2002) Nutritional evaluation of terrestrial invertebrates as traditional food in Amazonia. Biotropica 34:273–280

Martay B, Pearce-Higgins JW (2020) Opening a can of worms: Can the availability of soil invertebrates be indicated by birds? Ecol Indic 113:106222

McCay TS, Storm GL (1997) Masked shrew (*Sorex cinereus*) abundance, diet, and prey selection in an irrigated forest. Am Midl Nat 138:268–275

Medina-Sauza RM, Álvarez-Jiménez M, Delhal A, Reverchon F, Blouin M, Guerrero-Analco JA, Cerdán CR, Guevara R, Villain L, Barois I (2019) Earthworms building up soil microbiota, a review. Front Environ Sci 7:81

Miles HB (1963) Soil protozoa and earthworm nutrition. Soil Sci 95:407–409

Moeed A (1980) Diets of adult and nestling starlings (*Sturnus vulgaris*) in Hawke's Bay, New Zealand. NZ J Zool 7:247–256

Monroy F, Aira M, Domínguez J (2008) Changes in density of nematodes, protozoa and total coliforms after transit through the gut of four epigeic earthworms (Oligochaeta). Appl Soil Ecol 39:127–132

Morgan JE, Morgan AJ (1999) The accumulation of metals (Cd, Cu, Pb, Zn and Ca) by two ecologically contrasting earthworm species (Lumbricus rubellus and Aporrectodea caliginosa): implications for ecotoxicological testing. Appl Soil Ecol 13(1):9–20

Murchie AK, Gordon AW (2013) The impact of the 'New Zealand flatworm', *Arthurdendyus triangulatus*, on earthworm populations in the field. Biol Invasions 15(3):569–586

Nana PA, Fokam Z, Ntoungwa Ebague GM, Ngassam P, Bricheux G et al (2012) Description of five new species of hysterocinetian ciliates (Ciliophora, Hysterocinetidae), endocommensal in a terrestrial oligochaete of the genus Alma from Cameroon. Protistology 7(2):97–107

Obert T, Rurik I, Vd'ačný P. (2021) Diversity and eco-evolutionary associations of endosymbiotic astome ciliates with their lumbricid earthworm hosts. Front Microbiol 12:689987

Owen RB Jr, Galbraith WJ (1989) Earthworm biomass in relation to forest types, soil, and land use: implications for woodcock management. Wildl Soc Bull 17:130–136

Parkin TB, Berry EC (1999) Microbial nitrogen transformations in earthworm burrows. Soil Biol Biochem 31(13):1765–1771

Pass DA (2015) The earthworm microbiome (Doctoral dissertation, Cardiff University).

Pennycott TW, Grant D, Nager RG (2020) Earthworms in the diet of Herring Gulls *Larus argentatus* breeding on an off-shore island. Bird Study 67:131–134

Petersen KE, Yates TL (1980) Condylura cristata. Mamm Species 129:1–4

Picelli AM, Masseli GS, Kaefer IL (2019) Predation on giant earthworms (Clitellata: Rhinodrilidae) by the Diving Lizard *Uranoscodon superciliosus* (Linnaeus, 1758) (Squamata: Tropiduridae). Herpetol Notes 12:1203–1205

Piearce TG, and Phillips MJ, (1980) Fate of ciliates in the earthworm gut: An in vitro study. Microbial ecology, 5, pp. 313–319.

Pinnegar JK, Blanchard JL, Mackinson S, Scott RD, Duplisea DE (2005) Aggregation and removal of weak-links in food-web models: system stability and recovery from disturbance. Ecol Model 184(2–4):229–248

Pokarzhevskii AD, Zaboyev DP, Ganin GN, Gordienko SA (1997) Amino acids in earthworms: Are earthworms ecosystemivorous? Soil Biol Biochem 29:559–567

Pollierer MM, Langel R, Scheu S, Maraun M (2009) Compartmentalization of the soil animal food web as indicated by dual analysis of stable isotope ratios (15N/14N and 13C/12C). Soil Biol Biochem 41(6):1221–1226

Potapov AM, Tiunov AV, Scheu S, Larsen T, Pollierer MM (2019) Combining bulk and amino acid stable isotope analyses to quantify trophic level and basal resources of detritivores: a case study on earthworms. Oecologia 189:447–460

Ransom TS (2012) Comparison of direct, indirect, and ecosystem engineering effects of an earthworm on the red-backed salamander. Ecology 93(10):2198–2207

Reguera S, Santos X, Feriche M, Mociño-Deloya E, Setser K, Pleguezuelos JM (2011) Diet and energetic constraints of an earthworm specialist, the Mesa Central Blotched Garter Snake (*Thamnophis scaliger*). Can J Zool 89:1178–1187

Reynolds JW (1977) Earthworms utilized by the American woodcock. Proc Woodcock Symp 6:161–169

Richardson JB, Görres JH, Jackson BP, Friedland AJ (2015) Trace metals and metalloids in forest soils and exotic earthworms in northern New England, USA. Soil Biol Biochem 85:190–198

Roots BI (1956) The water relations of earthworms: II. Resistance to desiccation and immersion, and behaviour when submerged and when allowed a choice of environment. J Exp Biol 33:29–44

Sabine JR (1983) Earthworms as a source of food and drugs. In: Satchell JE (ed) Earthworm ecology: from Darwin to Vermiculture. Chapman and Hall, London, pp 285–296

Saim N (1985) The odor components of the defense mechanism of the "green worm", *Amynthas hupeiensis*, and wheat insect repellent in bay leaves (Laurus nobilis L.). M.S. Thesis, Kansas State University, Manhattan, KS

Salmon S, Geoffroy JJ, Ponge JF (2005) Earthworms and collembola relationships: effects of predatory centipedes and humus forms. Soil Biol Biochem 37(3):487–495

Sample BE, Schlekat C, Spurgeon DJ, Menzie C, Rauscher J, Adams B (2014) Recommendations to improve wildlife exposure estimation for development of soil screening and cleanup values. Integr Environ Assess Manag 10(3):372–387

Savin MC, Görres JH, Amador JA (2004) Microbial and microfaunal community dynamics in artificial and Lumbricus terrestris (L.) burrows. Soil Sci Soc Am J 68(1):116–124

Schall JJ (2021) Stomatocystis goerresi, a new species of gregarine parasite (Apicomplexa, Monocystidae) from the invasive Japanese earthworm Amynthas tokioensis (Megascolecidae), with a description of the parasite's life cycle. Folia Parasitol 68:1–7

Singh S, Singh J, Vig AP, Verma F, Suthar S (2020) Earthworm communities and soil structural properties. In: Bhat et al (eds) Earthworm assisted remediation of effluents and wastes. Springer Nature, Singapore, pp 333–350

Snyder BA, Callaham Jr MA, Lowe CN, and Hendrix PF (2013) Earthworm invasion in North America: Food resource competition affects native millipede survival and invasive earthworm reproduction. Soil Biology and Biochemistry, 57, pp. 212–216.

Strüssmann C, Brito ES, Marques OA (2013) What do water snakes eat? First report of predation by a Neotropical Hydropsini snake on giant earthworms (Glossoscolecidae). Salamandra 49:48–50

Sun Z, Jiang H (2017) Nutritive evaluation of earthworms as human food. In: Mikkola H (ed) Future foods. InTech, Croatia, pp 127–142

Swift MJ, Heal OW, Anderson JM, Anderson JM (1979) Decomposition in terrestrial ecosystems, vol 5. University of California Press

Taboga L (1980) The nutritional value of earthworms for chickens. Br Poult Sci 21:405–410

van den Brink N, Lammertsma D, Dimmers W, Boerwinkel MC, van der Hout A (2010) Effects of soil properties on food web accumulation of heavy metals to the wood mouse (Apodemus sylvaticus). Environ Pollut 158(1):245–251

Van Groenigen JW, Lubbers IM, Vos HM, Brown GG, De Deyn GB, Van Groenigen KJ (2014) Earthworms increase plant production: a meta-analysis. Sci Rep 4(1):6365

Vargas R, Hattori T (1986) Protozoan predation of bacterial cells in soil aggregates. FEMS Microbiol Ecol 2:233–242

Vermeulen F, Covaci A, d'Havé H, Van den Brink NW, Blust R, De Coen W, Bervoets L (2010) Accumulation of background levels of persistent organochlorine and organobromine pollutants through the soil–earthworm–hedgehog food chain. Environ Int 36(7):721–727

Wang T, Goyal A, Dubinkina V, Maslov S (2019) Evidence for a multi-level trophic organization of the human gut microbiome. PLoS Comput Biol 15:e1007524

Wang Y, Smith HK, Goossens E, Hertzog L, Bletz MC, Bonte D, Verheyen K, Lens L, Vences M, Pasmans F, Martel A (2021) Diet diversity and environment determine the intestinal microbiome and bacterial pathogen load of fire salamanders. Nat Sci Rep 11:20493

Weller S (2013) Ökologie und Paarungsverhalten des Regenwurms Lumbricus terrestris in Zusammenhang mit Parasitenbefall durch Monocystis sp (Doctoral dissertation, Universität Tübingen)

West JA (1910) A study of the food of moles of Illinois. Bull Illinois State Lab Nat Hist 9:14–22

Yakkou L, Houida S, Bilen S, Kaya LO, Raouane M, Amghar S, El Harti A (2022) Assessment of earthworm (Aporrectodea molleri)'s coelomic fluid-associated bacteria on different plant growth-promoting traits and maize germination and seedling growth. Biocatal Agric Biotechnol 42:102341

Yalden DW (1976) The food of the hedgehog in England. Acta Theriol 21:401–424

Yeates GW (1977) Soil nematodes in New Zealand pastures. Soil Sci 123(6):415–422

Yeates GW (2007) Abundance, diversity, and resilience of nematode assemblages in forest soils. Can J For Res 37(2):216–225

Zaborski ER (2002) Observations on feeding behavior by the terrestrial flatworm *Bipalium adventitium* (Platyhelminthes: Tricladida: Terricola) from Illinois. Am Midl Nat 148(2):401–408

Zhang W, Hendrix PF, Snyder BA, Molina M, Li J, Rao X, Siemann E, Fu S (2010) Dietary flexibility aids Asian earthworm invasion in North American forests. Ecology 91:2070–2079

Ziemba JL, Cameron AC, Peterson K, Hickerson CAM, Anthony CD (2015) Invasive Asian earthworms of the genus Amynthas alter microhabitat use by terrestrial salamanders. Can J Zool 93(10):805–811

Ziemba JL, Hickerson CAM, Anthony CD (2016) Invasive Asian earthworms negatively impact keystone terrestrial salamanders. PLoS One 11(5):e0151591

Earthworms as Soil Ecosystem Engineers

Patrick Lavelle and Alister V. Spain

Abstract In soils that are the basis of 16 of the 24 ecosystem services recognized by the Millennium Ecosystem Assessment, earthworms play a very important role as soon as conditions allow them to get established. They carry out three types of engineering, physical, community and biochemical. With the energy that they obtain from their mutualistic digestion in association with soil microbial communities, they create a large part of the physical structure of the soil. They create pores of very diverse nature, from the large sub vertical tubular galleries of anecics to porous spaces of diverse shapes and sizes, left between casts deposited in the soil by endogeics. They are also responsible for the aggregation of a significant proportion of soil of the upper horizons, by accumulating casts, in particular those of endogeic communities. The aggregation of several hundred tons of soil per ha in the soils where they are active, greatly influences the provision of hydrological services and C storage in these compact structures. Physical engineering is accompanied by community engineering resulting from the mutualistic mode of digestion which involves a significant selective stimulation of the soil microbial community ingested with soil. In addition, the creation of more or less connected porous spaces affects root penetration and populations of smaller invertebrates which find habitats and nutritional resources there. Finally, the biochemical engineering associated with earthworm activities, the least known of all, allows them, through the release of hormones and other signal molecules produced by the microorganisms under their control, to stimulate the growth of plants and defend them against certain parasites and diseases. Among the various mechanisms involved in biochemical engineering, the modification of the expression of plant genes is the most spectacular. A more in-depth knowledge of these three types of engineering will give stronger arguments to those who want to defend their communities affected by the unsustainable methods of certain agriculture practices and to use them optimally in modern agroecological practices.

P. Lavelle (✉)
Sorbonne Université, Institute of Ecological and Environmental Studies, Paris, France

A. V. Spain
Waihi Beach, New Zealand

Y. Kooch, Y. Kuzyakov (eds.), *Earthworms and Ecological Processes*,
https://doi.org/10.1007/978-3-031-64510-5_18

1 Introduction

The Millennium Ecosystem Assessment clearly emphasized the outstanding role of soils in providing ecosystem services (MEA 2005). Sixteen of the 24 services enumerated in this document are directly dependent on soil function (Dominati et al. 2010; Smith et al. 2015; Adhikari and Hartemink 2016). Climate regulation linked to C storage in soils, primary production dependent on nutrient cycling, infiltration and storage of water and its release to water tables and rivers, protection against erosion and flooding, and conservation of biodiversity are among the most cited contributions of soils to the maintenance of sustainable environment conditions. The contribution of soil organisms has been largely emphasized (Lavelle et al. 2006; Barrios 2007; FAO 2020), with special mention of earthworms as key actors of these processes (Blouin et al. 2013).

Soils are also part of the human sustainable development goals of the United Nations (UNDP 2015; Bouma 2014) and considered in the planetary boundary paradigm developed by Rockström et al. (2009). Large reports produced by international agencies give detailed assessments of soil degradation worldwide and expected trends (UNCCD 2016; UNEP 2016; OECD 2019; IPBES 2018).

In this chapter, we first recall the conceptual framework in which the ecological engineering of earthworms fits. We then recall essential features of their interactions with other soil organisms and summarize the various biochemical, community, and physical components of the engineering resulting from these interactions.

2 Conceptual Framework

2.1 *Ecosystem Engineering: A Multiscale Key Process of Soil Function*

Soils are self-organized systems in which interactions among physical, chemical, and biological processes, across a suite of nested scales, determine the formation and conservation of all basic properties and functions (Lavelle et al. 2016). A suite of five discrete scales has been proposed to describe the general organization in soil ecological processes (Fig. 1). At each scale, organisms interact with the physical elements of the environment and create structures that give the system emergent properties which in turn improve ecosystem functions and ultimately the ability of soils to provide ecosystem services. At the lowest level of resolution of micrometers, microorganisms form colonies (bacteria) and/or networks (fungi) that get associated with soil mineral and organic particles forming a specific type of microaggregates (Tisdall and Oades 1982; Tötsche et al. 2018) or rather large although fragile macroaggregates (Plante) resulting from the entanglement of soil particles in fungal networks.

Although all estimates show that this compartment is responsible for the largest part of soil heterotrophic respiration and overall mineralization activities (Lavelle

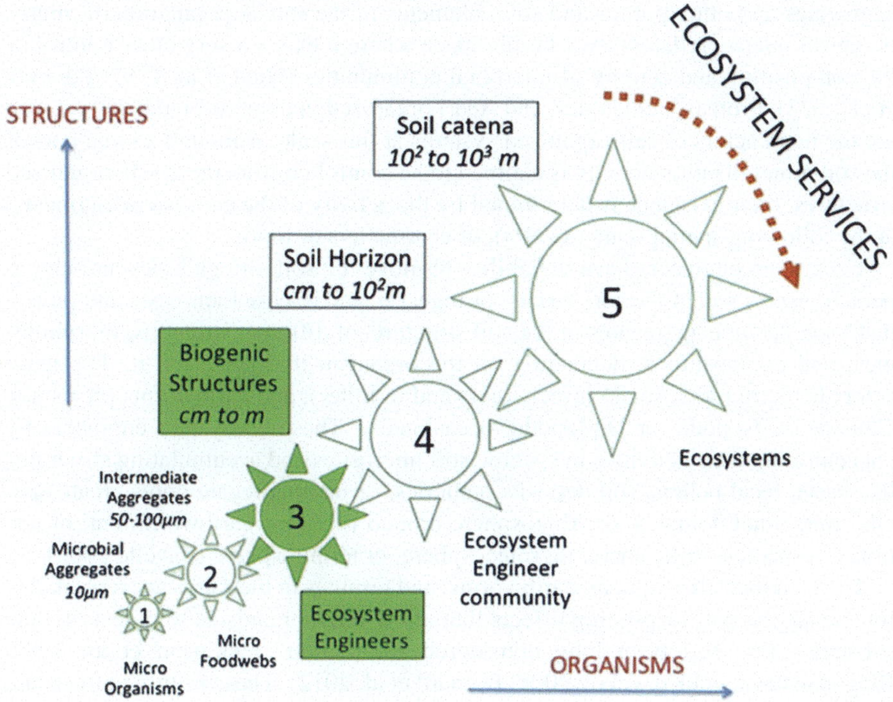

Fig. 1 Soil as a self-organized system of biological entities. Organisms of increasing sizes, from microorganisms to micro-invertebrates (micro-food webs), ecosystem engineers and ecosystems (horizontal axis) create structures of different sizes (vertical axis) and organize soil at different nested scales from 1 (microbial aggregates inhabited by microbial communities) to 5 (landscape viewed as a mosaic of different ecosystems). Most soil-based ecosystem services are perceived and delivered at the landscape (5) scale (Lavelle et al. 2006)

and Spain 2001; Bond-Lamberty et al. 2004; Sacca et al. 2017), microorganisms are inactive great part of the time due to their total (bacteria) or highly limited (fungi) ability to move toward new feeding sources once the one they were feeding on is exhausted. Indeed, at the scale of microbes, organic matter is not a homogeneously distributed source, but rather a set of fragments or local sources (root exudates) dispersed in the volume of mineral soil, randomly in the soil space and/or sporadic in time (Lavelle and Spain 2001).

This inactivity is suggested by visual evidences like dormant bacterial colonies or non-germinated fungal spores and quantified by the great difference observed between estimates of microbial biomass turnover times of several months to more than 1 year and the generation time measured in laboratory tests, when microbes are given optimum conditions for their growth (Spohn et al. 2016).

The composition and activity of microbial communities are then largely dependent on the activities of organisms that operate at larger temporal and spatial scales. The next scale-up (2), that of soil micro-food webs, is observed inpores and aggregate structures around 50 to 100 microns large, formed by the union of microbial aggregates and other porous and solid elements of the soil. A population of micro-organism micro-predators lives in this space where it exerts a very close control of the composition and activity of microbial communities (Hunt et al. 1987; Potapov et al. 2023). Protists, nematodes, and Acari organized in a micro-predator food web are the main actors of self-organized systems at this scale. Although a large part of the soil mineralization activity is claimed to be controlled from these self-organized structures, their existence is determined by the activity of the ecosystem engineers, at the following spatial scale (level 3), of centimeters or more.

Ecosystem engineers have the ability to move the soil, dig galleries and chambers, transport soil at the soil surface or ingest it, and release their casts and pellets that soon become aggregates in the soil structure, of different strengths, morphologies, and compositions, depending on the organism that produced it. The main actors here are roots, social insects (ants and termites), and earthworms plus some Coleoptera, Isopoda, or Diplopoda, occasionally. These ecosystem engineers by concentrating their activities in specific soil microsites and accumulating structures like casts, fecal pellets, soil deposits, galleries, or interaggregate pores create specific functional domains, the rhizosphere around root tips, earthworm, ant, or termites' respective drilosphere, myrmecosphere, or termitosphere (Lavelle 2002).

Roots in their rhizosphere, earthworms, and termites in their guts are responsible for the largest part of priming effects that allow microorganisms to digest organic substrates that had been long considered recalcitrant (Jenkinson et al. 1985; Blagodatskaya and Kuzyakov 2008; Bernard et al. 2012). These priming effects are triggered by the provision of highly energetic substrates that allow microbes to resume their activities and soon restore their enzymatic capabilities and become able to digest any kind of organic resource. Since the selection and activation of microbial communities occur in the gut of invertebrate engineers or in the immediate vicinity of root tips, part of the digestion products is recuperated by the engineer, in an unknown proportion, in what is considered a mutualist digestion system (Lavelle et al. 1995; Trigo et al. 1999). The energy captured this way is used by the engineers for their mechanical activities, moving in the compact soil structure and digging and/or ingesting soil and moving it through the intestinal tract.

At the next level (level 4), in the hierarchy of self-organized systems, that of a soil horizon or profile, communities of ecosystem engineers get organized making soil horizon mosaics of mixed functional domains, defined as the soil volume units associated dominated by the activity of a given engineer.

Finally at the landscape scale (5), soils with different communities get spatially organized in a mosaic of ecosystem types created by topography, changes of soil types along soil catenas, and/or human-driven land uses and plant covers.

Earthworms, as ecosystem engineers, operate at a strategic intermediate scale and are key actors in soil function. Their function as physical engineers has been largely acknowledged in a number of papers and reviews (Lavelle et al. 2006, 2007; Blouin et al. 2013; Bertrand et al. 2015). Less documented is their role as community engineers as they shape microbial, invertebrate, and even plant communities. Finally, their paper as biochemical engineers able to influence genetic expression of plants through complex interactions with microorganisms is a recent and promising field of investigation and application (Blouin 2018) (Fig. 2).

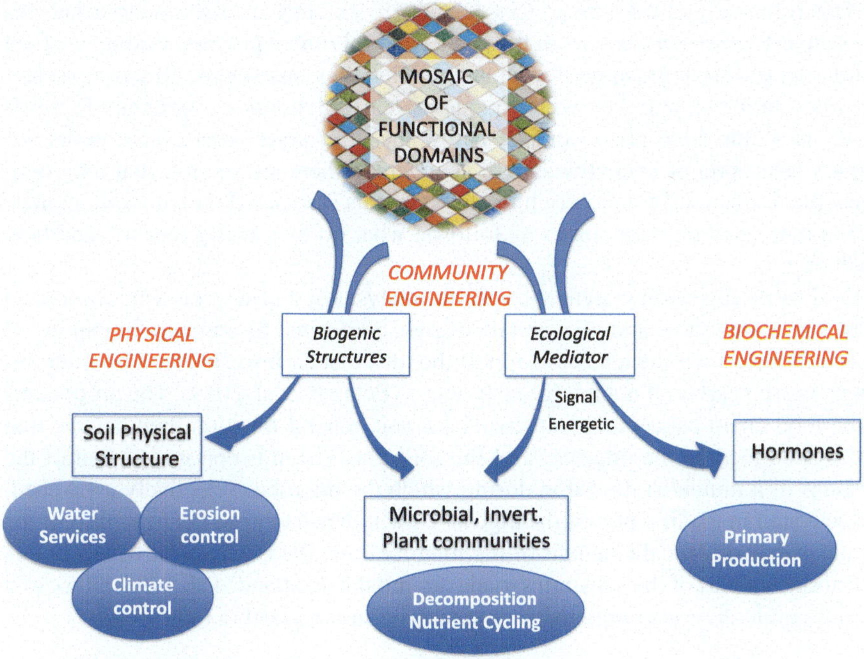

Fig. 2 Soil ecosystem engineers and the delivery of ecosystem services: three different processes (in red). (1) Physical engineering, the organization of soil physical structure and associated ecosystem services. (2) Community engineering, the selection of plant, microbial, and smaller invertebrate communities that determine decomposition and nutrient cycling processes. (3) Biochemical engineering: the release of hormones and signaling molecules that regulate primary production (Lavelle et al. 2016)

2.2 *Where Do Earthworms Get Their Energy From?*

Although the answer seems obvious, behind this apparently naïve question hides a great unknown (Potapov et al. 2022). Soil ecosystem engineers are the only organisms in soils able to move in this compact environment and to arrange it to allow the movement of other organisms and the construction of their habitats. Such a function requires a huge amount of energy. For example, it has been calculated that the African endogeic earthworm *Millsonia anomala* digests daily up to 30 times its own weight of soil when a recently born immature uses 95% of the assimilated energy in respiratory activities, mostly related to the huge effort needed to detach particles from the soil matrix and have them transit in a gut that may be more than 15 cm long in the largest adults (Lavelle 1978).

Depending on their ecological category, earthworms have different feeding behaviors identified by direct observations and confirmed by isotopic analyzes (Martin et al. 1992; Hyodo et al. 2008; Potapov et al. 2019). Epigeics eat leaf litter in the litter system, and anecics drag leaf litter into their burrows and eat them mixed to some soil, sometimes after it has been fermented during some time in an external rumen digestion strategy (Swift et al. 1979). They are also said to occasionally reingest their own casts, with the same external rumen process. Endogeics feed on mineral soil from the upper few cm of soil that they ingest with little or no selection (mesohumics), enriched with organic matter (polyhumics), for example, feeding in the plant rhizosphere, or just below the litter layer where fecal pellets of primary arthropod or enchytraeids are abundant. Others are so slim that they only ingest the thinner soil fraction rich in clay minerals associated with organic matter.

The real question then comes as follows: what do they really digest? And how much of it?

Earthworm digestion system and specific analyses of digestive enzymes produced in their guts are scarce and, to a certain degree, confusing. Strong enrichment in ^{13}C observed in earthworms tends to support the idea that earthworms feed on microbially processed carbon if not microbial biomass (Potapov et al. 2019). The proponents of the food chain conceptual framework for soil general function then believe that earthworms feed on the microflora of the soil. This vision is opposed to that of the partisans of a mutualist digestion during which the microbes selectively awakened, and activated in the first part of the digestive tract, then carry out the digestion of the organic substrates for the benefit of the microbes and the earthworm, which in this case absorbs a part of the substrates made assimilable (Lattaud et al. 1998). Research provides quite diverse complements that may be summarized in three points:

1. Earthworms generally have weak digestive enzymatic equipments. Cellulase is not produced in gut tissue cultures of some species (*Pontoscolex corethrurus*, *Millsonia anomala*, *Hormogaster elisae*) although *Polypheretima elongata*, *Hyperiodrilus africanus*, or *Dichogaster terrae-nigrae* can produce them (Lattaud et al. 1998; Trigo et al. 1999).
2. Earthworms can digest microorganisms, rather fungi than bacteria. Direct evidence however is only proposed so far for microorganisms that are not present in soil (Shan et al. 2013). On the other hand, decreases in microbial biomass interpreted as signs of a digestion (Aira et al. 2009) may be misleading. A few experiments actually show that the mutualist digestion system that consists in an activation of the ingested soil microflora by intestinal mucus may not be activated when the ingested soil has enough assimilable organic compounds to feed the earthworm (Lavelle et al. 1989).
3. In any case, the assimilation rate (of 8 to 29% as measured by Shan et al. (2013) of a microbial biomass that represents itself a low proportion of the organic matter (less than 5% of the soil organic contents) does not seem to meet the amount of OM actually assimilated, and around 9% in tropical endogeic earthworms (Lavelle 1978) and probably more in species fed more organic substrates. In addition, the amount (5 to 18% of the ingested soil) and energy cost of the intestinal mucus added to the ingested soil do not seem to be justified by a function that would be only lubricating the gut duct.

However, there is no doubt much remains to be discovered in order to fully account for the digestive process of earthworms and a number of intriguing unknowns remain. For example, Pokarzhevskii et al. (1997) show that the amino acid concentration of earthworms is an order of magnitude, or more, than the average soil concentration. These authors suggest that the source of the amino acids is the microbial biomass and/or that of the microfauna ingested with the soil.

In conclusion, although earthworms may digest some part of the ingested microbial biomass, especially fungal biomass in an opportunistic way, most of the assimilated material seems to be released by the mutualistic digestion system first described by Barois and Lavelle (1986). In this system, the activity of the dormant bacteria ingested with the soil is awakened and stimulated by the secretion of intestinal mucus and the addition of a volume of water and the intense mixing that takes place in the gizzard and the foregut. The dominance of this system over direct digestion by enzymes produced by the earthworm gut wall and/or digestion of living microbial biomass probably varies across ecological categories. One would expect that the mutualist system is all the more active and predominant as the earthworms ingest a soil with low organic matter content and low chemical quality (Lavelle et al. 1995; Trigo et al. 1999).

2.3 *Earthworm Ecological Categories, Ecosystem Engineering, and Ecosystem Services*

There are many ways to be an earthworm, and this diversity must be considered when discussing their impact on soil-borne ecosystem services (Table 1).

Table 1 From earthworms to ecosystem services. A suite of processes that link earthworm adaptive strategies in ecological categories to their feeding processes and the production of stable physical structures through different engineering processes and the emergent properties as ecosystem services

Ecological categories	Feeding processes	Drilosphere structures	Ecosystem engineering	Ecosystem emergent process	Ecosystem services
Epigeic Anecic Polyhumic endogeic Mesohumic endogeic Oligohumic endogeic	Food selection + ingestion	Casts	Biochemical	Decomposition Nutrient + signal molecule release	Nutrient cycling
					Primary production
			Community	Shaping micro-food webs and smaller invert. communities	Biodiversity
	Digestion (host-mediated + mutualist)		Physical	Soil aggregation Carbon storage GHG gas emissions	Climate control
	Comminution and mixing				Erosion control
	Activation selection of soil microbial communities	Pores		Water infiltration	Water services
				Water storage and transfer	Flooding control

Earthworm communities exhibit a continuum of ecological categories that can be ranked according to the relative proportions and qualities of the ingested organic matter and mineral soil and corresponding vertical distribution in the soil profile. Litter-feeding epigeics may ingest up to 100% organic residues, and their main function in the ecosystem is natural composting of fresh leaf litter. Polyhumic endogeics, the soil-living earthworms that ingest a soil rich in organic matter, may feed on concentrations of organic matter at the bottom of the litter system, root litter, and root exudates mixed with soil of the rhizosphere. Rhizophagy has been rarely documented, and experiments show that root litter at different stages of decomposition, in aerobic or anaerobic conditions, is a poor food for the African endogeic *Millsonia anomala* (Lavelle et al. 1989). Isotopic (Spain et al. 1990) or NIR signals (Zangerle et al. 2011) measured on earthworm bodies or casts however indicate feeding links with the rhizosphere system, possibly in the form of root exudates.

At the other end of the continuum, mesohumic and oligohumic endogeics that live on soil from the surface or deeper soil horizons with very low organic contents are mainly physical engineers. They daily ingest several times their own weight of soil, digging their way through the soil and leaving behind pores and casts of different sizes and shapes that comprise their respective drilospheres.

Anecic earthworms that ingest a mixture of leaf litter with mineral soil represent an intermediate situation.

All categories participate in community engineering at shaping the composition and activity of microbial communities, micro-food webs, and other invertebrates and vertebrates. All of them also contribute to the biochemical engineering effect at releasing hormones and other signal molecules with effects on plant growth and gene expression.

Intermediate ecosystem processes and emergent properties like nutrient release, soil macro- and microaggregation, water infiltration, storage in the porous space, and transfer to water tables form the basis of soil-based ecosystem services.

3 Earthworm Ecosystem Engineering

3.1 Biochemical Engineering

Biochemical engineering, the lesser-known mode of ecosystem engineering, profoundly affects the relationship of earthworms with the rest of the soil community. Earthworms produce a number of very specific molecules, here called ecological mediators, that may profoundly affect the activities of other organisms, microorganisms, and plants principally. Biochemical engineering is defined here as the outcome of these interactions, at the ecosystem level.

Ecological mediators of two kinds, energetic and signaling, are released in the drilosphere. Intestinal and cutaneous mucus—apart from their long-alleged function as internal and external lubricants—are key energetic mediators that trigger priming effects on microbial activities.

Signaling molecules produced by plants and soil organisms act as hormones or initiators of signaling pathways leading to the production of phytohormones (Medina-Sauza et al. 2019). These molecules produced in the drilospheres may activate plant growth or influence the expression of plant genes (Blouin et al. 2013).

Intestinal Mucus Extraction with hot water (60 °C) of the water-soluble compounds present in the anterior part of the digestive tract of earthworms produces a transparent yellowish solution which, once freeze-dried, takes on the appearance of a white foam (Martin et al. 1987). Chromatographic analysis of this extract taken in African endogeic earthworms showed three peaks: one between 40 and 60 K d comprised of a glycoprotein; another peak purely proteic around 25Kd, probably comprised of enzymes; and the last peak of low-molecular-weight osides, amino acids, or peptides around 180 d (Fig. 3). This mucus may comprise 5 to 18% of the anterior gut content of endogeic tropical earthworms and up to 25–40% in endogeic earthworms of temperate regions (Fig. 4). Concentrations tend to increase when the organic contents of the ingested soil decrease and/or as temperature decreases, with greater contents in peregrine than endemic species (Lavelle et al. 1995). Such results suggest earthworms invest all the more in mucus addition as low temperature, low contents of organic residues, and/or their relative recalcitrance to mineralization oppose constraints to the mutualistic digestive process (Lavelle et al. 1995). Only present in the anterior part of the earthworm gut, mucus then disappears, and soluble extracts from posterior gut parts present a number of small peaks. They correspond to assimilable organic matter temporarily available before its probable absorption by the gut wall or assimilation by microorganisms.

This mucus represents surprisingly high proportions, from 4 to up to 40% of the dry weight of the fore gut content. The energy represented by the production of this mucus is much higher than what the worm will extract from its digestion of the organic matter of the soil, limited to 3 to 19% of the ingested organic matter. No

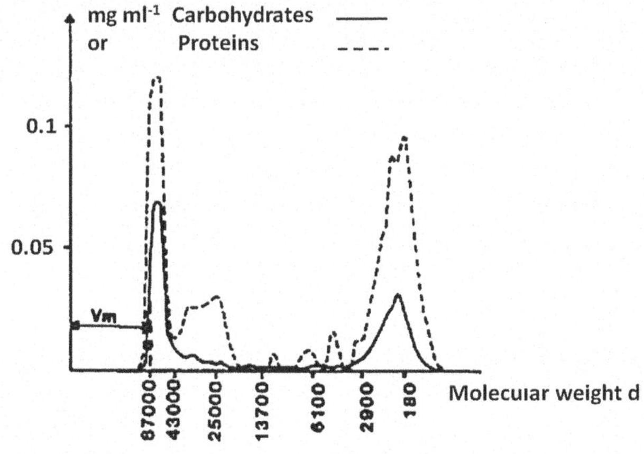

Fig. 3 Gel chromatography of water-soluble extracts from the anterior gut part of *Millsonia anomala* (now renamed *Reginaldia omodeoi*) (Martin et al. 1987)

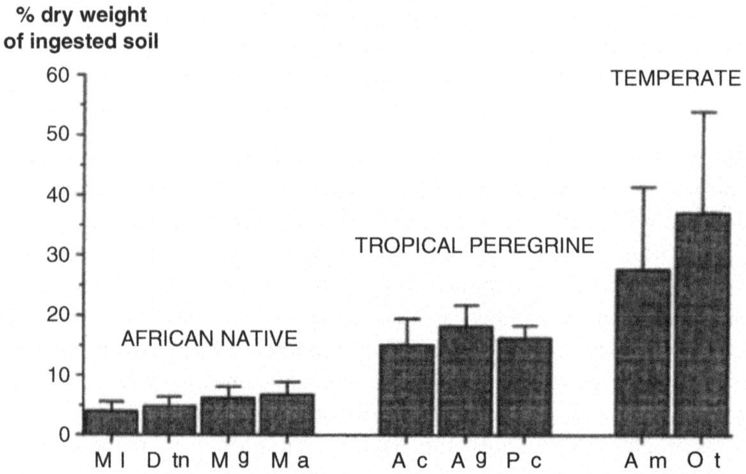

Fig. 4 Proportion of intestinal mucus mixed with the soil ingested by different earthworm species. Ml, *Millsonia lamtoiana*; Dtn, *Dichogaster terrae-nigrae*; Mg, *Millsonia ghanensis*; Ma, *Millsonia anomala*; Ac, *Amynthas corticis*; Ag, *A. gracilis*; Pc, *Pontoscolex corethrurus*; Am, *Allolobophora molleri*; Ot, *Octolasium tyrtaeum* (Lavelle et al. 1995)

longer found in the medium and posterior parts of the digestive tract, the part of the intestinal mucus that has not been metabolized by bacteria is probably re-absorbed.

Cutaneous mucus Cutaneous mucus is a complex mixture of carbohydrates with a high aliphatic/lipid content, acetylated carbohydrates, peptides, and low protein contents (Cortez and Bouché 1987; Shutenko et al. 2020). The amounts produced are not known. They probably vary according to ecological categories and are expected to have been selected by different evolutionary constraints like eliminating excess C from a nutrient poor diet (stoichiometric constraint). Seen from a stoichiometric perspective, the production of intestinal mucus, in addition to the high respiration cost of mechanical activities, is a process that allows eliminating C in excess and adjusts the stoichiometry of organic compounds assimilated during digestion to that of earthworm biomass. Cutaneous mucus is also considered as a mediator in community engineering that attracts Collembola (Salmon and Ponge 2001) and stimulates the development of specific microbial communities on burrow walls where it has been deposited (Loquet et al. 1977; Furlong et al. 2002). Cutaneous mucus also acts as a primer accelerating the turnover of soil organic (Hoang et al. 2016).

Hormones involved in plant growth have been found in earthworm casts of mainly epigeic forms (Medina-Sauza et al. 2019). Production of indole acetic acid, jasmonic acid, salicylic acid, amino-cyclopropane-1-carboxylate (ACC), auxin and auxin-like molecules, ethylene or ethylene precursors, and humic acids observed in drilospheres is mostly considered a result of the activation of specific bacteria, like plant growth-promoting bacteria, in the gut contents and casts of earthworms.

Celomic fluid of various earthworm species has exhibited significant effects as plant growth promoters and pesticidal and antifungal properties (Nadana et al. 2020; Ecimovic et al. 2021).

The outstanding effect of vermicomposts on plant growth is likely due to the release in these products of plant growth-promoting molecules (Pathma and Sakthivel 2012; Zhang et al. 2015).

Biochemical engineering also involves effects on the expression of plant genes, allowed by the release of specific ecological mediators. Modification of the expression of upland rice genes in the presence of earthworm has been reported to render these plants tolerant to plant parasitic nematode attacks (Blouin et al. 2005) Fig. 5. In this case, the overexpression of a leaf lipoxygenase gene and underexpression of a cysteine protease gene, responsible for protein degradation in leaves and their transfer to damaged roots, prevented the inhibition of photosynthesis and allowed a normal survival and growth of the plants. Other signaling molecules produced in drilospheres have the ability to influence the expression of a large number of plant genes (Jana et al. 2010; Puga-Freitas et al. 2012).

Nutrients Another outstanding case of biogeochemical engineering is the release of assimilable nutrients observed in earthworm casts. Fresh casts generally exhibit high concentrations of mineral N and assimilable P (Lavelle and Spain 2001). These nutrients, often produced within reach of growing roots, are available during a limited amount of time, ca. 24 h in the case of fresh casts of *Pontoscolex corethrurus* (Lavelle et al. 1992) before they are either expelled outside the cast and released in the external environment or immobilized in the microbial biomass (Decaëns et al.

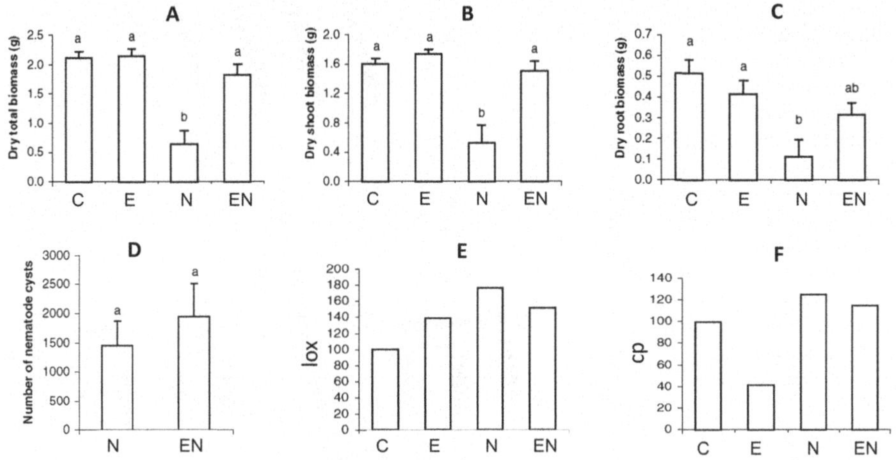

Fig. 5 Rice biomass in different treatments. (**A**) Total dry biomass, (**B**) shoot dry biomass, (**C**) root dry biomass, (**D**) number of nematode cysts per unit of root biomass, (**E**) gene expression (number of transcripts after RT-PCR on total RNAs) of lipoxygenase-1 (lox) and (**F**) cysteine protease (cp). *C* control, *E* earthworms, *N* nematodes, *EN* earthworms and nematodes. Significant differences between means (± 1 SE) are marked by different letters (least square means, Tukey-Kramer test, $P < 0.05$) (Blouin et al. 2005)

1999; Fig. 6). The synchrony between nutrient release and plant uptake realized by this spatiotemporal pattern coincidence prevents nutrient loss to the deeper soil or to the atmosphere (Swift and Anderson 1994). However, an analysis of the mineral contents of dry castings shows higher contents during the lifetime of these structures (Van Groenigen et al. 2019).

An interesting interface with climate regulation is the release of NOx gas, a powerful greenhouse gas, that naturally contributes to the regulation of the earth ozone layer. Claims that earthworms might significantly alter climate via the production of these gases based on the extrapolation of laboratory tests are most probably exaggerated (Lubbers et al. 2013). Excessive earthworm biomasses that create anaerobic conditions favorable to denitrification and the absence of plants to extract mineral N in most experiments have engendered conditions that must be rare if at all possible in nature.

3.2 *Community Engineering*

Earthworms greatly influence the composition and activities of communities of soil-dwelling microorganisms; micro-, meso-, and macrofauna; and even plants by their feeding and bioturbation activities.

Fig. 6 N dynamics in earthworm casts and the nearby soil. The arrows indicate the direction of water flow, and the numbers refer to a chronological order: (1) high NH4+ release in fresh casts, (2) vertical drainage in the gallery, (3) lateral runoff from the cast and the soil surface, (4) vertical leaching in the soil profile, (5) denitrification plus root and microbial uptake, and (6) N immobilization in dry casts (Decaëns et al. 1999)

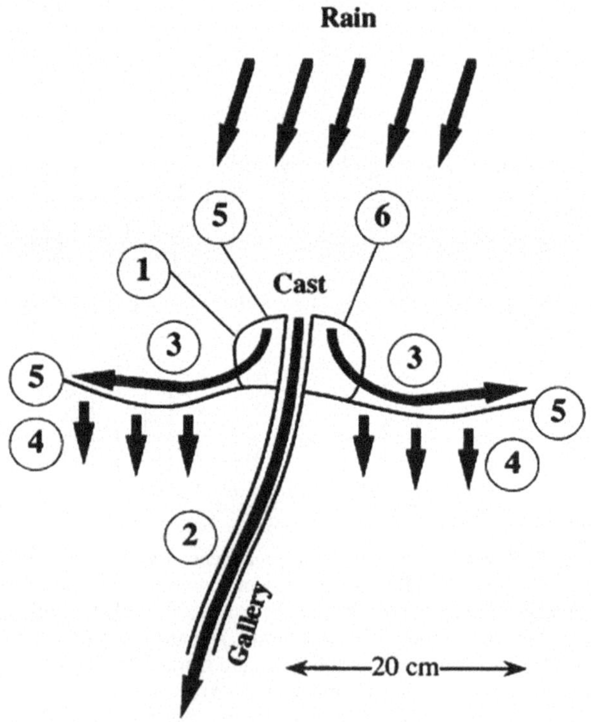

3.2.1 Microbial Communities

Earthworm gut represents a very special set of microsites with highly favorable environmental conditions for microbial activities: high water (up to +150% the content of the surrounding soil) and organic contents, neutral pH, and large amounts (4 to 40%) of intestinal mucus added as a primer in the anterior part of the gut (Barois and Lavelle 1986; Trigo et al. 1999; Lavelle and Spain 2001).

Oxygenation varies from microaerophilic to anoxic conditions, while soil progresses from the foregut to the midgut and hindgut parts (Barois and Lavelle 1986; Trigo et al. 1999; Meier et al. 2018). Although this condition may favor microaerobic microorganisms such as denitrifying bacteria, for which earthworm gut is a "transient heaven" (Drake and Horn 2006), soil microorganisms of a wide range of biological profiles may get stimulated during their stay in this succession of highly favorable microenvironments, from the gizzard and crop to fore mid- and hindgut section. The digested soil and/or litter expelled as fresh casts pass through another sequence of states from freshly deposited casts to dry casts and their final belonging to the soil macroaggregate structure (Lavelle et al. 1992; Zangerle et al. 2014).

Analyses of microbial communities in soil, the different parts of earthworm guts, and casts suggest that the specific conditions observed in the different gut parts and casts select and activate alternatively different groups of microorganisms.

While the existence of specific symbiotic bacteria in the earthworm gut is not discarded, there is a general consensus to consider that most microorganisms ingested with soil and litter transit unharmed through the gut and constitute the greatest part of the gut microbiome (Wolter and Scheu 1999; Furlong et al. 2002; Byzov et al. 2009; Medina-Sauza et al. 2019). Although specific bacteria and fungi can be isolated from cleaned gut walls, there is no proof so far that they are symbiotic species of the gut. Part of the microbial community ingested with soil and organic debris is selected and strongly activated (Parle 1963; Brown 1995; Barois and Lavelle 1986; Furlong et al. 2002). Another part seems to be destroyed by midgut intestinal fluid (Byzov et al. 2007). The selected community seems to be composed mostly of generalist bacteria (Huang et al. 2022) that become dominant, and microbial diversity in the gut is lower than in the bulk soil (Aira et al. 2022). A likely consequence of specific activation of different parts of the soil-born microbiota in different sections of the gut and in casts is the large dissimilarities observed in all these sections (Sampedro and Whalen 2007). The negative effect of gut transit on *E. coli* from pig slurry indicated by Monroy et al. (2008) suggests that some bacteria may not find suitable conditions for their survival in earthworm gut contents.

Comparing microbiome of soil, gut contents, and casts of the common endogeic Lumbricidae *Aporrectodea caliginosa*, Aira et al. (2022) found that a large proportion of bacteria identified with 16S rRNA sequencing was unique to each microbiome: 82% in soil, 89% in gut, and 75% in casts (75%). Alpha diversity was maximum during gut transit and decreased in casts. This decrease of overall bacterial richness and diversity in the gut content as compared with the soil has been confirmed in other studies based on different identification techniques (Xia et al. 2022). The effect of earthworm species and soil composition on this selection has also been noted. In an experiment where endogeic earthworms of the genus *Drawida* were fed a soil enriched with manure (HMs) or manure and lime (HMLs), only 520 OTUs (2.9% of a total of 17,654) were common to all soil and earthworm gut (HMe and HMLe) contents (Fig. 7). Between soil and earthworms of a same treatment, the percentage was 15.7% in the organic-amended HM soil and only 7.8% in the same soil with lime (Zhu et al. 2022) (Fig. 2a). Such differences in the composition of detectable microbial communities were significantly related with expected carbon metabolic functions of the community assessed with PICRUSt2 and FAPROTAX techniques.

In general, soil and organic matter inputs, on the one hand, earthworm species and, on the other hand, may promote different microbial communities. While fungi are commonly identified from earthworm guts and clean gut walls, their participation in the digestive process has not been indicated so far. Microaerophilic to anoxic conditions observed in the earthworm gut seem to limit their activity (Swaby 1950).

3.2.2 Mycorrhizae and Fungal Diseases

While earthworms seem to be closely associated with bacteria in their interactions, interactions with fungi are less well known. Fungi do not seem to participate in the mutualist digestion process, and cases of control by earthworms of fungal plant diseases have been documented. A rather large number of studies on earthworm: arbuscular mycorrhizae interactions show rather diverse outcomes (Reddell and Spain 1991). More recently, in a study by Meng et al. (2021), 60% of records showed enhancement of root colonization in the presence of earthworms and 20%

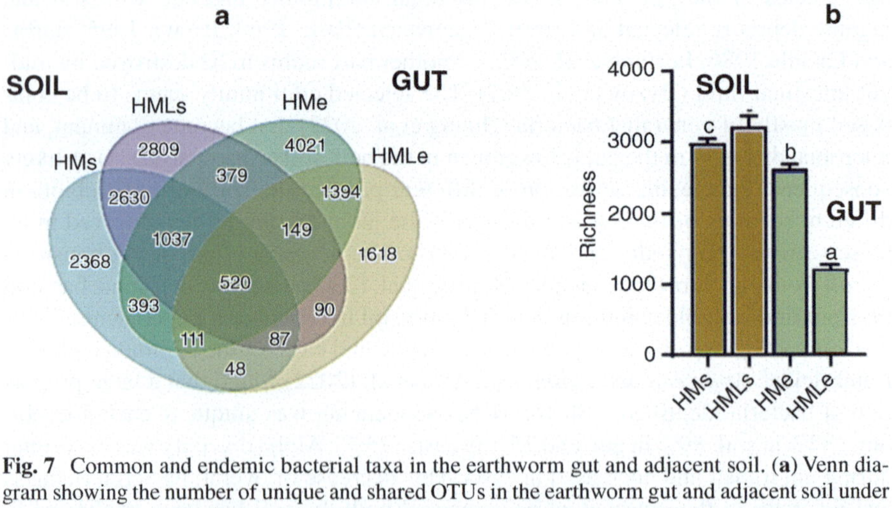

Fig. 7 Common and endemic bacterial taxa in the earthworm gut and adjacent soil. (**a**) Venn diagram showing the number of unique and shared OTUs in the earthworm gut and adjacent soil under manure treatments. (**b**) Richness in soil and gut contents. HMs, control soil with manure treatment; HMLs, control soil with manure and lime treatments; HMe, earthworm gut content fed soil with manure; HMLe, earthworm gut content fed soil with manure and lime

an inhibition, although different effects were observed depending on the ecological categories involved. Epigeic and anecic species generally stimulate root mycorrhizal infection in proportions of 3:1 and 4:1, respectively, while an inverse ratio of 1:2 was observed with endogeic species (Blouin et al. 2013). The higher occurrence of negative cases with endogeic species is attributed to the damage that they do to fungal hyphal networks and fine roots (Gange and Brown 2002) or possible fungal feeding.

Redell and Spain (1991a) reported the presence of putative ectomycorrhizal fungal spores and ectomycorrhizal root fragments in the casts of *Pontoscolex corethrurus* from Australian sites with sclerophyllous vegetation. Spore populations were low, and no concentration effects were noted, as with AM fungi. Few studies of the transmission of the infective propagules of ectomycorrhizal fungi by earthworms have been conducted.

3.2.3 Microfauna Communities of the Soil Micro-food Web

As earthworms ingest several hundred Mg of surface soil each year, and up to 1200 Mg in the humid savannahs of Ivory Coast (Lavelle and Spain 2001), it is likely that all the microsites in which the soil microfauna lives are digested at regular intervals. Earthworms have even been said to be ecosystemivorous, not saprophagous (Pokarzhevskii et al. 1997). Contrasted results have been observed with protozoa. While Rouelle et al. (1985) and Pokarzhevskii et al. (1997) claimed that earthworms might digest them, detailed observations of Monroy et al. (2008) show that they are either unaffected (testate amebae) or stimulated (flagellates and naked amoebae), with different effects of the three epigeic species tested. Astomatous ciliates have been indicated as permanent residents of the gut of anecic earthworms (De Puytorac 1954; Piearce and Phillips 1980). The effects of earthworms on nematodes have been much thoroughly studied with different kinds of results. In general, detritivorous nematodes are not affected by gut transit, while plant parasites often are severely reduced (Loranger-Merciris et al. 2012). While some of them may be directly digested during gut transit, others lose the chemoreceptors that drive them to roots after proteases released by bacteria during gut transit destroyed them. In other cases, eggs hatch in fresh earthworm casts, and larvae become trapped in the compact structure of the cast (Boyer et al. 2013).

3.2.4 Meso- and Macrofauna

Mesofauna and the smallest macroinvertebrates are also influenced by earthworm community engineering. While relationships between microarthropods and epigeics are variable, endogeics have mainly negative effects due to competition for food (Eisenhauer et al. 2010).

Anecics may locally favor microarthropods at providing specific habitats like middens or galleries. Collembola attracted by earthworm casts are often found in earthworm galleries (Salmon and Ponge 2001; Marinissen and Bok 1988). Earthworm galleries generally have positive effects providing habitat to smaller invertebrates. Higher densities of both Collembola and Acari have been reported in patches of high density of the endogeic *Polypheretima elongata* in pastures of Martinique Island (Loranger et al. 1998). In forest soils of Madagascar, a giant earthworm digs sub-horizontal galleries below the A horizon, all of the same diameter. These galleries seem to serve as ways for the smaller individuals that do not dig galleries (Figs. 8, and 9).

Earthworms may also have indirect negative effects on litter-feeding fauna when large populations of anecics incorporate the greatest part of leaf litter into the soil. This example of amensalism as indicated by Eisenhauer et al. (2010) is observed at the Butte Montceau site near Paris, along a natural gradient of pH created by the outcropping of sand in the upper parts and calcareous sediments on the lower part of a slope (Lavelle and Spain 2001; Fig. 9). With increasing pH, earthworm populations greatly increase, and litter-feeding arthropods decrease as their food resource decreases.

Similar effects on arthropod (Snyder et al. 2009; McCay and Scull 2019) and even woodland salamander populations (Maerz et al. 2009) have been observed when litter accumulation decreases after the invasion of exotic earthworms in North America.

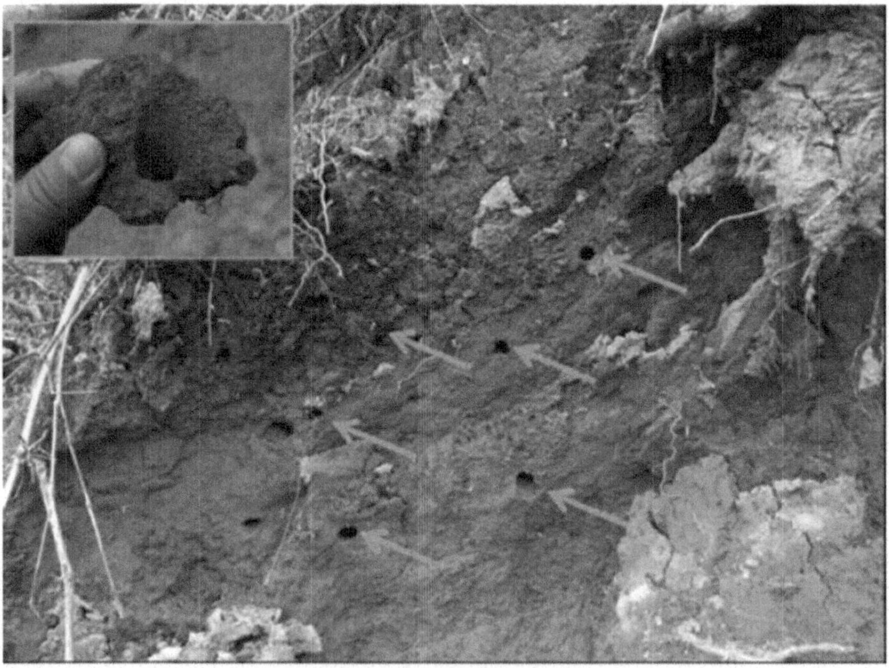

Fig. 8 Large burrow system built by a giant earthworm in the Didy forest (Madagascar). Galleries have the same diameter and orientation. Young worms of the same species and cocoon envelops have been found in these galleries (Photo: P. Lavelle)

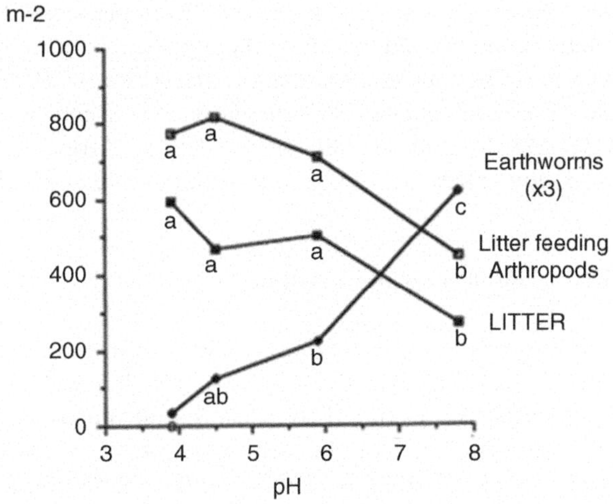

Fig. 9 Correlated changes in soil pH, litter mass, population densities of litter-feeding macroarthropods, and earthworms at Butte Montceau (France) on a hill slope with alternating sand and calcareous bedrock (Lavelle and Spain 2001)

3.2.5 Plant Communities

Earthworms affect plant communities via the selective ingestion and dispersion of seeds and their effects on their germination (Milcu et al. 2006). Seed ingestion by earthworms is common, and gut transit seems to enhance their germination. According to Decaëns et al. (2003) from 0.65% of the seed bank in savanna to 16.7% in old pastures is annually incorporated in earthworm casts in savannahs of the Colombian Orinoquia). At ingesting and egesting viable seeds in their casts, earthworms allow seedlings to germinate in a nutrient-rich medium, protected from seed predators. However, this positive effect may be offset by deep burial of the seeds below emergence depth limits (Regnier et al. 2008) and the digestion of seeds and seedlings observed in *Lumbricus terrestris* (Eisenhauer et al. 2010).

Finally, through a combination of effects on seeds and seedling emergence and differential stimulation effects, earthworms can modify plant communities (Laossi et al. 2009) and suppress weed occurrence in earthworm-corn-associated cropping systems (Li et al. 2020).

3.3 *Physical Engineering*

Much information has been collected on this effect of endogeic and anecic earthworms, the result of burrowing and casting processes (Blouin et al. 2013; Lavelle et al. 2007).

Although precise data on soil ingestion rates are relatively scarce, some general figures can be proposed. Active endogeic earthworms may daily ingest from 1–2 (*Aporrectodea caliginosa*; Zhang et al) to 20–30 (*Millsonia anomala*, Lavelle 1978) times their own weight of soil. Capowiez et al. (2014) estimated that on average

Allolobophora chlorotica burrowed 22 cm per week. Ingestion rates depend on the relative contents and quality of the organic matter in the ingested soil, temperature, and soil moisture and earthworm size (Lavelle 1975; Lavelle et al. 1989). This means several hundred Mg in most soils where earthworms are present, up to 1200 Mg dry soil annually ingested in an African savanna and transformed into casts, equivalent to several cm of soil, in some cases virtually all the A horizon.

3.3.1 Soil Creep and Erosion

Soil bioturbation and digging by earthworms have significant although contrasted effects as a great diversity of behaviors is observed among species. Casts can either limit surface water runoff (or cause significant erosion, depending on the firmness of their structure and the configuration of the surface, slope, and roughness) (Bertrand et al. 2015). A 75% decrease in runoff on a 40° slope was observed in a Vietnamese soil covered with casts of *Amynthas khami*. Middens of *L. terrestris* have the same effect at increasing soil roughness (Pitkanen and Nuutinen 1998; Le Bayon et al. 2002). Earthworms are also held responsible of part of the biotic soil flux when their casts are carried away by runoff water and accumulate in low-lying areas (Wilkinson et al. 2009). In this process, soil creeping is faster for fine soil particles and decreases with depth (Nye 1955; Nooren et al. 1995).

3.3.2 Earthworm-Induced Aggregation Processes

The ingested soil is first thoroughly dispersed by the intense mixing operated in the gizzard and crop segments and addition to the ingested soil of mucus and water (Shipitalo and Protz 1988; Barois et al. 1993). Digested soil is then defecated in soil or at the soil surface. While anecics tend to deposit a large part of their casts at the soil surface, endogeics deposit very little surface casts (Lavelle 1978; Lavelle and Spain 2001).

A large diversity of cast types has been observed and described (Lavelle and Spain 2001). Two opposite types of casts deposited inside soils can be defined for their either loose or compact structure (Blanchart et al. 1997). Compact casts have round shapes and very low porosities—with bulk densities of up to 1.6 g cm^{-3}. Very fragile and dispersable structures when fresh, they harden tremendously after the first drying event and become highly compact structures covered with a thin, 15-μm-thick crust, made of clay and colloidal molecules with hydrophobic properties (Blanchart et al. 1993; Fig. 10). Inside these casts that have become soil macroaggregates, microaggregates form associating organic matter particles with clay and polysaccharides produced by microorganisms (Fig. 11; Shipitalo and Protz 1989; Six et al. 2000; Bossuyt et al. 2004; Fonte et al. 2007). This protection of organic matter in microaggregates within macroaggregates is a powerful mechanism of C

Fig. 10 Scanning electron microscope view of a broken cast of *Millsonia anomala* showing a 15-µm-thick cortex made of fine particles that give the cast a smooth and close aspect (Blanchart et al. 1993)

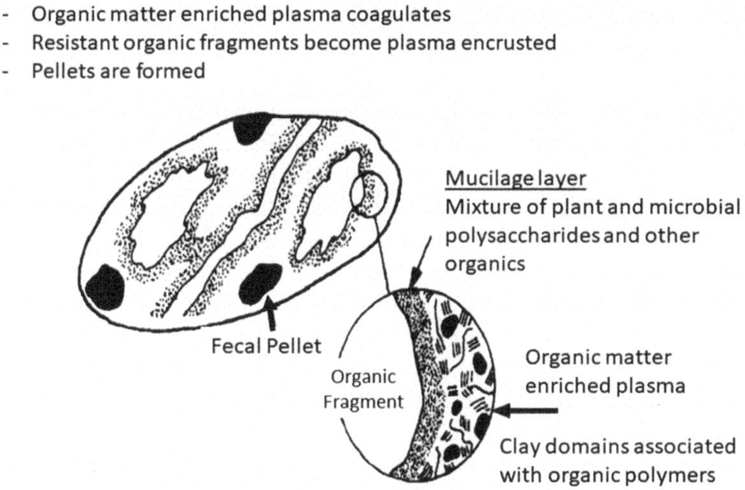

Fig. 11 Diagrammatic representation of aggregate formation in earthworm casts (Shipitalo and Protz 1989)

storage and protection in soils referred to as "the earthworm C trap" by Zhang et al. (2013). Martin (1991) shows that the loss of C due to the process of ingestion, transit, and digestion of soil organic matter by the endogeic *Millsonia anomala* is compensated for 100 days by the protection induced with the protection in the aggregates. If the aggregate persists beyond this time, the effect of the earthworm becomes a net protection favoring the increase of C stocks in the soil.

These casts may persist for years if no mechanical effect destroys them. In the Lamto savanna, for example, Blanchart et al. (1997) observed virtually no change in soil macroaggregation done by endogeic earthworms after the worms had been removed by flooding soil blocks 20*20 cm and 40 cm deep and replacing the blocks in soil, isolated from further colonization by a metal mesh.

In natural conditions, soil compaction is prevented by a parallel production of lose earthworm casts of the decompacting species or destruction of these compact structures by ant and termites in their digging of galleries and chambers to house their populations and their eventual mushroom cultures. Observation of a soil thin section from an Amazonian oxisol illustrates the participation of both types of macroaggregates to soil physical properties (Barros et al. 2001; Fig. 12). Diversity in soil ecosystem engineers seems to produce soil structures that allow resistance to erosion and C storage in compact macroaggregates, while the accumulation of loose macroaggregated structures, galleries, and interaggregate macro-porosities allows circulation of water and air.

Although earthworms are expected to help maintain stable but perhaps less important carbon stocks in their drilosphere as would occur in their absence, many factors may interfere. Blanchart et al. (2004), for example, show that the effect of earthworms on soil aggregation is maximal in kaolinite soils but negligible compared to the root-aggregating effect in smectite vertisols. Moreover, in soils richer in organic matter, earthworms ingest less soil and therefore produce less macroaggregates.

The overall effect of earthworms on soil aggregation and C storage and conservation in soils is thus the result of an equilibrium between compacting and decompacting effects, all other conditions being equal. The quality of the organic resources and variations in the microstructure of casts according to species and their feeding regimes probably adds their effects. In agricultural soils of Peruvian Amazonia inoculated with the endogeic *Pontoscolex corethrurus*, the proportion of macroaggregates (>1 cm) increased from 25.1 to 32.7% as compared with control with no earthworms, while the proportion of small macroaggregates (<2 cm) decreased from 33.2 to 26.1%, bulk density increased from 1.12 to 1.23 Mg m(−3), and porosity decreased from 58 to 53%. (Alegre et al. 1996; Blanchart et al. 1999).

In other conditions, the overaccumulation of compact casts where the peregrine endogeic species *Pontoscolex corethrurus* is abundant—and decompacting species are absent—may promote severe soil compaction (Chauvel et al. 1999) (Figs. 12 and 13).

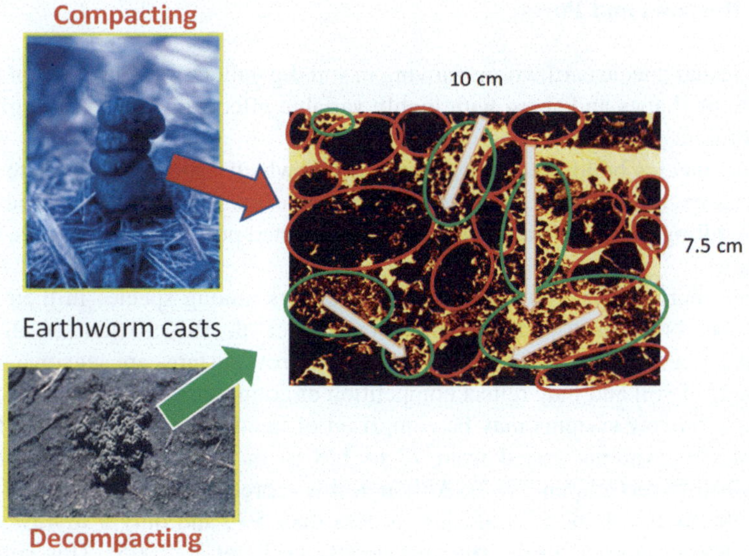

Fig. 12 Distribution of earthworm casts deposited by compacting (red) and decompacting (green) earthworm and other invertebrate species on a thin section of an Amazonian soil. Pale blue arrows suggest water flows in this structure

Fig. 13 Effect of earthworm (*P. corethrurus*) introduction and crop residue application on the proportion of aggregates (macroaggregates and microaggregates) and porosity (macro- and microporosity) in thin sections of soil from Yurimaguas (Loreto, Peru). (A) Surface strata (0–1.5 cm) and (B) deeper strata (1.5–4 cm) (Blanchart et al. 1999)

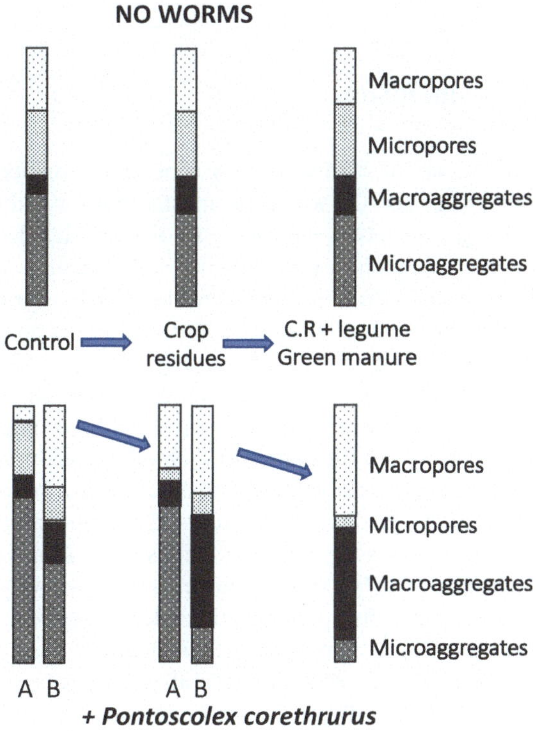

3.3.3 Burrows and Pores

Endogeic and anecic earthworms moving in soil dig galleries and burrows of a great diversity of shapes and sizes, with highly variable effects on soil water infiltration and hydraulic properties.

While anecics build open burrow systems in which water may circulate freely, endogeics excrete the soil they have digested into the gallery they have just dug, partially filling it up and leaving only interaggregated pores of very variable shapes and sizes.

Anecic burrow systems exhibit great variations among species in their depth, orientation, branching, and diameter (Joschko et al. 1993; Jegou et al. 2001; Bastardie et al. 2005). Experiments show that burrow systems are sensitive to local conditions of soil and may reflect competition or commensalism among species.

Anecic burrow systems may be comprised of hundreds of m per m^2. In French pastures, observations varied from 77 to 128 m m^{-3} (Bastardie et al. 2005) to 142–890 m (Kretzschmar 1982). Although this represents total surfaces of 0.7 to 12 m^2, the total volume is equivalent to less than 5%, and only 9 to 43% of this volume was connected to the soil surface (Bastardie et al. 2005). This volume is enough to significantly affect water infiltration, and hydraulic conductivity can be doubled in their presence (Joschko et al. 1993; Bertrand et al. 2015), while increases in infiltration rates of up to 282 mm h-1 per 100 g fw of anecics per m^2 have been measured by Bouché and Al-Addan (1997) in Mediterranean soils.

4 Conclusion

The effects of earthworms as ecosystem engineers have been long studied, with Aristoteles and Darwin as the most famous initial promotors. Their effects as physical engineers have been largely studied and detailed in dozens of papers that describe in great details how they organize soils, creating habitats for a wide range of organisms, from micro to macro, contribute to the storage and conservation of water and organic matter, and interact with roots in this process. The short summary given in this chapter only aims at identifying the major effects and indicates how diverse they are depending on soil types, climates, and ecological strategies. Beyond the wealth of information available, research should now engage toward understanding behaviors that explain the patterns observed. Which part is explained by the genetic heritage, which by interactions with other earthworms of the same or other species, other invertebrates and with plant roots. Communication among earthworms in the compact and supposedly still and silent soil environment is still a total mystery. Do they have territories? Or do they move on a random way? How do the African oligohumic endogeics that live permanently at depths between 30 and 60 cm, and never come to the soil surface, at low densities (sometimes less than one individual per m^2), achieve reproduction? How can they localize a sexual partner at that distance and reach it? Each individual digging its part of a tunnel toward the other one?

Community engineering has benefited enormously of the technical progresses of microbiology. Large amounts of detailed data have been published on the microbiome of the gut and casts of earthworms. However, the vision it gives of the interactions among microorganisms and earthworms is still mostly static, and some results are difficult to interpret. How can microbiomes be so different between the bulk soil, different parts of the gut content, and casts of different ages? Are there autochthonous microbes of the gut, and if it is the case, how can they stay when the gut is traversed by a constant and fast flux of a semi-liquid mixture of water, mineral soil, organic matter, and microorganisms and microfauna ingested with the soil? Another important unknown is the consequence for soil communities of the arrival of peregrine invasive earthworm species. Do they manage to change the microbial community of the soil at ingesting the whole volume of surface soil in a few years? And if so, which consequences does it suppose for native species that developed with a rather specific microbiome, to survive and/or recolonize the soil when suitable environmental conditions have been restored? The example of primary forests in Costa Rica where earthworm community is totally dominated by the invasive *Pontoscolex corethrurus* challenges (Lapied and Lavelle 2003). How can we discard the hypothesis that these forests that were replaced by cropping systems at the Maya time had before the kind of highly diverse and endemic earthworm community as found in the nearby Lacandon forest of Mexico (Fragoso and Lavelle 1987)?

Biochemical engineering is the less known part of earthworm ecosystem engineering. While knowledge of the composition and possible roles of intestinal and cutaneous mucus is available, the physiological bases for their production, metabolism, and possible recuperation in the case of intestinal mucus are not known. Their production has a very energetic cost that needs to be compensated inside the energy budget of earthworms. Energy budgets are only available for a few African earthworms and a single species from temperate areas. Much needs to be done here to better understand the whole system. The other fascinating part of biochemical engineering is the production of signal molecules, hormones, and other molecules involved in signaling pathways that allow earthworms and their drilosphere associates to act upon the expression of genes, in interactions that in the end will benefit the plant.

Beyond such mysteries research should also explore and refine the large conceptual framework and the logics that organizes all these processes. The self-organization theory seems to be a suitable basis to use. Possibly more needs to be imagined to account for all the mysteries of the drilosphere.

References

Adhikari K, Hartemink AE (2016) Linking soils to ecosystem services - a global review. Geoderma 262:101–111

Aira M, Monroy F, Dominguez J (2009) Changes in bacterial numbers and microbial activity of pig slurry during gut transit of epigeic and anecic earthworms. J Hazard Mater 162(2-3):1404–1407

Aira M, Perez-Losada M, Crandall KA, Dominguez J (2022) Composition, structure and diversity of soil bacterial communities before, during and after transit through the gut of the earthworm *Aporrectodea caliginosa*. Microorganisms 10:5

Alegre JC, Pashanasi B, Lavelle P (1996) Dynamics of soil physical properties in Amazonian agroecosystems inoculated with earthworms. Soil Sci Soc Am J 60(5):1522–1529

Barois I, Lavelle P (1986) Changes in respiration rate and somephysicochemical properties of a tropical soil during transit through Pontoscolex corethrurus (Glossoscolecidae, Oligochaeta). Soil Biol Biochem 18(5):539–541

Barois I, Villemin G, Lavelle P, Toutain F (1993) Transformation of the soil structure through Pontoscolex corethrurus (Oligochaeta) intestinal tract. Geoderma 56(1-4):57–66

Barrios E (2007) Soil biota, ecosystem services and land productivity. Ecol Econ 64(2):269–285

Barros E, Curmi P, Hallaire V, Chauvel A, Lavelle P (2001) The role of macrofauna in the transformation and reversibility of soil structure of an oxisol in the process of forest to pasture conversion. Geoderma 100(1-2):193–213

Bastardie F, Capowiez Y, Cluzeau D (2005) 3D characterisation of earthworm burrow systems in natural soil cores collected from a 12-year-old pasture. Appl Soil Ecol 30(1):34–46

Bernard L, Chapuis-Lardy L, Razafimbelo T, Razafindrakoto M, Pablo A-L, Legname E, Poulain J, Bruels T, O'Donohue M, Brauman A, Chotte J-L, Blanchart E (2012) Endogeic earthworms shape bacterial functional communities and affect organic matter mineralization in a tropical soil. ISME J 6(1):213–222

Bertrand M, Barot S, Blouin M, Whalen J, de Oliveira T, Roger-Estrade J (2015) Earthworm services for cropping systems. A review. Agron Sustain Dev 35(2):553–567

Blagodatskaya E, Kuzyakov Y (2008) Mechanisms of real and apparent priming effects and their dependence on soil microbial biomass and community structure: critical review. Biol Fertil Soils 45(2):115–131

Blanchart E, Bruand A, Lavelle P (1993) The physical structure of casts of Millsonia-anomala (Oligochaeta, Megascolecidae) in shrub savanna soils (Cote-divoire). Geoderma 56(1-4):119–132

Blanchart E, Lavelle P, Braudeau E, Le Bissonais Y, Valentin C (1997) Regulation of soil structure by geophagous earthworm activities in humid savannas of Cote d'Ivoire. Soil Biol Biochem 29(3/4):431–439

Blanchart E, Albrecht A, Alegre J, Duboisset A, Pashanasi B, Lavelle P, Brussaard L (1999) In: Lavelle P, Brussaard L, Hendrix P (eds) Effects of earthworms on soil structure and physical properties. Earthworm management in tropical. CAB International, Wallingford, pp 139–162

Blanchart E, Albrecht A, Brown G, Decaens T, Duboisset A, Lavelle P, Mariani L, Roose E (2004) Effects of tropical endogeic earthworms on soil erosion. Agric Ecosyst Environ 104(2):303–315

Blouin M (2018) Chemical communication: an evidence for co-evolution between plants and soil organisms. Appl Soil Ecol 123:409–415

Blouin M, Zuily-Fodil Y, Pham-Thi AT, Laffray D, Reversat G, Pando A, Tondoh J, Lavelle P (2005) Belowground organism activities affect plant aboveground phenotype, inducing plant tolerance to parasites. Ecol Lett 8(2):202–208

Blouin M, Hodson ME, Delgado EA, Baker G, Brussaard L, Butt KR, Dai J, Dendooven L, Peres G, Tondoh JE, Cluzeau D, Brun JJ (2013) A review of earthworm impact on soil function and ecosystem services. Eur J Soil Sci 64(2):161–182

Bossuyt H, Six J, Hendrix PF (2004) Rapid incorporation of carbon from fresh residues into newly formed stable microaggregates within earthworm casts. Eur J Soil Sci 55(2):393–399

Bouche MB, Al-Addan F (1997) Earthworms, water infiltration and soil stability: some new assessments. Soil Biol Biochem 29(3-4):441–452

Bouma J (2014) Soil science contributions towards Sustainable Development Goals and their implementation: linking soil functions with ecosystem services. J Plant Nutr Soil Sci 177(2):111–120

Boyer J, Reversat G, Lavelle P, Chabanne A (2013) Interactions between earthworms, and plant-parasitic nematodes. Eur J Soil Biol 59:43–47

Brown GG (1995) How do earthworms affect microfloral and faunal community diversity. Plant Soil 170(1):209–231

Byzov BA, Khomyakov NV, Kharin SA, Kurakov AV (2007) Fate of soil bacteria and fungi in the gut of earthworms. Eur J Soil Biol 43:S149–S156

Byzov BA, Nechitaylo TY, Bumazhkin BK, Kurakov AV, Golyshin PN, Zvyagintsev DG (2009) Culturable microorganisms from the earthworm digestive tract. Microbiology 78(3):360

Capowiez Y, Bottinelli N, Jouquet P (2014) Quantitative estimates of burrow construction and destruction, by anecic and endogeic earthworms in repacked soil cores. Appl Soil Ecol 74:46–50

Chauvel A, Grimaldi M, Barros E, Blanchart E, Desjardins T, Sarrazin M, Lavelle P (1999) Pasture damage by an Amazonian earthworm. Nature 398(6722):32–33

Cortez J, Bouché MB (1987) Composition chimique du mucus cutané de *Allolobophora chaetophora chaetophora* (Oligochaeta: Lumbricidae). C R Acad Sci Paris 305. III:207–210

De Puytorac P (1954) Contribution a l'etude cytologique et taxonomique des Infusoires. Annales des Sciences Naturelles. Zoologie 11:82–270

Decaëns T, Jimenez JJ, Lavelle P (1999a) Effect of exclusion of the anecic earthworm *Martiodrilus carimaguensis* Jimenez and Moreno on soil properties and plant growth in grasslands of the eastern plains of Colombia. Pedobiologia 43(6):835–841

Decaëns T, Rangel AF, Asakawa N, Thomas RJ (1999b) Carbon and nitrogen dynamics in ageing earthworm casts in grasslands of the eastern plains of Colombia. Biol Fertil Soils 30(1-2):20–28

Decaëns T, Mariani L, Betancourt N, Jimenez JJ (2003) Seed dispersion by surface casting activities of earthworms in Colombian grasslands. Acta Oecol 24(4):175–185

Dominati E, Patterson M, Mackay A (2010) A framework for classifying and quantifying the natural capital and ecosystem services of soils. Ecol Econ 69(9):1858–1868

Drake HL, Horn MA (2006) Earthworms as a transient heaven for terrestrial denitrifying microbes: a review. Eng Life Sci 6(3):261–265

Ecimovic S, Vrandecic K, Kujavec M, Zulj M, Cosic J, Velki M (2021) antifungal activity of earthworm coelomic fluid obtained from Eisenia andrei, Dendrobaena veneta and Allolobophora chlorotica on six species of phytopathogenic fungi. Environments 8:1–11

Eisenhauer N, Butenschoen O, Radsick S, Scheu S (2010) Earthworms as seedling predators: importance of seeds and seedlings for earthworm nutrition. Soil Biol Biochem 42(8):1245–1252

FAO (2020) State of knowledge of soil biodiversity. Status, challenges and potentialities. https://www.fao.org/policy-support/tools-and-publications/resources-details/es/c/1363310/

Fonte SJ, Kong AYY, van Kessel C, Hendrix PF, Six J (2007) Influence of earthworm activity on aggregate-associated carbon and nitrogen dynamics differs with agroecosystem management. Soil Biol Biochem 39(5):1014–1022

Fragoso C, Lavelle P (1987) The earthworm community of a Mexican tropical rainforest (Chajul, Chiapas). In: Bonvicini AM, Omodeo O (eds) On earthworms. Modena, Mucchi, pp 281–297

Furlong MA, Singleton DR, Coleman DC, Whitman WB (2002) Molecular and culture-based analyses of prokaryotic communities from an agricultural soil and the burrows and casts of the earthworm Lumbricus rubellus. Appl Environ Microbiol 68(3):1265–1279

Gange AC, Brown VK (2002) Soil food web components affect plant community structure during early succession. Ecol Res 17(2):217–227

Hoang DTT, Pausch J, Razavi BS, Kuzyakova I, Banfield CC, Kuzyakov Y (2016) Hotspots of microbial activity induced by earthworm burrows, old root channels, and their combination in subsoil. Biol Fertil Soils 52(8):1105–1119

Huang BC, Long J, Li J, Li Y, Wu D, Yang J, Ai YW (2022) Bacterial generalists in earthworm gut had stronger environmental adaptation potential and higher network vulnerability under antimony stress. J Clean Prod 380

Hunt HW, Coleman DC, Ingham ER, Elliott ET, Moore JC, Rose L, Reid CPP, Morley CR (1987) The detrital food web in a shortgrass prairie. Biol Fertil Soils 3:3. https://doi.org/10.1007/bf00260580

Hyodo F, Tayasu I, Konate S, Tondoh J, Lavelle P, Wada E (2008) Gradual enrichment of N-15 with humification of diets in a below-ground food web: relationship between N-15 and diet age determined using C-14. Funct Ecol 22(3):516–522

IPBES (2018) The IPBES assessment report on land degradation and restoration. In: Montanarella L, Scholes R, Brainich A (Eds.) Secretariat of the intergovernmental science-policy platform on biodiversity and ecosystem services, Bonn. 744 pages. doi:https://doi.org/10.5281/zenodo.3237392

Jana U, Barot S, Blouin M, Lavelle P, Laffray D, Repellin A (2010) Earthworms influence the production of above- and belowground biomass and the expression of genes involved in cell proliferation and stress responses in Arabidopsis thaliana. Soil Biol Biochem 42(2):244–252

Jegou D, Schrader S, Diestel H, Cluzeau D (2001) Morphological, physical and biochemical characteristics of burrow walls formed by earthworms. Appl Soil Ecol 17(2):165–174

Jenkinson DS, Fox RH, Rayner JH (1985) Interactions between fertilizer nitrogen and soil nitrogen, The so-called priming effect. J Soil Sci 36(3):425–444

Joschko M, Müller PC, Kotzke K, Döhring W, Larink O (1993) Earthworm burrow system development assessed by means of X-ray computed tomography. Geoderma 56:209–221

Kretzschmar A (1982) Description des galeries de vers de terre et variations saisonnières des réseaux (observations en conditions naturelles). Rev Ecol Biol Sol 19:579–591

Laossi KR, Noguera DC, Bartolome-Lasa A, Mathieu J, Blouin M, Barot S (2009) Effects of an endogeic and an anecic earthworm on the competition between four annual plants and their relative fecundity. Soil Biol Biochem 41(8):1668–1673

Lapied E, Lavelle P (2003) The peregrine earthworm Pontoscolex corethrurus in the East coast of Costa Rica. Pedobiologia 47(5-6):471–474

Lattaud C, Locati S, Mora P, Rouland C, Lavelle P (1998) The diversity of digestive systems in tropical geophagous earthworms. Appl Soil Ecol 9:189–195

Lavelle P (1975) Yearly earth consumption by natural-population of earthworms (Millsonia-anomala Omodeo, Acanthodrilidae, Oligochaeta) in Savanna of Lamto, Ivory-Coast. Revue D Ecologie Et De Biologie Du Sol 12(1):11–24

Lavelle P (1978) Les vers de terre de l savane de Lamto, Peuplements, populations et fonctions dans l'écosystème. Publications du Laboratoire de Zoologie de l'ENS, 12, 301p

Lavelle P (2002) Functional domains in soils. Ecol Res 17(4):441–450

Lavelle P, Spain AV (2001) Soil ecology. Kluwer Scientific Publications, Amsterdam

Lavelle P, Schaefer R, Zaidi Z (1989) Soil ingestion and growth in *Millsonia anomala*, a tropical earthworm, as influenced by the quality of the organic matter ingested. Pedobiologia 33:379–388

Lavelle P, Melendez G, Pashanasi B, Schaefer R (1992) Nitrogen mineralization and reorganization in casts of the geophagous tropical earthworm Pontoscolex corethrurus (Glossoscolecidae). Biol Fertil Soils 14(1):49–53

Lavelle P, Lattaud C, Trigo D, Barois I (1995) Mutualism and biodiversity in soils. Plant Soil 170(1):23–33

Lavelle P, Decaens T, Aubert M, Barot S, Blouin M, Bureau F, Margerie P, Mora P, Rossi JP (2006) Soil invertebrates and ecosystem services. Eur J Soil Biol 42:S3–S15

Lavelle P, Barot S, Blouin M, Decaëns T, Jimenez JJ, Jouquet P (2007) Earthworms as key actors in self organized soilspp 77-100. In: Cuddington K, Byers JE, Wilson WG, Hastings A (eds) Ecosystem engineers. Plants to protists. Academic

Lavelle P, Spain A, Blouin M, Brown G, Decaëns T, Grimaldi M, Jimenez JJ, McKey D, Mathieu J, Velasquez E, Zangerle A (2016) Ecosystem engineers in a self-organized soil: a review of concepts and future research questions. Soil Sci 181(3-4):91–109

Le Bayon RC, Moreau S, Gascuel-Odoux C, Binet F (2002) Annual variations in earthworm surface-casting activity and soil transport by water runoff under a temperate maize agroecosytem. Geoderma 106(1-2):121–135

Li T, Fan J, Qian Z, Yuan G, Meng D, Guo S, Lv W (2020) Predation on weed seeds and seedlings by Pheretima guillelmi and its potential for weed biocontrol. Weed Sci 68(6):639–645. https://doi.org/10.1017/wsc.2020.65
Loquet M, Bhatnagar T, Bouché MB, Rouelle J (1977) Essai d'estimation de l'influence écologique des lombriciens sur les microorganismes. Pedobiologia 17(6):400–417
Loranger G, Ponge JF, Blanchart E, Lavelle P (1998) Impact of earthworms on the diversity of microarthropods in a vertisol (Martinique). Biol Fertil Soils 27:21–26
Loranger-Merciris G, Cabidoche YM, Delone B, Queneherve P, Ozier-Lafontaine H (2012) How earthworm activities affect banana plant response to nematodes parasitism. Appl Soil Ecol 52:1–8
Lubbers IM, van Groenigen KJ, Fonte SJ, Six J, Brussaard L, van Groenigen JW (2013) Greenhouse-gas emissions from soils increased by earthworms. Nat Clim Chang 3(3):187–194
Maerz JC, Nuzzo VA, Blossey B (2009) Declines in woodland salamander abundance associated with non-native earthworm and plant invasions. Conserv Biol 23(4):975–981
Marinissen JCY, Bok J (1988) Earthworm-amended soil structure: its influence on Collembola populations in grassland. Pedobiologia 32:243–252
Martin A (1991) Short- and long-temr effects of the endogeic earthworm *Millsonia anomala* (Omodeo) (Megascolescidae, Oligochaeta) of tropical savannas, on soil organic matter. Biol Fertil Soils 11:134–138
Martin A, Cortez J, Barois I, Lavelle P (1987) The production of intestinal mucus by earthworms. A key process in their interactions with the soil microflora. Revue d'Ecologie et de Biologie du Sol 24(4):549–558
Martin A, Mariotti A, Balesdent J, Lavelle P (1992) Soil organic matter assimilaion by a geophagous tropical earthworm based on $\delta^{13}C$ measurments. Ecology 73(1):118–128
McCay TS, Scull P (2019) Invasive lumbricid earthworms in northeastern North American forests and consequences for leaf-litter fauna. Biol Invasions 21(6):2081–2093
MEA (2005) Millennium ecosystem assessment. www.millenniumassessment.org
Medina-Sauza RM, Alvarez-Jimenez M, Delhal A, Reverchon F, Blouin M, Guerrero-Analco JA, Cerdan CR, Guevara R, Villain L, Barois I (2019) Earthworms building up soil microbiota, a review. Front Environ Sci 7
Meier AB, Hunger S, Drake HL (2018) Differential engagement of fermentative taxa in gut contents of the earthworm Lumbricus terrestris. Appl Environ Microbiol 84:5
Meng LL, Srivastava AK, Kuča K, Bhoopander G, Rahman MM, Wu QS (2021) Interaction between earthworms and arbuscular mycorrhizal fungi in plants: a review. Phyton 90(3):687–699
Milcu A, Schumacher J, Scheu S (2006) Earthworms (Lumbricus terrestris) affect plant seedling recruitment and microhabitat heterogeneity. Funct Ecol 20(2):261–268
Monroy F, Aira M, Dominguez J (2008) Changes in density of nematodes, protozoa and total coliforms after transit through the gut of four epigeic earthworms (Oligochaeta). Appl Soil Ecol 39(2):127–132
Nadana GRV, Rajesh C, Kavitha A, Sivakumar P, Sridevi G, Palanichelvam K (2020) Induction of growth and defense mechanism in rice plants towards fungal pathogen by eco-friendly coelomic fluid of earthworm. Environ Technol Innov 19:101011
Nooren CAM, van Breemen N, Stoorvogel JJ, Jongmans AG (1995) The role of earthworms in the formation of sandy surface soils in a tropical forest in Ivory Coast. Geoderma 65:135–148
Nye PH (1955) Some soil-forming processes in the humid tropics. IV - the action of the soil fauna. J Soil Sci 6. 1:73–83
OECD (2019) Trends and drivers of agri-environmental performance in OECD countries. Éditions
Parle JN (1963) Micro-organisms in the intestines of earthworms. J Gen Microbiol 31:1–11
Piearce TG, Phillips MJ (1980) The fate of ciliates in the earthworm gut. An in vitro study. Microb Ecol 5:313–319
Pitkanen J, Nuutinen V (1998) Earthworm contribution to infiltration and surface runoff after 15 years of different soil management. Appl Soil Ecol 9(1-3):411–415

Pokarzhevskii AD, Zaboyev DP, Ganin GN, Gordienko SA (1997) Amino acids in earthworms: Are earthworms ecosystemivorous? Soil Biol Biochem 29(3-4):559–567

Potapov AM, Tiunov AV, Scheu S (2019) Uncovering trophic positions and food resources of soil animals using bulk natural stable isotope composition. Biol Rev 94(1):37–59

Potapov A, Lindo Z, Buchkowski R, Geisen S (2022) Multiple dimensions of soil food-web research: history and prospects. Eur J Soil Biol 117:103494

Potapov M, Buchkowski R, Geisen S, Lindo Z (2023) Back and better: Soil food-web researchers integrate empirical data and develop novel tools. Eur J Soil Biol, 119:103562.

Puga-Freitas R, Barot S, Taconnat L, Renou JP, Blouin M (2012) Signal molecules mediate the impact of the earthworm Aporrectodea caliginosa on growth, development and defence of the plant Arabidopsis thaliana. PLoS One 7:12

Regnier E, Harrison SK, Liu J, Schmoll JT, Edwards CA, Arancon N, Holloman C (2008) Impact of an exotic earthworm on seed dispersal of an indigenous US weed. J Appl Ecol 45(6):1621–1629

Reddell P, Spain AV (1991) Earthworms as vectors of viable propagules of mycorrhizal fungi. Soil Biology and Biochemistry 23 (8):767–774.

Rockström J, Steffen W, Noone K, Persson Å, Chapin FS III, Lambin E, Lenton TM, Scheffer M, Folke C, Schellnhuber H, Nykvist B, De Wit CA, Hughes T, van der Leeuw S, Rodhe H, Sörlin S, Snyder PK, Costanza R, Svedin U, Falkenmark M, Karlberg L, Corell RW, Fabry VJ, Hansen J, Walker B, Liverman D, Richardson K, Crutzen P, Foley J (2009) Planetary boundaries: exploring the safe operating space for humanity. Ecol Soc 14(2):32

Rouelle J, Pussard M, Randriamamonjizaka JL (1985) Interactions microbiennes (Bactéries, Protozoaires), alimentation des vers de terre et minéralisation de la matiere organique. Bulletin d¨Écologie 16, 1:83–88

Sacca ML, Caracciolo AB, Di Lenola M, Grenni P (2017) eEcosystem services provided by soil microorganisms. 3rd annual meeting of the cost-action-fp1305-biolink on linking belowground biodiversity and ecosystem function in European forests. Soil Biological Communities and Ecosystem resilience, pp 9–24

Salmon S, Ponge JF (2001) Earthworm excreta attract soil springtails: laboratory experiments on Heteromurus nitidus (Collembola: Entomobryidae). Soil Biol Biochem 33(14):1959–1969

Sampedro L, Whalen JK (2007) Changes in the fatty acid profiles through the digestive tract of the earthworm Lumbricus terrestris L. Appl Soil Ecol 35(1):226–236

Shan J, Liu J, Wang YF, Yan XY, Guo HY, Li XZ, Ji R (2013) Digestion and residue stabilization of bacterial and fungal cells, protein, peptidoglycan, and chitin by geophagous earthworm Metaphire guillelmi. Soil Biol Biochem 64:9–17

Shipitalo MJ, Protz R (1988) Factors influencing the dispersibility of clay in worm casts. Soil Sci Soc Am J 52(3):764–769

Shipitalo MJ, Protz R (1989) Chemistry and micromorphology of aggregation in earthworm casts. Geoderma 45:357–374

Shutenko GS, Kelleher BP, Simpson AJ, Soong R, Mobarhan YL, Schmidt O (2020) Evidence for substantial acetate presence in cutaneous earthworm mucus. J Soils Sediments 20(10):3627–3632

Six J, Elliott ET, Paustian K (2000) Soil macroaggregate turnover and microaggregate formation: a mechanism for C sequestration under no-tillage agriculture. Soil Biol Biochem 32(14):2099–2103

Smith P, Cotrufo ME, Rumpel C, Paustian K, Kuikman PJ, Elliott JA, McDowell R, Griffiths RI, Asakawa S, Bustamante M, House JI, Sobocka J, Harper R, Pan G, West PC, Gerber JS, Clark JM, Adhya T, Scholes RJ, Scholes MC (2015) Biogeochemical cycles and biodiversity as key drivers of ecosystem services provided by soils. Soil 1(2):665–685

Snyder BA, Boots B, Hendrix PF (2009) Competition between invasive earthworms (Amynthas corticis, Megascolecidae) and native North American millipedes (Pseudopolydesmus erasus, Polydesmidae). effects on carbon cycling and soil structure. Soil Biol Biochem 41(7):1442–1449

Spain AV, Safigna PG, Wood AW (1990) Tissue carbon sources for *Pontoscolex corethrurus* (Oligochaeta, Glossoscolecidae) in a sugarcane ecosystem. Soil Biol Biochem 22(5):703–706

Spohn M, Klaus K, Wanek W, Richter A (2016) Microbial carbon use efficiency and biomass turnover times depending on soil depth - implications for carbon cycling. Soil Biol Biochem 96:74–81. https://doi.org/10.1016/j.soilbio.2016.01.016

Swaby RJ (1950) The influence of earthworms on soil aggregation. J Soil Sci 1(2):195

Swift MJ, Anderson JM (1994) Biodiversity and ecosystem function in agricultural systems. In: Schulze ED, Mooney HA (eds) Biodiversity and ecosystem function. Springer, Berlin, pp 15–41

Swift MJ, Heal OW, Anderson GM (1979) Decomposition in terrestrial ecosystems. Blackwell, Scientific Publications, Oxford, 372p

Tisdall JM, Oades JM (1982) Organic matter and water-stable aggregates in soils. J Soil Sci 33:141–163

Tötsche KU, Amelung W, Gerzabek MH, Guggenberger G, Klump E, Knief C, Lehndorff E, Mikutta R, Peth S, Prechtel A, Ray N, Kogel-Knabner I (2018) Micro-aggregates in soils. J Plant Nutr Soil Sci 181(1):104–136

Trigo D, Barois I, Garvin MH, Huerta E, Irisson S, Lavelle P (1999) Mutualism between earthworms and soil microflora. Pedobiologia 43(6):866–873

UNCCD (2016). https://www.unccd.int/resources/global-land-outlook/overview

UNDP (2015) UN sustainable development goals. https://www.undp.org/sustainable-development-goals

UNEP (2016) Frontiers 2016: Emerging issues of environmental concern. www.unep.org › resources › frontiers-2016

Van Groenigen JW, Van Groenigen KJ, Koopmans GF, Stokkermans L, Vos HMJ, Lubbers IM (2019) How fertile are earthworm casts? A meta-analysis. Geoderma 338:525–535

Wilkinson MT, Richards PJ, Humphreys GS (2009) Breaking ground: Pedological, geological, and ecological implications of soil bioturbation. Earth Sci Rev 97(1-4):257–272

Wolter C, Scheu S (1999) Changes in bacterial numbers and hyphal lengths during the gut passage through Lumbricus terrestris (Lumbricidae, Oligochaeta). Pedobiologia 43(6):891–900

Xia R, Shi Y, Wang XW, Wu YL, Sun MM, Hu F (2022) Metagenomic sequencing reveals that the assembly of functional genes and taxa varied highly and lacked redundancy in the earthworm gut compared with soil under vanadium stress. Msystems 7:1

Zangerle A, Pando A, Lavelle P (2011) Do earthworms and roots cooperate to build soil macroaggregates? A microcosm experiment. Geoderma 167-68:303–309

Zangerle A, Hissler C, Blouin M, Lavelle P (2014) Near infrared spectroscopy (NIRS) to estimate earthworm cast age. Soil Biol Biochem 70:47–53

Zhang WX, Hendrix PF, Dame LE, Burke RA, Wu JP, Neher DA, Li JX, Shao YH, Fu SL (2013) Earthworms facilitate carbon sequestration through unequal amplification of carbon stabilization compared with mineralization. Nat Commun 4

Zhang H, Tan SN, Teo CH, Yew YR, Ge LY, Chen X, Yong JWH (2015) Analysis of phytohormones in vermicompost using a novel combinative sample preparation strategy of ultrasound-assisted extraction and solid-phase extraction coupled with liquid chromatography-tandem mass spectrometry. Talanta 139:189–197

Zhu GF, Schmidt O, Luan L, Xue JR, Fan JB, Geisen S, Sun B, Jiang YJ (2022) Bacterial keystone taxa regulate carbon metabolism in the earthworm gut. Microbiol Spect

Advances in Earthworms as Biological Remediators of Polluted Soils

Juan C. Sanchez-Hernandez, Angie M. Pedraza Torres, Natividad I. Navarro Pacheco, and Mallavarapu Megharaj

Abstract Agricultural soils receive many environmental contaminants through organic fertilization (e.g., composts, animal manure, and biosolids), irrigation with wastewater, and chemical control of pests. Thus, it is not surprising to detect a wide range of pollutants in soil, such as metals, nanoparticles, microplastics, pesticides, polycyclic aromatic hydrocarbons, polybrominated diphenyl ethers, phthalates, pharmaceuticals, personal care products, or perfluorinated substances. Furthermore, predictive models indicate that soil pollution will be exacerbated by global climate change. Therefore, transitioning to an organic agricultural system and environmentally friendly, bio-based soil remediation systems can be suitable measures to improve soil health and resilience. Accordingly, inoculating agricultural soils with earthworms or stimulating the native population of these organisms could be seen as an opportunity to increase the natural attenuation of soil against contaminant input and as a bioremediation strategy. To propose the use of earthworms in the remediation of polluted agricultural soils (vermiremediation), this chapter provides an overview of the main physicochemical and biological impacts of earthworms in agricultural soils. It then moves on to the drilosphere concept and the relationship between soil microorganisms and earthworm gut symbionts for removing pollutants. Vermiremediation depends on earthworm sensitivity to pollutants and bioaccumulation capacity, which are discussed in the context of earthworm ecotypes. A final section includes new integrative approaches in which earthworms interact with other biotic factors (exogenous microorganisms and mesofauna) and abiotic factors (biochar or hydrochar) to improve vermiremediation.

J. C. Sanchez-Hernandez (✉) · A. M. Pedraza Torres · N. I. Navarro Pacheco
Laboratory of Ecotoxicology, Institute of Environmental Sciences, University of Castilla-La Mancha, Toledo, Spain
e-mail: juancarlos.sanchez@uclm.es

M. Megharaj
Global Centre for Environmental Remediation, College of Engineering, Science and Environment, University of Newcastle, Callaghan, NSW, Australia

Y. Kooch, Y. Kuzyakov (eds.), *Earthworms and Ecological Processes*,
https://doi.org/10.1007/978-3-031-64510-5_19

1 Introduction

Intensive agriculture is characterized by short-rotation crops receiving high synthetic agrochemicals to maximize crop yields and profitability (Shennan et al. 2017). However, this agricultural system leads to a loss of soil fertility and biodiversity in the long term (Tsiafouli et al. 2015; Kopittke et al. 2019). Likewise, predictive models point out that climate change exacerbates the environmental impact of pesticides because of a higher geographical incidence of pesticide-resistant species (Ma et al. 2021) and, consequently, increased pesticide inputs (Noyes et al. 2009). Furthermore, new situations of environmental and food safety emerge in which agricultural soils receive a cocktail of pollutants via crop irrigation with reclaimed wastewater and soil fertilization with composts, biosolids, and animal manure. Thus, it is not surprising to detect residues of a wide range of pollutants in soil other than pesticides, such as metals, polycyclic aromatic hydrocarbons, polybrominated flame retardants, phthalates, pharmaceuticals, perfluorinated substances, or microplastics (<5 mm plastic fragments). For example, a pesticide survey study performed on agricultural soils from the European Union found that over 80% of soil samples contained more than one pesticide residue, glyphosate (and its metabolites), boscalid, epoxiconazole, and tebuconazole the most frequently detected pesticides (Silva et al. 2019). More recently, a field study performed in agricultural soils from the Atlantic Macaronesia archipelagos, including the Azores Islands, Canary Islands, Cape Verde, and Madeira, revealed that agricultural soils from the Canary Islands contained up to 109 pesticide residues and 4 pharmaceuticals (eprinomectin, sulfadiazine, fenbendazole, and oxfendazole) from a total of 310 organic contaminants including pesticides, persistent organic pollutants, and active pharmaceutical compounds (Acosta-Dacal et al. 2022). Likewise, the use of plastic materials in agriculture (e.g., mulch films, drip irrigation tapes, low tunnels, soil solarization films, or greenhouse covers) is a significant source of microplastics in the agroecosystems (Ng et al. 2018) and further contaminates the soil with chemical additives such as plasticizers or flame retardants (Scopetani et al. 2022). Currently, emphasis is placed on transitioning to an organic agricultural system marked, for instance, by integrated pest management programs and other environmentally friendly measures (e.g., replacing inorganic fertilizers with organic amendments) that improve soil health and reduce contaminant inputs (Padilla-Sánchez et al. 2015). Accordingly, bio-based strategies are gaining increasing attention in the scheme of sustainable agriculture transition (Plaas et al. 2019; Ning et al. 2022). In this context, inoculating agricultural soils with earthworms or stimulating native populations of these organisms could be an opportunity to increase the natural attenuation of soil against contaminant input.

This chapter proposes earthworms as biological remediators of polluted soils (so-called vermiremediation) and drivers for promoting soil resilience against chemical stressors. Some past review articles have already addressed the topic (Hickman and Reid 2008; Rodriguez-Campos et al. 2014), and four main highlights can be drawn: (i) earthworm activity increases the bioavailability of organic pollutants for biodegradation via the destruction of soil microaggregates during the gastrointestinal passage of soil; (ii) earthworm casts are a sink for organic pollutants because of the high organic matter content and fine-sized particles they contain; (iii) ecological characteristics of earthworms, particularly feeding habits and burrowing ability, are critical for removing pollutants; and (iv) environmental variables such as temperature, soil moisture, soil texture, and food availability are important drivers in earthworm activity and survival, being decisive in vermiremediation viability. This chapter does not attempt a comprehensive systematic review of vermiremediation, but based on the knowledge gained so far, it gives an updated and integrated perspective of how earthworms may remove or immobilize a group of environmental contaminants called "emerging pollutants" (i.e., chemicals detected in the environment but currently not included in regulatory monitoring programs and whose fate and biological impacts are poorly understood) (Hutchinson et al. 2013). The first section describes the soil ecosystem services in which earthworms have a significant contribution (Sect. 2) and then moves to the concept of drilosphere and its main structural and functional elements (Sect. 3). Next, we describe how this soil domain contributes to pollutant fate through the interaction of earthworms with microorganisms (Sect. 4) and the bioaccumulation and detoxification of emerging pollutants by earthworms (Sect. 5). We introduce the concept of integrated bioremediation considering two approaches: a multiple bio-based strategy and a mixed biotic-abiotic strategy (Sect. 6). Finally, a set of concluding remarks highlight the main knowledge gaps and technical issues that should be considered for the field application of vermiremediation.

2 Earthworm-Derived Ecosystem Services

Earthworms impact multiple physical, chemical, and biological parameters of soil function, such as pedogenesis, development and maintenance of soil structure, water regulation, nutrient cycling, biomass production, climate regulation, pollution remediation, and biodiversity (Blouin et al. 2013), which have direct and indirect implications on soil ecosystem services (Fig. 1). These impacts are the consequence of the earthworm bioturbation, i.e., "the reworking of soils by all kinds of organisms, including microbes, rooting plants, and burrowing animals" (Meysman et al. 2006).

Geophagous and detritivorous diets of earthworms contribute to humus formation in topsoil, transfer of organic matter from the soil surface into the subsoil, carbonate supply, and mineral and mechanical weathering (Georgiadis et al. 2019; Cunha et al. 2016; Zamanian et al. 2016; Andriuzzi et al. 2016); all these processes contribute to soil formation (Guhra et al. 2020). Moreover, earthworm burrowing and feeding activities change the aggregate distribution altering the soil structure (Shipitalo and Le Bayon 2004). In general, these organisms modify the proportion of the existing soil micro- and macroaggregates (Blouin et al. 2013) and create biogenic aggregates (i.e., formed under the influence of biological processes, including

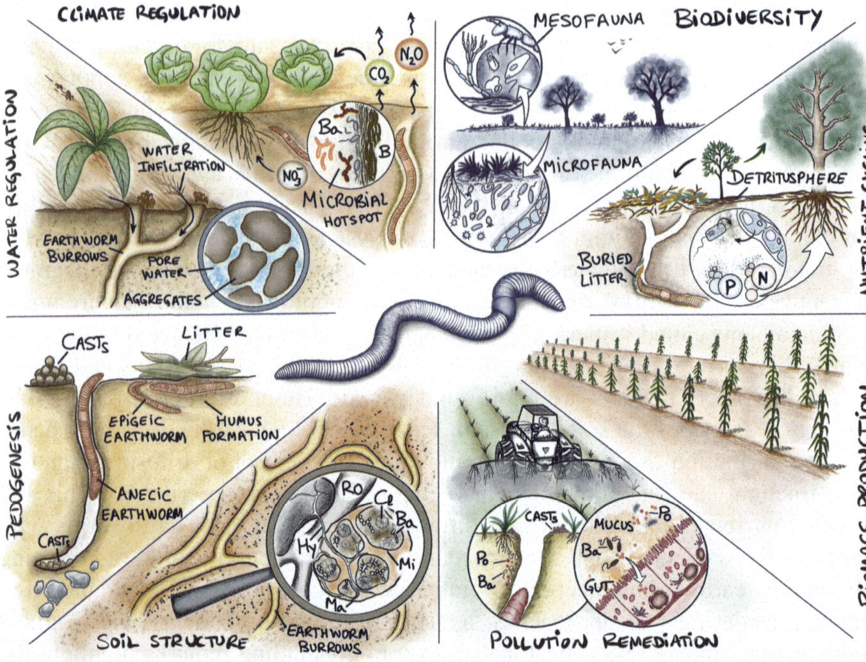

Fig. 1 An illustrative depiction of ecosystem services derived from the soil, highlighting the significant contributions of earthworms. *B* burrow wall, *Ba* bacteria, *Cl* clays; *Hy* hyphae, *Ma* macroaggregates, *Mi* microaggregates, *N* nitrogen, *Ro* root, *P* phosphorus, *Po* pollutants. Illustration elaborated by Juan C. Sanchez-Hernandez

earthworm activity and root and fungal exudates together to organism-induced mechanical stress (Guhra et al. 2022)). Earthworm-induced macroporosity in soil and biogenic aggregates indirectly affects soil aeration, hydraulic properties, and plant root development (Capowiez et al. 2014; Xiao et al. 2018). Water dynamic in the soil is significantly affected by earthworms, particularly those species that construct permanent, open surface, vertical burrows (anecic earthworms) (Blouin et al. 2013). From a chemical viewpoint, the feeding activity of earthworms enhances organic matter decomposition, alters nutrient cycling dynamics, stabilizes carbon, and facilitates nutrient uptake by plants and microorganisms (Rämert et al. 2001). Indeed, earthworms generally increase soil microbial activity and change microbial community structure, which is a consequence of casts (feces) rich in mucus and labile carbon molecules, the secretion of skin mucus, the selective foraging of earthworms on microorganisms, and the dispersion of microorganisms in the soil (Andriuzzi et al. 2016). Some studies have also proven that earthworm activity, via its influence on soil microbiota, contributes to the dynamic of greenhouse gas emissions from soil (Lubbers et al. 2013). Furthermore, earthworms influence soil meso- and macrofauna diversity (Stromberger et al. 2012; Stroud et al. 2016; Butt and Lowe 2007) and indirectly impact aboveground organisms (Wurst 2010). Even there is solid evidence that earthworms decrease soil-borne pathogens such as nematodes (Demetrio et al. 2019). These physicochemical and biological impacts collectively benefit plant growth and development (Van Groenigen et al. 2019).

Nevertheless, the impact of earthworms on soil function largely depends on factors such as earthworm species, population density, soil type, available food, climatic conditions, and soil mechanical state (Ruiz et al. 2021). The classical view of soil ecology classifies earthworms into three ecotypes: epigeic, anecic, and endogeic (Fig. 2). Endogeic (live in the subsoil creating temporary horizontal burrows and feeding on mineral soil to obtain nutrients) and anecic earthworms (live in the subsoil creating permanent vertical burrows and feed from the litter that they collect from the soil surface and drag into the burrow) have a higher impact on soil structure than epigeic (inhabit the soil surface and feed on litter) (Medina-Sauza et al. 2019). However, the latter significantly alters organic matter decomposition at the topsoil, whereas anecics have a prominent role in C translocation from the topsoil to the subsoil (Hoang et al. 2020). While earthworms have species-specific impacts on soil, they generally offer multiple benefits to soil-derived ecosystem functions and services (Fig. 1).

One of the main threats to soil health is chemical contamination (Montanarella and Panagos 2021). Pollutants contribute to soil degradation and compromise many

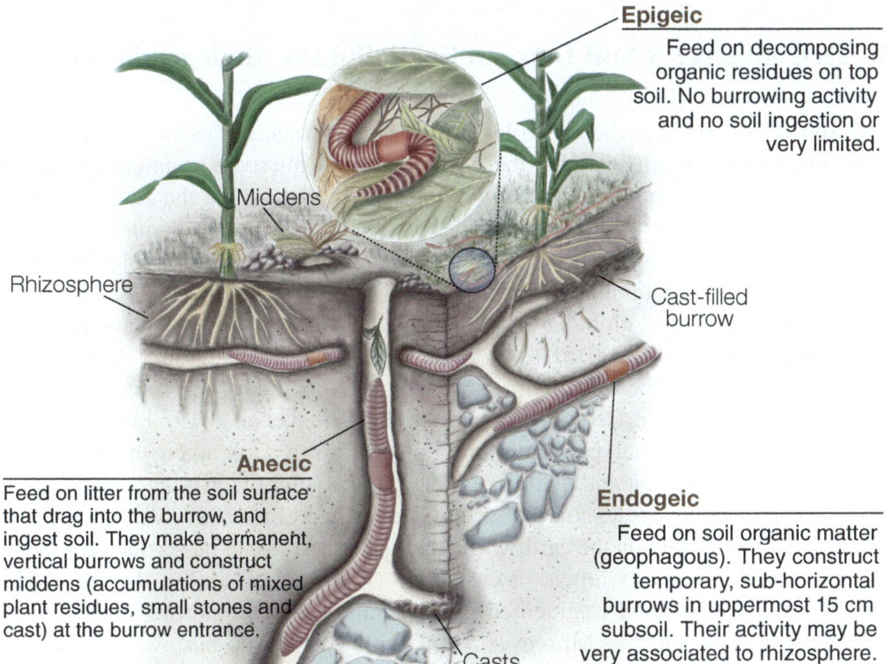

Fig. 2 Main features of earthworm ecotypes: epigeic, endogeic, and anecic. Illustration adapted with permission from Sanchez-Hernandez et al. (2023)

soil-associated ecosystem services such as biomass production, biodiversity conservation, pest and disease control, water quality and supply, and climate regulation (Bünemann et al. 2018). To fight against soil pollution, in situ bioremediation is a viable option to recover soil health or increase its resilience potential. Various engineered and biological remediation methodologies are used to clean up contaminated soils (Megharaj et al. 2011); however, bio-based strategies are highly recommended due to their lower adverse effects than those of engineered remediation. Soil bioremediation uses a variety of biological strategies to remove, degrade, or immobilize pollutants, such as the application of nutrient-rich amendments (e.g., composts) to activate native soil microbial communities (*biostimulation*), soil inoculation with lab-scale cultivated microorganisms with enhanced catalytic potential for contaminant degradation (*bioaugmentation*), plants (*phytoremediation*), free contaminant-degrading enzymes stabilized on physical supports (*enzymatic* remediation), and soil macrofauna as earthworms (*vermiremediation*) (Morillo and Villaverde 2017; Megharaj et al. 2011). Particularly, vermiremediation is gaining research interest because of the earthworm benefits in soil function other than pollutant removal (Zeb et al. 2020). In the following sections, the direct and indirect impacts of earthworms on pollutant degradation and immobilization are described with some studies as examples.

3 Drilosphere: A Soil Domain for Pollutant Biodegradation

Drilosphere is defined as the soil volume under the influence of earthworm activity. This functional domain includes the gastrointestinal microenvironment of earthworms and the biogenic structures they create in soil, which are the casts excreted on the soil surface and in the burrow, the burrow walls, and the middens (Brown et al. 2000). Middens are deposits of litter and small stones mixed with casts that anecic earthworms accumulate around the burrow's entrance (Stroud et al. 2016).

Casts are expelled as globular or granular pellets on the soil surface and the burrow wall (Brown et al. 2000). They contain high fresh organic matter content and nutrients compared to the surrounding soil, which favors microbial proliferation and increases organic matter decomposition processes (Vidal et al. 2019). Indeed, earthworm casts are potential plant-nutrient hotspots (Van Groenigen et al. 2019). Nevertheless, the amount and quality of the organic matter of casts and their microbiota largely depend on the soil and food types (Abail et al. 2017; Van Groenigen et al. 2019). Depending on the earthworm ecotype, burrows can be vertical, horizontal, permanent, temporary, open, or cast-filled (Brown et al. 2000). Burrow walls contain casts, mucus, and fragmented plant residues that earthworms (anecic species) drag into the burrow. Moreover, plant roots can invade burrows, increasing microbial activity and fostering nutrient cycling (Ning et al. 2022), which is concentrated in the first few millimeters from the burrow edge (Hoang et al. 2020). Middens are also biologically active structures that may develop a community of microorganisms and mesofauna (Stroud et al. 2016). Because middens are formed by plant litter that anecic earthworms accumulate, they can be considered part of the detritusphere (i.e., organic matter decomposition from litter accumulated on the soil surface and around plant root death) (Kuzyakov and Blagodatskaya 2015).

The earthworm gut microenvironment is another active component of the drilosphere. Earthworms ingest large amounts of organic matter and mineral soil, so there is a permanent mucus production to facilitate the transit of the ingested material. This intestinal mucus promotes symbiont activity, triggering a digestive mutualism between earthworms and gut microbiota (Brown et al. 2000). Microorganisms participate in the digestive processes of earthworms (Brown et al. 2000; Drake and Horn 2007), and, in turn, earthworms feed on certain soil microorganisms such as fungi (Song et al. 2020; Bonkowski et al. 2000). This particular trophic relationship modifies the microbial community structure in the gut and the soil (casts and burrow walls).

Figure 3 illustrates a conceptual model of the drilosphere elements, where the biological activity is more intense, and the decomposition and mineralization processes take place at a high rate. This leads us to hypothesize whether these hotspots of biological activity are drivers of pollutant degradation and immobilization. In

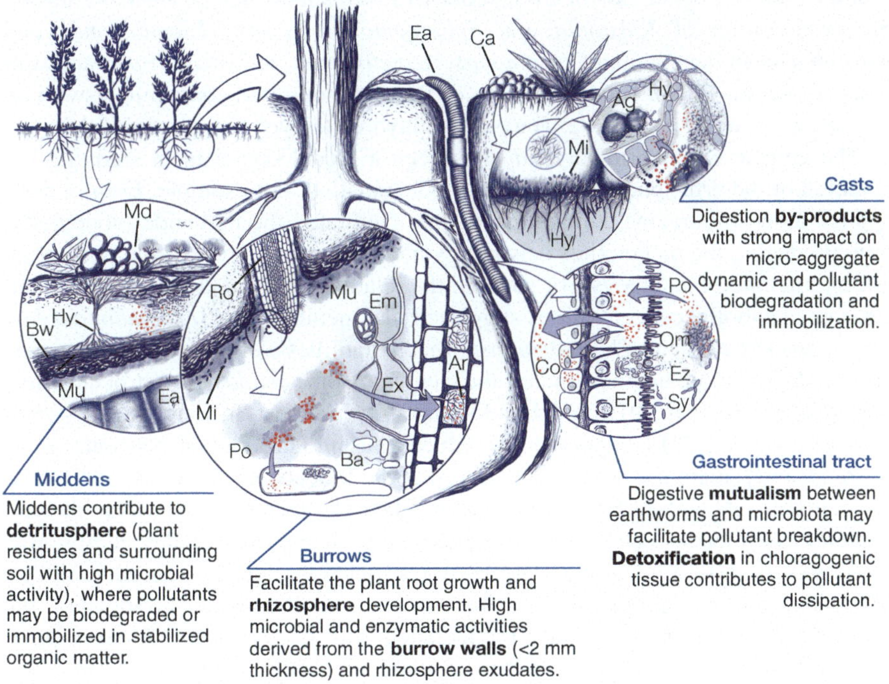

Fig. 3 Drilosphere hotspots of microbial activity and pollutant fate. *Ag* biogenic aggregates, *Ar* arbuscule, *Ba* bacteria, *Bw* burrow wall, *Ca* casts, *Co* coelomic cavity, *Ea* earthworm, *Em* endomycorrhiza, *En* enterocytes, *Ex* exudates, *Ez* enzymes, *Hy* hyphae, *Md* midden, *Mi* microorganisms, *Mu* mucus, *Om* organic matter, *Po* pollutants, *Ro* root, *Sy* symbionts. Pollutants are represented by red spots. Illustration elaborated by Juan C. Sanchez-Hernandez

fact, some studies strengthen the functional importance of the drilosphere in bioremediation. Although knowledge of the mechanisms underpinning the pollutant breakdown in the drilosphere is still scant, it seems that pollutant biodegradation between earthworm gut and air-exposed biogenic structure is markedly different. For example, the herbicide 2-methyl-4-chlorophenoxyacetic acid degraded faster in the burrow walls (<2 mm thickness) created by *Aporrectodea caliginosa* than in the surrounding soil, involving the alphaproteobacteria *Sphingomonas* and *Bradyrhizobium* in the aerobic degradation of the herbicide (Liu et al. 2011a). However, the anoxic environment of the earthworm gut likely impedes herbicide biodegradation, making it doubtful for the gut to contribute to the breakdown of these substances (Liu et al. 2011b). Similarly, the metabolite profile derived from l,l,l-trichloro-2,2-bis(4-chloropenyl)ethane (DDT) biodegradation by the endogeic earthworm *Amynthas robustus* differed depending on the drilosphere element. Aerobic dehydrochlorination of DDT into l,l-dichloro-2,2-bis(4-chlorophenyl)ethylene (DDE) mainly occurred in burrow walls. In contrast, DDT and DDE were anaerobically dechlorinated to l,l-dichloro-2,2-bis(4-chlorophenyl)ethane and 1-chloro-2,2-bis(4-chlorophenyl)ethylene (Xu et al. 2019). Microorganisms play an important role in this marked difference in pollutant biodegradation pathways. For example, the biodegradation of tetracycline by *Eisenia fetida* and *A. robustus* needed the proliferation of *Ralstonia* and *Sphingomonas* in soil, *Pseudomonas* and *Arthrobacter* in the gut, and *Comamonas*, *Acinetobacter*, and *Stenotrophomonas* in casts (Lin et al. 2021). Biodegradation of this antibiotic was concomitant with an increase in its main metabolites (i.e., 4-epianhydrotetracycline hydrochloride).

The epigeic earthworm *E. fetida* has been a model organism in studying the potential of the drilosphere in pollutant biodegradation. For example, *Eisenia* casts applied alone to atrazine-contaminated soils facilitated the herbicide biodegradation, producing the metabolites deisopropylatrazine and deethylatrazine (Luo et al. 2022). The microbial analysis of soil and the relationships found with soil pH, and atrazine metabolite concentrations suggested that members of *Pseudomonadaceae*, *Streptomycetaceae*, and *Thermomonosporaceae* families were responsible for atrazine biodegradation. In other studies, the incubation of *E. fetida* adults in acetochlor-spiked soils (Hao et al. 2018) and the addition of *E. fetida* casts in nonylphenol-spiked soils (Yang et al. 2014) significantly reduced the initial pollutant concentrations. More recently, a similar study evidenced that the inoculation of acetochlor-spiked agricultural soils with *E. fetida* increased the degradation rate of the herbicide with respect to sterilized or earthworm-free soils, and such enhanced biodegradation was progressively more evident after successive pesticide treatments (Han et al. 2021). Moreover, the biodegradation of this herbicide was enantiomer-specific, the *S*-enantiomer being preferentially biodegraded with respect to the *R*-enantiomer. The earthworm stimulated the growth of acetochlor microbial degraders (*Lysobacter*, *Kaistobacter*, *Flavobacterium*, *Arenimonas*, *Aquicell*, *Aeromonas*, and *Algoriphagus*), which caused a generalized increase in soil microbial activity linked to using small molecular weight compounds (Han et al. 2021). However, quantifying metabolites to confirm pollutant breakdown was not performed in those studies, leaving open the possibility of chemically low availability of pollutants instead of true biodegradation. Indeed, *E. fetida* casts have a high retention capacity against nonylphenol, thus reducing the bioavailability and biodegradation of this surfactant (Jiang et al. 2018).

Vermiremediation should not be considered as a mere method of removing pollutants but as a bio-based approach that promotes biological processes to boost soil resilience. This concept is shown in the study by Sun et al. (2020), which found that *E. fetida* played a key role in the biodegradation of the herbicide metolachlor. The researchers also observed that earthworms stimulated the growth of fungi, including *Fusarium*, *Metarhizium*, *Trichoderma*, and *Stachybotrys*, which positively correlated with soil enzyme activities. Moreover, *E. fetida* was shown to alleviate the toxicity of metolachlor on soil microorganisms and restore the soil microbial network to the levels of the control treatments. Similar findings were previously found in chlorpyrifos-spiked agricultural soils in the presence of *Lumbricus terrestris* (Sanchez-Hernandez et al. 2018). The burrowing activity of this species recovered the activity of esterases, phosphatases, and β-glucosidases which were previously inhibited by the organophosphorus pesticides. In another study, the interaction between *E. fetida* and the arbuscular mycorrhizal fungi *Rhizophagus intraradices* increased the biodegradation rate of the veterinary antibiotic oxytetracycline (Cao et al. 2015). The earthworms not only favored the development and colonization capacity of the fungi but also stimulated the biomass and activity of potential oxytetracycline microbial degraders (Cao et al. 2018). Moreover, the role of earthworms in pollutant degradation is not limited to stimulating potential soil microbial degraders; their gut microbiota may also degrade organic pollutants (Lin et al. 2018).

Despite these empirical data, the impact of earthworms on the bioremediation of polluted soils requires more research. The main reason for such a limited understanding is that most studies have used *Eisenia* sp. as a model organism. However, a comparable body of knowledge with earthworm species inhabiting the subsoil still needs to be explored. The vermiremediation of polluted soils is widely believed to require two essential processes: (i) the presence of microorganisms capable of breaking down target pollutants or enhancing their bioavailability through the secretion of surfactants (Liu et al. 2011a, b; Rodriguez-Campos et al. 2014; Megharaj et al. 2011) and (ii) the ability of earthworms to accumulate and detoxify environmental pollutants, which enables them to survive in contaminated soils while simultaneously aiding in pollutant removal (Sanchez-Hernandez et al. 2014; Katagi and Ose 2015). While this section has presented a general overview of the intricate interplay between earthworms and microorganisms in a polluted environment, the following section will outline the bioaccumulation and detoxification processes of emerging pollutants.

4 Pollutant Bioaccumulation and Toxicity: Potential Drivers for Vermiremediation

The sensitivity of earthworms to pollutants is a key factor in determining the effectiveness of vermiremediation. For instance, Pelosi et al. (2013) found through a comparative analysis of LC_{50} values that the epigeic earthworms *E. fetida* and *Eisenia andrei* are more resistant to pesticides than other species such as *A. caliginosa* (endogeic) or *L. terrestris* (anecic). Similarly, Chao et al. (2022) found that earthworm sensitivity to organic pollutants has a greater impact on vermiremediation than environmental factors like soil organic matter and nitrogen contents, soil texture, or pH. Moreover, these researchers confirmed that *Eisenia* species are more tolerant to pesticides than anecic and endogeic species. Collectively, these two studies highlight that earthworm sensitivity to pollutants and detoxification ability should be considered before using these organisms for contaminant removal. For example, levels of detoxifying enzymes such as cytochrome P450-dependent monooxygenases (Lu et al. 2017), carboxylesterases, and phosphotriesterases (Sanchez-Hernandez et al. 2023) markedly differ between ecological groups of earthworms. Because these enzymes actively participate in detoxifying environmental pollutants, their baseline activity levels and reactivity against pollutant exposure (e.g., induction and affinity of the active site of the enzyme for the pollutant) could be determinant factors in vermiremediation.

Emerging pollutants frequently detected in agricultural soils include active pharmaceutical ingredients, personnel care products, brominated flame retardants, per- and polyfluoroalkyl substances (PFAS), engineered nanoparticles, and microplastics. Many of these chemicals can be accumulated in plants, increasing the risk of human exposure (Fu et al. 2019; Ghisi et al. 2019) and altering the dynamics of soil microbial communities with adverse effects on soil functions (Bartrons and Peñuelas 2017; Kinney and Heuvel 2020). Despite the high risk that emerging pollutants pose for contaminating agricultural products, little attention has been paid to remediating soils contaminated by these chemicals. Previous studies examining the interaction between emerging pollutants and earthworms have largely focused on assessing environmental risks, including bioaccumulation and toxicity. Understanding the bioaccumulation and toxicity of emerging pollutants on earthworms is crucial for developing effective vermiremediation strategies, either alone or in combination with other remediation methods.

4.1 Pharmaceuticals and Personal Care Products

Pharmaceuticals and personal care products are commonly found in biosolids, reclaimed wastewater, and animal manure (veterinary drugs). Indeed, these materials serve as the primary routes for these emerging pollutants to enter agricultural soils (de Souza and Guimarães 2022). Although these chemicals have a low environmental persistence (Fu et al. 2019), their ongoing discharge through practices such as irrigation with reclaimed water results in a persistent presence in soil. This constant input has led to some pharmaceutical residues and personal care products being classified as pseudo-persistent (Qin et al. 2015). Environmental risks of pharmaceutical compounds in soil are the emergence of antibiotic-resistant microorganisms and plant uptake (Malchi et al. 2003; Kuppusamy et al. 2018); both risks compromise food safety.

Some studies have demonstrated that earthworms can accumulate pharmaceuticals. For example, *E. fetida* accumulated carbamazepine, diclofenac, fluoxetine, and orlistat, but the earthworms quickly eliminated the four pharmaceuticals when transferred to clean soil; 20–60% of accumulated drugs remained in earthworm tissues (Carter et al. 2016a). In addition, no metabolites were detected in the earthworms, suggesting a limited capacity of this species to degrade pharmaceuticals. Likewise, bioaccumulation of some of these pollutants depends on soil organic matter content, having a strong effect on neutral pharmaceuticals (orlistat and carbamazepine) with respect to ionizable ones (diclofenac and fluoxetine) (Carter et al. 2016b). Pharmaceutical bioaccumulation also depends on the earthworm ecotype. By comparing two earthworm species (*E. fetida* and *L. terrestris*), Carter et al. (2016a) observed significant differences in pharmaceutical bioaccumulation between them, which could be due to differences in feeding habits, earthworm size, temperature of testing assays, and lipid content of earthworms. Nevertheless, the gut microbiota is essential for removing active pharmaceutical ingredients. For example, the earthworm *Metaphire vulgaris* significantly degraded the antibiotic ciprofloxacin with the intervention of gut symbionts such as bacteria belonging to *Flavobacterium* and *Turicibacter* (Pu et al. 2020). Similarly, treatment of valsartan-spiked soils with *L. terrestris* increased the soil microbial and enzymatic activities, and members of *Proteobacteria*, *Bacteroidetes*, *Actinobacteria*, and *Firmicutes* were suggested as potential degraders of the antihypertensive drug (Gallego et al. 2021). Also, Zhang et al. (2022) conducted a study highlighting the crucial role of the drilosphere concept in evaluating the biodegradability of pharmaceuticals by earthworms. They found that *Pheretima guillelmi* accelerated the biodegradation of sulfamethoxazole by stimulating gut symbionts, such as *Microvirga*, *Sphingomonas*, *Methylobacterium*, *Bacillus*, and *Tumebacillus*. These microorganisms were found to colonize the soil via earthworm castings.

Personal care products may contain antimicrobial agents such as parabens or triclosan. These substances pose an environmental risk due to their potential to alter soil microbial communities, which can negatively impact the natural attenuation capacity of soil to resist chemical pollution. However, the scant information regarding the interaction between personal care products and earthworms does not allow us to draw conclusions on the vermiremediation of these chemicals. For example, research has shown that exposure to high concentrations of methyl-, propyl-, and butylparaben did not result in lethality or adverse effects on the reproduction of *E. fetida* (Arachchige Chamila Samarasinghe et al. 2021). However, it is important to note that parabens have a low level of soil persistence, which could contribute to their low toxicity. Similarly, no toxic effects were observed in *E. fetida* incubated in soils treated with triclosan-contaminated biosolids (0.05–1 mg/kg, range of triclosan concentrations in soil), indicating that triclosan does not present an environmental risk to the soil at environmentally realistic concentrations (Pannu et al. 2012).

4.2 Brominated Flame Retardants

Brominated flame retardants are additives in many consumer goods such as plastics, textiles, and furniture to reduce fire ignition and spread. Among them, hexabromocyclododecane (HBCD), polybrominated diphenyl ether (PBDE), and tetrabromobisphenol A (TBBPA) are the most studied flame retardants in environmental chemistry and toxicology (Smythe et al. 2022). Some lab-scale studies examining bioaccumulation and metabolism suggest that earthworms may contribute to removing flame retardants from soils. For example, *E. fetida* efficiently accumulated different polybrominated flame retardants with no toxicity signs (Nyholm et al. 2010). This species also accumulated PBDE after 28 days of incubation to diverse PBDE-contaminated substrates such as soil, biosolids, and compost (Gaylor et al. 2013). In another study, the toxicity of tris(1,3-dichloro-2-propyl)phosphate was very low for *E. fetida*. High concentrations (5 mg/kg) of this flame retardant caused a slight decrease (<10% of controls) in the mean body weight and 25% decrease in reproduction rate (Zhu et al. 2019). The transcriptomic and metabolic analysis of earthworms showed some hormone biosynthesis and metabolic routes altered by the flame retardant, which provides evidence of the ability of earthworms to metabolize this chemical.

Bioaccumulation of brominated flame retardants is species-specific. The TBBPA bioaccumulation rate is higher in *Metaphire guillelmi* than in *E. fetida* (Chen et al. 2017); even the bioaccumulation factor of this flame retardant in *E. fetida* decreased as the concentration in soil increased, probably due to saturation of uptake mechanisms (Rothenbacher and Pecquet 2018). These studies suggest that, for this type of flame retardants, the contribution of earthworms in removing them from the soil is minimal. Similarly, the bioaccumulation of HBCD isomers differed between *E. fetida* and *M. guillelmi* (Li et al. 2016). The epigeic species exhibited a higher accumulation rate than the anecic ones, probably because of the former's higher lipid and protein content. However, other studies describe a rapid excretion of PBDEs in *E. fetida* when transferred from PBDE-contaminated to clean soils, suggesting limited sequestering and metabolic capacity of PBDEs in this species (Jiang et al. 2021;

Qiao et al. 2022). Indeed, the study by Chen et al. (2017) showed that, although *E. fetida* and *M. guillelmi* could metabolize TBBPA, *M. guillelmi* broke down faster and almost entirely the TBBPA into dimethyl ether metabolite. These results suggest that earthworms, particularly anecic species, can probably remove and biodegrade TBBPA with the intervention of gut symbionts. Likewise, the occurrence of earthworms such as *M. guillelmi* in soil contaminated by flame retardants may inactivate the potential toxicity of these chemicals on soil microorganisms, as suggested by Feng et al. (2016). These researchers found an earthworm-dependent recovery of soil urease, catalase, and alkaline phosphatase activities after their inhibition by decabromodiphenyl ether.

4.3 Per- and Polyfluoroalkyl Substances

Per- and polyfluoroalkyl substances are synthetic organofluorine chemicals with many industrial and commercial applications (e.g., food packaging, waterproof textiles, cosmetics, paints, non-stick cookware, papermaking, oil production, mining, metal plating, electronics, and fire-fighting foams) (Sharifan et al. 2021). Because of their high environmental persistence, bioaccumulation potential, and toxicity, PFAS are currently emerging pollutants of global concern, being detected even in remote areas (Joerss et al. 2020). In the agroecosystem, PFAS reach the soil via crop irrigation with reclaimed wastewater and applying biosolids and composts as fertilizers (Sharifan et al. 2021). For example, Navarro et al. (2016) experimentally demonstrated that applying biosolids not only contaminated the soil with PFAS and PBDEs but also transferred these pollutants from the biosolid-soil system into earthworms (*E. andrei*). PFAS had a higher bioaccumulation potential in earthworms than PBDEs (Navarro et al. 2016). Nevertheless, PFAS bioaccumulation in earthworms seems to depend on co-contaminants in the environment, as demonstrated by Wang et al. (2022), who found that perfluorooctanoic acid (PFOA) bioaccumulation in *E. fetida* decreased in the presence of arsenic, whereas the contrary effect was found for the metalloid. In general, PFAS bioaccumulation in earthworms, particularly PFOA and perfluorooctane sulfonic acid (PFOS), decrease with the increase of PFAS concentrations in soil (Burkhard and Votava 2023), which could limit the potential of earthworms to remove these chemicals. While much attention has been given to PFAS bioaccumulation by earthworms, less research has focused on whether earthworms may facilitate PFAS degradation in soil.

Biodegradation of PFAS is limited or has little probability of occurring in the natural environment because of the C-F bond strength (Naidu et al. 2020). However, PFAS precursors such as perfluorooctane sulfonamides may be biodegraded to the more persistent PFOS and PFOA. Some studies with different earthworm species have demonstrated such biotransformation. For example, *L. terrestris* metabolized several zwitterionic/cationic perfluoroalkyl sulfonamides into PFOA and PFOS in 10 days (Jin et al. 2020). Similarly, the endogeic earthworm *M. guillelmi* also produced highly persistent PFAS during the metabolism of telomer-based polyfluoroalkyl phosphate esters (Zhu et al. 2021). *Eisenia fetida* also could transform PFAS precursors into PFOS (Zhao et al. 2018), and xenobiotic-detoxifying enzymes such as CYP450-dependent monooxygenases or glutathione *S*-transferase seemed to be involved in such detoxification (Zhao et al. 2020). Together these studies question the use of earthworms in removing certain PFAS because of their contribution to forming the environmentally persistent PFOA and PFOS, thus increasing their risk of being bioaccumulated through the food web. From this viewpoint, the most reasonable PFAS remediation methods in agricultural soils would be those based on PFAS immobilization with the minimum impact on soil properties (Sleep and Juhasz 2021). Among them, using carbonaceous materials (e.g., biochar and hydrochar) could mitigate the potential toxicity of PFAS in soil and their risk of mobility and plant uptake (Sørmo et al. 2021).

4.4 Engineered Nanoparticles

Engineered nanoparticles are mainly carbon- or metal-containing particles of <100 nm in size manufactured for a wide range of products (e.g., electronics, cosmetics, medicine, and textiles, among others), which are now found in soil as emerging pollutants. They can reach agricultural soils by direct application of agrochemicals bound to engineered nanoparticles (nanopesticides and nanofertilizers) (Santos et al. 2022) or indirectly through nanoparticle-contaminated biosolids (Kah et al. 2018; Thiagarajan and Ramasubbu 2021). Nanoparticle fate and toxicity largely depend on the physicochemical properties of soils, such as pH, organic matter content, salinity, clay content, or ionic strength (Lahive et al. 2023; Cornelis et al. 2014). Some studies suggest that earthworms may accumulate engineered nanoparticles, alter the chemical speciation of the nanoparticle-associated metal, and trigger a cascade of cellular and molecular responses to nanoparticle exposure. These biological responses allow earthworms to immobilize nanoparticles, thus reducing their toxicity. For example, *E. andrei* significantly accumulated metal-based nanoparticles after 14 days of exposure to nanoparticle-spiked soils (Pacheco et al. 2022). Coelomocytes play an important role in trapping the nanoparticles via the encapsulation process (Navarro Pacheco et al. 2021; Gupta et al. 2014). This is a cellular immune response whereby nanoparticles and other non-recognized materials are immobilized by forming a cluster of coelomocytes named brown bodies, which may be expelled through skin pores (Valembois et al. 1994). Combining nanoparticle bioaccumulation and the cellular immune response could be considered a biological process to remove nanoparticles from the soil. However, there are still many unknowns around the bioremediation meaning of these detoxification processes. For example, it is needed to confirm that such sequestering mechanisms also operate in soil-dwelling earthworms, and whether the secretion of brown bodies by earthworms or their release when the animal is dead means an input of bioavailable nanoparticles again into the soil.

4.5 Microplastics and Additives

In the last decade, soil microplastic pollution has attracted the attention of the scientific community. Excessive accumulation of microplastics in soil may cause adverse effects on plant growth, altering root development and soil physicochemical properties (de Souza Machado et al. 2018; Qi et al. 2018). Likewise, plants may uptake nanoplastics (< 100 nm) and translocate them to aerial parts, thus increasing the risk of human consumption of these particles (Rillig 2019). Furthermore, the degradation of plastic fragments in soil may release chemical additives (e.g., phthalate esters and flame retardants) that contribute to plastic toxicity (Ng et al. 2018). Most research efforts on microplastic pollution have dealt with extraction methods and analytical determination of microplastics, their fate and main input routes, global distribution of microplastics in soil, and their potential toxicity on soil function, including plants, soil organisms, and trophic transfer. However, little attention has been paid to plastic biodegradation in soil and how we can facilitate microplastic degradation (Sanchez-Hernandez et al. 2020). Inoculating plastic-contaminated soils with earthworms could be a viable option to mitigate microplastic-derived adverse effects on plants. The study by Qi et al. (2018) is an example of such a bio-based system. These researchers found that *L. terrestris* in soils previously contaminated with 1% (w/w) of macro- or microplastics derived from starch-based mulch films inactivated the adverse effects of these plastic fragments on wheat growth (*Triticum aestivum*). Some studies have also shown that *L. terrestris* strongly interacts with plastics. This anecic species can transport plastic fragments into their burrows via three processes, thus contributing to microplastic transport and accumulation in the subsoil (Rillig et al. 2017): (i) deposition of microplastic-containing casts in deep burrows, (ii) the drag of litter mixed with microplastics, and (iii) microplastics adhered on earthworm skin. Equally, this species may translocate nanoplastics to deeper soil by ingesting nanoplastic-contaminated food and further deposition of nanoplastic-contaminated casts (Heinze et al. 2021; Chao et al. 2022). Since burrow walls and casts are considered hotspots of microbial proliferation and exoenzyme production (Kuzyakov and Blagodatskaya 2015), it has been proposed that buried microplastics by earthworm burrowing activity could lead to microbial colonization of microplastics and further biodegradation, at least for biodegradable plastic polymers (Sanchez-Hernandez et al. 2020). Likewise, the ingestion of microplastics by earthworms has led some researchers to explore potential gut symbiont plastic degraders. For example, the gut luminal content of *L. terrestris* exposed to polyethylene microplastics contained potential microbial degraders that could depolymerize polyethylene (Huerta Lwanga et al. 2018). Thus, that study opens an exciting field in biotechnology for plastic polymer degradation and the bioremediation of plastics in soil.

In summary, while great progress has been made over the last decade in identifying the main detoxification route and parameters driving the bioaccumulation of emerging pollutants, more is needed to know about the real contribution of earthworms to the dissipation of these pollutants in soil. Using *E. fetida* or *E. andrei* for such a purpose is justified because these are recommended earthworm species in soil standardized toxicity and bioaccumulation tests. Therefore, an extensive body of knowledge on earthworm ecotoxicology is based on these epigeic species. However, it is important to note that factors related to species-specific biological, ecological, and toxicological traits should be considered in the vermiremediation of soil contaminated by emerging pollutants.

5 An Integrative Approach to Vermiremediation: Combining Multiple Strategies for Effective Environmental Cleanup

Restoration of degraded or contaminated soils with earthworms is not new, as some past field trials and case studies have demonstrated the viability of this restoring strategy (Baker et al. 2006; Butt 2008). Vermiremediation largely depends on environmental (climatic conditions, soil type, soil hydromechanical state, food availability), biological (earthworm ecology and sensitivity to pollutants), logistic (earthworm availability and survival in ex situ culture), and regulation factors (risk of exotic species spreading) (Ferlian et al. 2018). The most critical technical issue is probably to achieve good environmental conditions for earthworm survival and activity, implying food availability, adequate soil temperature and moisture, low soil compaction, and low predation rate (Ducasse et al. 2021; Morillo and Villaverde 2017). Adding green manure or organic mulching facilitates earthworm survival (Baldivieso-Freitas et al. 2018). Despite the environmental constraints, the available evidence suggests that inoculating contaminated soils with earthworms is a viable remediation strategy (Auclerc et al. 2022). However, an effective vermiremediation plan should consider two strategic principles: Firstly, the function-directed remediation proposed by Megharaj et al. (2011) recognizes that agricultural soils need not necessarily entail restoring all-natural soil functions or cleanup of large areas of land. This principle makes vermiremediation an attractive option for more efficient and cost-effective soil cleanup. Secondly, a combination of multiple bio-based and abiotic strategies can ensure earthworm survival and the removal or inactivation of pollutants in soil. Implementing the first principle requires a survey of contaminated land to identify hotspots of high pollution and restore soil functions, while the second explores the co-application of earthworms with various soil organisms and exogenous materials. This section focuses on the second principle, which examines the main advantages and limitations of two strategies, such as a multiple bio-based strategy and a mixed biotic-abiotic strategy, with earthworms as the central component.

5.1 A Multiple Bio-based Strategy

Currently, it is increasingly recognized that effective soil bioremediation needs the interplay of multiple biological actors in soil. This is particularly true if the goal of bioremediation is not only to remove or deactivate pollutants but also to restore soil functions. In vermiremediation, earthworms cooperate with native soil and symbiont microorganisms, which can co-metabolize pollutants. However, exogenous microorganisms can be co-applied with earthworms to enhance the biodegradation of pollutants or facilitate their bioavailability to be accumulated or degraded by others. For example, the co-application of *E. fetida*, two bacteria strains with superior catalytic potential, and alfalfa (*Medicago sativa*) was the best option to remove 50–70% of dieldrin and 20–25% of benzo(a)pyrene and heavy metals from historically contaminated soils (Urionabarrenetxea et al. 2021). Co-applying *L. terrestris* with the isopod *Porcellio scaber* also resulted in an excellent combination to remove oil derivatives from soil experimentally contaminated with petroleum products (Gospodarek et al. 2022). Although *L. terrestris* reduced 9% of total petroleum hydrocarbons, both species contributed to 29% of hydrocarbon removal, which suggested functional cooperation for soil bioremediation.

The remediation of agricultural soils contaminated by heavy metals may have two opposite objectives: the immobilization of metals to reduce their uptake and accumulation by crop plants or the increase of metal availability to be removed from the soil via phytoremediation during non-cropping seasons. The latter option is less risky in ensuring food security. Many studies agree that earthworms increase heavy metal availability in soil, thus facilitating the absorption by plants (Sizmur and Hodson 2009; Sizmur et al. 2011). Additionally, the co-application of earthworms and microorganisms, particularly arbuscular mycorrhizal fungi (AMF), is emerging as a cooperative strategy for improving the metal phytoextraction (Meng et al. 2021). For example, the co-application of *E. fetida* and the AMF *Rhizophagus intraradices* increased the Cd accumulation by *Solanum nigrum* (Wang et al. 2020). Interestingly, the plants grown in soils with earthworms and AMF developed high root and shoot biomass compared to plants from earthworm- or AMF-free soils. Furthermore, the combination of earthworms and AMF increased the capacity of *S. nigrum* to mitigate phytotoxicity derived from high tissular concentrations of Cd, which was attributed to enhanced antioxidant mechanisms of plants (Wang et al. 2021). The phytoremediation of Pb-contaminated soils using the Bermuda grass (*Cynodon dactylon*) was significantly higher when soils were inoculated with *E. fetida*, AMF, and rhizobacteria (plant growth promoters) compared to earthworm-free soils, suggesting that earthworms had a critical role in Pb mobilization (Mahohi and Raiesi 2021). Similarly, Cu phytoextraction by *Canavalia ensiformis* increased in the presence of *E. andrei* and the AMF *Rhizoglomus clarum* (Santana et al. 2019). In that study, the co-application of earthworms and AMF not only improved the plant tolerance to Cu and increased plant biomass but also facilitated the metal translocation from roots to shoot. The synthesis of low-molecular-weight organic acids induced by earthworms could act as organic transporters in the Cu translocation (Santana et al. 2019). However, the co-application of earthworms and AMF in metal-contaminated soils may also result in metal immobilization in the rhizosphere (so-called phytostabilization) due to earthworm mucus and hyphae exudates, which bind metals (Wang et al. 2022; Sizmur et al. 2011). These contrasting effects reveal that the impact of earthworms in heavy metal fate may be unpredictable. Thus a predictive environmental risk assessment of vermiremediation of metal-contaminated soils is highly recommended. Despite this limitation, the studies discussed above clearly suggest that an integrated bioremediation scheme combining earthworms, microorganisms (AMF), and plants is the best option for increasing phytoremediation and improving soil health.

5.2 *A Mixed Biotic-Abiotic Strategy*

Soil health and resilience are threatened by many environmental stressors, such as climate change-derived impacts, declining organic matter, loss of biodiversity, or contamination. The increase in soil resilience potential could be a fighting measure against soil degradation. Because soil resilience measures "the ability of soil to resist or recover from an anthropogenic or natural perturbation" (Lal 1997), it should be theoretically possible to increase it by acting on soil physicochemical and biological properties. In the case of pollution, soil resilience and health can be promoted via long-term stimulation of potential biological degraders or increasing the soil organo-mineral complexes to favor pollutant immobilization. In this context, biochar/hydrochar technology emerges as an attractive option because of its high capacity to adsorb environmental pollutants, stimulate soil microorganisms, and be a compatible material with soil meso- and macrofauna (Beesley et al. 2011; Han et al. 2016).

Biochar is formed during the pyrolysis or thermochemical conversion of biomass at high temperatures (>300 °C) and anoxic atmosphere, whereas hydrochar is obtained by hydrothermal carbonization of biomass at temperatures between 180 °C and 260 °C, in the presence of water and at high pressures (2–6 MPa) (Kambo and Dutta 2015). Production of both materials is considered a carbon sequestration strategy with direct applications on soil fertilization and pollutant removal. The co-application of biochar and earthworms has been proposed as a promising bioremediation strategy for several reasons. First, earthworms disperse biochar particles throughout the bulk soil via their feeding and burrowing activities, thereby promoting the beneficial effects of biochar in soil. However, the degree of impact depends on the particle size and type of biochar (Elmer et al. 2015). Second, earthworms may enhance the retention of enzymes and microorganisms on the surface of biochar, potentially through the action of their mucus, which acts as a cross-linking agent and increases the adsorption capacity of the biochar (Sanchez-Hernandez et al. 2019a). Third, biochar can alter the soil microbial communities and the gut microbiota of earthworms (Jin et al. 2022). Fourth, while biochar can cause oxidative stress in earthworms (Sanchez-Hernandez et al. 2019b), they do not avoid it, making the co-application of earthworms and biochar a feasible option for improving soil resilience. Finally, the functional earthworm-biochar tandem can increase soil biological parameters, such as enzyme activities with detoxification potential (Garau et al. 2022; Sanchez-Hernandez et al. 2019c).

The potential of hydrochar and earthworms for remediation has received little attention. To date, only one study has investigated the interaction between *A. caliginosa* and hydrochar derived from beetroot chips, a sugar-processing residue, on the growth rate of *Plantago lanceolata* and the performance of AMF (Salem et al. 2013). The researchers found that when 10% (w/w) hydrochar was co-applied with earthworms, it induced plant root colonization by AMF, increased plant biomass, and elevated the P content in leaves, compared to treatments without earthworms or hydrochar. These findings suggest that co-applying hydrochar and earthworms may yield similar synergistic effects in contaminated soils as those observed with biochar and earthworms.

Figure 4 illustrates a conceptual model of how co-applying earthworms and biochar work to remove soil pollutants and simultaneously improve soil health. Biochar would significantly impact three soil domains: rhizosphere, drilosphere, and charosphere. Root and microbial exudates may contribute to biogenic aggregate formation (Guhra et al. 2022), which can incorporate fine biochar particles. Biochar may indeed stimulate fungal hyphae growth by supplying physical support and nutrients (Hammer et al. 2014). Interaction of biochar with the drilosphere may occur at multiple levels. Earthworms may fragment ingested biochar particles at the gizzard and crop segment, so finer particles may be later deposited with casts. Entrapped biochar particles in casts may change the chemical and microbial features of casts (Ferreira et al. 2021). Anecic earthworms may translocate biochar particles from the topsoil to subsoil, triggering the chemical-biological interactions associated with biochar in the deep soil (Sanchez-Hernandez et al. 2019a, c; Elmer et al. 2015). The charosphere, which refers to the soil immediately surrounding biochar and directly affected by its physicochemical properties (Quilliam et al. 2013), may harbor a distinct microbial community compared to the rest of the soil (Yu et al. 2021). This means that the ability of the charosphere to break down or sequester pollutants may differ from that of the bulk soil. In summary, the combined use of earthworms and biochar has the potential to not only remediate polluted soils but also promote their overall health within the context of ecological transition. This approach represents a promising opportunity to address environmental concerns associated with intensive agriculture and support sustainable practices.

6 Concluding Remarks

Earthworms can directly and indirectly interact with environmental pollutants, promoting their biodegradation or reducing their bioavailability. However, their bioremediation potential is largely influenced by several environmental and biological factors. Firstly, earthworms' sensitivity to pollutants is a critical variable, as it can

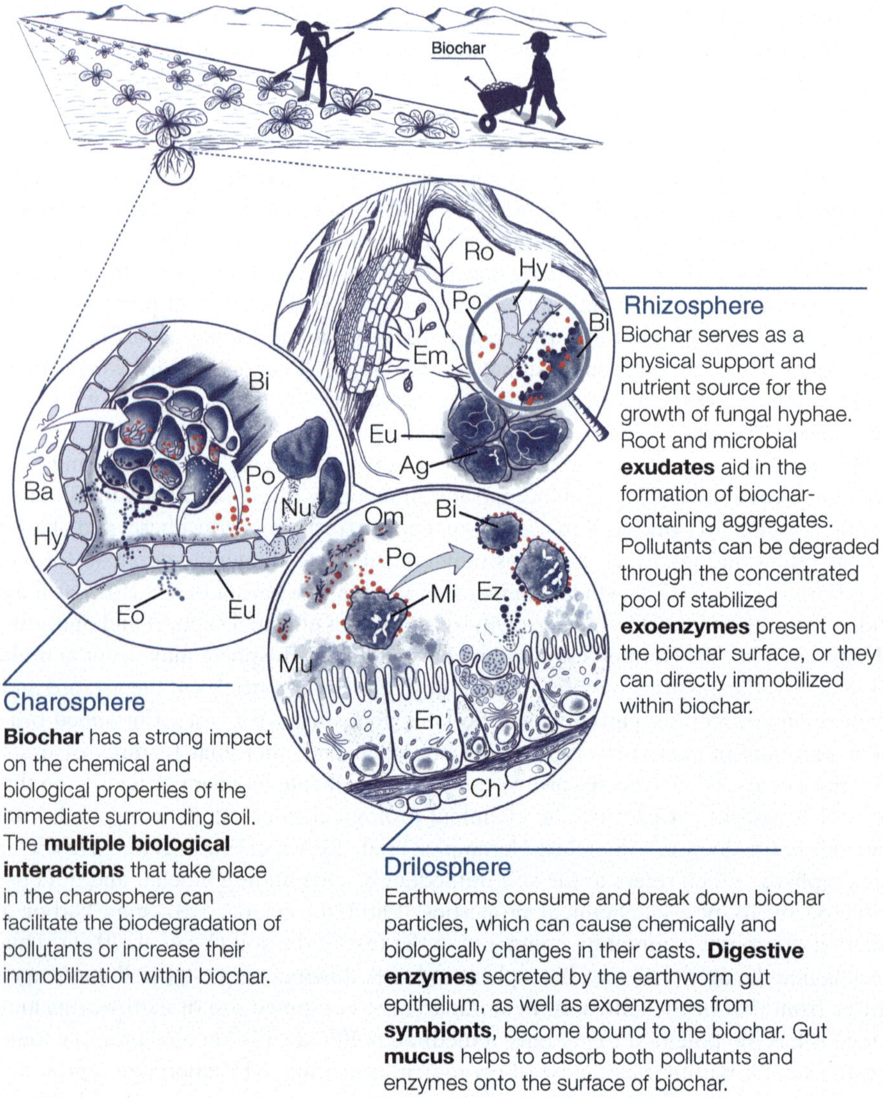

Fig. 4 Main effects of co-applying biochar and earthworms on pollutant fate. *Ag* biogenic aggregates, *Ba* bacteria, *Bi* biochar particles, *Ch* chloragogen cells, *Em* endomycorrhiza, *En* enterocytes, *Eo* exoenzyme, *Eu* exudates, *Ez* enzymes, *Hy* hyphae, *Mi* microorganisms, *Mu* mucus, *Om* organic matter, *Nu* nutrients, *Po* pollutants, and *Ro* root. Pollutants are represented by red spots. Illustration elaborated by Juan C. Sanchez-Hernandez

provoke an avoidance response or vermiremediation failure due to high earthworm mortality. Secondly, the survival of earthworms in polluted areas depends on favorable climatic conditions, such as soil temperature, moisture, and food availability. To encourage the growth of natural earthworm populations or to introduce exogenous earthworm species, extra effort is therefore required in the form of periodic nutrient supply to the soil and maintaining the soil moisture at a suitable level for earthworm activity. Alternatively, vermiremediation can be intentionally carried out through multiple episodes that coincide with the seasonality of wet periods. Thirdly, the earthworm ecotype is another determinant of vermiremediation success. Most studies have been performed on *Eisenia* species in laboratory settings. Although this species appears more resistant to organic pollutants than other soil-dwelling earthworms, their limited presence in agricultural soils and inability to burrow and ingest soil could limit their potential for bioremediation. In contrast, endogeic and anecic earthworms are better candidates for bioremediation due to the significant modifications these species cause in soil structure and chemical-biological properties. Finally, earthworm-assisted bioremediation should consider the integrated biological universe of soil and the interactions among multiple taxa. In this context, using different earthworm ecotypes could increase the soil volume for cleanup and promote soil biodiversity. Additionally, carbonaceous materials like biochar and hydrochar can improve earthworm-assisted bioremediation. Despite promising lab-scale results, it is crucial to conduct field studies to understand the long-term effects of co-applying earthworms and carbonaceous materials due to their strong adsorption capacity, which could potentially release pollutants into the soil during biochar particle degradation and fragmentation.

Acknowledgments We thank the Ministerio de Ciencia e Innovación, Spanish Government (grant no. PGC2018-098851-B-I00), Ministerio de Universidades (Salvador de Madariaga program, grant no. PRX21/00611), and the Junta de Castilla-La Mancha (grant no. SBPLY/19/180501/000259) for the financial support of this research.

References

Abail Z, Sampedro L, Whalen JK (2017) Short-term carbon mineralization from endogeic earthworm casts as influenced by properties of the ingested soil material. Appl Soil Ecol 116:79–86

Acosta-Dacal A, Hernández-Marrero ME, Rial-Berriel C, Díaz-Díaz R, Bernal-Suárez MDM, Zumbado M, Henríquez-Hernández LA, Boada LD, Luzardo OP (2022) Comparative study of organic contaminants in agricultural soils at the archipelagos of the Macaronesia. Environ Pollut 301:118979

Andriuzzi WS, Ngo P-T, Geisen S, Keith AM, Dumack K, Bolger T, Bonkowski M, Brussaard L, Faber JH, Chabbi A (2016) Organic matter composition and the protist and nematode communities around anecic earthworm burrows. Biol Fert Soils 52:91–100

Arachchige Chamila Samarasinghe SV, Krishnan K, Aitken RJ, Naidu R, Megharaj M (2021) Persistence of the parabens in soil and their potential toxicity to earthworms. Environ Toxicol Pharmacol 83:103574

Auclerc A, Beaumelle L, Barantal S, Chauvat M, Cortet J, De Almeida T, Dulaurent A-M, Dutoit T, Joimel S, Séré G, Blight O (2022) Fostering the use of soil invertebrate traits to restore ecosystem functioning. Geoderma 424:116019

Baker GH, Brown G, Butt K, Curry JP, Scullion J (2006) Introduced earthworms in agricultural and reclaimed land: their ecology and influences on soil properties, plant production and other soil biota. Biol Invasions 8:1301–1316

Baldivieso-Freitas P, Blanco-Moreno JM, Gutiérrez-López M, Peigné J, Pérez-Ferrer A, Trigo-Aza D, Sans FX (2018) Earthworm abundance response to conservation agriculture practices in organic arable farming under Mediterranean climate. Pedobiologia 66:58–64

Bartrons M, Peñuelas J (2017) Pharmaceuticals and personal-care products in plants. Trends Plant Sci 22:194–203

Beesley L, Moreno-Jiménez E, Gomez-Eyles JL, Harris E, Robinson B, Sizmur T (2011) A review of biochars' potential role in the remediation, revegetation and restoration of contaminated soils. Environ Pollut 159:3269–3282

Blouin M, Hodson ME, Delgado EA, Baker G, Brussaard L, Butt KR, Dai J, Dendooven L, Peres G, Tondoh JE, Cluzeau D, Brun J-J (2013) A review of earthworm impact on soil function and ecosystem services. Eur J Soil Sci 64:161–182

Bonkowski M, Griffiths BS, Ritz K (2000) Food preferences of earthworms for soil fungi. Pedobiologia 44:666–676

Brown GG, Barois I, Lavelle P (2000) Regulation of soil organic matter dynamics and microbial activityin the drilosphere and the role of interactionswith other edaphic functional domains. Eur J Soil Biol 36:177–198

Bünemann EK, Bongiorno G, Bai Z, Creamer RE, De Deyn G, de Goede R, Fleskens L, Geissen V, Kuyper TW, Mäder P, Pulleman M, Sukkel W, van Groenigen JW, Brussaard L (2018) Soil quality – a critical review. Soil Biol Biochem 120:105–125

Burkhard LP, Votava LK (2023) Review of per- and polyfluoroalkyl substances (PFAS) bioaccumulation in earthworms. Environ Adv 11:100335

Butt KR (2008) Earthworms in soil restoration: lessons learned from United Kingdom case studies of land reclamation. Restor Ecol 16:637–641

Butt KR, Lowe CN (2007) Presence of earthworm species within and beneath *Lumbricus terrestris* (L.) middens. Eur J Soil Biol 43:S57–S60

Cao J, Ji D, Wang C (2015) Interaction between earthworms and arbuscular mycorrhizal fungi on the degradation of oxytetracycline in soils. Soil Biol Biochem 90:283–292

Cao J, Wang C, Dou Z, Liu M, Ji D (2018) Hyphospheric impacts of earthworms and arbuscular mycorrhizal fungus on soil bacterial community to promote oxytetracycline degradation. J Hazard Mater 341:346–354

Capowiez Y, Sammartino S, Michel E (2014) Burrow systems of endogeic earthworms: effects of earthworm abundance and consequences for soil water infiltration. Pedobiologia 57:303–309

Carter LJ, Ryan JJ, Boxall AB (2016a) Does uptake of pharmaceuticals vary across earthworm species. Bull Environ Contam Toxicol 97:316–322

Carter LJ, Ryan JJ, Boxall ABA (2016b) Effects of soil properties on the uptake of pharmaceuticals into earthworms. Environ Pollut 213:922–931

Chao H, Sun M, Wu Y, Xia R, Yuan S, Hu F (2022) Quantitative relationship between earthworms' sensitivity to organic pollutants and the contaminants' degradation in soil: a meta-analysis. J Hazard Mater 429:128286

Chen X, Gu J, Wang Y, Gu X, Zhao X, Wang X, Ji R (2017) Fate and O-methylating detoxification of Tetrabromobisphenol A (TBBPA) in two earthworms (*Metaphire guillelmi* and *Eisenia fetida*). Environ Pollut 227:526–533

Cornelis G, Hund-Rinke K, Kuhlbusch T, van den Brink N, Nickel C (2014) Fate and bioavailability of engineered nanoparticles in soils: a review. Crit Rev Environ Sci Technol 44:2720–2764

Cunha L, Brown GG, Stanton DWG, Da Silva E, Hansel FA, Jorge G, McKey D, Vidal-Torrado P, Macedo RS, Velasquez E, James SW, Lavelle P, Kille P (2016) Soil animals and pedogenesis. Soil Sci 181:110–125

de Souza RB, Guimarães JR (2022) Effects of avermectins on the environment based on its toxicity to plants and soil invertebrates-a review. Water Air Soil Pollut 233:259

de Souza Machado AA, Lau CW, Till J, Kloas W, Lehmann A, Becker R, Rillig MC (2018) Impacts of microplastics on the soil biophysical environment. Environ Sci Technol 52:9656–9665

Demetrio WC, Dionísio JA, Maceda A (2019) Negative effects of earthworms on soil nematodes are dependent on earthworm density, ecological category and experimental conditions. Pedobiologia 76:150568

Drake HL, Horn MA (2007) As the worm turns: the earthworm gut as a transient habitat for soil microbial biomes. Ann Rev Microbiol 61:169–189

Ducasse V, Darboux F, Auclerc A, Legout A, Ranger J, Capowiez Y (2021) Can Lumbricus terrestris be released in forest soils degraded by compaction? Preliminary results from laboratory and field experiments. Appl Soil Ecol 168:104131

Elmer WH, Lattao CV, Pignatello JJ (2015) Active removal of biochar by earthworms (*Lumbricus terrestris*). Pedobiologia 58:1–6

Feng M, He Q, Shi J, Qin L, Zhang X, Sun P, Wang Z (2016) Effect of decabromodiphenyl ether (BDE-209) on a soil-biota system: Role of earthworms and ryegrass. Environ Toxicol Chem 35:1349–1357

Ferlian O, Eisenhauer N, Aguirrebengoa M, Camara M, Ramirez-Rojas I, Santos F, Tanalgo K, Thakur MP (2018) Invasive earthworms erode soil biodiversity: a meta-analysis. J Anim Ecol 87:162–172

Ferreira T, Hansel FA, Maia CMBF, Guiotoku M, Cunha L, Brown GG (2021) Earthworm-biochar interactions: a laboratory trial using *Pontoscolex corethrurus*. Sci Total Environ 777:146147

Fu Q, Malchi T, Carter LJ, Li H, Gan J, Chefetz B (2019) Pharmaceutical and personal care products: from wastewater treatment into agro-food systems. Environ Sci Technol 53:14083–14090

Gallego S, Nos D, Montemurro N, Sanchez-Hernandez JC, Pérez S, Solé M, Martin-Laurent F (2021) Ecotoxicological impact of the antihypertensive valsartan on earthworms, extracellular enzymes and soil bacterial communities. Environ Pollut 275:116647

Garau M, Sizmur T, Coole S, Castaldi P, Garau G (2022) Impact of *Eisenia fetida* earthworms and biochar on potentially toxic element mobility and health of a contaminated soil. Sci Total Environ 806:151255

Gaylor MO, Harvey E, Hale RC (2013) Polybrominated diphenyl ether (PBDE) accumulation by earthworms (*Eisenia fetida*) exposed to biosolids-, polyurethane foam microparticle-, and Penta-BDE-amended soils. Environ Sci Technol 47:13831–13839

Georgiadis A, Marhan S, Lattacher A, Mäder P, Rennert T (2019) Do earthworms affect the fractionation of silicon in soil. Pedobiologia 75:1–7

Ghisi R, Vamerali T, Manzetti S (2019) Accumulation of perfluorinated alkyl substances (PFAS) in agricultural plants: a review. Environ Res 169:326–341

Gospodarek J, Petryszak P, Kafel A, Paśmionka IB (2022) *Porcellio scaber* Latr. and *Lumbricus terrestris* L.—PAHs content and remediation of long-term aging soil contamination with petroleum products during a single- and two-species experiment. Energies 15:7835

Guhra T, Stolze K, Schweizer S, Totsche KU (2020) Earthworm mucus contributes to the formation of organo-mineral associations in soil. Soil Biol Biochem 145:107785

Guhra T, Stolze K, Totsche KU (2022) Pathways of biogenically excreted organic matter into soil aggregates. Soil Biol Biochem 164:108483

Gupta S, Kushwah T, Yadav S (2014) Earthworm coelomocytes as nanoscavenger of ZnO NPs. Nanoscale Res Lett 9:259

Hammer EC, Balogh-Brunstad Z, Jakobsen I, Olsson PA, Stipp SLS, Rillig MC (2014) A mycorrhizal fungus grows on biochar and captures phosphorus from its surfaces. Soil Biol Biochem 77:252–260

Han L, Ro KS, Sun K, Sun H, Wang Z, Libra JA, Xing B (2016) New evidence for high sorption capacity of hydrochar for hydrophobic organic pollutants. Environ Sci Technol 50:13274–13282

Han L, Fang K, Liu Y, Fang J, Wang F, Wang X (2021) Earthworms accelerated the degradation of the highly toxic acetochlor S-enantiomer by stimulating soil microbiota in repeatedly treated soils. J Hazard Mater 420:126669

Hao Y, Zhao L, Sun Y, Li X, Weng L, Xu H, Li Y (2018) Enhancement effect of earthworm (*Eisenia fetida*) on acetochlor biodegradation in soil and possible mechanisms. Environ Pollut 242:728–737

Heinze WM, Mitrano DM, Lahive E, Koestel J, Cornelis G (2021) Nanoplastic transport in soil via bioturbation by *Lumbricus terrestris*. Environ Sci Technol 55:16423–16433

Hickman ZA, Reid BJ (2008) Earthworm assisted bioremediation of organic contaminants. Environ Int 34:1072–1081

Hoang DTT, Maranguit D, Kuzyakov Y, Razavi BS (2020) Accelerated microbial activity, turnover and efficiency in the drilosphere is depth dependent. Soil Biol Biochem 147:107852

Huerta Lwanga E, Thapa B, Yang X, Gertsen H, Salánki T, Geissen V, Garbeva P (2018) Decay of low-density polyethylene by bacteria extracted from earthworm's guts: a potential for soil restoration. Sci Total Environ 624:753–757

Hutchinson TH, Lyons BP, Thain JE, Law RJ (2013) Evaluating legacy contaminants and emerging chemicals in marine environments using adverse outcome pathways and biological effects-directed analysis. Mar Pollut Bull 74:517–525

Jiang L, Yang Y, Jia LX, Liu Y, Pan B, Lin Y (2018) Effects of earthworm casts on sorption-desorption, degradation, and bioavailability of nonylphenol in soil. Environ Sci Pollut Res Int 25:7968–7977

Jiang L, Ling S, Fu M, Peng C, Zhang W, Lin K, Zhou B (2021) Bioaccumulation, elimination and metabolism in earthworms and microbial indices responses after exposure to decabromodiphenyl ethane in a soil-earthworm-microbe system. Environ Pollut 289:117965

Jin B, Mallula S, Golovko SA, Golovko MY, Xiao F (2020) In vivo generation of PFOA, PFOS, and other compounds from cationic and zwitterionic per- and polyfluoroalkyl substances in a terrestrial invertebrate (*Lumbricus terrestris*). Environ Sci Technol 54:7378–7387

Jin B-J, Liu X-P, Roux XL, Bi Q-F, Li K-J, Wu C-Y, Sun C-L, Zhu Y-G, Lin X-Y (2022) Biochar addition regulates soil and earthworm gut microbiome and multifunctionality. Soil Biol Biochem 173:108810

Joerss H, Xie Z, Wagner CC, von Appen WJ, Sunderland EM, Ebinghaus R (2020) Transport of legacy perfluoroalkyl substances and the replacement compound HFPO-DA through the Atlantic gateway to the Arctic ocean-is the Arctic a sink or a source. Environ Sci Technol 54:9958–9967

Kah M, Kookana RS, Gogos A, Bucheli TD (2018) A critical evaluation of nanopesticides and nanofertilizers against their conventional analogues. Nat Nanotechnol 13:677–684

Kambo HS, Dutta A (2015) A comparative review of biochar and hydrochar in terms of production, physico-chemical properties and applications. Renew Sust Energ Rev 45:359–378

Katagi T, Ose K (2015) Toxicity, bioaccumulation and metabolism of pesticides in the earthworm. J Pest Sci 40:69–81

Kinney CA, Heuvel BV (2020) Translocation of pharmaceuticals and personal care products after land application of biosolids. Curr Opin Environ Sci Health 14:23–30

Kopittke PM, Menzies NW, Wang P, McKenna BA, Lombi E (2019) Soil and the intensification of agriculture for global food security. Environ Int 132:105078

Kuppusamy S, Kakarla D, Venkateswarlu K, Megharaj M, Yoon Y-E, Lee YB (2018) Veterinary antibiotics (VAs) contamination as a global agro-ecological issue: a critical view. Agric Ecosyst Environ 257:47–59

Kuzyakov Y, Blagodatskaya E (2015) Microbial hotspots and hot moments in soil: concept & review. Soil Biol Biochem 83:184–199

Lahive E, Matzke M, Svendsen C, Spurgeon DJ, Pouran H, Zhang H, Lawlor A, Glória Pereira M, Lofts S (2023) Soil properties influence the toxicity and availability of Zn from ZnO nanoparticles to earthworms. Environ Pollut 319:120907

Lal R (1997) Degradation and resilience of soils. Philos Trans R Soc Lond Ser B Biol Sci 352:997–1010

Li B, Yao T, Sun H, Zhang Y, Yang J (2016) Diastereomer- and enantiomer-specific accumulation, depuration, bioisomerization, and metabolism of hexabromocyclododecanes (HBCDs) in two ecologically different species of earthworms. Sci Total Environ 542:427–434

Lin Z, Zhen Z, Ren L, Yang J, Luo C, Zhong L, Hu H, Liang Y, Li Y, Zhang D (2018) Effects of two ecological earthworm species on atrazine degradation performance and bacterial community structure in red soil. Chemosphere 196:467–475

Lin Z, Zhen Z, Luo S, Ren L, Chen Y, Wu W, Zhang W, Liang YQ, Song Z, Li Y, Zhang D (2021) Effects of two ecological earthworm species on tetracycline degradation performance, pathway and bacterial community structure in laterite soil. J Hazard Mater 412:125212

Liu YJ, Liu SJ, Drake HL, Horn MA (2011a) Alphaproteobacteria dominate active 2-methyl-4-chlorophenoxyacetic acid herbicide degraders in agricultural soil and drilosphere. Environ Microbiol 13:991–1009

Liu YJ, Zaprasis A, Liu SJ, Drake HL, Horn MA (2011b) The earthworm *Aporrectodea caliginosa* stimulates abundance and activity of phenoxyalkanoic acid herbicide degraders. ISME J 5:473–485

Lu X, Li Y, Thunders M, Cavanagh J, Matthew C, Wang X, Zhou X, Qiu J (2017) Differential protein expression and localization of CYP450 enzymes in three species of earthworm; is this a reflection of environmental adaptation. Chemosphere 171:485–490

Lubbers IM, van Groenigen KJ, Fonte SJ, Six J, Brussaard L, van Groenigen JW (2013) Greenhouse-gas emissions from soils increased by earthworms. Nat Clim Chang 3:187–194

Luo S, Ren L, Wu W, Chen Y, Li G, Zhang W, Wei T, Liang YQ, Zhang D, Wang X, Zhen Z, Lin Z (2022) Impacts of earthworm casts on atrazine catabolism and bacterial community structure in laterite soil. J Hazard Mater 425:127778

Ma CS, Zhang W, Peng Y, Zhao F, Chang XQ, Xing K, Zhu L, Ma G, Yang HP, Rudolf VHW (2021) Climate warming promotes pesticide resistance through expanding overwintering range of a global pest. Nat Commun 12:5351

Mahohi A, Raiesi F (2021) The performance of mycorrhizae, rhizobacteria, and earthworms to improve Bermuda grass (*Cynodon dactylon*) growth and Pb uptake in a Pb-contaminated soil. Environ Sci Pollut Res Int 28:3019–3034

Malchi T, Maor Y, Tadmor G, Shenker M, Chefetz B (2003) Irrigation of root vegetables with treated wastewater: evaluating uptake of pharmaceuticals and the associated human health risks. Environ Sci Technol 166:145–167

Medina-Sauza RM, Álvarez-Jiménez M, Delhal A, Reverchon F, Blouin M, Guerrero-Analco JA, Cerdán CR, Guevara R, Villain L, Barois I (2019) Earthworms building up soil microbiota, a review. Front Environ Sci 7

Megharaj M, Ramakrishnan B, Venkateswarlu K, Sethunathan N, Naidu R (2011) Bioremediation approaches for organic pollutants: a critical perspective. Environ Int 37:1362–1375

Meng L, Srivastava AK, Kuča K, Giri B, Mahabubur Rahman M, Wu Q (2021) Interaction between earthworms and arbuscular mycorrhizal fungi in plants: a review. Phyton 90:687–699

Meysman FJR, Middelburg JJ, Heip CHR (2006) Bioturbation: a fresh look at Darwin's last idea. Trends Ecol Evol 21:688–695

Montanarella L, Panagos P (2021) The relevance of sustainable soil management within the European Green Deal. Land Use Policy 100:104950

Morillo E, Villaverde J (2017) Advanced technologies for the remediation of pesticide-contaminated soils. Sci Total Environ 586:576–597

Naidu R, Nadebaum P, Fang C, Cousins I, Pennell K, Conder J, Newell CJ, Longpré D, Warner S, Crosbie ND, Surapaneni A, Bekele D, Spiese R, Bradshaw T, Slee D, Liu Y, Qi F, Mallavarapu M, Duan L, McLeod L, Bowman M, Richmond B, Srivastava P, Chadalavada S, Umeh A, Biswas B, Barclay A, Simon J, Nathanail P (2020) Per- and poly-fluoroalkyl substances (PFAS): current status and research needs. Environ Technol Inn 19:100915

Navarro Pacheco NI, Roubalova R, Dvorak J, Benada O, Pinkas D, Kofronova O, Semerad J, Pivokonsky M, Cajthaml T, Bilej M, Prochazkova P (2021) Understanding the toxicity mechanism of CuO nanoparticles: the intracellular view of exposed earthworm cells. Environ Sci Nano 8:2464–2477

Navarro I, Torre ADL, Sanz P, Pro J, Carbonell G, Martínez MA (2016) Bioaccumulation of emerging organic compounds (perfluoroalkyl substances and halogenated flame retardants) by earthworm in biosolid amended soils. Environ Res 149:32–39

Ng E-L, Huerta Lwanga E, Eldridge SM, Johnston P, Hu H-W, Geissen V, Chen D (2018) An overview of microplastic and nanoplastic pollution in agroecosystems. Sci Total Environ 627:1377–1388

Ning T, Liu Z, Hu H, Li G, Kuzyakov Y (2022) Physical, chemical and biological subsoiling for sustainable agriculture. Soil Tillage Res 223:105490

Noyes PD, McElwee MK, Miller HD, Clark BW, Van Tiem LA, Walcott KC, Erwin KN, Levin ED (2009) The toxicology of climate change: environmental contaminants in a warming world. Environ Int 35:971–986

Nyholm JR, Asamoah RK, van der Wal L, Danielsson C, Andersson PL (2010) Accumulation of polybrominated diphenyl ethers, hexabromobenzene, and 1,2-dibromo-4-(1,2-dibromoethyl) cyclohexane in earthworm (*Eisenia fetida*). Effects of soil type and aging. Environ Sci Technol 44:9189–9194

Pacheco NIN, Semerad J, Pivokonsky M, Cajthaml T, Filip J, Busquets-Fité M, Dvorak J, Rico A, Prochazkova P (2022) Effects of silver sulfide nanoparticles on the earthworm *Eisenia andrei*. Comp Biochem Physiol 257C:109355

Padilla-Sánchez JA, Romero-González R, Plaza-Bolaños P, Garrido Frenich A, Martínez Vidal JL (2015) Residues and organic contaminants in agricultural soils in intensive agricultural areas of Spain: a three years survey. CLEAN–Soil Air Water 43:746–753

Pannu MW, O'Connor GA, Toor GS (2012) Toxicity and bioaccumulation of biosolids-borne triclosan in terrestrial organisms. Environ Toxicol Chem 31:646–653

Pelosi C, Joimel S, Makowski D (2013) Searching for a more sensitive earthworm species to be used in pesticide homologation tests – a meta-analysis. Chemosphere 90:895–900

Plaas E, Meyer-Wolfarth F, Banse M, Bengtsson J, Bergmann H, Faber J, Potthoff M, Runge T, Schrader S, Taylor A (2019) Towards valuation of biodiversity in agricultural soils: a case for earthworms. Ecol Econ 159:291–300

Pu Q, Wang HT, Pan T, Li H, Su JQ (2020) Enhanced removal of ciprofloxacin and reduction of antibiotic resistance genes by earthworm Metaphire vulgaris in soil. Sci Total Environ 742:140409

Qi Y, Yang X, Pelaez AM, Huerta Lwanga E, Beriot N, Gertsen H, Garbeva P, Geissen V (2018) Macro- and micro- plastics in soil-plant system: Effects of plastic mulch film residues on wheat (*Triticum aestivum*) growth. Sci Total Environ 645:1048–1056

Qiao Z, Lu C, Han Y, Luo K, Fu M, Zhou S, Peng C, Zhang W (2022) Enrichment and removal of five brominated flame retardants in the presence of co-exposure in a soil-earthworm system. Environ Pollut 310:119877

Qin Q, Chen X, Zhuang J (2015) The fate and impact of pharmaceuticals and personal care products in agricultural soils irrigated with reclaimed water. Crit Rev Environ Sci Technol 45:1379–1408

Quilliam RS, Glanville HC, Wade SC, Jones DL (2013) Life in the 'charosphere' – Does biochar in agricultural soil provide a significant habitat for microorganisms. Soil Biol Biochem 65:287–293

Rämert B, Bugg RL, Clark MS, Werner MR, McGuinn RP, Poudel DD, Berry AM (2001) Influence of Lumbricus terrestris inoculation on green manure disappearance and the decomposer community in a walnut orchard. Soil Biol Biochem 33:1509–1516

Rillig MC (2019) Microplastic effects on plants. New Phytol 223:1066–1070

Rillig MC, Ziersch L, Hempel S (2017) Microplastic transport in soil by earthworms. Sci Rep 7:1362

Rodriguez-Campos J, Dendooven L, Alvarez-Bernal D, Contreras-Ramos SM (2014) Potential of earthworms to accelerate removal of organic contaminants from soil: a review. Appl Soil Ecol 79:10–25

Rothenbacher KP, Pecquet AM (2018) Summary of historical terrestrial toxicity data for the brominated flame retardant tetrabromobisphenol A (TBBPA): effects on soil microorganisms, earthworms, and seedling emergence. Environ Sci Pollut Res Int 25:17268–17277

Ruiz SA, Bickel S, Or D (2021) Global earthworm distribution and activity windows based on soil hydromechanical constraints. Commun Biol 4:612

Salem M, Kohler J, Wurst S, Rillig MC (2013) Earthworms can modify effects of hydrochar on growth of *Plantago lanceolata* and performance of arbuscular mycorrhizal fungi. Pedobiologia 56:219–224

Sanchez-Hernandez JC, Narvaez C, Sabat P, Martínez Mocillo S (2014) Integrated biomarker analysis of chlorpyrifos metabolism and toxicity in the earthworm *Aporrectodea caliginosa*. Sci Total Environ 490:445–455

Sanchez-Hernandez JC, Notario Del Pino J, Capowiez Y, Mazzia C, Rault M (2018) Soil enzyme dynamics in chlorpyrifos-treated soils under the influence of earthworms. Sci Total Environ 612:1407–1416

Sanchez-Hernandez JC, Cares XA, Pérez MA, del Pino JN (2019a) Biochar increases pesticide-detoxifying carboxylesterases along earthworm burrows. Sci Total Environ 667:761–768

Sanchez-Hernandez JC, Ríos JM, Attademo AM, Malcevschi A, Andrade Cares X (2019b) Assessing biochar impact on earthworms: Implications for soil quality promotion. J Hazard Mater 366:582–591

Sanchez-Hernandez JC, Ro KS, Díaz FJ (2019c) Biochar and earthworms working in tandem: Research opportunities for soil bioremediation. Sci Total Environ 688:574–583

Sanchez-Hernandez JC, Capowiez Y, Ro KS (2020) Potential use of earthworms to enhance decaying of biodegradable plastics. ACS Sustain Chem Eng 8:4292–4316

Sanchez-Hernandez JC, Narváez C, Cares XA, Sabat P, Naidu R (2023) Predicting the bioremediation potential of earthworms of different ecotypes through a multi-biomarker approach. Sci Total Environ 862:160547

Santana NA, Ferreira PAA, Tarouco CP, Schardong IS, Antoniolli ZI, Nicoloso FT, Jacques RJS (2019) Earthworms and mycorrhization increase copper phytoextraction by *Canavalia ensiformis* in sandy soil. Ecotoxicol Environ Saf 182:109383

Santos E, Montanha GS, Gomes MHF, Duran NM, Corrêa CG, Romeu SLZ, Pereira AES, Oliveira JL, Almeida E, Pérez-de-Luque A, Ghoshal S, Santaella C, de Lima R, Fraceto LF, Carvalho HWP (2022) Are nanomaterials leading to more efficient agriculture? Outputs from 2009 to 2022 research metadata analysis. Environ Sci Nano 9:3711–3724

Scopetani C, Chelazzi D, Cincinelli A, Martellini T, Leiniö V, Pellinen J (2022) Hazardous contaminants in plastics contained in compost and agricultural soil. Chemosphere 293:133645

Sharifan H, Bagheri M, Wang D, Burken JG, Higgins CP, Liang Y, Liu J, Schaefer CE, Blotevogel J (2021) Fate and transport of per- and polyfluoroalkyl substances (PFASs) in the vadose zone. Sci Total Environ 771:145427

Shennan C, Krupnik TJ, Baird G, Cohen H, Forbush K, Lovell RJ, Olimpi E (2017) Organic and conventional agriculture: a useful framing? Annu Rev Environ Res 42:317–346

Shipitalo M, Le Bayon R-C (2004) Quantifying the effects of earthworms on soil aggregation and porosity. In: Earthworm ecology. CRC Press

Silva V, Mol HGJ, Zomer P, Tienstra M, Ritsema CJ, Geissen V (2019) Pesticide residues in European agricultural soils – a hidden reality unfolded. Sci Total Environ 653:1532–1545

Sizmur T, Hodson ME (2009) Do earthworms impact metal mobility and availability in soil? – a review. Environ Pollut 157:1981–1989

Sizmur T, Tilston EL, Charnock J, Palumbo-Roe B, Watts MJ, Hodson ME (2011) Impacts of epigeic, anecic and endogeic earthworms on metal and metalloid mobility and availability. J Environ Monit 13:266–273

Sleep JA, Juhasz AL (2021) A review of immobilisation-based remediation of per- and polyfluoroalkyl substances (PFAS) in soils. Curr Pollut Rep 7:524–539

Smythe TA, Su G, Bergman Å, Letcher RJ (2022) Metabolic transformation of environmentally-relevant brominated flame retardants in fauna: a review. Environ Int 161:107097

Song K, Sun Y, Qin Q, Sun L, Zheng X, Terzaghi W, Lv W, Xue Y (2020) The effects of earthworms on fungal diversity and community structure in farmland soil with returned straw. Front Microbiol 11:594265

Sørmo E, Silvani L, Bjerkli N, Hagemann N, Zimmerman AR, Hale SE, Hansen CB, Hartnik T, Cornelissen G (2021) Stabilization of PFAS-contaminated soil with activated biochar. Sci Total Environ 763:144034

Stromberger ME, Keith AM, Schmidt O (2012) Distinct microbial and faunal communities and translocated carbon in *Lumbricus terrestris* drilospheres. Soil Biol Biochem 46:155–162

Stroud JL, Irons DE, Carter JE, Watts CW, Murray PJ, Norris SL, Whitmore AP (2016) *Lumbricus terrestris* middens are biological and chemical hotspots in a minimum tillage arable ecosystem. Appl Soil Ecol 105:31–35

Sun Y, Zhao L, Li X, Xu H, Weng L, Yang L, Li Y (2020) Response of soil bacterial and fungal community structure succession to earthworm addition for bioremediation of metolachlor. Ecotoxicol Environ Saf 189:109926

Thiagarajan V, Ramasubbu S (2021) Fate and behaviour of TiO2 nanoparticles in the soil: their impact on staple food crops. Water Air Soil Pollut 232:274

Tsiafouli MA, Thébault E, Sgardelis SP, de Ruiter PC, van der Putten WH, Birkhofer K, Hemerik L, de Vries FT, Bardgett RD, Brady MV, Bjornlund L, Jørgensen HB, Christensen S, Hertefeldt TD, Hotes S, Gera Hol WH, Frouz J, Liiri M, Mortimer SR, Setälä H, Tzanopoulos J, Uteseny K, Pižl V, Stary J, Wolters V, Hedlund K (2015) Intensive agriculture reduces soil biodiversity across Europe. Glob Chang Biol 21:973–985

Urionabarrenetxea E, Garcia-Velasco N, Anza M, Artetxe U, Lacalle R, Garbisu C, Becerril T, Soto M (2021) Application of in situ bioremediation strategies in soils amended with sewage sludges. Sci Total Environ 766:144099

Valembois P, Seymour J, Lassègues M (1994) Evidence of lipofuscin and melanin in the brown body of the earthworm *Eisenia fetida andrei*. Cell Tissue Res 277:183–188

Van Groenigen JW, Van Groenigen KJ, Koopmans GF, Stokkermans L, Vos HMJ, Lubbers IM (2019) How fertile are earthworm casts? A meta-analysis. Geoderma 338:525–535

Vidal A, Watteau F, Remusat L, Mueller CW, Nguyen Tu T-T, Buegger F, Derenne S, Quenea K (2019) Earthworm cast formation and development: a shift from plant litter to mineral associated organic matter. Front Environ Sci 7

Wang G, Wang L, Ma F, You Y, Wang Y, Yang D (2020) Integration of earthworms and arbuscular mycorrhizal fungi into phytoremediation of cadmium-contaminated soil by *Solanum nigrum* L. J Hazard Mater 389:121873

Wang G, Wang L, Ma F, Yang D, You Y (2021) Earthworm and arbuscular mycorrhiza interactions: Strategies to motivate antioxidant responses and improve soil functionality. Environ Pollut 272:115980

Wang L, Yang D, Chen R, Ma F, Wang G (2022) How a functional soil animal-earthworm affect arbuscular mycorrhizae-assisted phytoremediation in metals contaminated soil. J Hazard Mater 435:128991

Wurst S (2010) Effects of earthworms on above-and belowground herbivores. Appl Soil Ecol 45:123–130

Xiao Z, Wang X, Koricheva J, Kergunteuil A, Le Bayon R-C, Liu M, Hu F, Rasmann S (2018) Earthworms affect plant growth and resistance against herbivores: a meta-analysis. Funct Ecol 32:150–160

Xu HJ, Bai J, Li WY, Zhao LX, Li YT (2019) Removal of persistent DDT residues from soils by earthworms: a mechanistic study. J Hazard Mater 365:622–631

Yang C-W, Tang S-L, Chen L-Y, Chang B-V (2014) Removal of nonylphenol by earthworms and bacterial community change. Int Biodeterior Biodegradation 96:9–17

Yu M, Su W-q, Huang L, Parikh SJ, Tang C, Dahlgren RA, Xu J (2021) Bacterial community structure and putative nitrogen-cycling functional traits along a charosphere gradient under waterlogged conditions. Soil Biol Biochem 162:108420

Zamanian K, Pustovoytov K, Kuzyakov Y (2016) Pedogenic carbonates: forms and formation processes. Earth-Sci Rev 157:1–17

Zeb A, Li S, Wu J, Lian J, Liu W, Sun Y (2020) Insights into the mechanisms underlying the remediation potential of earthworms in contaminated soil: a critical review of research progress and prospects. Sci Total Environ 740:140145

Zhang Y, Song K, Zhang J, Xu X, Ye G, Cao H, Chen M, Cai S, Cao X, Zheng X, Lv W (2022) Removal of sulfamethoxazole and antibiotic resistance genes in paddy soil by earthworms (*Pheretima guillelmi*): intestinal detoxification and stimulation of indigenous soil bacteria. Sci Total Environ 851:158075

Zhao S, Wang B, Zhu L, Liang T, Chen M, Yang L, Lv J, Liu L (2018) Uptake, elimination and biotransformation of N-ethyl perfluorooctane sulfonamide (N-EtFOSA) by the earthworms (*Eisenia fetida*) after *in vivo* and *in vitro* exposure. Environ Pollut 241:19–25

Zhao S, Wang B, Zhong Z, Liu T, Liang T, Zhan J (2020) Contributions of enzymes and gut microbes to biotransformation of perfluorooctane sulfonamide in earthworms (*Eisenia fetida*). Chemosphere 238:124619

Zhu Y, Zhang J, Liu Y, Su G, Zhu L, Lin D (2019) Environmentally relevant concentrations of the flame retardant tris(1,3-dichloro-2-propyl) phosphate inhibit the growth and reproduction of earthworms in soil. Environ Sci Technol Lett 6:277–282

Zhu Y, Jia Y, Liu M, Yang L, Yi S, Feng X, Zhu L (2021) Mechanisms for tissue-specific accumulation and phase I/II transformation of 6:2 fluorotelomer phosphate diester in earthworm (*M. guillelmi*). Environ Int 151:106451

Perspectives on the Study of Earthworms: The Global Impact of Earthworms on Soil Health and Plant Ecology

Amy Rowe and Felicity Crotty

Abstract Earthworms are often considered to be the emblem of soil health due to the large impact agricultural management can have on their wellbeing and habitat, leading to changes in earthworm abundance and diversity. Earthworms also actively improve soil health through soil formation, nutrient cycling and soil aeration. In fact, earthworms have a wider influence than just soil health, affecting seed germination and distribution, alongside microbial stimulation and water infiltration. Earthworms are ecosystem engineers and interact with the soil and the wider ecology through a diversity of interrelationships; promoting crop growth through enhancing microbial activity, cast production and mobilising plant-available nutrients. Thus, they have the potential to increase aboveground biodiversity as well as beneficially changing the habitat belowground.

1 Introduction

Soil health is a current topic among soil scientists, stakeholders and policymakers, trying to promote sustainable agriculture with metrics that are measurable and easy to understand and change with improvements in soil management. However, there are several debates about how to quantify soil health as a whole and what factors need to be included. Does the use of the vague term "soil health" contribute to advancing academic research, or is it primarily beneficial for facilitating comprehension among individuals not involved in soil research, as discussed by Lehmann et al. (2021) and Powlson (2021). Only recently, has the primary term in discussions about soil shifted from "soil quality" to "soil health"? While soil quality

A. Rowe
School of Agriculture, Food and Environment, Royal Agricultural University, Cirencester, UK
e-mail: Amy.Rowe@student.rau.ac.uk

F. Crotty (✉)
Agriculture and Land Team, Ricardo Energy and Environment, Harwell, Didcot, UK
e-mail: felicity.crotty@ricardo.com

Y. Kooch, Y. Kuzyakov (eds.), *Earthworms and Ecological Processes*,
https://doi.org/10.1007/978-3-031-64510-5_20

traditionally emphasised the measurement of soil chemistry and physical metrics, soil health also includes a biological component—like earthworms, as a key indicator. Research within the field of soil health and soil quality have been growing exponentially, although there is still a limited inclusion of earthworm research within these publications (Fig. 1).

Current researchers are suggesting changing the paradigm of soil science to a more holistic approach that encompasses key measurements and the ecosystem functions they provide within a soil system rather than focusing solely on individual measurements (Harris et al. 2022). The Department for Environment, Food and Rural Affairs (DEFRA, a UK government agency) has announced a Soil Health Action Plan (SHAPE) to allow for a greater focus and to establish legislation around soil conservation within the UK (DEFRA 2022). The plan is being developed with input from academics as well as agri-businesses, farmers, ministers, charities and environmental non-governmental organisations. However, specific details regarding the plan have still not been clarified, and it will possibly be incorporated into "Environmental Improvement Plans" instead, rather than being solely focused on soil. The EU has established its own soil strategy for 2030, with the goal of safeguarding and rejuvenating soils and promoting their sustainable utilisation. In addition, the EU plans to formulate a new Soil Health Law to guarantee the protection of soil (EC 2023). With more than a third of the world's soils being partially degraded on a global scale (FAO and ITPS 2015), it is crucial to prioritise the development of methods to monitor, maintain and enhance soil health.

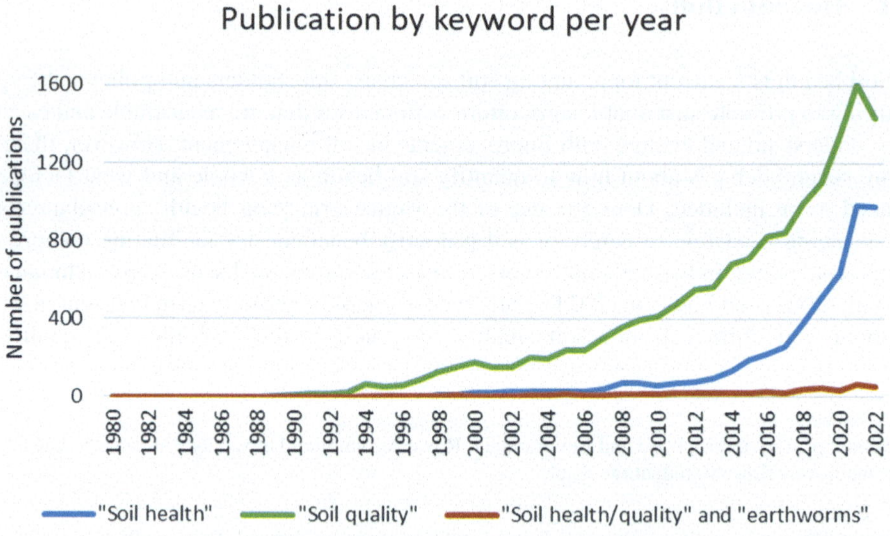

Fig. 1 Growth in the numbers of scientific publications by year 1980–2022, using the keywords "soil health" and/or "soil quality" and "earthworms" (Data from Web of Science. Accessed 9th January 2023)

2 Earthworms Impact on Soil Health and Crop Growth

Earthworms are often credited as being ecosystem engineers, benefitting both food security and ecosystem services (Stroud 2019). As ecosystem engineers they provide numerous benefits to soil (Miglani and Bisht 2019). These benefits include a positive impact on plant growth, water retention, soil microbe biodiversity, soil microbial biomass and the soil fungal community (Al-Maliki et al. 2021). Earthworms have been shown to accelerate the change in biological, physical and chemical properties of soil (Fig. 2), thereby affecting the soil structure, aeration, nutrient availability, productivity and drainage (Siddiqui 2020). Current studies are investigating the impact that earthworms can have on removing toxic elements from contaminated soil (Xiao et al. 2022). Often soil can be polluted from the use of chemical fertilisers, herbicides, fungicides and pesticides in agriculture. Chemicals from industrial processes can also be a risk to soil health, soil biodiversity, food security and human health (Lionetto et al. 2011). Research into the numerous species and diverse functions of earthworms has led to extensive and ongoing exploration of their positive effects on soil and, consequently, agricultural production.

Due to this significant positive impact, earthworms are often associated with increased crop yields and improved crop germination (Plaas et al. 2019). The impact earthworms have on germination has been referred to as a "maternal effect" due to the effect their presence has on seeds (Laossi et al. 2010); however, different earthworms and plant species interactions have yielded different results. When compared to lime and phosphorus-based fertilisers, earthworms alone were observed to have more of a significant impact on plant growth (Fonte et al. 2019). A meta-analysis of crop growth showed that the presence of earthworms could increase crop yields by 25% and overall aboveground biomass by 23% in low N environments (Groenigen et al. 2015). The earthworm's ability to fix nitrogen through secretion of hormones known to affect plant growth (Ahmed and Al-Mutairi 2022) has been credited as the reason for these increases in crop yields. The hormones can be found within the earthworm casts and have been attributed to crop growth and yield in both natural and managed agricultural settings (Singh 2018). The hormones produced are a by-product of increased microbial activity of earthworms that directly impact plant growth and yields (Rodge et al. 2018). The effect of the hormone activity from the casts depends on dosage and plant species (Tomati and Grappelli 1988).

Earthworms also secrete a cutaneous mucus that is metabolised by other soil organisms and becomes a source of plant-available nutrients (Guhra et al. 2020). The cutaneous mucus is thought to be composed of proteins, peptides and carbohydrates, is secreted through the epidermis and helps facilitate respiration and movement through soil with high compaction (Shutenko et al. 2020). However, there is currently little research on the earthworm cutaneous mucus, so a detailed composition is not known and is likely to differ between species.

There is a concern that some earthworms consume live plants (Hoeffner et al. 2022 and the references therein); however, this impact is insignificant compared with other organisms that damage crops (Edwards and Arancon 2022a). It also

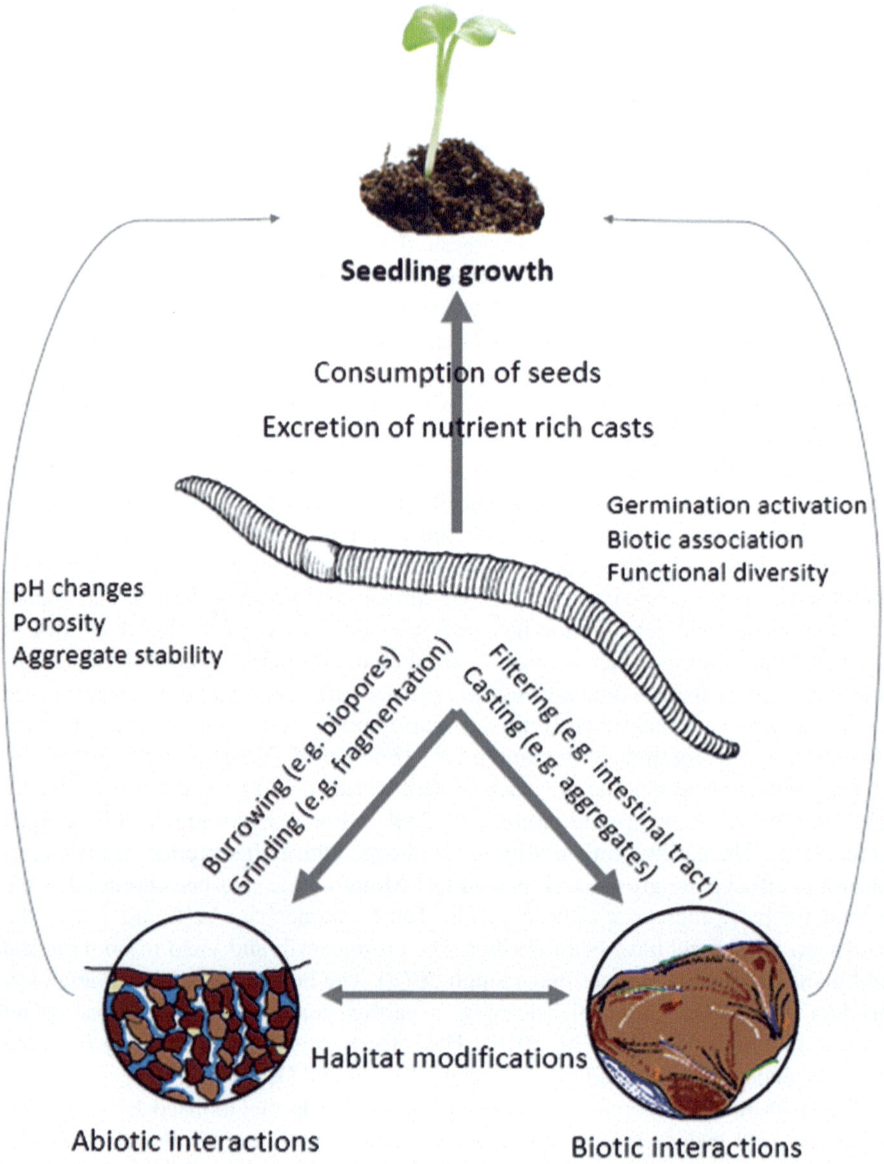

Fig. 2 Conceptual diagram showing pathways of earthworm contributions to seedling germination and plant growth (Inspired by Liu et al. (2019))

varies, depending on earthworm species, plant and parts of plant (roots, shoots or seeds). Thus, the well-documented positive effects of earthworms to soil health and crop growth negate any potential damage to crops.

3 Earthworm Cast-Producing Species

There are around 7000 described species of earthworms globally (Orgiazzi et al. 2016). However, within the UK, there are only 27 known native species of earthworms (Sherlock and Carpenter 2009). Earthworm species can be classified into three main ecological groups: epigeic, endogeic and anecic; this concept was first introduced by Bouché (1977). They are categorised based on their location within the soil profile, their feeding habitats and the earthworm's morphology (Hsu et al. 2023). Epigeic earthworms dwell closer to the surface of the soil profile and within the leaf litter. Their presence within the soil and leaf litter is known to increase the speed in which organic matter is decomposed through digestion and ingestion of the organic matter (University of Buffalo 2022; Gomez-Brandon et al. 2011). Endogeic earthworms live and feed within the soil profile and create transitory burrows through either ingesting or manoeuvring soil (Capoweiz et al. 2021). They are the dominant group of earthworms, found within most soil profiles, and are little affected by agricultural management (Hallam et al. 2021). Anecic earthworms live within the soil in permanent vertical burrows which they create through the ingesting and manoeuvring of the soil. They also take plant material and seeds from the litter layer of the soil and bring this food/organic matter down into their burrows (Hoeffner et al. 2018). Thus, seed dispersal by anecic earthworms may have benefits to seed germination, seedling recruitment and plant community composition (Milcu et al. 2006).

Anecic earthworms have the strongest effect on litter mass loss when compared to the other functional groups (Huang et al. 2020). This litter mass is ingested and transported throughout their vertical burrows. The consumption of litter mass by earthworms is important for nutrients, alongside seeds and seedlings. Another way anecic earthworms can ensure a good diet is by re-ingestion of casts (Curry and Schmidt 2007). As earthworm casts are known to be high in nutrients, the re-ingestion can ensure that the earthworms receive and assimilate the optimum amount of nutrients and reduce any nutrient wastage, it is possible that this is something that occurs more frequently in soil that is in poor health—an area which requires further investigation.

Anecic earthworms are the only functional group of earthworms to produce casts found on the soil surface (Li et al. 2021). These nutrient-rich casts enhance the soil fertility directly alongside the other benefits anecic earthworms provide within the soil habitat. Their vertical burrows reduce compaction and increase porosity and aeration of the soil. The casts left on the surface, also known as soil middens, increase the surface roughness and overall reduce soil runoff and erosion from rainfall (Wen et al. 2022; Stroud et al. 2022). The burrowing and manoeuvring of the soil contribute to the generation of topsoil (Bottinelli et al. 2021) and therefore negate some of the impact of soil erosion. Organic matter is also transported down into the burrow formations; this then becomes a habitat for microorganisms and can increase soil microorganism biomass, abundance and biodiversity (Hoang et al. 2017).

In the UK various species of anecic earthworms such as *Aporrectodea longa*, *Aporrectodea nocturna*, *Lumbricus friendi* and *Lumbricus terrestris* can be found (Earthworm Society of Britain 2022). However, although the two families of earthworm are both anecic, they have been noted to have some behavioural and physical differences. *Aporrectodea* species prefer to feed on aged and native soil organic matter, whereas *Lumbricus* species prefer fresher plant material on the soil surface (Hoeffner et al. 2022). The different species of earthworms also differ in their burrowing habits. *Aporrectodea* species make burrows that are more complex than those of *Lumbricus* species (Arrazola-Vasques et al. 2022). This highlights the importance of not only earthworm abundance but also diversity of species. Having a higher biodiversity of earthworms within a soil allows for increased positive functions and impacts. Studies have found that higher biodiversity of earthworm species increase earthworm functions when compared to singular or fewer earthworm species (Schon et al. 2021).

Anecic earthworms are vulnerable to conventional tillage practices. Studies have found a reduced abundance of anecic earthworms in over-worked soils (Stroud 2019). Toxic residues from pesticides also pose a risk to earthworms and can alter biodiversity and reduce ecosystem function (Pelosi et al. 2021). Gowri and Thangaraj (2018) found that the increased use of an organophosphate insecticide had a negative impact on earthworms' survival and physiology. As the insecticide concentration increased, so did earthworm mortality rates and the number of defective cocoons. Continued use of pesticides also damaged the epidermis and muscular cells of the earthworms (Gowri and Thangaraj 2018). Anecic earthworms have been shown to have higher levels of sensitivity to fungicides when compared with epigeic and endogeic species (Amosse et al. 2020). Numerous studies have shown earthworms to be impacted by tillage (Crotty 2020 and references therein), and consensus has now been reached that no-till agricultural systems prove better than other methods in regards to improving earthworm abundance, biodiversity and biomass (Stroud 2020). Earthworms have been found to be impacted by herbicides (even though they are not the target of the application). A study using glyphosate found the casting activity of anecic earthworms ceased for 3 weeks after herbicide application (Gaupp-Berghausen et al. 2015). However, mechanical weeding has been found to have no impact on the abundance of earthworms and may be better for earthworm biodiversity when compared to pesticides (Fishkis and Koch 2023).

4 Earthworm Casts

Earthworm casts are known for their fertilising properties and their ability to affect plant growth (Van Groenigen et al. 2019). The casts are created by anecic earthworms digesting soil and organic matter (Koren and Bisesi 2016). After ingestion of the soil and organic matter, the soil microstructures are broken down (Fig. 3), so that the organic particles become mixed with the mineral particles and are bound by mucus to form casts (Vidal et al. 2019). Earthworm casts are found to contain

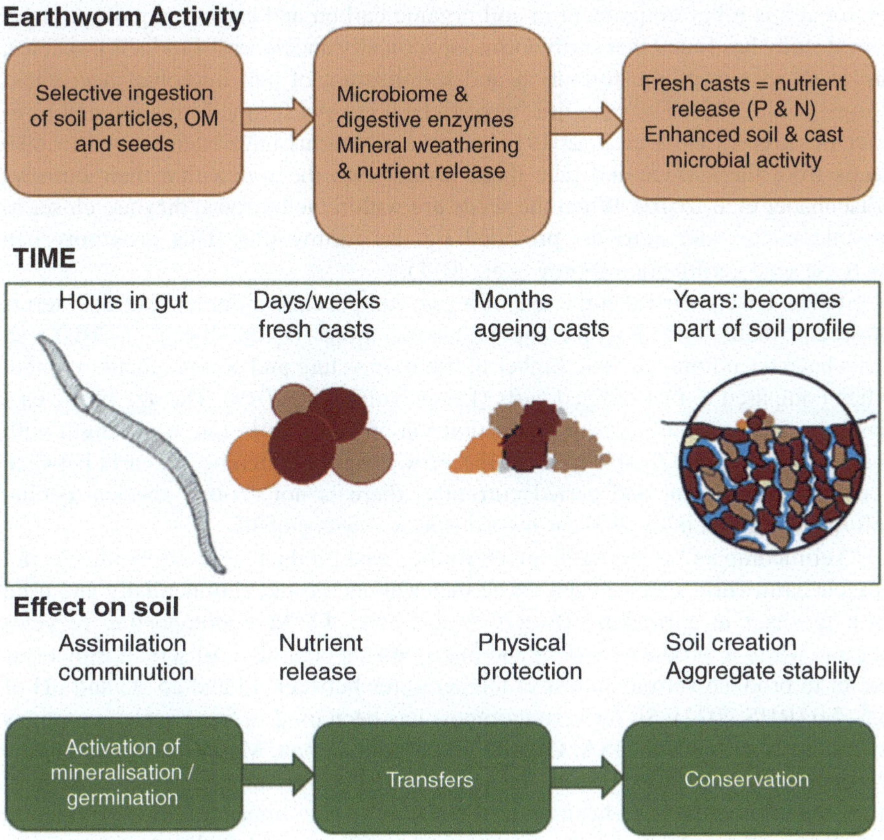

Fig. 3 Diagrammatic representation of the feeding ecology of earthworms and cast breakdown over time (Inspired by Cunha et al. (2016))

important nutrients such as nitrogen, phosphorus, potassium and magnesium, all of which are key for plant growth and therefore valued in agriculture. The casts contain microorganisms that work with the high, and varied, nutrient content to facilitate plant growth (Arunachalama and Entoori 2022). Once casts are produced, they can then improve microbial activities, nutrient cycling, aggregate formation and soil stability (Kumar et al. 2020). Soil type, vegetation, organic matter availability, land use and the surrounding environment all have an impact on the production of casts by earthworms and their nutrient and microbial content (Edwards and Arancon 2022b). Although these factors have been found to have an impact on the casts, studies show that different species of earthworms produce casts with a different microbiome that is not specific to that of the diet (Aira et al. 2022). This means that the microbiome of the earthworm's gut will also be a contributing factor to the content of the cast.

There are also studies to suggest that the earthworm casts have a significant potential to aid with carbon sequestration (Thomas et al. 2020). Microbial necromass is a key component in soil organic carbon and carbon sequestration. A recent study has found that earthworms, specifically anecic, cast-producing species, are at the centre of the formation and stabilisation of soil microbial necromass (Angst et al. 2022). The casts are found to be a hotspot for the formation and turn-over of carbon (Vidal et al. 2019). Anecic earthworms have been known to take seeds from the surface and take them further into the soil within their burrows (Eisenhauer et al. 2010). When the seeds are within the burrows, they are closer to organic matter and nutrients provided by the earthworms. This proximity can increase seed germination (Forey et al. 2011).

Studies have shown that the age of a cast can produce different results in terms of organic matter and mineral compound associations (Bottinelli et al. 2020). Fresh casts have been found to have higher nutrient, moisture and organic matter content when compared to that of aged casts (Boonchami et al. 2019). The age of the cast has been shown to affect root adaptation via an increase in fine root length with maturation of casts (Agapit et al. 2018). However, the difference in casts between species is significant and varied currently; there is not enough species-specific information about individual earthworm species casts globally.

"Vermicompost" is the name given to the "cast product" created by anecic and epigeic earthworm species from waste material and is sold commercially and used as a fertiliser in agriculture (Singh et al. 2020a, b). Vermicomposting recycles organic waste to produce nutrient and microbe-rich organic fertilisers. Earthworms prefer to produce vermicompost in temperatures between 18 and 25 °C and pH of 6.5–7.0 (RHS 2022). So far vermicompost has been used in large-scale operations in countries such as the USA, Canada, India, Italy, Japan, Malaysia, Nepal and the Philippines with positive results (Ansari et al. 2019). Vermicompost and the functions of earthworms have been praised not only for the impact on crop production but also for the ecological and economic impact (Singh et al. 2020a, b). The vermicompost has fewer negative effects on the environment in terms of less runoff and nutrient leaching than artificial fertiliser. Vermicompost also lowers the cost of food production in comparison to chemical fertilisers (Mistry 2015).

5 Earthworms and Their Casts in Arable Production

Earthworm casts are reduced in land managed for agricultural production and have been found to be of smaller biomass and contribute less to nutrient mobilisation when compared to casts in undisturbed land (Singh et al. 2017). This is likely due to the reduction in earthworm abundance as a result of conventional agricultural methods (e.g. tillage/inversion ploughing). In recent years the UK government has put a significant focus on sustainable agriculture and the negative effects of current practices on the environment, including soil health. A short-term, 5-year experiment showed that conservation tillage methods alone, such as direct drilling and shallow non-inversion tillage, did not significantly impact the overall soil health when compared to conventional ploughing in arable production (Cooper et al. 2020). This research suggests that other agricultural methods used in arable production such as the use of chemical fertilisers, pesticides, herbicides and fungicides have more of a significant impact on soil health and earthworm numbers. This agrees with the study by Crotty et al. (2016) that used a split-plot experiment, to assess the legacy effect of previous forage crops as part of an arable rotation. Earthworm abundance declined across all plots when changed from permanent forage cover to a spring cereal unrelated to whether the plots were direct drilled or ploughed, although the ploughed plots had a significantly larger decline (Crotty et al. 2016). However, using reduced tillage still has benefits such as reducing production costs and mitigating carbon emissions (Jordon et al. 2022). However, while overall soil health may not be significantly influenced by tillage systems, earthworms generally are. A global meta-analysis showed that when soil was disturbed less, it had a significant impact on earthworm abundance and biomass when compared to conventional ploughing methods (Briones and Schmidt 2017). When using conservational tillage methods in rice-wheat cropping systems, it was found that the reduced tilling enhanced micronutrient output from earthworm casts (Sharma and Dhaliwal 2021). The study suggests that it is the tillage methods that have the significant impact and not the chosen crop.

One way to help reduce the negative impact of tillage and further promote soil health, and therefore earthworm communities, is to diversify crop rotations within arable farming (Wang et al. 2022). Prolonged monoculture and the increased chemical intervention needed within the system can cause soil degradation which negatively impacts belowground communities. A diverse crop rotation with reduced or no tillage systems has been proven to promote anecic earthworms and their activities within the soil (Torppa and Taylor 2022).

Vermicompost, as mentioned previously, has a significant impact on arable production. Analysis of vermicompost has shown that the earthworm produce can often elicit an improved crop response when compared with standard fertilisers. This is likely due to the abundance of nutrients, with high nitrogen content as well as the high biomass of microbes (Broz et al. 2016). Further analysis highlights the significant effect on crop growth. A meta-analysis by Blouin et al. (2019) found that on average vermicompost increased commercial crop yield by 26%, shoot biomass by 78% and total biomass by 13%. Maximum potential for crop growth was reached when vermicompost composed between 30 and 50% of the soil total volume (Blouin et al. 2019). This positive impact from earthworm casts is not only analysed with agricultural crops but also other plant species. Studies have shown that flower nectar volume was increased which therefore improves pollinating functions (Jusselme et al. 2019). Flower diameter, number of leaves, stem diameter, fresh and dry weight of flowers has also been found to increase with the use of earthworm casts (Karagoz et al. 2019).

Focusing on in situ earthworm casts, a study collected earthworm casts from oilseed rape fields and used these as a fertiliser to increase rice crop yields. This study found that the use of earthworm casts not only improved the growth of the rice plants but also increased uptake from nitrogen fertilisers (Huang et al. 2016). The increased nitrogen uptake allows for a reduction in fertiliser waste. This has a positive impact not only on reducing environmental risk from fertiliser runoff and leaching but also on economic value as less fertiliser is wasted overall.

6 Earthworms and Their Casts in Grassland Production

Earthworms have been found to increase aboveground grass biomass within grassland settings while also increasing the phosphorus uptake (Vos et al. 2022). This was found to be due to anecic species functions, including the production of casts. Pot trials have shown that earthworm casts are acting as a slow-release nutrient source for grass leys aiding establishment (Abernethy 2017). Earthworm communities have been found to prefer more biodiverse grasslands rather than monocultures (Schorpp and Schrader 2016). When earthworm density is increased in grassland, earthworm function also increases, including the production of earthworm casts from anecic species (Fig. 4). It was found that from this increase in function, the grass root system was significantly increased in length and density (Arnone and Zaller 2014). Earthworms can also promote greater richness and abundance of different plant species within grasslands, through the translocation of seeds through burrows, across the grassland (Clause et al. 2016). This is positive for biodiversity within farmland habitats, which overall is better for the environment and agricultural sustainability. However, some farmers are unaware of the benefits of earthworms and their casts and their impact on grassland biodiversity.

In intensively managed grasslands, earthworm community activity can decrease. This is likely due to regular grass cutting and heavy chemical fertiliser use impacting earthworm function. However, sheep grazing has been found to actively increase earthworm activity (Singh et al. 2020a, b). In grazed pastures, livestock can still negatively affect earthworm communities through compaction of the habitable pore space (Schon et al. 2012). However, positive connections can be found between the cattle and soil macrofauna such as increased grassland sward productivity and cattle manure as a food source (Bacher et al. 2018). As the grass sward is increased, food for grazing livestock, such as cattle and sheep, is also increased. The earthworms and livestock provide food for each other as part of a cycle (Schon and Dominati 2020). This production of food for livestock is considered to be a valuable ecosystem service as it improves soil health and lowers costs for feeding livestock in terms of buying less feed and fertiliser for the grass.

7 Growth of Casts Over a Period of Time

Historically, the production of casts by earthworms over time was first monitored in experiments at Rothamsted Research UK in 1947 and 1948 (Evans and Guild 1947; Evans 1948). Recent studies are relatively similar, for example, Wang et al. (2021) monitored the same plantation, daily, for a year and measured the dry weight of

Fig. 4 Example of earthworm casts on soil surface (Photographs taken as part of the STFC-funded Remote Worming project (PI's Crotty and Castagna))

casts. They found that annual surface cast production was around 0.83 kg m^{-2} for the area studied (Wang et al. 2021). Wang et al. (2021) also showed that cast production was higher in months with high rainfall and high temperatures, agreeing with the results of Evans and Guild (1947) although these results were obtained from permanent pasture. Le Bayon et al. (2002) also found that earthworm cast abundance increased when rainfall increased. However, this study was performed on arable land and found that earthworm cast production was "visibly absent" for several days after harvest. Another study in arable crops found an annual cast production rate of between 2.5 and 3.2 kg m^{-2}, far higher in comparison to that of the plantation (Binet and Le Bayon 1998). However, the original work by Evans (1948) was within a similar range of both studies, with an average annual earthworm cast production across eight permanent pasture fields equating to 1.71 kg m^{-2}. These previous studies highlighted rainfall and temperature as important factors in the cast production period. A study from 1977 attributes the reduction in earthworm cast production to reduction in food supply (Sharpley and Syers 1977). In the colder and drier months, organic matter, microbial activity and leaf litter are reduced, so the food sources for earthworms are also reduced. The reduction in food supply means earthworms will have to conserve more of their energy and therefore will be less active and produce less casts overall, and those that are produced may be re-ingested by the earthworms. This reduction in food availability could also be why earthworm casts were "visibly absent" after crop harvest (Le Bayon et al. 2002).

8 Earthworm Seed Ingestion, Egestion and Digestion

The varied diet of an earthworm has been found to include living roots, fungi, algae, dead plant matter, soil and seeds (Potapov et al. 2022). Earthworms have also been described as "granivores" due to the habit of consuming seeds and seedlings (Eisenhauer et al. 2010). Research has shown that different earthworm species have different preferences for which seeds they choose to ingest (Piearce et al. 1994). This was further investigated through trials of choice and no-choice. For example, in one study, *L. terrestris* was able to pick specific seeds, when given the choice from a large pool of seeds that differed in species and appearance (Burilo et al. 2020). The ingestion of seeds depends on several different factors, earthworm species, seed appearance and previous egestion. It has been found that the length, width, weight and oil content can influence ingestion (Clause et al. 2011). Seed width and oil content were found to be most influential factors for earthworm ingestion. This is likely due to the width of the seed determining if the earthworm could physically ingest the seed and the oil content determining the nutritional factor for the earthworm.

As earthworms are granivores, it is important the seeds they do ingest have nutritional value (Griffith et al. 2013). Some evidence suggests that earthworms prefer seeds that have higher nitrogen content such as legume seeds (Eisenhauer and Scheu 2008). White clover seeds are small and are epigeal germinators that germinate close to the surface and in the top layer of the soil profile (Gibson and Cope 1985). A number of studies have found earthworm abundance to increase in legume-rich crops (e.g. Crotty et al. 2015; Roarty et al. 2017). In particular, anecic earthworms have been found to become the dominant functional group in grass/white clover mixes as the sward ages (Krogh et al. 2021). It is thought that the introduction of clover into a system increases the organic matter input, therefore providing a good quantity and quality of food for earthworms (Schmidt et al. 2003). When compared to other legumes and grasses, white clover seems to be the preferred species for earthworms, most likely due to the clover's ability to promote healthy soil environments (Hill et al. 2021), but could also be due to increased food availability via seed ingestion and organic matter increases.

Earthworms also tend to not ingest seeds that have been previously egested by other earthworms (McTavish and Murphey 2019) or if colonised by fungus (Rowan and Latch 1994). It is thought this is because after egestion, the nutritional value of the seed will have been reduced, so the earthworms would prefer to ingest "fresh" seeds. This will allow for the earthworm to achieve the maximum benefit from the seed once it has been digested and prolong their survival. Partial digestion happens within the earthworm's gut, where the seeds are not entirely digested and are later egested. The earthworm can absorb some of the nutrients needed from the seed without entirely degrading and destroying them. Earthworms' guts are host to a species-specific, rich microbiome (Sapkota et al. 2020). The microbes found within the gut could have been digested from the soil habitat and are transient (Zeibich et al. 2019). However, the earthworm gut has been found to have ideal conditions for anaerobic bacteria (Sun et al. 2020); anaerobic conditions are not usually found within a healthy soil and are therefore unlikely to be transient microbes taken up from the surrounding soil.

Once the seed has been digested and egested, the seeds' viability may have been impacted. Some studies show that the ingestion of the seeds may negatively impact the germination rates or have no impact on the germination (Melampy et al. 2019). This is likely due to nutrients being lost when absorbed through the earthworm's gut. The positive impact on germination has been attributed to post-digestion and the seed being within an earthworm cast (Clause et al. 2015). As the seed is within, or close to, a cast, it will also benefit from the nutrients that the earthworm has excreted. In habitats where earthworm communities are present, it has been found that there are more seeds within earthworm casts, likely from egestion, rather than in the habitat seed bank (Clause et al. 2016). Anecic earthworms are known to take seeds from soil seed banks and transport them throughout their network of burrows and within the soil profile (McTavish and Murphey 2021). When transporting the seeds, they may distribute them further from the surface, which can have a negative effect on germination as some species need to be close to the surface for effective germination. However, when seeds are transported closer to earthworm casts, this positively influenced seed germination and therefore plant growth (Clause et al. 2015). Earthworms have also been found to disperse endangered plant species within the agri-environment (Türke et al. 2013). The translocation of seeds via ingestion has been found to mitigate weed populations in agricultural settings allowing for higher germination rates of crop species (Li et al. 2023). Even though the journey through the earthworms' gut can negatively impact the seed morphology and, potentially, its ability to germinate, the overwhelmingly positive impact of the cast negates the impact of the digestive action. This digestive action could also have a positive impact where the seeds have a hard seed coat that needs to be degraded prior to germination.

9 Earthworms and Arable Plants

Agricultural intensification has led to a decline in arable plant species, with 582 species considered to have "threatened status" across the European Union (Richner et al. 2015), and many are classified as "rare arable plants". An example of this is the Red Hemp Nettle (*Galeopsis angustifolia*) (Fig. 5), a critically endangered species of arable weed that is currently found in only a few dozen places in the UK after its decline in population over the last 60 years (Plantlife 2022).

Red Hemp Nettle's population decline has been attributed to an increase in the use of herbicides within agriculture and also due to the time of year it sets seeds (Gibson et al. 2006). However, the reasons for its decline (along with other rare arable plants) may be due to a wider degradation of the ecosystem and the interactions between plants and soil biology. Many of the endangered/rare species of arable plants, like *G. angustifolia*, have a hard seed coat which imposes a mechanical constriction on plant emergence until the seed coat has been broken down (Plantlife

Fig. 5 Photograph of Red Hemp Nettle (*Galeopsis angustifolia*) taken by Dr. K Hemmings as part of the Royal Agricultural University's Red Hemp Nettle Back from the Brink project in collaboration with Plantlife

2022). Laboratory dormancy alleviation trials for *G. angustifolia* have found only after seed coat and endosperm removal surrounding the embryo: will germination occur (Chapman et al. 2018)—mimicking consumption, digestion and egestion by earthworms? More examples of mechanical scarification to alleviate seed dormancy can be found within the literature (e.g. Basaran et al. 2012; Teketay 1998). Many forage legumes have hard seed coats impeding germination, as a preferred seed for consumption by earthworms: could this be a sign of co-evolution between legumes and earthworms? As previously discussed, when seeds are ingested by earthworms, variable damage can occur to the seed coat dependent on soil and seed type (Gan et al. 2023), with a large number of seedlings germinating after consumption. Declines in rare arable plants with hard seed coats could be occurring in parallel with earthworm declines. A recent study assessing the abundance of earthworm data over the last 100 years has found earthworms to have declined across farmland within the UK and could be contributing to declines in ecosystem function (Barnes et al. 2023).

10 In Summary

Earthworms are widely considered to be ecosystem engineers and the emblem of soil health, but as well as their direct activity improving the soil environment (burrowing, redistribution of organic matter and breakdown of plant material), their indirect activity is also crucial. The cycling of key nutrients through earthworms via cast production enhances plant growth, aggregate stability, erosion reduction and runoff. Alongside this consumption of seeds by earthworms has been shown to potentially increase germination rates as well as aiding dispersal of rare plants. Overall, we still do not fully understand the complex web of interactions occurring within the soil habitat between earthworms and plants, but it is imperative that we restore soil health globally and try to increase the abundance of earthworms within the soil.

References

Abernethy P (2017) Agronomic effectiveness of vermicompost in grassland systems. [online] Available from: https://researcharchive.lincoln.ac.nz/bitstream/handle/10182/8997/Abernethy_BAgrSc%28Hons%29.pdf?sequence=3&isAllowed=y (Date Accessed: 24/11/2022)

Agapit C, Gigon A, Puga-Freitas R, Zeller B, Blouin M (2018) Plant-earthworm interactions: influence of age and proportion of casts in the soil on plant growth, morphology and nitrogen uptake. Plant Soil 424 49–61. Available from: https://link.springer.com/article/10.1007/s11104-017-3544-y (Date Accessed: 26/11/2022)

Ahmed N, Al-Mutairi KA (2022) Earthworms effect on microbial population and soil fertility as well as their interaction with agriculture practices. Sustainability 2022(13):14. https://doi.org/10.3390/su14137803

Aira M, Perez-Losada M, Crandall KA, Dominguez J (2022) Host taxonomy determines the composition, structure, and diversity of the earthworm cast microbiome under homogenous feeding conditions. FEMS Microbiol Ecol 98:9. https://doi.org/10.1093/femsec/fiac093

Al-Maliki S, Al-Taey DKA, Al-Mommori HZ (2021) Earthworms and eco-consequences: Considerations to soil biological indicators and plant function: a review. Acta Ecol Sinica 41(6):512–523. https://doi.org/10.1016/j.chnaes.2021.02.003

Amosse J, Bart S, Brulle F, Tebby C, Beaudouin R, Nelieu S, Lamy I, Pery ARR, Pelosi C (2020) A two years field experiment to assess the impact of two fungicides on earthworm communities and their recovery. Ecotoxicol Environ Saf 203. https://doi.org/10.1016/j.ecoenv.2020.110979

Angst G, Frouz J, Groenigen JWV, Scheu S, Kogel-Knaber I, Eisenhauer N (2022) Earthworms as catalysts in the formation and stabilization of soil microbial necromass. Glob Chang Biol. https://doi.org/10.1111/gcb.16208

Ansari AA, Ori L, Ramnarain YI (2019) An effective organic waste recycling through vermicompost technology for soil health restoration. In: Meena, R. (eds) Soil Health Restoration and Management. Springer, Singapore. https://doi.org/10.1007/978-981-13-8570-4_3 (Date Accessed: 24/11/2022)

Arnone JA, Zaller JG (2014) Earthworm effects on native grassland root system dynamics under natural and increased rainfall. Funct Plant Ecol. Available from: https://www.frontiersin.org/articles/10.3389/fpls.2014.00152/full (Date Accessed: 25/11/2022)

Arrazola-Vasques E, Larbosa M, Capowiez Y, Taylor A, Sandin M, Iseskog D, Keller T (2022) Earthworm burrowing modes and rates depend on earthworm species and soil mechanical resistance. Appl Soil Ecol 178:104568. https://doi.org/10.1016/j.apsoil.2022.104568

Arunachalama A, Entoori K (2022) Role of earthworms in soil fertility and its impact on agriculture: a review. Int J Fauna Biol Stud 9(3) 55–63. Available from: https://www.faunajournal.com/archives/2022/vol9issue3/PartA/9-3-7-702.pdf (Date Accessed: 19/11/2022)

Bacher MG, Fenton O, Bondi G, Creamer RE, Karmarker M, Schmidt O (2018) The impact of cattle dung pats on earthworm distribution in grazed pastures. BMC Ecol 18(59). Available from: https://bmcecol.biomedcentral.com/articles/10.1186/s12898-018-0216-6 (Date Accessed: 25/11/2022)

Barnes AE, Robinson RA, Pearce-Higgins JW (2023) Collation of a century of soil invertebrate abundance data suggests long-term declines in earthworms but not tipulids. PLoS One 18(4):e0282069. https://doi.org/10.1371/journal.pone.0282069

Basaran U, Mut H, Asci OO, Ayan I, Acar Z (2012) Germination pattern of naturally grown Lathyrus and Vicia species to different methods and seedbeds. Int J Plant Prod 6(3):325–335

Binet F, Le Bayon RC (1998) Space-time dynamics in situ of earthworm casts under temperate cultivated soils. Soil Biol Biochem 31(1):85–93. https://doi.org/10.1016/S0038-0717(98)00109-6

Blouin M, Barrere J, Meyer N, Lartigue S, Barot S, Mathieu J (2019) Vermicompost significantly affects plant growth. A meta-analysis. Agron Sustain Dev 39. Available from: https://link.springer.com/article/10.1007/s13593-019-0579-x (Date Accessed: 24/11/2022)

Boonchami C, Iwai CB, Ta-Oun M (2019) Physical-chemical properties of earthworm casts in different earthworm species. Available from: https://www.jstage.jst.go.jp/article/ijerd/10/1/10_1/_article/-char/ja/ (Date Accessed: 18/11/2020)

Bottinelli N, Kaupenjohann M, Marten M, Jouquet P, Soucemarianadin L, Baudin F, Tran TM, Rumpel C (2020) Age matters: fate of soil organic matter during ageing of earthworm casts produced by the anecic earthworm Amynthas khami. Soil Biol Biochem 148. https://doi.org/10.1016/j.soilbio.2020.107906

Bottinelli N, Maeght JL, Pham RD, Valentin C, Rumpel C, Pham QV, Nguyen TT, Lam DH, Nguyen AD, Tran TM, Zaiss R, Jouquet P (2021) Anecic earthworms generate more topsoil than they contribute to erosion – evidence at catchment scale in northern Vietnam. Catena 201. https://doi.org/10.1016/j.catena.2021.105186

Bouché MB (1977) Strategies lombriciennes. Ecol Bull 25:122–132. http://www.jstor.org/stable/20112572
Briones MJI, Schmidt O (2017) Conventional tillage decreases the abundance and biomass of earthworms and alters their community structure in a global meta-analysis. Glob Chang Biol. https://doi.org/10.1111/gcb.13744
Broz AP, Verma PO, Appel C (2016) Nitrogen dynamics of vermicompost use in sustainable agriculture. J Soil Sci Environ Manag 7(11):173–183. https://doi.org/10.5897/JSSEM2016.0587
Burilo S, McTavish MJ, Murphey SD (2020) Consistent seed ingestion preferences of earthworms (Lumbricus terrestris) between no-choice and choice feeding experiments. Pedobiolgia 80. https://doi.org/10.1016/j.pedobi.2020.150643
Capoweiz Y, Sammartino S, Keller T, Bottinelli N (2021) Decreased burrowing activity of endogeic earthworms and effects on water infiltration in response to an increase in soil bulk density. Pedobiologia 85–86. https://doi.org/10.1016/j.pedobi.2021.150728
Chapman T, Pocock S, Davies R (2018) Propagation protocols for ten threatened cornfield annuals. https://brahmsonline.kew.org/Content/Projects/msbp/resources/Training/Colour_in_the_margins_Report.pdf (Date Accessed: 31/05/23)
Clause J, Margerie P, Langlois E, Decaens T, Forey E (2011) Fat but slim: criteria of seed attractiveness for earthworms. Pedobiolgia:159–165. https://doi.org/10.1016/j.pedobi.2011.08.007
Clause J, Barot S, Forey E (2015) Effects of cast properties and passage through the earthworm gut on seed germination and seedling growth. Appl Soil Ecol 96:108–113. https://doi.org/10.1016/j.apsoil.2015.07.007
Clause J, Barot S, Forey E (2016) Earthworms promote greater richness and abundance in the emergence of plant species across a grassland-forest ecotone. J Plant Ecol 9(1):703–711. https://doi.org/10.1093/jpe/rtw008
Cooper RJ, Hama-Aziz ZQ, Hiscock KM, Lovett AA, Vrain E, Dugdale SJ, Sunnenburg G, Dockerty T, Hovesen P, Noble L (2020) Conservation tillage and soil health: Lessons from a 5-year UK farm trial (2013–2018). Soil Tillage Res 202. https://doi.org/10.1016/j.still.2020.104648
Crotty FV (2020) Soil organisms within arable habitats. In: Hurford C, Wilson P, Storkey J (eds) The changing status of arable habitats in Europe. Springer, Cham
Crotty FV, Fychan R, Scullion J, Sanderson R, Marley CL (2015) Assessing the impact of agricultural forage crops on soil biodiversity and abundance. Soil Biol Biochem 91:119–126. https://doi.org/10.1016/j.soilbio.2015.08.036
Crotty FV, Fychan R, Sanderson R, Rhymes JR, Bourdin F, Scullion J, Marley CL (2016) Understanding the legacy effect of previous forage crop and tillage management on soil biology, after conversion to an arable crop rotation. Soil Biol Biochem 103:241–252. https://doi.org/10.1016/j.soilbio.2016.08.018
Cunha L, Brown GG, Stanton DWG, Da Silva E, Hansel FA, Jorge G et al (2016) Soil animals and pedogenesis: the role of earthworms in anthropogenic soils. Soil Sci 181(3/4). https://doi.org/10.1097/SS.0000000000000144
Curry JP, Schmidt O (2007) The feeding ecology of earthworms - A review. Pedobiologia 50:463–477
DEFRA. Soil health: a UK-wide discussion. Available from: https://defrafarming.blog.gov.uk/2022/06/28/soil-health-a-uk-wide-discussion/ (Date Accessed:04/12/2022)
Earthworm Society of Britain (2022) Earthworm ecology. Available from: https://www.earthwormsoc.org.uk/earthworm-ecology#:~:text=Anecic-species-are-the-largest-species-of-earthworms,A-porrectodea-nocturna%2C-Lumbricus-friendi-and-Lumbricus-terrestris (Date Accessed: 13/11/2022)
Edwards CA, Arancon NQ (2022a) Earthworms as pests and benefactors. Biol Ecol Earthworms 335–370. Available from: https://link.springer.com/chapter/10.1007/978-0-387-74943-3_11 (Date Accessed: 24/11/2022)
Edwards CA, Arancon NQ (2022b) The role of earthworms in organic matter and nutrient cycles. Biol Ecol Earthworms 233–271. Available from: https://link.springer.com/chapter/10.1007/978-0-387-74943-3_8 (Date Accessed: 18/11/2022)

Eisenhauer N, Scheu S (2008) Invasibility of experimental grassland communities: the role of earthworms, plant functional group identity and seed size. Oikos 117(7):1026–1036. https://doi.org/10.1111/j.0030-1299.2008.16812.x

Eisenhauer N, Butenschoen O, Radsick S, Scheu S (2010) Earthworms as seedling predators: importance of seeds and seedlings for earthworm nutrition. Soil Biol Biochem 42(8):1245–1252. https://doi.org/10.1016/j.soilbio.2010.04.012

European Commission (EC) (2023) Soil strategy for 2030. Available from: https://environment.ec.europa.eu/topics/soil-and-land/soil-strategy_en#:~:text=It-also-announces-a-new,of-the-European-Green-Deal (Date Accessed: 09/01/2023)

Evans AC (1948) Studies on the relationships between earthworms and soil fertility. Ann Appl Biol 35(1):1–13. https://doi.org/10.1111/j.1744-7348.1948.tb07344.x

Evans AC, Guild WJM (1947) Studies on the relationships between earthworms and soil fertility: I. Biological studies in the field. Ann Appl Biol 34(3):307–330. https://doi.org/10.1111/j.1744-7348.1947.tb06367.x

FAO & ITPS (2015) Status of the world's soil resources – main report. Good and Agricultural Organization of the United Nations and Intergovernmental Technical Panel on Soils, Rome

Fishkis O, Koch H-J (2023) Effect of mechanical weeding on soil erosion and earthworm abundance in sugar beet (Beta vulgaris L.). Soil Tillage Res 225. https://doi.org/10.1016/j.still.2022.105548

Fonte SJ, Botero C, Quintero DC, Lavelle P, Kessel C (2019) Earthworms regulate plant productivity and the efficacy of soil fertility amendments in acid soils of the Colombian Llanos. Soil Biol Biochem 129:136–143. https://doi.org/10.1016/j.soilbio.2018.11.016

Forey E, Barot S, Decaens T, Langlois E, Laossi KR, Margerie P, Scheu S, Eisenhauer N (2011) Importance of earthworm–seed interactions for the composition and structure of plant communities: a review. Acta Oecolgia 37(6):594–603. https://doi.org/10.1016/j.actao.2011.03.001

Gan SH, McTavish MJ, Bourchier RS, Smith SM (2023) Soil grit influences the effects of earthworms (*Lumbricus terrestris*) on seed damage. Appl Soil Ecol 185. https://doi.org/10.1016/j.apsoil.2023.104807

Gaupp-Berghausen M, Hofer M, Rewald B, Zaller JG (2015) Glyphosate-based herbicides reduce the activity and reproduction of earthworms and lead to increased soil nutrient concentrations. Sci Rep 5:12886. https://doi.org/10.1038/srep12886

Gibson PB, Cope WA (1985) White clover. Clover Sci Technol 25. https://doi.org/10.2134/agronmonogr25.c20

Gibson RH, Nelson IL, Hopkins GW, Hamlett BJ, Memmott J (2006) Pollinator webs, plant communities and the conservation of rare plants: arable weeds as a case study. J Appl Ecol 43(2):246–257. https://doi.org/10.1111/j.1365-2664.2006.01130.x

Gomez-Brandon M, Aira M, Lores M, Dominguez J (2011) Epigeic earthworms exert a bottleneck effect on microbial communities through gut associated processes. [online]. https://doi.org/10.1371/journal.pone.0024786

Gowri S, Thangaraj R (2018) Studies on the toxic effects of agrochemical pesticide (Monocrotophos) on physiological and reproductive behavior of indigenous and exotic earthworm species. Int J Environ Health Res 30:2. Available from: https://www.tandfonline.com/doi/abs/10.1080/09603123.2019.1590538 (Date Accessed: 19/11/2022)

Griffith B, Turke M, Weisser WW, Eisenhauer N (2013) Herbivore behavior in the anecic earthworm species Lumbricus terrestris L.? Eur J Soil Biol 55:62–65. https://doi.org/10.1016/j.ejsobi.2012.12.002

Groenigen JW, Lubbers IM, Vos HMJ, Brown GG, Deyn GB, Groenigen KJ (2015) Earthworms increase plant production: a meta-analysis. Sci Rep 4. https://doi.org/10.1038/srep06365

Guhra T, Stolze K, Schweizer S, Totsche KU (2020) Earthworm mucus contributes to the formation of organo-mineral associations in soil. Soil Biol Biochem 145. https://doi.org/10.1016/j.soilbio.2020.107785

Hallam J, Holden J, Robinson DA, Hodson ME (2021) Effects of winter wheat and endogeic earthworms on soil physical and hydraulic properties. Geoderma 400. https://doi.org/10.1016/j.geoderma.2021.115126

Harris JA, Evans DL, Mooney SJ (2022) A new theory for soil health. Eur J Soil Health 73(4). https://doi.org/10.1111/ejss.13292

Hill NS, Levi M, Basinger N, Thompson A, Cabrera M, Wallace J, Saikawa E, Avramov A, Mullican J (2021) White clover living mulch enhances soil health vs. annual cover crops. Agron J 113:3697–3707. https://doi.org/10.1002/agj2.20768

Hoang DTT, Bauke S, Kuzyakov Y, Pausch J (2017) Rolling in the deep: Priming effects in earthworm biopores in topsoil and subsoil. Soil Biol Biochem 114:59–71. https://doi.org/10.1016/j.soilbio.2017.06.021

Hoeffner K, Monard C, Santonja M, Cluzeau D (2018) Feeding behaviour of epi-anecic earthworm species and their impacts on soil microbial communities. Soil Biol Biochem 125:1–9. https://doi.org/10.1016/j.soilbio.2018.06.017

Hoeffner K, Butt KR, Monard C, Frazão J, Pérès G, Cluzeau D (2022) Two distinct ecological behaviours within anecic earthworm species in temperate climates. Eur J Soil Biol 113:103446. https://doi.org/10.1016/j.ejsobi.2022.103446

Hsu G-C, Szlavecz K, Csuzdi C, Bernard M, Change C-H (2023) Ecological groups and isotopic niches of earthworms. Appl Soil Ecol 181. https://doi.org/10.1016/j.apsoil.2022.104655

Huang M, Zhou X, Xie X, Zhoa C, Chen J, Cao F, Zou Y (2016) Rice yield and the fate of fertilizer nitrogen as affected by addition of earthworm casts collected from oilseed rape fields: a pot experiment. PLoS One. https://doi.org/10.1371/journal.pone.0167152

Huang W, Gonzalez G, Zou X (2020) Earthworm abundance and functional group diversity regulate plant litter decay and soil organic carbon level: a global meta-analysis. Appl Soi Ecol 150. https://doi.org/10.1016/j.apsoil.2019.103473

Jordon MW, Smith P, Long PR, Burkner P-C, Petrokofsky G, Willis KJ (2022) Can regenerative agriculture increase national soil carbon stocks? Simulated country-scale adoption of reduced tillage, cover cropping, and ley-arable integration using RothC. Sci Total Environ 825. https://doi.org/10.1016/j.scitotenv.2022.153955

Jusselme MD, Pruvost C, Motard E, Giussti-Miller S, Frechault S, Alphonse V, Balland-Bolou-Bi C, Dajoz I, Mora P (2019) Increasing the ability of a green roof to provide ecosystem services by adding organic matter and earthworms. Appl Soil Ecol 143. https://doi.org/10.1016/j.apsoil.2019.05.028

Karagoz FP, Dursen A, Tekiner N, Kul R, Kotan R (2019) Efficacy of vermicompost and/or plant growth promoting bacteria on the plant growth and development in gladiolus. Ornam Hort 25(2). https://doi.org/10.14295/oh.v25i2.2023

Koren H, Bisesi MS (2016) Private and public sewage disposals of soil. In: Handbook of Environmental Health, Volume II: Pollutant Interactions in Air, Water, and Soil (4th ed.). CRC Press. https://doi.org/10.1201/9780849378003 (Date accessed: 17/11/2022)

Krogh PH, Lamande M, Holmstrup M, Eriksen J (2021) Earthworm burrow number and vertical distribution are affected by the crop sequence of a grass-clover rotation system. Eur J Soil Biol 103. https://doi.org/10.1016/j.ejsobi.2021.103294

Kumar R, Sharma P, Gupta RK, Kumar S, Sharma MMM, Singh S, Pradhan G (2020) Earthworms for eco-friendly resource efficient agriculture. Resourc Use Effic Agricult 47–84. Available from: https://link.springer.com/chapter/10.1007/978-981-15-6953-1_2 (Date Accessed: 17/11/2022)

Laossi K-R, Noguera D-C, Barot S (2010) Earthworm-mediated maternal effects on seed germination and seedling growth in three annual plants. Soil Biol Biochem 42(2):319–323. https://doi.org/10.1016/j.soilbio.2009.11.010

Le Bayon RC, Moreau S, Gascuel-Ordoux C, Binet F (2002) Annual variations in earthworm surface-casting activity and soil transport by water runoff under a temperate maize agroecosystem. Geoderma 106(1-2):121–135. https://doi.org/10.1016/S0016-7061(01)00121-5

Lehmann P, Williams J, Smith R (2021). Soil health: a new buzzword or a new paradigm? Available from: https://uksoils.org/soiltalks-04 (Date Accessed: 04/12/2022)

Li Y, Wang J, Shao M (2021) Application of earthworm cast improves soil aggregation and aggregate-associated carbon stability in typical soils from Loess Plateau. J Environ Manag 278. https://doi.org/10.1016/j.jenvman.2020.111504

Li T, Fan J-Q, Qian H-W, Wei J-HW, Qian Z-G, Guo S, Lv W (2023). Earthworm activities enhance taro production by reducing weed infestation through taro–earthworm coculture. Agriculture, Ecosystems & Environment, 352, 108533. https://doi.org/10.1016/j.agee.2023.108533

Lionetto MG, Calisi A, Schettino T (2011) Earthworm biomarkers as tools for soil pollution assessments. In M. C. Hernandez-Soriano (Ed.), Soil Health and Land Use Management (pp. Ch. 16). Rijeka: IntechOpen. https://doi.org/10.5772/28265

Liu T, Chen X, Gong X, Lubbers IM, Jiang Y, Feng W et al (2019) Earthworms coordinate soil biota to improve multiple ecosystem functions. Curr Biol 29(20):3420–3429.e3425. https://doi.org/10.1016/j.cub.2019.08.045

McTavish MJ, Murphey SD (2019) Seed density and previous egestion affect earthworm seed ingestion: preliminary observations using granivory of Alliaria petiolata by Lumbricus terrestris. Appl Soil Ecol 139:29–31. https://doi.org/10.1016/j.apsoil.2019.01.019

McTavish MJ, Murphey SD (2021) Three-dimensional mapping of earthworm (Lumbricus terrestris) seed transport. Pedobiologia 87–88. doi:https://doi.org/10.1016/j.pedobi.2021.150752

Melampy MN, Mansbach SN, Durkin GK (2019) Assessing the impact of lumbricid earthworms on forest seeds in Northeastern Ohio. Am Midl Nat 181(1). https://doi.org/10.1674/0003-0031-181.1.92

Miglani R, Bisht SS (2019) [online]. World of earthworms with pesticides and insecticides. Interdiscip Toxicol 12(2):17–82. https://doi.org/10.2478/intox-2019-0008

Milcu A, Schumacher J, Scheu S (2006) Earthworms (Lumbricus terrestris) affect plant seedling recruitment and microhabitat heterogeneity. Funct Ecol 20(2):261–268. https://doi.org/10.1111/j.1365-2435.2006.01098.x

Mistry J (2015) Vermicompost, a best superlative for organic farming: a review. J Adv Stud Agricult Biol Environ Sci 2(3). Available from: http://jabe.in/2.3.15/38-46-JAYANTA-MISTRY.pdf (Date Accessed: 24/11/2022)

Orgiazzi A, Bardgett RD, Barrios E, Behan-Pelletier V, Briones MJI, Chotte J-L et al (2016) Global soil biodiversity atlas. Publications Office of the European Union, Luxembourg

Pelosi C, Bertrand C, Daniele G, Coeurdassier M, Benoit P, Nelieu S, Lafay F, Bretagnolle V, Gaba S, Vulliet E, Fritsch C (2021) Residues of currently used pesticides in soils and earthworms: a silent threat? Agricult Ecosyst Environ 305. https://doi.org/10.1016/j.agee.2020.107167

Pierce TG, Roggero N, Tipping R (1994) Earthworms and seeds. J Biol Educ 28(3):195–202. https://doi.org/10.1080/00219266.1994.9655392

Plaas E, Meyer-Wolfarth F, Banse M, Bengsston J, Bergmann H, Faber J, Potthoff M, Runge T, Schrader S, Taylor A (2019) [online]. Towards valuation of biodiversity in agricultural soils: a case for earthworms. Ecol Econ 159:291–300. https://doi.org/10.1016/j.ecolecon.2019.02.003

PlantLife (2022) Red Hemp-Nettle. Available from: https://naturebftb.co.uk/wp-content/uploads/2021/04/Red_Hemp-nettle_FINAL-LORES-01.04.21.pdf (Date Accessed: 31/05/23)

Potapov AM, Beaulieu F, Birkhofer K, Bluhm SL, Degtyarev MI, Devtter M, Goncharov AA, Gongalsky KB, Klarner B, Korobushkin DL, Liebke DF, Maraun M, McDonnell RJ, Pollierer MM, Schaefer I, Shrubovych J, Semenyuk II, Sendra A, Tuma J, Tůmová M, Vassilieva AB, Chen T-W, Geisen S, Schmidt O, Tiunov AV, Scheu S (2022) Feeding habits and multifunctional classification of soil-associated consumers from protists to vertebrates. Biol Rev 97:1057–1117. https://doi.org/10.1111/brv.12832

Powlson DS (2021) Is 'soil health' meaningful as a scientific concept or as terminology? Soil Use Manag 37:403–405. https://doi.org/10.1111/sum.12721

Richner N, Holderegger R, Linder HP, Walter T (2015) Reviewing change in the arable flora of Europe: a meta-analysis. Weed Res 55:1–13. https://doi.org/10.1111/wre.12123

Roarty S, Hackett RA, Schmidt O (2017) Earthworm populations in twelve cover crop and weed management combinations. Appl Soil Ecol 114:142–151. https://doi.org/10.1016/j.apsoil.2017.02.001

Rodge PB, Pandhare RA, Khyade VB (2018) Indole acetic acid and radish biomass productive influence of Rhizobacteria, Pseudomonas trivialias (L) and Earthworm, Eudrilus eugenae (L). Sci Res Centre 6(4):17–37. Available from: https://scientificrc.com/dl/journals/2-IJBEMS/v6-i4-oct-dec2018/paper2.pdf (Date Accessed: 06/11/2022)

Rowan DD, Latch GCM (1994) Utilization of endophyte-Infected perennial ryegrasses for increased insect resistance. Biotechnology of endophytic fungi of grasses. Available from: https://www.taylorfrancis.com/chapters/edit/10.1201/9781351070324-12/utilization-endophyte-Infected-perennial-ryegrasses-increased-insect-resistance-daryl-rowan-garrick-latch (Date Accessed: 12/12/2022)

Royal Horticultural Society (RHS) (2022) Worm composting. Available from: https://www.rhs.org.uk/soil-composts-mulches/worm-composting (Date Accessed: 25/11/2022)

Sapkota R, Santos S, Farias P, Krogh PH, Winding A (2020) Insights into the earthworm gut multi-kingdom microbial communities. Sci Total Environ 727. https://doi.org/10.1016/j.scitotenv.2020.138301

Schmidt O, Clements RO, Donaldson G (2003) Why do cereal–legume intercrops support large earthworm populations? Appl Soil Ecol 22(2):181–190. https://doi.org/10.1016/S0929-1393(02)00131-2

Schon NL, Dominati EJ (2020) Valuing earthworm contribution to ecosystem services delivery. Ecosyst Serv 43. https://doi.org/10.1016/j.ecoser.2020.101092

Schon NL, Mackay AD, Minor MA (2012) Relationship between food resource, soil physical condition, and invertebrates in pastoral soils. Soil Sci Soc Am J 76(5):1644–1654. https://doi.org/10.2136/sssaj2011.0375

Schon NL, Fraser PM, Mackay AD (2021) Relationship between earthworm abundance, ecological diversity and soil function in pastures. Soil Res 59(8):767–777. https://doi.org/10.1071/SR20273

Schorpp Q, Schrader S (2016) Earthworm functional groups respond to the perennial energy cropping system of the cup plant (Silphium perfoliatum L.). Biomass Bioenergy 87:61–68. https://doi.org/10.1016/j.biombioe.2016.02.009

Sharma S, Dhaliwal SS (2021) Conservation agriculture based practices enhanced micronutrients transformation in earthworm cast soil under rice-wheat cropping system. Ecol Eng 163. https://doi.org/10.1016/j.ecoleng.2021.106195

Sharpley AN, Syers JK (1977) Seasonal variation in casting activity and in the amounts and release to solution of phosphorus forms in earthworm casts. Soil Biol Biochem 9(4):227–231. https://doi.org/10.1016/0038-0717(77)90026-8

Sherlock E, Carpenter D (2009) An updated earthworm list for the British Isles and two new 'exotic' species to Britain from Kew Gardens. Eur J Soil Biol 45(5–6):431–435. https://doi.org/10.1016/j.ejsobi.2009.07.002

Shutenko GS, Kelleher BP, Simpson AJ, Soong R, Mobarhan YL, Schmidt O (2020) Evidence for substantial acetate presence in cutaneous earthworm mucus. J Soils Sediment 20:3627–3632. Available from: https://link.springer.com/article/10.1007/s11368-020-02674-4 (Date Accessed: 18/11/2022)

Siddiqui S (2020) Interaction of earthworm activity with soil structure and enzymes. Biol Compost Soil Biol 58. https://doi.org/10.1007/978-3-030-39173-7_5

Singh J (2018) 3 - Role of earthworm in sustainable agriculture. Sustain Food Syst Agric Industry:83–122. doi:https://doi.org/10.1016/B978-0-12-811935-8.00003-2

Singh P, Mitra S, Majumdar D, Bhattacharyya P, Prakash A, Borah P, Paul A, Rangan L (2017) Nutrient and enzyme mobilization in earthworm casts: a comparative study with addition of selective amendments in undisturbed and agricultural soils of a mountain ecosystem. Int Biodeterior Biodegradation 119:437–447. https://doi.org/10.1016/j.ibiod.2016.09.008

Singh A, Karmegam N, Singh GS, Bhadauria T, Chang SW, Awasthi MK, Sudhakar S, Arunachalam KD, Biruntha M, Ravindran B (2020a) Earthworms and vermicompost: an eco-friendly approach for repaying nature's debt. Environ Geochem Health 42:1617–1642. [online] Available from: https://link.springer.com/article/10.1007/s10653-019-00510-4 (Date Accessed: 24/11/2022)

Singh J, Cameron E, Reitz T, Schadler M, Eisenhauer N (2020b) Grassland management effects on earthworm communities under ambient and future climatic conditions. Eur J Soil Sci. https://doi.org/10.1111/ejss.12942

Stroud JL (2019) Soil health pilot study in England: outcomes from an on-farm earthworm survey. PLoS One 14(2). https://doi.org/10.1371/journal.pone.0203909

Stroud JL (2020) Earthworms in no-till: the key to soil biological farming. No-till farming systems for sustainable agriculture: 267–280. Available from: https://link.springer.com/chapter/10.1007/978-3-030-46409-7_16 (Date Accessed: 17/11/2022)

Stroud JL, Dummett I, Kemp SJ, Sturrock CJ (2022) Working with UK farmers to investigate anecic earthworm middens and soil biophysical properties. Ann Appl Biol. https://doi.org/10.1111/aab.12795

Sun M, Chao H, Zheng X, Deng S, Ye M, Hu F (2020) Ecological role of earthworm intestinal bacteria in terrestrial environments: a review. Sci Total Environ 740. https://doi.org/10.1016/j.scitotenv.2020.140008

Teketay D (1998) Germination of Acacia origena, A-pilispina and Pterolobium stellatum in response to different pre-sowing seed treatments, temperature and light. J Arid Environ 38(4):551–560. https://doi.org/10.1006/jare.1997.0332

Thomas E, Prabha VS, Kurien VT, Thomas AP (2020) The potential of earthworms in soil carbon storage: a review. Environ Exp Biol 18:61–75. Available from: http://eeb.lu.lv/EEB/202006/EEB_XVIII_Thomas.pdf (Date Accessed: 17/11/2022)

Tomati U, Grappelli A (1988) The hormone-like effect of earthworm casts on plant growth. Biol Fertil Soils 5(4):288–294. https://doi.org/10.1007/BF00262133

Torppa KA, Taylor AR (2022) Alternative combinations of tillage practices and crop rotations can foster earthworm density and bioturbation. Appl Soil Ecol 175

Türke M, Blattmann T, Knop E, Kindermann A, Prestele J, Marquez L et al (2013) Weeds and endangered herbs have unforeseen dispersal helpers in the agri-environment: gastropods and earthworms. Renew Agricult Food Syst 28(4):380–383. https://doi.org/10.1017/S1742170512000415

University of Buffalo (2022) Epigeic earthworms. [online]. Available from: https://soil.evs.buffalo.edu/index.php/Epigeic_Earthworms (Date Accessed: 13/11/2022)

Van Groenigen JW, Van Groenigen KJ, Koopmans GF, Stokkermans L, Vos HMJ, Lubbers IM (2019) How fertile are earthworm casts? A meta-analysis. Geoderma 338:525–535. https://doi.org/10.1016/j.geoderma.2018.11.001

Vidal A, Watteau F, Remusat L, Mueller CW, Tu TTN, Buegger F, Derenne S, Quenea K (2019) Earthworm cast formation and development: a shift from plant litter to mineral associated organic matter. Front Environ Sci. https://doi.org/10.3389/fenvs.2019.00055

Vos HMJ, Koopmans GF, Ferron LME, Oenma O, Groenigen JWV (2022) Do earthworms increase grass biomass production and phosphorus uptake under field conditions? Appl Soil Ecol 180. https://doi.org/10.1016/j.apsoil.2022.104598

Wang X, Fu S, Wang X, Li Z, Li J, Zhang W (2021) One-year monitoring of daily earthworm cast production: surface cast contribution to soil fertility in a subtropical forest. Forests 12(7):865. https://doi.org/10.3390/f12070865

Wang G, Li X, Xi X, Cong W-F (2022) Crop diversification reinforces soil microbiome functions and soil health. Plant Soil 476:375–383. Available from: https://link.springer.com/article/10.1007/s11104-022-05436-y (Date Accessed: 20/11/2022)

Wen S, Wang J, Li Y, Shao M (2022) Effects of anecic earthworms on runoff and erosion on the slope with soil from the Loess Plateau under a rainfall simulation experiment. Agric Water Manag 259. https://doi.org/10.1016/j.agwat.2021.107230

Xiao R, Ali A, Xu Y, Abdelrahman H, Li R, Lin Y, Bolan N, Shaheen SM, Rinklebe J, Zhang Z (2022) Earthworms as candidates for remediation of potentially toxic elements contaminated soils and mitigating the environmental and human health risks: a review. Environ Int 158. https://doi.org/10.1016/j.envint.2021.106924

Zeibich L, Schmidt O, Drake HL (2019) Fermenters in the earthworm gut: do transients matter? FEMS Microbiol Ecol 95(2). https://doi.org/10.1093/femsec/fiy221

Part IV
Conclusion

Historical Perspectives on the Study of Earthworms: A Global Issue

John Warren Reynolds and Mohammad Kazem Parsapour

Abstract Earthworms are often cited to be the emblem of soil health and have become a global focus of research initiatives within agricultural science, with large monitoring programs and citizen science projects occurring worldwide. There are over 6000 species of identified earthworms, and they can be found in the majority of ecosystems globally, with invasive earthworms also found in unexpected regions such as the arctic. Earthworms do not always have a central role within soil health, and the diversity of interrelationships with the soil and wider ecology is discussed. In this chapter, an overview of the global perspectives of earthworm ecology and research will be presented under the headings: "Training of Taxonomists," "Taxonomic Tools," "Parthenogenesis in Taxonomy," "Earthworm Surveys," "Life Histories," "Modern Techniques," "Earthworms in Waste Management," and "Plain Language and Less Esotery."

1 Introduction

Nearly six decades have passed since Reynolds first started to write on the earthworms' biogeography, taxonomy, and diversity in North America (Reynolds 1995, 1998, 2004). For these papers, Reynolds utilized a range of headings including the following: "Taxonomist Training," "The Role of Parthenogenesis in Taxonomy," "Surveys on Earthworms," "Exploration of Life Histories," "Utilization of Modern

J. W. Reynolds (✉)
Oligochaetology Laboratory, Kitchener, ON, Canada

New Brunswick Museum, Saint John, NB, Canada

M. K. Parsapour
Research Division of Natural Resources, Chaharmahal and Bakhtiari Agricultural and Natural Resources Research and Education Center, AREEO, Shahrekord, Iran

Y. Kooch, Y. Kuzyakov (eds.), *Earthworms and Ecological Processes*,
https://doi.org/10.1007/978-3-031-64510-5_21

Techniques," "The Earthworms' Significance in Waste Management," "The Potential of Earthworms for Tracking Environmental Conditions," and "The Importance of Using Plain Language and Avoiding Overly Technical Terminology." This chapter is an opportunity to reflect back and describes how our understanding has changed over this time. Much has happened since the turn of the century, and we will discuss the advances in these areas since then.

A significant challenge in earthworm study in North America has been the scarcity of earthworm taxonomists, who were all living in isolation. When Reynolds visited earthworm conferences, this situation was not restricted to North America. Many of the previous specialists in the late twentieth century had died or retired, and some of the young emerging scientists in the field died early, while they were still actively engaged in research (William Murchie (48), Richard Tandy (55), Per-Olof Ljungström (24), Narcis Mršić (46), Randy Baker (28), Gilbert Right (62)—there may be others, and their omission does not diminish their contribution to the field). Even though these former colleagues left us too soon, they made significant contributions to the study of earthworms in various geographical and taxonomic areas.

2 Training of Taxonomists

Prior to 2000, there were a number of symposia and colloquia held every couple of years, where colleagues could gather and exchange ideas. Since the turn of the century, there have been more and varied meetings, the following is just a sample: Advances en taxonomy de Lumbricus de Tierra (Advancement in taxonomy of earthworms) (Spain, 2004), Minhocas na América Latina: Biodiversidade e Ecologia (Earthworms in Latin America: Biodiversity and ecology) (Brazil, 2007), 9th International Symposium on Earthworm Ecology (Mexico, 2010), 6th National Symposium on Earthworm Ecology (India, 2010), 10th International Symposium on Earthworm Ecology (USA, 2014), 5ᵉ Encontro Latino-Americano de Ecologia e Taxonomia de Oligoquetas (Brazil, 2015), 16th Vermiculture Conference USA, 2015), XVII International Colloquium on Soil Zoology (Japan, 2016), 7th International Oligochaete Taxonomy Meeting (France, 2016), European Worm Meeting (Austria, 2022), and 12th International Symposium on Earthworm Ecology (France, 2022). It is even possible to take a training course online—Earthworm Course—Summer 2021 Online, sponsored by the Seismological Facility for the Advancement of Geoscience. There have also been individual small training sessions such as the ones Reynolds gave at Sir Sandford Fleming College (Canada) and the Universidad Nacional de Córdoba (Argentina).

3 Taxonomic Tools

Nomenclatura Oligochaetologica, a compilation of type specimens, names, and descriptions of the Oligochaeta, was developed by Reynolds and Cook (1976) to help stabilize earthworm taxonomy and nomenclature. It lists all species by their original binomen along with the author (s), year, journal description citation, and type specimens. The original book went through three supplements, before Reynolds and Wetzel (2024) integrated the original book in 2013 with the three supplements and what was to be the fourth supplement into the second edition as an electronic source which can be accessed by anyone anywhere at any time. In addition to listing the species, subgenera, and genera, the project has instructions in ten languages as well as a dedication with biographies of deceased specialists (see Wetzel and Reynolds 2023).

A user-friendly key, devised by Emma Sherlock (2012) in the United Kingdom, is particularly beneficial to users who have no prior experience in identifying earthworms. It is a full-color publication containing a multitude of newly added photographs that help in identification. In addition to sampling and preserving earthworms, it provides insightful information, making it a valuable resource for earthworm enthusiasts. This latest revised publication provides valuable information on the taxonomy, nomenclature, morphology, and general biological and ecological properties of earthworms across Canada and the northern part of the United States. It is beneficial for both scientists and those who are not specialists. This book (in English and French versions) also includes an illustrated glossary, detailed drawings, color photographs, and a new key (Reynolds 2022).

Another beneficial effort started in 2013—taxo.drilobase.org "Drilobase is a database that has been created through a partnership among experts of the global earthworm scientific community." The data is provided for each taxonomic unit, and all identified taxa are contained. It is imperative to bear in mind that this is an in progress effort and will be continuously revised. The earthworms' identification is a highly intricate process that demands a substantial amount of specialized knowledge. Certain taxa necessitate DNA sequencing. We do not offer a service of this type.

Their stated goals are:

- Arranging seminars and scientific courses focused on DriloBASE and the study of earthworm ecology, biology, and taxonomy.
- Producing and distributing publications and methods of communication to spread the latest scientific information on earthworms.
- Providing financial support or sharing the financial burden of field trips aimed at contributing to the compilation of a comprehensive inventory of earthworm species worldwide.

There are other databases, but one worth mentioning is WoRMS (World Register of Marine Species) that was started in 1999, but in May 2017 WoRMS added earthworms to their register. In a recent publication by Mısırlıoğlu et al. (2023), an international group has provided a modified checklist of available species and families, along with valuable insights into their distribution.

4 Parthenogenesis in Taxonomy

The presence of parthenogenesis in earthworms caused potential problems in early classification of earthworms. This was succinctly stated by Gates (1972: 16–17):

> Basic in classical thinking was a belief that the only evolutionary modifications of an ancestral holandric-hologynous state, during all of the eons since megadriles arose … were eliminations of one or the other pair of ovaries along with, or without, deletion of one or another of the ancestral pairs of testes. Implicit, though perhaps never so stated before, were beliefs that the sex of each pair in the octagonal battery was immutable and that reproduction was obligatorily biparental. So deeply entrenched were those ideas that evidence to the contrary was ignored, explained as the result of massive parasitic infestations, or attributed to evolution of dioecism.

This situation over the years, during the time before the mid-twentieth century, when sperm was absent during breeding season or spermathecae were lacking, gave rise to the contributions regarding some studies in chromosome numbers and polyploidy (Muldal 1952; Omodeo 1955). This was the first step in helping eliminate the description of synonymous species. The next advance was the use of electrophoresis. Bøgh (1993) was able to illustrate differences in species.

Since the late 1990s, we are employing DNA techniques to help identify species. The following is summarized by Decaëns et al. (2013):

For discrimination of species, DNA barcoding employs a standardized region of 658 bp from the mitochondrial gene cytochrome c oxidase I (COI) (Hebert et al. 2003). The method has a number of benefits (Rougerie et al. 2009): (1) it is a verifiable and reproducible technique as it maintains a connection between a voucher sample and a barcode; (2) for regular identifications on a large scale, it is generally quicker and more cost-effective than conventional morphological identifications; and (3) it is applicable for species identification, detecting hidden diversity, and conducting alpha taxonomy.

Decaëns et al. (2013) reviewed some of the achievements:

Although significant research has been carried out on earthworm families such as Rhinodrilidae and Megascolecidae, numerous other families have been largely neglected, despite their significant contribution to the overall variety of earthworms and the geographical areas they inhabit. South and North America together with Africa have the most pronounced lack of molecular phylogenetic study. While molecular phylogenetic investigations have been conducted for the majority of those families, they often represent independent and disconnected endeavors aimed at addressing very particular evolutionary inquiries.

The study conducted by Decaëns et al. (2013) aimed to promote the use of molecular phylogenetic tools among earthworm specialists. The objective was to develop reliable genus systems, spatio-temporal evolutionary structures, evaluate cryptic diversity, and compare the geographical distribution of different species using biogeographical reconstructions and time-calibrated phylogenies. In circumstances when molecular biology resources are unavailable, international cooperation should give sufficient responses. The worrying reduction in the number of earthworm taxonomic experts in recent years is an issue affecting each continent identically. Without taxonomists with competence in the target species, even the most powerful molecular phylogenetic methods would be useless. Hence, it is crucial for earthworm taxonomists to prioritize the training of a new cohort of investigators who can effectively combine their understanding of the complex classification of earthworms with the use of advanced molecular techniques.

There are numerous papers on the use of DNA in earthworm taxonomy and systematics as currently it is a very active area of research. As techniques improve, this will become more important in the future. However, those using DNA analysis rarely use classical taxonomy alongside it. Currently, the technique is not infallible, as results do occur at times with inconclusive or inaccurate determinations. We will also always need classical taxonomists to assist with the description of new and old species.

5 Earthworm Surveys

There were early country earthworm surveys prior to the mid-twentieth century (e.g., Černosvitov, 1935 for Czechoslovakia), but changes in taxonomy and nomenclature may give problems to those new to the science. In relatively recent years, there have been a number of excellent examples, and here are just a few: Bouché (1972) for France, Omodeo and Rota (1989) for Turkey, Mršić (1991) for the Balkans, Fragoso (2001) for Mexico, Pižl (2002) for Czechia, Csuzdi and Zicsi (2003) for Hungary, Reynolds and Fragoso (2004) for Bermuda, Xiao (2019) for China, Reynolds (2020) for the United States, Reynolds (2022) for Canada, and Singh and Singh (2023) for India.

There have been attempts to survey countries where no earthworm records existed, e.g., the United Arab Emirates (Reynolds 2017, 2018, 2019; Reynolds and Reynolds 2010). There are still many areas in the world where little or no information is known on which earthworms are present.

6 Life Histories

Out of the approximately 6000 species of earthworms, only a small fraction, maybe about 20 species (0.0033%), has been the subject of recent and updated life history research. Several factors may be the cause: (1) some species are known only from a very few individuals, in some cases only the holotype; (2) some species, e.g., *Diplocardia*, where attempts to culture them have been unsuccessful (Vail 1972);

and (3) some life histories, particularly in the Lumbricidae, were done at a time when species lumping occurred, i.e., factors attributed to a complex which included several species and thus accounted for the range of data, vis-à-vis a specific value of other species (Reynolds 1995). We believe in time that many obstacles to determining life histories will be overcome.

7 Modern Techniques

This was partially covered in parthenogenesis in earthworms, above. The following is from Reynolds (2022: 5–7).

7.1 Sampling Techniques

Various techniques for sampling earthworm communities have been reported in the literature. These techniques could be categorized into four groups: *vibration*, *chemical extraction process*, *electrical extraction process*, and *excavation and manual sorting*. The efficacy of these strategies depends on the specific environment and earthworm species observed at an investigation location. It should be emphasized that no specific strategy is universally applicable to all species across all ecosystems.

The excavation and manual sorting by hand is the most dependable approach for sampling. Despite being laborious, this approach has been extensively employed for sampling earthworms and evaluating the efficacy of other techniques. When digging for earthworms, it is important to take into account the presence of humidity and organic material. The chances of finding earthworms are higher when digging focuses in spots where both of these factors are prominent. Different implements such as shovel, garden fork, trowel, or soil cores could be utilized for the purpose of excavation. Afterward, the soil could be compressed and sifted through the fingers, or sieves can be used. This approach has two distinct benefits. Firstly, it allows for the collection of active individuals, aestivating individuals, and cocoons within a specific sample area. Secondly, it enables the selection of a properly specified sampling region, which facilitates the collection of quantitative data. There are many drawbacks related to digging techniques. Firstly, digging requires a lot of physical effort and takes up a significant amount of time. Secondly, while collecting samples, those that are smaller than 2 cm might manage to evade capture. Lastly, if digging is limited to the upper layers of soil, larger individuals can be able to get away into lower levels of soil. Significant habitat degradation and sample damage are additional potential consequences.

Chemical extraction process is a well-established technique for collecting earthworms. Evans and Guild (1947) conducted preliminary research on chemical extraction, applying a solution of potassium permanganate (K_2PO_4) to stimulate earthworms and drive them to the surface for easier collection. Raw (1959) and

Waters (1955) performed further studies using chemical the extraction process, particularly using formalin. For years, Reynolds has used an established sampling technique to quantitatively extract samples. This involves using one or more 0.25 m^2 quadrats of the soil's surface at a particular location. Each quadrat is treated with a solution containing 25 mL of undiluted formalin (37% formaldehyde solution, USP) mixed with 4.5 L of water. The solution is applied in a way that ensures it fully penetrates the soil without any runoff. The earthworms that appear within 10 minutes after the effective treatment are gathered. When collecting samples for non-scientific goals, such as bait, it is essential to consider the timing, concentration, and frequency of application. However, it is worth noting that using a solution more powerful than 15 mL of formalin per liter of water could damage the grass. Additionally, if you intend to keep the samples alive for an extended period of time, it is crucial to rinse them in fresh water as soon as they are brought to the surface, as formalin can function as a vermicide. Formalin, a well-established carcinogen, has lost favor among many organizations in the past few years.

An approach that has proven to be highly effective, according to my recent quantitative sampling, was implemented by Dr. Catherine Fox of Agriculture Canada using a mustard solution. Cindy Hale, in her current study on introduced and native earthworm populations in forested regions of Minnesota (Hale 2013), also utilized this technique (Reynolds et al. 2002; Byron 2003). Chan and Munro (2001) and Bartlett et al. (2006) conducted more analysis on this approach. By using a chemical extraction technique, the amount of time and effort required for sampling is decreased. Additionally, this approach allows for the selection of a specific sample region and minimizes any disruption to the natural environment. There are a few drawbacks to this method. It only collects active individuals and not cocoons or individuals in a state of aestivation or hibernation. Additionally, it is only effective for species that reside in shallow areas or have burrow mechanisms. The success of the vermicide depends on the condition of the soil, and various species might react variously to the treatment. This strategy is often efficient for gathering lumbricids but less successful for species in other families. Starking and Roloff (2019) have provided an assessment of the mustard extraction method.

Electrical extraction: Fishermen have historically obtained lure through electrical extraction, a process that has been documented by numerous researchers (Walton 1933; Johnstone-Wallace 1937; Doeksen 1950; Satchell 1961). One to three electrodes and a generator are required for the procedure. The current that flows through the soil serves as an exceptional mechanism. Limited habitat disturbance is a benefit of this technique. The drawbacks include the prolonged time needed for each specimen, the challenge of precisely determining the boundaries of the treated soil volume, and the inconsistency in the physical and chemical characteristics of the soil. For instance, when the soil is hydrated, earthworms that typically reside deep in the soil may emerge to the surface, whereas in dry conditions, they may burrow even deeper. Excessive electrical current results in the death of earthworms in close proximity to the electrodes, and the reaction to electricity differs among various species. This approach underwent recent scrutiny by Bartlett et al. (2010), Lardo et al. (2012), and Mazur-Pączka et al. (2020).

Vibration techniques, often referred to as *mechanical extraction*, commonly known as “fiddling” or “grunting,” are mostly restricted to the southeastern region of the United States. Fish bait catchers use several adaptations of this method to get earthworms in remarkable numbers (Reynolds 1972, 1977). Earthworm extraction via vibrational mechanical stimulation exhibits remarkable efficiency for certain earthworm families, particularly Acanthodrilidae and Megascolecidae. However, this method appears to have minimal influence on Lumbricidae earthworms. Interestingly, Acanthodrilidae and Megascolecidae are less prevalent in regions like Europe and Canada. This ecological disparity might explain the scarcity of mentions regarding vibration-based extraction techniques within European discussions on earthworm collection practices. The benefits of mechanical extraction include less habitat disruption and decreased time needed for each specimen. The drawbacks include the challenge of precisely determining the treated soil’s amount, the impact of the soil’s physicochemical parameters’ fluctuation, and the varying reaction of various species.

The literature also discusses a number of additional sampling methods, including pitfall traps, moist sieving, floatation, and heat extraction through diverse funnels. However, the utilization of these methods has been limited to a few research endeavors due to the frequent production of earthworm specimens in the degraded physical state, which impedes or obstructs accurate identification. Further knowledge and published references for these methods were provided by Reynolds (1977) and Fender (1985). Extraction methods have been thoroughly examined in recent reviews by Singh et al. (2015) and González et al. (2020).

7.2 *Preservation Techniques*

The appropriate fixation and conservation of samples for the purpose of identification, transportation, and storage have been a persistent issue. There is a scarcity of reliable accounts regarding preservation and fix methods readily accessible to individuals desiring to submit materials to experts for analysis. A 10–15% formalin solution stands out as a preferred choice for earthworm preservation by virtue of its tissue fixation properties. This process induces a hardening effect, facilitating subsequent handling and manipulation during morphological examinations. Diluted alcohol solutions are inadequate for preserving specimens as they can leave them soft and pliable. Conversely, excessively concentrated alcohol solutions can render specimens undesirably brittle. Alcohol induces a phenomenon called “alcohol browning” in both situations. Consequently, the assessment of color in conserved alcohol samples has limited diagnostic significance. Overall, formalin does not drastically change the color. As previously said, we suggest that collectors document the color of living samples *before* they are fixed or preserved. Buffered formalin provides a clear benefit by inhibiting the production of unwanted acid-formalin-hematin pigment in tissue slices. Furthermore, it aids in reducing the breakdown of calcium-based structures present in the samples. Borax, which is commonly used as a laundry detergent enhancer (sodium borate, sodium tetra decahydrate: $Na_2B_4O_7 \cdot 10H_2O$), can be employed to neutralize formalin. Additionally, magnesium carbonate [$MgCO_3$] and calcium carbonate [$CaCO_3$] frequently serve as formalin buffers. To neutralize formalin, a 1:15 ratio of borax, calcium carbonate, or magnesium carbonate to formalin is recommended. The mixture should be shaken vigorously and allowed to settle prior to apply. Hexamine [$(CH_2)_6N_4$], with a concentration of 8 mg per liter of a 2% formalin solution, has been successfully utilized as a buffer for formalin. However, it can be challenging to acquire.

An efficient and straightforward method for preparing samples for investigation involves killing the worms by submerging them in a solution of ethyl alcohol with a concentration of 70%. Once motion ceases, the specimens ought to be positioned in a straight line on absorbent paper and left to dry for a couple of minutes. In order to ensure appropriate preservation of tissues and maintain their integrity over a long period of time, it is necessary to relocate the specimens to a container holding a solution of 10–15% buffered formalin. The samples then go through a hardening process while maintaining the position they were originally placed in. It is preferable for the samples to be straight, as curled or twisted ones can pose challenges during internal inspection and dissection. After allowing the specimens to rest in this container for the night, they can be transferred to bottles or vials containing a fresh 10–15% formalin solution lacking a significant risk of curling. The formalin solution should be replaced every week, or more frequently if it rapidly loses its color, in order to achieve optimal outcomes. Typically, it is recommended to replace the formalin solution every week until it becomes transparent.

8 Earthworms in Waste Management

The utilization of earthworms for vermicomposting and organic e-waste recycling has experienced a surge in popularity since 1990, and this trend is continuing to spread swiftly in both developed and developing nations. According to a report by Edwards (2004, cited in Edwards and Arancon 2006), commercial vermicomposting initiatives have been established in numerous nations as of that year. These include the United States, Japan, China, India, the Philippines, and various Southeast Asian countries; the United Kingdom, Germany, France, Italy, Spain, the Netherlands, and Poland; and New Zealand, Australia, Hawaii, American Samoa, and numerous countries in South America. Vermicompost and vermimeal production in the Philippines began in 1979 (Guerrero III 2009). However, the African night crawler (ANC), a kind of composting earthworm, was not introduced into the nation until 1982 (Guerrero III et al. 1984). The earthworm species that have been most successful in vermicomposting are as follows: *Dendrobaena veneta* (= *Eisenia hortensis*), *Dendrodrilus rubidus*, *Eisenia andrei*, *E. fetida*, *Eudrilus eugeniae*, *Lumbricus rubellus*, *Perionyx excavatus*, and *Polypheretima elongata*. In the future, there may be a species currently known or yet to be discovered which will be a

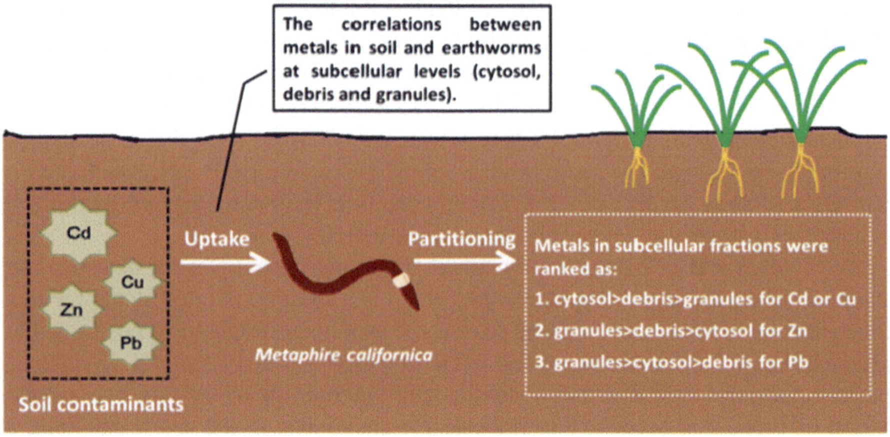

Fig. 1 The uptake of heavy metals (cadmium, copper, lead, and zinc) in earthworms (from Zhang et al. 2020)

prime candidate for vermicomposting. India has been a very active center for vermicomposting research (Sreenivasan 2020).

9 Earthworms for Environmental Monitoring

One of the earliest studies was by Gish and Christianson (1973) who examined heavy metals along the Washington-Baltimore Parkway where they found cadmium, nickel, lead, and zinc in earthworms. The second early study was by Van Hook (1974) who looked at biological accumulation in earthworms. Kumar et al. (2008) examined heavy metal effects on earthworms in vermicomposting municipal solid waste. Some of the recent studies on heavy metal effects and tolerance to earthworms found the same heavy metals in other countries (Le Roux et al. 2016; Wang et al. 2018; Zhang et al. 2020; Fig. 1). It is our belief that the effects and toxicity of various metals and substances, like pesticides, on earthworms worldwide will be important in the coming decades.

10 Plain Language and Less Esotery

Reynolds has always emphasized the need of making scientific knowledge more readily available to the general public, who may not have a scientific background. The progress of entomologists and ornithologists has been significantly accelerated by the collaborative efforts of citizen scientists "amateurs" and ordinary collectors. A pioneer in the field of earthworm biology during the beginning of the twentieth century was the Reverend Hilderic Friend, who encountered resistance and rejection from contemporary earthworm scientists. He had excellent skills as a naturalist; nevertheless, a significant number of his theories were subjected to derision, dismissal, and assimilation into similar ideas. Had he had more encouragement, his study could have had a more significant influence on the field of oligochaetology as it is understood in the present day. In the mid-twentieth century, another biologist was discouraged (Jim Joyner) when he learned that dissection was essential for many of the diplocardian earthworms he encountered. Thus, there is only one publication to his credit.

Citizen scientist programs became popular around the year 2000. Citizen Scientists was formed with three main goals: (1) to educate volunteers, (2) to foster local stewardship, and (3) to monitor local habitats. Several of the early programs are Toronto and Region Conservation Authority, Rouge Valley Conservation Centre, and British Ecological Society—Science Hub.

The purpose of the OPAL Soil and Earthworm Survey is to gather information on soil and earthworms and explore the connections between various earthworm species, ecosystems, and soil types. A grand number of 5510 online responds were received for the OPAL Soil and Earthworm Survey.

The survey comprises the following five activities:

1. Excavating a soil pit to extract a soil sample
2. Analyzing the characteristics of the soil sample (such as firmness, moisture content, pH level, and texture)
3. Documenting various earthworm species
4. Performing a search for earthworms in different environments
5. Documenting the presence of additional organisms discovered in the soil specimen

The OPAL project spawned the following programs:

The Earthworm Society of Great Britain and Northern Ireland (ESB) is a program dedicated only to earthworms. Its main objective is to advance and facilitate scientific study in order to enhance our understanding of earthworms and their habitat. The Society's objective is to promote the preservation of earthworms and their natural environments, as well as to inform and motivate the public, ensuring the continued enjoyment of these wonderful organisms in the future.

WormWatch is an entertaining science-based education program that is available in both the United States and Canada. It is a component of *NatureWatch*, a nationwide initiative where volunteers monitor and detect ecological alterations that might be influencing our ecosystem. This site will:

- Elucidate on the significance of earthworms as a bioindicator of soil biodiversity.
- Introduce the anatomy and ecology of worms.
- Provide the resources and instruments necessary for identifying and tracking worms.

The *WormWatch* site offers distinct initiatives that encourage volunteers to gather data on various earthworm species and their environments. We extend an invitation to all individuals to engage in the process of collecting and categorizing earthworms.

iNaturalist, among the most well-known nature applications in the world, assists users in identifying familiar flora and animals (Mathieu et al. 2023). iNaturalist.org was initiated in 2008 as the Master's Culminating Project at the School of Information, UC Berkeley, by Ken-ichi Ueda, Nate Agrin, and Jessica Kline. *iNaturalist* was incorporated by the California Academy of Sciences in 2014 and the National Geographic Society in 2017 as a joint project. On an international basis, *iNaturalist* receives support from various organizations via the *iNaturalist* Network.

11 Glimpses into the Future

Previously, Reynolds had the belief that all earthworms were beneficial and that their activities in the soil had no detrimental effects. Their proliferation as invasive species and the impact of climate change have compelled me to reconsider my initial stance in recent years. In the near future, climate change will pose serious risks to biodiversity, putting a majority of ecosystem functioning and services at jeopardy. Earthworms have a disadvantage due to the porosity of their outer covering and their need on humid conditions for their existence (Siebert et al. 2019).

Scientists have observed significant summer melting rates over the Greenland Ice Sheet, providing more proof of the alarming Arctic warming. Canada's last ice shelf, located in Nunavut, disintegrated and fell into the ocean at the beginning of August 2020. In the summer of 2020, several areas of Arctic Siberia and Svalbard, which are a cluster of Norwegian islands in the Arctic Ocean, had very high temperatures, reaching a record-breaking 38°C (100°F) (Brigham-Grette and Petsch 2020; Labe et al. 2020).

Glacial melt causing flooding submerges habitats off shore, low flood plains, and small low island countries, such as Maldives, Republic of the Marshall Islands, Cayman Islands, Cocos (Keeling) Islands, Tuvalu, and a great number of islands in the Indian and Pacific Ocean. The increase in the sea level has much exceeded the forecast made by the Intergovernmental Panel on Climate Change in 2007 and has the potential to reach a height of 2 m during the current century. This will eliminate terrestrial earthworms from these places.

Global warming has greatly increased the number of forest and grassland fires, destroying the habitat for corticole, epigeic, and anecic species in particular. They lose their food source, e.g., corticoles since the deadwood is consumed, epigeics as the litter layer is destroyed, and anecics although they live below the surface they require a litter layer as a food source.

Earthworms are of critical importance in the global determination of soil greenhouse gas balances, and their impact is anticipated to increase in the coming decades. It is believed that they promote carbon sequestration in soil aggregates but concurrently contribute to elevated emissions of carbon dioxide (CO_2) and nitrous oxide

(N_2O), which are the primary greenhouse gases. Therefore, the extent to which earthworms influence soils as a net source or sink of greenhouse gases continues to be a subject of intense debate (Reynolds 2022).

The Arctic is now experiencing unprecedented warmth not seen in the last three million years. Geoscientists are very concerned about the implications of this phenomenon in the Arctic and its potential impact on the global scale. In just 200 years, the mankind has successfully reversed the course that started 50 million years ago and brought the world back to CO_2 levels that have not been seen for millions of years. It is anticipated that by 2040, for instance, the Arctic will be devoid of summer ice entirely (Brigham-Grette and Petsch 2020). In our viewpoint, the earthworms' abundance and diversity in Arctic soils will significantly increase by 2050 compared to the present.

We feel confident that these important issues will be solved in the future because of the new young scientists throughout the world who are concentrating on earthworm research. When Reynolds began his biological research, Watson and Crick (1953) had recently published their research on DNA. These young scientists are familiar with DNA and a number of new techniques available for use in their research. They will also develop as yet unknown techniques and instruments to help answer the challenges in the future.

It has taken oligochaetologists a long time to accept the assistance of citizen scientists which have been a great asset to other branches of biology, e.g., entomologists and ornithologists. Recently, *iNaturalist*—citizen scientists have reached out to the public for help (Mathieu et al. 2023). Therefore, I (JWR) feel confident that when my contributions to earthworm research have ended, there will be many new scientists ready to pick up the gauntlet.

Acknowledgment We wish to thank Wilma M. Reynolds of the Oligochaetology Laboratory, Kitchener, Canada, for reviewing the manuscript and her comments and suggestions which have greatly improved the final version.

References

Bartlett MD, Harris JA, James IT, Ritz K (2006) Inefficiency of mustard extraction technique for assessing size and structure of earthworm communities in UK pasture. Soil Biol Biochem 38:2990–2992

Bartlett MD, Briones MJI, Neilson R, Schmidt O, Spurgeon D, Creamer RE (2010) A critical review of current methods in earthworm ecology: from individuals to populations. Eur J Soil Biol 46(2):67–73

Bøgh PS (1993) Identification of earthworms (Lumbricidae): choice of method and dstinction criteria. Megadrilogica 4(10):163–174

Bouché MB (1972) Lombriciens de France, écologie et systématique. Institut national de la recherche agronomique, 671p

Brigham-Grette J, Petsch S (2020) The Arctic hasn't been this warm for 3 million years – and that foreshadows big changes for the rest of the planet. The Conversation, October 5th, 2020

Byron E (2003) The worm turns: off home turf. It's down to no good. Wall Street J CCXLII(4): A1; A6. (July 7th issue)
Chan K-Y, Munro K (2001) Evaluating mustard extracts for earthworm sampling. Pedobiologia 45(3):272–278
Csuzdi C, Zicsi A (2003) Earthworms of Hungary (Annelida: Oligochaeta, Lumbricidae). Pedozoologica Hungarica 1:1–272
Decaëns T, Porco D, Rougerie R, Brown GG, James SW (2013) Potential of DNA barcoding for earthworm research in taxonomy and ecology. Appl Soil Ecol 65:35–42
Doeksen J (1950) An electrical method of sampling soil for earthworms. Transactions of the 4th international congress of soil science, pp 129–131
Edwards CA, Arancon NQ (2006) The science of vermiculture: the use of earthworms in organic waste management. In: Guerrero III RD, Guerrero-del Castillo MRA (Eds) Vermi Technologies for Developing Countries. Proceedings of the international symposium-workshop on vermi technologies for developing countries. Nov 16–18, 2005, Philippines: Los Baños, Laguna. Philippine Fisheries Association, Inc., pp 1–30
Evans AC, Guild WJML (1947) Studies on the relationships between earthworms and soil fertility. I. Biological studies in the field. Annu Appl Biol 34(3):307–330
Fender WM (1985) Earthworms of the western United States. Part I. Lumbricidae. Megadrilogica 4(5):93–129
Fragoso C (2001) Las lombrices de tierra de Mexico (Annelida, Oligochaeta): diversidad, ecologia y manejo. Acta Zoologica Mexicana (n.s.). Numero especial 1:131–171
Gates GE (1972) Burmese earthworms. An introduction to the systematics of megadrile oligochaetes with special references to southeast Asia. Trans Am Philos Soc (new series) 62(7):1–326
Gish CD, Christianson RE (1973) Cadmium, nickel, lead and zinc in earthworms from roadside soil. Environ Sci Technol 7:1060–1062
González G, Barberena-Arias MF, Huang W, Ospina-Sánchez CM (2020) Sampling methods for soil and litter fauna. Springer: Measuring Arthropod Biodiversity, pp 495–522
Guerrero III RD (2009) Vermicompost and vermimeal production. MARID Agribusiness Technology Guide. 22p
Guerrero RD III, Guerrrero LA, Carggado AU (1984) Studies on the culture of the earthworm, *Eudrilus euginae*, and its use as feed for Macrobracchium Idella and fertilizer source for Brassica compensis. Trans Natl Acad Sci 6:33–40
Hale C (2013) Earthworms of the Great Lakes. Kollath+Stensaas Publications, Duluth, MN, 47p
Hebert PDN, Cywinska A, Ball SL, deWaard JR (2003) Biological identifications through DNA barcodes. Proc R Soc Biol Sci 270(1512):313–321
Johnstone-Wallace DB (1937) The influence of wild white clover on the seasonal production and chemical composition of pasture herbage and upon soil temperatures. Soil moistures and erosion control. 4th International Grassland Congress Report, pp 188–196
Kumar S, Sharma V, Bhoyar RV, Bhattacharyya JK, Chakrabarti T (2008) Effect of heavy metals on earthworm activities during vermicomposting of municipal solid waste. Water Environ Res 80(2):154–161
Labe ZM, Peings Y, Magnusdottir G (2020) Warm Arctic, cold Siberia pattern: role of full Arcticamplification versus sea ice loss alone. Geophysical Res Lett. https://doi.org/10.1029/2020GL088583
Lardo L, Coll P, Le Cadre E, Palese AM, Villenave C, Xiloyannis C, Celano G (2012) Electromagnetic induction (EMI) measurements as a proxy of earthworm presence in southern French vineyards. Appl Soil Ecol 61:76–84
Le Roux S, Baker P, Crouch A (2016) Bioaccumulation of total mercury in the earthworm *Eisenia andrei*. SpringerPlus 5:681. https://doi.org/10.1186/s40064-016-2282-6
Mathieu J, Bartz M, Brown G, Cameron E, Decaëns T, Dupont L, Fernandez Marchan D, Fragoso C, Ganault P, Geraskina A, Golovanova E, Hedde M, Keith A, Reynolds J, and others (2023) Collaboration call: building an image collection for computer vision identifica-

tion of worldwide earthworm species in iNaturalist. https://www.inaturalist.org/projects/curated-collection-of-earthworms-id
Mazur-Pączka A, Pączka G, Kostecka J, Podolak A, Garczyńska M (2020) Effectiveness of Lumbricidae extracting with an environmentally friendly method. J Ecol Eng 21(5):114–119
Mısırlıoğlu M, Reynolds JW, Stojanović M, Trakić T, Sekulić J, James SW, Decaëns T, Lapied E, Phillips H, Cameron E, Brown GG (2023) Earthworms (Clitellata, Megadrili) of the world: an updated checklist of valid species and families, with notes on their distribution. Zootaxa 5255(1):417–438
Mršić N (1991) Monograph on earthworms (Lumbricidae) of the Balkans I–II. Slovenska Akademija Znanosti in Umetnosti, Ljubljana, 757p
Muldal S (1952) The chromosomes of the earthworms. 1. The evolution of poloyploidy. Heredity 6:55–76
Omodeo P (1955) Cariologia dei Lumbricdae. II Contributo. Caryologia 8:137–178
Omodeo P, Rota E (1989) Earthworms of Turkey. Ital J Zool 56(2):167–198
Pižl V (2002) Žižaly Česke republiky. Earthworms of the Czech Republic. Sbornik prirodovedneho klubu v Uherském Hradišti 9:5–154
Raw F (1959) Estimating earthworm populations by using formalin. Nature (London) 184:1661–1662
Reynolds JW (1972) Réparation et biomasse des lombricidés (Oligochaeta: Lumbricidae) dans trois bois de la France (Département d'Essonne). Megadrilogica 1(4):1–6
Reynolds JW, Cook DG (1976) Nomenclatura Oligochaetologica: a catalogue of the names, descriptions and type specimens of the Oligochaeta. University of New Brunswick, Fredericton, *x* + 217 pp
Reynolds JW (1977) The earthworms (Lumbricidae and Sparganophilidae) of Ontario. Life Sciences Miscellaneous Publication, Royal Ontario Museum, *xi* + 141 pp
Reynolds JW (1995) The status of exotic earthworm systematics and biogeography in North America. In: Hendrix PF (ed) Ecology and biogeography of earthworms in North America. Lewis, Boca Raton, FL, pp 1, 244p–27
Reynolds JW (1998) The status of earthworm biogeography, diversity and taxonomy in North America revisited with glimpses into the future. In: Edwards CA (ed) Earthworm ecology. St. Lucie Press, Delray Beach, FL, pp 15, 389 p–34
Reynolds JW (2004) The status of earthworm biogeography, diversity, and taxonomy in North America revisited with glimpses into the future. In: Edwards CA (ed) Earthworm ecology, 2nd edn. CRC Press, St. Lucia, FL, pp 63, 441p–74
Reynolds JW (2017) *Lampito mauritii* (Oligochaeta: Megascolecidae) first earthworm record for Sharjah, U.A.E. Megadrilogica 22(5):131–134
Reynolds JW (2018) New earthworm (Oligochaeta: Megascolecidae and Octochaetidae) records from the Emirate of Dubai. UAE Megadrilogica 23(9):121–128
Reynolds JW (2019) First earthworm (Oligochaeta: Octochaetidae) records from Ras al Khaimah and new record for Dubai. United Arabic Emirates Megadrilogica 24(10):129–136
Reynolds JW (2020) Earthworms in American Ecoregions. OmniScriptum Publishing Group, Beau Bassin, p 430
Reynolds JW (2022) The Earthworms (Lumbricidae, Megascolecidae and Sparganophilidae) in Canada. Canada Food Inspection Agency, Ottawa, *i* + 179 p
Reynolds JW, Fragoso C (2004) The earthworms (Oligochaeta: Acanthodrilidae, Eudrilidae, Glossoscolecidae, Lumbricidae, Megascolecidae and Ocnerodrilidae) of Bermuda. Megadrilogica 10(4):17–26
Reynolds JW, Reynolds DW (2010) The great global earthworm race goes to Dubai, U.A.E. Megadrilogica 14(2):49–52
Reynolds JW, Wetzel MJ (2024) Nomenclatura Oligochaetologica – a catalogue of names, descriptions and type specimens of the Oligochaeta. Editio Secunda. https://nomenclatura-oligochaetologica.inhs.illinois.edu (date accessed: 04-28-2024)

Reynolds JW, Linden DR, Hale CM (2002) The earthworms of Minnesota (Oligochaeta: Acanthodrilidae, Lumbricidae and Megascolecidae). Megadrilogica 8(12):85–99

Rougerie R, Decaëns T, Deharveng L, Porco D, James SW, Chang C-H, Richard B, Potapov M, Suhardjono Y, Hebert PDN (2009) DNA barcodes for soil animal taxonomy. Pesquisa Agropecuaria Brasileira 44(8):789–801

Satchell JE (1961) An electrical method of sampling earthworm populations. pp. 89–97. In: Rodale R (ed) The challenge of earthworm research. Soil and Health Foundation, Emmas, PA, 102p

Siebert J, Eisenhauer N, Poll C, Marhan S, Bonkowski M, Hines J, Koller R, Ruess L, Thakur MP (2019) Earthworms modulate the effects of climate warming on the taxon richness of soil meso- and macrofauna in an agricultural system. Agric Ecosyst Environ 278:72–80

Singh PK, Singh K (2023) Updated checklist of earthworms belonging to the family Megascolecidae (Annelida: Clitellata: Oligochaeta) in India. Munis Entomol Zool 18(2):1423–1447

Singh J, Singh S, Vig AP (2015) Extraction of earthworm from soil by different sampling methods: a review. Environ Dev Sustain. https://doi.org/10.1007/s10668-015-9703-5

Sreenivasan E (2020) Solid waste management and its role in mitigating environmental degradationwith special reference to recycling of wood wastes, pp 166–182. In: Kumar V, Singh J, Kumar P (eds) Environmental degradation: causes and remediation strategies. Springer Nature Switzerland AG, xvii + 544 p

Starking MD, Roloff GJ (2019) Evaluating a mustard extraction technique for sampling earthworms. Wildl Soc Bull 43(4):760–765

Vail VA (1972) Natural history and reproduction in *Diplocardia mississippiensis* (Oligochaeta). Bulletin of Tall Timbers Research Station, No. 11, pp 1–39

Van Hook RI (1974) Cadmium, lead and zinc distributions between earthworms and soils: potentials for biological accumulation. Bull Environ Contam Toxicol 12:509–511

Walton WR (1933) The reaction of earthworms to alternating currents of electricity in the soil. Proc Entomol Soc Wash 35(2):24–27

Wang K, Qiao Y, Zhang H, Yue S, Li H, Ji X, Liu L (2018) Influence of metal-contamination on distribution in subcellular fractions of the earthworm (*Metaphire californica*) from Hunan Province, China. J Environ Sci 73:127–137

Waters RAS (1955) Numbers and weights of earthworms under a highly productive pastures. N Z J Sci Technol Sect A Agric Res 36:516–525

Watson JD, Crick FHC (1953) A structure for deoxyribose nucleic acid. Nature 171:737–738

Wetzel MJ, Reynolds JW (2023) *Nomenclatura Oligochaetologica* – a catalogue of names, descriptions, and type specimens of the Oligochaeta. *Editio Secunda*~~. (pp. xx-xx)~~ In: Martin P, Pinder A, Wetzel MJ (eds) Proceedings of the 15th international symposium of aquatic oligochaeta (20–24 September 2022, Brussels, Belgium). Zoosymposia 23: 98–100

Xiao N (2019) Terrestrial earthworms (Oligochaeta: Opisthopora) of China. Academic, London, 418p

Zhang D, Caiping Y, Abbas N, Mao Z, Zhang Y (2020) Multiple heavy metal tolerance and removal by an earthworm gut fungus *Trichoderma brevicompactum* QYCD-6. Sci Rep 10(6940):1–9

The manufacturer's authorised representative in the EU is Springer Nature Customer Service Centre GmbH, Europaplatz 3, 69115 Heidelberg, Germany. If you have any concerns regarding our products, please contact ProductSafety@springernature.com

Printed and bound by CPI Group (UK) Ltd, Croydon, CR0 4YY
01/10/2026
02225910-0018